T0140060

Shai Avidan · Gabriel Brostow ·
Moustapha Cissé · Giovanni Maria Farinella ·
Tal Hassner (Eds.)

Computer Vision – ECCV 2022

17th European Conference
Tel Aviv, Israel, October 23–27, 2022
Proceedings, Part I

 Springer

Editors
Shai Avidan
Tel Aviv University
Tel Aviv, Israel

Gabriel Brostow 🆔
University College London
London, UK

Moustapha Cissé
Google AI
Accra, Ghana

Giovanni Maria Farinella 🆔
University of Catania
Catania, Italy

Tal Hassner 🆔
Facebook (United States)
Menlo Park, CA, USA

ISSN 0302-9743   ISSN 1611-3349 (electronic)
Lecture Notes in Computer Science
ISBN 978-3-031-19768-0   ISBN 978-3-031-19769-7 (eBook)
https://doi.org/10.1007/978-3-031-19769-7

This Springer imprint is published by the registered company Springer Nature Switzerland AG
The registered company address is: Gewerbestrasse 11, 6330 Cham, Switzerland

Foreword

Organizing the European Conference on Computer Vision (ECCV 2022) in Tel-Aviv during a global pandemic was no easy feat. The uncertainty level was extremely high, and decisions had to be postponed to the last minute. Still, we managed to plan things just in time for ECCV 2022 to be held in person. Participation in physical events is crucial to stimulating collaborations and nurturing the culture of the Computer Vision community.

There were many people who worked hard to ensure attendees enjoyed the best science at the 16th edition of ECCV. We are grateful to the Program Chairs Gabriel Brostow and Tal Hassner, who went above and beyond to ensure the ECCV reviewing process ran smoothly. The scientific program includes dozens of workshops and tutorials in addition to the main conference and we would like to thank Leonid Karlinsky and Tomer Michaeli for their hard work. Finally, special thanks to the web chairs Lorenzo Baraldi and Kosta Derpanis, who put in extra hours to transfer information fast and efficiently to the ECCV community.

We would like to express gratitude to our generous sponsors and the Industry Chairs, Dimosthenis Karatzas and Chen Sagiv, who oversaw industry relations and proposed new ways for academia-industry collaboration and technology transfer. It's great to see so much industrial interest in what we're doing!

Authors' draft versions of the papers appeared online with open access on both the Computer Vision Foundation (CVF) and the European Computer Vision Association (ECVA) websites as with previous ECCVs. Springer, the publisher of the proceedings, has arranged for archival publication. The final version of the papers is hosted by SpringerLink, with active references and supplementary materials. It benefits all potential readers that we offer both a free and citeable version for all researchers, as well as an authoritative, citeable version for SpringerLink readers. Our thanks go to Ronan Nugent from Springer, who helped us negotiate this agreement. Last but not least, we wish to thank Eric Mortensen, our publication chair, whose expertise made the process smooth.

October 2022

Rita Cucchiara
Jiří Matas
Amnon Shashua
Lihi Zelnik-Manor

Preface

Welcome to the proceedings of the European Conference on Computer Vision (ECCV 2022). This was a hybrid edition of ECCV as we made our way out of the COVID-19 pandemic. The conference received 5804 valid paper submissions, compared to 5150 submissions to ECCV 2020 (a 12.7% increase) and 2439 in ECCV 2018. 1645 submissions were accepted for publication (28%) and, of those, 157 (2.7% overall) as orals.

846 of the submissions were desk-rejected for various reasons. Many of them because they revealed author identity, thus violating the double-blind policy. This violation came in many forms: some had author names with the title, others added acknowledgments to specific grants, yet others had links to their github account where their name was visible. Tampering with the LaTeX template was another reason for automatic desk rejection.

ECCV 2022 used the traditional CMT system to manage the entire double-blind reviewing process. Authors did not know the names of the reviewers and vice versa. Each paper received at least 3 reviews (except 6 papers that received only 2 reviews), totalling more than 15,000 reviews.

Handling the review process at this scale was a significant challenge. To ensure that each submission received as fair and high-quality reviews as possible, we recruited more than 4719 reviewers (in the end, 4719 reviewers did at least one review). Similarly we recruited more than 276 area chairs (eventually, only 276 area chairs handled a batch of papers). The area chairs were selected based on their technical expertise and reputation, largely among people who served as area chairs in previous top computer vision and machine learning conferences (ECCV, ICCV, CVPR, NeurIPS, etc.).

Reviewers were similarly invited from previous conferences, and also from the pool of authors. We also encouraged experienced area chairs to suggest additional chairs and reviewers in the initial phase of recruiting. The median reviewer load was five papers per reviewer, while the average load was about four papers, because of the emergency reviewers. The area chair load was 35 papers, on average.

Conflicts of interest between authors, area chairs, and reviewers were handled largely automatically by the CMT platform, with some manual help from the Program Chairs. Reviewers were allowed to describe themselves as senior reviewer (load of 8 papers to review) or junior reviewers (load of 4 papers). Papers were matched to area chairs based on a subject-area affinity score computed in CMT and an affinity score computed by the Toronto Paper Matching System (TPMS). TPMS is based on the paper's full text. An area chair handling each submission would bid for preferred expert reviewers, and we balanced load and prevented conflicts.

The assignment of submissions to area chairs was relatively smooth, as was the assignment of submissions to reviewers. A small percentage of reviewers were not happy with their assignments in terms of subjects and self-reported expertise. This is an area for improvement, although it's interesting that many of these cases were reviewers hand-picked by AC's. We made a later round of reviewer recruiting, targeted at the list of authors of papers submitted to the conference, and had an excellent response which helped provide enough emergency reviewers. In the end, all but six papers received at least 3 reviews.

The challenges of the reviewing process are in line with past experiences at ECCV 2020. As the community grows, and the number of submissions increases, it becomes ever more challenging to recruit enough reviewers and ensure a high enough quality of reviews. Enlisting authors by default as reviewers might be one step to address this challenge.

Authors were given a week to rebut the initial reviews, and address reviewers' concerns. Each rebuttal was limited to a single pdf page with a fixed template.

The Area Chairs then led discussions with the reviewers on the merits of each submission. The goal was to reach consensus, but, ultimately, it was up to the Area Chair to make a decision. The decision was then discussed with a buddy Area Chair to make sure decisions were fair and informative. The entire process was conducted virtually with no in-person meetings taking place.

The Program Chairs were informed in cases where the Area Chairs overturned a decisive consensus reached by the reviewers, and pushed for the meta-reviews to contain details that explained the reasoning for such decisions. Obviously these were the most contentious cases, where reviewer inexperience was the most common reported factor.

Once the list of accepted papers was finalized and released, we went through the laborious process of plagiarism (including self-plagiarism) detection. A total of 4 accepted papers were rejected because of that.

Finally, we would like to thank our Technical Program Chair, Pavel Lifshits, who did tremendous work behind the scenes, and we thank the tireless CMT team.

October 2022

Gabriel Brostow
Giovanni Maria Farinella
Moustapha Cissé
Shai Avidan
Tal Hassner

Organization

General Chairs

Rita Cucchiara   University of Modena and Reggio Emilia, Italy
Jiří Matas   Czech Technical University in Prague, Czech Republic
Amnon Shashua   Hebrew University of Jerusalem, Israel
Lihi Zelnik-Manor   Technion – Israel Institute of Technology, Israel

Program Chairs

Shai Avidan   Tel-Aviv University, Israel
Gabriel Brostow   University College London, UK
Moustapha Cissé   Google AI, Ghana
Giovanni Maria Farinella   University of Catania, Italy
Tal Hassner   Facebook AI, USA

Program Technical Chair

Pavel Lifshits   Technion – Israel Institute of Technology, Israel

Workshops Chairs

Leonid Karlinsky   IBM Research, Israel
Tomer Michaeli   Technion – Israel Institute of Technology, Israel
Ko Nishino   Kyoto University, Japan

Tutorial Chairs

Thomas Pock   Graz University of Technology, Austria
Natalia Neverova   Facebook AI Research, UK

Demo Chair

Bohyung Han   Seoul National University, Korea

Social and Student Activities Chairs

Tatiana Tommasi   Italian Institute of Technology, Italy
Sagie Benaim   University of Copenhagen, Denmark

Diversity and Inclusion Chairs

Xi Yin   Facebook AI Research, USA
Bryan Russell   Adobe, USA

Communications Chairs

Lorenzo Baraldi   University of Modena and Reggio Emilia, Italy
Kosta Derpanis   York University & Samsung AI Centre Toronto,
                                        Canada

Industrial Liaison Chairs

Dimosthenis Karatzas   Universitat Autònoma de Barcelona, Spain
Chen Sagiv   SagivTech, Israel

Finance Chair

Gerard Medioni   University of Southern California & Amazon,
                                        USA

Publication Chair

Eric Mortensen   MiCROTEC, USA

Area Chairs

Lourdes Agapito   University College London, UK
Zeynep Akata   University of Tübingen, Germany
Naveed Akhtar   University of Western Australia, Australia
Karteek Alahari   Inria Grenoble Rhône-Alpes, France
Alexandre Alahi   École polytechnique fédérale de Lausanne,
                                        Switzerland
Pablo Arbelaez   Universidad de Los Andes, Columbia
Antonis A. Argyros   University of Crete & Foundation for Research
                                        and Technology-Hellas, Crete
Yuki M. Asano   University of Amsterdam, The Netherlands
Kalle Åström   Lund University, Sweden
Hadar Averbuch-Elor   Cornell University, USA

Hossein Azizpour   KTH Royal Institute of Technology, Sweden
Vineeth N. Balasubramanian   Indian Institute of Technology, Hyderabad, India
Lamberto Ballan   University of Padova, Italy
Adrien Bartoli   Université Clermont Auvergne, France
Horst Bischof   Graz University of Technology, Austria
Matthew B. Blaschko   KU Leuven, Belgium
Federica Bogo   Meta Reality Labs Research, Switzerland
Katherine Bouman   California Institute of Technology, USA
Edmond Boyer   Inria Grenoble Rhône-Alpes, France
Michael S. Brown   York University, Canada
Vittorio Caggiano   Meta AI Research, USA
Neill Campbell   University of Bath, UK
Octavia Camps   Northeastern University, USA
Duygu Ceylan   Adobe Research, USA
Ayan Chakrabarti   Google Research, USA
Tat-Jen Cham   Nanyang Technological University, Singapore
Antoni Chan   City University of Hong Kong, Hong Kong, China
Manmohan Chandraker   NEC Labs America, USA
Xinlei Chen   Facebook AI Research, USA
Xilin Chen   Institute of Computing Technology, Chinese
                                   Academy of Sciences, China
Dongdong Chen   Microsoft Cloud AI, USA
Chen Chen   University of Central Florida, USA
Ondrej Chum   Vision Recognition Group, Czech Technical
                                   University in Prague, Czech Republic
John Collomosse   Adobe Research & University of Surrey, UK
Camille Couprie   Facebook, France
David Crandall   Indiana University, USA
Daniel Cremers   Technical University of Munich, Germany
Marco Cristani   University of Verona, Italy
Canton Cristian   Facebook AI Research, USA
Dengxin Dai   ETH Zurich, Switzerland
Dima Damen   University of Bristol, UK
Kostas Daniilidis   University of Pennsylvania, USA
Trevor Darrell   University of California, Berkeley, USA
Andrew Davison   Imperial College London, UK
Tali Dekel   Weizmann Institute of Science, Israel
Alessio Del Bue   Istituto Italiano di Tecnologia, Italy
Weihong Deng   Beijing University of Posts and
                                   Telecommunications, China
Konstantinos Derpanis   Ryerson University, Canada
Carl Doersch   DeepMind, UK

Matthijs Douze	Facebook AI Research, USA
Mohamed Elhoseiny	King Abdullah University of Science and Technology, Saudi Arabia
Sergio Escalera	University of Barcelona, Spain
Yi Fang	New York University, USA
Ryan Farrell	Brigham Young University, USA
Alireza Fathi	Google, USA
Christoph Feichtenhofer	Facebook AI Research, USA
Basura Fernando	Agency for Science, Technology and Research (A*STAR), Singapore
Vittorio Ferrari	Google Research, Switzerland
Andrew W. Fitzgibbon	Graphcore, UK
David J. Fleet	University of Toronto, Canada
David Forsyth	University of Illinois at Urbana-Champaign, USA
David Fouhey	University of Michigan, USA
Katerina Fragkiadaki	Carnegie Mellon University, USA
Friedrich Fraundorfer	Graz University of Technology, Austria
Oren Freifeld	Ben-Gurion University, Israel
Thomas Funkhouser	Google Research & Princeton University, USA
Yasutaka Furukawa	Simon Fraser University, Canada
Fabio Galasso	Sapienza University of Rome, Italy
Jürgen Gall	University of Bonn, Germany
Chuang Gan	Massachusetts Institute of Technology, USA
Zhe Gan	Microsoft, USA
Animesh Garg	University of Toronto, Vector Institute, Nvidia, Canada
Efstratios Gavves	University of Amsterdam, The Netherlands
Peter Gehler	Amazon, Germany
Theo Gevers	University of Amsterdam, The Netherlands
Bernard Ghanem	King Abdullah University of Science and Technology, Saudi Arabia
Ross B. Girshick	Facebook AI Research, USA
Georgia Gkioxari	Facebook AI Research, USA
Albert Gordo	Facebook, USA
Stephen Gould	Australian National University, Australia
Venu Madhav Govindu	Indian Institute of Science, India
Kristen Grauman	Facebook AI Research & UT Austin, USA
Abhinav Gupta	Carnegie Mellon University & Facebook AI Research, USA
Mohit Gupta	University of Wisconsin-Madison, USA
Hu Han	Institute of Computing Technology, Chinese Academy of Sciences, China

Bohyung Han	Seoul National University, Korea
Tian Han	Stevens Institute of Technology, USA
Emily Hand	University of Nevada, Reno, USA
Bharath Hariharan	Cornell University, USA
Ran He	Institute of Automation, Chinese Academy of Sciences, China
Otmar Hilliges	ETH Zurich, Switzerland
Adrian Hilton	University of Surrey, UK
Minh Hoai	Stony Brook University, USA
Yedid Hoshen	Hebrew University of Jerusalem, Israel
Timothy Hospedales	University of Edinburgh, UK
Gang Hua	Wormpex AI Research, USA
Di Huang	Beihang University, China
Jing Huang	Facebook, USA
Jia-Bin Huang	Facebook, USA
Nathan Jacobs	Washington University in St. Louis, USA
C.V. Jawahar	International Institute of Information Technology, Hyderabad, India
Herve Jegou	Facebook AI Research, France
Neel Joshi	Microsoft Research, USA
Armand Joulin	Facebook AI Research, France
Frederic Jurie	University of Caen Normandie, France
Fredrik Kahl	Chalmers University of Technology, Sweden
Yannis Kalantidis	NAVER LABS Europe, France
Evangelos Kalogerakis	University of Massachusetts, Amherst, USA
Sing Bing Kang	Zillow Group, USA
Yosi Keller	Bar Ilan University, Israel
Margret Keuper	University of Mannheim, Germany
Tae-Kyun Kim	Imperial College London, UK
Benjamin Kimia	Brown University, USA
Alexander Kirillov	Facebook AI Research, USA
Kris Kitani	Carnegie Mellon University, USA
Iasonas Kokkinos	Snap Inc. & University College London, UK
Vladlen Koltun	Apple, USA
Nikos Komodakis	University of Crete, Crete
Piotr Koniusz	Australian National University, Australia
Philipp Kraehenbuehl	University of Texas at Austin, USA
Dilip Krishnan	Google, USA
Ajay Kumar	Hong Kong Polytechnic University, Hong Kong, China
Junseok Kwon	Chung-Ang University, Korea
Jean-Francois Lalonde	Université Laval, Canada

Ivan Laptev   Inria Paris, France
Laura Leal-Taixé   Technical University of Munich, Germany
Erik Learned-Miller   University of Massachusetts, Amherst, USA
Gim Hee Lee   National University of Singapore, Singapore
Seungyong Lee   Pohang University of Science and Technology,
                                       Korea
Zhen Lei   Institute of Automation, Chinese Academy of
                                       Sciences, China
Bastian Leibe   RWTH Aachen University, Germany
Hongdong Li   Australian National University, Australia
Fuxin Li   Oregon State University, USA
Bo Li   University of Illinois at Urbana-Champaign, USA
Yin Li   University of Wisconsin-Madison, USA
Ser-Nam Lim   Meta AI Research, USA
Joseph Lim   University of Southern California, USA
Stephen Lin   Microsoft Research Asia, China
Dahua Lin   The Chinese University of Hong Kong,
                                       Hong Kong, China
Si Liu   Beihang University, China
Xiaoming Liu   Michigan State University, USA
Ce Liu   Microsoft, USA
Zicheng Liu   Microsoft, USA
Yanxi Liu   Pennsylvania State University, USA
Feng Liu   Portland State University, USA
Yebin Liu   Tsinghua University, China
Chen Change Loy   Nanyang Technological University, Singapore
Huchuan Lu   Dalian University of Technology, China
Cewu Lu   Shanghai Jiao Tong University, China
Oisin Mac Aodha   University of Edinburgh, UK
Dhruv Mahajan   Facebook, USA
Subhransu Maji   University of Massachusetts, Amherst, USA
Atsuto Maki   KTH Royal Institute of Technology, Sweden
Arun Mallya   NVIDIA, USA
R. Manmatha   Amazon, USA
Iacopo Masi   Sapienza University of Rome, Italy
Dimitris N. Metaxas   Rutgers University, USA
Ajmal Mian   University of Western Australia, Australia
Christian Micheloni   University of Udine, Italy
Krystian Mikolajczyk   Imperial College London, UK
Anurag Mittal   Indian Institute of Technology, Madras, India
Philippos Mordohai   Stevens Institute of Technology, USA
Greg Mori   Simon Fraser University & Borealis AI, Canada

Vittorio Murino	Istituto Italiano di Tecnologia, Italy
P. J. Narayanan	International Institute of Information Technology, Hyderabad, India
Ram Nevatia	University of Southern California, USA
Natalia Neverova	Facebook AI Research, UK
Richard Newcombe	Facebook, USA
Cuong V. Nguyen	Florida International University, USA
Bingbing Ni	Shanghai Jiao Tong University, China
Juan Carlos Niebles	Salesforce & Stanford University, USA
Ko Nishino	Kyoto University, Japan
Jean-Marc Odobez	Idiap Research Institute, École polytechnique fédérale de Lausanne, Switzerland
Francesca Odone	University of Genova, Italy
Takayuki Okatani	Tohoku University & RIKEN Center for Advanced Intelligence Project, Japan
Manohar Paluri	Facebook, USA
Guan Pang	Facebook, USA
Maja Pantic	Imperial College London, UK
Sylvain Paris	Adobe Research, USA
Jaesik Park	Pohang University of Science and Technology, Korea
Hyun Soo Park	The University of Minnesota, USA
Omkar M. Parkhi	Facebook, USA
Deepak Pathak	Carnegie Mellon University, USA
Georgios Pavlakos	University of California, Berkeley, USA
Marcello Pelillo	University of Venice, Italy
Marc Pollefeys	ETH Zurich & Microsoft, Switzerland
Jean Ponce	Inria, France
Gerard Pons-Moll	University of Tübingen, Germany
Fatih Porikli	Qualcomm, USA
Victor Adrian Prisacariu	University of Oxford, UK
Petia Radeva	University of Barcelona, Spain
Ravi Ramamoorthi	University of California, San Diego, USA
Deva Ramanan	Carnegie Mellon University, USA
Vignesh Ramanathan	Facebook, USA
Nalini Ratha	State University of New York at Buffalo, USA
Tammy Riklin Raviv	Ben-Gurion University, Israel
Tobias Ritschel	University College London, UK
Emanuele Rodola	Sapienza University of Rome, Italy
Amit K. Roy-Chowdhury	University of California, Riverside, USA
Michael Rubinstein	Google, USA
Olga Russakovsky	Princeton University, USA

Mathieu Salzmann	École polytechnique fédérale de Lausanne, Switzerland
Dimitris Samaras	Stony Brook University, USA
Aswin Sankaranarayanan	Carnegie Mellon University, USA
Imari Sato	National Institute of Informatics, Japan
Yoichi Sato	University of Tokyo, Japan
Shin'ichi Satoh	National Institute of Informatics, Japan
Walter Scheirer	University of Notre Dame, USA
Bernt Schiele	Max Planck Institute for Informatics, Germany
Konrad Schindler	ETH Zurich, Switzerland
Cordelia Schmid	Inria & Google, France
Alexander Schwing	University of Illinois at Urbana-Champaign, USA
Nicu Sebe	University of Trento, Italy
Greg Shakhnarovich	Toyota Technological Institute at Chicago, USA
Eli Shechtman	Adobe Research, USA
Humphrey Shi	University of Oregon & University of Illinois at Urbana-Champaign & Picsart AI Research, USA
Jianbo Shi	University of Pennsylvania, USA
Roy Shilkrot	Massachusetts Institute of Technology, USA
Mike Zheng Shou	National University of Singapore, Singapore
Kaleem Siddiqi	McGill University, Canada
Richa Singh	Indian Institute of Technology Jodhpur, India
Greg Slabaugh	Queen Mary University of London, UK
Cees Snoek	University of Amsterdam, The Netherlands
Yale Song	Facebook AI Research, USA
Yi-Zhe Song	University of Surrey, UK
Bjorn Stenger	Rakuten Institute of Technology
Abby Stylianou	Saint Louis University, USA
Akihiro Sugimoto	National Institute of Informatics, Japan
Chen Sun	Brown University, USA
Deqing Sun	Google, USA
Kalyan Sunkavalli	Adobe Research, USA
Ying Tai	Tencent YouTu Lab, China
Ayellet Tal	Technion – Israel Institute of Technology, Israel
Ping Tan	Simon Fraser University, Canada
Siyu Tang	ETH Zurich, Switzerland
Chi-Keung Tang	Hong Kong University of Science and Technology, Hong Kong, China
Radu Timofte	University of Würzburg, Germany & ETH Zurich, Switzerland
Federico Tombari	Google, Switzerland & Technical University of Munich, Germany

James Tompkin	Brown University, USA
Lorenzo Torresani	Dartmouth College, USA
Alexander Toshev	Apple, USA
Du Tran	Facebook AI Research, USA
Anh T. Tran	VinAI, Vietnam
Zhuowen Tu	University of California, San Diego, USA
Georgios Tzimiropoulos	Queen Mary University of London, UK
Jasper Uijlings	Google Research, Switzerland
Jan C. van Gemert	Delft University of Technology, The Netherlands
Gul Varol	Ecole des Ponts ParisTech, France
Nuno Vasconcelos	University of California, San Diego, USA
Mayank Vatsa	Indian Institute of Technology Jodhpur, India
Ashok Veeraraghavan	Rice University, USA
Jakob Verbeek	Facebook AI Research, France
Carl Vondrick	Columbia University, USA
Ruiping Wang	Institute of Computing Technology, Chinese Academy of Sciences, China
Xinchao Wang	National University of Singapore, Singapore
Liwei Wang	The Chinese University of Hong Kong, Hong Kong, China
Chaohui Wang	Université Paris-Est, France
Xiaolong Wang	University of California, San Diego, USA
Christian Wolf	NAVER LABS Europe, France
Tao Xiang	University of Surrey, UK
Saining Xie	Facebook AI Research, USA
Cihang Xie	University of California, Santa Cruz, USA
Zeki Yalniz	Facebook, USA
Ming-Hsuan Yang	University of California, Merced, USA
Angela Yao	National University of Singapore, Singapore
Shaodi You	University of Amsterdam, The Netherlands
Stella X. Yu	University of California, Berkeley, USA
Junsong Yuan	State University of New York at Buffalo, USA
Stefanos Zafeiriou	Imperial College London, UK
Amir Zamir	École polytechnique fédérale de Lausanne, Switzerland
Lei Zhang	Alibaba & Hong Kong Polytechnic University, Hong Kong, China
Lei Zhang	International Digital Economy Academy (IDEA), China
Pengchuan Zhang	Meta AI, USA
Bolei Zhou	University of California, Los Angeles, USA
Yuke Zhu	University of Texas at Austin, USA

Todd Zickler   Harvard University, USA
Wangmeng Zuo   Harbin Institute of Technology, China

Technical Program Committee

Davide Abati
Soroush Abbasi
　Koohpayegani
Amos L. Abbott
Rameen Abdal
Rabab Abdelfattah
Sahar Abdelnabi
Hassan Abu Alhaija
Abulikemu Abuduweili
Ron Abutbul
Hanno Ackermann
Aikaterini Adam
Kamil Adamczewski
Ehsan Adeli
Vida Adeli
Donald Adjeroh
Arman Afrasiyabi
Akshay Agarwal
Sameer Agarwal
Abhinav Agarwalla
Vaibhav Aggarwal
Sara Aghajanzadeh
Susmit Agrawal
Antonio Agudo
Touqeer Ahmad
Sk Miraj Ahmed
Chaitanya Ahuja
Nilesh A. Ahuja
Abhishek Aich
Shubhra Aich
Noam Aigerman
Arash Akbarinia
Peri Akiva
Derya Akkaynak
Emre Aksan
Arjun R. Akula
Yuval Alaluf
Stephan Alaniz
Paul Albert
Cenek Albl

Filippo Aleotti
Konstantinos P.
　Alexandridis
Motasem Alfarra
Mohsen Ali
Thiemo Alldieck
Hadi Alzayer
Liang An
Shan An
Yi An
Zhulin An
Dongsheng An
Jie An
Xiang An
Saket Anand
Cosmin Ancuti
Juan Andrade-Cetto
Alexander Andreopoulos
Bjoern Andres
Jerone T. A. Andrews
Shivangi Aneja
Anelia Angelova
Dragomir Anguelov
Rushil Anirudh
Oron Anschel
Rao Muhammad Anwer
Djamila Aouada
Evlampios Apostolidis
Srikar Appalaraju
Nikita Araslanov
Andre Araujo
Eric Arazo
Dawit Mureja Argaw
Anurag Arnab
Aditya Arora
Chetan Arora
Sunpreet S. Arora
Alexey Artemov
Muhammad Asad
Kumar Ashutosh

Sinem Aslan
Vishal Asnani
Mahmoud Assran
Amir Atapour-Abarghouei
Nikos Athanasiou
Ali Athar
ShahRukh Athar
Sara Atito
Souhaib Attaiki
Matan Atzmon
Mathieu Aubry
Nicolas Audebert
Tristan T.
　Aumentado-Armstrong
Melinos Averkiou
Yannis Avrithis
Stephane Ayache
Mehmet Aygün
Seyed Mehdi
　Ayyoubzadeh
Hossein Azizpour
George Azzopardi
Mallikarjun B. R.
Yunhao Ba
Abhishek Badki
Seung-Hwan Bae
Seung-Hwan Baek
Seungryul Baek
Piyush Nitin Bagad
Shai Bagon
Gaetan Bahl
Shikhar Bahl
Sherwin Bahmani
Haoran Bai
Lei Bai
Jiawang Bai
Haoyue Bai
Jinbin Bai
Xiang Bai
Xuyang Bai

Yang Bai
Yuanchao Bai
Ziqian Bai
Sungyong Baik
Kevin Bailly
Max Bain
Federico Baldassarre
Wele Gedara Chaminda
    Bandara
Biplab Banerjee
Pratyay Banerjee
Sandipan Banerjee
Jihwan Bang
Antyanta Bangunharcana
Aayush Bansal
Ankan Bansal
Siddhant Bansal
Wentao Bao
Zhipeng Bao
Amir Bar
Manel Baradad Jurjo
Lorenzo Baraldi
Danny Barash
Daniel Barath
Connelly Barnes
Ioan Andrei Bârsan
Steven Basart
Dina Bashkirova
Chaim Baskin
Peyman Bateni
Anil Batra
Sebastiano Battiato
Ardhendu Behera
Harkirat Behl
Jens Behley
Vasileios Belagiannis
Boulbaba Ben Amor
Emanuel Ben Baruch
Abdessamad Ben Hamza
Gil Ben-Artzi
Assia Benbihi
Fabian Benitez-Quiroz
Guy Ben-Yosef
Philipp Benz
Alexander W. Bergman

Urs Bergmann
Jesus Bermudez-Cameo
Stefano Berretti
Gedas Bertasius
Zachary Bessinger
Petra Bevandić
Matthew Beveridge
Lucas Beyer
Yash Bhalgat
Suvaansh Bhambri
Samarth Bharadwaj
Gaurav Bharaj
Aparna Bharati
Bharat Lal Bhatnagar
Uttaran Bhattacharya
Apratim Bhattacharyya
Brojeshwar Bhowmick
Ankan Kumar Bhunia
Ayan Kumar Bhunia
Qi Bi
Sai Bi
Michael Bi Mi
Gui-Bin Bian
Jia-Wang Bian
Shaojun Bian
Pia Bideau
Mario Bijelic
Hakan Bilen
Guillaume-Alexandre
    Bilodeau
Alexander Binder
Tolga Birdal
Vighnesh N. Birodkar
Sandika Biswas
Andreas Blattmann
Janusz Bobulski
Giuseppe Boccignone
Vishnu Boddeti
Navaneeth Bodla
Moritz Böhle
Aleksei Bokhovkin
Sam Bond-Taylor
Vivek Boominathan
Shubhankar Borse
Mark Boss

Andrea Bottino
Adnane Boukhayma
Fadi Boutros
Nicolas C. Boutry
Richard S. Bowen
Ivaylo Boyadzhiev
Aidan Boyd
Yuri Boykov
Aljaz Bozic
Behzad Bozorgtabar
Eric Brachmann
Samarth Brahmbhatt
Gustav Bredell
Francois Bremond
Joel Brogan
Andrew Brown
Thomas Brox
Marcus A. Brubaker
Robert-Jan Bruintjes
Yuqi Bu
Anders G. Buch
Himanshu Buckchash
Mateusz Buda
Ignas Budvytis
José M. Buenaposada
Marcel C. Bühler
Tu Bui
Adrian Bulat
Hannah Bull
Evgeny Burnaev
Andrei Bursuc
Benjamin Busam
Sergey N. Buzykanov
Wonmin Byeon
Fabian Caba
Martin Cadik
Guanyu Cai
Minjie Cai
Qing Cai
Zhongang Cai
Qi Cai
Yancheng Cai
Shen Cai
Han Cai
Jiarui Cai

Bowen Cai
Mu Cai
Qin Cai
Ruojin Cai
Weidong Cai
Weiwei Cai
Yi Cai
Yujun Cai
Zhiping Cai
Akin Caliskan
Lilian Calvet
Baris Can Cam
Necati Cihan Camgoz
Tommaso Campari
Dylan Campbell
Ziang Cao
Ang Cao
Xu Cao
Zhiwen Cao
Shengcao Cao
Song Cao
Weipeng Cao
Xiangyong Cao
Xiaochun Cao
Yue Cao
Yunhao Cao
Zhangjie Cao
Jiale Cao
Yang Cao
Jiajiong Cao
Jie Cao
Jinkun Cao
Lele Cao
Yulong Cao
Zhiguo Cao
Chen Cao
Razvan Caramalau
Marlène Careil
Gustavo Carneiro
Joao Carreira
Dan Casas
Paola Cascante-Bonilla
Angela Castillo
Francisco M. Castro
Pedro Castro

Luca Cavalli
George J. Cazenavette
Oya Celiktutan
Hakan Cevikalp
Sri Harsha C. H.
Sungmin Cha
Geonho Cha
Menglei Chai
Lucy Chai
Yuning Chai
Zenghao Chai
Anirban Chakraborty
Deep Chakraborty
Rudrasis Chakraborty
Souradeep Chakraborty
Kelvin C. K. Chan
Chee Seng Chan
Paramanand Chandramouli
Arjun Chandrasekaran
Kenneth Chaney
Dongliang Chang
Huiwen Chang
Peng Chang
Xiaojun Chang
Jia-Ren Chang
Hyung Jin Chang
Hyun Sung Chang
Ju Yong Chang
Li-Jen Chang
Qi Chang
Wei-Yi Chang
Yi Chang
Nadine Chang
Hanqing Chao
Pradyumna Chari
Dibyadip Chatterjee
Chiranjoy Chattopadhyay
Siddhartha Chaudhuri
Zhengping Che
Gal Chechik
Lianggangxu Chen
Qi Alfred Chen
Brian Chen
Bor-Chun Chen
Bo-Hao Chen

Bohong Chen
Bin Chen
Ziliang Chen
Cheng Chen
Chen Chen
Chaofeng Chen
Xi Chen
Haoyu Chen
Xuanhong Chen
Wei Chen
Qiang Chen
Shi Chen
Xianyu Chen
Chang Chen
Changhuai Chen
Hao Chen
Jie Chen
Jianbo Chen
Jingjing Chen
Jun Chen
Kejiang Chen
Mingcai Chen
Nenglun Chen
Qifeng Chen
Ruoyu Chen
Shu-Yu Chen
Weidong Chen
Weijie Chen
Weikai Chen
Xiang Chen
Xiuyi Chen
Xingyu Chen
Yaofo Chen
Yueting Chen
Yu Chen
Yunjin Chen
Yuntao Chen
Yun Chen
Zhenfang Chen
Zhuangzhuang Chen
Chu-Song Chen
Xiangyu Chen
Zhuo Chen
Chaoqi Chen
Shizhe Chen

xxi

Organization

Xiaotong Chen
Xiaozhi Chen
Dian Chen
Defang Chen
Dingfan Chen
Ding-Jie Chen
Ee Heng Chen
Tao Chen
Yixin Chen
Wei-Ting Chen
Lin Chen
Guang Chen
Guangyi Chen
Guanying Chen
Guangyao Chen
Hwann-Tzong Chen
Junwen Chen
Jiacheng Chen
Jianxu Chen
Hui Chen
Kai Chen
Kan Chen
Kevin Chen
Kuan-Wen Chen
Weihua Chen
Zhang Chen
Liang-Chieh Chen
Lele Chen
Liang Chen
Fanglin Chen
Zehui Chen
Minghui Chen
Minghao Chen
Xiaokang Chen
Qian Chen
Jun-Cheng Chen
Qi Chen
Qingcai Chen
Richard J. Chen
Runnan Chen
Rui Chen
Shuo Chen
Sentao Chen
Shaoyu Chen
Shixing Chen

Shuai Chen
Shuya Chen
Sizhe Chen
Simin Chen
Shaoxiang Chen
Zitian Chen
Tianlong Chen
Tianshui Chen
Min-Hung Chen
Xiangning Chen
Xin Chen
Xinghao Chen
Xuejin Chen
Xu Chen
Xuxi Chen
Yunlu Chen
Yanbei Chen
Yuxiao Chen
Yun-Chun Chen
Yi-Ting Chen
Yi-Wen Chen
Yinbo Chen
Yiran Chen
Yuanhong Chen
Yubei Chen
Yuefeng Chen
Yuhua Chen
Yukang Chen
Zerui Chen
Zhaoyu Chen
Zhen Chen
Zhenyu Chen
Zhi Chen
Zhiwei Chen
Zhixiang Chen
Long Chen
Bowen Cheng
Jun Cheng
Yi Cheng
Jingchun Cheng
Lechao Cheng
Xi Cheng
Yuan Cheng
Ho Kei Cheng
Kevin Ho Man Cheng

Jiacheng Cheng
Kelvin B. Cheng
Li Cheng
Mengjun Cheng
Zhen Cheng
Qingrong Cheng
Tianheng Cheng
Harry Cheng
Yihua Cheng
Yu Cheng
Ziheng Cheng
Soon Yau Cheong
Anoop Cherian
Manuela Chessa
Zhixiang Chi
Naoki Chiba
Julian Chibane
Kashyap Chitta
Tai-Yin Chiu
Hsu-kuang Chiu
Wei-Chen Chiu
Sungmin Cho
Donghyeon Cho
Hyeon Cho
Yooshin Cho
Gyusang Cho
Jang Hyun Cho
Seungju Cho
Nam Ik Cho
Sunghyun Cho
Hanbyel Cho
Jaesung Choe
Jooyoung Choi
Chiho Choi
Changwoon Choi
Jongwon Choi
Myungsub Choi
Dooseop Choi
Jonghyun Choi
Jinwoo Choi
Jun Won Choi
Min-Kook Choi
Hongsuk Choi
Janghoon Choi
Yoon-Ho Choi

Yukyung Choi
Jaegul Choo
Ayush Chopra
Siddharth Choudhary
Subhabrata Choudhury
Vasileios Choutas
Ka-Ho Chow
Pinaki Nath Chowdhury
Sammy Christen
Anders Christensen
Grigorios Chrysos
Hang Chu
Wen-Hsuan Chu
Peng Chu
Qi Chu
Ruihang Chu
Wei-Ta Chu
Yung-Yu Chuang
Sanghyuk Chun
Se Young Chun
Antonio Cinà
Ramazan Gokberk Cinbis
Javier Civera
Albert Clapés
Ronald Clark
Brian S. Clipp
Felipe Codevilla
Daniel Coelho de Castro
Niv Cohen
Forrester Cole
Maxwell D. Collins
Robert T. Collins
Marc Comino Trinidad
Runmin Cong
Wenyan Cong
Maxime Cordy
Marcella Cornia
Enric Corona
Huseyin Coskun
Luca Cosmo
Dragos Costea
Davide Cozzolino
Arun C. S. Kumar
Aiyu Cui
Qiongjie Cui

Quan Cui
Shuhao Cui
Yiming Cui
Ying Cui
Zijun Cui
Jiali Cui
Jiequan Cui
Yawen Cui
Zhen Cui
Zhaopeng Cui
Jack Culpepper
Xiaodong Cun
Ross Cutler
Adam Czajka
Ali Dabouei
Konstantinos M. Dafnis
Manuel Dahnert
Tao Dai
Yuchao Dai
Bo Dai
Mengyu Dai
Hang Dai
Haixing Dai
Peng Dai
Pingyang Dai
Qi Dai
Qiyu Dai
Yutong Dai
Naser Damer
Zhiyuan Dang
Mohamed Daoudi
Ayan Das
Abir Das
Debasmit Das
Deepayan Das
Partha Das
Sagnik Das
Soumi Das
Srijan Das
Swagatam Das
Avijit Dasgupta
Jim Davis
Adrian K. Davison
Homa Davoudi
Laura Daza

Matthias De Lange
Shalini De Mello
Marco De Nadai
Christophe De
  Vleeschouwer
Alp Dener
Boyang Deng
Congyue Deng
Bailin Deng
Yong Deng
Ye Deng
Zhuo Deng
Zhijie Deng
Xiaoming Deng
Jiankang Deng
Jinhong Deng
Jingjing Deng
Liang-Jian Deng
Siqi Deng
Xiang Deng
Xueqing Deng
Zhongying Deng
Karan Desai
Jean-Emmanuel Deschaud
Aniket Anand Deshmukh
Neel Dey
Helisa Dhamo
Prithviraj Dhar
Amaya Dharmasiri
Yan Di
Xing Di
Ousmane A. Dia
Haiwen Diao
Xiaolei Diao
Gonçalo José Dias Pais
Abdallah Dib
Anastasios Dimou
Changxing Ding
Henghui Ding
Guodong Ding
Yaqing Ding
Shuangrui Ding
Yuhang Ding
Yikang Ding
Shouhong Ding

Haisong Ding
Hui Ding
Jiahao Ding
Jian Ding
Jian-Jiun Ding
Shuxiao Ding
Tianyu Ding
Wenhao Ding
Yuqi Ding
Yi Ding
Yuzhen Ding
Zhengming Ding
Tan Minh Dinh
Vu Dinh
Christos Diou
Mandar Dixit
Bao Gia Doan
Khoa D. Doan
Dzung Anh Doan
Debi Prosad Dogra
Nehal Doiphode
Chengdong Dong
Bowen Dong
Zhenxing Dong
Hang Dong
Xiaoyi Dong
Haoye Dong
Jiangxin Dong
Shichao Dong
Xuan Dong
Zhen Dong
Shuting Dong
Jing Dong
Li Dong
Ming Dong
Nanqing Dong
Qiulei Dong
Runpei Dong
Siyan Dong
Tian Dong
Wei Dong
Xiaomeng Dong
Xin Dong
Xingbo Dong
Yuan Dong

Samuel Dooley
Gianfranco Doretto
Michael Dorkenwald
Keval Doshi
Zhaopeng Dou
Xiaotian Dou
Hazel Doughty
Ahmad Droby
Iddo Drori
Jie Du
Yong Du
Dawei Du
Dong Du
Ruoyi Du
Yuntao Du
Xuefeng Du
Yilun Du
Yuming Du
Radhika Dua
Haodong Duan
Jiafei Duan
Kaiwen Duan
Peiqi Duan
Ye Duan
Haoran Duan
Jiali Duan
Amanda Duarte
Abhimanyu Dubey
Shiv Ram Dubey
Florian Dubost
Lukasz Dudziak
Shivam Duggal
Justin M. Dulay
Matteo Dunnhofer
Chi Nhan Duong
Thibaut Durand
Mihai Dusmanu
Ujjal Kr Dutta
Debidatta Dwibedi
Isht Dwivedi
Sai Kumar Dwivedi
Takeharu Eda
Mark Edmonds
Alexei A. Efros
Thibaud Ehret

Max Ehrlich
Mahsa Ehsanpour
Iván Eichhardt
Farshad Einabadi
Marvin Eisenberger
Hazim Kemal Ekenel
Mohamed El Banani
Ismail Elezi
Moshe Eliasof
Alaa El-Nouby
Ian Endres
Francis Engelmann
Deniz Engin
Chanho Eom
Dave Epstein
Maria C. Escobar
Victor A. Escorcia
Carlos Esteves
Sungmin Eum
Bernard J. E. Evans
Ivan Evtimov
Fevziye Irem Eyiokur
  Yaman
Matteo Fabbri
Sébastien Fabbro
Gabriele Facciolo
Masud Fahim
Bin Fan
Hehe Fan
Deng-Ping Fan
Aoxiang Fan
Chen-Chen Fan
Qi Fan
Zhaoxin Fan
Haoqi Fan
Heng Fan
Hongyi Fan
Linxi Fan
Baojie Fan
Jiayuan Fan
Lei Fan
Quanfu Fan
Yonghui Fan
Yingruo Fan
Zhiwen Fan

Zicong Fan
Sean Fanello
Jiansheng Fang
Chaowei Fang
Yuming Fang
Jianwu Fang
Jin Fang
Qi Fang
Shancheng Fang
Tian Fang
Xianyong Fang
Gongfan Fang
Zhen Fang
Hui Fang
Jiemin Fang
Le Fang
Pengfei Fang
Xiaolin Fang
Yuxin Fang
Zhaoyuan Fang
Ammarah Farooq
Azade Farshad
Zhengcong Fei
Michael Felsberg
Wei Feng
Chen Feng
Fan Feng
Andrew Feng
Xin Feng
Zheyun Feng
Ruicheng Feng
Mingtao Feng
Qianyu Feng
Shangbin Feng
Chun-Mei Feng
Zunlei Feng
Zhiyong Feng
Martin Fergie
Mustansar Fiaz
Marco Fiorucci
Michael Firman
Hamed Firooz
Volker Fischer
Corneliu O. Florea
Georgios Floros

Wolfgang Foerstner
Gianni Franchi
Jean-Sebastien Franco
Simone Frintrop
Anna Fruehstueck
Changhong Fu
Chaoyou Fu
Cheng-Yang Fu
Chi-Wing Fu
Deqing Fu
Huan Fu
Jun Fu
Kexue Fu
Ying Fu
Jianlong Fu
Jingjing Fu
Qichen Fu
Tsu-Jui Fu
Xueyang Fu
Yang Fu
Yanwei Fu
Yonggan Fu
Wolfgang Fuhl
Yasuhisa Fujii
Kent Fujiwara
Marco Fumero
Takuya Funatomi
Isabel Funke
Dario Fuoli
Antonino Furnari
Matheus A. Gadelha
Akshay Gadi Patil
Adrian Galdran
Guillermo Gallego
Silvano Galliani
Orazio Gallo
Leonardo Galteri
Matteo Gamba
Yiming Gan
Sujoy Ganguly
Harald Ganster
Boyan Gao
Changxin Gao
Daiheng Gao
Difei Gao

Chen Gao
Fei Gao
Lin Gao
Wei Gao
Yiming Gao
Junyu Gao
Guangyu Ryan Gao
Haichang Gao
Hongchang Gao
Jialin Gao
Jin Gao
Jun Gao
Katelyn Gao
Mingchen Gao
Mingfei Gao
Pan Gao
Shangqian Gao
Shanghua Gao
Xitong Gao
Yunhe Gao
Zhanning Gao
Elena Garces
Nuno Cruz Garcia
Noa Garcia
Guillermo
    Garcia-Hernando
Isha Garg
Rahul Garg
Sourav Garg
Quentin Garrido
Stefano Gasperini
Kent Gauen
Chandan Gautam
Shivam Gautam
Paul Gay
Chunjiang Ge
Shiming Ge
Wenhang Ge
Yanhao Ge
Zheng Ge
Songwei Ge
Weifeng Ge
Yixiao Ge
Yuying Ge
Shijie Geng

Zhengyang Geng
Kyle A. Genova
Georgios Georgakis
Markos Georgopoulos
Marcel Geppert
Shabnam Ghadar
Mina Ghadimi Atigh
Deepti Ghadiyaram
Maani Ghaffari Jadidi
Sedigh Ghamari
Zahra Gharaee
Michaël Gharbi
Golnaz Ghiasi
Reza Ghoddoosian
Soumya Suvra Ghosal
Adhiraj Ghosh
Arthita Ghosh
Pallabi Ghosh
Soumyadeep Ghosh
Andrew Gilbert
Igor Gilitschenski
Jhony H. Giraldo
Andreu Girbau Xalabarder
Rohit Girdhar
Sharath Girish
Xavier Giro-i-Nieto
Raja Giryes
Thomas Gittings
Nikolaos Gkanatsios
Ioannis Gkioulekas
Abhiram
  Gnanasambandam
Aurele T. Gnanha
Clement L. J. C. Godard
Arushi Goel
Vidit Goel
Shubham Goel
Zan Gojcic
Aaron K. Gokaslan
Tejas Gokhale
S. Alireza Golestaneh
Thiago L. Gomes
Nuno Goncalves
Boqing Gong
Chen Gong

Yuanhao Gong
Guoqiang Gong
Jingyu Gong
Rui Gong
Yu Gong
Mingming Gong
Neil Zhenqiang Gong
Xun Gong
Yunye Gong
Yihong Gong
Cristina I. González
Nithin Gopalakrishnan
  Nair
Gaurav Goswami
Jianping Gou
Shreyank N. Gowda
Ankit Goyal
Helmut Grabner
Patrick L. Grady
Ben Graham
Eric Granger
Douglas R. Gray
Matej Grcić
David Griffiths
Jinjin Gu
Yun Gu
Shuyang Gu
Jianyang Gu
Fuqiang Gu
Jiatao Gu
Jindong Gu
Jiaqi Gu
Jinwei Gu
Jiaxin Gu
Geonmo Gu
Xiao Gu
Xinqian Gu
Xiuye Gu
Yuming Gu
Zhangxuan Gu
Dayan Guan
Junfeng Guan
Qingji Guan
Tianrui Guan
Shanyan Guan

Denis A. Gudovskiy
Ricardo Guerrero
Pierre-Louis Guhur
Jie Gui
Liangyan Gui
Liangke Gui
Benoit Guillard
Erhan Gundogdu
Manuel Günther
Jingcai Guo
Yuanfang Guo
Junfeng Guo
Chenqi Guo
Dan Guo
Hongji Guo
Jia Guo
Jie Guo
Minghao Guo
Shi Guo
Yanhui Guo
Yangyang Guo
Yuan-Chen Guo
Yilu Guo
Yiluan Guo
Yong Guo
Guangyu Guo
Haiyun Guo
Jinyang Guo
Jianyuan Guo
Pengsheng Guo
Pengfei Guo
Shuxuan Guo
Song Guo
Tianyu Guo
Qing Guo
Qiushan Guo
Wen Guo
Xiefan Guo
Xiaohu Guo
Xiaoqing Guo
Yufei Guo
Yuhui Guo
Yuliang Guo
Yunhui Guo
Yanwen Guo

Akshita Gupta
Ankush Gupta
Kamal Gupta
Kartik Gupta
Ritwik Gupta
Rohit Gupta
Siddharth Gururani
Fredrik K. Gustafsson
Abner Guzman Rivera
Vladimir Guzov
Matthew A. Gwilliam
Jung-Woo Ha
Marc Habermann
Isma Hadji
Christian Haene
Martin Hahner
Levente Hajder
Alexandros Haliassos
Emanuela Haller
Bumsub Ham
Abdullah J. Hamdi
Shreyas Hampali
Dongyoon Han
Chunrui Han
Dong-Jun Han
Dong-Sig Han
Guangxing Han
Zhizhong Han
Ruize Han
Jiaming Han
Jin Han
Ligong Han
Xian-Hua Han
Xiaoguang Han
Yizeng Han
Zhi Han
Zhenjun Han
Zhongyi Han
Jungong Han
Junlin Han
Kai Han
Kun Han
Sungwon Han
Songfang Han
Wei Han

Xiao Han
Xintong Han
Xinzhe Han
Yahong Han
Yan Han
Zongbo Han
Nicolai Hani
Rana Hanocka
Niklas Hanselmann
Nicklas A. Hansen
Hong Hanyu
Fusheng Hao
Yanbin Hao
Shijie Hao
Udith Haputhanthri
Mehrtash Harandi
Josh Harguess
Adam Harley
David M. Hart
Atsushi Hashimoto
Ali Hassani
Mohammed Hassanin
Yana Hasson
Joakim Bruslund Haurum
Bo He
Kun He
Chen He
Xin He
Fazhi He
Gaoqi He
Hao He
Haoyu He
Jiangpeng He
Hongliang He
Qian He
Xiangteng He
Xuming He
Yannan He
Yuhang He
Yang He
Xiangyu He
Nanjun He
Pan He
Sen He
Shengfeng He

Songtao He
Tao He
Tong He
Wei He
Xuehai He
Xiaoxiao He
Ying He
Yisheng He
Ziwen He
Peter Hedman
Felix Heide
Yacov Hel-Or
Paul Henderson
Philipp Henzler
Byeongho Heo
Jae-Pil Heo
Miran Heo
Sachini A. Herath
Stephane Herbin
Pedro Hermosilla Casajus
Monica Hernandez
Charles Herrmann
Roei Herzig
Mauricio Hess-Flores
Carlos Hinojosa
Tobias Hinz
Tsubasa Hirakawa
Chih-Hui Ho
Lam Si Tung Ho
Jennifer Hobbs
Derek Hoiem
Yannick Hold-Geoffroy
Aleksander Holynski
Cheeun Hong
Fa-Ting Hong
Hanbin Hong
Guan Zhe Hong
Danfeng Hong
Lanqing Hong
Xiaopeng Hong
Xin Hong
Jie Hong
Seungbum Hong
Cheng-Yao Hong
Seunghoon Hong

Yi Hong
Yuan Hong
Yuchen Hong
Anthony Hoogs
Maxwell C. Horton
Kazuhiro Hotta
Qibin Hou
Tingbo Hou
Junhui Hou
Ji Hou
Qiqi Hou
Rui Hou
Ruibing Hou
Zhi Hou
Henry Howard-Jenkins
Lukas Hoyer
Wei-Lin Hsiao
Chiou-Ting Hsu
Anthony Hu
Brian Hu
Yusong Hu
Hexiang Hu
Haoji Hu
Di Hu
Hengtong Hu
Haigen Hu
Lianyu Hu
Hanzhe Hu
Jie Hu
Junlin Hu
Shizhe Hu
Jian Hu
Zhiming Hu
Juhua Hu
Peng Hu
Ping Hu
Ronghang Hu
MengShun Hu
Tao Hu
Vincent Tao Hu
Xiaoling Hu
Xinting Hu
Xiaolin Hu
Xuefeng Hu
Xiaowei Hu

Yang Hu
Yueyu Hu
Zeyu Hu
Zhongyun Hu
Binh-Son Hua
Guoliang Hua
Yi Hua
Linzhi Huang
Qiusheng Huang
Bo Huang
Chen Huang
Hsin-Ping Huang
Ye Huang
Shuangping Huang
Zeng Huang
Buzhen Huang
Cong Huang
Heng Huang
Hao Huang
Qidong Huang
Huaibo Huang
Chaoqin Huang
Feihu Huang
Jiahui Huang
Jingjia Huang
Kun Huang
Lei Huang
Sheng Huang
Shuaiyi Huang
Siyu Huang
Xiaoshui Huang
Xiaoyang Huang
Yan Huang
Yihao Huang
Ying Huang
Ziling Huang
Xiaoke Huang
Yifei Huang
Haiyang Huang
Zhewei Huang
Jin Huang
Haibin Huang
Jiaxing Huang
Junjie Huang
Keli Huang

Lang Huang
Lin Huang
Luojie Huang
Mingzhen Huang
Shijia Huang
Shengyu Huang
Siyuan Huang
He Huang
Xiuyu Huang
Lianghua Huang
Yue Huang
Yaping Huang
Yuge Huang
Zehao Huang
Zeyi Huang
Zhiqi Huang
Zhongzhan Huang
Zilong Huang
Ziyuan Huang
Tianrui Hui
Zhuo Hui
Le Hui
Jing Huo
Junhwa Hur
Shehzeen S. Hussain
Chuong Minh Huynh
Seunghyun Hwang
Jaehui Hwang
Jyh-Jing Hwang
Sukjun Hwang
Soonmin Hwang
Wonjun Hwang
Rakib Hyder
Sangeek Hyun
Sarah Ibrahimi
Tomoki Ichikawa
Yerlan Idelbayev
A. S. M. Iftekhar
Masaaki Iiyama
Satoshi Ikehata
Sunghoon Im
Atul N. Ingle
Eldar Insafutdinov
Yani A. Ioannou
Radu Tudor Ionescu

Umar Iqbal
Go Irie
Muhammad Zubair Irshad
Ahmet Iscen
Berivan Isik
Ashraful Islam
Md Amirul Islam
Syed Islam
Mariko Isogawa
Vamsi Krishna K. Ithapu
Boris Ivanovic
Darshan Iyer
Sarah Jabbour
Ayush Jain
Nishant Jain
Samyak Jain
Vidit Jain
Vineet Jain
Priyank Jaini
Tomas Jakab
Mohammad A. A. K.
  Jalwana
Muhammad Abdullah
  Jamal
Hadi Jamali-Rad
Stuart James
Varun Jampani
Young Kyun Jang
YeongJun Jang
Yunseok Jang
Ronnachai Jaroensri
Bhavan Jasani
Krishna Murthy
  Jatavallabhula
Mojan Javaheripi
Syed A. Javed
Guillaume Jeanneret
Pranav Jeevan
Herve Jegou
Rohit Jena
Tomas Jenicek
Porter Jenkins
Simon Jenni
Hae-Gon Jeon
Sangryul Jeon

Boseung Jeong
Yoonwoo Jeong
Seong-Gyun Jeong
Jisoo Jeong
Allan D. Jepson
Ankit Jha
Sumit K. Jha
I-Hong Jhuo
Ge-Peng Ji
Chaonan Ji
Deyi Ji
Jingwei Ji
Wei Ji
Zhong Ji
Jiayi Ji
Pengliang Ji
Hui Ji
Mingi Ji
Xiaopeng Ji
Yuzhu Ji
Baoxiong Jia
Songhao Jia
Dan Jia
Shan Jia
Xiaojun Jia
Xiuyi Jia
Xu Jia
Menglin Jia
Wenqi Jia
Boyuan Jiang
Wenhao Jiang
Huaizu Jiang
Hanwen Jiang
Haiyong Jiang
Hao Jiang
Huajie Jiang
Huiqin Jiang
Haojun Jiang
Haobo Jiang
Junjun Jiang
Xingyu Jiang
Yangbangyan Jiang
Yu Jiang
Jianmin Jiang
Jiaxi Jiang

Jing Jiang
Kui Jiang
Li Jiang
Liming Jiang
Chiyu Jiang
Meirui Jiang
Chen Jiang
Peng Jiang
Tai-Xiang Jiang
Wen Jiang
Xinyang Jiang
Yifan Jiang
Yuming Jiang
Yingying Jiang
Zeren Jiang
ZhengKai Jiang
Zhenyu Jiang
Shuming Jiao
Jianbo Jiao
Licheng Jiao
Dongkwon Jin
Yeying Jin
Cheng Jin
Linyi Jin
Qing Jin
Taisong Jin
Xiao Jin
Xin Jin
Sheng Jin
Kyong Hwan Jin
Ruibing Jin
SouYoung Jin
Yueming Jin
Chenchen Jing
Longlong Jing
Taotao Jing
Yongcheng Jing
Younghyun Jo
Joakim Johnander
Jeff Johnson
Michael J. Jones
R. Kenny Jones
Rico Jonschkowski
Ameya Joshi
Sunghun Joung

Felix Juefei-Xu
Claudio R. Jung
Steffen Jung
Hari Chandana K.
Rahul Vigneswaran K.
Prajwal K. R.
Abhishek Kadian
Jhony Kaesemodel Pontes
Kumara Kahatapitiya
Anmol Kalia
Sinan Kalkan
Tarun Kalluri
Jaewon Kam
Sandesh Kamath
Meina Kan
Menelaos Kanakis
Takuhiro Kaneko
Di Kang
Guoliang Kang
Hao Kang
Jaeyeon Kang
Kyoungkook Kang
Li-Wei Kang
MinGuk Kang
Suk-Ju Kang
Zhao Kang
Yash Mukund Kant
Yueying Kao
Aupendu Kar
Konstantinos Karantzalos
Sezer Karaoglu
Navid Kardan
Sanjay Kariyappa
Leonid Karlinsky
Animesh Karnewar
Shyamgopal Karthik
Hirak J. Kashyap
Marc A. Kastner
Hirokatsu Kataoka
Angelos Katharopoulos
Hiroharu Kato
Kai Katsumata
Manuel Kaufmann
Chaitanya Kaul
Prakhar Kaushik

Yuki Kawana
Lei Ke
Lipeng Ke
Tsung-Wei Ke
Wei Ke
Petr Kellnhofer
Aniruddha Kembhavi
John Kender
Corentin Kervadec
Leonid Keselman
Daniel Keysers
Nima Khademi Kalantari
Taras Khakhulin
Samir Khaki
Muhammad Haris Khan
Qadeer Khan
Salman Khan
Subash Khanal
Vaishnavi M. Khindkar
Rawal Khirodkar
Saeed Khorram
Pirazh Khorramshahi
Kourosh Khoshelham
Ansh Khurana
Benjamin Kiefer
Jae Myung Kim
Junho Kim
Boah Kim
Hyeonseong Kim
Dong-Jin Kim
Dongwan Kim
Donghyun Kim
Doyeon Kim
Yonghyun Kim
Hyung-Il Kim
Hyunwoo Kim
Hyeongwoo Kim
Hyo Jin Kim
Hyunwoo J. Kim
Taehoon Kim
Jaeha Kim
Jiwon Kim
Jung Uk Kim
Kangyeol Kim
Eunji Kim

Daeha Kim
Dongwon Kim
Kunhee Kim
Kyungmin Kim
Junsik Kim
Min H. Kim
Namil Kim
Kookhoi Kim
Sanghyun Kim
Seongyeop Kim
Seungryong Kim
Saehoon Kim
Euyoung Kim
Guisik Kim
Sungyeon Kim
Sunnie S. Y. Kim
Taehun Kim
Tae Oh Kim
Won Hwa Kim
Seungwook Kim
YoungBin Kim
Youngeun Kim
Akisato Kimura
Furkan Osman Kınlı
Zsolt Kira
Hedvig Kjellström
Florian Kleber
Jan P. Klopp
Florian Kluger
Laurent Kneip
Byungsoo Ko
Muhammed Kocabas
A. Sophia Koepke
Kevin Koeser
Nick Kolkin
Nikos Kolotouros
Wai-Kin Adams Kong
Deying Kong
Caihua Kong
Youyong Kong
Shuyu Kong
Shu Kong
Tao Kong
Yajing Kong
Yu Kong

Zishang Kong
Theodora Kontogianni
Anton S. Konushin
Julian F. P. Kooij
Bruno Korbar
Giorgos Kordopatis-Zilos
Jari Korhonen
Adam Kortylewski
Denis Korzhenkov
Divya Kothandaraman
Suraj Kothawade
Iuliia Kotseruba
Satwik Kottur
Shashank Kotyan
Alexandros Kouris
Petros Koutras
Anna Kreshuk
Ranjay Krishna
Dilip Krishnan
Andrey Kuehlkamp
Hilde Kuehne
Jason Kuen
David Kügler
Arjan Kuijper
Anna Kukleva
Sumith Kulal
Viveka Kulharia
Akshay R. Kulkarni
Nilesh Kulkarni
Dominik Kulon
Abhinav Kumar
Akash Kumar
Suryansh Kumar
B. V. K. Vijaya Kumar
Pulkit Kumar
Ratnesh Kumar
Sateesh Kumar
Satish Kumar
Vijay Kumar B. G.
Nupur Kumari
Sudhakar Kumawat
Jogendra Nath Kundu
Hsien-Kai Kuo
Meng-Yu Jennifer Kuo
Vinod Kumar Kurmi

Yusuke Kurose
Keerthy Kusumam
Alina Kuznetsova
Henry Kvinge
Ho Man Kwan
Hyeokjun Kweon
Heeseung Kwon
Gihyun Kwon
Myung-Joon Kwon
Taesung Kwon
YoungJoong Kwon
Christos Kyrkou
Jorma Laaksonen
Yann Labbe
Zorah Laehner
Florent Lafarge
Hamid Laga
Manuel Lagunas
Shenqi Lai
Jian-Huang Lai
Zihang Lai
Mohamed I. Lakhal
Mohit Lamba
Meng Lan
Loic Landrieu
Zhiqiang Lang
Natalie Lang
Dong Lao
Yizhen Lao
Yingjie Lao
Issam Hadj Laradji
Gustav Larsson
Viktor Larsson
Zakaria Laskar
Stéphane Lathuilière
Chun Pong Lau
Rynson W. H. Lau
Hei Law
Justin Lazarow
Verica Lazova
Eric-Tuan Le
Hieu Le
Trung-Nghia Le
Mathias Lechner
Byeong-Uk Lee

Chen-Yu Lee
Che-Rung Lee
Chul Lee
Hong Joo Lee
Dongsoo Lee
Jiyoung Lee
Eugene Eu Tzuan Lee
Daeun Lee
Saehyung Lee
Jewook Lee
Hyungtae Lee
Hyunmin Lee
Jungbeom Lee
Joon-Young Lee
Jong-Seok Lee
Joonseok Lee
Junha Lee
Kibok Lee
Byung-Kwan Lee
Jangwon Lee
Jinho Lee
Jongmin Lee
Seunghyun Lee
Sohyun Lee
Minsik Lee
Dogyoon Lee
Seungmin Lee
Min Jun Lee
Sangho Lee
Sangmin Lee
Seungeun Lee
Seon-Ho Lee
Sungmin Lee
Sungho Lee
Sangyoun Lee
Vincent C. S. S. Lee
Jaeseong Lee
Yong Jae Lee
Chenyang Lei
Chenyi Lei
Jiahui Lei
Xinyu Lei
Yinjie Lei
Jiaxu Leng
Luziwei Leng

Jan E. Lenssen
Vincent Lepetit
Thomas Leung
María Leyva-Vallina
Xin Li
Yikang Li
Baoxin Li
Bin Li
Bing Li
Bowen Li
Changlin Li
Chao Li
Chongyi Li
Guanyue Li
Shuai Li
Jin Li
Dingquan Li
Dongxu Li
Yiting Li
Gang Li
Dian Li
Guohao Li
Haoang Li
Haoliang Li
Haoran Li
Hengduo Li
Huafeng Li
Xiaoming Li
Hanao Li
Hongwei Li
Ziqiang Li
Jisheng Li
Jiacheng Li
Jia Li
Jiachen Li
Jiahao Li
Jianwei Li
Jiazhi Li
Jie Li
Jing Li
Jingjing Li
Jingtao Li
Jun Li
Junxuan Li
Kai Li

Kailin Li
Kenneth Li
Kun Li
Kunpeng Li
Aoxue Li
Chenglong Li
Chenglin Li
Changsheng Li
Zhichao Li
Qiang Li
Yanyu Li
Zuoyue Li
Xiang Li
Xuelong Li
Fangda Li
Ailin Li
Liang Li
Chun-Guang Li
Daiqing Li
Dong Li
Guanbin Li
Guorong Li
Haifeng Li
Jianan Li
Jianing Li
Jiaxin Li
Ke Li
Lei Li
Lincheng Li
Liulei Li
Lujun Li
Linjie Li
Lin Li
Pengyu Li
Ping Li
Qiufu Li
Qingyong Li
Rui Li
Siyuan Li
Wei Li
Wenbin Li
Xiangyang Li
Xinyu Li
Xiujun Li
Xiu Li

Xu Li
Ya-Li Li
Yao Li
Yongjie Li
Yijun Li
Yiming Li
Yuezun Li
Yu Li
Yunheng Li
Yuqi Li
Zhe Li
Zeming Li
Zhen Li
Zhengqin Li
Zhimin Li
Jiefeng Li
Jinpeng Li
Chengze Li
Jianwu Li
Lerenhan Li
Shan Li
Suichan Li
Xiangtai Li
Yanjie Li
Yandong Li
Zhuoling Li
Zhenqiang Li
Manyi Li
Maosen Li
Ji Li
Minjun Li
Mingrui Li
Mengtian Li
Junyi Li
Nianyi Li
Bo Li
Xiao Li
Peihua Li
Peike Li
Peizhao Li
Peiliang Li
Qi Li
Ren Li
Runze Li
Shile Li

Sheng Li
Shigang Li
Shiyu Li
Shuang Li
Shasha Li
Shichao Li
Tianye Li
Yuexiang Li
Wei-Hong Li
Wanhua Li
Weihao Li
Weiming Li
Weixin Li
Wenbo Li
Wenshuo Li
Weijian Li
Yunan Li
Xirong Li
Xianhang Li
Xiaoyu Li
Xueqian Li
Xuanlin Li
Xianzhi Li
Yunqiang Li
Yanjing Li
Yansheng Li
Yawei Li
Yi Li
Yong Li
Yong-Lu Li
Yuhang Li
Yu-Jhe Li
Yuxi Li
Yunsheng Li
Yanwei Li
Zechao Li
Zejian Li
Zeju Li
Zekun Li
Zhaowen Li
Zheng Li
Zhenyu Li
Zhiheng Li
Zhi Li
Zhong Li

Zhuowei Li
Zhuowan Li
Zhuohang Li
Zizhang Li
Chen Li
Yuan-Fang Li
Dongze Lian
Xiaochen Lian
Zhouhui Lian
Long Lian
Qing Lian
Jin Lianbao
Jinxiu S. Liang
Dingkang Liang
Jiahao Liang
Jianming Liang
Jingyun Liang
Kevin J. Liang
Kaizhao Liang
Chen Liang
Jie Liang
Senwei Liang
Ding Liang
Jiajun Liang
Jian Liang
Kongming Liang
Siyuan Liang
Yuanzhi Liang
Zhengfa Liang
Mingfu Liang
Xiaodan Liang
Xuefeng Liang
Yuxuan Liang
Kang Liao
Liang Liao
Hong-Yuan Mark Liao
Wentong Liao
Haofu Liao
Yue Liao
Minghui Liao
Shengcai Liao
Ting-Hsuan Liao
Xin Liao
Yinghong Liao
Teck Yian Lim

Che-Tsung Lin
Chung-Ching Lin
Chen-Hsuan Lin
Cheng Lin
Chuming Lin
Chunyu Lin
Dahua Lin
Wei Lin
Zheng Lin
Huaijia Lin
Jason Lin
Jierui Lin
Jiaying Lin
Jie Lin
Kai-En Lin
Kevin Lin
Guangfeng Lin
Jiehong Lin
Feng Lin
Hang Lin
Kwan-Yee Lin
Ke Lin
Luojun Lin
Qinghong Lin
Xiangbo Lin
Yi Lin
Zudi Lin
Shijie Lin
Yiqun Lin
Tzu-Heng Lin
Ming Lin
Shaohui Lin
SongNan Lin
Ji Lin
Tsung-Yu Lin
Xudong Lin
Yancong Lin
Yen-Chen Lin
Yiming Lin
Yuewei Lin
Zhiqiu Lin
Zinan Lin
Zhe Lin
David B. Lindell
Zhixin Ling

Zhan Ling
Alexander Liniger
Venice Erin B. Liong
Joey Litalien
Or Litany
Roee Litman
Ron Litman
Jim Little
Dor Litvak
Shaoteng Liu
Shuaicheng Liu
Andrew Liu
Xian Liu
Shaohui Liu
Bei Liu
Bo Liu
Yong Liu
Ming Liu
Yanbin Liu
Chenxi Liu
Daqi Liu
Di Liu
Difan Liu
Dong Liu
Dongfang Liu
Daizong Liu
Xiao Liu
Fangyi Liu
Fengbei Liu
Fenglin Liu
Bin Liu
Yuang Liu
Ao Liu
Hong Liu
Hongfu Liu
Huidong Liu
Ziyi Liu
Feng Liu
Hao Liu
Jie Liu
Jialun Liu
Jiang Liu
Jing Liu
Jingya Liu
Jiaming Liu

Jun Liu
Juncheng Liu
Jiawei Liu
Hongyu Liu
Chuanbin Liu
Haotian Liu
Lingqiao Liu
Chang Liu
Han Liu
Liu Liu
Min Liu
Yingqi Liu
Aishan Liu
Bingyu Liu
Benlin Liu
Boxiao Liu
Chenchen Liu
Chuanjian Liu
Daqing Liu
Huan Liu
Haozhe Liu
Jiaheng Liu
Wei Liu
Jingzhou Liu
Jiyuan Liu
Lingbo Liu
Nian Liu
Peiye Liu
Qiankun Liu
Shenglan Liu
Shilong Liu
Wen Liu
Wenyu Liu
Weifeng Liu
Wu Liu
Xiaolong Liu
Yang Liu
Yanwei Liu
Yingcheng Liu
Yongfei Liu
Yihao Liu
Yu Liu
Yunze Liu
Ze Liu
Zhenhua Liu

Zhenguang Liu
Lin Liu
Lihao Liu
Pengju Liu
Xinhai Liu
Yunfei Liu
Meng Liu
Minghua Liu
Mingyuan Liu
Miao Liu
Peirong Liu
Ping Liu
Qingjie Liu
Ruoshi Liu
Risheng Liu
Songtao Liu
Xing Liu
Shikun Liu
Shuming Liu
Sheng Liu
Songhua Liu
Tongliang Liu
Weibo Liu
Weide Liu
Weizhe Liu
Wenxi Liu
Weiyang Liu
Xin Liu
Xiaobin Liu
Xudong Liu
Xiaoyi Liu
Xihui Liu
Xinchen Liu
Xingtong Liu
Xinpeng Liu
Xinyu Liu
Xianpeng Liu
Xu Liu
Xingyu Liu
Yongtuo Liu
Yahui Liu
Yangxin Liu
Yaoyao Liu
Yaojie Liu
Yuliang Liu

Yongcheng Liu
Yuan Liu
Yufan Liu
Yu-Lun Liu
Yun Liu
Yunfan Liu
Yuanzhong Liu
Zhuoran Liu
Zhen Liu
Zheng Liu
Zhijian Liu
Zhisong Liu
Ziquan Liu
Ziyu Liu
Zhihua Liu
Zechun Liu
Zhaoyang Liu
Zhengzhe Liu
Stephan Liwicki
Shao-Yuan Lo
Sylvain Lobry
Suhas Lohit
Vishnu Suresh Lokhande
Vincenzo Lomonaco
Chengjiang Long
Guodong Long
Fuchen Long
Shangbang Long
Yang Long
Zijun Long
Vasco Lopes
Antonio M. Lopez
Roberto Javier
   Lopez-Sastre
Tobias Lorenz
Javier Lorenzo-Navarro
Yujing Lou
Qian Lou
Xiankai Lu
Changsheng Lu
Huimin Lu
Yongxi Lu
Hao Lu
Hong Lu
Jiasen Lu

Juwei Lu
Fan Lu
Guangming Lu
Jiwen Lu
Shun Lu
Tao Lu
Xiaonan Lu
Yang Lu
Yao Lu
Yongchun Lu
Zhiwu Lu
Cheng Lu
Liying Lu
Guo Lu
Xuequan Lu
Yanye Lu
Yantao Lu
Yuhang Lu
Fujun Luan
Jonathon Luiten
Jovita Lukasik
Alan Lukezic
Jonathan Samuel Lumentut
Mayank Lunayach
Ao Luo
Canjie Luo
Chong Luo
Xu Luo
Grace Luo
Jun Luo
Katie Z. Luo
Tao Luo
Cheng Luo
Fangzhou Luo
Gen Luo
Lei Luo
Sihui Luo
Weixin Luo
Yan Luo
Xiaoyan Luo
Yong Luo
Yadan Luo
Hao Luo
Ruotian Luo
Mi Luo

Tiange Luo
Wenjie Luo
Wenhan Luo
Xiao Luo
Zhiming Luo
Zhipeng Luo
Zhengyi Luo
Diogo C. Luvizon
Zhaoyang Lv
Gengyu Lyu
Lingjuan Lyu
Jun Lyu
Yuanyuan Lyu
Youwei Lyu
Yueming Lyu
Bingpeng Ma
Chao Ma
Chongyang Ma
Congbo Ma
Chih-Yao Ma
Fan Ma
Lin Ma
Haoyu Ma
Hengbo Ma
Jianqi Ma
Jiawei Ma
Jiayi Ma
Kede Ma
Kai Ma
Lingni Ma
Lei Ma
Xu Ma
Ning Ma
Benteng Ma
Cheng Ma
Andy J. Ma
Long Ma
Zhanyu Ma
Zhiheng Ma
Qianli Ma
Shiqiang Ma
Sizhuo Ma
Shiqing Ma
Xiaolong Ma
Xinzhu Ma

Gautam B. Machiraju
Spandan Madan
Mathew Magimai-Doss
Luca Magri
Behrooz Mahasseni
Upal Mahbub
Siddharth Mahendran
Paridhi Maheshwari
Rishabh Maheshwary
Mohammed Mahmoud
Shishira R. R. Maiya
Sylwia Majchrowska
Arjun Majumdar
Puspita Majumdar
Orchid Majumder
Sagnik Majumder
Ilya Makarov
Farkhod F. Makhmudkhujaev
Yasushi Makihara
Ankur Mali
Mateusz Malinowski
Utkarsh Mall
Srikanth Malla
Clement Mallet
Dimitrios Mallis
Yunze Man
Dipu Manandhar
Massimiliano Mancini
Murari Mandal
Raunak Manekar
Karttikeya Mangalam
Puneet Mangla
Fabian Manhardt
Sivabalan Manivasagam
Fahim Mannan
Chengzhi Mao
Hanzi Mao
Jiayuan Mao
Junhua Mao
Zhiyuan Mao
Jiageng Mao
Yunyao Mao
Zhendong Mao
Alberto Marchisio

Diego Marcos
Riccardo Marin
Aram Markosyan
Renaud Marlet
Ricardo Marques
Miquel Martí i Rabadán
Diego Martin Arroyo
Niki Martinel
Brais Martinez
Julieta Martinez
Marc Masana
Tomohiro Mashita
Timothée Masquelier
Minesh Mathew
Tetsu Matsukawa
Marwan Mattar
Bruce A. Maxwell
Christoph Mayer
Mantas Mazeika
Pratik Mazumder
Scott McCloskey
Steven McDonagh
Ishit Mehta
Jie Mei
Kangfu Mei
Jieru Mei
Xiaoguang Mei
Givi Meishvili
Luke Melas-Kyriazi
Iaroslav Melekhov
Andres Mendez-Vazquez
Heydi Mendez-Vazquez
Matias Mendieta
Ricardo A. Mendoza-León
Chenlin Meng
Depu Meng
Rang Meng
Zibo Meng
Qingjie Meng
Qier Meng
Yanda Meng
Zihang Meng
Thomas Mensink
Fabian Mentzer
Christopher Metzler

Gregory P. Meyer
Vasileios Mezaris
Liang Mi
Lu Mi
Bo Miao
Changtao Miao
Zichen Miao
Qiguang Miao
Xin Miao
Zhongqi Miao
Frank Michel
Simone Milani
Ben Mildenhall
Roy V. Miles
Juhong Min
Kyle Min
Hyun-Seok Min
Weiqing Min
Yuecong Min
Zhixiang Min
Qi Ming
David Minnen
Aymen Mir
Deepak Mishra
Anand Mishra
Shlok K. Mishra
Niluthpol Mithun
Gaurav Mittal
Trisha Mittal
Daisuke Miyazaki
Kaichun Mo
Hong Mo
Zhipeng Mo
Davide Modolo
Abduallah A. Mohamed
Mohamed Afham
Mohamed Aflal
Ron Mokady
Pavlo Molchanov
Davide Moltisanti
Liliane Momeni
Gianluca Monaci
Pascal Monasse
Ajoy Mondal
Tom Monnier

Aron Monszpart
Gyeongsik Moon
Suhong Moon
Taesup Moon
Sean Moran
Daniel Moreira
Pietro Morerio
Alexandre Morgand
Lia Morra
Ali Mosleh
Inbar Mosseri
Sayed Mohammad
    Mostafavi Isfahani
Saman Motamed
Ramy A. Mounir
Fangzhou Mu
Jiteng Mu
Norman Mu
Yasuhiro Mukaigawa
Ryan Mukherjee
Tanmoy Mukherjee
Yusuke Mukuta
Ravi Teja Mullapudi
Lea Müller
Matthias Müller
Martin Mundt
Nils Murrugarra-Llerena
Damien Muselet
Armin Mustafa
Muhammad Ferjad Naeem
Sauradip Nag
Hajime Nagahara
Pravin Nagar
Rajendra Nagar
Naveen Shankar Nagaraja
Varun Nagaraja
Tushar Nagarajan
Seungjun Nah
Gaku Nakano
Yuta Nakashima
Giljoo Nam
Seonghyeon Nam
Liangliang Nan
Yuesong Nan
Yeshwanth Napolean

Dinesh Reddy
    Narapureddy
Medhini Narasimhan
Supreeth
    Narasimhaswamy
Sriram Narayanan
Erickson R. Nascimento
Varun Nasery
K. L. Navaneet
Pablo Navarrete Michelini
Shant Navasardyan
Shah Nawaz
Nihal Nayak
Farhood Negin
Lukáš Neumann
Alejandro Newell
Evonne Ng
Kam Woh Ng
Tony Ng
Anh Nguyen
Tuan Anh Nguyen
Cuong Cao Nguyen
Ngoc Cuong Nguyen
Thanh Nguyen
Khoi Nguyen
Phi Le Nguyen
Phong Ha Nguyen
Tam Nguyen
Truong Nguyen
Anh Tuan Nguyen
Rang Nguyen
Thao Thi Phuong Nguyen
Van Nguyen Nguyen
Zhen-Liang Ni
Yao Ni
Shijie Nie
Xuecheng Nie
Yongwei Nie
Weizhi Nie
Ying Nie
Yinyu Nie
Kshitij N. Nikhal
Simon Niklaus
Xuefei Ning
Jifeng Ning

Yotam Nitzan
Di Niu
Shuaicheng Niu
Li Niu
Wei Niu
Yulei Niu
Zhenxing Niu
Albert No
Shohei Nobuhara
Nicoletta Noceti
Junhyug Noh
Sotiris Nousias
Slawomir Nowaczyk
Ewa M. Nowara
Valsamis Ntouskos
Gilberto Ochoa-Ruiz
Ferda Ofli
Jihyong Oh
Sangyun Oh
Youngtaek Oh
Hiroki Ohashi
Takahiro Okabe
Kemal Oksuz
Fumio Okura
Daniel Olmeda Reino
Matthew Olson
Carl Olsson
Roy Or-El
Alessandro Ortis
Guillermo Ortiz-Jimenez
Magnus Oskarsson
Ahmed A. A. Osman
Martin R. Oswald
Mayu Otani
Naima Otberdout
Cheng Ouyang
Jiahong Ouyang
Wanli Ouyang
Andrew Owens
Poojan B. Oza
Mete Ozay
A. Cengiz Oztireli
Gautam Pai
Tomas Pajdla
Umapada Pal

Simone Palazzo
Luca Palmieri
Bowen Pan
Hao Pan
Lili Pan
Tai-Yu Pan
Liang Pan
Chengwei Pan
Yingwei Pan
Xuran Pan
Jinshan Pan
Xinyu Pan
Liyuan Pan
Xingang Pan
Xingjia Pan
Zhihong Pan
Zizheng Pan
Priyadarshini Panda
Rameswar Panda
Rohit Pandey
Kaiyue Pang
Bo Pang
Guansong Pang
Jiangmiao Pang
Meng Pang
Tianyu Pang
Ziqi Pang
Omiros Pantazis
Andreas Panteli
Maja Pantic
Marina Paolanti
Joao P. Papa
Samuele Papa
Mike Papadakis
Dim P. Papadopoulos
George Papandreou
Constantin Pape
Toufiq Parag
Chethan Parameshwara
Shaifali Parashar
Alejandro Pardo
Rishubh Parihar
Sarah Parisot
JaeYoo Park
Gyeong-Moon Park

Hyojin Park
Hyoungseob Park
Jongchan Park
Jae Sung Park
Kiru Park
Chunghyun Park
Kwanyong Park
Sunghyun Park
Sungrae Park
Seongsik Park
Sanghyun Park
Sungjune Park
Taesung Park
Gaurav Parmar
Paritosh Parmar
Alvaro Parra
Despoina Paschalidou
Or Patashnik
Shivansh Patel
Pushpak Pati
Prashant W. Patil
Vaishakh Patil
Suvam Patra
Jay Patravali
Badri Narayana Patro
Angshuman Paul
Sudipta Paul
Rémi Pautrat
Nick E. Pears
Adithya Pediredla
Wenjie Pei
Shmuel Peleg
Latha Pemula
Bo Peng
Houwen Peng
Yue Peng
Liangzu Peng
Baoyun Peng
Jun Peng
Pai Peng
Sida Peng
Xi Peng
Yuxin Peng
Songyou Peng
Wei Peng

Weiqi Peng
Wen-Hsiao Peng
Pramuditha Perera
Juan C. Perez
Eduardo Pérez Pellitero
Juan-Manuel Perez-Rua
Federico Pernici
Marco Pesavento
Stavros Petridis
Ilya A. Petrov
Vladan Petrovic
Mathis Petrovich
Suzanne Petryk
Hieu Pham
Quang Pham
Khoi Pham
Tung Pham
Huy Phan
Stephen Phillips
Cheng Perng Phoo
David Picard
Marco Piccirilli
Georg Pichler
A. J. Piergiovanni
Vipin Pillai
Silvia L. Pintea
Giovanni Pintore
Robinson Piramuthu
Fiora Pirri
Theodoros Pissas
Fabio Pizzati
Benjamin Planche
Bryan Plummer
Matteo Poggi
Ashwini Pokle
Georgy E. Ponimatkin
Adrian Popescu
Stefan Popov
Nikola Popović
Ronald Poppe
Angelo Porrello
Michael Potter
Charalambos Poullis
Hadi Pouransari
Omid Poursaeed

Shraman Pramanick
Mantini Pranav
Dilip K. Prasad
Meghshyam Prasad
B. H. Pawan Prasad
Shitala Prasad
Prateek Prasanna
Ekta Prashnani
Derek S. Prijatelj
Luke Y. Prince
Véronique Prinet
Victor Adrian Prisacariu
James Pritts
Thomas Probst
Sergey Prokudin
Rita Pucci
Chi-Man Pun
Matthew Purri
Haozhi Qi
Lu Qi
Lei Qi
Xianbiao Qi
Yonggang Qi
Yuankai Qi
Siyuan Qi
Guocheng Qian
Hangwei Qian
Qi Qian
Deheng Qian
Shengsheng Qian
Wen Qian
Rui Qian
Yiming Qian
Shengju Qian
Shengyi Qian
Xuelin Qian
Zhenxing Qian
Nan Qiao
Xiaotian Qiao
Jing Qin
Can Qin
Siyang Qin
Hongwei Qin
Jie Qin
Minghai Qin

Yipeng Qin
Yongqiang Qin
Wenda Qin
Xuebin Qin
Yuzhe Qin
Yao Qin
Zhenyue Qin
Zhiwu Qing
Heqian Qiu
Jiayan Qiu
Jielin Qiu
Yue Qiu
Jiaxiong Qiu
Zhongxi Qiu
Shi Qiu
Zhaofan Qiu
Zhongnan Qu
Yanyun Qu
Kha Gia Quach
Yuhui Quan
Ruijie Quan
Mike Rabbat
Rahul Shekhar Rade
Filip Radenovic
Gorjan Radevski
Bogdan Raducanu
Francesco Ragusa
Shafin Rahman
Md Mahfuzur Rahman
   Siddiquee
Hossein Rahmani
Kiran Raja
Sivaramakrishnan
   Rajaraman
Jathushan Rajasegaran
Adnan Siraj Rakin
Michaël Ramamonjisoa
Chirag A. Raman
Shanmuganathan Raman
Vignesh Ramanathan
Vasili Ramanishka
Vikram V. Ramaswamy
Merey Ramazanova
Jason Rambach
Sai Saketh Rambhatla

Clément Rambour
Ashwin Ramesh Babu
Adín Ramírez Rivera
Arianna Rampini
Haoxi Ran
Aakanksha Rana
Aayush Jung Bahadur
   Rana
Kanchana N. Ranasinghe
Aneesh Rangnekar
Samrudhdhi B. Rangrej
Harsh Rangwani
Viresh Ranjan
Anyi Rao
Yongming Rao
Carolina Raposo
Michalis Raptis
Amir Rasouli
Vivek Rathod
Adepu Ravi Sankar
Avinash Ravichandran
Bharadwaj Ravichandran
Dripta S. Raychaudhuri
Adria Recasens
Simon Reiß
Davis Rempe
Daxuan Ren
Jiawei Ren
Jimmy Ren
Sucheng Ren
Dayong Ren
Zhile Ren
Dongwei Ren
Qibing Ren
Pengfei Ren
Zhenwen Ren
Xuqian Ren
Yixuan Ren
Zhongzheng Ren
Ambareesh Revanur
Hamed Rezazadegan
   Tavakoli
Rafael S. Rezende
Wonjong Rhee
Alexander Richard

Christian Richardt
Stephan R. Richter
Benjamin Riggan
Dominik Rivoir
Mamshad Nayeem Rizve
Joshua D. Robinson
Joseph Robinson
Chris Rockwell
Ranga Rodrigo
Andres C. Rodriguez
Carlos Rodriguez-Pardo
Marcus Rohrbach
Gemma Roig
Yu Rong
David A. Ross
Mohammad Rostami
Edward Rosten
Karsten Roth
Anirban Roy
Debaditya Roy
Shuvendu Roy
Ahana Roy Choudhury
Aruni Roy Chowdhury
Denys Rozumnyi
Shulan Ruan
Wenjie Ruan
Patrick Ruhkamp
Danila Rukhovich
Anian Ruoss
Chris Russell
Dan Ruta
Dawid Damian Rymarczyk
DongHun Ryu
Hyeonggon Ryu
Kwonyoung Ryu
Balasubramanian S.
Alexandre Sablayrolles
Mohammad Sabokrou
Arka Sadhu
Aniruddha Saha
Oindrila Saha
Pritish Sahu
Aneeshan Sain
Nirat Saini
Saurabh Saini

Takeshi Saitoh
Christos Sakaridis
Fumihiko Sakaue
Dimitrios Sakkos
Ken Sakurada
Parikshit V. Sakurikar
Rohit Saluja
Nermin Samet
Leo Sampaio Ferraz
    Ribeiro
Jorge Sanchez
Enrique Sanchez
Shengtian Sang
Anush Sankaran
Soubhik Sanyal
Nikolaos Sarafianos
Vishwanath Saragadam
István Sárándi
Saquib Sarfraz
Mert Bulent Sariyildiz
Anindya Sarkar
Pritam Sarkar
Paul-Edouard Sarlin
Hiroshi Sasaki
Takami Sato
Torsten Sattler
Ravi Kumar Satzoda
Axel Sauer
Stefano Savian
Artem Savkin
Manolis Savva
Gerald Schaefer
Simone Schaub-Meyer
Yoni Schirris
Samuel Schulter
Katja Schwarz
Jesse Scott
Sinisa Segvic
Constantin Marc Seibold
Lorenzo Seidenari
Matan Sela
Fadime Sener
Paul Hongsuck Seo
Kwanggyoon Seo
Hongje Seong

Dario Serez
Francesco Setti
Bryan Seybold
Mohamad Shahbazi
Shima Shahfar
Xinxin Shan
Caifeng Shan
Dandan Shan
Shawn Shan
Wei Shang
Jinghuan Shang
Jiaxiang Shang
Lei Shang
Sukrit Shankar
Ken Shao
Rui Shao
Jie Shao
Mingwen Shao
Aashish Sharma
Gaurav Sharma
Vivek Sharma
Abhishek Sharma
Yoli Shavit
Shashank Shekhar
Sumit Shekhar
Zhijie Shen
Fengyi Shen
Furao Shen
Jialie Shen
Jingjing Shen
Ziyi Shen
Linlin Shen
Guangyu Shen
Biluo Shen
Falong Shen
Jiajun Shen
Qiu Shen
Qiuhong Shen
Shuai Shen
Wang Shen
Yiqing Shen
Yunhang Shen
Siqi Shen
Bin Shen
Tianwei Shen

Xi Shen
Yilin Shen
Yuming Shen
Yucong Shen
Zhiqiang Shen
Lu Sheng
Yichen Sheng
Shivanand Venkanna
    Sheshappanavar
Shelly Sheynin
Baifeng Shi
Ruoxi Shi
Botian Shi
Hailin Shi
Jia Shi
Jing Shi
Shaoshuai Shi
Baoguang Shi
Boxin Shi
Hengcan Shi
Tianyang Shi
Xiaodan Shi
Yongjie Shi
Zhensheng Shi
Yinghuan Shi
Weiqi Shi
Wu Shi
Xuepeng Shi
Xiaoshuang Shi
Yujiao Shi
Zenglin Shi
Zhenmei Shi
Takashi Shibata
Meng-Li Shih
Yichang Shih
Hyunjung Shim
Dongseok Shim
Soshi Shimada
Inkyu Shin
Jinwoo Shin
Seungjoo Shin
Seungjae Shin
Koichi Shinoda
Suprosanna Shit

Palaiahnakote
    Shivakumara
Eli Shlizerman
Gaurav Shrivastava
Xiao Shu
Xiangbo Shu
Xiujun Shu
Yang Shu
Tianmin Shu
Jun Shu
Zhixin Shu
Bing Shuai
Maria Shugrina
Ivan Shugurov
Satya Narayan Shukla
Pranjay Shyam
Jianlou Si
Yawar Siddiqui
Alberto Signoroni
Pedro Silva
Jae-Young Sim
Oriane Siméoni
Martin Simon
Andrea Simonelli
Abhishek Singh
Ashish Singh
Dinesh Singh
Gurkirt Singh
Krishna Kumar Singh
Mannat Singh
Pravendra Singh
Rajat Vikram Singh
Utkarsh Singhal
Dipika Singhania
Vasu Singla
Harsh Sinha
Sudipta Sinha
Josef Sivic
Elena Sizikova
Geri Skenderi
Ivan Skorokhodov
Dmitriy Smirnov
Cameron Y. Smith
James S. Smith
Patrick Snape

Mattia Soldan
Hyeongseok Son
Sanghyun Son
Chuanbiao Song
Chen Song
Chunfeng Song
Dan Song
Dongjin Song
Hwanjun Song
Guoxian Song
Jiaming Song
Jie Song
Liangchen Song
Ran Song
Luchuan Song
Xibin Song
Li Song
Fenglong Song
Guoli Song
Guanglu Song
Zhenbo Song
Lin Song
Xinhang Song
Yang Song
Yibing Song
Rajiv Soundararajan
Hossein Souri
Cristovao Sousa
Riccardo Spezialetti
Leonidas Spinoulas
Michael W. Spratling
Deepak Sridhar
Srinath Sridhar
Gaurang Sriramanan
Vinkle Kumar Srivastav
Themos Stafylakis
Serban Stan
Anastasis Stathopoulos
Markus Steinberger
Jan Steinbrener
Sinisa Stekovic
Alexandros Stergiou
Gleb Sterkin
Rainer Stiefelhagen
Pierre Stock

Organization   xli

Ombretta Strafforello
Julian Straub
Yannick Strümpler
Joerg Stueckler
Hang Su
Weijie Su
Jong-Chyi Su
Bing Su
Haisheng Su
Jinming Su
Yiyang Su
Yukun Su
Yuxin Su
Zhuo Su
Zhaoqi Su
Xiu Su
Yu-Chuan Su
Zhixun Su
Arulkumar Subramaniam
Akshayvarun Subramanya
A. Subramanyam
Swathikiran Sudhakaran
Yusuke Sugano
Masanori Suganuma
Yumin Suh
Yang Sui
Baochen Sun
Cheng Sun
Long Sun
Guolei Sun
Haoliang Sun
Haomiao Sun
He Sun
Hanqing Sun
Hao Sun
Lichao Sun
Jiachen Sun
Jiaming Sun
Jian Sun
Jin Sun
Jennifer J. Sun
Tiancheng Sun
Libo Sun
Peize Sun
Qianru Sun

Shanlin Sun
Yu Sun
Zhun Sun
Che Sun
Lin Sun
Tao Sun
Yiyou Sun
Chunyi Sun
Chong Sun
Weiwei Sun
Weixuan Sun
Xiuyu Sun
Yanan Sun
Zeren Sun
Zhaodong Sun
Zhiqing Sun
Minhyuk Sung
Jinli Suo
Simon Suo
Abhijit Suprem
Anshuman Suri
Saksham Suri
Joshua M. Susskind
Roman Suvorov
Gurumurthy Swaminathan
Robin Swanson
Paul Swoboda
Tabish A. Syed
Richard Szeliski
Fariborz Taherkhani
Yu-Wing Tai
Keita Takahashi
Walter Talbott
Gary Tam
Masato Tamura
Feitong Tan
Fuwen Tan
Shuhan Tan
Andong Tan
Bin Tan
Cheng Tan
Jianchao Tan
Lei Tan
Mingxing Tan
Xin Tan

Zichang Tan
Zhentao Tan
Kenichiro Tanaka
Masayuki Tanaka
Yushun Tang
Hao Tang
Jingqun Tang
Jinhui Tang
Kaihua Tang
Luming Tang
Lv Tang
Sheyang Tang
Shitao Tang
Siliang Tang
Shixiang Tang
Yansong Tang
Keke Tang
Chang Tang
Chenwei Tang
Jie Tang
Junshu Tang
Ming Tang
Peng Tang
Xu Tang
Yao Tang
Chen Tang
Fan Tang
Haoran Tang
Shengeng Tang
Yehui Tang
Zhipeng Tang
Ugo Tanielian
Chaofan Tao
Jiale Tao
Junli Tao
Renshuai Tao
An Tao
Guanhong Tao
Zhiqiang Tao
Makarand Tapaswi
Jean-Philippe G. Tarel
Juan J. Tarrio
Enzo Tartaglione
Keisuke Tateno
Zachary Teed

Ajinkya B. Tejankar
Bugra Tekin
Purva Tendulkar
Damien Teney
Minggui Teng
Chris Tensmeyer
Andrew Beng Jin Teoh
Philipp Terhörst
Kartik Thakral
Nupur Thakur
Kevin Thandiackal
Spyridon Thermos
Diego Thomas
William Thong
Yuesong Tian
Guanzhong Tian
Lin Tian
Shiqi Tian
Kai Tian
Meng Tian
Tai-Peng Tian
Zhuotao Tian
Shangxuan Tian
Tian Tian
Yapeng Tian
Yu Tian
Yuxin Tian
Leslie Ching Ow Tiong
Praveen Tirupattur
Garvita Tiwari
George Toderici
Antoine Toisoul
Aysim Toker
Tatiana Tommasi
Zhan Tong
Alessio Tonioni
Alessandro Torcinovich
Fabio Tosi
Matteo Toso
Hugo Touvron
Quan Hung Tran
Son Tran
Hung Tran
Ngoc-Trung Tran
Vinh Tran

Phong Tran
Giovanni Trappolini
Edith Tretschk
Subarna Tripathi
Shubhendu Trivedi
Eduard Trulls
Prune Truong
Thanh-Dat Truong
Tomasz Trzcinski
Sam Tsai
Yi-Hsuan Tsai
Ethan Tseng
Yu-Chee Tseng
Shahar Tsiper
Stavros Tsogkas
Shikui Tu
Zhigang Tu
Zhengzhong Tu
Richard Tucker
Sergey Tulyakov
Cigdem Turan
Daniyar Turmukhambetov
Victor G. Turrisi da Costa
Bartlomiej Twardowski
Christopher D. Twigg
Radim Tylecek
Mostofa Rafid Uddin
Md. Zasim Uddin
Kohei Uehara
Nicolas Ugrinovic
Youngjung Uh
Norimichi Ukita
Anwaar Ulhaq
Devesh Upadhyay
Paul Upchurch
Yoshitaka Ushiku
Yuzuko Utsumi
Mikaela Angelina Uy
Mohit Vaishnav
Pratik Vaishnavi
Jeya Maria Jose Valanarasu
Matias A. Valdenegro Toro
Diego Valsesia
Wouter Van Gansbeke
Nanne van Noord

Simon Vandenhende
Farshid Varno
Cristina Vasconcelos
Francisco Vasconcelos
Alex Vasilescu
Subeesh Vasu
Arun Balajee Vasudevan
Kanav Vats
Vaibhav S. Vavilala
Sagar Vaze
Javier Vazquez-Corral
Andrea Vedaldi
Olga Veksler
Andreas Velten
Sai H. Vemprala
Raviteja Vemulapalli
Shashanka
    Venkataramanan
Dor Verbin
Luisa Verdoliva
Manisha Verma
Yashaswi Verma
Constantin Vertan
Eli Verwimp
Deepak Vijaykeerthy
Pablo Villanueva
Ruben Villegas
Markus Vincze
Vibhav Vineet
Minh P. Vo
Huy V. Vo
Duc Minh Vo
Tomas Vojir
Igor Vozniak
Nicholas Vretos
Vibashan VS
Tuan-Anh Vu
Thang Vu
Mårten Wadenbäck
Neal Wadhwa
Aaron T. Walsman
Steven Walton
Jin Wan
Alvin Wan
Jia Wan

Jun Wan
Xiaoyue Wan
Fang Wan
Guowei Wan
Renjie Wan
Zhiqiang Wan
Ziyu Wan
Bastian Wandt
Dongdong Wang
Limin Wang
Haiyang Wang
Xiaobing Wang
Angtian Wang
Angelina Wang
Bing Wang
Bo Wang
Boyu Wang
Binghui Wang
Chen Wang
Chien-Yi Wang
Congli Wang
Qi Wang
Chengrui Wang
Rui Wang
Yiqun Wang
Cong Wang
Wenjing Wang
Dongkai Wang
Di Wang
Xiaogang Wang
Kai Wang
Zhizhong Wang
Fangjinhua Wang
Feng Wang
Hang Wang
Gaoang Wang
Guoqing Wang
Guangcong Wang
Guangzhi Wang
Hanqing Wang
Hao Wang
Haohan Wang
Haoran Wang
Hong Wang
Haotao Wang

Hu Wang
Huan Wang
Hua Wang
Hui-Po Wang
Hengli Wang
Hanyu Wang
Hongxing Wang
Jingwen Wang
Jialiang Wang
Jian Wang
Jianyi Wang
Jiashun Wang
Jiahao Wang
Tsun-Hsuan Wang
Xiaoqian Wang
Jinqiao Wang
Jun Wang
Jianzong Wang
Kaihong Wang
Ke Wang
Lei Wang
Lingjing Wang
Linnan Wang
Lin Wang
Liansheng Wang
Mengjiao Wang
Manning Wang
Nannan Wang
Peihao Wang
Jiayun Wang
Pu Wang
Qiang Wang
Qiufeng Wang
Qilong Wang
Qiangchang Wang
Qin Wang
Qing Wang
Ruocheng Wang
Ruibin Wang
Ruisheng Wang
Ruizhe Wang
Runqi Wang
Runzhong Wang
Wenxuan Wang
Sen Wang

Shangfei Wang
Shaofei Wang
Shijie Wang
Shiqi Wang
Zhibo Wang
Song Wang
Xinjiang Wang
Tai Wang
Tao Wang
Teng Wang
Xiang Wang
Tianren Wang
Tiantian Wang
Tianyi Wang
Fengjiao Wang
Wei Wang
Miaohui Wang
Suchen Wang
Siyue Wang
Yaoming Wang
Xiao Wang
Ze Wang
Biao Wang
Chaofei Wang
Dong Wang
Gu Wang
Guangrun Wang
Guangming Wang
Guo-Hua Wang
Haoqing Wang
Hesheng Wang
Huafeng Wang
Jinghua Wang
Jingdong Wang
Jingjing Wang
Jingya Wang
Jingkang Wang
Jiakai Wang
Junke Wang
Kuo Wang
Lichen Wang
Lizhi Wang
Longguang Wang
Mang Wang
Mei Wang

Min Wang
Peng-Shuai Wang
Run Wang
Shaoru Wang
Shuhui Wang
Tan Wang
Tiancai Wang
Tianqi Wang
Wenhai Wang
Wenzhe Wang
Xiaobo Wang
Xiudong Wang
Xu Wang
Yajie Wang
Yan Wang
Yuan-Gen Wang
Yingqian Wang
Yizhi Wang
Yulin Wang
Yu Wang
Yujie Wang
Yunhe Wang
Yuxi Wang
Yaowei Wang
Yiwei Wang
Zezheng Wang
Hongzhi Wang
Zhiqiang Wang
Ziteng Wang
Ziwei Wang
Zheng Wang
Zhenyu Wang
Binglu Wang
Zhongdao Wang
Ce Wang
Weining Wang
Weiyao Wang
Wenbin Wang
Wenguan Wang
Guangting Wang
Haolin Wang
Haiyan Wang
Huiyu Wang
Naiyan Wang
Jingbo Wang

Jinpeng Wang
Jiaqi Wang
Liyuan Wang
Lizhen Wang
Ning Wang
Wenqian Wang
Sheng-Yu Wang
Weimin Wang
Xiaohan Wang
Yifan Wang
Yi Wang
Yongtao Wang
Yizhou Wang
Zhuo Wang
Zhe Wang
Xudong Wang
Xiaofang Wang
Xinggang Wang
Xiaosen Wang
Xiaosong Wang
Xiaoyang Wang
Lijun Wang
Xinlong Wang
Xuan Wang
Xue Wang
Yangang Wang
Yaohui Wang
Yu-Chiang Frank Wang
Yida Wang
Yilin Wang
Yi Ru Wang
Yali Wang
Yinglong Wang
Yufu Wang
Yujiang Wang
Yuwang Wang
Yuting Wang
Yang Wang
Yu-Xiong Wang
Yixu Wang
Ziqi Wang
Zhicheng Wang
Zeyu Wang
Zhaowen Wang
Zhenyi Wang

Zhenzhi Wang
Zhijie Wang
Zhiyong Wang
Zhongling Wang
Zhuowei Wang
Zian Wang
Zifu Wang
Zihao Wang
Zirui Wang
Ziyan Wang
Wenxiao Wang
Zhen Wang
Zhepeng Wang
Zi Wang
Zihao W. Wang
Steven L. Waslander
Olivia Watkins
Daniel Watson
Silvan Weder
Dongyoon Wee
Dongming Wei
Tianyi Wei
Jia Wei
Dong Wei
Fangyun Wei
Longhui Wei
Mingqiang Wei
Xinyue Wei
Chen Wei
Donglai Wei
Pengxu Wei
Xing Wei
Xiu-Shen Wei
Wenqi Wei
Guoqiang Wei
Wei Wei
XingKui Wei
Xian Wei
Xingxing Wei
Yake Wei
Yuxiang Wei
Yi Wei
Luca Weihs
Michael Weinmann
Martin Weinmann

Congcong Wen
Chuan Wen
Jie Wen
Sijia Wen
Song Wen
Chao Wen
Xiang Wen
Zeyi Wen
Xin Wen
Yilin Wen
Yijia Weng
Shuchen Weng
Junwu Weng
Wenming Weng
Renliang Weng
Zhenyu Weng
Xinshuo Weng
Nicholas J. Westlake
Gordon Wetzstein
Lena M. Widin Klasén
Rick Wildes
Bryan M. Williams
Williem Williem
Ole Winther
Scott Wisdom
Alex Wong
Chau-Wai Wong
Kwan-Yee K. Wong
Yongkang Wong
Scott Workman
Marcel Worring
Michael Wray
Safwan Wshah
Xiang Wu
Aming Wu
Chongruo Wu
Cho-Ying Wu
Chunpeng Wu
Chenyan Wu
Ziyi Wu
Fuxiang Wu
Gang Wu
Haiping Wu
Huisi Wu
Jane Wu

Jialian Wu
Jing Wu
Jinjian Wu
Jianlong Wu
Xian Wu
Lifang Wu
Lifan Wu
Minye Wu
Qianyi Wu
Rongliang Wu
Rui Wu
Shiqian Wu
Shuzhe Wu
Shangzhe Wu
Tsung-Han Wu
Tz-Ying Wu
Ting-Wei Wu
Jiannan Wu
Zhiliang Wu
Yu Wu
Chenyun Wu
Dayan Wu
Dongxian Wu
Fei Wu
Hefeng Wu
Jianxin Wu
Weibin Wu
Wenxuan Wu
Wenhao Wu
Xiao Wu
Yicheng Wu
Yuanwei Wu
Yu-Huan Wu
Zhenxin Wu
Zhenyu Wu
Wei Wu
Peng Wu
Xiaohe Wu
Xindi Wu
Xinxing Wu
Xinyi Wu
Xingjiao Wu
Xiongwei Wu
Yangzheng Wu
Yanzhao Wu

Yawen Wu
Yong Wu
Yi Wu
Ying Nian Wu
Zhenyao Wu
Zhonghua Wu
Zongze Wu
Zuxuan Wu
Stefanie Wuhrer
Teng Xi
Jianing Xi
Fei Xia
Haifeng Xia
Menghan Xia
Yuanqing Xia
Zhihua Xia
Xiaobo Xia
Weihao Xia
Shihong Xia
Yan Xia
Yong Xia
Zhaoyang Xia
Zhihao Xia
Chuhua Xian
Yongqin Xian
Wangmeng Xiang
Fanbo Xiang
Tiange Xiang
Tao Xiang
Liuyu Xiang
Xiaoyu Xiang
Zhiyu Xiang
Aoran Xiao
Chunxia Xiao
Fanyi Xiao
Jimin Xiao
Jun Xiao
Taihong Xiao
Anqi Xiao
Junfei Xiao
Jing Xiao
Liang Xiao
Yang Xiao
Yuting Xiao
Yijun Xiao

Yao Xiao

Zeyu Xiao

Zhisheng Xiao

Zihao Xiao

Binhui Xie

Christopher Xie

Haozhe Xie

Jin Xie

Guo-Sen Xie

Hongtao Xie

Ming-Kun Xie

Tingting Xie

Chaohao Xie

Weicheng Xie

Xudong Xie

Jiyang Xie

Xiaohua Xie

Yuan Xie

Zhenyu Xie

Ning Xie

Xianghui Xie

Xiufeng Xie

You Xie

Yutong Xie

Fuyong Xing

Yifan Xing

Zhen Xing

Yuanjun Xiong

Jinhui Xiong

Weihua Xiong

Hongkai Xiong

Zhitong Xiong

Yuanhao Xiong

Yunyang Xiong

Yuwen Xiong

Zhiwei Xiong

Yuliang Xiu

An Xu

Chang Xu

Chenliang Xu

Chengming Xu

Chenshu Xu

Xiang Xu

Huijuan Xu

Zhe Xu

Jie Xu

Jingyi Xu

Jiarui Xu

Yinghao Xu

Kele Xu

Ke Xu

Li Xu

Linchuan Xu

Linning Xu

Mengde Xu

Mengmeng Frost Xu

Min Xu

Mingye Xu

Jun Xu

Ning Xu

Peng Xu

Runsheng Xu

Sheng Xu

Wenqiang Xu

Xiaogang Xu

Renzhe Xu

Kaidi Xu

Yi Xu

Chi Xu

Qiuling Xu

Baobei Xu

Feng Xu

Haohang Xu

Haofei Xu

Lan Xu

Mingze Xu

Songcen Xu

Weipeng Xu

Wenjia Xu

Wenju Xu

Xiangyu Xu

Xin Xu

Yinshuang Xu

Yixing Xu

Yuting Xu

Yanyu Xu

Zhenbo Xu

Zhiliang Xu

Zhiyuan Xu

Xiaohao Xu

Yanwu Xu

Yan Xu

Yiran Xu

Yifan Xu

Yufei Xu

Yong Xu

Zichuan Xu

Zenglin Xu

Zexiang Xu

Zhan Xu

Zheng Xu

Zhiwei Xu

Ziyue Xu

Shiyu Xuan

Hanyu Xuan

Fei Xue

Jianru Xue

Mingfu Xue

Qinghan Xue

Tianfan Xue

Chao Xue

Chuhui Xue

Nan Xue

Zhou Xue

Xiangyang Xue

Yuan Xue

Abhay Yadav

Ravindra Yadav

Kota Yamaguchi

Toshihiko Yamasaki

Kohei Yamashita

Chaochao Yan

Feng Yan

Kun Yan

Qingsen Yan

Qixin Yan

Rui Yan

Siming Yan

Xinchen Yan

Yaping Yan

Bin Yan

Qingan Yan

Shen Yan

Shipeng Yan

Xu Yan

Yan Yan
Yichao Yan
Zhaoyi Yan
Zike Yan
Zhiqiang Yan
Hongliang Yan
Zizheng Yan
Jiewen Yang
Anqi Joyce Yang
Shan Yang
Anqi Yang
Antoine Yang
Bo Yang
Baoyao Yang
Chenhongyi Yang
Dingkang Yang
De-Nian Yang
Dong Yang
David Yang
Fan Yang
Fengyu Yang
Fengting Yang
Fei Yang
Gengshan Yang
Heng Yang
Han Yang
Huan Yang
Yibo Yang
Jiancheng Yang
Jihan Yang
Jiawei Yang
Jiayu Yang
Jie Yang
Jinfa Yang
Jingkang Yang
Jinyu Yang
Cheng-Fu Yang
Ji Yang
Jianyu Yang
Kailun Yang
Tian Yang
Luyu Yang
Liang Yang
Li Yang
Michael Ying Yang

Yang Yang
Muli Yang
Le Yang
Qiushi Yang
Ren Yang
Ruihan Yang
Shuang Yang
Siyuan Yang
Su Yang
Shiqi Yang
Taojiannan Yang
Tianyu Yang
Lei Yang
Wanzhao Yang
Shuai Yang
William Yang
Wei Yang
Xiaofeng Yang
Xiaoshan Yang
Xin Yang
Xuan Yang
Xu Yang
Xingyi Yang
Xitong Yang
Jing Yang
Yanchao Yang
Wenming Yang
Yujiu Yang
Herb Yang
Jianfei Yang
Jinhui Yang
Chuanguang Yang
Guanglei Yang
Haitao Yang
Kewei Yang
Linlin Yang
Lijin Yang
Longrong Yang
Meng Yang
MingKun Yang
Sibei Yang
Shicai Yang
Tong Yang
Wen Yang
Xi Yang

Xiaolong Yang
Xue Yang
Yubin Yang
Ze Yang
Ziyi Yang
Yi Yang
Linjie Yang
Yuzhe Yang
Yiding Yang
Zhenpei Yang
Zhaohui Yang
Zhengyuan Yang
Zhibo Yang
Zongxin Yang
Hantao Yao
Mingde Yao
Rui Yao
Taiping Yao
Ting Yao
Cong Yao
Qingsong Yao
Quanming Yao
Xu Yao
Yuan Yao
Yao Yao
Yazhou Yao
Jiawen Yao
Shunyu Yao
Pew-Thian Yap
Sudhir Yarram
Rajeev Yasarla
Peng Ye
Botao Ye
Mao Ye
Fei Ye
Hanrong Ye
Jingwen Ye
Jinwei Ye
Jiarong Ye
Mang Ye
Meng Ye
Qi Ye
Qian Ye
Qixiang Ye
Junjie Ye

Sheng Ye
Nanyang Ye
Yufei Ye
Xiaoqing Ye
Ruolin Ye
Yousef Yeganeh
Chun-Hsiao Yeh
Raymond A. Yeh
Yu-Ying Yeh
Kai Yi
Chang Yi
Renjiao Yi
Xinping Yi
Peng Yi
Alper Yilmaz
Junho Yim
Hui Yin
Bangjie Yin
Jia-Li Yin
Miao Yin
Wenzhe Yin
Xuwang Yin
Ming Yin
Yu Yin
Aoxiong Yin
Kangxue Yin
Tianwei Yin
Wei Yin
Xianghua Ying
Rio Yokota
Tatsuya Yokota
Naoto Yokoya
Ryo Yonetani
Ki Yoon Yoo
Jinsu Yoo
Sunjae Yoon
Jae Shin Yoon
Jihun Yoon
Sung-Hoon Yoon
Ryota Yoshihashi
Yusuke Yoshiyasu
Chenyu You
Haoran You
Haoxuan You
Yang You

Quanzeng You
Tackgeun You
Kaichao You
Shan You
Xinge You
Yurong You
Baosheng Yu
Bei Yu
Haichao Yu
Hao Yu
Chaohui Yu
Fisher Yu
Jin-Gang Yu
Jiyang Yu
Jason J. Yu
Jiashuo Yu
Hong-Xing Yu
Lei Yu
Mulin Yu
Ning Yu
Peilin Yu
Qi Yu
Qian Yu
Rui Yu
Shuzhi Yu
Gang Yu
Tan Yu
Weijiang Yu
Xin Yu
Bingyao Yu
Ye Yu
Hanchao Yu
Yingchen Yu
Tao Yu
Xiaotian Yu
Qing Yu
Houjian Yu
Changqian Yu
Jing Yu
Jun Yu
Shujian Yu
Xiang Yu
Zhaofei Yu
Zhenbo Yu
Yinfeng Yu

Zhuoran Yu
Zitong Yu
Bo Yuan
Jiangbo Yuan
Liangzhe Yuan
Weihao Yuan
Jianbo Yuan
Xiaoyun Yuan
Ye Yuan
Li Yuan
Geng Yuan
Jialin Yuan
Maoxun Yuan
Peng Yuan
Xin Yuan
Yuan Yuan
Yuhui Yuan
Yixuan Yuan
Zheng Yuan
Mehmet Kerim Yücel
Kaiyu Yue
Haixiao Yue
Heeseung Yun
Sangdoo Yun
Tian Yun
Mahmut Yurt
Ekim Yurtsever
Ahmet Yüzügüler
Edouard Yvinec
Eloi Zablocki
Christopher Zach
Muhammad Zaigham
  Zaheer
Pierluigi Zama Ramirez
Yuhang Zang
Pietro Zanuttigh
Alexey Zaytsev
Bernhard Zeisl
Haitian Zeng
Pengpeng Zeng
Jiabei Zeng
Runhao Zeng
Wei Zeng
Yawen Zeng
Yi Zeng

Yiming Zeng
Tieyong Zeng
Huanqiang Zeng
Dan Zeng
Yu Zeng
Wei Zhai
Yuanhao Zhai
Fangneng Zhan
Kun Zhan
Xiong Zhang
Jingdong Zhang
Jiangning Zhang
Zhilu Zhang
Gengwei Zhang
Dongsu Zhang
Hui Zhang
Binjie Zhang
Bo Zhang
Tianhao Zhang
Cecilia Zhang
Jing Zhang
Chaoning Zhang
Chenxu Zhang
Chi Zhang
Chris Zhang
Yabin Zhang
Zhao Zhang
Rufeng Zhang
Chaoyi Zhang
Zheng Zhang
Da Zhang
Yi Zhang
Edward Zhang
Xin Zhang
Feifei Zhang
Feilong Zhang
Yuqi Zhang
GuiXuan Zhang
Hanlin Zhang
Hanwang Zhang
Hanzhen Zhang
Haotian Zhang
He Zhang
Haokui Zhang
Hongyuan Zhang

Hengrui Zhang
Hongming Zhang
Mingfang Zhang
Jianpeng Zhang
Jiaming Zhang
Jichao Zhang
Jie Zhang
Jingfeng Zhang
Jingyi Zhang
Jinnian Zhang
David Junhao Zhang
Junjie Zhang
Junzhe Zhang
Jiawan Zhang
Jingyang Zhang
Kai Zhang
Lei Zhang
Lihua Zhang
Lu Zhang
Miao Zhang
Minjia Zhang
Mingjin Zhang
Qi Zhang
Qian Zhang
Qilong Zhang
Qiming Zhang
Qiang Zhang
Richard Zhang
Ruimao Zhang
Ruisi Zhang
Ruixin Zhang
Runze Zhang
Qilin Zhang
Shan Zhang
Shanshan Zhang
Xi Sheryl Zhang
Song-Hai Zhang
Chongyang Zhang
Kaihao Zhang
Songyang Zhang
Shu Zhang
Siwei Zhang
Shujian Zhang
Tianyun Zhang
Tong Zhang

Tao Zhang
Wenwei Zhang
Wenqiang Zhang
Wen Zhang
Xiaolin Zhang
Xingchen Zhang
Xingxuan Zhang
Xiuming Zhang
Xiaoshuai Zhang
Xuanmeng Zhang
Xuanyang Zhang
Xucong Zhang
Xingxing Zhang
Xikun Zhang
Xiaohan Zhang
Yahui Zhang
Yunhua Zhang
Yan Zhang
Yanghao Zhang
Yifei Zhang
Yifan Zhang
Yi-Fan Zhang
Yihao Zhang
Yingliang Zhang
Youshan Zhang
Yulun Zhang
Yushu Zhang
Yixiao Zhang
Yide Zhang
Zhongwen Zhang
Bowen Zhang
Chen-Lin Zhang
Zehua Zhang
Zekun Zhang
Zeyu Zhang
Xiaowei Zhang
Yifeng Zhang
Cheng Zhang
Hongguang Zhang
Yuexi Zhang
Fa Zhang
Guofeng Zhang
Hao Zhang
Haofeng Zhang
Hongwen Zhang

Hua Zhang
Jiaxin Zhang
Zhenyu Zhang
Jian Zhang
Jianfeng Zhang
Jiao Zhang
Jiakai Zhang
Lefei Zhang
Le Zhang
Mi Zhang
Min Zhang
Ning Zhang
Pan Zhang
Pu Zhang
Qing Zhang
Renrui Zhang
Shifeng Zhang
Shuo Zhang
Shaoxiong Zhang
Weizhong Zhang
Xi Zhang
Xiaomei Zhang
Xinyu Zhang
Yin Zhang
Zicheng Zhang
Zihao Zhang
Ziqi Zhang
Zhaoxiang Zhang
Zhen Zhang
Zhipeng Zhang
Zhixing Zhang
Zhizheng Zhang
Jiawei Zhang
Zhong Zhang
Pingping Zhang
Yixin Zhang
Kui Zhang
Lingzhi Zhang
Huaiwen Zhang
Quanshi Zhang
Zhoutong Zhang
Yuhang Zhang
Yuting Zhang
Zhang Zhang
Ziming Zhang

Zhizhong Zhang
Qilong Zhangli
Bingyin Zhao
Bin Zhao
Chenglong Zhao
Lei Zhao
Feng Zhao
Gangming Zhao
Haiyan Zhao
Hao Zhao
Handong Zhao
Hengshuang Zhao
Yinan Zhao
Jiaojiao Zhao
Jiaqi Zhao
Jing Zhao
Kaili Zhao
Haojie Zhao
Yucheng Zhao
Longjiao Zhao
Long Zhao
Qingsong Zhao
Qingyu Zhao
Rui Zhao
Rui-Wei Zhao
Sicheng Zhao
Shuang Zhao
Siyan Zhao
Zelin Zhao
Shiyu Zhao
Wang Zhao
Tiesong Zhao
Qian Zhao
Wangbo Zhao
Xi-Le Zhao
Xu Zhao
Yajie Zhao
Yang Zhao
Ying Zhao
Yin Zhao
Yizhou Zhao
Yunhan Zhao
Yuyang Zhao
Yue Zhao
Yuzhi Zhao

Bowen Zhao
Pu Zhao
Bingchen Zhao
Borui Zhao
Fuqiang Zhao
Hanbin Zhao
Jian Zhao
Mingyang Zhao
Na Zhao
Rongchang Zhao
Ruiqi Zhao
Shuai Zhao
Wenda Zhao
Wenliang Zhao
Xiangyun Zhao
Yifan Zhao
Yaping Zhao
Zhou Zhao
He Zhao
Jie Zhao
Xibin Zhao
Xiaoqi Zhao
Zhengyu Zhao
Jin Zhe
Chuanxia Zheng
Huan Zheng
Hao Zheng
Jia Zheng
Jian-Qing Zheng
Shuai Zheng
Meng Zheng
Mingkai Zheng
Qian Zheng
Qi Zheng
Wu Zheng
Yinqiang Zheng
Yufeng Zheng
Yutong Zheng
Yalin Zheng
Yu Zheng
Feng Zheng
Zhaoheng Zheng
Haitian Zheng
Kang Zheng
Bolun Zheng

Haiyong Zheng
Mingwu Zheng
Sipeng Zheng
Tu Zheng
Wenzhao Zheng
Xiawu Zheng
Yinglin Zheng
Zhuo Zheng
Zilong Zheng
Kecheng Zheng
Zerong Zheng
Shuaifeng Zhi
Tiancheng Zhi
Jia-Xing Zhong
Yiwu Zhong
Fangwei Zhong
Zhihang Zhong
Yaoyao Zhong
Yiran Zhong
Zhun Zhong
Zichun Zhong
Bo Zhou
Boyao Zhou
Brady Zhou
Mo Zhou
Chunluan Zhou
Dingfu Zhou
Fan Zhou
Jingkai Zhou
Honglu Zhou
Jiaming Zhou
Jiahuan Zhou
Jun Zhou
Kaiyang Zhou
Keyang Zhou
Kuangqi Zhou
Lei Zhou
Lihua Zhou
Man Zhou
Mingyi Zhou
Mingyuan Zhou
Ning Zhou
Peng Zhou
Penghao Zhou
Qianyi Zhou

Shuigeng Zhou
Shangchen Zhou
Huayi Zhou
Zhize Zhou
Sanping Zhou
Qin Zhou
Tao Zhou
Wenbo Zhou
Xiangdong Zhou
Xiao-Yun Zhou
Xiao Zhou
Yang Zhou
Yipin Zhou
Zhenyu Zhou
Hao Zhou
Chu Zhou
Daquan Zhou
Da-Wei Zhou
Hang Zhou
Kang Zhou
Qianyu Zhou
Sheng Zhou
Wenhui Zhou
Xingyi Zhou
Yan-Jie Zhou
Yiyi Zhou
Yu Zhou
Yuan Zhou
Yuqian Zhou
Yuxuan Zhou
Zixiang Zhou
Wengang Zhou
Shuchang Zhou
Tianfei Zhou
Yichao Zhou
Alex Zhu
Chenchen Zhu
Deyao Zhu
Xiatian Zhu
Guibo Zhu
Haidong Zhu
Hao Zhu
Hongzi Zhu
Rui Zhu
Jing Zhu

Jianke Zhu
Junchen Zhu
Lei Zhu
Lingyu Zhu
Luyang Zhu
Menglong Zhu
Peihao Zhu
Hui Zhu
Xiaofeng Zhu
Tyler (Lixuan) Zhu
Wentao Zhu
Xiangyu Zhu
Xinqi Zhu
Xinxin Zhu
Xinliang Zhu
Yangguang Zhu
Yichen Zhu
Yixin Zhu
Yanjun Zhu
Yousong Zhu
Yuhao Zhu
Ye Zhu
Feng Zhu
Zhen Zhu
Fangrui Zhu
Jinjing Zhu
Linchao Zhu
Pengfei Zhu
Sijie Zhu
Xiaobin Zhu
Xiaoguang Zhu
Zezhou Zhu
Zhenyao Zhu
Kai Zhu
Pengkai Zhu
Bingbing Zhuang
Chengyuan Zhuang
Liansheng Zhuang
Peiye Zhuang
Yixin Zhuang
Yihong Zhuang
Junbao Zhuo
Andrea Ziani
Bartosz Zieliński
Primo Zingaretti

Nikolaos Zioulis
Andrew Zisserman
Yael Ziv
Liu Ziyin
Xingxing Zou
Danping Zou
Qi Zou

Shihao Zou
Xueyan Zou
Yang Zou
Yuliang Zou
Zihang Zou
Chuhang Zou
Dongqing Zou

Xu Zou
Zhiming Zou
Maria A. Zuluaga
Xinxin Zuo
Zhiwen Zuo
Reyer Zwiggelaar

Contents – Part I

Learning Depth from Focus in the Wild

Changyeon Won⬤ and Hae-Gon Jeon$^{(\boxtimes)}$⬤

Gwangju Institute of Science and Technology, Gwangju, South Korea
cywon1997@gm.gist.ac.kr, haegonj@gist.ac.kr

Abstract. For better photography, most recent commercial cameras including smartphones have either adopted large-aperture lens to collect more light or used a burst mode to take multiple images within short times. These interesting features lead us to examine depth from focus/defocus. In this work, we present a convolutional neural network-based depth estimation from single focal stacks. Our method differs from relevant state-of-the-art works with three unique features. First, our method allows depth maps to be inferred in an end-to-end manner even with image alignment. Second, we propose a sharp region detection module to reduce blur ambiguities in subtle focus changes and weakly texture-less regions. Third, we design an effective downsampling module to ease flows of focal information in feature extractions. In addition, for the generalization of the proposed network, we develop a simulator to realistically reproduce the features of commercial cameras, such as changes in field of view, focal length and principal points. By effectively incorporating these three unique features, our network achieves the top rank in the DDFF 12-Scene benchmark on most metrics. We also demonstrate the effectiveness of the proposed method on various quantitative evaluations and real-world images taken from various off-the-shelf cameras compared with state-of-the-art methods. Our source code is publicly available at https://github.com/wcy199705/DfFintheWild.

Keywords: Depth from focus · Image alignment · Sharp region detection and simulated focal stack dataset

1 Introduction

As commercial demand for high-quality photographic applications increases, images have been increasingly utilized in scene depth computation. Most commercial cameras, including smartphone and DSLR cameras have two interesting configurations: large-aperture lens and a dual-pixel (DP) sensor. Both are reasonable choices to collect more light and to quickly sweep the focus through multiple depths. Because of this, images appear to have a shallow depth of field (DoF) and are formed as focal stacks with corresponding meta-data such as focal

Supplementary Information The online version contains supplementary material available at https://doi.org/10.1007/978-3-031-19769-7_1.

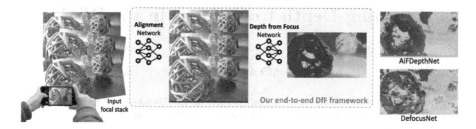

Fig. 1. Results of our true end-to-end DfF framework with comparisons to state-of-the-art methods.

length and principal points. One method to accomplish this is to use single dual-pixel (DP) images which have left and right sub-images with narrow baselines and limited DoFs. A straightforward way is to find correspondences between the left and right sub-images [7,36,41]. Despite an abundance of research, such methods are heavily dependent on the accurate retrieval of correspondences due to the inherent characteristics of DP images. Pixel disparities between the two sub-images result in blurred regions, and the amount of spatial shifts is proportional to the degree of blurrings. Another group of approaches solves this problem using different angles. The out-of-focus regions make it possible to use depth-from-defocus (DfD) techniques to estimate scene depths [2,24,40]. Since there is a strong physical relationship between scene depths and the amount of defocus blurs, the DfD methods account for it in data-driven manners by learning to directly regress depth values. However, there is a potential limitation to these works [2,24,40]. A classic issue, an aperture effect, makes an analysis of defocus blur in a local window difficult. In addition, some of them recover deblurred images from input, but image deblurring also belongs to a class of ill-posed inverse problems for which the uniqueness of the solution cannot be established [19].

These shortcomings motivate us to examine depth from focus (DfF) as an alternative. DfF takes in a focal stack to a depth map during focus sweeping, which is available in most off-the-shelf cameras, and determines the focus in the input focal stack. In particular, the inherent operations of convolutional neural networks (CNNs), convolution and maxpooling, are suitable for measuring the values obtained from derivatives of the image/feature map based on the assumption that focused images contain sharper edges [12,21,37]. Nevertheless, there is still room for improvements with respect to model generalization, due to the domain gap between public datasets and real-world focal stack images, and an alignment issue that we will discuss.

In this work, we achieve a high-quality and well-generalized depth prediction from single focal stacks. Our contributions are threefold (see Fig. 1): First, we compensate the change in image appearance due to magnification during the focus change, and the slight translations from principal point changes. Compared to most CNN-based DfD/DfF works [12,21,37] which either assume that input sequential images are perfectly aligned or use hand-crafted feature-based alignment techniques, we design a learnable context-based image alignment, which works well in defocusing blurred images. Second, the proposed sharp region detection (SRD) module addresses blur ambiguities resulting from subtle defocus changes in weakly-textured regions. SRD consists of convolution layers and a residual block, and allows the extraction of more powerful feature representations for image sharpness. Third, we also propose an efficient downsampling (EFD) module for the DfF framework. The proposed EFD combines output feature maps from upper scales using a stride convolution and a 3D convolution with maxpooling and incorporates them to both keep the feature representation of the original input and to ease the flow of informative features for focused regions. To optimize and generalize our network, we develop a high performance simulator to produce photo-realistic focal stack images with corresponding meta-data such as camera intrinsic parameters.

With this depth from focus network, we achieve state-of-the-art results over various public datasets as well as the top rank in the DDFF benchmark [12]. Ablation studies indicate that each of these technical contributions appreciably improves depth prediction accuracy.

2   Related Work

The mainstream approaches for depth prediction such as monocular depth estimation [6,8,9], stereo matching [3,32] and multiview stereo [10,16] use all-in-focus images. As mentioned above, they overlook the functional properties of off-the-shelf cameras and are out-of-scope for this work. In this section, we review depth from defocus blur images, which are closely related to our work.

Depth from Defocus. Some unsupervised monocular depth estimation [11,33] approaches utilize a defocus blur cue as a supervisory signal. A work in [33] proposes differentiable aperture rendering functions to train a depth prediction network which generates defocused images from input all-in-focus images. The network is trained by minimizing distances between ground truth defocused images and output defocused images based on an estimated depth map. Inspired by [33], a work in [11] introduces a fast differentiable aperture rendering layer from hypothesis of defocus blur. In spite of depth-guided defocus blur, both these works need all-in-focus images as input during an inference time. Anwar *et al.* [2] formulate a reblur loss based on circular blur kernels to regularize depth estimation, and design a CNN architecture to minimize input blurry images and reblurred images from output deblurring images as well. Zhang and Sun [40] propose a regularization term to impose a consistency between depth and defocus maps from single out-of-focus images.

Depth from DP Images. Starting with the use of traditional stereo matching, CNN-based approaches have been adopted for depth from DP images [36]. A work in [7] introduces that an affine ambiguity exists between a scene depth and its disparity from DP data, and then alleviates it using both novel 3D assisted loss and folded loss. In [41], a dual-camera with DP sensors is proposed to take

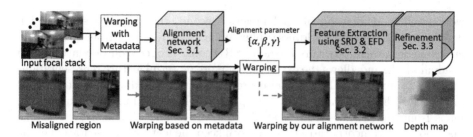

Fig. 2. An overview of the proposed network.

advantage of both stereo matching and depth from DP images. In [27], unsupervised depth estimation by modeling a point spread function of DP cameras. The work in [24] proposes an end-to-end CNN for depth from single DP images using both defocus blur and correspondence cues. In addition, they provide a simulator that makes a synthetic DP dataset from all-in-focus images and the corresponding depth map. In [39], single DP images are represented via multi-plane images [35] with a calibrated point spread function for a certain DP camera model. The representation is used for both unsupervised defocus map and all-in-focus image generation.

Depth from Focus. DfF accounts for changes in blur sizes in the focal stack and determines scene depths according to regions adjacent to the focus [19, 21,26]. In particular, conventional DfF methods infer depth values from a focal stack by comparing the sharpness of a local window at each pixel [17,28,34]. The research in [12] introduces a CNN-based DfF by leveraging focal stack datasets made with light field and RGB-D cameras. In [21], domain invariant defocus blur is used for dealing with the domain gap. The defocus blur is supervised to train data-driven models for DfF as an intermediate step, and is then utilized for permutation-invariant networks to achieve a better generalization from synthetic datasets to real photos. In addition, the work uses a recurrent auto-encoder to handle scene movements which occur during focal sweeps[1]. In [37], a CNN learns to make an intermediate attention map which is shared to predict scene depth prediction and all-in-focus images reconstruction from focal stack images.

3 Methodology

Our network is composed of two major components: One is an image alignment model for sequential defocused images. It is a prerequisite that we should first address the non-alignment issue on images captured with smartphones whose focus is relayed to focus motors adjusting locations of camera lenses. Another component is a focused feature representation, which encodes the depth information of scenes. To be sensitive to subtle focus changes, it requires two consecutive

---

[1] Unfortunately, both the source codes for training/test and its pre-trained weight are not available in public.

feature maps of the corresponding modules from our sharp region detector (SRD) and an effective downsampling module for defocused images (EFD). The overall procedure is depicted in Fig. 2.

3.1   A Network for Defocus Image Alignment

Since camera field of views (FoVs) vary according to the focus distance, a zoom-like effect is induced during a focal sweep [13], called focal breathing. Because of the focal breathing, an image sharpness cannot be accurately measured on the same pixel coordinates across focal slices. As a result, traditional DfF methods perform feature-based defocus image alignment to compensate this, prior to depth computations. However, recent CNN-based approaches disregard the focal breathing because either all public synthetic datasets for DfF/DfD, whose scale is enough to generalize CNNs well, provide well-aligned focal stacks, or are generated by single RGB-D images. Because of this gap between real-world imagery and easy to use datasets, their generality is limited. Therefore, as a first step to implementing a comprehensive, all-in-one solution to DfF, we introduce a defocus image alignment network.

Field of View. Scene FoVs are calculated by work distances, focus distances, and the focal length of cameras in Eq. (1). Since the work distances are fixed during a focal sweep, relative values of FoVs (Relative FoVs) are the same as the inverse distance between sensor and lens. We thus perform an initial alignment of a focal stack using these relative FoVs as follows:

$$FoV_n = W \times \frac{A}{s_n} \tag{1}$$

$$Relative\ FoV_n = \frac{FoV_n}{FoV_{min}} = \frac{s_{min}}{s_n} \quad (s_n = \frac{F_n \times f}{F_n - f}),$$

where s_n is the distance between the lens and the sensor in an n-th focal slice. A is the sensor size, and W is the working distance. f and F_n are the focal length of the lens and a focus distance, respectively. min denotes an index of a focal slice whose FoV is the smallest among focal slices. In this paper, we call this focal slice with the min index as the target focal slice. We note that the values are available by accessing the metadata information in cameras without any user calibration.

Nevertheless, the alignment step is not perfectly appropriate for focal stack images due to hardware limitations, as described in [13]. Most smartphone cameras control their focus distances by spring-installed voice coil motors (VCMs). The VCMs adjust the positions of the camera lens by applying voltages to a nearby electromagnet which induces spring movements. Since the elasticity of the spring can be changed by temperature and usage, there will be an error between real focus distances and values in the metadata. In addition, the principal point of cameras also changes during a focal sweep because the camera lens is

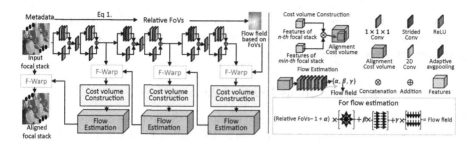

Fig. 3. An illustration of our alignment network. Given initially-aligned images with camera metadata, this network produces an aligned focal stack. In the flow estimation, we use three basis functions to model radial, horizontal and vertical motions of VCMs.

not perfectly parallel to the image sensor, due to some manufacturing imperfections. Therefore, we propose an alignment network to adjust this mis-alignment and a useful simulator to ensure realistic focal stack acquisition.

Alignment Network. As shown in Fig. 3, our alignment network has 3-level encoder-decoder structures, similar to the previous optical flow network [15]. The encoder extracts multi-scale features, and multi-scale optical flow volumes are constructed by concatenating the features of a reference and a target focal slice. The decoder refines the multi-scale optical flow volumes in a coarse-to-fine manner using feature warping (F-warp). However, we cannot directly use the existing optical flow framework for alignment because defocus blur breaks the brightness constancy assumption [34].

To address this issue, we constrain the flow using three basis vectors with corresponding coefficients (α, β, γ) for each scene motion. To compute the coefficients instead of the direct estimation of the flow field, we add an adaptive average pooling layer to each layer of the decoder. The first basis vector accounts for an image crop which reduces errors in the FoVs. We elaborate the image crop as a flow that spreads out from the center. The remaining two vectors represent $x-$ and $y-$axis translations, which compensate for errors in the principal point of the cameras. These parametric constraints of flow induce the network to train geometric features which are not damaged by defocus blur. We optimize this alignment network using a robust loss function L_{align}, proposed in [20], as follows:

$$L_{align} = \sum_{n=0}^{N} \rho(I_n(\Gamma + D(\Gamma)) - I_{min}(\Gamma)), \qquad (2)$$

where $\rho(\cdot) = (|\cdot| + \varepsilon)^q$. q and ε are set to 0.4 and 0.01, respectively. I_n is a focal slice of a reference image, and I_{min} is the target focal slice. $D(\Gamma)$ is an output flow of the alignment network at a pixel position, Γ.

We note that the first basis might be insufficient to describe the zooming effects with spatially-varying motions. However, our design for the image crop

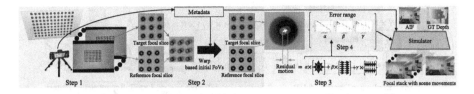

Fig. 4. A pipeline of our simulator. Red and green dots mean a center of a circle. The misalignment error occurs due to inaccurate intrinsic parameters. Our simulator produces misaligned focal stack images because of the hardware limitations for autofocus. (Color figure online)

shows consistently promising results for the alignment, thanks to the combination of the three basis functions that compensate for a variety of motions in real-world.

Simulator. Because public datasets do not describe changes in FoVs or hardware limitations in off-the-shelf cameras, we propose a useful simulator to render realistic sequential defocus images for training our alignment network. Here, the most important part is to determine the error ranges of the intrinsic camera parameters, such as principal points and focal distances. We estimate them as the following process in Fig. 4: (1) We capture circle patterns on a flat surface using various smartphone models by changing focus distances. (2) We initially align focal stacks with the recorded focus distances. (3) After the initial alignment, we decompose the residual motions of the captured circles using 3 basis vectors, image crop and $x-$ and $y-$axis translations. (4) We statistically calculate the error ranges of the principal points and focus distances from the three parameters of the basis vectors. Given metadata of cameras used, our simulator renders focal stacks induced from blur scales based on the focus distance and the error ranges of the basis vector.

3.2 Focal Stack-oriented Feature Extraction

For high-quality depth prediction, we consider three requirements that must be imposed on our network. First, to robustly measure focus in the feature space, it is effective to place a gap space in the convolution operations, as proved in [17]. In addition, even though feature downsampling such as a convolution with strides and pooling layers is necessary to reduce the computations in low-level computer vision tasks like stereo matching [22], such downsampling operations can make a defocused image and its feature map sharper. This fails to accurately determine the focus within the DfF framework. Lastly, feature representations for DfF need to identify subtle distinctions in blur magnitudes between input images.

Initial Feature Extraction. In an initial feature extraction step, we utilize a dilated convolution to extract focus features. After the dilated convolution, we extract feature pyramids to refine the focal volumes in the refinement step. Given an input focal stack $S \in R^{H*W*N*3}$ where H, W and N denote the height, width

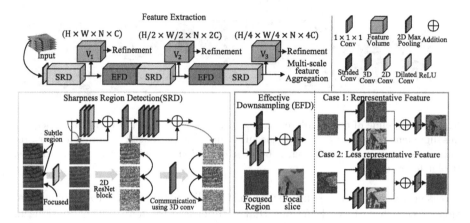

Fig. 5. An architecture of our feature extraction. If feature maps from neighbor focal slices have similar values, our SRD gives an attention score to the sharpest focal slice. Our EFD preserves informative defocus feature representation during downsampling.

and the number of focal slices respectively, we extract three pyramidal feature volumes whose size is $H/2^L \times W/2^L \times N \times C * 2^L$ where $L \in \{0, 1, 2\}$ and C is the number of channels in the focal volume. This pyramidal feature extraction consists of three structures in which SRD and EFD are iteratively performed, as described in Fig. 5. Each pyramidal feature volume is then used as the input to the next EFD module. The last one is utilized as an input of the multi-scale feature aggregation step in Sect. 3.3.

Sharp Region Detector. The initial feature of each focal slice is needed to communicate with other neighboring focal slices, to measure the focus of the pixel of interest. A work in [21] extracts focus features using a global pooling layer as a communication tool across a stack dimension. However, we observe that the global pooling layer causes a loss of important information due to its inherent limitation that all values across focal slices become single values.

Using our SRD module consisting of both a 2D convolution and a 3D convolution, we overcome the limitation. In Fig. 5 (left), we extract features using a 2D ResNet block and add an attention score which is computed from them by 3D convolutions and a ReLU activation. The 3D convolution enables the detection of subtle defocus variations in weakly texture-less regions by communicating the features with neighbor focal slices. With this module, our network encodes more informative features for the regions than previous works [21,37].

EFfective Downsampling. Unlike stereo matching networks that use convolutions with strides for downsampling features [3,32], the stride of a convolution causes a loss in spatial information because most of the focused regions may not be selected. As a solution to this issue, one previous DfF work [21] uses a combination of maxpooling and average pooling with the feature extraction step.

Inspired by [21], we propose a EFD module leveraging a well-known fact that a feature has higher activation in a focused region than weakly textured regions in Fig. 5 (right). The EFD module employs a 2D max-pooling as a downsampling operation and applies a 3D convolution to its output. Through our EFD module, our network can both take representative values of focused regions in a local window and communicate the focal feature with neighbor focal slices.

3.3 Aggregation and Refinement

Our network produces a final depth map after multi-scale feature aggregation and refinement steps.

Multi-scale Feature Aggregation. The receptive field of our feature extraction module might be too small to learn non-local features. Therefore, we propose a multi-scale feature aggregation module using one hour-glass module to expand the receptive field, which is similar to the stereo matching network in [32]. At an initial step, we use three different sizes of kernels (2×2, 4×4, 8×8) in the average pooling layer. Unlike [32], the reason for using average pooling is to avoid a memory consumption issue because DfF requires more input images. We then apply a ResBlock on each output of average pooling in order to extract multi-scale features. These features are embedded into the encoder and aggregated by the decoder of the hour-glass module. The aggregated feature volume is utilized as an input in the refinement step.

Refinement and Regression. The refinement module has three hour-glass modules with skip-connections like [3]. Here, we add transposed convolutions to resize the output of each hourglass whose size is the same as each level of a pyramidal feature volume from the feature extraction module. We construct an input focal volume of each hourglass by concatenating pyramidal feature volumes of the feature extraction module with the output focal volume of the previous hourglass. As each hourglass handles increasingly higher resolutions with pyramidal feature volumes, the focal volumes are refined in a coarse to fine manner. To obtain a depth map from the output focal volumes, we multiply a focus distance value and the probability of each focus distance leading to maximal sharpness. The probability is computed by applying a normalized soft-plus in the output focal volumes in a manner similar to [37]. The whole depth prediction network is optimized using a weighted loss function L_{depth} from scratch as follows:

$$L_{depth} = \sum_{i=1}^{4} w_i * ||D_i - D_{gt}||_2 \qquad (3)$$

where $|| \cdot ||_2$ means a l_2 loss and D_{gt} indicates a ground truth depth map. $i \in \{1, 2, 3, 4\}$ means the scale level of the hour-glass module. In our implementation, we set w_i to 0.3, 0.5, 0.7 and 1.0, respectively.

Implementation Details. We train our network using the following strategy: (1) We first train the alignment network in Sect. 3.1 during 100 epochs using the alignment loss in Eq. (2). (2) We freeze the alignment network and merge it

Table 1. Quantitative evaluation on DDFF 12-Scene [12]. We directly refer to the results from [37]. Since the result of DefocusNet [21] is not uploaded in the official benchmark, we only bring the MAE value from [21]. **bold**: Best, <u>Underline</u>: Second best. Unit: pixel.

Method	MSE ↓	RMSE log ↓	AbsRel ↓	SqRel ↓	Bump ↓	$\delta = 1.25$ ↑	$\delta = 1.25^2$ ↑	$\delta = 1.25^3$ ↑
Lytro	$2.1e^{-3}$	0.31	0.26	**0.01**	1.0	55.65	82.00	93.09
VDFF [23]	$7.3e^{-3}$	1.39	0.62	0.05	0.8	8.42	19.95	32.68
PSP-LF [42]	$2.7e^{-3}$	0.45	0.46	<u>0.03</u>	**0.5**	39.70	65.56	82.46
PSPNet [42]	$9.4e^{-4}$	<u>0.29</u>	0.27	**0.01**	<u>0.6</u>	62.66	85.90	<u>94.42</u>
DFLF [12]	$4.8e^{-3}$	0.59	0.72	0.07	0.7	28.64	53.55	71.61
DDFF [12]	$9.7e^{-4}$	0.32	0.29	**0.01**	<u>0.6</u>	61.95	85.14	92.98
DefocusNet [21]	$9.1e^{-4}$	–	–	–	–	–	–	–
AiFDepthNet [37]	<u>$8.6e^{-4}$</u>	<u>0.29</u>	<u>0.25</u>	**0.01**	<u>0.6</u>	<u>68.33</u>	<u>87.40</u>	93.96
Ours	**$5.7e^{-4}$**	0.21	**0.17**	**0.01**	<u>0.6</u>	**77.96**	**93.72**	**97.94**

with the depth prediction network. (3) We train the merged network during 1500 epochs with the depth loss in Eq. (3). (4) In an inference step, we can estimate the depth map from the misaligned focal stack in an end-to-end manner. We note that our network is able to use an arbitrary number of images as input, like the previous CNN-based DfF/DfD [21,37]. The number of parameters of our alignment network and feature extraction module is 0.195M and 0.067M, respectively. And, the multi-scale feature aggregation module and the refinement module have 2.883M and 1.067M learnable parameters, respectively. That's, the total parameters of our network is 4.212M. We implement our network using a public PyTorch framework [25], and optimize it using Adam optimizer [18] ($\beta_1 = 0.9$, $\beta_2 = 0.99$) with a learning rate 10^{-3}. Our model is trained on a single NVIDIA RTX 2080Ti GPU with 4 mini-batches, which usually takes three days. For data augmentation, we apply random spatial transforms (rotation, flipping and cropping) and color jittering (brightness, contrast and gamma correction).

4   Evaluation

We compare the proposed network with state-of-the-art methods related to DfD, DfF and depth from light field images. We also conduct extensive ablation studies to demonstrate the effectiveness of each component of the proposed network. For quantitative evaluation, we use standard metrics as follows: mean absolute error (MAE), mean squared error (MSE), absolute relative error (AbsRel), square relative error (SqRel), root mean square error (RMSE), log root-mean-squared error (RMSE log), bumpiness (Bump), inference time (Secs) and accuracy metric $\delta_i = 1.25^i$ for $i \in \{1, 2, 3\}$. Following [37], we exclude pixels whose depth ranges are out of focus distance at test time.

4.1   Comparisons to State-of-the-art Methods

We validate the robustness of the proposed network by showing experimental results on various public datasets: DDFF 12-Scene [12], DefocusNet Dataset [21],

Table 2. Quantitative evaluation on DefocusNet dataset [21] (unit: meter), 4D Light Field dataset [14] (unit: pixel) and Smartphone dataset [13] (unit: meter). For DefocusNet dataset and 4D Light Field dataset, we directly refer to the results from [37]. For Smartphone dataset [13], we multiply confidence scores on metrics ('MAE' and 'MSE') which are respectively denoted as 'MAE*' and 'MSE*'. **bold**: Best.

Method	DefocusNet dataset [21]			4D light field [14]			Smartphone [13]		
	MAE ↓	MSE ↓	AbsRel ↓	MSE ↓	RMSE ↓	Bump ↓	MAE* ↓	MSE* ↓	Secs ↓
DefocusNet [21]	0.0637	0.0175	0.1386	0.0593	0.2355	2.69	0.1650	0.0800	0.1598
AiFDepthNet [37]	0.0549	0.0127	0.1115	0.0472	0.2014	1.58	0.1568	0.0764	0.1387
Ours	**0.0403**	**0.0087**	**0.0809**	**0.0230**	**0.1288**	**1.29**	**0.1394**	**0.0723**	**0.1269**

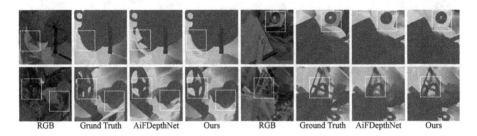

RGB   Grund Truth   AiFDepthNet   Ours   RGB   Ground Truth   AiFDepthNet   Ours

Fig. 6. Examples of depth prediction from AiFDepthNet and ours on DefocusNet dataset.

4D Light Field Dataset [14], Smartphone [13] as well as focal stack images generated from our simulator. The datasets provide pre-aligned defocused images. We use the training split of each dataset to build our depth estimation network in both Sect. 3.2 and Sect. 3.3 from scratch, and validate it on the test split.

DDFF 12-Scene [12]. DDFF 12-Scene dataset provides focal stack images and its ground truth depth maps captured by a light-field camera and a RGB-D sensor, respectively. The images have shallow DoFs and show texture-less regions. Our method shows the better performance than those of recent published works in Table 1 and achieves the top rank in almost evaluation metrics of the benchmark site[2].

DefocusNet Dataset [21]. This dataset is rendered in a virtual space and generated using Blender Cycles renderer [4]. Focal stack images consist of only five defocused images whose focus distances are randomly sampled in an inverse depth space. The quantitative results are shown in Table 2. As shown in Fig. 6, our method successfully reconstructs the smooth surface and the sharp depth discontinuity rather than previous methods.

4D Light Field Dataset [14]. This synthetic dataset has 10 focal slices with shallow DoFs for each focal stack. The number of focal stacks in training and test split is 20 and 4, respectively. For fair comparison on this dataset, we fol-

---

[2] https://competitions.codalab.org/competitions/17807#results.

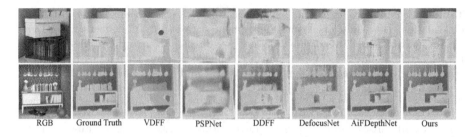

RGB   Ground Truth   VDFF   PSPNet   DDFF   DefocusNet   AiFDepthNet   Ours

Fig. 7. Qualitative comparison on 4D Light Field dataset.

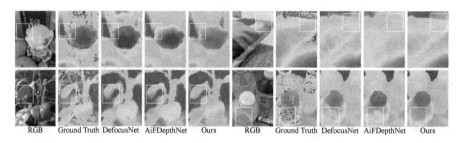

RGB   Ground Truth DefocusNet AiFDepthNet   Ours   RGB   Ground Truth DefocusNet AiFDepthNet   Ours

Fig. 8. Qualitative results on Smartphone dataset.

low the evaluation protocol in the relevant work [37]. In qualitative comparisons Fig. 7, our SRD and EFD enable to capture sharp object boundaries like the box and fine details like lines hanging from the ceiling. In quantitative evaluation of Table 2 the MSE and RMSE are half of them from the comparison methods [1, 38].

Smartphone [13]. This dataset shows real-world scenes captured from Pixel 3 smartphones. Unlike previous datasets, ground truth depth maps are obtained by multiview stereo [30,31] and its depth holes are not considered in the evaluation. As expected, our network achieves the promising performance over the state-of-the-art methods, whose results are reported in Table 2 and Fig. 8. We note that our method consistently yields the best quality depth maps from focal stack images regardless of dataset, thanks to our powerful defocused feature representations using both SRD and EFD.

Generalization Across Different Datasets. Like [37], we demonstrate the generality of the proposed network. For this, we train our network on Flyingthings3D [22] which is a large-scale synthetic dataset, and test it on two datasets including Middlebury Stereo [29] and DefocusNet dataset [21]. As shown in Table 3 and Fig. 9, our network still shows impressive results on both datasets.

4.2   Ablation Studies

We carry out extensive ablation studies to demonstrate the effectiveness of each module of the proposed network.

Table 3. Quantitative result across different datasets for generalization of the state-of-the-art methods and ours. We train our depth prediction model on FlyingThings3D and test them on Middlebury stereo (unit: pixel) and DefocusNet dataset (unit: meter). For fair comparison, we directly refer the results of the works [21,37] from [37].

Method	Train dataset	Test dataset	MAE ↓	MSE ↓	RMSE ↓	AbsRel ↓	SqRel ↓
DefocusNet [21]	FlyingThings3d	Middlebury	7.408	157.440	9.079	0.231	4.245
AiFDepthNet [37]			3.825	58.570	5.936	0.165	3.039
Ours			**1.645**	**9.178**	**2.930**	**0.068**	**0.376**
DefocusNet [21]	FlyingThings3d	DefocusNet	0.320	0.148	0.372	1.383	0.700
AiFDepthNet [37]			0.183	0.080	0.261	0.725	0.404
Ours			**0.163**	**0.076**	**0.259**	**0.590**	**0.360**

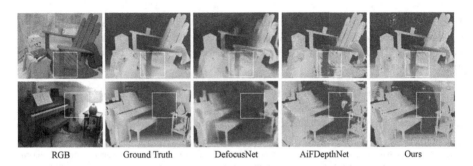

| RGB | Ground Truth | DefocusNet | AiFDepthNet | Ours |

Fig. 9. Qualitative results on Middlebury dataset.

Alignment Network. We first evaluate our alignment network. To do this, we render focal stacks using our simulator which generates defocused images based on a camera metadata. We test our alignment network in consideration of four cases: 1) without any warping, 2) with only initial FoVs in (1), 3) a classical homography method [5], 4) our alignment network with initial FoVs. The quantitative results are reported in Table 4, whose example is displayed in Fig. 10. The results demonstrate that our alignment network achieves much faster and competable performance with the classic homography-based method.

SRD and EFD. We compare our modules with other feature extraction modules depicted in Fig. 11. We conduct this ablation study on DefocusNet dataset [21] because it has more diverse DoF values than other datasets. The quantitative result is reported in Table 5.

When we replace our SRD module with either 3D ResNet block or 2D ResNet block only, there are performance drops, even with more learnable parameters for the 3D ResNet block. We also compare our EFD module with four replaceable modules: max-pooling+3D Conv, average pooling+3D Conv, Stride convolution and 3D pooling layer. As expected, our EFD module achieves the best performance because it allows better gradient flows preserving defocus property.

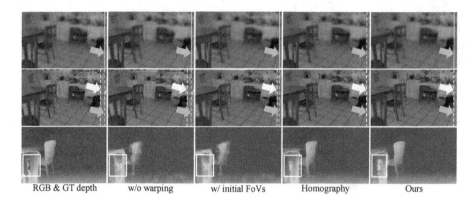

| RGB & GT depth | w/o warping | w/ initial FoVs | Homography | Ours |

Fig. 10. Ablation study on our alignment network. The first and second row refer a target and reference focal slice whose FoVs have the smallest and the biggest values, respectively. The third row shows depth estimation results in accordance to the alignment methods.

Table 4. Ablation study for alignment network. Unit: meter

Module	MAE ↓	MSE ↓	RMSE log ↓	AbsRel ↓	SqRel ↓	$\delta = 1.25$ ↑	$\delta = 1.25^2$ ↑	$\delta = 1.25^3$ ↑	Secs ↓	GPU
w/o alignment	0.0247	0.0014	0.0915	0.0067	0.0034	0.9707	0.9970	0.9995	**0.0107**	2080Ti
w/ initial FoVs	0.0165	0.0009	0.0636	0.0400	0.0019	0.9867	0.9976	0.9994	0.0358	2080Ti
Homography-based	**0.0151**	**0.0007**	**0.0570**	0.0369	**0.0015**	**0.9907**	**0.9986**	**0.9997**	0.8708	R3600
Ours	**0.0151**	**0.0007**	0.0578	**0.0365**	0.0016	0.9898	0.9984	0.9996	0.0923	2080Ti

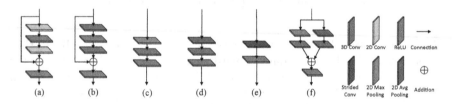

Fig. 11. Candidate modules of our SRD and EFD. (a) 2D ResNet block, (b) 3D ResNet block, (c) Max pooling + 3D Conv, (d) Average pooling + 3D Conv, (e) Strided Conv and (f) 3D pooling layer.

Number of Focal Slices. Like previous DfF networks [21,37], our network can handle an arbitrary number of focal slices by the virtue of 3D convolutions. Following the relevant work [37], we train our network from three different ways, whose result is reported in Fig. 12: The '5' means a model trained using five focal slices; The 'Same' denotes that the number of focal slices in training and test phase is same; The 'Random' is a model trained using an arbitrary number of focal slices.

The '5' case performs poorly when the different number of focal slices is used in the test phase, and the 'Same' case shows promising performances. Neverthe-

Table 5. Ablation studies for SRD and EFD. Unit: meter

Module	MAE ↓	MSE ↓	RMSE log ↓	AbsRel ↓	SqRel ↓	$\delta = 1.25$ ↑	$\delta = 1.25^2$ ↑	$\delta = 1.25^3$ ↑
SRD → 2D ResNet block	0.0421	0.0095	0.1614	0.0842	0.0142	0.9082	0.9722	0.9873
SRD → 3D ResNet block	0.0409	0.0088	0.1576	0.0818	**0.0128**	0.9123	0.9725	0.9891
EFD → Maxpooling + 3D Conv	0.0421	0.0094	0.1622	0.0845	0.0143	0.9125	0.9712	0.9849
EFD → Avgpooling + 3D Conv	0.0422	0.0097	0.1628	0.0830	0.0141	0.9126	0.9718	0.9860
EFD → Strided Conv	0.0419	0.0091	0.1630	0.0842	0.0135	0.9144	0.9725	0.9867
EFD → 3D Poolying Layer	0.0414	0.0089	0.1594	0.0843	0.0132	0.9088	0.9747	0.9886
Ours	**0.0403**	**0.0087**	**0.1534**	**0.0809**	0.0130	**0.9137**	**0.9761**	**0.9900**

(a)   (b) Focal Slice (c) Ground Truth (d) Random 2   (e) Random 3   (f) Random 4   (g) Random 5

Fig. 12. (a) The performance change according to the number of focal slices in training and test phase. (b) One of focal slices and (c) its ground truth depth map. (d) to (g) output depth maps on the random number of input focal slices in training phase.

less, the 'Random' case consistently achieves good performances regardless of the number of focal slices

5 Conclusion

In this paper, we have presented a novel and true end-to-end DfF architecture. To do this, we first propose a trainable alignment network for sequential defocused images. We then introduce a novel feature extraction and an efficient downsampling module for robust DfF tasks. The proposed network achieves the best performance in the public DfF/DfD benchmark and various evaluations.

Limitation. There are still rooms for improvements. A more sophisticated model for flow fields in the alignment network would enhance depth prediction results. More parameters can be useful for extreme rotations. Another direction is to make depth prediction better by employing focal slice selection like defocus channel attention in the aggregation process.

Acknowledgement. This work is in part supported by the Institute of Information & communications Technology Planning & Evaluation (IITP) (No.2021-0-02068, Artificial Intelligence Innovation Hub), Vehicles AI Convergence Research & Development Program through the National IT Industry Promotion Agency of Korea (NIPA), 'Project for Science and Technology Opens the Future of the Region' program through the INNOPOLIS FOUNDATION (Project Number: 2022-DD-UP-0312) funded by the Ministry of Science and ICT (No. S1602-20-1001), the National Research Foundation of Korea (NRF) (No. 2020R1C1C1012635) grant funded by the Korea government (MSIT), the Ministry of Trade, Industry and Energy (MOTIE) and Korea Institute for Advancement of Technology (KIAT) through the International Cooperative R&D

program (P0019797), and the GIST-MIT Collaboration grant funded by the GIST in 2022.

References

1. Abuolaim, A., Brown, M.S.: Defocus deblurring using dual-pixel data. In: Vedaldi, A., Bischof, H., Brox, T., Frahm, J.-M. (eds.) ECCV 2020. LNCS, vol. 12355, pp. 111–126. Springer, Cham (2020). https://doi.org/10.1007/978-3-030-58607-2_7
2. Anwar, S., Hayder, Z., Porikli, F.: Depth estimation and blur removal from a single out-of-focus image. In: BMVC (2017)
3. Chang, J.R., Chen, Y.S.: Pyramid stereo matching network. In: Proceedings of IEEE Conference on Computer Vision and Pattern Recognition (CVPR) (2018)
4. Community, B.O.: Blender-a 3D modelling and rendering package. Blender Foundation (2018)
5. Evangelidis, G.D., Psarakis, E.Z.: Parametric image alignment using enhanced correlation coefficient maximization. IEEE Trans. Pattern Anal. Mach. Intell. **30**(10), 1858–1865 (2008)
6. Fu, H., Gong, M., Wang, C., Batmanghelich, K., Tao, D.: Deep ordinal regression network for monocular depth estimation. In: Proceedings of IEEE Conference on Computer Vision and Pattern Recognition (CVPR) (2018)
7. Garg, R., Wadhwa, N., Ansari, S., Barron, J.T.: Learning single camera depth estimation using dual-pixels. In: Proceedings of International Conference on Computer Vision (ICCV) (2019)
8. Godard, C., Mac Aodha, O., Brostow, G.J.: Unsupervised monocular depth estimation with left-right consistency. In: Proceedings of IEEE Conference on Computer Vision and Pattern Recognition (CVPR) (2017)
9. Godard, C., Mac Aodha, O., Firman, M., Brostow, G.J.: Digging into self-supervised monocular depth estimation. In: Proceedings of International Conference on Computer Vision (ICCV) (2019)
10. Gu, X., Fan, Z., Zhu, S., Dai, Z., Tan, F., Tan, P.: Cascade cost volume for high-resolution multi-view stereo and stereo matching. In: Proceedings of IEEE Conference on Computer Vision and Pattern Recognition (CVPR) (2020)
11. Gur, S., Wolf, L.: Single image depth estimation trained via depth from defocus cues. In: Proceedings of IEEE Conference on Computer Vision and Pattern Recognition (CVPR) (2019)
12. Hazirbas, C., Soyer, S.G., Staab, M.C., Leal-Taixé, L., Cremers, D.: Deep depth from focus. In: Proceedings of Asian Conference on Computer Vision (ACCV) (2018)
13. Herrmann, C., et al.: Learning to autofocus. In: Proceedings of IEEE Conference on Computer Vision and Pattern Recognition (CVPR) (2020)
14. Honauer, K., Johannsen, O., Kondermann, D., Goldluecke, B.: A dataset and evaluation methodology for depth estimation on 4D light fields. In: Proceedings of Asian Conference on Computer Vision (ACCV) (2016)
15. Hui, T.W., Tang, X., Loy, C.C.: Liteflownet: a lightweight convolutional neural network for optical flow estimation. In: Proceedings of IEEE Conference on Computer Vision and Pattern Recognition (CVPR) (2018)
16. Im, S., Jeon, H.G., Lin, S., Kweon, I.S.: Dpsnet: end-to-end deep plane sweep stereo. arXiv preprint arXiv:1905.00538 (2019)
17. Jeon, H.G., Surh, J., Im, S., Kweon, I.S.: Ring difference filter for fast and noise robust depth from focus. IEEE Trans. Image Process. **29**, 1045–1060 (2019)

18. Kingma, D.P., Ba, J.: Adam: a method for stochastic optimization. arXiv preprint arXiv:1412.6980 (2014)
19. Levin, A., Fergus, R., Durand, F., Freeman, W.T.: Image and depth from a conventional camera with a coded aperture. ACM Trans. Graph. (TOG) **26**(3), 70-es (2007)
20. Liu, P., King, I., Lyu, M.R., Xu, J.: Ddflow: learning optical flow with unlabeled data distillation. In: Proceedings of the AAAI Conference on Artificial Intelligence (AAAI) (2019)
21. Maximov, M., Galim, K., Leal-Taixé, L.: Focus on defocus: bridging the synthetic to real domain gap for depth estimation. In: Proceedings of IEEE Conference on Computer Vision and Pattern Recognition (CVPR) (2020)
22. Mayer, N., et al.: A large dataset to train convolutional networks for disparity, optical flow, and scene flow estimation. In: Proceedings of IEEE Conference on Computer Vision and Pattern Recognition (CVPR) (2016)
23. Moeller, M., Benning, M., Schönlieb, C., Cremers, D.: Variational depth from focus reconstruction. IEEE Trans. Image Process. **24**(12), 5369–5378 (2015)
24. Pan, L., Chowdhury, S., Hartley, R., Liu, M., Zhang, H., Li, H.: Dual pixel exploration: simultaneous depth estimation and image restoration. In: Proceedings of IEEE Conference on Computer Vision and Pattern Recognition (CVPR) (2021)
25. Paszke, A., et al.: Pytorch: an imperative style, high-performance deep learning library. Adv. Neural Inf. Process. Syst. **32**, 8026–8037 (2019)
26. Pertuz, S., Puig, D., Garcia, M.A.: Analysis of focus measure operators for shape-from-focus. Pattern Recogn. **46**(5), 1415–1432 (2013)
27. Punnappurath, A., Abuolaim, A., Afifi, M., Brown, M.S.: Modeling defocus-disparity in dual-pixel sensors. In: 2020 IEEE International Conference on Computational Photography (ICCP), pp. 1–12. IEEE (2020)
28. Sakurikar, P., Narayanan, P.: Composite focus measure for high quality depth maps. In: Proceedings of International Conference on Computer Vision (ICCV) (2017)
29. Scharstein, D., et al.: High-resolution stereo datasets with subpixel-accurate ground truth. In: Jiang, X., Hornegger, J., Koch, R. (eds.) GCPR 2014. LNCS, vol. 8753, pp. 31–42. Springer, Cham (2014). https://doi.org/10.1007/978-3-319-11752-2_3
30. Schönberger, J.L., Frahm, J.M.: Structure-from-motion revisited. In: Proceedings of IEEE Conference on Computer Vision and Pattern Recognition (CVPR) (2016)
31. Schönberger, J.L., Zheng, E., Frahm, J.-M., Pollefeys, M.: Pixelwise view selection for unstructured multi-view stereo. In: Leibe, B., Matas, J., Sebe, N., Welling, M. (eds.) ECCV 2016. LNCS, vol. 9907, pp. 501–518. Springer, Cham (2016). https://doi.org/10.1007/978-3-319-46487-9_31
32. Shen, Z., Dai, Y., Rao, Z.: Cfnet: cascade and fused cost volume for robust stereo matching. In: Proceedings of IEEE Conference on Computer Vision and Pattern Recognition (CVPR) (2021)
33. Srinivasan, P.P., Garg, R., Wadhwa, N., Ng, R., Barron, J.T.: Aperture supervision for monocular depth estimation. In: Proceedings of IEEE Conference on Computer Vision and Pattern Recognition (CVPR) (2018)
34. Suwajanakorn, S., Hernandez, C., Seitz, S.M.: Depth from focus with your mobile phone. In: Proceedings of IEEE Conference on Computer Vision and Pattern Recognition (CVPR) (2015)
35. Tucker, R., Snavely, N.: Single-view view synthesis with multiplane images. In: Proceedings of IEEE Conference on Computer Vision and Pattern Recognition (CVPR) (2020)

36. Wadhwa, N., et al.: Synthetic depth-of-field with a single-camera mobile phone. ACM Trans. Graph. (ToG) **37**(4), 1–13 (2018)
37. Wang, N.H., et al.: Bridging unsupervised and supervised depth from focus via all-in-focus supervision. In: Proceedings of International Conference on Computer Vision (ICCV) (2021)
38. Wanner, S., Goldluecke, B.: Globally consistent depth labeling of 4D light fields. In: 2012 IEEE Conference on Computer Vision and Pattern Recognition. IEEE (2012)
39. Xin, S., et al.: Defocus map estimation and deblurring from a single dual-pixel image. In: Proceedings of International Conference on Computer Vision (ICCV) (2021)
40. Zhang, A., Sun, J.: Joint depth and defocus estimation from a single image using physical consistency. IEEE Trans. Image Process. **30**, 3419–3433 (2021)
41. Zhang, Y., Wadhwa, N., Orts-Escolano, S., Häne, C., Fanello, S., Garg, R.: Du^{2}net: learning depth estimation from dual-cameras and dual-pixels. In: Vedaldi, A., Bischof, H., Brox, T., Frahm, J.-M. (eds.) ECCV 2020. LNCS, vol. 12346, pp. 582–598. Springer, Cham (2020). https://doi.org/10.1007/978-3-030-58452-8_34
42. Zhao, H., Shi, J., Qi, X., Wang, X., Jia, J.: Pyramid scene parsing network. In: Proceedings of IEEE Conference on Computer Vision and Pattern Recognition (CVPR) (2017)

Learning-Based Point Cloud Registration for 6D Object Pose Estimation in the Real World

Zheng Dang[1]([✉])([iD]), Lizhou Wang[2]([iD]), Yu Guo[2]([iD]), and Mathieu Salzmann[1,3]([iD])

[1] CVLab, EPFL, Lausanne, Switzerland
{zheng.dang,mathieu.salzmann}@epfl.ch
[2] Xi'an Jiaotong University, Xi'an, Shaanxi, China
dzyxwanglizhou@stu.xjtu.edu.cn, yu.guo@xjtu.edu.cn
[3] Clearspace, Renens, Switzerland

Abstract. In this work, we tackle the task of estimating the 6D pose of an object from point cloud data. While recent learning-based approaches to addressing this task have shown great success on synthetic datasets, we have observed them to fail in the presence of real-world data. We thus analyze the causes of these failures, which we trace back to the difference between the feature distributions of the source and target point clouds, and the sensitivity of the widely-used SVD-based loss function to the range of rotation between the two point clouds. We address the first challenge by introducing a new normalization strategy, Match Normalization, and the second via the use of a loss function based on the negative log likelihood of point correspondences. Our two contributions are general and can be applied to many existing learning-based 3D object registration frameworks, which we illustrate by implementing them in two of them, DCP and IDAM. Our experiments on the real-scene TUD-L [26], LINEMOD [23] and Occluded-LINEMOD [7] datasets evidence the benefits of our strategies. They allow for the first time learning-based 3D object registration methods to achieve meaningful results on real-world data. We therefore expect them to be key to the future development of point cloud registration methods. Our source code can be found at https://github.com/Dangzheng/MatchNorm.

Keywords: 6D object pose estimation · Point cloud registration

1 Introduction

Estimating the 6D pose, i.e., 3D rotation and 3D translation, of an object has many applications in various domains, such as robotics grasping, simultaneous localization and mapping (SLAM), and augmented reality. In this context, great progress has been made by learning-based methods operating on RGB(D) images [32,36,42,44,48,57,58,61,62,74]. In particular, these methods achieve impressive results on real-world images.

© The Author(s), under exclusive license to Springer Nature Switzerland AG 2022
S. Avidan et al. (Eds.): ECCV 2022, LNCS 13661, pp. 19–37, 2022.
https://doi.org/10.1007/978-3-031-19769-7_2

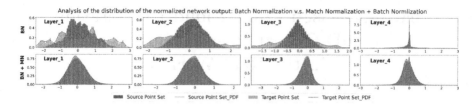

Fig. 1. Feature distributions at different network layers with real-world data. **Top:** With Batch Normalization, the distributions of the features extracted from the model (source) and input (target) point clouds differ significantly. **Bottom:** Our Match Normalization strategy makes these distributions much more similar.

In parallel to this line of research, and thanks to the development of point cloud processing networks [4,35,46,47,56,64,73], several learning-based 3D object registration algorithms [3,65,66,71,72] have emerged to estimate 6D object poses from 3D measurements only, such as those obtained with a LiDAR. Since they focus solely on 3D information, discarding any RGB appearance, these methods have demonstrated excellent generalization to previously unseen objects. However, in contrast with *scene-level* registration methods [11–13,15,16,27], these *object-level* learning-based techniques are typically evaluated only on synthetic datasets, and virtually never on real-world data, such as the TUD-L [26], LineMod [23] and LineMod-Occluded [7] datasets.

In our experiments with the state-of-the-art learning-based object-level registration frameworks, we observed them to struggle with the following challenges. First, in contrast with synthetic datasets where all objects have been normalized to a common scale, the size of different objects in the real world varies widely. The fact that the sensor depicts only an unknown portion the object precludes a simple re-scaling of the target point cloud. In synthetic data, the target point cloud is typically sampled from the normalized model, thus ignoring this difficulty. Second, while synthetic datasets typically limit the rotation between the source and target point clouds in the 45° range, real-world sensors may observe the target object from any viewpoint, covering the full rotation range.

As shown in the top row of Fig. 1, the first above-mentioned challenge translates to a significant gap between the feature distributions of the source and target point clouds in the inner layers of the network. The greater the difference between the two distributions, the fewer correct inlier matches will be found, which then yields a decrease in performance. To address this, we propose a new normalization method, which we refer to as Match Normalization. Match Normalization exploits an instance-level scale parameter in each layer of the feature extraction network. This parameter is shared by the source and target point clouds, thus providing robustness to partial observations and outliers. This makes the distributions of the two point clouds more concentrated and similar, as shown in the bottom row of Fig. 1, and yields increase the number of inlier matches, as evidenced by our experiments.

Furthermore, we observed the second above-mentioned challenge to cause instabilities in the network convergence when relying on the widely-used SVD-based loss function [65, 66, 71] for training. To address this, we exploit a simple negative log-likelihood (NLL) loss function, which we show to improve convergence and lead to better object-level pose estimation accuracy.

Altogether, our contributions have the following advantages: (i) The proposed Match Normalization is applicable to many point cloud registration network architectures; (ii) Both contributions only involve small changes to the network and yet substantially improve its performance on real object-level pose estimation datasets; (iii) They allow for the first time a learning-based point cloud registration method to achieve meaningful results on real-world 6D object pose estimation datasets, such as the TUD-L [26], LINEMOD [23] and Occluded-LINEMOD [7] datasets. We release our code to facilitate reproducibility and future research.

2 Related Work

Traditional Point Cloud Registration Methods. The Iterative Closest Point (ICP) [5] is the best-known local registration methods.Several variants, such as Generalized-ICP [55] and Sparse ICP [6], have been proposed to improve robustness to noise and mismatches, and we refer the reader to [45,51] for a complete review of ICP-based strategies. The main drawback of these methods is their requirement for a reasonable initialization to converge to a good solution. Only relatively recently has this weakness been addressed by the globally-optimal registration method Go-ICP [70]. In essence, this approach follows a branch-and-bound strategy to search the entire 3D motion space $SE(3)$. A similar strategy is employed by the sampling-based global registration algorithm Super4PCS [40]. While globally optimal, Go-ICP come at a much higher computational cost than vanilla ICP. This was, to some degree, addressed by the Fast Global Registration (FGR) algorithm [75], which leverages a local refinement strategy to speed up computation. While effective, FGR still suffers from the presence of noise and outliers in the point sets, particularly because, as vanilla ICP, it relies on 3D point-to-point distance to establish correspondences. In principle, this can be addressed by designing point descriptors that can be more robustly matched. For example, [60] relies on generating pose hypotheses via feature matching, followed by a RANSAC-inspired method to choose the candidate pose with the largest number of support matches. Similarly, TEASER [68] and its improved version TEASER++ [69] take putative correspondences obtained via feature matching as input and remove the outlier ones by an adaptive voting scheme. In addition to the above, there are many algorithms [1, 2, 9, 10, 18–20, 22, 24, 29, 30, 33, 37, 39–41, 49, 50, 52, 53, 68] that have contributed to this direction.

Learning-Based Object Point Cloud Registration Methods. A key requirement to enable end-to-end learning-based registration was the design of deep networks acting on unstructured sets. Deep sets [73] and PointNet [46] constitute the pioneering works in this direction. In particular, PointNetLK [3] combines the PointNet backbone with the traditional, iterative Lucas-Kanade

(LK) algorithm [38] so as to form an end-to-end registration network; DCP [65] exploits DGCNN [64] backbones followed by Transformers [59] to establish 3D-3D correspondences, which are then passed through an SVD layer to obtain the final rigid transformation. While effective, PointNetLK and DCP cannot tackle the partial-to-partial registration scenario. That is, they assume that both point sets are fully observed, during both training and test time. This was addressed by PRNet [66] via a deep network designed to extract keypoints from each input set and match these keypoints. This network is then applied in an iterative manner, so as to increasingly refine the resulting transformation. IDAM [34] builds on the same idea as PRNet, using a two-stage pipeline and a hybrid point elimination strategy to select keypoints. By contrast, RPM-Net [71] builds on DCP and adopts a different strategy, replacing the softmax layer with an optimal transport one so as to handle outliers. Nevertheless, as PRNet, RPM-Net relies on an iterative strategy to refine the computed transformation. DeepGMR [72] leverages mixtures of Gaussians and formulates registration as the minimization of the KL-divergence between two probability distributions to handle outliers. In any event, the methods discussed above were designed to handle point-clouds in full 3D, and were thus neither demonstrated for registration from 2.5D measurements, nor evaluated on real scene datasets, such as TUD-L, LINEMOD and Occluded-LINEMOD. In this work, we identify and solve the issues that prevent the existing learning-based methods from working on real-world data, and, as a result, develop the first learning-based point cloud registration method able to get reasonable result on the real-world 6D pose estimation datasets.

3   Methodology

3.1   Problem Formulation

Let us now introduce our approach to object-level 3D registration. We consider the problem of partial-to-whole registration between two point clouds $\mathcal{X} = \{\mathbf{x}_1, \cdots, \mathbf{x}_M\}$ and $\mathcal{Y} = \{\mathbf{y}_1, \cdots, \mathbf{y}_N\}$, which are two sets of 3D points sampled from the same object surface, with $\mathbf{x}_i, \mathbf{y}_j \in \mathbb{R}^3$. We typically refer to $\mathcal{X}$ as the source point set, which represents the whole object, and to $\mathcal{Y}$ as the target point set, which only contains a partial view of the object. We obtain the source point set $\mathcal{X}$ by uniform sampling from the mesh model, and the target one $\mathcal{Y}$ from a depth map I_{depth} acquired by a depth sensor, assuming known camera intrinsic parameters. Our goal is to estimate the rotation matrix $\mathbf{R} \in SO(3)$ and translation vector $\mathbf{t} \in \mathbb{R}^3$ that align $\mathcal{X}$ and $\mathcal{Y}$. The transformation $\mathbf{R}, \mathbf{t}$ can be estimated by solving

$$\min_{\mathbf{R}, \mathbf{t}} = \sum_{\mathbf{x} \in \mathcal{X}_s} \|\mathbf{R}\mathbf{x} + \mathbf{t} - \mathcal{Q}(\mathbf{x})\|_2^2, \tag{1}$$

where $\mathcal{Q} : \mathcal{X}_s \rightarrow \mathcal{Y}_s$ denotes the function that returns the best matches from set $\mathcal{X}_s$ to set $\mathcal{Y}_s$, with $\mathcal{X}_s$ and $\mathcal{Y}_s$ the selected inlier point sets. Given the matches, Eq. 1 can be solved via SVD [5,21]. The challenging task therefore is to estimate the matching function $\mathcal{Q}$, with only $\mathcal{X}$ and $\mathcal{Y}$ as input.

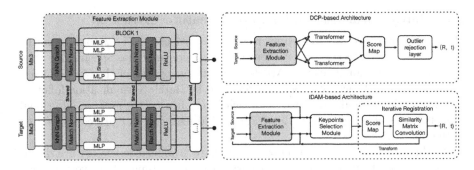

Fig. 2. Architecture of our DCP-based and IDAM-based frameworks. We integrate Match Normalization in every block of the feature extraction module, sharing the scale parameter between the corresponding source and target point sets. The feature extraction module is central to the success of 3D registration architectures, such as DCP and IDAM. Its outputs greatly affect the subsequent score maps, and thus the construction of matches and ultimately the pose estimates. Our Match Normalization strategy yields robust features in the presence of real-world data.

3.2   Method Overview

Most learning-based 3D registration methods rely on an architecture composed of two modules: the feature extraction module and the point matching module. The feature extraction module takes the two point sets as input, and outputs a feature vector $\mathbf{f}_x^{(i)}$, resp. $\mathbf{f}_y^{(j)}$, for either each point [65,71] in $\mathcal{X}$, resp. $\mathcal{Y}$, or each keypoint [34,66] extracted from $\mathcal{X}$, resp. $\mathcal{Y}$.

Given these feature vectors, a score map $\mathcal{S} \in \mathbb{R}^{M \times N}$ is formed by computing the similarity between each source-target pair of descriptors. That is, the (i,j)-th element of $\mathcal{S}$ is computed as

$$\mathcal{S}_{i,j} = <\mathbf{f}_x^{(i)}, \mathbf{f}_y^{(j)}>, \quad \forall (i,j) \in [1,M] \times [1,N] , \tag{2}$$

where $< \cdot, \cdot >$ is the inner product, and $\mathbf{f}_x^{(i)}, \mathbf{f}_y^{(j)} \in \mathbb{R}^P$. This matrix is then given to the point matching module, whose goal is to find the correct inlier matches between the two point sets while rejecting the outliers.

The parameters of the network are typically trained by minimizing the difference between the ground-truth rotation and translation and those obtained by solving Eq. 1. Because, given predicted matches, the solution to Eq. 1 can be obtained by SVD, training can be achieved by backpropagating the gradient through the SVD operator, following the derivations in [28].

In the remainder of this section, we first introduce our approach to robustifying the feature extraction process, and then discuss the loss function we use to stabilize the training process in the presence of a full rotation range. Finally, we provide the details of the two models, DCPv2 [65] and IDAM [34], in which we implemented our strategies.

3.3   Match Normalization for Robust Feature Extraction

Most learning-based 3D registration methods [34,65,66,71,72] build on the PointNet [46] architecture for the feature extraction module. Specifically, they use either convolutional layers or MLPs operating on a graph or the raw point set, with the output of each layer being normalized by Batch Normalization followed by ReLU. The normalized features of each layer then go through an additional subnetwork, which aggregates them into global features that are concatenated to the point-wise ones.

In this process, Batch Normalization aims to standardize the feature distributions to speed up training. Batch normalization assumes that every sample follows the same global statistics. In synthetic datasets, where all point clouds have been normalized to a common size, this assumption is typically satisfied. However, real-world data obtained by capturing objects with a depth sensor often includes objects of highly diverse sizes and only depicts unknown portions of these objects. Therefore, the data does not meet the Batch Normalization assumption, and using the same normalization values for all samples within a mini-batch yields a large gap in the distributions of the features extracted by the network. This is illustrated in the top row of Fig. 1, where we show the histogram and corresponding probability density function of the output of each layer of a network trained with Batch Normalization. Specifically, for each layer, we show a histogram encompassing all the feature channels for all the points in the source point cloud, and a similar histogram for the corresponding target point cloud. Additional examples are provided in the supplementary material. These plots clearly highlights the differences between the distributions of the corresponding source and target point clouds. In practice, these differences then affect the number of matched points, ultimately leading to low registration accuracy.

To overcome this, we introduce Match Normalization. The central idea between Match Normalization is to force the two point sets to have similar distributions. Specifically, we achieve this by centering the source and target point sets separately but by scaling them with the same parameter. The resulting normalized features then satisfy the Batch Normalization assumption, and, as shown in Fig. 2, we then feed them into a Batch Normalization layer to retain the benefit of fast training.

Formally, Match Normalization can be expressed as follows. For a layer with C output channels, let $\mathbf{o}_x \in \mathbb{R}^{C \times M}$ and $\mathbf{o}_y \in \mathbb{R}^{C \times N}$ be the features obtained by processing $\mathcal{X}$ and $\mathcal{Y}$, respectively. We then normalize the features for each point i as

$$\hat{\mathbf{o}}_x^{(i)} = \frac{1}{\beta}(\mathbf{o}_x^{(i)} - \mu_x), \qquad \hat{\mathbf{o}}_y^{(i)} = \frac{1}{\beta}(\mathbf{o}_y^{(i)} - \mu_y), \qquad (3)$$

where $\mu_x = \frac{1}{M}\sum_{i=1}^{M} \mathbf{o}_x^{(i)}$, and similarly for μ_y, are calculated separately for $\mathbf{o}_x$ and $\mathbf{o}_y$, but the scale

$$\beta = \max_{(i,j)=(1,1)}^{(M,C)} |\mathbf{o}_x^{(i,j)}|, \qquad (4)$$

where $\mathbf{o}_x^{(i,j)}$ denotes the j-th feature of the i-th point, is shared by the corresponding source and target point sets. This scale parameter is computed from

Table 1. Influence of the loss function. The models are evaluated on the partial-to-partial registration task on ModelNet40 (clean) as in [34,66,71], with either a 45° rotation range, or a full one. For a given rotation range, the NLL loss yields better results than the SVD-based one. In the full rotation range scenario, the SVD-based loss fails completely, while the NLL loss still yields reasonably accurate pose estimates.

Method	Rotation mAP						Translation mAP					
	5°	10°	15°	20°	25°	30°	0.001	0.005	0.01	0.05	0.1	0.5
DCP(v2)+SVD (45°)	0.36	0.75	0.90	0.96	0.98	0.99	0.51	0.92	0.98	1.00	1.00	1.00
DCP(v2)+NLL (45°)	0.72	0.97	1.00	1.00	1.00	1.00	0.54	0.99	1.00	1.00	1.00	1.00
DCP(v2)+SVD (Full)	0.00	0.01	0.02	0.03	0.04	0.05	0.04	0.23	0.42	0.96	1.00	1.00
DCP(v2)+NLL (Full)	0.35	0.67	0.86	0.94	0.97	0.98	0.19	0.57	0.85	1.00	1.00	1.00

the source features, which are not subject to partial observations as the source points are sampled from the object model.

One advantage of using the same scale parameter for both point sets is robustness to partial observations and to outliers. Indeed, if the target point cloud had its own scaling parameter, the presence of partial observations, respectively outlier measurements, in the target point cloud might lead to stretching, respectively squeezing, it. By contrast, the source point cloud is complete and does not contain outliers. We thus leverage the intuition that the source and target point sets should be geometrically similar, and use the same scaling parameters in the Match Normalization process.

3.4 NLL Loss Function for Stable Training

In the commonly-used synthetic setting, the relative rotation between the two point clouds is limited to the $[0°, 45°]$ range. By contrast, in real data, the objects' pose may cover the full rotation range. To mimic this, we use the synthetic ModelNet40 dataset and modify the augmentation so as to generate samples in the full rotation range. As shown in Table 1, our DCPv2 baseline, although effective for a limited rotation range, fails in the full range setting. Via a detailed analysis of the training behavior, we traced the reason for this failure back to the choice of loss function. Specifically, the use of an SVD-based loss function yields instabilities in the gradient computation. This is due to the fact that, as can be seen from the mathematical expression of the SVD derivatives in [28], when two singular values are close to each other in magnitude, the derivatives explode.

To cope with this problem, inspired by [54], we propose to use the negative log likelihood loss to impose a direct supervision on the score map. To this end, let $\mathcal{M} \in \{0,1\}^{M \times N}$ be the matrix of ground-truth correspondences, with a 1 indicating a correspondence between a pair of points. To build the ground-truth assignment matrix $\mathcal{M}$, we transform $\mathcal{X}$ using the ground-truth transformation $\mathcal{T}$, giving us $\tilde{\mathcal{X}}$. We then compute the pairwise Euclidean distance matrix between $\tilde{\mathcal{X}}$ and $\mathcal{Y}$, which we threshold to obtain a correspondence matrix $\mathcal{M} \in \{0,1\}$.

We augment $\mathcal{M}$ with an extra row and column acting as outlier bins to obtain $\bar{\mathcal{M}}$. The points without any correspondence are treated as outliers, and the corresponding positions in $\bar{\mathcal{M}}$ are set to one. This strategy does not guarantee a bipartite matching, which we address using a forward-backward check.

We then express our loss function as the negative log-likelihood

$$\mathcal{L}(\bar{\mathcal{P}}, \bar{\mathcal{M}}) = \frac{-\sum_{i=1}^{M}\sum_{j=1}^{N}(\log \bar{\mathcal{P}}_{i,j})\bar{\mathcal{M}}_{i,j}}{\sum_{i=1}^{M}\sum_{j=1}^{N}\bar{\mathcal{M}}_{i,j}}, \tag{5}$$

where $\bar{\mathcal{P}}$ is the estimated score map, and where the denominator normalizes the loss value so that different training samples containing different number of correspondences have the same influence in the overall empirical risk.

3.5   Network Architectures

In this section, we present the two architectures in which we implemented our strategies. One of them relies on point-wise features whereas the other first extract keypoints, thus illustrating the generality of our contributions.

DCP-Based Architecture. In DCPv2, the feature extraction module, denoted by $\Theta(\cdot, \cdot)$, takes two point sets as input, and outputs the feature matrix θ_x, resp. θ_y, i.e., one P-dimensional feature vector per 3D point for $\mathcal{X}$, resp. $\mathcal{Y}$. Then two feature matrices be passed to a transformer, which learns a function $\phi : \mathbb{R}^{M \times P} \times \mathbb{R}^{N \times P} \to \mathbb{R}^{M \times P}$, that combines the information of the two point sets. Ultimately, this produces the descriptor matrix $\mathbf{f}_x$, resp. $\mathbf{f}_y$, for $\mathcal{X}$, resp. $\mathcal{Y}$, written as

$$\mathbf{f}_x = \theta_x + \phi(\theta_x, \theta_y), \quad \mathbf{f}_y = \theta_y + \phi(\theta_y, \theta_x) . \tag{6}$$

For our architecture, we integrate our Match Normalization strategy to the layers of the feature extractor $\Theta(\cdot, \cdot)$, while keeping the transformer architecture unchanged.

Inspired by previous work [54,71], we choose to use a Sinkhorn layer to handle the outliers. Specifically, we extend the score matrix $\mathcal{S}$ of Eq. 2 by one row and one column to form an augmented score matrix $\bar{\mathcal{S}}$. The values at the newly-created positions in $\bar{\mathcal{S}}$ are set to

$$\bar{\mathcal{S}}_{i,N+1} = \bar{\mathcal{S}}_{M+1,j} = \bar{\mathcal{S}}_{M+1,N+1} = \alpha, \tag{7}$$

$\forall i \in [1, M]$, $\forall j \in [1, N]$, where $\alpha \in \mathbb{R}$ is a fixed parameter, which we set to 1 in practice. The values at the other indices directly come from $\mathcal{S}$. Given the augmented score map $\bar{\mathcal{S}}$, we aim to find a partial assignment $\bar{\mathcal{P}} \in \mathbb{R}^{(M+1) \times (N+1)}$, defining correspondences between the two point sets, extended with the outlier bins. This assignment is obtained by a differentiable version of the Sinkhorn algorithm [54,71], and is used in the calculation of the loss function of Eq. 2.

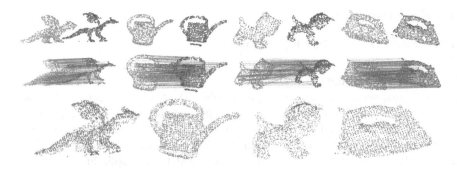

Fig. 3. Qualitative results of Ours-DCP+ICP. Top: Input source (orange) and target (blue) point clouds. We show objects from the TUD-L (dragon,watering can), LINEMOD (kitten) and Occluded-LINEMOD (iron) datasets. **Middle:** Matches found by Ours-DCP+ICP, with the true inlier matches in green, and the outlier matches in red. **Bottom:** Registration results, showing that the source and target sets are correctly aligned.

At test time, we use the output of the Sinkhorn layer to find the best set of corresponding points between the two point clouds. In addition to the points found as outliers, we also discard those whose value in the score map are below a threshold.

IDAM-Based Architecture. The main difference compared to the DCP-based architecture is that this architecture builds the score map upon selected keypoints instead of all the input points. Similarly to the DCP-based architecture, the IDAM-based one uses a feature extraction module $\Theta(\cdot, \cdot)$. This module can be a traditional local descriptor, FPFH [52], or a learning-based method. We therefore integrate our Match Normalization to the learning-based feature extraction network. The extracted features θ_x and θ_y are then passed to a keypoint selection module, corresponding to a second network that outputs a significance score. The significance score is used to obtain a fixed number of keypoints. We denote the features of the reduced keypoint sets as $\tilde{\theta}_x \in \mathbb{R}^{M' \times P}$ and $\tilde{\theta}_y \in \mathbb{R}^{N' \times P}$. These features, combined with their corresponding original coordinates, are used to calculate a reduced score map $\mathcal{S}' \in \mathbb{R}^{M' \times N' \times (2P+4)}$, which is processed by a similarity matrix convolutional neural network to obtain the final score map used to find the matches. IDAM further incorporates an iterative registration loop to this process to refine the results, as illustrated in Fig. 2. Please refer to [34] for more detail.

For IDAM, the loss function is composed of three terms: One to supervise the score matrix as in the DCP-based architecture, and two to supervise the keypoint selection network. The framework assumes that the outliers have been eliminated in the keypoint selection process. Therefore, at test time, we simply compute the argmax of each row to find the best matches. More details can be found in the paper [34].

4   Experiments

4.1   Datasets and Training Parameters

We evaluate our method on three object-level pose estimation real scene datasets: TUD-L [26], LINEMOD [23], Occluded-LINEMOD [7]. For TUD-L, we use the provided real scene training data for training. As there are only 1214 testing images and no explicit training data in Occluded-LINEMOD, we train our network based on the LINEMOD training data. To be specific, we use the PBR dataset provided by the BOP Benchmark [26]. For testing, we follow the BOP 2019 challenge instructions and use the provided testing split for testing. We implement our DCP-based pose estimation network in Pytorch [43] and train it from scratch. We use the Adam optimizer [31] with a learning rate of 10^{-3} and mini-batches of size 32, and train the network for 30,000 iterations. For the OT layer, we use $k = 50$ iterations and set $\lambda = 0.5$. For our IDAM-based architecture, we train the model with the Adam optimizer [31] until convergence, using a learning rate of 10^{-4} and mini-batches of size 32. We use the FPFH implementation from the Open3D [76] library and our custom DGCNN for feature extraction. We set the number of refinement iterations for both the FPFH-based and DGCNN-based versions to 3. For both frameworks, we set the number of points for $\mathcal{X}$ and $\mathcal{Y}$ to be 1024 and 768, respectively, encoding the fact that $\mathcal{Y}$ only contains a visible portion of $\mathcal{X}$. To obtain the target point clouds from the depth maps, we use the masks provided with the datasets. Training was performed on one NVIDIA RTX8000 GPU.

4.2   Evaluation Metrics

For evaluation, in addition to the three metrics used by the BOP benchmark, Visible Surface Discrepancy (VSD) [25,26], Maximum Symmetry-Aware Surface Distance (MSSD) [17] and Maximum Symmetry-Aware Projection Distance (MSPD) [8], we also report the rotation and translation error between the predictions $\hat{R}, \hat{\mathbf{t}}$ and the ground truth $R_{gt}, \mathbf{t}_{gt}$. These errors are computed as

$$E_{rot}(\hat{R}, R_{gt}) = arccos\frac{trace(\hat{R}^\top R_{gt}) - 1}{2} \;, E_{trans}(\hat{\mathbf{t}}, \mathbf{t}_{gt}) = \left\|\hat{\mathbf{t}} - \mathbf{t}_{gt}\right\|_2^2 \;. \quad (8)$$

We summarize the results in terms of mean average precision (mAP) of the estimated relative pose under varying accuracy thresholds, as in [14]. We keep the rotation error unchanged. Furthermore, we also report the ADD metric [67], which measures the average distance between the 3D model points transformed using the predicted pose and those obtained with the ground-truth one. We set the threshold to be 10% of the model diameter, as commonly done in 6D pose estimation (Fig. 4).

4.3   Comparison with Existing Methods

We compare our method to both traditional techniques and learning-based ones. Specifically, for the traditional methods, we used the Open3D [76] implementations of ICP [5] and FGR [75], the official implementation of TEASER++ [68],

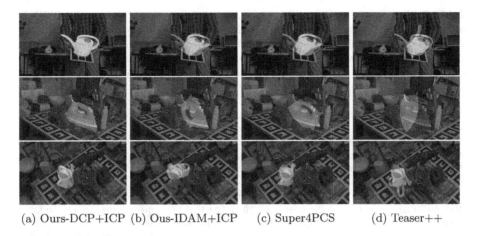

(a) Ours-DCP+ICP (b) Ous-IDAM+ICP   (c) Super4PCS   (d) Teaser++

Fig. 4. Qualitative results on the TUD-L (top), LINEMOD (middle) and Occluded-LINEMOD (bottom) datasets.

and the author's binary file for Super4PCS [40]. For DCP, we used the default DCPv2 training settings. However, because of instabilities caused by the SVD computation, we had to train the model several times to eventually find a point before a crash achieving reasonable accuracy. Note that this problem was also reported in [63]. We also tried to train PRNet and RPMNet, but failed to get reasonable results because of similar SVD-related crashes, as also observed in [12]; specifically, SVD always crashes at the beginning of training when processing real data. For IDAM, we report results using both traditional FPFH features and features extracted with a DGCNN. Ours-DCP denotes our approach implemented in the DCPv2 architecture, where we replace the SVD-based loss with the NLL one, replace the softmax layer with a Sinkhorn layer, and integrate Match Normalization with the DGCNN. Ours-IDAM denotes our approach within the IDAM architecture, incorporating Match Normalization in the DGCNN. Since the IDAM does not use an SVD-based loss function, we retained its original loss.

TUD-L Dataset. The results of all methods for the TUD-L dataset are summarized in Table 2. Note that the traditional methods based on FPFH features yield poor results, because FPFH yields unreliable features in the presence of many smooth areas on the objects. Vanilla DCPv2 and IDAM also struggle with such real-world data. However, these baselines are significantly improved by our Match Normalization strategy, Ours-DCP and Ours-IDAM, both of which outperform Super4PCS. Our results can further be boosted by the use of ICP as a post-processing step.

In the table, we also provide the results of 'Vidal-Sensors18' and 'Drost-CVPR10-3D-Edges', the two best depth-only performers in the BOP leaderboard. Note that these are traditional methods whose results were obtained without using a mask to segment the target point cloud, which makes the com-

Table 2. Quantitative comparison of our method with previous work on the **TUD-L** real scene dataset. The results for Vidal-Sensor18 [60] and Drost (Drost-CVPR10-3D-Edges) [18] were directly taken from the BOP leaderboard, and, in contrast with all the other results, were obtained without using a mask for the target point cloud. Our contributions allow existing learning-based methods, such as IDAM and DCP, to successfully register real-world data.

Method	Rotation mAP			Translation mAP			ADD	BOP Benchmark			
	5°	10°	20°	1 cm	2 cm	5 cm	0.1d	VSD	MSSD	MSPD	AR
ICP	0.02	0.02	0.02	0.01	0.14	0.57	0.02	0.117	0.023	0.027	0.056
FGR(FPFH)	0.00	0.01	0.01	0.04	0.25	0.63	0.01	0.071	0.007	0.008	0.029
TEASER++(FPFH)	0.13	0.17	0.19	0.03	0.22	0.56	0.17	0.175	0.196	0.193	0.188
Super4PCS	0.30	0.50	0.56	0.05	0.40	0.92	0.54	0.265	0.500	0.488	0.418
DCP(v2)	0.00	0.01	0.02	0.02	0.07	0.55	0.01	0.0253	0.051	0.039	0.038
IDAM(FPFH)	0.05	0.12	0.20	0.03	0.17	0.63	0.13	0.100	0.194	0.166	0.153
IDAM	0.03	0.05	0.10	0.02	0.08	0.49	0.05	0.067	0.108	0.099	0.091
⋆Vidal-Sensors18	–	–	–	–	–	–	–	0.811	0.910	0.907	0.876
⋆Drost	–	–	–	–	–	–	–	0.809	0.875	0.872	0.852
Ours-IDAM	0.36	0.46	0.53	0.23	0.47	0.75	0.46	0.339	0.502	0.492	0.444
Ours-IDAM+ICP	0.56	0.58	0.61	0.55	0.66	0.81	0.58	0.580	0.604	0.618	0.601
Ours-DCP	0.70	0.81	0.87	0.71	0.86	0.97	0.85	0.700	0.853	0.852	0.801
Ours-DCP+ICP	**0.91**	**0.92**	**0.93**	**0.86**	**0.95**	**0.99**	**0.93**	**0.859**	**0.914**	**0.935**	**0.903**

parison favorable to our approach. Nevertheless, our results evidence that our contributions can make learning-based 3D object registration applicable to real-world data, which we believe to be a significant progress in the field.

LINEMOD Dataset. In contrast with TUD-L, the LINEMOD dataset contains symmetrical objects and small occlusions at the object boundaries, which increase the difficulty of this dataset. As shown in Table 3, even Super4PCS is therefore unable to yield meaningful results on this dataset. Furthermore, as the LINEMOD training data does not contain any real-world measurements, the training-testing domain gap further complicates the task for learning-based methods. Nevertheless, our approach improves the results of both DCP and IDAM, allowing them to produce reasonably accurate pose estimates.

Occluded-LINEMOD Dataset. The Occluded-LINEMOD dataset further increases the challenge compared to LINEMOD by adding severe occlusions, in addition to the still-existing domain gap. As such, as shown in Table 4, the results of all the methods deteriorate. Nevertheless, our approach still allows DCP and IDAM to yield meaningful pose estimates.

Table 3. Quantitative comparison of our method with previous work on the **LINEMOD** real scene dataset. PPF_3D_ICP [18] and Drost (Drost-CVPR10-3D-Only) [18] are traditional methods and represent the best depth-only performers from the BOP leaderboard.

Method	Rotation mAP			Translation mAP			ADD	BOP Benchmark			
	5°	10°	20°	1 cm	2 cm	5 cm	0.1 d	VSD	MSSD	MSPD	AR
ICP	0.00	0.01	0.01	0.04	0.27	0.82	0.01	0.092	0.014	0.027	0.044
FGR(FPFH)	0.00	0.00	0.00	0.05	0.31	0.89	0.00	0.068	0.000	0.010	0.026
TEASER++(FPFH)	0.01	0.03	0.05	0.03	0.21	0.73	0.03	0.108	0.076	0.098	0.094
Super4PCS	0.02	0.09	0.15	0.04	0.31	0.89	0.10	0.117	0.178	0.201	0.165
DCP(v2)	0.00	0.00	0.01	0.05	0.24	0.83	0.00	0.057	0.025	0.049	0.044
IDAM(FPFH)	0.00	0.01	0.03	0.03	0.16	0.67	0.01	0.053	0.055	0.069	0.059
IDAM	0.00	0.01	0.05	0.03	0.16	0.71	0.02	0.050	0.070	0.081	0.067
⋆PPF_3D_ICP	–	–	–	–	–	–	–	**0.719**	**0.856**	**0.866**	**0.814**
⋆Drost	–	–	–	–	–	–	–	0.678	0.786	0.789	0.751
Ours-IDAM	0.01	0.07	0.15	0.13	0.38	0.87	0.11	0.148	0.194	0.209	0.184
Ours-IDAM+ICP	0.15	0.23	0.27	0.25	0.54	0.91	0.23	0.352	0.311	0.345	0.336
Ours-DCP	0.10	0.27	0.49	0.26	0.60	0.95	0.37	0.319	0.490	0.529	0.446
Ours-DCP+ICP	**0.43**	**0.59**	**0.67**	**0.49**	**0.83**	**0.97**	**0.60**	0.616	0.680	0.737	0.678

Table 4. Quantitative comparison of our method with previous work on the **Occluded-LINEMOD** real scene dataset. Vidal-Sensors18 [60] and PPF_3D_ICP [18] are traditional methods and represent the best depth-only performers from the BOP leaderboard.

Method	Rotation mAP			Translation mAP			ADD	BOP Benchmark			
	5°	10°	20°	1 cm	2 cm	5 cm	0.1 d	VSD	MSSD	MSPD	AR
ICP	0.01	0.01	0.01	0.07	0.36	0.85	0.01	0.085	0.014	0.032	0.044
FGR(FPFH)	0.00	0.00	0.00	0.08	0.43	0.85	0.00	0.055	0.000	0.009	0.021
TEASER++(FPFH)	0.01	0.02	0.05	0.04	0.26	0.77	0.02	0.096	0.060	0.093	0.083
Super4PCS	0.01	0.03	0.06	0.06	0.31	0.83	0.03	0.054	0.072	0.113	0.080
DCP(v2)	0.00	0.00	0.01	0.03	0.30	0.83	0.00	0.055	0.018	0.059	0.044
IDAM(FPFH)	0.00	0.00	0.02	0.04	0.18	0.73	0.00	0.044	0.033	0.066	0.048
IDAM	0.00	0.02	0.06	0.07	0.26	0.76	0.02	0.063	0.088	0.119	0.090
⋆Vidal-Sensors18	–	–	–	–	–	–	–	0.473	0.625	0.647	0.582
⋆PPF_3D_ICP	–	–	–	–	–	–	–	**0.523**	**0.669**	**0.716**	**0.636**
Ours-IDAM	0.02	0.08	0.18	0.15	0.44	0.84	0.12	0.155	0.204	0.248	0.202
Ours-IDAM+ICP	0.15	0.22	0.32	0.23	0.58	0.88	0.25	0.349	0.320	0.374	0.348
Ours-DCP	0.07	0.19	0.36	0.24	0.57	0.88	0.28	0.263	0.384	0.450	0.365
Ours-DCP+ICP	**0.31**	**0.46**	**0.56**	**0.37**	**0.70**	**0.91**	**0.47**	0.478	0.542	0.612	0.544

4.4 Ablation Study

In this section, we conduct an ablation study to justify the effectiveness of the proposed Match Normalization. Specifically, we evaluate our proposed DCP-based architecture with and without Match Normalization on the TUD-L dataset. In addition to the metrics used in the previous section, we report the number of matches predicted by the network and the number of real inliers within these predicted matches. The predicted matches are those extracted directly from

Table 5. Ablation Study. We evaluate the influence of our Match Normalization strategy in our DCP-based baseline not only on the same metrics as before but also on the number of matches found by the network.

Method	Rotation mAP			Translation mAP			ADD	BOP Benchmark				Matches	
	5°	10°	20°	1 cm	2 cm	5 cm	0.1 d	VSD	MSSD	MSPD	AR	pred	true
Ours w/o MN	0.21	0.22	0.24	0.22	0.32	0.61	0.23	0.27	0.27	0.27	0.27	24.84	10.44
Ours	0.91	0.92	0.93	0.86	0.95	0.99	0.93	0.86	0.91	0.94	0.90	276.10	262.96

the predicted score map $\bar{P}$. We set the threshold to identify the true inliers to 0.02.

As shown in Table 5, the number of matches significantly increases with our Match Normalization, and a vast majority of them are true inliers. This evidences that normalizing the source and target point sets with Match Normalization indeed helps the network to find correct matches, and eventually improves the pose estimation performance. Qualitative results obtained with our method are shown in Fig. 3. Importantly, Match Normalization (MN) does not affect the training efficiency as it adds no new learnable parameters.

5   Conclusion

We have identified two factors that prevent the existing learning-based 3D object registration methods from working on real-world data. One is the gap between the feature distributions of the source and target point sets. The larger the difference, the fewer inlier matches are found, which deteriorates the performance. Another is the instability in gradient computations when using an SVD-based loss function, which complicates the network's convergence when the data can undergo a full rotation range. To tackle the first issue, we have proposed a new normalization method, Match Normalization, which encourages the two point sets to have similar feature distributions by sharing the same scaling parameter. For the second problem, we have replaced the SVD-based loss function with a simple yet robust NLL loss function that imposes direct supervision on the score map. Our two solutions are simple, effective and can be easily applied to many existing learning-based 3D registration frameworks. We have integrated them into a DCP-based and an IDAM-based architecture, and have proven the effectiveness of our method on three real-world 6D object pose estimation datasets, TUD-L, LINEMOD and Occluded-LINEMOD. To the best of our knowledge, this is the first time that a learning-based 3D object registration method achieves meaningful results on real-world data.

Acknowledgements. Zheng Dang would like to thank to H. Chen for the highly-valuable discussions and for her encouragement. This work was funded in part by the Swiss Innovation Agency (Innosuisse).

References

1. Agamennoni, G., Fontana, S., Siegwart, R.Y., Sorrenti, D.G.: Point clouds registration with probabilistic data association. In: 2016 IEEE/RSJ International Conference on Intelligent Robots and Systems (IROS), pp. 4092–4098. IEEE (2016)
2. Aiger, D., Mitra, N.J., Cohen-Or, D.: 4-points congruent sets for robust pairwise surface registration. In: ACM SIGGRAPH 2008 papers, pp. 1–10 (2008)
3. Aoki, Y., Goforth, H., Srivatsan, R.A., Lucey, S.: Pointnetlk: robust & efficient point cloud registration using pointnet. In: Conference on Computer Vision and Pattern Recognition, Long Beach, California, pp. 7163–7172 (2019)
4. Atzmon, M., Maron, H., Lipman, Y.: Point convolutional neural networks by extension operators. ACM Trans. Graph. (TOG) (2018)
5. Besl, P., Mckay, N.: A method for registration of 3D shapes. IEEE Trans. Pattern Anal. Mach. Intell. **14**(2), 239–256 (1992)
6. Bouaziz, S., Tagliasacchi, A., Pauly, M.: Sparse iterative closest point. In: Computer Graphics Forum, vol. 32, pp. 113–123. Wiley Online Library, Hoboken (2013)
7. Brachmann, E., Krull, A., Michel, F., Gumhold, S., Shotton, J., Rother, C.: Learning 6D object pose estimation Using 3D object coordinates. In: Fleet, D., Pajdla, T., Schiele, B., Tuytelaars, T. (eds.) ECCV 2014. LNCS, vol. 8690, pp. 536–551. Springer, Cham (2014). https://doi.org/10.1007/978-3-319-10605-2_35
8. Brachmann, E., Michel, F., Krull, A., Yang, M.Y., Gumhold, S., et al.: Uncertainty-driven 6D pose estimation of objects and scenes from a single RGB image. In: Proceedings of the IEEE Conference on Computer Vision and Pattern Recognition, pp. 3364–3372 (2016)
9. Bronstein, A.M., Bronstein, M.M.: Regularized partial matching of rigid shapes. In: Forsyth, D., Torr, P., Zisserman, A. (eds.) ECCV 2008. LNCS, vol. 5303, pp. 143–154. Springer, Heidelberg (2008). https://doi.org/10.1007/978-3-540-88688-4_11
10. Bronstein, A.M., Bronstein, M.M., Bruckstein, A.M., Kimmel, R.: Partial similarity of objects, or how to compare a centaur to a horse. Int. J. Comput. Vision **84**(2), 163–183 (2009)
11. Cao, A.Q., Puy, G., Boulch, A., Marlet, R.: Pcam: product of cross-attention matrices for rigid registration of point clouds. In: Proceedings of the IEEE/CVF International Conference on Computer Vision, pp. 13229–13238 (2021)
12. Choy, C., Dong, W., Koltun, V.: Deep global registration. In: Proceedings of the IEEE Conference on Computer Vision and Pattern Recognition (2020)
13. Choy, C., Park, J., Koltun, V.: Fully convolutional geometric features. In: Proceedings of the IEEE/CVF International Conference on Computer Vision, pp. 8958–8966 (2019)
14. Dang, Z., Moo Yi, K., Hu, Y., Wang, F., Fua, P., Salzmann, M.: Eigendecomposition-free training of deep networks with zero eigenvalue-based losses. In: European Conference on Computer Vision, Munich, Germany, pp. 768–783 (2018)
15. Deng, H., Birdal, T., Ilic, S.: Ppf-foldnet: unsupervised learning of rotation invariant 3D local descriptors. In: Proceedings of the European Conference on Computer Vision (ECCV), pp. 602–618 (2018)
16. Deng, H., Birdal, T., Ilic, S.: Ppfnet: global context aware local features for robust 3D point matching. In: Conference on Computer Vision and Pattern Recognition, Salt Lake City, Utah, pp. 195–205 (2018)

17. Drost, B., Ulrich, M., Bergmann, P., Hartinger, P., Steger, C.: Introducing mvtec itodd-a dataset for 3D object recognition in industry. In: Proceedings of the IEEE International Conference on Computer Vision Workshops, Venice, Italy, pp. 2200–2208 (2017)
18. Drost, B., Ulrich, M., Navab, N., Ilic, S.: Model globally, match locally: efficient and robust 3D object recognition. In: 2010 IEEE Computer Society Conference on Computer Vision and Pattern Recognition, pp. 998–1005 (2010)
19. Fitzgibbon, A.W.: Robust registration of 2D and 3D point sets. Image Vision Comput. **21**(13–14), 1145–1153 (2003)
20. Gelfand, N., Mitra, N.J., Guibas, L.J., Pottmann, H.: Robust global registration. In: Symposium on geometry processing, Vienna, Austria, p. 5 (2005)
21. Gower, J.C.: Generalized procrustes analysis. Psychometrika **40**(1), 33–51 (1975)
22. Hähnel, D., Burgard, W.: Probabilistic matching for 3D scan registration. In: Proceedings of the VDI-Conference Robotik, vol. 2002. Citeseer (2002)
23. Hinterstoisser, S., et al.: Model based training, detection and pose estimation of texture-less 3D objects in heavily cluttered scenes. In: Lee, K.M., Matsushita, Y., Rehg, J.M., Hu, Z. (eds.) ACCV 2012. LNCS, vol. 7724, pp. 548–562. Springer, Heidelberg (2013). https://doi.org/10.1007/978-3-642-37331-2_42
24. Hinzmann, T., et al.: Collaborative 3D reconstruction using heterogeneous UAVs: system and experiments. In: Kulić, D., Nakamura, Y., Khatib, O., Venture, G. (eds.) ISER 2016. SPAR, vol. 1, pp. 43–56. Springer, Cham (2017). https://doi.org/10.1007/978-3-319-50115-4_5
25. Hodaň, T., Matas, J., Obdržálek, Š: On evaluation of 6D object pose estimation. In: Hua, G., Jégou, H. (eds.) ECCV 2016. LNCS, vol. 9915, pp. 606–619. Springer, Cham (2016). https://doi.org/10.1007/978-3-319-49409-8_52
26. Hodan, T., et al.: Bop: benchmark for 6D object pose estimation. In: European Conference on Computer Vision, Munich, Germany, pp. 19–34 (2018)
27. Huang, S., Gojcic, Z., Usvyatsov, M., Wieser, A., Schindler, K.: Predator: registration of 3D point clouds with low overlap. In: Proceedings of the IEEE/CVF Conference on Computer Vision and Pattern Recognition, pp. 4267–4276 (2021)
28. Ionescu, C., Vantzos, O., Sminchisescu, C.: Matrix backpropagation for deep networks with structured layers. In: Conference on Computer Vision and Pattern Recognition, Boston, MA, USA (2015)
29. Izatt, G., Dai, H., Tedrake, R.: Globally optimal object pose estimation in point clouds with mixed-integer programming. In: Amato, N.M., Hager, G., Thomas, S., Torres-Torriti, M. (eds.) Robotics Research. SPAR, vol. 10, pp. 695–710. Springer, Cham (2020). https://doi.org/10.1007/978-3-030-28619-4_49
30. Johnson, A.E., Hebert, M.: Using spin images for efficient object recognition in cluttered 3d scenes. TPAMI **21**(5), 433–449 (1999)
31. Kingma, D.P., Ba, J.: Adam: a method for stochastic optimization. In: International Conference on Learning Representations, San Diego, CA, USA (2015)
32. Labbé, Y., Carpentier, J., Aubry, M., Sivic, J.: CosyPose: consistent multi-view multi-object 6D pose estimation. In: Vedaldi, A., Bischof, H., Brox, T., Frahm, J.-M. (eds.) ECCV 2020. LNCS, vol. 12362, pp. 574–591. Springer, Cham (2020). https://doi.org/10.1007/978-3-030-58520-4_34
33. Le, H.M., Do, T.T., Hoang, T., Cheung, N.M.: Sdrsac: semidefinite-based randomized approach for robust point cloud registration without correspondences. In: Proceedings of the IEEE/CVF Conference on Computer Vision and Pattern Recognition, pp. 124–133 (2019)

34. Li, J., Zhang, C., Xu, Z., Zhou, H., Zhang, C.: Iterative distance-aware similarity matrix convolution with mutual-supervised point elimination for efficient point cloud registration. In: Vedaldi, A., Bischof, H., Brox, T., Frahm, J.-M. (eds.) ECCV 2020. LNCS, vol. 12369, pp. 378–394. Springer, Cham (2020). https://doi.org/10.1007/978-3-030-58586-0_23

35. Li, Y., Bu, R., Sun, M., Wu, W., Di, X., Chen, B.: Pointcnn: convolution on x-transformed points. In: Advances in Neural Information Processing Systems, Montréal, Quebec, Canada, pp. 820–830 (2018)

36. Li, Y., Wang, G., Ji, X., Xiang, Y., Fox, D.: DeepIM: deep iterative matching for 6D pose estimation. In: European Conference on Computer Vision, Munich, Germany, pp. 683–698 (2018)

37. Litany, O., Bronstein, A.M., Bronstein, M.M.: Putting the pieces together: regularized multi-part shape matching. In: Fusiello, A., Murino, V., Cucchiara, R. (eds.) ECCV 2012. LNCS, vol. 7583, pp. 1–11. Springer, Heidelberg (2012). https://doi.org/10.1007/978-3-642-33863-2_1

38. Lucas, B.D., Kanade, T., et al.: An iterative image registration technique with an application to stereo vision. In: International Joint Conference on Artificial Intelligence. Vancouver, British Columb (1981)

39. Maron, H., Dym, N., Kezurer, I., Kovalsky, S., Lipman, Y.: Point registration via efficient convex relaxation. ACM Trans. Graph. (TOG) **35**(4), 1–12 (2016)

40. Mellado, N., Aiger, D., Mitra, N.J.: Super 4 pcs fast global pointcloud registration via smart indexing. In: Computer Graphics Forum, vol. 33, pp. 205–215. Wiley Online Library (2014)

41. Mohamad, M., Ahmed, M.T., Rappaport, D., Greenspan, M.: Super generalized 4pcs for 3D registration. In: 2015 International Conference on 3D Vision, pp. 598–606. IEEE (2015)

42. Park, K., Patten, T., Vincze, M.: Pix2pose: pixel-wise coordinate regression of objects for 6D pose estimation. In: International Conference on Computer Vision, Seoul, Korea, pp. 7668–7677 (2019)

43. Paszke, A., et al.: Automatic differentiation in pytorch. In: International Conference on Learning Representations, Toulon, France (2017)

44. Peng, S., Liu, Y., Huang, Q., Zhou, X., Bao, H.: PVNet: pixel-wise voting network for 6DoF pose estimation. In: Conference on Computer Vision and Pattern Recognition, Long Beach, California, pp. 4561–4570 (2019)

45. Pomerleau, F., Colas, F., Siegwart, R., et al.: A review of point cloud registration algorithms for mobile robotics. Found. Trends® Rob. **4**(1), 1–104 (2015)

46. Qi, C., Su, H., Mo, K., Guibas, L.: Pointnet: deep learning on point sets for 3D classification and segmentation. In: Conference on Computer Vision and Pattern Recognition, Honolulu, Hawaii (2017)

47. Qi, C., Yi, L., Su, H., Guibas, L.: Pointnet++: deep hierarchical feature learning on point sets in a metric space. In: Advances in Neural Information Processing Systems, Long Beach, California, United States (2017)

48. Rad, M., Lepetit, V.: Bb8: a scalable, accurate, robust to partial occlusion method for predicting the 3D poses of challenging objects without using depth. In: International Conference on Computer Vision, Venice, Italy, pp. 3828–3836 (2017)

49. Raposo, C., Barreto, J.P.: Using 2 point+ normal sets for fast registration of point clouds with small overlap. In: 2017 IEEE International Conference on Robotics and Automation (ICRA), pp. 5652–5658. IEEE (2017)

50. Rosen, D.M., Carlone, L., Bandeira, A.S., Leonard, J.J.: Se-sync: a certifiably correct algorithm for synchronization over the special euclidean group. Int. J. Rob. Res. **38**(2–3), 95–125 (2019)

51. Rusinkiewicz, S., Levoy, M.: Efficient variants of the icp algorithm. In: Proceedings Third International Conference on 3-D Digital Imaging and Modeling, pp. 145–152. IEEE, Quebec City (2001)
52. Rusu, R.B., Blodow, N., Beetz, M.: Fast point feature histograms (fpfh) for 3D registration. In: International Conference on Robotics and Automation, pp. 3212–3217. IEEE, Kobe (2009)
53. Rusu, R.B., Blodow, N., Marton, Z.C., Beetz, M.: Aligning point cloud views using persistent feature histograms. In: International Conference on Intelligent Robots and Systems, pp. 3384–3391. IEEE, Nice (2008)
54. Sarlin, P.E., DeTone, D., Malisiewicz, T., Rabinovich, A.: Superglue: learning feature matching with graph neural networks. In: Conference on Computer Vision and Pattern Recognition. IEEE, Long Beach (2019)
55. Segal, A., Haehnel, D., Thrun, S.: Generalized-icp. In: In Robotics: Science and Systems, Cambridge (2009)
56. Su, H., et al.: Splatnet: sparse lattice networks for point cloud processing. In: Proceedings of the IEEE Conference on Computer Vision and Pattern Recognition, pp. 2530–2539 (2018)
57. Sundermeyer, M., Marton, Z.C., Durner, M., Brucker, M., Triebel, R.: Implicit 3D orientation learning for 6D object detection from RGB images. In: European Conference on Computer Vision, Munich, Germany, pp. 699–715 (2018)
58. Tremblay, J., To, T., Sundaralingam, B., Xiang, Y., Fox, D., Birchfield, S.: Deep object pose estimation for semantic robotic grasping of household objects. arXiv preprint arXiv:1809.10790 (2018)
59. Vaswani, A., et al.: Attention is all you need. In: Advances in Neural Information Processing Systems, Long Beach, California, United States, pp. 5998–6008 (2017)
60. Vidal, J., Lin, C.Y., Lladó, X., Martí, R.: A method for 6D pose estimation of free-form rigid objects using point pair features on range data. Sensors $18(8)$, 2678 (2018)
61. Wang, C., et al.: Densefusion: 6D object pose estimation by iterative dense fusion. In: Conference on Computer Vision and Pattern Recognition, Long Beach, California, pp. 3343–3352 (2019)
62. Wang, H., Sridhar, S., Huang, J., Valentin, J., Song, S., Guibas, L.J.: Normalized object coordinate space for category-level 6D object pose and size estimation. In: International Conference on Computer Vision, Seoul, Korea, pp. 2642–2651 (2019)
63. Wang, W., Dang, Z., Hu, Y., Fua, P., Salzmann, M.: Backpropagation-friendly eigendecomposition. In: Advances in Neural Information Processing Systems, Vancouver, British Columbia, Canada, pp. 3156–3164 (2019)
64. Wang, Y., Sun, Y., Liu, Z., Sarma, S., Bronstein, M., Solomon, J.: Dynamic graph cnn for learning on point clouds. ACM Trans. Graph. (TOG) (2019)
65. Wang, Y., Solomon, J.M.: Deep closest point: learning representations for point cloud registration. In: International Conference on Computer Vision, Seoul, Korea, pp. 3523–3532 (2019)
66. Wang, Y., Solomon, J.M.: Prnet: Self-supervised learning for partial-to-partial registration. In: Advances in Neural Information Processing Systems, Vancouver, British Columbia, Canada, pp. 8812–8824 (2019)
67. Xiang, Y., Schmidt, T., Narayanan, V., Fox, D.: Posecnn: a convolutional neural network for 6D object pose estimation in cluttered scenes. In: Robotics: Science and Systems Conference, Pittsburgh, PA, USA (2018)
68. Yang, H., Carlone, L.: A polynomial-time solution for robust registration with extreme outlier rates. In: Robotics: Science and Systems Conference, Freiburg im Breisgau, Germany (2019)

69. Yang, H., Shi, J., Carlone, L.: Teaser: fast and certifiable point cloud registration. arXiv Preprint (2020)
70. Yang, J., Li, H., Campbell, D., Jia, Y.: Go-icp: a globally optimal solution to 3D icp point-set registration. TPAMI **38**(11), 2241–2254 (2015)
71. Yew, Z.J., Lee, G.H.: Rpm-net: robust point matching using learned features. In: Conference on Computer Vision and Pattern Recognition. Online (2020)
72. Yuan, W., Eckart, B., Kim, K., Jampani, V., Fox, D., Kautz, J.: DeepGMR: learning latent gaussian mixture models for registration. In: Vedaldi, A., Bischof, H., Brox, T., Frahm, J.-M. (eds.) ECCV 2020. LNCS, vol. 12350, pp. 733–750. Springer, Cham (2020). https://doi.org/10.1007/978-3-030-58558-7_43
73. Zaheer, M., Kottur, S., Ravanbakhsh, S., Poczos, B., Salakhutdinov, R.R., Smola, A.J.: Deep sets. In: Advances in Neural Information Processing Systems, Long Beach, California, United States, pp. 3391–3401 (2017)
74. Zakharov, S., Shugurov, I., Ilic, S.: DPOD: 6D pose object detector and refiner. In: International Conference on Computer Vision, Seoul, Korea (2019)
75. Zhou, Q.-Y., Park, J., Koltun, V.: Fast global registration. In: Leibe, B., Matas, J., Sebe, N., Welling, M. (eds.) ECCV 2016. LNCS, vol. 9906, pp. 766–782. Springer, Cham (2016). https://doi.org/10.1007/978-3-319-46475-6_47
76. Zhou, Q.Y., Park, J., Koltun, V.: Open3D: a modern library for 3D data processing. arXiv Preprint (2018)

An End-to-End Transformer Model for Crowd Localization

Dingkang Liang[1], Wei Xu[2], and Xiang Bai[1(✉)]

[1] Huazhong University of Science and Technology, Wuhan 430074, China
{dkliang,xbai}@hust.edu.cn
[2] Beijing University of Posts and Telecommunications, Beijing 100876, China
xuwei2020@bupt.edu.cn

Abstract. Crowd localization, predicting head positions, is a more practical and high-level task than simply counting. Existing methods employ pseudo-bounding boxes or pre-designed localization maps, relying on complex post-processing to obtain the head positions. In this paper, we propose an elegant, end-to-end **C**rowd **L**ocalization **TR**ansformer named CLTR that solves the task in the regression-based paradigm. The proposed method views the crowd localization as a direct set prediction problem, taking extracted features and trainable embeddings as input of the transformer-decoder. To reduce the ambiguous points and generate more reasonable matching results, we introduce a KMO-based Hungarian matcher, which adopts the nearby context as the auxiliary matching cost. Extensive experiments conducted on five datasets in various data settings show the effectiveness of our method. In particular, the proposed method achieves the best localization performance on the NWPU-Crowd, UCF-QNRF, and ShanghaiTech Part A datasets.

Keywords: Crowd localization · Crowd counting · Transformer

1 Introduction

Crowd localization, a fundamental subtask of crowd analysis, aims to provide the location of each instance. Here, the location means the center points of heads because annotating the bounding box for each head is expensive and laborious in dense scenes. Thus, most crowd datasets only provide point-level annotations. A powerful crowd localization algorithm can give great potential for similar tasks, *e.g.*, crowd tracking [45], object counting [5,16], and object localization [3,46].

The mainstream crowd localization methods can be generally categorized into detection-based (Fig. 1(a)) and map-based (Fig. 1(b)) methods. The detection-based methods [24,31] utilize nearest-neighbor head distances to initialize the

---

Project page at https://dk-liang.github.io/CLTR/.

Supplementary Information The online version contains supplementary material available at https://doi.org/10.1007/978-3-031-19769-7_3.

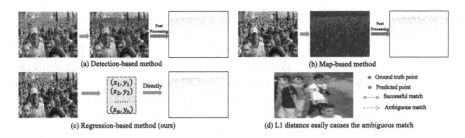

Fig. 1. (a) Detection-based methods, using the predefined pseudo GT bounding boxes. (b) Map-based methods, using high-resolution representation and complicated post-processing. (c) Regression-based methods, mapping the input images to the point coordinates directly. (d) Due to lack of context, the $L1$ distance easily causes the ambiguous match pair.

pseudo ground truth (GT) bounding boxes. However, these detection-based methods can not report satisfactory performance. Moreover, the heuristic non-maximum suppression (NMS) is used to remove the negative predictions. Most crowd localization methods [1,46] are map-based because it has relatively higher localization accuracy. Nevertheless, the map-based methods still suffer some inevitable problems. For instance, complex multi-scale representation is necessary to generate sharp maps. Besides, they adopt non-differentiable post-processing (*e.g.,* find-maxima) to extract the location, which precludes end-to-end training.

In contrast, the regression-based methods, directly regressing the coordinates, are more straightforward than the detection-based and map-based methods, as shown in Fig. 1(c). The benefits of regression-based can be summarized as two folds. (1) It is simple, elegant, and end-to-end trainable since it does not need pre-processing (*e.g.,* pseudo GT boxes or maps generation) and post-processing (*e.g.,* NMS or find-maxima). (2) It does not rely on complex multi-scale fusion mechanisms to generate high-quality feature maps.

Recently, we have witnessed the rise of Transformer [2] in computer vision. A pioneer is DETR [2], an end-to-end trainable detector without NMS, which models the relations of the object queries and context via Transformer and achieves competitive performance only using a single-level feature map. This simple and effective detection method gives rise to a question: *can crowd localization be solved with such a simple model as well?*

Our answer is: "Yes, such a framework can be applied to crowd localization." Indeed, it is nothing special to directly apply the DETR-based pipeline in crowd localization. However, crowd localization is quite different from object detection. DETR shows terrible performance in the crowd localization task, attributed to the intrinsic limitation of the matcher. Specifically, the key component in DETR is the $L1$-based Hungarian matcher, which measures $L1$ distance and bounding box IoU to match the prediction-GT bounding box pairs, showing superior performance in object detection. However, no bounding box is given in crowd datasets, and more importantly, $L1$ distance easily gives rise to ambiguous matching in the point-to-point pairs (*i.e.*, a point that can belong to multiple *gts* simultaneously as shown in Fig. 1(d)). The main reasons are two-fold: (1) The $L1$-based Hungarian is a local view without context. (2) Different from the object detection, the crowd images only contain one category (heads), and the dense heads usually have similar textures, reporting close confidence, confusing the matcher. To this end, we introduce the k-nearest neighbors (KNN) matching objective named KMO as an auxiliary matching cost. The KMO-based Hungarian considers the context from nearby heads, which helps to reduce the ambiguous points and generate more reasonable matching results.

In summary, the main contributions of this paper are two-fold: 1) We propose an end-to-end **C**rowd **L**ocalization **TR**ansformer framework named CLTR, which formulates the crowd localization as a point set prediction task. CLTR significantly simplifies the crowd localization pipeline by removing pre-processing and post-processing. 2) We introduce the KMO-based Hungarian bipartite matching, which takes the context from nearby heads as an auxiliary matching cost. As a result, the matcher can effectively reduce the ambiguous points and generate more reasonable matching results.

Extensive experiments are carried out on five challenge datasets, and significant improvements from KMO-based Hungarian matcher indicate its effectiveness. In particular, just with a single-scale and low-resolution ($\frac{1}{32}$ of input images) feature map, CLTR can achieve state-of-the-art or highly competitive localization performance.

2   Related Works

2.1   Detection-Based Methods

The detection-based methods [24,31,44] mainly follow the pipeline of Faster RCNN [29]. Specifically, PSDDN [24] utilizes the nearest neighbor distance to initialize the pseudo bounding boxes and update the pseudo boxes by choosing smaller predicted boxes in the training phase. LSC-CNN [31] also uses a similar mechanism to generate the pseudo bounding boxes and propose a new winner-take-all loss for better training at higher resolutions. These methods [24,31,44] usually use NMS to filter the predicted boxes, which is not end-to-end trainable.

2.2   Map-Based Methods

Map-based methods are the mainstream of the crowd localization task. Idress *et al.* [13], and Gao *et al.* [7] utilize small Gaussian kernel density maps, and the head locations are equal to the maxima of density maps. Even though using the small kernel can generate sharp density maps, it still exists overlaps in the extremely dense region, making the head location undistinguishable. To solve this, some methods [1,8,18,46] focus on designing new maps to handle the extremely dense region, such as the distance label map [46], Focal Inverse Distance Transform Map (FIDTM) [18] and Independent Instance Map (IIM) [8]. These methods can effectively avoid overlap in the dense region, but they need post-processing ("find-maxima") to extract the instance location and rely on multi-scale feature maps, which is not simple and elegant.

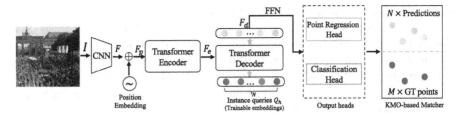

Fig. 2. The overview of our CLTR. First, the input image I is fed to the CNN-based backbone to extract the features F. Second, the features F are added position embedding, resulting in F_p, fed to the transformer-encoder layers, outputting F_e. Third, we define $N\times$ trainable embeddings Q_h as query, F_e as key, and transformer decoder takes the Q_h and F_e as input to generate the decoded feature F_d. Finally, the F_d can be decoupled to the point coordinate and corresponding confidence score.

2.3 Regression-Based Methods

Just a few research works focus on regression-based. We note a recent paper [34], P2PNet, also a regression-based framework for crowd localization. But this is a concurrent work that has appeared while this manuscript is under preparation. P2PNet [34] defines surrogate regression on a large set of proposals, and the model relies on pre-processing, such as producing $8 \times W \times H$ point proposals (anchors). In contrast, our method replaces massive fixed point proposals with a few trainable instance queries, which is more elegant and unified.

2.4 Visual Transformer

Recently, visual transformers [2,4,23,28,37] have gone viral in computer vision. In particular, DETR [2] utilizes the Transformer-decoder to model object detection in an end-to-end pipeline, successfully removing the need for post-processing. Based on DETR [2], Conditional DETR [28] further adopts the spatial queries and keys to a band containing the object extremity or a region, accelerating the convergence of DETR [2]. In the crowd analysis, Liang *et al.* [17,26] propose TransCrowd, which reformulates the weakly-supervised counting problem from a sequence-to-count perspective. Several methods [35,36] demonstrate the power of transformers in point-supervised crowd counting setup. Method [6] adopts the IIM [8] in the swin transformer [25] to implement crowd localization.

3   Our Method

The overview of our method is shown in Fig. 2. The proposed method is an end-to-end network, directly predicting all instances at once without additional pre-processing and post-processing. The approach consists of a CNN-based backbone, a transformer encoder, a transformer decoder, and a KMO-based matcher. Given an image by $I \in \mathbb{R}^{H \times W \times 3}$, where H, W are the height and width of the image,

- The CNN-based backbone first extracts the feature maps $F \in \mathbb{R}^{\frac{H}{32} \times \frac{W}{32} \times C}$ from the input image I. To verify the effectiveness of our method, the F is only a single-scale feature map without feature aggregation.
- The feature maps F are then flattened into a 1D sequence with positional embedding, and the channel dimension is reduced from C to c, which results in $F_p \in \mathbb{R}^{\frac{HW}{32^2} \times c}$. The transformer-encoder layers take the F_p as input and output encoded features F_e.
- Next, the transformer-decoder layers take the trainable head queries Q_h and encoded features F_e as input and interact with each other via cross attention to generate the decoded embedding F_d, which contains the point (person's head) and category information.
- Finally, the decoded embeddings F_d are decoupled to the point coordinates and confidence scores by a point regression head and a classification head, respectively.

3.1   Transformer Encoder

We use a 1×1 convolution to reduce the channel dimension of the extracted feature maps F from $\mathbb{R}^{\frac{H}{32} \times \frac{W}{32} \times C}$ to $\mathbb{R}^{\frac{H}{32} \times \frac{W}{32} \times c}$ (c set as 256). Due to the transformer-encoder adopt a $1D$ sequence as input, we reshape the extracted features F and add position embedding, resulting in $F_p \in \mathbb{R}^{\frac{HW}{32^2} \times c}$. The F_p are then fed into the transformer-encoder layers to generate the encoded features F_e. Here the encoder contains many encoder layers, and each layer consists of a self-attention (SA) layer and a feed-forward (FC) layer. The SA consists of three inputs, including query (Q), key (K), and value (V), defined as follow:

$$SA(Q, K, V) = softmax(\frac{QK^T}{\sqrt{c}})V, \tag{1}$$

where Q, K and V are obtained from the same input Z (e.g., $Q = ZW_Q$). In particular, we use the multi-head self-attention (MSA) to model the complex feature relation, which is an extension with several independent SA modules: $MSA = [SA_1; SA_2; \cdots ; SA_m]W_O$, where W_O is a re-projection matrix and m is the number of attention heads set as 8.

3.2   Transformer Decoder

The transformer-decoder consists of many decoder layers, and each layer is composed of three sub-layers: (1) a self-attention (SA) layer. (b) a cross attention (CA) layer. (3) a feed-forward (FC) layer. The SA and FC are the same as the encoder. The CA module takes two different embeddings as input instead of the same inputs in SA. Let us denote two embeddings as X and Z, and the CA can be written as $CA = SA(Q = XW_Q, K = ZW_K, V = ZW_V)$. Following [28], each decoder layer takes a set of trainable embedding ($Q_h \in \mathbb{R}^{N \times c}$) as query and the features from the last encoder layers as key and value. The decoder output the decoded features F_d, which are used to predict the point coordinates (regression head) and confidence scores (classification head).

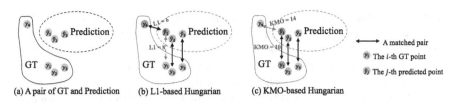

(a) A pair of GT and Prediction   (b) L1-based Hungarian   (c) KMO-based Hungarian

Fig. 3. (a) A pair of GT and predictions. (b) The $L1$-based Hungarian generate unsatisfactory matching results. (c) The proposed KMO-based Hungarian models the context as the matching cost, generating more reasonable matching results.

3.3   KMO-Based Matcher

To train the model, we need to match the predictions and GT by one-to-one, and the unmatched predicted points are considered to the "background" class.

Let us denote the predicted points set as $\hat{y} = \{\hat{y}_j\}_{j=1}^N$ and GT points set as $y = \{y_i\}_{i=1}^M$. N and M refer to the number of predicted heads and GT, respectively. N is larger than M to ensure each GT matches a prediction, and the rest of the predictions match failed can be classified as negative. Next, we need to find a bipartite matching between these two sets with the lowest cost. A straightforward way is to take the $L1$ distance and confidence as matching cost:

$$L_m(y_i, \hat{y}_j) = ||y_i^p - \hat{y}_j^p||_1 - \hat{C}_j, i \in M, j \in N, \qquad (2)$$

where $||*||_1$ means the $L1$ distance and $\hat{C}_j$ is the confidence of the j-th predicted point. y_i^p is a vector that defines the i-th GT point coordinates. Accordingly, $\hat{y}_j^p$ is formed as the point coordinates of j-th predicted head. Based on the L_m, we can utilize the Hungarian [14] to implement one-to-one matching. However, we find that merely taking the $L1$ with confidence maybe generate unsatisfactory matching results (seen Fig. 1(d)). Another toy example is shown in Fig. 3, given a pair of GT and prediction set (Fig. 3(a)), from the whole perspective, the $\hat{y}_1$ should match the y_1 ideally (just like $\hat{y}_2$ match y_2). Using Eq. 2 for matching cost[1], it will match the $\hat{y}_1$ and y_4 since the $L1$-based Hungarian lack of context information. Thus, we introduce the KMO-based Hungarian, which effectively utilizes the context as auxiliary matching cost, formulated as L_m^k:

$$L_m^k(y_i, \hat{y}_j) = ||y_i^p - \hat{y}_j^p||_1 + ||y_i^k - \hat{y}_j^k||_1 - \hat{C}_j,$$

$$y_i^k = \frac{1}{k}\sum_{k=1}^{k} d_i^k, \qquad \hat{y}_j^k = \frac{1}{k}\sum_{k=1}^{k} \hat{d}_j^k, \qquad (3)$$

where d_i^k means the distance between i-th GT point and its k-th neighbour. y_i^k refer to the average neighbour-distance of the i-th GT point. $\hat{d}_j^k$ and $\hat{y}_j^k$ have similar definitions as d_i^k and y_i^k, respectively. Taking inspirations from [24,31],

---

[1] Here, we ignore the $\hat{C}_j$ for simply illustrating since heads usually report similar confidence score.

in our experiments, $\hat{y}_j^k$ is predicted by the network. The proposed L_m^k, revisiting the label assignment from a context view, turns to find the whole optimum. As shown in Fig. 3(c), the proposed KMO-based Hungarian makes sure the $\hat{y}_1$ successfully matches the y_1. Regarding the predictions as a point set containing the geometric relationships. Assignment on Fig. 3(b) is not wrong, but it is a local view without context, and it will break the internal geometric relationships of the point set. Assignment on Fig. 3(c) considers the context information from the nearby heads, pursuing the whole optimum and maintaining the geometric relationships of the point set, making the model easier to be optimized, which is more reasonable. For the case in Fig. 1(d), when multiple *gts* tend to match the same predicted point, the KMO-based Hungarian will resolve their conflicts by using the context information. Note that the matcher is just used in the training phase.

3.4   Loss Function

After obtaining the one-to-one matching results, we calculate the loss for back-propagate. We make point predictions directly. The loss consists of point regression and classification. For the point regression, we employ the commonly-used L_1 loss, defined as:

$$L_{loc} = ||y_i^p - \hat{y}_{\sigma(i)}^p||_1, \tag{4}$$

where $\hat{y}_{\sigma(i)}^p$ is the matched subset from y_i^p by using the proposed KMO-based Hungarian. It is noteworthy that we normalize all ground truth point range to $[0, 1]$ for scale invariance. We utilize the focal loss as the classification loss L_{cls}, and the final loss L is defined as:

$$L = L_{cls} + \lambda L_{loc}, \tag{5}$$

where λ is a balance weight, set as 2.5. These two losses are normalized by the number of instances inside the batch.

4   Experiments

4.1   Implementation Details

We use the ResNet50 [10] as the backbone. The number of transformer encoder layers and decoder layers are both set to 6. The N is set to 500 (number of instance queries Q_h). We augment the training data using random cropping, random scaling, and horizontal flipping with a 0.5 probability. The crop size is set as 128×128 for ShanghaiTech Part A, 256×256 for the rest datasets. We use Adam with the learning rate of 1e-4 to optimize the model. For the large-scale datasets (*i.e.*, UCF-QNRF, JHU-Crowd++, and NWPU-Crowd), we ensure the longer size is less than 2048, keeping the original aspect ratio. The batch size is set to 16. k is set as 4 for all datasets. During the testing phase, each image is split into non-overlapped patches (size same as training phase). Zero padding is adopted if a cropped patch is smaller than the predefined size. And a simple confidence threshold (set to 0.35) is used to filter the "background" class.

Table 1. Localization performance on NWPU-Crowd dataset. * means the methods rely on box-level instead of point-level annotations.

Method	Validation set			Test set		
	P(%)	R(%)	F(%)	P(%)	R(%)	F(%)
Faster RCNN* [29]	**96.4%**	3.8%	7.3%	**95.8%**	3.5%	6.7%
TinyFaces* [11]	54.3%	**66.6%**	**59.8%**	52.9%	61.1%	56.7%
TopoCount* [1]	–	–	–	69.5%	**68.7%**	**69.1%**
GPR [7]	61.0%	52.2%	56.3%	55.8%	49.6%	52.5%
RAZ_Loc [19]	69.2%	56.9%	62.5%	66.6%	54.3%	59.8%
AutoScale_loc [46]	70.1%	63.8%	66.8%	67.3%	57.4%	62.0%
Crowd-SDNet [44]	–	–	–	65.1%	62.4%	63.7%
GL [39]	–	–	–	80.0%	56.2%	66.0%
CLTR (**ours**)	**73.9%**	**71.3%**	**72.6%**	69.4%	**67.6%**	**68.5%**

Table 2. Localization performance on the UCF-QNRF dataset. We report the Average Precision, Recall, and F1-measure at different thresholds σ: $(1, 2, 3, \ldots, 100)$ pixels.

Method	Av. precision	Av. recall	F1-measure
CL [13]	75.80%	59.75%	66.82%
LCFCN [15]	77.89%	52.40%	62.65%
Method in [30]	75.46%	49.87%	60.05%
LSC-CNN [31]	75.84%	74.69%	75.26%
AutoScale_loc [46]	81.31%	75.75%	78.43%
GL [39]	78.20%	74.80%	76.40%
TopoCount [1]	81.77%	78.96%	80.34%
CLTR (**ours**)	**82.22%**	**79.75%**	**80.97%**

4.2 Dataset

We evaluate our method on five challenging public datasets, each being elaborated below.

NWPU-Crowd [42] is a large-scale dataset collected from various scenes, consisting of 5,109 images. The images are randomly split into training, validation, and test sets, which contain 3109, 500, and 1500 images, respectively. This dataset provides point-level and box-level annotations.

JHU-Crowd++ [33] is a challenging dataset containing 4372 crowd images. This dataset consists of 2272 training images, 500 validation images, and 1600 test images. And the total number of people in each image ranges from 0 to 25791.

UCF-QNRF [13], a dense dataset, contains 1535 images (1201 for training and 334 for testing) and about one million annotations. The average number of pedestrians per image is 815, and the maximum number reaches 12865.

ShanghaiTech [48] is divided into Part A and Part B. Part A consists of 300 training images and 182 test images. Part B consists of 400 training images and 316 test images.

Table 3. Comparison of the localization performance on the Part A dataset.

Method	$\sigma = 4$			$\sigma = 8$		
	P (%)	R (%)	F (%)	P (%)	R (%)	F (%)
LCFCN [15]	43.3%	26.0%	32.5%	**75.1%**	45.1%	56.3%
Method in [30]	34.9%	20.7%	25.9%	67.7%	44.8%	53.9%
LSC-CNN [31]	33.4%	31.9%	32.6%	63.9%	61.0%	62.4%
TopoCount [1]	41.7%	40.6%	41.1%	74.6%	72.7%	73.6%
CLTR (ours)	**43.6%**	**42.7%**	**43.2%**	74.9%	**73.5%**	**74.2%**

4.3 Evaluation Metrics

This paper mainly focuses on crowd localization, and counting is an incidental task, *i.e.*, the total count is equal to the number of predicted points.

Localization Metrics. In this work, we utilize the Precision, Recall, and F1-measure as the localization metrics, following [13, 42]. If the distance between a predicted point and GT point is less than the predefined distance threshold σ, this predicted point will be treated as True Positive (TP). For the NWPU-Crowd dataset [42], containing the box-level annotations, we set σ to $\sqrt{w^2 + h^2}/2$, where w and h are the width and height of each head, respectively. For the ShanghaiTech dataset, we utilize two fixed thresholds, including $\sigma = 4$ and $\sigma = 8$. For the UCF-QNRF, we use various threshold ranges from [1, 100], following CL [13].

Counting Metrics. The Mean Absolute Error (MAE) and Mean Square Error (MSE) are used as counting metrics, defined as: $MAE = \frac{1}{N_c} \sum_{i=1}^{N_c} |P_i - G_i|$, $MSE = \sqrt{\frac{1}{N_c} \sum_{i=1}^{N_c} |P_i - G_i|^2}$, where N_c is the total number of images, P_i and G_i are the predicted and GT count of the i-th image, respectively.

5 Results and Analysis

5.1 Crowd Localization

We first evaluate the localization performance with some state-of-the-art localization methods [1, 39, 46], as shown in Table 1, Table 2, and Table 3. For the NWPU-Crowd (see Table 1), a large-scale dataset, our CLTR outperforms GL [39] and AutoScale [46] at least 5.8% (*resp.* 2.5%) for F1-measure on the validation set (*resp.* test set). It is noteworthy that this dataset provides precise box-level annotations, and the TopoCount [1] relies on the labeled box in the training phase instead of using the point-level annotations. Even though our method is just based on point-annotation, a more weak label mechanism, it can still achieve competitive performance against the TopoCount [1] on the NWPU-Crowd (test set). For the dense dataset, UCF-QNRF (see Table 2), our method achieves the best Average Precision, Average Recall and F1-measure. For the

Table 4. Counting results of various methods on the NWPU validation and test sets.

Method	Output position information	Validation set		Test set	
		MAE	MSE	MAE	MSE
MCNN [48]	✗	218.5	700.6	232.5	714.6
CSRNet [16]	✗	104.8	433.4	121.3	387.8
CAN [22]	✗	93.5	489.9	106.3	**386.5**
SCAR [9]	✗	81.5	397.9	110.0	495.3
BL [27]	✗	93.6	470.3	105.4	454.2
SFCN [43]	✗	95.4	608.3	105.4	424.1
DM-Count [41]	✗	**70.5**	**357.6**	**88.4**	388.6
RAZ_loc [19]	✔	128.7	665.4	151.4	634.6
AutoScale_loc [46]	✔	97.3	571.2	123.9	515.5
TopoCount [1]	✔	–	–	107.8	438.5
GL [39]	✔	–	–	79.3	346.1
CLTR (ours)	✔	**61.9**	**246.3**	**74.3**	**333.8**

Table 5. Comparison of the counting performance on the UCF-QNRF, ShanghaiTech Part A, and Part B datasets.

Method	Output position information	QNRF		Part A		Part B	
		MAE	MSE	MAE	MSE	MAE	MSE
CSRNet [16]	✗	–	–	68.2	115.0	10.6	16.0
L2SM [47]	✗	104.7	173.6	64.2	98.4	7.2	11.1
DSSI-Net [21]	✗	99.1	159.2	60.6	96.0	6.9	10.3
MBTTBF [32]	✗	97.5	165.2	60.2	94.1	8.0	15.5
BL [27]	✗	88.7	154.8	62.8	101.8	7.7	12.7
AMSNet [12]	✗	101.8	163.2	56.7	**93.4**	**6.7**	**10.2**
LibraNet [20]	✗	88.1	**143.7**	**55.9**	97.1	7.3	11.3
KDMG [40]	✗	99.5	173.0	63.8	99.2	7.8	12.7
NoisyCC [38]	✗	85.8	150.6	61.9	99.6	7.4	11.3
DM-Count [41]	✗	**85.6**	148.3	59.7	95.7	7.4	11.8
CL [13]	✔	132.0	191.0	–	–	–	–
RAZ_loc+ [19]	✔	118.0	198.0	71.6	120.1	9.9	15.6
PSDDN [24]	✔	–	–	65.9	112.3	9.1	14.2
LSC-CNN [31]	✔	120.5	218.2	66.4	117.0	8.1	12.7
TopoCount [1]	✔	89.0	159.0	61.2	104.6	7.8	13.7
AutoScale_loc [46]	✔	104.4	174.2	65.8	112.1	8.6	13.9
GL [39]	✔	**84.3**	147.5	61.3	95.4	7.3	11.7
CLTR (ours)	✔	85.8	**141.3**	**56.9**	95.2	**6.5**	**10.6**

ShanghaiTech Part A (see Table 3), a sparse dataset, our CLTR outperforms the state-of-the-art method TopoCount [1] by 2.1% F1-measure for the strict setting ($\sigma = 4$), and still get ahead for the less strict settings ($\sigma = 8$). These results demonstrate that the proposed method can cope with various scenes,

Table 6. Categorical counting results on JHU-Crowd++ dataset. Low, Medium, and High respectively indicate three categories based on different ranges: [0, 50], (50, 500], and (500, +∞). Weather means the degraded images (*e.g.*, haze, snow, rain).

Methods	Output position information	Low		Medium		High		Weather		Overall	
		MAE	MSE	MAE	MSE	MAE	MSE	MAE	MSE	MAE	MSE
MCNN [48]	✗	97.1	192.3	121.4	191.3	618.6	1,166.7	330.6	852.1	188.9	483.4
CSRNET [16]	✗	27.1	64.9	43.9	71.2	356.2	784.4	141.4	640.1	85.9	309.2
JHU++ [33]	✗	14.0	42.8	35.0	**53.7**	**314.7**	712.3	**120.0**	580.8	71.0	278.6
LSC-CNN [31]	✗	10.6	**31.8**	34.9	55.6	601.9	1,172.2	178.0	744.3	112.7	454.4
BL [27]	✗	**10.1**	32.7	**34.2**	54.5	352.0	768.7	140.1	675.7	75.0	299.9
AutoScale_loc [46]	✔	13.2	30.2	32.3	52.8	425.6	916.5	–	–	85.6	356.1
TopoCount [1]	✔	**8.2**	**20.5**	**28.9**	**50.0**	282.0	685.8	120.4	635.1	60.9	267.4
GL [39]	✔	–	–	–	–	–	–	–	–	59.9	259.5
CLTR (**ours**)	✔	8.3	21.8	30.7	53.8	**265.2**	**614.0**	**109.5**	**568.5**	**59.5**	**240.6**

Table 7. Effect of transformer size (the number of layers and the number of trainable queries Q_h) on UCF-QNRF dataset.

Layers	N (Queries number)	Localization			Counting	
		P (%)	R (%)	F (%)	MAE	MSE
3	500	80.60%	79.44%	80.02%	88.4	149.9
6	500	**82.22%**	**79.75%**	**80.97%**	**85.8**	**141.3**
12	500	80.82%	79.41%	80.11%	87.7	150.3
6	300	80.61%	79.18%	79.89%	89.9	153.6
6	700	81.32%	79.38%	80.34%	86.8	146.4

including large-scale, dense and sparse scenes. Note that all of other localization methods [1,39,46] adopt multi-scale or higher-resolution feature that potentially benefit our approach, which is currently not our focus and left as our future work.

5.2   Crowd Counting

In this section, we compare the counting performance with various methods (including density map regression-based and localization-based), as shown in Table 4, Table 5 and Table 6. Although our method only uses a single-scale and low-resolution ($\frac{1}{32}$ of input image) feature map, it can achieve state-of-the-art or highly competitive performance in all experiments. Specifically, our method achieves the first MAE and MSE on the NWPU-Crowd test set (see Table 4). Compared with the localization-based counting methods (the bottom part of Table 5), which can provide the position information, our method achieves the best counting performance in MAE and MSE on ShanghaiTech Part A and Part B datasets. On the UCF-QNRF dataset, our method achieves the best MSE and reports comparable MAE. On the JHU-Crowd++ dataset (Table 6), our method outperforms the state-of-the-art method GL [39] by a significant margin of 18.9 MSE. Furthermore, CLTR has superior performance on the extremely dense (the "High" part) and degraded set (the "Weather" part).

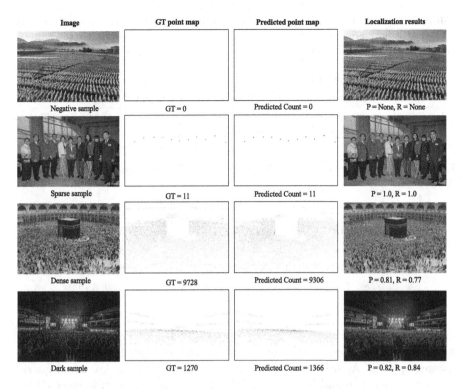

Image	GT point map	Predicted point map	Localization results
Negative sample	GT = 0	Predicted Count = 0	P = None, R = None
Sparse sample	GT = 11	Predicted Count = 11	P = 1.0, R = 1.0
Dense sample	GT = 9728	Predicted Count = 9306	P = 0.81, R = 0.77
Dark sample	GT = 1270	Predicted Count = 1366	P = 0.82, R = 0.84

Fig. 4. Some examples from the NWPU-Crowd dataset (validation set). From left to right, there are images, GT points, predicted points, and localization results. Row 1, row 2, row 3, and row 4 refer to the negative, sparse, dense and dark samples, respectively. In the last column, P and R are the Precision and Recall, respectively. The green, red and magenta points denote true positive (TP), false negative (FN) and false positive (FP), respectively. It can be seen that the proposed method can effectively handle various scenes. (Color figure online)

5.3   Visualizations

We further give some qualitative visualizations to analyze the effectiveness of our method, as shown in Fig. 4. The samples are selected from some typical scenes on the NWPU-Crowd dataset (validation set), including negative, sparse, extremely dense and dark scenes. In the first row, CLTR shows a strong robust on the negative sample ("dense fake humans"). CLTR performs well in different congested scenes, such as the sparse scene (row 2) and extremely dense scene (row 3). Additionally, we find that CLTR can also make promising localization results in dark scenes (row 4). These impressive visualizations demonstrate the effectiveness of our method in crowd localization and counting.

5.4   Ablation Studies

The ablation studies are carried out on the UCF-QNRF dataset, a large and dense dataset, which can effectively avoid overfitting.

Table 8. The effectiveness of the proposed KMO-based Hungarian on UCF-QNRF dataset. L_m only adopts the $L1$ distance with confidence as matching cost, and L_m^k contains the proposed KMO.

Matching cost	Localization			Counting	
	Av. precision (%)	Av. recall (%)	F1-measure (%)	MAE	MSE
L_m	80.89%	79.17%	80.02%	91.3	157.4
L_m^k (ours)	**82.22%**	**79.75%**	**80.97%**	**85.8**	**141.3**

Table 9. The influence of using different numbers of nearest-neighbour on the UCF-QNRF dataset.

k	Localization			Counting	
	Av. precision (%)	Av. recall (%)	F1-measure (%)	MAE	MSE
3	81.46%	79.19%	80.31%	87.1	146.8
4	**82.22%**	**79.75%**	**80.97%**	**85.8**	**141.3**
5	81.52%	79.34%	80.42%	86.9	148.1

Effect of Transformer. We first study the influence by changing the size of the transformer, including the number of encoder/decoder layers and trainable instance queries. As listed in Table 7, we find that when the layer and queries number are set to 6 and 500, the CLTR achieves the best performance. When the number of queries changes to 700 (resp. 300), the performance of MAE drops from 85.8 to 86.8 (resp. 89.9). We hypothesize that, by using a small number of queries, CLTR may lose potential heads, while using a large number of queries, CLTR may generate massive negative samples. We empirically find that all the pre-defined non-overlap patches contain less than 500 persons. The following ablation studies are organized using 6 transformer layers and 500 queries.

Effect of Matching Cost. We next study the impact of the proposed KMO, as shown in Table 8. When removing the KMO, we observe a significant performance drop for the counting (MAE from 85.8 to 91.3) and localization as well. We hypothesize that the $L1$ with classification can not provide a strong matching indicator, while the proposed KMO gives a direct signal to achieve great one-to-one matching based on whole-optimal.

Effect of K. We then study the effect of using different k (the number of nearest-neighbor), listed in Table 9. The proposed CLTR with different k consistently achieves improvement compared with the baseline, demonstrating the proposed KMO-based Hungarian's effectiveness. When the k is set to 4, we find that the result achieves the best on the UCF-QNRF dataset. We then set the same k in all datasets without further fine-tuning, which works well. We also try to use a fixed radius around each point and take as many NN as they fall within that circle. However, the training time is unacceptable because calculating dynamic KNN in each circle is time-consuming.

(a) Large scale head, image size: 2048×1536   (b) "Fake" people in the background

Fig. 5. Failure cases analysis. (a) Large-scale head, significantly larger than the crop-size. (b) Confusing regions that do not need to localize. The green, red, and magenta denote true positive (TP), false negative (FN), and false positive (FP), respectively. (Color figure online)

Table 10. The comparisons of complexity. The experiments are conducted on a 3090 GPU, and the size of the input image is 1024 × 768.

Method	Parameters (M)	MACs (G)
LSC-CNN [31]	35.0	1244.3
AutoScale [46]	24.9	1074.6
TopoCount [1]	25.8	797.2
GL [39]	**21.5**	324.6
CLTR (**ours**)	43.4	**157.2**

The Computational Statistics. Finally, we report the Multiply-Accumulate Operations (MACs) and parameters, as listed in Table 10. Although the proposed method has the largest parameters (mainly from the transformer part), it still reports the smallest MACs. Speeding up our model is a future work that is worthy of being studied.

5.5 Limitations

Our method has some limitations. For instance, due to the CLTR crop a fixed-size (*i.e.*, 256 × 256) sub-image for training and testing, it may fail on extremely large heads, which are significantly larger than the crop size, as shown in Fig. 5(a). This problem can be solved by resizing the image into a small resolution. Another case of unsatisfied localization is shown in Fig. 5(b), where there are some confused background regions (containing "fake" people that do not need localization). This failure case can be solved using more modalities, such as thermal images.

6 Conclusion

In this work, we propose an end-to-end crowd localization framework named CLTR, solving the task in the regression-based paradigm. The proposed method follows a one-to-one matching mechanism during the training phase. To achieve a good matching result, we propose the KMO-based Hungarian matcher, using the context information as an auxiliary matching cost. Our approach is simple yet effective. Experiments on five challenge datasets demonstrate the effectiveness of our methods. We hope our method can provide a new perspective for the crowd localization task.

Acknowledgment. This work was supported by National Key R&D Program of China (Grant No. 2018YFB1004602).

References

1. Abousamra, S., Hoai, M., Samaras, D., Chen, C.: Localization in the crowd with topological constraints. In: Proceedings of the AAAI Conference on Artificial Intelligence (2021)
2. Carion, N., Massa, F., Synnaeve, G., Usunier, N., Kirillov, A., Zagoruyko, S.: End-to-end object detection with transformers. In: Vedaldi, A., Bischof, H., Brox, T., Frahm, J.-M. (eds.) ECCV 2020. LNCS, vol. 12346, pp. 213–229. Springer, Cham (2020). https://doi.org/10.1007/978-3-030-58452-8_13
3. Chen, Y., Liang, D., Bai, X., Xu, Y., Yang, X.: Cell localization and counting using direction field map. IEEE J. Biomed. Health Inf. **26**(1), 359–368 (2021)
4. Dosovitskiy, A., et al.: An image is worth 16×16 words: transformers for image recognition at scale. In: Proceedings of International Conference on Learning Representations (2020)
5. Du, D., et al.: VisDrone-CC2020: the vision meets drone crowd counting challenge results. In: Bartoli, A., Fusiello, A. (eds.) ECCV 2020. LNCS, vol. 12538, pp. 675–691. Springer, Cham (2020). https://doi.org/10.1007/978-3-030-66823-5_41
6. Gao, J., Gong, M., Li, X.: Congested crowd instance localization with dilated convolutional swin transformer. arXiv preprint arXiv:2108.00584 (2021)
7. Gao, J., Han, T., Wang, Q., Yuan, Y.: Domain-adaptive crowd counting via inter-domain features segregation and gaussian-prior reconstruction. arXiv preprint arXiv:1912.03677 (2019)
8. Gao, J., Han, T., Yuan, Y., Wang, Q.: Learning independent instance maps for crowd localization. arXiv preprint arXiv:2012.04164 (2020)
9. Gao, J., Wang, Q., Yuan, Y.: Scar: spatial-/channel-wise attention regression networks for crowd counting. Neurocomputing **363**, 1–8 (2019)
10. He, K., Zhang, X., Ren, S., Sun, J.: Deep residual learning for image recognition. In: Proceedings of IEEE International Conference on Computer Vision and Pattern Recognition (2016)
11. Hu, P., Ramanan, D.: Finding tiny faces. In: Proceedings of IEEE International Conference on Computer Vision and Pattern Recognition (2017)
12. Hu, Y., et al.: NAS-count: counting-by-density with neural architecture search. In: Vedaldi, A., Bischof, H., Brox, T., Frahm, J.-M. (eds.) ECCV 2020. LNCS, vol. 12367, pp. 747–766. Springer, Cham (2020). https://doi.org/10.1007/978-3-030-58542-6_45
13. Idrees, H., et al.: Composition loss for counting, density map estimation and localization in dense crowds. In: Proceedings of European Conference on Computer Vision (2018)

14. Kuhn, H.W.: The Hungarian method for the assignment problem. Naval Res. Logist. Q. **2**(1–2), 83–97 (1955)
15. Laradji, I.H., Rostamzadeh, N., Pinheiro, P.O., Vazquez, D., Schmidt, M.: Where are the blobs: counting by localization with point supervision. In: Proceedings of European Conference on Computer Vision (2018)
16. Li, Y., Zhang, X., Chen, D.: CSRNet: dilated convolutional neural networks for understanding the highly congested scenes. In: Proceedings of IEEE International Conference on Computer Vision and Pattern Recognition (2018)
17. Liang, D., Chen, X., Xu, W., Zhou, Y., Bai, X.: Transcrowd: weakly-supervised crowd counting with transformers. Sci. China Inf. Sci. **65**(6), 1–14 (2022)
18. Liang, D., Xu, W., Zhu, Y., Zhou, Y.: Focal inverse distance transform maps for crowd localization and counting in dense crowd. arXiv preprint arXiv:2102.07925 (2021)
19. Liu, C., Weng, X., Mu, Y.: Recurrent attentive zooming for joint crowd counting and precise localization. In: Proceedings of IEEE International Conference on Computer Vision and Pattern Recognition (2019)
20. Liu, L., Lu, H., Zou, H., Xiong, H., Cao, Z., Shen, C.: Weighing counts: sequential crowd counting by reinforcement learning. In: Proceedings of European Conference on Computer Vision (2020)
21. Liu, L., Qiu, Z., Li, G., Liu, S., Ouyang, W., Lin, L.: Crowd counting with deep structured scale integration network. In: Proceedings of IEEE International Conference on Computer Vision (2019)
22. Liu, W., Salzmann, M., Fua, P.: Context-aware crowd counting. In: Proceedings of IEEE International Conference on Computer Vision and Pattern Recognition (2019)
23. Liu, X., et al.: End-to-end temporal action detection with transformer. IEEE Trans. Image Process. **31**, 5427–5441 (2022)
24. Liu, Y., Shi, M., Zhao, Q., Wang, X.: Point in, box out: beyond counting persons in crowds. In: Proceedings of IEEE International Conference on Computer Vision and Pattern Recognition (2019)
25. Liu, Z., et al.: Swin transformer: hierarchical vision transformer using shifted windows. In: Proceedings of IEEE International Conference on Computer Vision, pp. 10012–10022 (2021)
26. Liu, Z., et al.: Visdrone-cc2021: the vision meets drone crowd counting challenge results. In: Proceedings of IEEE International Conference on Computer Vision, pp. 2830–2838 (2021)
27. Ma, Z., Wei, X., Hong, X., Gong, Y.: Bayesian loss for crowd count estimation with point supervision. In: Proceedings of IEEE International Conference on Computer Vision (2019)
28. Meng, D., et al.: Conditional detr for fast training convergence. In: Proceedings of IEEE International Conference on Computer Vision, pp. 3651–3660 (2021)
29. Ren, S., He, K., Girshick, R., Sun, J.: Faster r-cnn: towards real-time object detection with region proposal networks. In: Proceedings of Advances in Neural Information Processing Systems (2015)
30. Ribera, J., Güera, D., Chen, Y., Delp, E.J.: Locating objects without bounding boxes. In: Proceedings of IEEE International Conference on Computer Vision and Pattern Recognition (2019)
31. Sam, D.B., Peri, S.V., Sundararaman, M.N., Kamath, A., Radhakrishnan, V.B.: Locate, size and count: accurately resolving people in dense crowds via detection. IEEE Trans. Pattern Anal. Mach. Intell. **43**, 2739–2751 (2020)

32. Sindagi, V.A., Patel, V.M.: Multi-level bottom-top and top-bottom feature fusion for crowd counting. In: Proceedings of IEEE International Conference on Computer Vision (2019)
33. Sindagi, V.A., Yasarla, R., Patel, V.M.: Jhu-crowd++: large-scale crowd counting dataset and a benchmark method. IEEE Trans. Pattern Anal. Mach. Intell. **44**, 2594–2609 (2020)
34. Song, Q., et al.: Rethinking counting and localization in crowds: a purely point-based framework. In: Proceedings of IEEE International Conference on Computer Vision, pp. 3365–3374 (2021)
35. Sun, G., Liu, Y., Probst, T., Paudel, D.P., Popovic, N., Van Gool, L.: Boosting crowd counting with transformers. arXiv preprint arXiv:2105.10926 (2021)
36. Tian, Y., Chu, X., Wang, H.: Cctrans: simplifying and improving crowd counting with transformer. arXiv preprint arXiv:2109.14483 (2021)
37. Touvron, H., Cord, M., Douze, M., Massa, F., Sablayrolles, A., Jégou, H.: Training data-efficient image transformers & distillation through attention. In: Proceedings of International Conference on Machine Learning, pp. 10347–10357. PMLR (2021)
38. Wan, J., Chan, A.: Modeling noisy annotations for crowd counting. Adv. Neural Inf. Process. Syst. **33**, 3386–3396 (2020)
39. Wan, J., Liu, Z., Chan, A.B.: A generalized loss function for crowd counting and localization. In: Proceedings of IEEE International Conference on Computer Vision and Pattern Recognition, pp. 1974–1983 (2021)
40. Wan, J., Wang, Q., Chan, A.B.: Kernel-based density map generation for dense object counting. IEEE Trans. Pattern Anal. Mach. Intell. **44**, 1357–1370 (2020)
41. Wang, B., Liu, H., Samaras, D., Hoai, M.: Distribution matching for crowd counting. In: Proceedings of Advances in Neural Information Processing Systems (2020)
42. Wang, Q., Gao, J., Lin, W., Li, X.: Nwpu-crowd: a large-scale benchmark for crowd counting and localization. IEEE Trans. Pattern Anal. Mach. Intell. **43**, 2141–2149 (2020)
43. Wang, Q., Gao, J., Lin, W., Yuan, Y.: Learning from synthetic data for crowd counting in the wild. In: Proceedings of IEEE International Conference on Computer Vision and Pattern Recognition (2019)
44. Wang, Y., Hou, J., Hou, X., Chau, L.P.: A self-training approach for point-supervised object detection and counting in crowds. IEEE Trans. Image Process. **30**, 2876–2887 (2021)
45. Wen, L., et al.: Detection, tracking, and counting meets drones in crowds: a benchmark. In: Proceedings of IEEE International Conference on Computer Vision and Pattern Recognition, pp. 7812–7821 (2021)
46. Xu, C., et al.: Autoscale: learning to scale for crowd counting. Int. J. Comput. Vision **130**, 1–30 (2022)
47. Xu, C., Qiu, K., Fu, J., Bai, S., Xu, Y., Bai, X.: Learn to scale: generating multi-polar normalized density map for crowd counting. In: Proceedings of IEEE International Conference on Computer Vision (2019)
48. Zhang, Y., Zhou, D., Chen, S., Gao, S., Ma, Y.: Single-image crowd counting via multi-column convolutional neural network. In: Proceedings of IEEE International Conference on Computer Vision and Pattern Recognition (2016)

Few-Shot Single-View 3D Reconstruction with Memory Prior Contrastive Network

Zhen Xing[1] , Yijiang Chen[1] , Zhixin Ling[1] , Xiangdong Zhou[1(✉)] ,
and Yu Xiang[2]

[1] School of Computer Science, Fudan University, Shanghai 200433, China
{zxing20,chenyj20,2021201005,xdzhou}@fudan.edu.cn
[2] The University of Texas at Dallas, Richardson, USA
yu.xiang@utdallas.edu

Abstract. 3D reconstruction of novel categories based on few-shot learning is appealing in real-world applications and attracts increasing research interests. Previous approaches mainly focus on how to design shape prior models for different categories. Their performance on unseen categories is not very competitive. In this paper, we present a Memory Prior Contrastive Network (MPCN) that can store shape prior knowledge in a few-shot learning based 3D reconstruction framework. With the shape memory, a multi-head attention module is proposed to capture different parts of a candidate shape prior and fuse these parts together to guide 3D reconstruction of novel categories. Besides, we introduce a 3D-aware contrastive learning method, which can not only complement the retrieval accuracy of memory network, but also better organize image features for downstream tasks. Compared with previous few-shot 3D reconstruction methods, MPCN can handle the inter-class variability without category annotations. Experimental results on a benchmark synthetic dataset and the Pascal3D+ real-world dataset show that our model outperforms the current state-of-the-art methods significantly.

Keywords: Few-shot learning · 3D reconstruction · Memory network

1 Introduction

Reconstructing 3D shapes from RGB images is valuable in many real-world applications such as autonomous driving, virtual reality, CAD and robotics. Traditional methods for 3D reconstruction such as Structure From Motion (SFM) [29] and Simultaneous Localization and Mapping (SLAM) [2] often require significant efforts in data acquisition. For example, a large number of images need to be captured and the camera parameters need to be calibrated, which limit the applications of these traditional methods.

Supplementary Information The online version contains supplementary material available at https://doi.org/10.1007/978-3-031-19769-7_4.

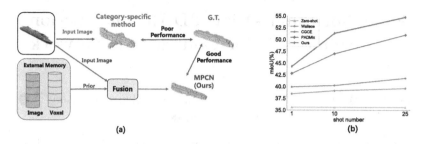

Fig. 1. (a) Novel category 3D reconstruction results on category-specific method and our method combination of prior. (b) The mIoU(%) of current methods against the number of shots. Our MPCN outperforms the SOTA approaches with different shot.

In recent years, 3D reconstruction from single image based on deep neural networks attracts great research interests. However, previous methods of single-view 3D reconstruction are mostly category-specific [8,42]. Therefore, they only perform well in the specific training categories. These methods also require a large number of labeled training images, which is time consuming and costly to obtain.

Notably, Tatarchenko et al. [31] shows that single-view 3D reconstruction of specific categories is closely related to image recognition for shape matching. Several simple image retrieval baseline methods can even outperform state-of-the-art 3D reconstruction methods.

In this paper, we propose a novel category-agnostic model for single-view 3D reconstruction in the few-shot learning settings. In our method, the network parameters are first optimized on some specific object categories. Then given novel categories, the model can be quickly transferred to 3D reconstruction of these categories based on few-shot learning techniques.

To the best of our knowledge, there are mainly three previous works focus on unseen category 3D reconstruction. Wallace et al. [34] present a network for single-view 3D reconstruction with additional shape prior of object categories. However, their shape prior model cannot catch the diversity of shapes within an object category. The authors of Compositional Global Class Embeddings (CGCE) [22] adopt a solution to optimize the shape prior codebooks for the reconstruction of unseen classes. Their model depends on finetuning with additional codebooks, which makes the training process complicated and makes the performance unstable. Pose Adaptive Dual Mixup (PADMix) [6] proposes to apply mixup [46] at the feature level and introduce pose level information, which reaches a new state-of-the-art performance in this task.

In addition, all the works rely on shape prior of specific categories. As a result, additional category annotation is needed to recognize the category of the input image. Then these methods can construct shape prior according to the category annotation, which is not very suitable for category-agnostic 3D reconstruction with novel categories.

The previous works of exploring shape prior for 3D reconstruction of novel categories are insightful and reasonable. However, there still exits a challenge on how to handle shape variety within a novel category in the context of few-shot learning. In this paper, we present a novel deep network model with a memory that can store a shape and its corresponding image as a key-value pair for retrieval. When an image of a novel category is inputted to the network, our deep network can select and combine appropriate shapes retrieved from the memory without category annotation to guide a decoder for 3D reconstruction. In order to adaptively combine the stored shapes as a prior for the downstream 3D reconstruction task, a multi-head attention shape prior module is proposed. Figure 1(a) shows the example on novel watercraft category 3D reconstruction performance between traditionally category-specific method [8] and our method.

Besides, we propose a 3D-aware contrastive loss that pulls together the image features of objects with similar shape and pushes away the image features with different 3D shape in the latent feature space, which helps for both organizing the image feature and improving the retrieval accuracy of memory network. Our 3D-aware contrastive loss takes into account the difference of shape as a weighting term to reduce or stress the positiveness of a pair, regardless of the category or instance, as we aim at a category-agnostic 3D reconstruction network.

In Smmary, Our Contributions are as Follows: We propose a novel Memory Prior Contrastive Network (MPCN) that can retrieve shape prior as an intermediate representation to help the neural network to infer the shape of novel objects without any category annotations.

Our multi-head attention prior module can automatically learn the association between retrieved prior and pay attention to different part of shape prior. It can not only provide prior information between object categories, but also represent the differences within objects in the same category.

The network with both reconstruction loss and a contrastive loss works together for better result. Our improved 3D-aware contrastive loss takes into account of the difference of positive samples, which is more suitable for supervised 3D tasks.

Experimental results on ShapeNet [3] dataset show that our method greatly improves the state-of-the-art methods on the two mainstream evaluation metrics using Intersection over Union and F-score. The reconstruction results on the real-world Pascal3D+ [41] dataset also demonstrate the effectiveness of our method quantitatively and qualitatively.

2   Related Work

Deep Learning 3D Reconstruction. Recently, Convolutional Neural Network (CNN) based single-view 3D reconstruction methods become more and more popular. Using voxels to represent a 3D shape is suitable for 3D CNNs. In the early work 3D Recurrent Reconstruction Neural Network (3D-R2N2) [8], the encoder with a Recurrent Neural Network (RNN) structure is used to predict

3D voxels by a 3D decoder. A follow-up work, 3D-VAE-GAN [39], explores the generation ability of Variational Autoencoders (VAEs) and Generative Adversarial Networks (GANs) to infer 3D shapes. Marrnet [38] and ShapeHD [40] predict the 2.5D information such as depth, silhouette and surface normal of an input RGB image, and then use these intermediate information to reconstruct the 3D object. OGN [30] and Matryoshka [27] utilize octrees or nested shape layers to represent high resolution 3D volume. Pix2Vox [42] and Pix2Vox++ [43] mainly improve the fusion of multi-view 3D reconstruction. Mem3D [45] introduces external memory for category-specific 3D reconstruction. However, its performance relies on a large number of samples saved during the training process. More recently, SSP3D [47] propose a semi-supervised setting for 3D Reconstruction. In addition to voxels, 3D shapes can also be represented by point clouds [9,20,35], meshes [36,37] and signed distance fields [21,44].

Few-Shot Learning. Few-shot learning models can be roughly classified into two categories: metric-based methods and meta-based methods. Metric-based methods mainly utilize Siamese networks [19], match networks [7,33] or prototype networks [10] to model the distance distribution between samples such that similar samples are closer to each other and heterogeneous samples are far away from each other. Meta-based methods [26] [25] and meta-gradient based methods [4,15,28] are teaching models by using few unseen samples to quickly update the model parameters in order to achieve generalization.

Few-Shot 3D Reconstruction. Wallace et al. [34] introduce the first method for single-view 3D reconstruction in the few-shot settings. They propose to combine category-specific shape priors with an input image to guide the 3D reconstruction process. However, in their work, a random shape or a calculated average shape is selected for each category as the prior information, which cannot account for shape diversity among objects in a category. In addition, the method does not explicitly learn the inter-class concepts. Compositional Global Class Embeddings (CGCE) [22] adopts a solution to quickly optimize codebooks for 3D reconstruction of novel categories. Before testing on a novel category, the parameters of other modules are fixed. Only the weight vector of codebooks are optimized with a few support samples from the novel category. Therefore, given a new category, CGCE needs to add a new codebook vector for this category and finetune the weight parameters, which makes the whole process complicated and inefficient. Pose Adaptive Dual Mixup (PADMix) [6] proposes a pose adaptive procedure and a three-stage training method with mixup [46]. It tries to solve the pose-invariance issue by an autoencoder but its shape prior module is similar to Wallace [34] and has the drawbacks of complicated three-stage training strategy.

3   Method

Our aim is to design a category-agnostic model, which can achieve superior generalization ability of single-view 3D reconstruction for novel categories with

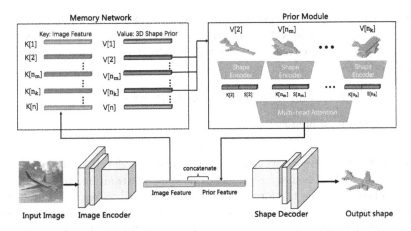

Fig. 2. An overview of the proposed MPCN. In the training stage, we only use base categories to train the model, and insert memory slots into the memory network as alternative prior according to the rules we set. In the test phase, the memory network saves few shot of support set of the novel category to reconstruct 3D volumes of the query set.

limited support data. Suppose there is a base 3D dataset, defined as $D_b = \{(I_i, V_i)\}_{i=1}^{K}$, where I_i and V_i denote the ith image and its corresponding 3D shape represented using voxels, respectively. K is the number of image and voxel pairs in the dataset. We denote the categories in the base dataset D_b as base categories. Let $D_s = \{(I_i, V_i)\}_{i=1}^{M}$ be another dataset of (image, voxel) pairs with M examples. The categories in D_s are defined as the novel categories, which are different from those in D_b. D_s is called the support set, where $M \ll K$. Meanwhile, there is a large test or query set $D_q = \{(I_i, V_i)\}_{i=1}^{N}$ with N examples. Examples in D_q and D_s are all within the novel categories. Note that we only use D_b and D_s for training. The support set D_s can be used as prior information. We hope that the model can be designed to be category-agnostic and achieve good performance in the query set D_q.

3.1 Memory Network

In most previous works on single-view 3D reconstruction, the shape prior information is learned from the model parameters, which leads to category collapse when transferring to novel categories. As mentioned in [31], such kind of model makes the problem degenerated into a classification task. To alleviate this issue, directly using 3D shapes as priors is an intuitive and effective way. As shown in Fig. 2, we adopt an explicit key-value memory network to store and calculate shape priors. In the training and testing stages, the CNN features of the input image is extracted by a 2D encoder. Then a retrieval task is performed, where the keys of the samples stored in the memory network are compared to the query vector and the corresponding Top-k retrieved shapes are sent to the

prior module for generating prior features. Specifically, as shown in Eq. (1), the
input image I_q is first encoded by a 2D encoder E_{2D}, then the image features
and shape prior features are concatenated. Finally, the 3D shape is inferred by
a 3D decoder D_{3D}.

$$pr = D_{3D}(\text{Concatenate}(E_{2D}(I_q), \text{prior feature})), \tag{1}$$

where pr is the final predict volume.

Memory Store. The external memory module is a database of experiences.
Each column in the memory represents one data distribution. In MPCN, two
columns of structures in the form of key-value are stored. A key is a deep feature
vector of an image, and its value is the corresponding 3D shape represented using
voxels. Each memory slot is composed of [image feature, voxel], and the memory
module database can be defined as $\mathcal{M} = \{(I_i, V_i)\}_{i=1}^m$, where m represents the
size of the external memory module. We use a simple but effective memory
storage strategy to store data in a limited memory size. During training stage,
when generating a target shape with MPCN, we calculate the distance $d(pr, gt)$
between all the samples' prediction and target shape of a batch as in Eq. (2):

$$d(pr, gt) = \frac{1}{r_v^3} \sum_{i,j,k} (pr_{(i,j,k)} - gt_{(i,j,k)})^2, \tag{2}$$

where r_v is the resolution of 3D volumes, gt is the ground truth volume. For a
sample (I_k, V_k), if $d(pr, gt)$ is greater than a specified threshold δ, we consider
that the current network parameters and the prior have poor reconstruction
performance on this shape. So we insert (I_k, V_k) into the external memory module
and store it as a memory slot in order to guide the reconstruction of similar
shapes in the future. We maintain an external memory module similar to the
memory bank (queue). When the memory is full, the memory slot initially added
to the queue will be replaced by later one. This makes sense because the later
image features are updated with iteration of the model training.

Memory Reader. Each row of the external memory database represents a
memory slot. The retrieval of the memory module is based on a k-nearest neighbor algorithm. When comparing the CNN features of the current query image
and the image features of all slots in the memory network, we use the Euclidean
metric to measure the differences. In order to obtain the distance between the
query matrix and the key matrix of the memory network conveniently, we use
the effective distance matrix computation to calculate the matrix of Euclidean
distance as shown in Eq. (3):

$$\text{Distance} = \|Q\| + \|K\| - 2 * QK^T, \tag{3}$$

where $Q \in \mathbb{R}^{b \times 2048}$ is query matrix, $K \in \mathbb{R}^{m \times 2048}$ is memory-key matrix, b is
the batch size, and m is the memory size.

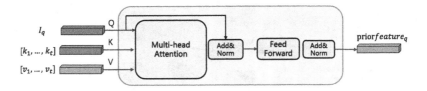

Fig. 3. An overview of the proposed Memory Prior Module.

After calculating the distance, we select the nearest k retrieval results as prior information, which is defined as $R = \{(I_i, V_i)\}_{i=1}^k$. $\{I_i\}_{i=1}^k$ represents k retrieved image features, $\{V_i\}_{i=1}^k$ represents k retrieved voxels. When the first batch is searched, the memory will be empty. However, we will take out k all-zero tensors, which increases the robustness of the model to some extent. Even without the shape prior as a guide, our model should reconstruct the 3D model according to the 2D features of the image.

3.2 Prior Module

The prior module can first obtain the set $\{(I_i, V_i)\}_{i=1}^k$ retrieved by the external memory module from the previous step. Note that shape volume is the original voxel saved at this stage, and its size is 32^3. So the model needs to extract shape features by a 3D shape encoder before the downstream processing.

$$k_i = I_i, v_i = \text{Encoder3D}(V_i), \tag{4}$$

$$Q = I_q W_q, K = k_i W_k, V = v_i W_v, \tag{5}$$

$$e_q = Q + \text{LayerNorm}(\text{MHA}(Q, K, V)), \tag{6}$$

$$prior\ feature_q = e_q + \text{LayerNorm}(\text{FFN}(e_q)), \tag{7}$$

In previous works, only 3D voxels are regarded as the prior features. In contrast, as shown in Fig. 3, we use the attention based architecture to extract shape prior features by exploring the association between image features and 3D shape. Concretely, we take the query image feature I_q as the query, the retrieved image feature $\{I_i\}_{i=1}^k$ as the key, and its corresponding 3D shape feature $\{v_i\}_{i=1}^k$ as the value. As shown in Eq. (5), we first use three separate linear layers parameterized by W_q, W_k and W_v to extract query, key, value embedding Q, K and V.

Then the embeddings are forwarded to the multi-head attention (MHA) [32] and layer normalization (LayerNorm) module [1] to perform cross-attention between the query and every key. The output of the attention is fused to the original input query embedding to get enhanced feature e_q. Afterward, the obtained features e_q are sent into feed-forward network (FFN) and layer normalization (LayerNorm). The output $prior feature_q$ is obtained by adding up the feed-forward module output with residual connection as in Eq. (6) and Eq. (7).

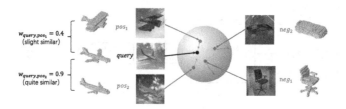

Fig. 4. An example of 3D-Aware Contrastive Loss, which pulls together the positive samples with similar 3D shape (e.g., pos_1 and pos_2) by different *weight*, and pushes apart the negative samples with different 3D shape (e.g., neg_1 and neg_2).

3.3 3D-Aware Contrastive Learning Method

We believe that Memory Prior Network can work effectively mainly based on the accuracy of 2D image embedding retrieval. In order to improve the retrieved prior accuracy provided by memory network, previous work generally used triple loss [13] to optimize encoder, which is effective for simple classification problems [16] [48]. However, for 3D reconstruction task, the triplets need to construct positive and negative samples according to the threshold of specific shape difference, which is an empirical and troublesome step [45].

Recently, contrastive learning method [5] [11] train an image encoder maximizing the similarity of different transformations of the same sample and minimizing the similarity with other samples. The proposed loss $\mathcal{L}_{infoNCE}$ [11] achieves great success in self-supervised representation learning. However, their success depends on the large batch size and memory occupation, and taking all of samples in the same batch as negative pairs may be wrong for supervised tasks. In addition, supervised contrastive loss [17] tries to solve this problem but it mainly focuses on simple classification problem.

We hope to design a loss that can adaptively pull together the image embedding pairs with similar 3D shape and push away the image embeddings with different 3D shapes. Therefore, as shown in Fig. 4. we introduce an improved 3D-aware contrastive loss, which considers the positivity of positive pairs. Concretely, for each image pair (q, k), we calculate the distance between the associated 3D shape $d(V_q, V_k) \in [0, 1]$, we take (q, k) as a positive pair if $d(V_q, V_k) < \delta$, then a weight is calculated by $w_{q,k} = (1 - d(V_q, V_k) \times \gamma)$, which is considered as the important weight of the positive pair in our 3D-aware contrastive loss:

$$\mathcal{L}_{3DNCE} = -log \frac{\sum_{p \in [1..M]} w_{q,p} \cdot exp(f_q \cdot f_p / \tau)}{M \cdot \sum_{k \in [1..N]} exp(f_q \cdot f_k / \tau).} \tag{8}$$

where $d(V_q, V_p)$ is the same as Eq (2), q is a query image, p is the positive samples of q according to $d(V_q, V_p) < \delta$, $\sum_{k \in [1..N]}$ mean the samples in the same batch with q, and f is image encoder. Intuitively, the more similar the 3D shape of two objects (q, p) is, the greater its weight $w_{q,p}$ is, and the closer the image features of the two objects are.

Algorithm 1. Training algorithm

1: **for** epoch in epochs **do**
2:      flush memory slots
3:      **for** batch_idx in range(max_episode): **do**
4:          Load query images, target shape from train set D_b
5:          embed2d = $Encoder2d$(query image)
6:          Key, Value, Distance = $Top\text{-}K$(embed2d)
7:          embedPrior = $Prior$(embed2d, Key, Value, Distance)
8:          embed = $concatenate$(Embed2d, embedPrior)
9:          predict = $Decoder3d$(embed)
10:         d = $computeDis$(predict, target shape)
11:         **if** $d > \delta$: **then**
12:             insert(image, voxel) to external memory
13:         **end if**
14:         Train on predict and target with backprop
15:     **end for**
16: **end for**

3.4 Training Procedure in Few-Shot Settings

We adopt a two-stage training strategy. In the first stage, we train the initialization model on the base category data D_b. In the second stage, we use few-shot novel category samples in support set D_s to finetune the network. Both stage adopt the training method based on episodes as shown in the Algorithm 1. At the beginning of each epoch, all slots of the memory are cleared to ensure that the new round of training can re-determine which samples are inserted into the memory module according to our memory store strategy. For test phase, we first insert samples in support set D_s to the memory module as candidate prior information according to the few-shot settings. Then it follows the same steps as training stage to predict 3D shape in query set D_q. Finally, the reconstruction results are evaluated by evaluation metric.

3.5 Architecture

Image Encoder. The 2D encoder shares the same ResNet backbone [12] as that of CGCE [22]. Then our model follows with three layers of convolution layer, batch normalization layer and Relu layer. The convolution kernel size of the three convolution layers is 3^2 with padding 1. The last two convolution layers are then followed by a max pooling layer. The kernel size of the pooling layer is 3^2 and 2^2, respectively. The output channels of the three convolution layers are 512, 256 and 128, respectively.

Prior Module. The 3D shape encoder of the prior module includes four convolution layers and two max-pooling layers. Each layer has a LeakyRelu activation layer, and the convolution kernel sizes are 5^3, 3^3, 3^3 and 3^3, respectively. The output Q,K,V embedding dimension of the Linear layer is 2048. The size of key

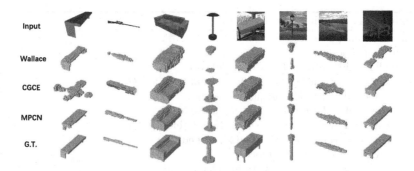

Fig. 5. Examples of single-view 3D Reconstruction on novel category of ShapeNet with shot-10. We show examples with clean background and with random background.

and value of the attention module are both 2048. This module has 2 heads in attention blocks. Finally, the prior feature dimension is 2048.

Shape Decoder. There are five 3D deconvolution layers in this module. The convolution kernel size is 4^3, stripe is 2^3, and padding is 1. The first four convolutions are followed by batch normalization and Relu, and the last is sigmoid function. The output channels of the five convolution layers are 256, 128, 32, 8 and 1, respectively. The final output is a voxel representation with size 32^3.

3.6   Loss Function

Reconstruction Loss. For the 3D reconstructions network, both the reconstruction prediction and the ground truth are based on voxel. We follow previous works [22, 34, 42, 43] that adopt the binary cross entropy loss as our loss function:

$$\mathcal{L}_{rec} = \frac{1}{r_v^3} \sum_{i=1}^{r_v^3} [gt_i \log(pr_i) + (1 - gt_i) \log(1 - pr_i)], \tag{9}$$

where r_v represents the resolution of the voxel space, pr and gt represent the predict and the ground truth volume.

Total Loss. The MPCN is trained end-to-end with the reconstruction loss and 3D-aware contrastive loss together as following:

$$\mathcal{L}_{total} = \mathcal{L}_{rec} + \lambda \mathcal{L}_{3DNCE} \tag{10}$$

where λ is hyperparameter, which is set to 0.001 in this work.

4 Experiment

4.1 Experimental Setup

Dataset. We experiment with the **ShapeNet** dataset [3]. The setting of this dataset is the same as [34]. Seven of 13 categories are designated as the base classes that are used for training: airplanes, cars, chairs, displays, phones, speakers and tables. The other categories are set as novel classes for testing. To fairly compare with the previous works, we use the same dataset split as in [34] and CGCE [22]. The datasets are provided by [8] which are composed with 137×137 RGB images and $32 \times 32 \times 32$ voxelized representations. **Pascal3D+** [41] dataset has 12 different categories. It provides approximate 3D annotations for Pascal VOC2012 and ImageNet [41]. Each category has about 10 CAD models, which are generally not used for training directly. We finetune our MPCN with 8 categories of Pascal3D+, and test it on four categories: bicycle, motorbike, bottle and train. For fair comparison, we use Binvox [23] tool to render the voxel representation from the CAD model, and the voxel resolution is also $32 \times 32 \times 32$.

Table 1. Comparison of single-view 3D object reconstruction on novel class of ShapeNet at 32^3 resolution with different available shot.We report the **mean IoU** per category. The best number for each category is highlighted in bold.

Category	0-shot	1-shot				10-shot				25-shot			
	B0	Wallace	CGCE	PADMix	MPCN	Wallace	CGCE	PADMix	MPCN	Wallace	CGCE	PADMix	MPCN
Cabinet	0.69	0.69	0.71	0.67	**0.72**	0.69	**0.71**	0.66	0.68	0.69	0.71	0.68	**0.74**
Sofa	0.52	0.54	0.54	0.54	**0.57**	0.54	0.54	0.57	**0.60**	0.54	0.55	0.59	**0.65**
Bench	0.37	0.37	0.37	0.37	**0.39**	0.36	0.37	0.41	**0.41**	0.36	0.38	0.42	**0.45**
Watercraft	0.28	0.33	0.39	**0.41**	**0.41**	0.36	0.41	0.46	**0.54**	0.37	0.43	0.52	**0.55**
Lamp	0.19	0.20	0.20	**0.29**	**0.29**	0.19	0.20	0.31	**0.32**	0.19	0.20	0.32	**0.37**
Firearm	0.13	0.21	0.23	**0.31**	0.24	0.24	0.23	0.39	**0.52**	0.26	0.28	0.50	**0.52**
Mean	0.36	0.38	0.40	0.43	**0.44**	0.40	0.41	0.47	**0.51**	0.41	0.43	0.51	**0.54**

Evaluation Metrics. For fair comparison, we follow previous work [34] [22] [42] using **Intersection over Union (IoU)** as the evaluation metrics. The IoU is defined as following:

$$\text{IoU} = \frac{\sum_{i,j,k} \mathcal{I}(\hat{p}_{(i,j,k)} > t)\mathcal{I}(p_{(i,j,k)})}{\sum_{i,j,k} \mathcal{I}[\mathcal{I}(\hat{p}_{(i,j,k)} > t) + \mathcal{I}(p_{(i,j,k)})]}, \tag{11}$$

where $\hat{p}_{(i,j,k)}$ and $p_{(i,j,k)}$ represent the predicted possibility and the value of ground truth at point (i, j, k), respectively. $\mathcal{I}$ is the function that is one when the requirements are satisfied. t represents a threshold of this point, which is set to 0.3 in our experiment.

Implementation Details. We used 224×224 RGB images as input to train the model with a batch size of 16. Our MPCN is implemented in PyTorch [24] and trained by the Adam optimizer [18]. We set the learning rate as $1e - 4$. The δ and γ are set to 0.1 and 10. The k of the retrieval samples Top-k is 5, the τ is set 0.1. The memory size m is set to 4000 in the training stage, and only 200 in the testing stage.

Baseline. We compare our proposed MPCN with three state-of-the-art methods: Wallace [34], CGCE [22] and PADMix [6]. We also follows the zero-shot lower baseline in CGCE [22]. Zero-shot refers to the result of training on the base categories with only single-image and testing directly on the novel class without any prior. Image-Finetune method refers to training on the base categories and finetuning the full network with few available novel categories samples.

4.2   Results on ShapeNet Dataset

We compare with the state-of-the-art methods on the ShapeNet novel categories. Table 1 shows the IoUs of our MPCN and other methods. Results in Table 1 from other models are taken from [6]. For few-shot settings, we follow the evaluation in CGCE [22] and PADMix [6] shown the results in the settings of 1-shot, 10-shot and 25-shot. It can be seen that our method is much better than the zero-shot baseline respectively, and greatly outperforms SOTA's methods. Experimental results show that MPCN has great advantages when the shot number increases, mainly because it can retrieve prior information more related to the target shape. Even in the results of 1-shot, there are some improvements, mainly because our model can select the most appropriate prior of shapes as well as image features, and use the differences of other shapes to exclude other impossible shapes. Our MPCN results are shown in Fig. 5. It can be seen that our model can obtain satisfactory reconstruction results for novel categories than any other SOTA methods even when the angles of input images are different.

Table 2. Comparison of single-view 3D object reconstruction on Pascal3D+ at 32^3 resolution. We report both the mean IoU of every novel category. The best number is highlighted in bold.

	Bicycle	Motorbike	Train	Bottle	Mean
Zero-shot	0.11	0.27	0.35	0.10	0.2074
Image-Finetune	0.20	0.28	0.35	0.32	0.2943
Wallace [34]	0.21	0.29	**0.40**	0.43	0.3324
CGCE [22]	0.23	0.33	0.37	0.35	0.3223
MPCN(ours)	**0.28**	**0.39**	0.37	**0.46**	**0.3748**

4.3   Results on Real-world Dataset

In order to compare with Wallace et al. [34] further, we also conducted experiments on the real-world Pascal3D+ dataset [41]. Firstly, the model is pre-trained on all 13 categories of the ShapeNet dataset [3]. Then the model is finetuned with Pascal3D+ base category dataset, and the final test set is selected from the four novel categories. Note that the experiment is set of 10-shot. The experimental results show that our method outperform zero-shot baseline by 16.74%, also it is the best compared with SOTA methods. Especially for bicycle and motorbike categories with large shape difference and variability, our model perform best. But for the category with subtle shape difference (*e.g.*, train), the reconstruction results tend to align to the average of prior shape, so Wallace [34] shows marginal improvement over ours. But note that our MPCN selects prior information by memory network automatically, while Wallace et al. [34] need the category annotations of the input images and choose shape priors manually. The experimental results are shown in the Table 2.

Table 3. The effectiveness of the different modules and losses. All the results are tested on novel categories of ShapeNet and Pascal3D+ with 5-shot. We report the mean IoU(%) of novel categories. The best number is highlighted in bold.

	MHA	LSTM	Random	Average	Finetune	$\mathcal{L}_{3DNCE}$	$\mathcal{L}_{infoNCE}$	ShapeNet	Pascal3D+
Zero-shot								35.68	20.74
ONN(5)								40.90	42.50
Image-Finetune					✓			40.82	27.49
MPCN-Top1	✓				✓	✓		39.95	31.65
w LSTM		✓			✓			40.52	25.63
w Random prior			✓		✓			35.52	22.05
w average				✓	✓			41.32	29.85
w $\mathcal{L}_{infoNCE}$	✓						✓	42.98	27.98
w/o $\mathcal{L}_{3DNCE}$	✓							43.05	28.76
w/o finetune	✓					✓		45.82	30.32
ours	✓				✓	✓		**47.53**	**33.54**

5 Ablation Study

In this part, we evaluate the effectiveness of proposed modules and the impact of different losses. We choose the setup of 5-shot on both ShapeNet and Pascal3D+ dataset for comparative experiment if not mentioned elsewhere.

Retrieval or Reconstruction. In order to prove that our method is superior to the retrieval method, we just take the highest similarity retrieved shape as the target shape. That is the result of MPCN-Top1 in Table 3. In addition, Image-Finetune method is also shown for comparison. The results in Table 3 shows that our MPCN outperforms any upper retrieval or finetune methods in the few-shot settings.

Analysis of Prior Module. We analyze the prior extraction module based on attention in MPCN. Because previous methods using external memory network adopt LSTM [14] in the shape prior fusion stage, we replace the attention part of MPCN with LSTM (w LSTM) for the purpose of comparison. We also compare the average-fusion (w average) of the retrieved Top-5 object volume features. In addition, the random initialization of prior vectors (w Random prior) is also compared in the experiment.

The experimental results in Table 3 show that our prior module plays an important role for guiding the reconstruction of 3D objects. The fusion of Top-5

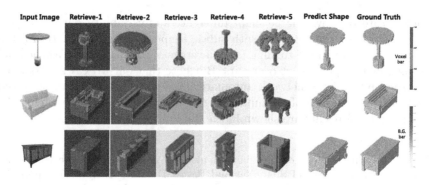

Fig. 6. Visualization of features maps from the retrieved 3D volumes and the corresponding reconstructions.

average method and random prior obviously cannot make full use of similar 3D volumes. Our attention module can capture the relevance of different 3D objects better than LSTM, and shows more powerful ability of inferring 3D shapes by using prior information in the few-shot settings. Figure 6 illustrates some shape priors selected by our model. We demonstrate that the multi-head attention module can adaptively detect the proper parts of the retrieved shapes for 3D reconstruction of novel categories.

Analysis of Loss and Finetune. In order to further prove the effectiveness of our proposed 3D-aware contrastive loss, we remove $\mathcal{L}_{3DNCE}$ in the training stage, as (MPCN w/o $\mathcal{L}_{3DNCE}$) shown in Table 3. Besides, we replace the improved $\mathcal{L}_{3DNCE}$ with the traditional contrastive loss, as shown (MPCN w $\mathcal{L}_{infoNCE}$) in Table 3. The results show that our comparison $\mathcal{L}_{3DNCE}$ has a great contribution to the improvement of experimental effect, mainly because it not only improves the retrieval accuracy of memory a prior module, but also makes the intent space of 2D representation more reasonable. In addition, using few-shot novel category samples in support set D_s to finetune the network (w finetune) in the second training stage is also important for the performance.

6   Conclusion

In this paper, we propose a novel category-agnostic model for 3D object reconstruction. Inspired by the novel 3D object recognizing ability by human-beings, we introduce an external memory network to assist in guiding the object to reconstruct the 3D model in few-shot settings. Compared with the existing methods, our method provides an advanced module to select shape priors, and fuses shape priors and image features to reconstruction 3D shapes. In addition, a 3D-aware contrastive method is proposed for encode 2D latent space, which may be used for other supervised tasks of 3D vision. The experimental results show that our MPCN can outperform existing methods on the ShapeNet dataset and the Pascal3D+ dataset under the settings of few-shot learning.

Acknowledgments. This work was supported by the National Key Research and Development Program of China, No.2018YFB1402600.

References

1. Ba, J.L., Kiros, J.R., Hinton, G.E.: Layer normalization. arXiv preprint arXiv:1607.06450 (2016)
2. Cadena, C., et al.: Past, present, and future of simultaneous localization and mapping: toward the robust-perception age. IEEE Trans. Robot. **32**(6), 1309–1332 (2016)
3. Chang, A.X., et al.: ShapeNet: an information-rich 3D model repository. arXiv preprint arXiv:1512.03012 (2015)
4. Chen, R., Chen, T., Hui, X., Wu, H., Li, G., Lin, L.: Knowledge graph transfer network for few-shot recognition. In: AAAI (2020)
5. Chen, T., Kornblith, S., Norouzi, M., Hinton, G.: A simple framework for contrastive learning of visual representations. In: ICML (2020)
6. Cheng, T.Y., Yang, H.R., Trigoni, N., Chen, H.T., Liu, T.L.: Pose adaptive dual Mixup for few-shot single-view 3D reconstruction. In: AAAI (2022)
7. Choi, J., Krishnamurthy, J., Kembhavi, A., Farhadi, A.: Structured set matching networks for one-shot part labeling. In: CVPR (2018)
8. Choy, C.B., Xu, D., Gwak, J., Chen, K., Savarese, S.: 3D-R2N2: a unified approach for single and multi-view 3D object reconstruction. In: ECCV (2016)
9. Fan, H., Su, H., Guibas, L.J.: A point set generation network for 3D object reconstruction from a single image. In: CVPR (2017)
10. Gao, T., Han, X., Liu, Z., Sun, M.: Hybrid attention-based prototypical networks for noisy few-shot relation classification. In: AAAI (2019)
11. He, K., Fan, H., Wu, Y., Xie, S., Girshick, R.: Momentum contrast for unsupervised visual representation learning. In: CVPR (2020)
12. He, K., Zhang, X., Ren, S., Sun, J.: Deep residual learning for image recognition. In: CVPR (2016)
13. Hermans, A., Beyer, L., Leibe, B.: In defense of the triplet loss for person re-identification. arXiv preprint arXiv:1703.07737 (2017)
14. Hochreiter, S., Schmidhuber, J.: Long short-term memory. Neural Comput. **9**(8), 1735–1780 (1997)
15. Jeong, M., Choi, S., Kim, C.: Few-shot open-set recognition by transformation consistency. In: CVPR (2021)
16. Kaiser, Ł., Nachum, O., Roy, A., Bengio, S.: Learning to remember rare events. In: ICLR (2017)
17. Khosla, P., et al.: Supervised contrastive learning. In: NeurIPS (2020)
18. Kingma, D.P., Ba, J.: Adam: a method for stochastic optimization. In: ICLR (2015)
19. Koch, G., Zemel, R., Salakhutdinov, R., et al.: Siamese neural networks for one-shot image recognition. In: ICMLW (2015)
20. Lin, Y., Wang, Y., Li, Y., Wang, Z., Gao, Y., Khan, L.: Single view point cloud generation via unified 3D prototype. In: AAAI (2021)
21. Mescheder, L., Oechsle, M., Niemeyer, M., Nowozin, S., Geiger, A.: Occupancy networks: learning 3D reconstruction in function space. In: CVPR (2019)
22. Michalkiewicz, M., Parisot, S., Tsogkas, S., Baktashmotlagh, M., Eriksson, A., Belilovsky, E.: Few-shot single-view 3-D object reconstruction with compositional priors. In: ECCV (2020)
23. Nooruddin, F.S., Turk, G.: Simplification and repair of polygonal models using volumetric techniques. IEEE Trans. Vis. Comput. Graph. **9**(2), 191–205 (2003)

24. Paszke, A., et al.: Pytorch: an imperative style, high-performance deep learning library. In: NeurIPS (2019)
25. Ramalho, T., Garnelo, M.: Adaptive posterior learning: few-shot learning with a surprise-based memory module. In: ICLR (2018)
26. Ravichandran, A., Bhotika, R., Soatto, S.: Few-shot learning with embedded class models and shot-free meta training. In: ICCV (2019)
27. Richter, S.R., Roth, S.: Matryoshka networks: predicting 3D geometry via nested shape layers. In: CVPR (2018)
28. Satorras, V.G., Estrach, J.B.: Few-shot learning with graph neural networks. In: ICLR (2018)
29. Schonberger, J.L., Frahm, J.M.: Structure-from-motion revisited. In: CVPR (2016)
30. Tatarchenko, M., Dosovitskiy, A., Brox, T.: Octree generating networks: efficient convolutional architectures for high-resolution 3D outputs. In: ICCV (2017)
31. Tatarchenko, M., Richter, S.R., Ranftl, R., Li, Z., Koltun, V., Brox, T.: What do single-view 3D reconstruction networks learn? In: CVPR (2019)
32. Vaswani, A., et al.: Attention is all you need. In: NeurIPS (2017)
33. Vinyals, O., Blundell, C., Lillicrap, T., Wierstra, D., et al.: Matching networks for one shot learning. In: NeurIPS (2016)
34. Wallace, B., Hariharan, B.: Few-shot generalization for single-image 3D reconstruction via priors. In: ICCV (2019)
35. Wang, J., Sun, B., Lu, Y.: MVPNet: multi-view point regression networks for 3D object reconstruction from a single image. In: AAAI (2019)
36. Wang, N., Zhang, Y., Li, Z., Fu, Y., Liu, W., Jiang, Y.G.: Pixel2mesh: generating 3D mesh models from single RGB images. In: ECCV (2018)
37. Wen, C., Zhang, Y., Li, Z., Fu, Y.: Pixel2mesh++: multi-view 3D mesh generation via deformation. In: ICCV (2019)
38. Wu, J., Wang, Y., Xue, T., Sun, X., Freeman, B., Tenenbaum, J.: MarrNet: 3D shape reconstruction via 2.5 D sketches. In: NeurIPS (2017)
39. Wu, J., Zhang, C., Xue, T., Freeman, W.T., Tenenbaum, J.B.: Learning a probabilistic latent space of object shapes via 3D generative-adversarial modeling. In: NeurIPS (2016)
40. Wu, J., Zhang, C., Zhang, X., Zhang, Z., Freeman, W.T., Tenenbaum, J.B.: Learning shape priors for single-view 3D completion and reconstruction. In: ECCV (2018)
41. Xiang, Y., Mottaghi, R., Savarese, S.: Beyond pascal: a benchmark for 3D object detection in the wild. In: WACV (2014)
42. Xie, H., Yao, H., Sun, X., Zhou, S., Zhang, S.: Pix2vox: context-aware 3D reconstruction from single and multi-view images. In: ICCV (2019)
43. Xie, H., Yao, H., Zhang, S., Zhou, S., Sun, W.: Pix2Vox++: multi-scale context-aware 3D object reconstruction from single and multiple images. Int. J. Comput. Vision **128**(12), 2919–2935 (2020). https://doi.org/10.1007/s11263-020-01347-6
44. Xu, Q., Wang, W., Ceylan, D., Mech, R., Neumann, U.: DISN: deep implicit surface network for high-quality single-view 3D reconstruction. In: NeurIPS (2019)
45. Yang, S., Xu, M., Xie, H., Perry, S., Xia, J.: Single-view 3D object reconstruction from shape priors in memory. In: CVPR (2021)
46. Zhang, H., Cisse, M., Dauphin, Y.N., Lopez-Paz, D.: Mixup: beyond empirical risk minimization. In: ICLR (2018)
47. Zhen Xing, Hengduo li, Z.W., Jiang, Y.G.: Semi-supervised single-view 3D reconstruction via prototype shape priors. In: ECCV (2022)
48. Zhu, L., Yang, Y.: Compound memory networks for few-shot video classification. In: ECCV, pp. 751–766 (2018)

DID-M3D: Decoupling Instance Depth for Monocular 3D Object Detection

Liang Peng[1,2], Xiaopei Wu[1], Zheng Yang[2], Haifeng Liu[1], and Deng Cai[1,2(✉)]

[1] State Key Lab of CAD&CG, Zhejiang University, Hangzhou, China
{pengliang,wuxiaopei,haifengliu}@zju.edu.cn, dengcai@cad.zju.edu.cn
[2] Fabu Inc., Hangzhou, China
yangzheng@fabu.ai

Abstract. Monocular 3D detection has drawn much attention from the community due to its low cost and setup simplicity. It takes an RGB image as input and predicts 3D boxes in the 3D space. The most challenging sub-task lies in the instance depth estimation. Previous works usually use a direct estimation method. However, in this paper we point out that the instance depth on the RGB image is non-intuitive. It is coupled by visual depth clues and instance attribute clues, making it hard to be directly learned in the network. Therefore, we propose to reformulate the instance depth to the combination of the instance visual surface depth (**visual depth**) and the instance attribute depth (**attribute depth**). The visual depth is related to objects' appearances and positions on the image. By contrast, the attribute depth relies on objects' inherent attributes, which are invariant to the object affine transformation on the image. Correspondingly, we decouple the 3D location uncertainty into visual depth uncertainty and attribute depth uncertainty. By combining different types of depths and associated uncertainties, we can obtain the final instance depth. Furthermore, data augmentation in monocular 3D detection is usually limited due to the physical nature, hindering the boost of performance. Based on the proposed instance depth disentanglement strategy, we can alleviate this problem. Evaluated on KITTI, our method achieves new state-of-the-art results, and extensive ablation studies validate the effectiveness of each component in our method. The codes are released at https://github.com/SPengLiang/DID-M3D.

Keywords: Monocular 3D detection · Instance depth estimation

1 Introduction

Monocular 3D object detection is an important topic in the self-driving and computer vision community. It is popular due to its low price and configuration simplicity. Rapid improvements [5,6,25,28,36,60] have been conducted in recent

Supplementary Information The online version contains supplementary material available at https://doi.org/10.1007/978-3-031-19769-7_5.

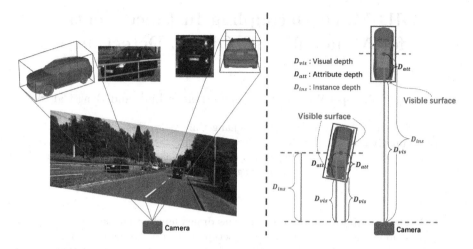

Fig. 1. We decouple the instance depth into visual depths and attribute depths due to the coupled nature of instance depth. Please refer to the text for more details. Best viewed in color with zoom in.

years. A well-known challenge in this task lies in instance depth estimation, which is the bottleneck towards boosting the performance since the depth information is lost after the camera projection process.

Many previous works [2,12,39] directly regress the instance depth. This manner does not consider the ambiguity brought by the instance depth itself. As shown in Fig. 1, for the right object, its instance depth is the sum of car tail depth and half-length of the car, where the car length is ambiguous since both car's left and right sides are invisible. For the left object, except for the intuitive visible surface depth, the instance depth further depends on the car dimension and orientation. We can observe that the instance depth is non-intuitive. It requires the network to additionally learn instance inherent attributes on the instance depth head. Previous direct estimation and mediate optimization methods do not fully consider this coupled nature. Thus they lead to suboptimal performance on the instance depth estimation, showing less accurate results.

Based on the analysis above, in this paper we propose to decouple the instance depth to **instance visual surface depth (visual depth)** and **instance attribute depth (attribute depth)**. We illustrate some examples in Fig. 1. For each point (or small patch) on the object, the visual depth denotes the absolute depth towards the agent (car/robot) camera, and we define the attribute depth as the relative depth offset from the point (or small patch) to the object's 3D center. This decoupled manner encourages the network to learn different feature patterns of the instance depth. Visual depth on monocular imagery depends on objects' appearances and positions [11] on the image, which is affine-sensitive. By contrast, attribute depth highly relies on object inherent attributes (*e.g.*, dimensions and orientations) of the object. It focuses on features inside the RoI, which is affine-invariant. (See Sect. 4.1 and 4.2 for detailed discussion). Thus the attribute depth is independent of the visual depth, and decoupled instance depth allows us to use separate heads to extract different types of features for different types of depths.

Specifically, for an object image patch, we divide it into $m \times n$ grids. Each grid can denote a small area on the object, with being assigned a visual depth and the corresponding attribute depth. Considering the occlusion and 3D location uncertainty, we use the uncertainty to denote the confidence of each depth prediction. At inference, every object can produce $m \times n$ instance depth predictions, thus we take advantage of them and associated uncertainties to adaptively obtain the final instance depth and confidence.

Furthermore, prior works usually are limited by the diversity of data augmentation, due to the complexity of keeping alignment between 2D and 3D objects when enforcing affine transformation in a 2D image. Based on the decoupled instance depth, we show that our method can effectively perform data augmentation, including the way using affine transformation. It is achieved by the affine-sensitive and affine-invariant nature of the visual depth and attribute depth, respectively (See Sect. 4.3). To demonstrate the effectiveness of our method, we perform experiments on the widely used KITTI dataset. The results suggest that our method outperforms prior works with a significant margin.

In summary, our contributions are listed as follows:

1. We point out the coupled nature of instance depth. Due to the entangled features, the previous way of directly predicting instance depth is suboptimal. Therefore, we propose to decouple the instance depth into attribute depths and visual depths, which are independently predicted.
2. We present two types of uncertainties to represent depth estimation confidence. Based on this, we propose to adaptively aggregate different types of depths into the final instance depth and correspondingly obtain the 3D localization confidence.
3. With the help of the proposed attribute depth and visual depth, we alleviate the limitation of using affine transformation in data augmentation for monocular 3D detection.
4. Evaluated on KITTI benchmark, our method sets the new state of the art (SOTA). Extensive ablation studies demonstrate the effectiveness of each component in our method.

2   Related Work

2.1   LiDAR-Based 3D Object Detection

LiDAR-based methods utilize precise LiDAR point clouds to achieve high performance. According to the representations usage, they can be categorized into point-based, voxel-based, hybrid, and range-view-based methods. Point-based methods [37,42,51] directly use raw point clouds to preserve fine-grained structures of objects. However, they usually suffer from high computational costs. Voxel-based methods [30,50,53,56,58] voxelize the unordered point clouds into regular grids so that the CNNs can be easily applied. These methods are more efficient, but voxelization introduces quantization errors, resulting in information loss. To explore advantages of different representations, some hybrid methods [8,32,41,43,52] are proposed. They validate that combining point-based and voxel-based methods can achieve a better trade-off between accuracy and efficiency. Range-view-based methods [1,4,13,19] organize point clouds in range view, which is a compact representation of point clouds. These methods are also computationally efficient but are under-explored.

2.2   Monocular 3D Object Detection

Due to the low cost and setup simplicity, monocular 3D object detection is popular in recent years. Previous monocular works can be roughly divided into image-only based and depth-map based methods. The pioneer method [6] integrates different types of information such as segmentation and scene priors for performing 3D detection. To learn spatial features, OFTNet [40] projects 2D image features to the 3D space. M3D-RPN [2] attempts to extract depth-aware features by using different convolution kernels on different image columns. Kinematic3D [3] uses multi-frames to capture the temporal information by employing a tracking module. GrooMeD-NMS [18] develops a trainable NMS-step to boost the final performance. At the same period of time, many works fully take advantage of the scene and geometry priors [23,44,48,59]. Due to the ill-posed nature of monocular imagery, some works resort to using the dense depth estimation [12,26,35,39,46,47,49]. With the help of estimated depth maps, RoI-10D [29] use CAD models to augment training samples. Pseudo-LiDAR [27,49] based methods are also popular. They convert estimated depth maps to point clouds, then well-designed LiDAR 3D detectors can be directly employed. CaDDN [39] predicts a categorical depth distribution, to precisely project depth-aware features to 3D space. In sum, benefiting from rapid developments of deep learning technologies, monocular 3D detection has conducted remarkable progress.

2.3   Estimation of Instance Depth

Most monocular works directly predict the instance depth. There are also some works that use auxiliary information to help the instance depth estimation in the post-processing. They usually take advantage of the geometry constraints and scene priors. The early work Deep3DBox [31] regresses object dimensions and orientations, the remaining 3D box parameters including the instance depth are estimated by 2D-3D box geometry constraints. This indirect way has poor performance because it does not fully use the supervisions. To use geometric projection constraints, RTM3D [21] predicts nine perspective key-points (center and corners) of a 3D bounding box in the image space. Then the initially estimated instance depth can be optimized by minimizing projection errors. KM3D [20] follows this line and integrates this optimization into a end-to-end training process. More recently, MonoFlex [55] also predicts the nine perspective key-points. In addition to the directly predicted instance depth, it uses the projection heights

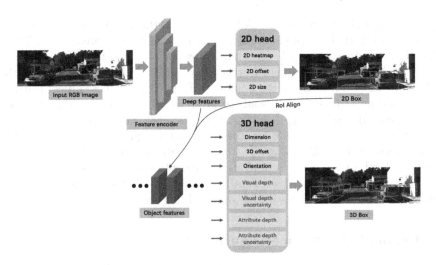

Fig. 2. Network framework. The overall design follows GUPNet [25]. The estimated 2D boxes are used to extract specific features for each object, followed by 3D box heads, which predict required 3D box parameters. The red parts in 3D heads denote the newly-proposed components. They are used to decouple the instance depth.

in pair key-points, using geometric relationships to produce new instance depths. MonoFlex develops an ensemble strategy to obtain the final instance depth. Differing from previous works, GUPNet [25] proposes an uncertainty propagation method for the instance depth. It uses estimated object 3D dimensions and 2D height to obtain initial instance depth, with additionally predicting a depth bias to refine the instance depth. GUPNet mainly focuses on tackling the error amplification problem in the geometry projection process. MonoRCNN [44] also introduces a distance decomposition based on the 2D height and 3D height. Such methods use geometric or auxiliary information to refine the estimated instance depth, while they do not fully use the coupled nature of the instance depth.

3 Overview and Framework

Preliminaries. Monocular 3D detection takes an image captured by an RGB camera as input, predicting amodal 3D bounding boxes of objects in 3D space. These 3D boxes are parameterized by the 3D center location (x, y, z), dimension (h, w, l), and the orientation (θ). Please note, in the self-driving scenario, the orientation usually refers to the yaw angle, and the roll and pitch angles are zeros by default. Also, the ego-car/robot has been calibrated.

In this paragraph, we provide an overview and describe the framework. The overall framework is shown in Fig. 2. First, the network takes an RGB image $\mathbf{I} \in \mathbb{R}^{H \times W \times 3}$ as the input. After feature encoding, we have deep features $\mathbf{F} \in \mathbb{R}^{\frac{H}{4} \times \frac{W}{4} \times C}$, where C is the channel number. Second, deep features $\mathbf{F}$ are fed into three 2D detection heads, namely, 2D heatmap $\mathbf{H} \in \mathbb{R}^{\frac{H}{4} \times \frac{W}{4} \times B}$, 2D offset $\mathbf{O}_{2d} \in$

$\mathbb{R}^{\frac{H}{4} \times \frac{W}{4} \times 2}$, and 2D size $\mathbf{S}_{2d} \in \mathbb{R}^{\frac{H}{4} \times \frac{W}{4} \times 2}$, where B is the number of categories. By combining such 2D head predictions, we can achieve 2D box predictions. Then, with 2D box estimates, single object features are obtained from deep features $\mathbf{F}$ by RoI Align. We have object features $\mathbf{F}_{obj} \in \mathbb{R}^{n \times 7 \times 7 \times C}$, where 7×7 is the RoI Align size and n refers to the number of RoIs. Finally, these object features $\mathbf{F}_{obj}$ are fed into 3D detection heads to produce 3D parameters. Therefore, we have 3D box dimension $\mathbf{S}_{3d} \in \mathbb{R}^{n \times 3}$, 3D center projection offset $\mathbf{O}_{3d} \in \mathbb{R}^{n \times 2}$, orientation $\Theta \in \mathbb{R}^{n \times k \times 2}$ (we follow the multi-bin design [31] where k is the bin number), visual depth $\mathbf{D}_{vis} \in \mathbb{R}^{n \times 7 \times 7}$, visual depth uncertainty $\mathbf{U}_{vis} \in \mathbb{R}^{n \times 7 \times 7}$, attribute depth $\mathbf{D}_{att} \in \mathbb{R}^{n \times 7 \times 7}$, and attribute depth uncertainty $\mathbf{U}_{att} \in \mathbb{R}^{n \times 7 \times 7}$. Using the parameters above, we can achieve the final 3D box predictions. In the following sections, we will detail the proposed method.

4   Decoupled Instance Depth

We divide the RoI image patch into 7×7 grids, assigning each grid a visual depth value and an attribute depth value. We provide the ablation on the grid size for visual and attribute depth in experiments (See Sect. 5.4 and Table 4). In the following, we first detail the two types of depths, followed by the decoupled-depth based data augmentation, then introduce the way of obtaining the final instance depth, and finally describe loss functions.

4.1   Visual Depth

The visual depth denotes the physical depth of the object surface on the small RoI image grid. For each grid, we define the visual depth as the average pixel-wise depth within the grid. If the grid is 1×1 pixel, the visual depth is equal to the pixel-wise depth. Given that a pixel denotes the quantified surface of the object, we can regard visual depths as the general extension of pixel-wise depths.

The visual depth in monocular imagery has an important property. For a monocular-based system, visual depth highly relies on the object's 2D box size (the faraway object appears small on the images and vice versa) and the position on the image (lower v coordinates under image coordinate system indicate larger depths) [11]. Therefore, if we perform an affine transformation to the image, the visual depth should be correspondingly transformed, where the depth value should be scaled. We call this nature the affine-sensitive.

4.2   Attribute Depth

The attribute depth refers to the depth offset from the visual surface to the object's 3D center. We call it attribute depth because it is more likely related to the object's inherent attributes. For example, when the car orientation is parallel to z-axis (depth direction) in 3D space, the attribute depth of the car tail is the car's half-length. By contrast, the attribute depth is car's half-width if the orientation is parallel to x-axis. We can see that the attribute depth depends on

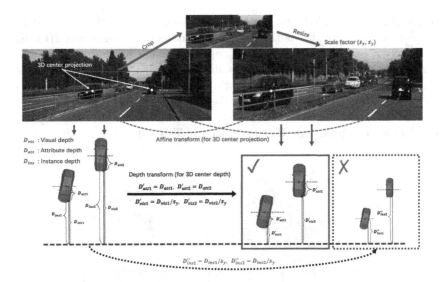

Fig. 3. Affine transformation based data augmentation. We do not change the object's inherent attributes, *i.e.*, attribute depths, dimensions, and observation angles. The visual depth is scaled according to the 2D height scale factor. The 3D center projection is transformed together with the image affine transformation.

the object semantics and its inherent attributes. In contrast to the affine-sensitive nature of visual depth, attribute depth is invariant to any affine transformation because object inherent characteristics will not change. We call this nature the affine-invariant.

As described above, we use two separate heads to estimate the visual depth and attribute depth, respectively. The disentanglement of instance depth has several advantages: (1) The object depth is decoupled in a reasonable and intuitive manner, thus we can more comprehensively and precisely represent the object; (2) The network is allowed to extract different types of features for different types of depths, which facilitates the learning; (3) Benefitting from the decoupled depth, our method can effectively perform affine transformation based data augmentation, which is usually limited in previous works.

4.3 Data Augmentation

In monocular 3D detection, many previous works are limited by data augmentation. Most of them only employ photometric distortion and flipping transformation. Data augmentation using affine transformation is hard to be adopted because the transformed instance depth is agnostic. Based on the decoupled depth, we point out that our method can alleviate this problem.

We illustrate an example in Fig. 3. Specifically, we add the random cropping and scaling strategy [57] in the data augmentation. The 3D center projection point on the image follows the same affine transformation process of the image.

78   L. Peng et al.

The visual depth is scaled by the scale factor along y-axis on the image because $d = \frac{f \cdot h_{3d}}{h_{2d}}$, where f, h_{3d}, h_{2d} is the focal length, object 3D height and 2D height, respectively. Conversely, the attribute depth keeps the same due to its affine-invariant nature. We do not directly scale the instance depth, as this manner will implicitly damage the attribute depth. Similarly, other inherent attributes of objects, *i.e.*, the observation angle and the dimension, keep the same as original values. We empirically show that the data augmentation works well. We provide the ablations in experiments (See Sect. 5.4 and Table 3).

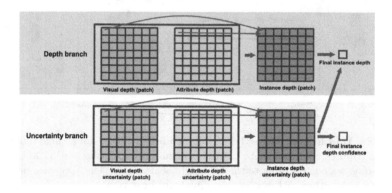

Fig. 4. Depth flow for an object. We use the visual depth, attribute depth, and the associated uncertainty to obtain the final instance depth.

4.4 Depth Uncertainty and Aggregation

The 2D classification score cannot fully express the confidence in monocular 3D detection because of the difficulty in 3D localization. Previous works [25,45] use the instance depth confidence or 3D IoU loss, integrating with 2D detection confidence to represent the final 3D detection confidence. Given that we have decoupled the instance depth into visual depth and attribute depth, we can further decouple the instance depth uncertainty. Only when an object has low visual uncertainty and low attribute depth uncertainty simultaneously, the instance depth can have high confidence.

Inspired by [16,25], we assume every depth prediction is a Laplace distribution. Specifically, for each visual depth d_{vis} in $\mathbf{D}_{vis} \in \mathbb{R}^{n \times 7 \times 7}$ and the corresponding uncertainty u_{vis} in $\mathbf{U}_{vis} \in \mathbb{R}^{n \times 7 \times 7}$, they follow the Laplace distribution $L(d_{vis}, u_{vis})$. Similarly, the attribute depth distribution is $L(d_{att}, u_{att})$, where d_{att} in $\mathbf{D}_{att} \in \mathbb{R}^{n \times 7 \times 7}$ and u_{att} in $\mathbf{U}_{att} \in \mathbb{R}^{n \times 7 \times 7}$. Therefore, the instance depth distribution derived by associated visual and attribute depth is $L(\tilde{d}_{ins}, \tilde{u}_{ins})$, where $\tilde{d}_{ins} = d_{vis} + d_{att}$ and $\tilde{u}_{ins} = \sqrt{u_{vis}^2 + u_{att}^2}$. Then we use $\tilde{\mathbf{D}}_{ins(patch)} \in \mathbb{R}^{n \times 7 \times 7}$ and $\tilde{\mathbf{U}}_{ins(patch)} \in \mathbb{R}^{n \times 7 \times 7}$ to denote the instance depth and the uncertainty on the RoI patch. We illustrate the depth flow process in Fig. 4.

To obtain the final instance depth, we first convert the uncertainty to probability [16,25], which can be written as $\mathbf{P}_{ins(patch)} = exp(-\tilde{\mathbf{U}}_{ins(patch)})$, where $\mathbf{P}_{ins(patch)} \in \mathbb{R}^{n \times 7 \times 7}$. Then we aggregate the instance depth on the patch to the final instance depth. For the ith object $(i = 1, ..., n)$, we have:

$$d_{ins} = \sum \frac{\tilde{\mathbf{D}}_{ins(patch)_i} \mathbf{P}_{ins(patch)_i}}{\sum \mathbf{P}_{ins(patch)_i}} \tag{1}$$

The corresponding final instance depth confidence is:

$$p_{ins} = \sum \left(\frac{\mathbf{P}_{ins(patch)_i}}{\sum \mathbf{P}_{ins(patch)_i}} \mathbf{P}_{ins(patch)_i} \right) \tag{2}$$

Therefore, the final 3D detection confidence is $p = p_{2d} \cdot p_{ins}$, where p_{2d} is the 2D detection confidence.

4.5   Loss Functions

2D Detection Part: As shown in Fig. 2, for the 2D object detection part, we follow the design in CenterNet [57]. The 2D heatmap $\mathbf{H}$ aims to indicate the rough object center on the image. The size is $\frac{H}{4} \times \frac{W}{4} \times B$, where H, W is the input image size and B is the number of categories. The 2D offset O_{2d} refers to the residual towards rough 2D centers, and the 2D size S_{2d} denotes the 2D box height and width. Following CenterNet [57], we have loss functions $\mathcal{L}_H, \mathcal{L}_{O_{2d}}, \mathcal{L}_{S_{2d}}$, respectively.

3D Detection Part: For the 3D object dimension, we follow the typical transformation and loss design [2] $\mathcal{L}_{S_{3d}}$. For the orientation, the network predicts the observation angle and uses the multi-bin loss [31] $\mathcal{L}_{\Theta}$. Also, we use the 3D center projection on the image plane and the instance depth to recover the object's 3D location. For the 3D center projection, we achieve it by predicting the 3D projection offset to the 2D center. The loss function is: $\mathcal{L}_{O_{3d}} = \text{Smooth}L_1(O_{3d}, O_{3d}^*)$. We use $*$ to denote corresponding labels. As mentioned above, we decouple the instance depth into visual depth and attribute depth. The visual depth labels are obtained by projecting LiDAR points onto the image and the attribute depth labels are obtained by subtracting instance depth labels with visual depth labels. Combing with the uncertainty [9,16], the visual depth loss is: $\mathcal{L}_{D_{vis}} = \frac{\sqrt{2}}{u_{vis}} \|d_{vis} - d_{vis}^*\| + log(u_{vis})$, where u_{vis} is the uncertainty. Similarly, we have attribute depth loss $\mathcal{L}_{D_{att}}$ and instance depth loss $\mathcal{L}_{D_{ins}}$. Among these loss terms, the losses concerning instance depth ($\mathcal{L}_{D_{vis}}$, $\mathcal{L}_{D_{att}}$, and $\mathcal{L}_{D_{ins}}$) play the most important role since they matter objects' localization in the 3D space. We empirically set 1.0 for weights of all loss terms, and the overall loss is:

$$\mathcal{L} = \mathcal{L}_H + \mathcal{L}_{O_{2d}} + \mathcal{L}_{S_{2d}} + \mathcal{L}_{S_{3d}} + \mathcal{L}_{\Theta} + \mathcal{L}_{O_{3d}} + \mathcal{L}_{D_{vis}} + \mathcal{L}_{D_{att}} + \mathcal{L}_{D_{ins}} \tag{3}$$

Table 1. Comparisons on KITTI testing set. The red refers to the highest result and blue is the second-highest result. Our method achieves state-of-the-art results. Note that DD3D [33] uses a large private dataset (DDAD15M), which includes 15M frames.

Approaches	Venue	Runtime	AP_{BEV} (IoU=0.7)$\mid R_{40}$			AP_{3D} (IoU=0.7)$\mid R_{40}$		
			Easy	Moderate	Hard	Easy	Moderate	Hard
MonoGRNet [38]	AAAI19	400 ms	18.19	11.17	8.73	15.74	9.61	4.25
ROI-10D [29]	CVPR19	200 ms	9.78	4.91	3.74	4.32	2.02	1.46
MonoPSR [17]	CVPR19	200 ms	18.33	12.58	9.91	10.76	7.25	5.85
M3D-RPN [2]	ICCV19	160 ms	21.02	13.67	10.23	14.76	9.71	7.42
AM3D [27]	ICCV19	400 ms	25.03	17.32	14.91	16.50	10.74	9.52
MonoPair [9]	CVPR20	60 ms	19.28	14.83	12.89	13.04	9.99	8.65
D4LCN [12]	CVPR20	200 ms	22.51	16.02	12.55	16.65	11.72	9.51
RTM3D [21]	ECCV20	40 ms	19.17	14.20	11.99	14.41	10.34	8.77
PatchNet [26]	ECCV20	400 ms	22.97	16.86	14.97	15.68	11.12	10.17
Kinematic3D [3]	ECCV20	120 ms	26.69	17.52	13.10	19.07	12.72	9.17
Neighbor-Vote [10]	MM21	100 ms	27.39	18.65	16.54	15.57	9.90	8.89
Ground-Aware [23]	RAL21	50 ms	29.81	17.98	13.08	21.65	13.25	9.91
MonoRUn [5]	CVPR21	70 ms	27.94	17.34	15.24	19.65	12.30	10.58
DDMP-3D [46]	CVPR21	180 ms	28.08	17.89	13.44	19.71	12.78	9.80
Monodle [28]	CVPR21	40 ms	24.79	18.89	16.00	17.23	12.26	10.29
CaDDN [39]	CVPR21	630 ms	27.94	18.91	17.19	19.17	13.41	11.46
GrooMeD-NMS [18]	CVPR21	120 ms	26.19	18.27	14.05	18.10	12.32	9.65
MonoEF [59]	CVPR21	30 ms	29.03	19.70	17.26	21.29	13.87	11.71
MonoFlex [55]	CVPR21	35 ms	28.23	19.75	16.89	19.94	13.89	12.07
MonoRCNN [44]	ICCV21	70 ms	25.48	18.11	14.10	18.36	12.65	10.03
AutoShape [24]	ICCV21	40 ms	30.66	20.08	15.95	22.47	14.17	11.36
GUPNet [25]	ICCV21	34 ms	30.29	21.19	18.20	22.26	15.02	13.12
PCT [47]	NeurIPS21	45 ms	29.65	19.03	15.92	21.00	13.37	11.31
MonoCon [22]	AAAI22	26 ms	**31.12**	**22.10**	**19.00**	**22.50**	16.46	13.95
DD3D [33]	*ICCV21*	-	*32.35*	*23.41*	*20.42*	*23.19*	*16.87*	*14.36*
DID-M3D (ours)	ECCV22	40 ms	32.95	22.76	19.83	24.40	16.29	**13.75**
Improvements			+1.83	+0.66	+0.83	+1.90	-0.17	−0.20

5   Experiments

5.1   Implementation Details

We conduct experiments on 2 NVIDIA RTX 3080Ti GPUs with batch size 16. We use the PyTorch framework [34]. We train the network with 200 epochs and employ the Hierarchical Task Learning (HTL) [25] training strategy. The Adam optimizer is used with the initial learning rate $1e-5$. The learning rate increases to $1e-3$ in the first 5 epochs by employing the linear warm-up strategy and decays in epoch 90 and 120 with rate 0.1. We set k 12 in the multi-bin orientation Θ. For the backbone and head, we follow the design in [25,54]. Inspired by CaDDN [39], we project LiDAR point clouds onto the image frame to create sparse depth maps and then depth completion [15] is performed to generate depth values at each pixel in the image. Considering the space limitation, we provide more experimental results and discussion in the supplementary material.

5.2   Dataset and Metrics

Following the commonly adopted setup in previous works [2,12,18,24], we perform experiments on the widely used KITTI [14] 3D detection dataset. The KITTI dataset provides 7,481 training samples and 7,518 testing samples, where training sample labels are publicly available and the labels of testing samples keep secret in the KITTI website, which are only used for online evaluation and ranking. To conduct ablations, previous work further divides the 7,481 samples into a new training set with 3,712 samples and a *val* set with 3,769 samples. This data split [7] is widely adopted by most previous works. Additionally, KITTI divides objects into the *easy*, *moderate*, and *hard* level according to the object 2D box height (related to the depth), occlusion and truncation levels. For evaluation metrics, we use the suggested AP_{40} metric [45] under the two core tasks, *i.e.*, 3D and bird's-eye-view (BEV) detection.

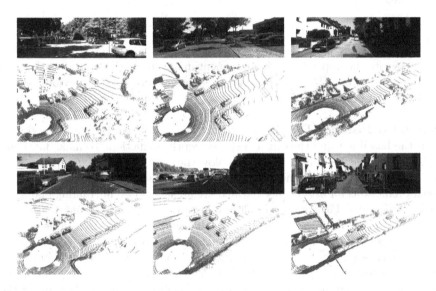

Fig. 5. Qualitative results on KITTI *val* set. Red: ground-truth 3D boxes; Green: our predictions. We can observe that the model conducts accurate 3D box predictions. Best viewed in color with zoom in.

5.3   Performance on KITTI Benchmark

We compare our method (DID-M3D) with other methods in KITTI test set, which is the official benchmark for monocular 3D object detection. The results are shown in Table 1. We can see that our method achieves a new state-of-the-art. For example, compared to GUPNet [25] (*ICCV21*), we boost the performance from 21.19/15.02 to 22.26/16.29 under the moderate setting. As for PCT [47] (*NeurIPS21*), we exceed it with 3.23/2.92 AP under the moderate setting, which is a significant improvement. When compared to the recent method MonoCon

[22] (*AAAI22*), our method still shows better performance on all BEV metrics and a 3D metric. Also, the runtime of our method is comparable to other real-time methods. Such results validate the superiority of our method. Additionally, to demonstrate the generalizability on other categories, we perform experiments on cyclist and pedestrian categories. As shown in Table 5, our method brings obvious improvements to the baseline (without employing the proposed components). The results suggest that our method works well for other categories.

Moreover, we provide qualitative results on the RGB image and 3D space in Fig. 5. We can observe that for most simple cases (*e.g.*, close objects without occlusion and truncation), the model predictions are quite precise. However, for the heavily occluded, truncated, or faraway objects, the orientation or instance depth is less accurate. This is a common dilemma for most monocular works due to the limited information in monocular imagery. In the supplementary material, we will provide more experimental results and make detailed discussions on failure cases.

5.4   Ablation Study

To investigate the impact of each component in our method, we conduct detailed ablation studies on KITTI. Following the common practice in previous works, all ablation studies are evaluated in the *val* set on the car category.

Decoupled Instance Depth. We report the results in Table 2. Experiment (a) is the baseline using the direct instance depth prediction. To make fair comparisons, for the baseline, we also employ the grid design (Experiment (b)). Similar to our method, it means that the network also produce 7×7 instance depth predictions for every object, which are all supervised in training and averaged at inference. Then, we decouple the instance depth into visual depths and attribute depths (Experiment (b) → (c)), this simple modification improves the accuracy significantly. This result indicates the network performs suboptimally due to the coupled nature of instance depth, demonstrating our viewpoint. From Experiment (c) → (d, e), we can see that the depth uncertainty brings improvements, because the uncertainty stabilizes the training of depth, benefiting the network learning. When simultaneously enforcing both two types of uncertainty, the performance is further boosted. Please note, the decoupled instance depth is the precondition of decoupled uncertainty. Given that the two types of depth uncertainty are achieved, we can obtain the final instance depth uncertainty (Experiment (f) → (g)). This can be regarded as the 3D location confidence. It is used to combine with the original 2D detection confidence, which brings obvious improvements. Finally, we can use the decoupled depth and corresponding uncertainties to adaptively obtain the final instance depth (Experiment (h)), while previous experiments use the average value on the patch. We can see that this design enhances the performance. In summary, by using the decoupled depth strategy, we improve the baseline performance from 16.79/11.24 to 22.76/16.12 (Experiment (b) → (h)). It is an impressive result. Overall, the ablations validate the effectiveness of our method.

Table 2. Ablation for decoupled instance depth. "Dec.": decoupled; "ID.": instance depth; "u_{vis}": visual depth uncertainty; "u_{att}": attribute depth uncertainty; "Conf.": confidence; "AA.": adaptive aggregation.

Experiments	Grid	Dec. ID.	u_{vis}	u_{att}	ID. Conf.	ID. AA.	AP_{BEV}/AP_{3D} (IoU=0.7)|R_{40}		
							Easy	Moderate	Hard
(a)							19.84/14.07	16.25/11.20	14.13/9.97
(b)	✓						21.01/14.67	16.79/11.24	15.41/10.18
(c)	✓	✓					22.98/16.95	18.72/13.24	16.57/11.23
(d)	✓	✓	✓				25.13/16.85	19.84/13.45	17.19/12.03
(e)	✓	✓		✓			24.83/17.29	19.66/13.50	17.06/11.48
(f)	✓	✓	✓	✓			25.23/18.14	20.06/13.91	17.63/12.52
(g)	✓	✓	✓	✓	✓		29.34/21.51	21.53/15.57	18.55/12.84
(h)	✓	✓	✓	✓	✓	✓	**31.10/22.98**	**22.76/16.12**	**19.50/14.03**

Affine Transformation Based Data Augmentation. In this paragraph we aim to understand the effect of affine transformation based data augmentation. The comparisons are shown in Table 3. We can see that the method obviously benefits from affine-based data augmentation. Note that the proper depth transformation is very important. When enforcing affine-based data augmentation, the visual depth should be scaled while the attribute depth should not be changed due to their affine-sensitive and affine-invariant nature, respectively. If we change the attribute depth without scaling visual depth, the detector even performs worse than the one without affine-based data augmentation (AP_{3D} downgrades from 12.76 to 12.65). It is because this manner misleads the network in the training with incorrect depth targets. After revising visual depth, the network can benefit from the augmented training samples, boosting the performance from 19.05/12.76 to 21.74/15.48 AP under the moderate setting. We can see that the improper visual depth can result in larger impacts compared to the improper attribute depth on the final performance, as the visual depth has a larger value range. Finally, we obtain the best performance when employing the proper visual depth and attribute depth transformation strategy.

Table 3. Ablation for affine transformation based data augmentation. "Aff. Aug." in the table denotes the affine-based data augmentation.

w/Aff. Aug.	Scaled d_{vis}	Scaled d_{att}	AP_{BEV}/AP_{3D} (IoU=0.7)|R_{40}		
			Easy	Moderate	Hard
			25.63/17.61	19.05/12.76	16.34/11.21
✓			26.97/18.98	19.80/14.33	17.71/12.00
✓		✓	22.23/15.27	18.98/12.65	16.52/10.75
✓	✓	✓	28.67/21.43	21.74/15.48	18.76/13.47
✓	✓		**31.10/22.98**	**22.76/16.12**	**19.50/14.03**

Grid Size for Visual and Attribute Depth. As described in Sect. 4, we divide the RoI image patch into $m \times m$ grids, where each grid has a visual depth and an attribute depth. This paragraph investigates the impact brought by the grid size. When increasing grid size m, visual depths and attribute depths are becoming fine-grained. This tendency makes visual depth more intuitive, which is close to the pixel-wise depth. However, the fine-grained grid will lead to sub-optimal performance in terms of learning object attributes since the attributes focus on the overall object. It indicates that there exists a trade-off. Therefore, we perform ablations on the grid size m, as shown in Table 4. We achieve the best performance when m is set to 7.

Table 4. Ablation for the grid size on visual depth and attribute depth.

Grid size	AP_{BEV}/AP_{3D} (IoU=0.7)$|_{R_{40}}$		
	Easy	Moderate	Hard
1×1	26.11/19.39	18.76/13.01	16.00/11.34
3×3	27.03/19.73	19.91/14.33	18.10/12.25
5×5	29.20/21.36	21.53/15.13	18.61/12.53
9×9	30.21/21.78	22.28/14.98	18.93/12.47
13×13	29.88/21.67	21.97/15.29	18.96/12.72
19×19	28.20/20.36	21.53/15.13	18.61/12.53
7×7	**31.10/22.98**	**22.76/16.12**	**19.50/14.03**

Table 5. Comparisons on pedestrian and cyclist categories on KITTI *val* set under IoU criterion 0.5. Our method brings obvious improvements to the baseline.

Approaches	Pedestrian, $AP_{BEV}/AP_{3D}|_{R_{40}}$			Cyclist, $AP_{BEV}/AP_{3D}|_{R_{40}}$		
	Easy	Moderate	Hard	Easy	Moderate	Hard
Baseline	5.97/5.13	4.75/3.88	3.87/3.05	5.11/4.27	2.68/2.46	2.50/2.17
Baseline+Ours	**8.86/7.27**	**7.01/5.87**	**5.46/4.89**	**6.13/5.54**	**3.09/2.59**	**2.67/2.49**

6   Conclusion

In this paper, we point out that the instance depth is coupled by visual depth clues and object inherent attributes. Its entangled nature makes it hard to be precisely estimated with the previous direct method. Therefore, we propose to decouple the instance depth into visual depths and attribute depths. This manner allows the network to learn different types of features for instance depth. At the inference stage, the instance depth is obtained by aggregating visual depth, attribute depth, and associated uncertainties. Using the decoupled depth, we can effectively perform affine transformation based data augmentation on the image, which is usually limited in previous works. Finally, extensive experiments demonstrate the effectiveness of our method.

Acknowledgments. This work was supported in part by The National Key Research and Development Program of China (Grant Nos: 2018AAA0101400), in part by The National Nature Science Foundation of China (Grant Nos: 62036009, U1909203, 61936006, 61973271), in part by Innovation Capability Support Program of Shaanxi (Program No. 2021TD-05).

References

1. Bewley, A., Sun, P., Mensink, T., Anguelov, D., Sminchisescu, C.: Range conditioned dilated convolutions for scale invariant 3D object detection. arXiv preprint arXiv:2005.09927 (2020)
2. Brazil, G., Liu, X.: M3D-RPN: monocular 3D region proposal network for object detection. In: Proceedings of the IEEE International Conference on Computer Vision, pp. 9287–9296 (2019)
3. Brazil, G., Pons-Moll, G., Liu, X., Schiele, B.: Kinematic 3D object detection in monocular video. In: Vedaldi, A., Bischof, H., Brox, T., Frahm, J.-M. (eds.) ECCV 2020. LNCS, vol. 12368, pp. 135–152. Springer, Cham (2020). https://doi.org/10.1007/978-3-030-58592-1_9
4. Chai, Y., et al.: To the point: efficient 3D object detection in the range image with graph convolution kernels. In: Proceedings of the IEEE/CVF Conference on Computer Vision and Pattern Recognition, pp. 16000–16009 (2021)
5. Chen, H., Huang, Y., Tian, W., Gao, Z., Xiong, L.: Monorun: monocular 3D object detection by reconstruction and uncertainty propagation. In: Proceedings of the IEEE/CVF Conference on Computer Vision and Pattern Recognition, pp. 10379–10388 (2021)
6. Chen, X., Kundu, K., Zhang, Z., Ma, H., Fidler, S., Urtasun, R.: Monocular 3D object detection for autonomous driving. In: Proceedings of the IEEE Conference on Computer Vision and Pattern Recognition, pp. 2147–2156 (2016)
7. Chen, X., Kundu, K., Zhu, Y., Ma, H., Fidler, S., Urtasun, R.: 3D object proposals using stereo imagery for accurate object class detection. IEEE Trans. Pattern Anal. Mach. Intell. **40**(5), 1259–1272 (2017)
8. Chen, Y., Liu, S., Shen, X., Jia, J.: Fast point R-CNN. In: ICCV (2019)
9. Chen, Y., Tai, L., Sun, K., Li, M.: Monopair: monocular 3D object detection using pairwise spatial relationships. In: Proceedings of the IEEE/CVF Conference on Computer Vision and Pattern Recognition, pp. 12093–12102 (2020)
10. Chu, X., et al.: Neighbor-vote: improving monocular 3D object detection through neighbor distance voting. arXiv preprint arXiv:2107.02493 (2021)
11. Dijk, T.V., Croon, G.D.: How do neural networks see depth in single images? In: Proceedings of the IEEE/CVF International Conference on Computer Vision, pp. 2183–2191 (2019)
12. Ding, M., Huo, Y., Yi, H., Wang, Z., Shi, J., Lu, Z., Luo, P.: Learning depth-guided convolutions for monocular 3D object detection. In: Proceedings of the IEEE/CVF Conference on Computer Vision and Pattern Recognition, pp. 11672–11681 (2020)
13. Fan, L., Xiong, X., Wang, F., Wang, N., Zhang, Z.: Rangedet: in defense of range view for lidar-based 3D object detection. arXiv preprint arXiv:2103.10039 (2021)
14. Geiger, A., Lenz, P., Urtasun, R.: Are we ready for autonomous driving? the kitti vision benchmark suite. In: 2012 IEEE Conference on Computer Vision and Pattern Recognition, pp. 3354–3361. IEEE (2012)

15. Hu, M., Wang, S., Li, B., Ning, S., Fan, L., Gong, X.: Penet: towards precise and efficient image guided depth completion. In: 2021 IEEE International Conference on Robotics and Automation (ICRA), pp. 13656–13662. IEEE (2021)
16. Kendall, A., Gal, Y.: What uncertainties do we need in Bayesian deep learning for computer vision? Adv. Neural Inf. Process. Syst. 30 (2017)
17. Ku, J., Pon, A.D., Waslander, S.L.: Monocular 3D object detection leveraging accurate proposals and shape reconstruction. In: Proceedings of the IEEE Conference on Computer Vision and Pattern Recognition, pp. 11867–11876 (2019)
18. Kumar, A., Brazil, G., Liu, X.: GrooMed-NMS: grouped mathematically differentiable NMS for monocular 3D object detection. In: Proceedings of the IEEE/CVF Conference on Computer Vision and Pattern Recognition, pp. 8973–8983 (2021)
19. Li, B., Zhang, T., Xia, T.: Vehicle detection from 3D lidar using fully convolutional network. arXiv preprint arXiv:1608.07916 (2016)
20. Li, P., Zhao, H.: Monocular 3D detection with geometric constraint embedding and semi-supervised training. IEEE Robot. Autom. Lett. **6**(3), 5565–5572 (2021)
21. Li, P., Zhao, H., Liu, P., Cao, F.: Rtm3d: real-time monocular 3D detection from object keypoints for autonomous driving. arXiv preprint arXiv:2001.03343 (2020)
22. Liu, X., Xue, N., Wu, T.: Learning auxiliary monocular contexts helps monocular 3D object detection. arXiv preprint arXiv:2112.04628 (2021)
23. Liu, Y., Yixuan, Y., Liu, M.: Ground-aware monocular 3D object detection for autonomous driving. IEEE Robot. Autom. Lett. **6**(2), 919–926 (2021)
24. Liu, Z., Zhou, D., Lu, F., Fang, J., Zhang, L.: Autoshape: real-time shape-aware monocular 3d object detection. In: Proceedings of the IEEE/CVF International Conference on Computer Vision, pp. 15641–15650 (2021)
25. Lu, Y., et al.: Geometry uncertainty projection network for monocular 3D object detection. In: Proceedings of the IEEE/CVF International Conference on Computer Vision, pp. 3111–3121 (2021)
26. Ma, X., Liu, S., Xia, Z., Zhang, H., Zeng, X., Ouyang, W.: Rethinking pseudo-lidar representation. arXiv preprint arXiv:2008.04582 (2020)
27. Ma, X., Wang, Z., Li, H., Zhang, P., Ouyang, W., Fan, X.: Accurate monocular 3D object detection via color-embedded 3D reconstruction for autonomous driving. In: Proceedings of the IEEE International Conference on Computer Vision, pp. 6851–6860 (2019)
28. Ma, X., et al.: Delving into localization errors for monocular 3D object detection. In: Proceedings of the IEEE/CVF Conference on Computer Vision and Pattern Recognition, pp. 4721–4730 (2021)
29. Manhardt, F., Kehl, W., Gaidon, A.: Roi-10d: monocular lifting of 2D detection to 6D pose and metric shape. In: Proceedings of the IEEE Conference on Computer Vision and Pattern Recognition, pp. 2069–2078 (2019)
30. Mao, J., et al.: Voxel transformer for 3D object detection. arXiv preprint arXiv:2109.02497 (2021)
31. Mousavian, A., Anguelov, D., Flynn, J., Kosecka, J.: 3D bounding box estimation using deep learning and geometry. In: Proceedings of the IEEE Conference on Computer Vision and Pattern Recognition, pp. 7074–7082 (2017)
32. Noh, J., Lee, S., Ham, B.: HVPR: hybrid voxel-point representation for single-stage 3d object detection. In: Proceedings of the IEEE/CVF Conference on Computer Vision and Pattern Recognition, pp. 14605–14614 (2021)
33. Park, D., Ambrus, R., Guizilini, V., Li, J., Gaidon, A.: Is pseudo-lidar needed for monocular 3D object detection? In: Proceedings of the IEEE/CVF International Conference on Computer Vision, pp. 3142–3152 (2021)

34. Paszke, A., et al.: PyTorch: an imperative style, high-performance deep learning library. In: Advances in Neural Information Processing Systems, pp. 8026–8037 (2019)
35. Peng, L., Liu, F., Yan, S., He, X., Cai, D.: OCM3D: object-centric monocular 3D object detection. arXiv preprint arXiv:2104.06041 (2021)
36. Peng, L., et al.: Lidar point cloud guided monocular 3D object detection. arXiv preprint arXiv:2104.09035 (2021)
37. Qi, C.R., Liu, W., Wu, C., Su, H., Guibas, L.J.: Frustum PointNets for 3D object detection from RGB-D data. In: Proceedings of the IEEE Conference on Computer Vision and Pattern Recognition, pp. 918–927 (2018)
38. Qin, Z., Wang, J., Lu, Y.: MonogrNet: a geometric reasoning network for monocular 3D object localization. In: Proceedings of the AAAI Conference on Artificial Intelligence. vol. 33, pp. 8851–8858 (2019)
39. Reading, C., Harakeh, A., Chae, J., Waslander, S.L.: Categorical depth distribution network for monocular 3D object detection. In: Proceedings of the IEEE/CVF Conference on Computer Vision and Pattern Recognition, pp. 8555–8564 (2021)
40. Roddick, T., Kendall, A., Cipolla, R.: Orthographic feature transform for monocular 3D object detection. arXiv preprint arXiv:1811.08188 (2018)
41. Shi, S., et al.: PV-RCNN: point-voxel feature set abstraction for 3D object detection. In: CVPR, pp. 10529–10538 (2020)
42. Shi, S., Wang, X., Li, H.: PointRCNN: 3D object proposal generation and detection from point cloud. In: CVPR, pp. 770–779 (2019)
43. Shi, S., Wang, Z., Shi, J., Wang, X., Li, H.: From points to parts: 3D object detection from point cloud with part-aware and part-aggregation network. IEEE Trans. Pattern Anal. Mach. Intell. **43**(8), 2647–2664 (2020)
44. Shi, X., Ye, Q., Chen, X., Chen, C., Chen, Z., Kim, T.K.: Geometry-based distance decomposition for monocular 3D object detection. In: Proceedings of the IEEE/CVF International Conference on Computer Vision, pp. 15172–15181 (2021)
45. Simonelli, A., Bulo, S.R., Porzi, L., López-Antequera, M., Kontschieder, P.: Disentangling monocular 3D object detection. In: Proceedings of the IEEE International Conference on Computer Vision, pp. 1991–1999 (2019)
46. Wang, L., et al.: Depth-conditioned dynamic message propagation for monocular 3D object detection. In: Proceedings of the IEEE/CVF Conference on Computer Vision and Pattern Recognition, pp. 454–463 (2021)
47. Wang, L., et al.: Progressive coordinate transforms for monocular 3D object detection. Adv. Neural. Inf. Process. Syst. **34**, 13364–13377 (2021)
48. Wang, T., Xinge, Z., Pang, J., Lin, D.: Probabilistic and geometric depth: detecting objects in perspective. In: Conference on Robot Learning, pp. 1475–1485. PMLR (2022)
49. Wang, Y., Chao, W.L., Garg, D., Hariharan, B., Campbell, M., Weinberger, K.Q.: Pseudo-lidar from visual depth estimation: bridging the gap in 3D object detection for autonomous driving. In: Proceedings of the IEEE Conference on Computer Vision and Pattern Recognition, pp. 8445–8453 (2019)
50. Yan, Y., Mao, Y., Li, B.: SECOND: sparsely embedded convolutional detection. Sensors **18**(10), 3337 (2018)
51. Yang, Z., Sun, Y., Liu, S., Jia, J.: 3DSSD: point-based 3D single stage object detector. In: CVPR, pp. 11040–11048 (2020)
52. Yang, Z., Sun, Y., Liu, S., Shen, X., Jia, J.: STD: sparse-to-dense 3D object detector for point cloud. In: ICCV, pp. 1951–1960 (2019)

53. Yin, T., Zhou, X., Krahenbuhl, P.: Center-based 3D object detection and tracking. In: Proceedings of the IEEE/CVF Conference on Computer Vision and Pattern Recognition (CVPR), pp. 11784–11793 (2021)
54. Yu, F., Wang, D., Shelhamer, E., Darrell, T.: Deep layer aggregation. In: Proceedings of the IEEE Conference on Computer Vision and Pattern Recognition, pp. 2403–2412 (2018)
55. Zhang, Y., Lu, J., Zhou, J.: Objects are different: flexible monocular 3D object detection. In: Proceedings of the IEEE/CVF Conference on Computer Vision and Pattern Recognition, pp. 3289–3298 (2021)
56. Zheng, W., Tang, W., Jiang, L., Fu, C.W.: SE-SSD: self-ensembling single-stage object detector from point cloud. In: Proceedings of the IEEE/CVF Conference on Computer Vision and Pattern Recognition, pp. 14494–14503 (2021)
57. Zhou, X., Wang, D., Krähenbühl, P.: Objects as points. arXiv preprint arXiv:1904.07850 (2019)
58. Zhou, Y., Tuzel, O.: VoxelNet: end-to-end learning for point cloud based 3D object detection. In: Proceedings of the IEEE Conference on Computer Vision and Pattern Recognition, pp. 4490–4499 (2018)
59. Zhou, Y., He, Y., Zhu, H., Wang, C., Li, H., Jiang, Q.: Monocular 3D object detection: an extrinsic parameter free approach. In: Proceedings of the IEEE/CVF Conference on Computer Vision and Pattern Recognition, pp. 7556–7566 (2021)
60. Zou, Z., Ye, X., Du, L., Cheng, X., Tan, X., Zhang, L., Feng, J., Xue, X., Ding, E.: The devil is in the task: exploiting reciprocal appearance-localization features for monocular 3D object detection. In: Proceedings of the IEEE/CVF International Conference on Computer Vision, pp. 2713–2722 (2021)

Adaptive Co-teaching for Unsupervised Monocular Depth Estimation

Weisong Ren[1], Lijun Wang[1(✉)], Yongri Piao[1], Miao Zhang[1], Huchuan Lu[1,2], and Ting Liu[3]

[1] Dalian University of Technology, Dalian, China
kalilia1102@mail.dlut.edu.cn,
{ljwang,yrpiao,miaozhang,lhchuan}@dlut.edu.cn
[2] Peng Cheng Laboratory, Shenzhen, China
[3] Alibaba Group, Hangzhou, China
brooks.lt@alibaba-inc.com

Abstract. Unsupervised depth estimation using photometric losses suffers from local minimum and training instability. We address this issue by proposing an adaptive co-teaching framework to distill the learned knowledge from unsupervised teacher networks to a student network. We design an ensemble architecture for our teacher networks, integrating a depth basis decoder with multiple depth coefficient decoders. Depth prediction can then be formulated as a combination of the predicted depth bases weighted by coefficients. By further constraining their correlations, multiple coefficient decoders can yield a diversity of depth predictions, serving as the ensemble teachers. During the co-teaching step, our method allows different supervision sources from not only ensemble teachers but also photometric losses to constantly compete with each other, and adaptively select the optimal ones to teach the student, which effectively improves the ability of the student to jump out of the local minimum. Our method is shown to significantly benefit unsupervised depth estimation and sets new state of the art on both KITTI and Nuscenes datasets.

Keywords: Unsupervised · Monocular depth estimation · Knowledge distillation · Ensemble learning

1 Introduction

Monocular depth estimation is a fundamental research task in computer vision, with a wide application ranging from navigation [32], 3D reconstruction [10] to simultaneous localization and mapping [2]. With the rapid development of deep learning techniques, monocular depth estimation based on deep convolutional

Supplementary Information The online version contains supplementary material available at https://doi.org/10.1007/978-3-031-19769-7_6.

S. Avidan et al. (Eds.): ECCV 2022, LNCS 13661, pp. 89–105, 2022.
https://doi.org/10.1007/978-3-031-19769-7_6

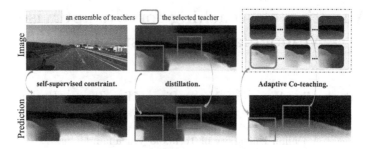

Fig. 1. Strength of our adaptive co-teaching framework. We conduct visualized comparison of our proposed adaptive co-teaching strategy (3rd column), photometric constraint based unsupervised training (1st column), and the straightforward distillation method (2nd column) on KITTI. Our scheme can transfer more accurate knowledge to the student maintaining fine-grained details.

networks has achieved significant progress in recent years. Nonetheless, training complex deep networks entails large-scale annotated depth data which are expensive to achieve and still very limited in terms of amount and diversity. To alleviate the need of depth annotations, unsupervised approaches have recently been investigated for learning monocular depth using either stereo images [12,14] or monocular videos [15,16], which have shown promising performance.

In this paper, we focus on unsupervised depth estimation using unlabeled monocular videos. Under the assumption of static scenes and Lambertian surfaces, both depth and pose networks can be jointly learned in an unsupervised manner by minimizing the photometric reconstruction losses of view synthesis. However, the above assumption may not hold in many scenarios, leading to unstable unsupervised learning and local minimum issues in dynamic regions and non-Lambertian or low-textured surfaces.

To alleviate this issue, recent works [7,17,24] propose to exploit semantic segmentation labels as prior knowledge to facilitate training. Though impressive performance has been achieved, these methods rely on additional manual annotations, and thus fail to maintain the advantages of unsupervised learning. An alternative idea to mitigate this challenge is to leverage the distillation techniques [27–30,37]. Although distillation learning can circumvent the unstable training issue, the teacher network is trained using conventional unsupervised framework. As shown in the second column of Fig. 1, it still suffers from the mentioned drawbacks, leading to low-quality pseudo labels and degraded final performance.

In light of the above issues, we propose an adaptive co-teaching framework for unsupervised depth estimation, which operates in a two-stage fashion. In the first stage, we train an ensemble of teacher networks for depth estimation in a unsupervised manner. By penalizing the correlation among teacher networks, we obtain a diversity of plausible solutions for depth estimation, providing a significantly higher chance to escape from local minimums of unsupervised learning compared to training a single teacher network. In the second stage, we transfer the learned knowledge from the teacher ensemble to the student network through a cost-aware co-teaching loss, which allows teacher networks in the ensemble to compete with each other and is able to adaptively select the optimal supervision source to teach the student network.

As another contribution of this work, we present the MUlti-STream ensemble network (MUSTNet) to facilitate more effective teacher ensemble learning. Our MUSTNet comprises a basis decoder and multiple coefficient decoders, where the basis decoder decomposes the depth map into a set of depth bases, and each coefficient decoder predicts a weight vector to linearly combine the depth bases, giving rise to one prediction of the depth map. By integrating both the basis and coefficient decoders in a single network, the MUSTNet serves as a more compact architecture for ensemble networks. In addition, the decomposition of the depth map into depth bases permits a more elegant and convenient way to enforce model diversity within the ensemble. As shown in our experiments, by incorporating the MUSTNet into our co-teaching framework, we achieve new state-of-the-art performance on popular benchmarks.

The contributions of our approach can be summarized into three-folds:

- An adaptive co-teaching framework for unsupervised depth estimation that enjoys the strengths of knowledge distillation and ensemble learning for more accurate depth estimation.
- A novel Multi-Stream Ensemble Network which decomposes depth maps into depth bases weighted by depth coefficients, providing a compact architecture for both the teach ensemble and the student model.
- A cost-aware co-teaching loss which leverages both the ensemble teachers and the photometric constraint to adaptively distill our student network.

Our model outperforms state-of-the-art monocular unsupervised approaches on the KITTI and Nuscenes datasets.

2   Related Work

2.1   Unsupervised Monocular Depth Estimation

Monocular depth estimation is an inherent ill-posed problem. Recently, with the help of Multi-view Stereo or Structure from Motion, some works [1,15,43] propose to tackle this problem within an unsupervised learning manner replacing the need of ground truth annotations.

Unsupervised Stereo Training. [12] proposes the first approach that estimates depth maps in an unsupervised manner with the help of multi-view synthesis. This work provides a basic paradigm for unsupervised depth estimation. After that, [14] employs a view synthesis loss and a depth smoothness loss to further improve the depth estimation performance. Very recently, [38] proposes a two-stage training strategy, which firstly generates pseudo depth labels from input images and secondly refine the network with a self-training strategy. It achieves better performance than state-of-the-art unsupervised methods.

Unsupervised Monocular Training. Different from multi-view based methods, monocular video sequence based methods require additional process to obtain camera poses. [43] jointly estimates depths and relative poses between adjacent frames, and indirectly supervise the depth and pose networks by computing the image re-projecting losses. However, this training strategy highly relies on the assumption that the adjacent frames comprise rigid scenes. Consequently, non-rigid parts caused by moving objects may seriously affect the performance. To solve this problem, [1] introduces an additional network to predict per-pixel invalid masks to ignore regions violating this assumption. [15] proposes a simple yet effective auto-masking and min re-projecting method to solve the problem of moving objects and occlusion. [24] decomposes the motion to the relative camera motion and instance-wise object motion to geometrically correct the projection process. [35] presents a novel tightly-coupled approach that leverages the interdependence of depth and ego motion at training and inference time. Although these methods have achieved matured performance, it is still an open question to solve the problem introduced by the photometric loss.

2.2 Knowledge Distillation

Knowledge distillation is originally proposed by [3] and popularized by [20]. The idea has been exploited for many computer vision tasks [6,18,26] for its ability to compressing a large network to a much smaller one. Recently, some works attempt to exploit distillation for unsupervised depth estimation.

Multi-view Training. [29] propose a self-distillation strategy for unsupervised multi-view depth estimation in which a sub-network of a bidirectional teacher is self-distilled to exploit the cycle inconsistency knowledge. [38] generates pseudo depth labels from the input images and secondly refine the network with a self-training strategy. More recently, [28] also apply a self-distillation method to unsupervised multi-view depth estimation and try to generate pseudo labels based on their proposed post-processing method. However, these multi-view images based methods fail to cope with monocular videos. It is hard for these methods to generate high quality pseudo labels for the camera poses keep unknown.

Monocular Training. For monocular videos, some approaches also attempt to employ knowledge distillation to unsupervised depth estimation. [27] train a complex teacher network in a unsupervised manner and distill the knowledge to a lightweight model to compress parameters while maintain high performance. [37] inference depth from multi-frame cost volume and generate depth prior information with a monocular unsupervised approach to teach the cost volume network in those potentially problematic regions. All the above introduced methods boost unsupervised depth estimation performance. However, those approaches leverage single depth teacher for distillation, which can not contribute to the student network to jump out of local minimal. Different from these methods, our work conducts an adaptive co-teaching strategy that leverages an ensemble of diverse teachers for better distillation.

3 Problem Formulation

During training, we consider a pair of input images: source image I_s and target image I_t of size $H \times W$. Two convolutional networks are leveraged to estimate the depth map of I_t and the relative camera pose $p_{s \to t} = [R|t]$ between I_s and I_t, respectively. After that, a synthesized target image $I_{t'}$ can be generated by rendering the source image I_s with predicted depth d_t, relative pose p_t, and the given camera intrinsic K. As a consequence, the depth and relative pose can be jointly optimized by minimizing the photometric loss given by

$$\mathcal{L}_p = \alpha \left(1 - SSIM\left(I_t, I_{t'}\right)\right) + (1 - \alpha) \left\| I_t - I_{t'} \right\|_1, \tag{1}$$

where $|| * ||_1$ measures the pixel-wise similarity and $SSIM$ indicates the structural similarity to measure the discrepancy between the synthesized and the real images structure. α is the hyper-parameter used to balance these two loss terms. Following [15], we set $\alpha = 0.85$.

Furthermore, in order to mitigate spatial fluctuation, we apply the edge-aware depth smoothness loss used in [15,21], which can be described as

$$\mathcal{L}_s\left(D_i\right) = \left|\delta_x D_i\right| e^{-\left|\delta_x I_i\right|} + \left|\delta_y D_i\right| e^{-\left|\delta_y I_i\right|}. \tag{2}$$

Following [15], for each pixel we optimize the loss for the best matching source image by selecting the per-pixel minimum over the reconstruction loss.

4 Methodology

Motivation and Overview. There may exist multiple feasible solutions to the photometric constraint (1), especially at low-texture regions, leading to training ambiguity and sub-optimal convergence. In this paper, we address this issue by proposing an adaptive co-teaching framework for unsupervised depth estimation. Our philosophy is to firstly learn an ensemble of depth estimation networks in the unsupervised manner, which can deliver a variety of depth prediction for an input image. We then treat these pre-trained networks as teachers and select their optimal predictions as pseudo labels to train the final student network using our cost-aware loss. As a result, training the student network under our co-teaching framework can circumvent the unstable issue of unsupervised learning. As verified in our experiment, through knowledge distillation from ensemble teachers, the student network outperforms each individual teacher network. In the following, we first introduce a new network architecture named MUSTNet for the teacher ensemble in Sect. 4.1. We then elaborate on the proposed adaptive co-teaching framework in Sect. 4.2. Finally, Sect. 4.3 presents the implementation details of our method.

4.1 MUSTNet: MUlit-STream Ensemble Network

A conventional approach to building network ensembles is to group a set of independent networks with similar architectures. The memory complexity and

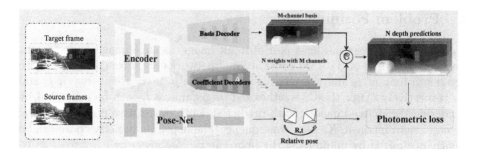

Fig. 2. Illustration of the proposed MUlti-STream ensemble Network.

computational overhead for training such an ensemble of networks will be linearly increased w.r.t. to the amount of ensemble members. Besides, the diversity of the learned network parameters can not be easily guaranteed. Inspired by the above observation, we design a more compact network structure named MUlti-Stream ensemble network (MUSTNet) for monocular depth estimation. In contrast to the above conventional approach, the proposed MUSTNet integrates a number of depth net into a single network. As shown in Fig. 2, MUSTNet contains one encoder followed by one basis decoder and N coefficient decoders in parallel. Given an input image, the basis decoder produces an output of M channels with each channel as one depth basis. Meanwhile, each coefficient decoder generates a coefficient vector of M dimensions. Each coefficient can be used as weight to linearly combine the depth basis, producing an estimation of depth map. The N coefficient decoders will then make N predictions of the depth map. As a result, the MUSTNet is equivalent to an ensemble of N depth estimation networks.

Basis Decoder. The basis decoder is designed motivated by prior methods [15,27]. It receives a multi-scale feature pyramid from the encoder, and the features are then progressively combined from coarse to fine level. Different from the basic structure, we extend the output channels of the last layer convolution to M. For each channel, we employ a sigmoid activation function to generate a normalized basis, representing the disparity values.

Coefficient Decoder. The coefficient decoder receives the coarsest output from the depth encoder. It consists of three 3×3 convolution layers followed by non-linear *ReLU* layers. An additional 3×3 convolution is add to compress feature channels to M, which keeps the same as the number of depth bases. Finally, we use a global average pooling layer to generate the coefficients.

Discussion. Our MUSTNet provides a compact structure for depth estimation, which enjoys more flexibility than conventional network ensembles from multiple aspects. As can be seen in Sect. 4.2, the diversity of the teacher ensemble is essential in our co-teaching framework. By decomposing depth estimation into depth bases and coefficient prediction, our MUSTNet permits an elegant and convenient way (introduced in Sect. 4.2) to enforce ensemble diversity. In addition, the MUSTNet structure is also scalable, i.e., it can not only serve as our

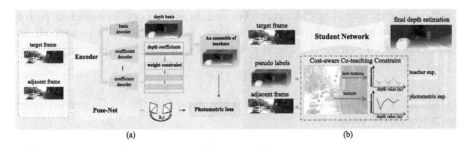

Fig. 3. Illustration of the proposed adaptive co-teaching framework. Specifically, **(a)**: In the first stage, we train the pose network (θ_{pose}) and the depth network (θ_{depth}) in an unsupervised manner. We employ an ensemble network for the depth estimation to generate a diversity of pseudo labels. **(b)**: In the second stage, we transfer the learned knowledge to a student monocular depth network leveraging the proposed cost aware co-teaching loss.

teacher ensemble, but is also suitable for the student depth network by using only one coefficient decoder. By using consistent architectures for teachers and student network, we expect the distillation learning to be more coherent, and thus more superior final performance.

4.2 Adaptive Co-teaching Framework

As shown in Fig. 3, our adaptive co-teaching framework operates in a two-stage fashion. In the first stage, we train a teacher ensemble through unsupervised learning. By penalizing their correlation, a diversity of teachers can be achieved. The second training stage is introduced to transfer the learned knowledge to the student network through a cost-aware co-teaching loss. It allows teacher networks in the ensemble to compete with each other and can adaptively select the optimal supervision source to teach the student network for much better predictions.

Unsupervised Teacher Ensemble Learning. We adopt the MUSTNet with $N > 1$ coefficient decoders as our teacher ensemble, where each coefficient decoder correspond to an individual teacher. We design the following two constraints to pursue model diversity.

Bases Diversity Constraint. We prefer the components of the bases have significant different distributions, which can help the network jump out of local minimal solutions. As suggested by [25], we assume the disparity values subject to a Gaussian distribution, and force the generated depth bases to have similar variances and different means for irrelevant outputs

$$L_v = \frac{\frac{1}{n}\sum_{i=1}^{n}\sigma_i^2 - \left(\frac{1}{n}\sum_{i=1}^{n}\sigma_i\right)^2}{\frac{1}{n}\sum_{i=1}^{n}\mu_i^2 - \left(\frac{1}{n}\sum_{i=1}^{n}\mu_i\right)^2}, \tag{3}$$

where μ_n and σ_n denote the mean and variance of the n-th depth basis, respectively.

Coefficients Orthogonality Constraint. For each predicted depth coefficient vector w, we compute its normalized version as $\overline{w} = \frac{w}{|w|}$, and stack all the normalized coefficients into a matrix W. The correlation of matrix W can be computed as

$$C_w = W \cdot W^T. \tag{4}$$

To penalize the relevance among coefficients, we define the coefficient orthogonality constraint as follows

$$L_w = ||C_w - E||, \tag{5}$$

where E represents the identity matrix.

The final loss function for training the teacher ensemble combines the photometric loss with the above two constraints and can be described as

$$L_{self} = L_p + \alpha L_s + \beta L_v + \gamma L_w, \tag{6}$$

where, α, β and γ are the hyper-parameters. In our work, we set $\alpha = 1e-3$, $\beta = 1e-3$, $\gamma = 1e-5$.

Student Learning via Adaptive Co-teaching. Given the pre-trained teacher ensemble, we learn a student depth network using a cost-aware co-teaching loss, which can not only identify the best teacher for distillation but also adaptively switch between distillation and unsupervised learning to select the optimal supervision sources.

Ensemble Distillation Loss. To aggregate multiple predictions from the teacher ensemble to synthesize satisfactory pseudo labels for distillation learning, we measure the accuracy of the predicted depth maps using the photometric reconstruction error (1). For each spatial location in the training image, we select the depth value predicted by the teacher ensemble with the minimum reconstruction error as the final pseudo label. The distillation loss is then defined as

$$\mathcal{L}_{distill} = ||d^s - d||_1 + \sum_i \left[(\nabla_x D_i)^2 + (\nabla_y D_i)^2 \right], \tag{7}$$

where d denotes the depth map predicted by the student network, d^s denotes the synthesized pseudo label, and $D_i = \log(d_i) - \log(d_i^s)$. The first term measures the difference between the predictions and pseudo labels, while the second term enforces the smoothness of the predicted depth maps.

Adaptive Switch Between Supervisions. In order to select the optimal supervision sources for learning the student, we propose to adaptively switch between the distillation learning (7) and unsupervised learning (1). Our basic idea is to estimate the quality of the potential solution yielded by the unsupervised learning. When unsupervised learning is likely to suffer from failure, we then switch to distillation learning. To this purpose, we introduce a cost volume [37,39] based masking mechanism. The cost volume V is constructed using the photometric reconstruction errors of K discrete depth planes. The planes are uniformly distributed over the disparity space. We set $K=16$ in our experiments. For each spatial location i, we compute the minimum error across all the depth planes as $e_i = \min_k(V_i[k])$. A smaller e_i mostly indicates a high-quality solution. However, this does not hold in low-texture regions, where multiple different solutions can produce similar errors. To address this issue, we further convert the cost volume into a confidence volume P as

$$P = \text{softmax}\left(\beta \frac{1}{V}\right), \tag{8}$$

where the softmax normalization is performed for each spatial location along the K depth planes, and β is a hyper-parameter that controls the height of local peaks. β is set as 0.05 in our work. The maximum confidence for location i is further computed as $c_i = \max_k(P_i[k])$. For low-texture regions with multiple local minimum errors, their maximum confidences are relatively low (See Fig. 3 (b) for an example). Therefore, unsupervised learning is only applicable to regions with low minimum errors and high maximum confidences. We define the selection mask M^u for unsupervised learning as

$$M_i^u = \begin{cases} 1, & \text{if } e_i < \tau_e \text{ and } c_i > \tau_c, \\ 0, & \text{otherwise,} \end{cases} \tag{9}$$

where τ_e and τ_c are two pre-defined thresholds. In our work, τ_e and τ_c are set as 0.6, 0.002, respectively. M_i^u indicates the mask value at position i. For regions with large minimum errors, we explore distillation from ensemble teachers for better supervision. However, the pseudo labels for distillation are generated based on photometric reconstruction losses, which are unreliable in low-texture regions. Therefore, the selection mask M^d for distillation learning also discards the low-texture regions and can be defined as

$$M_i^d = \begin{cases} 1, & \text{if } e_i \geq \tau_e \text{ and } c_i > \tau_c, \\ 0, & \text{otherwise.} \end{cases} \tag{10}$$

Finally, the cost aware co-teaching loss is formulated as a combination of the ensemble distillation loss and the unsupervised loss spatially weighted by their corresponding selection masks.

4.3 Implementation

Network Architecture. For our MUSTNet, we use the ResNet-18 [19] as the encoder. Furthermore, we employ the Redesigned Skip Connection block proposed in [27] to decrease the semantic gap between different scale features and obtain sharper depth details. For the relative pose, we employ the same design as [15], which predicts 6-DoF axis-angle representation. The first three dimensions represent translation vectors and the last three represent Euler angles.

Training Details. We implement our models on PyTorch and train on one Nvidia 2080Ti GPU. We use the Adam optimizer [23] with $\beta_1 = 0.9$, $\beta_2 =$

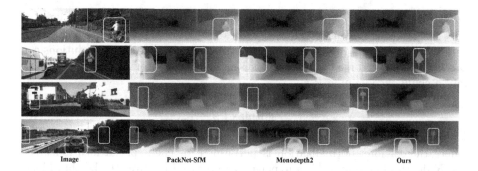

Fig. 4. Visual comparison of our method and recent works. The predicted depth maps of our method are perceptually more accurate with more details.

0.999. For the first training stage of the adaptive co-teaching framework, we employ a main-assistant training strategy to pursue the ability of the teachers. We firstly train a teacher network with only one coefficient decoder, namely main teacher, for 20 epochs, with a batch size of 12. As in [15], the initial learning rate for both depth and pose models is set to 1e−4 and decays after 15 epochs by factor 10. Then, we fix the parameters of the depth and pose networks, and train additional N−1 parallel coefficient decoders, namely assistant teachers, for another 5 epoch with a learning rate of 1e−5. The second stage of the adaptive co-teaching framework is trained for 15 epochs, with a batch size of 8. The learning rate is set to 1e−4. We adopt data augmentation strategies including random color jittering and horizontal flipping to improve generalization ability.

5   Experiments

In this section, we evaluate our proposed approach on two publicly available datasets, and perform ablation studies to validate the effectiveness of our design. Source code will be released at https://github.com/Mkalilia/MUSTNet.

5.1   Datasets

KITTI. The KITTI benchmark [13] is the most widely used dataset for training and test monocular depth methods. We employ the training and test split of [8]. As in [15], we use the pre-processing strategy to remove static frames. In particular, we use 39810 monocular triplets for training, 4424 for validation and 697 for evaluation. We follow [15], which uses the same intrinsics for all the images and sets the camera principal point to the image center. The focal length is set as the average of that of all the samples in KITTI.

Nuscenes. Nuscenes [4] is a recently released 3D dataset for multiple vision tasks. It carries the full autonomous vehicle sensor suite: 6 cameras, 5 radars and 1 lidar, all with full 360°C field of view. We only use this dataset for evaluation to assess the generalization capability of our approach. Based on [16], we evaluate

Table 1. Quantitative results on KITTI dataset for distance up to 80 m. **M** refers to methods supervised by monocular videos. **MS** refers to methods supervised by monocular videos and semantic information. **MF** refers to methods that jointly train depth and optical flow network. At test time, we scale depth with median ground-truth LiDAR information. Best results are marked **bold**.

	Methods	Sup.	Backbone.	Error metric↓				Accuracy metric↑		
				$AbsRel$	$SqRel$	$RMSE$	RMS_{lg}	$\delta < 1.25$	$\delta < 1.25^2$	$\delta < 1.25^3$
Lower	SfMLearner [43]	M	DispNet	0.198	1.836	6.565	0.275	0.718	0.901	0.960
	Struct2D [5]	M	ResNet18	0.141	1.026	5.291	0.215	0.816	0.945	0.979
	Geo-Net [40]	MF	ResNet50	0.155	1.296	5.857	0.233	0.793	0.931	0.973
Standard Resolution	SC-SfM [1]	M	DispNet	0.128	1.047	5.234	0.208	0.846	0.947	0.976
	MD2 [15]	M	ResNet18	0.115	0.903	4.863	0.193	0.877	0.959	0.981
	SAFE-Net [7]	MS	ResNet18	0.112	0.788	4.582	0.187	0.878	0.963	**0.983**
	HR-Depth [27]	M	ResNet18	0.109	0.792	4.632	0.185	0.884	0.962	**0.983**
	SCSI [36]	M	ResNet18	0.109	0.779	4.641	0.186	0.883	0.962	0.982
	Ours	M	ResNet18	**0.106**	**0.763**	**4.562**	**0.182**	**0.888**	**0.963**	**0.983**
Higher	Zhou et al. [42]	M	ResNet50	0.121	0.837	4.945	0.197	0.853	0.955	0.982
	PackNet [16]	M	PackNet	0.111	0.785	4.601	0.189	0.878	0.960	0.982
	Adrian et al. [22]	M	ResNet101	0.106	0.861	4.699	0.185	0.889	0.962	0.982
	Zhao et al. [41]	MF	ResNet18	0.113	**0.704**	4.581	0.184	0.871	0.961	0.984
	Lee et al. [24]	MS	ResNet18	0.112	0.777	4.772	0.191	0.872	0.959	0.982
	MD2 [15]	M	ResNet18	0.115	0.882	4.701	0.190	0.879	0.961	0.982
	Fang et al. [11]	M	ResNet18	0.110	0.806	4.681	0.187	0.881	0.961	0.982
	HR-Depth [27]	M	ResNet18	0.106	0.755	4.472	0.181	0.892	**0.966**	**0.984**
	Ours	M	ResNet18	**0.104**	0.750	**4.451**	**0.180**	**0.895**	**0.966**	**0.984**
	PackNet [16]	M	PackNet	0.107	0.802	4.538	0.186	0.889	0.962	0.981
	Chang et al. [33]	M	ResNet50	0.104	0.729	4.481	0.179	0.893	0.965	0.984

our network on the official NuScenes-mini train-val dataset, which contains 404 front-facing images with ground-truth depth.

5.2 Quantitative Evaluation

Results on KITTI Eigen Split. We report the performance of our network on KITTI raw data with the evaluation metrics described in [9]. We evaluated our model at standard resolution (192 × 640) and high resolution (1024 × 320). We compare our model with state-of-the-arts. Results in Table 1 show the superiority of our approach compared with all existing ResNet-18 based unsupervised approaches [11,15,27,36]. We also outperform recent models [16,22,33,42] with much larger backbones and recent works using additional semantic information [7,24] or optical flow information [24,41]. Furthermore, we showcase the qualitative comparison in Fig. 4. The depth predictions of our method are perceptually more accurate depth information with sharper depth details.

Generalization Capability. We also evaluate our approach on the recently proposed nuScenes dataset [4]. As shown in Table 2, our approach outperforms state-of-the-arts which confirms the generalization ability of our method across a large spectrum of vehicles and countries.

Table 2. Quantitative results of depth estimation on nuScenes dataset at the standard resolution. Best are marked **bold**.

Methods	Sup.	Error metric↓				Accuracy metric↑		
		$AbsRel$	$SqRel$	$RMSE$	$RMSE_{log}$	$\delta < 1.25$	$\delta < 1.25^2$	$\delta < 1.25^3$
SC-SfMLearner [43]	M	0.210	2.257	9.358	0.316	0.677	0.868	0.936
MD2 [15]	M	0.199	2.236	9.316	0.311	0.697	0.869	0.936
HR-Depth [27]	M	0.196	2.191	8.894	0.308	0.702	0.869	**0.937**
Ours	M	**0.192**	**2.143**	**8.888**	**0.305**	**0.716**	**0.870**	0.936

5.3   Ablation Studies

We conduct multiple ablative analysis on our approach, to further study the performance improvements provided by each component.

Ablation Study for MUSTNet. As one of our main contributions, MUSTNet serves as the architecture of both the teacher ensemble and the student model.

The Teacher Ensemble: We compare three variances of the model in an unsupervised manner, shown in L1-3 in Table 3. *Baseline* employs a single stream depth decoder outputting one channel prediction which keeps the same as [15,27]. *+BD* extends the decoder to the basis decoder with 16 output channels. The final outputs are generated by averaging all the channels. Compared with *Baseline*, both +BD and MUSTNet improve the performance. With 16 channel bases, more diverse representations of depth are embedded to guide the network out of local minimums and converge to a better solution. +BD is a special case of MUSTNet with a fixed weight $\frac{1}{N}[1,1,...,1]^T$ for all frames. MUSTNet learns more appropriate representations by co-adapting bases and coefficients. Higher quality pseudo-labels can be generated leveraging the MUSTNet as the teacher model. Meanwhile, as shown in L4-5, the student model learned from the MUSTNet also beats that from the *Baseline*.

The Student Model: As discussed in Sect. 4.1, the proposed MUSTNet is also suitable for student network by using one coefficient decoder. Compared with *baseline*, it provides an elegant way to adaptively learn feature channels and perform high quality predictions by dividing the single channel output into depth bases. As shown in L5-6 in Table 3, the results of a MUSTNet student are much better than the results of a *Baseline* student.

Ablation Study for the Adaptive Co-teaching Scheme. In this part we conduct 4 different training schemes on the MUSTNet to compare our adaptive co-teaching framework with the conventional unsupervised method and other distillation methods [27,31]. The results can be seen in L6-9 in Table 3.

Distill With a Single Teacher: L6 represents that we train only one teacher model for distillation. It is a straight forward distillation method, which is used in [27]. However, the improvement over the unsupervised baseline (L3) is marginal. The

Table 3. Ablation study of the Adaptive Co-teaching Framework. **P-Cons** denotes photometric constraint. **D-Cons** denotes distillation constraint with depth pseudo labels generated by **T-Model**. **AS** denotes adaptively switch between supervisions. **TS** denotes the total training stages. **NT** denotes number of teachers. **PM** denotes the teacher has N× parameters of the student network. Results for different variants of our method are trianed at 192 × 640.

Model	P-Cons.	D-Cons.	AS.	TS.	NT.	T-Model.	PM.	Error ↓		Accuracy ↑
								AbsRel	RMSE	δ < 1.25
Baseline	✓			1				0.113	4.795	0.880
† + BD	✓			1				0.112	4.808	0.881
MUSTNet	✓			1				0.109	4.656	0.883
Baseline		✓		2	1	Baseline	1×	0.113	4.801	0.879
Baseline		✓		2	1	MUSTNet	1×	0.112	4.744	0.882
MUSTNet		✓		2	1	MUSTNet	1×	0.110	4.718	0.883
MUSTNet		✓		2	4	MUSTNet	4×	0.108	4.622	0.885
MUSTNet	✓	✓		2	4	MUSTNet	1×	0.108	4.574	0.886
MUSTNet	✓	✓	✓	2	4	MUSTNet	1×	**0.106**	**4.562**	**0.888**

student just try to regress to pseudo labels of the teacher and can not learn more robust feature representations resulting in sub-optimal solutions.

Distill With Teacher Ensemble: Then we use the method provided by [31] to distill the student, shown in L7. Specifically, we train N randomly initialized network with the same architecture (baseline) and obtain empirical mean as the pseudo depth. The parameters of this bootstrapped ensemble architecture is N times of our compact ensemble, while the performance gain is still marginal.

Distill With Our Adaptive Co-teaching Framework: The teacher ensemble can be easily acquired by training a MUSTNet with a basis decoder and N coefficient decoders. We distill the student model with both photometric loss and pseudo label constraint, shown in L8. Then we further add the selective masks for the final cost-aware co-teaching loss, shown in L9. Compared with other distillation methods, the performance gain brought by our framework is significant with much lighter teacher ensemble and much simpler training process.

Visualized Comparison. For the sake of the pages limit, we report more visualized comparisons in the supplementary material. Specifically, i) the visualization of the depth bases. ii) the comparison of the outputs of the teacher ensemble and student model. iii) more quantitative and qualitative results.

5.4   Extension of Our Work

While our method significantly improves the performance of our proposed model, it also retains the advantages of being integrated into other models with two

Table 4. Quantitative results on [13]. **P** denotes photometric constraint. **TS** denotes the total training stages. **A** represents our adaptive co-teaching scheme.

	Methods	Sup.	TS.	Error metric↓				Accuracy metric↑		
				AbsRel	SqRel	RMSE	$RMSE_{log}$	$\delta < 1.25$	$\delta < 1.25^2$	$\delta < 1.25^3$
192 * 640	MD2 [15]	P	1	0.115	0.903	4.863	0.193	0.877	0.959	0.981
	Dual-stream	P	1	0.111	0.829	4.782	0.189	0.878	0.960	0.982
	Dual-stream	A	2	**0.108**	**0.795**	**4.649**	**0.185**	**0.885**	**0.962**	**0.983**
320 * 1024	MD2 [15]	P	1	0.115	0.882	4.701	0.190	0.879	0.961	0.982
	Dual-stream	P	1	0.108	0.818	4.591	0.185	0.890	0.963	0.982
	Dual-stream	A	2	**0.103**	**0.745**	**4.451**	**0.180**	**0.894**	**0.965**	**0.983**

simple modifications: (i) replace the single stream depth decoder with our multi-stream decoder format, and (ii) distill a dual-stream student model leveraging our adaptive co-teaching scheme.

We conduct experiments on one famous unsupervised method [15], and the quantitative results can be seen in Table 4. More details of this experiment are presented in our supplementary materials. Our framework improve the *AbsRel* from 0.115 to 0.103 at high resolution, which further verifies our contribution.

5.5 Discussion

Dynamic Mask. The cost volume based masking mechanism essentially follows the static scene assumption. It fails to mask out the dynamic parts. To solve this problem, we further conduct an extensive experiment to mask out those dynamic parts with the help of additional optical flow information generated by [34]. The detailed ablation study is shown in our supplementary material.

Moving Object. Despite good performance achieved by our adaptive co-teaching framework, one limitation appears on the moving objects that keep station relative to the camera. We have considered to mask out potential dynamic regions leveraging additional optical flow information. However, our framework essentially follows the static scene assumption which makes it fails to distinguish relative station parts and infinity. We support that, under the unsupervised learning pipeline, additional semantic guidance is essential to cope with this situation.

Video Inference. Another limitation appears on the scale inconsistency of video inferring. For the lack of pose ground truth, the scene scale keeps unknown. We tried to train our network in a scale consistent manner leveraging the geometric loss proposed in [43]. However, the joint depth-pose learning process is affected for it enforcing a consistent scale across all images.

6   Conclusion

The core design of our work is an adaptive co-teaching framework, which aims to solve the problem of training ambiguity and sub-optimal convergence for unsupervised depth estimation. We first design a compact ensemble architecture, namely MUlti-STream ensemble network, integrating a depth basis decoder with multiple depth coefficient decoders. Meanwhile, we propose a cost-aware co-teaching loss to transfer the learned knowledge from the teacher ensemble to a student network. As verified in our experiment, training the student network under our framework can circumvent the unstable issue of unsupervised learning. Our method sets new state-of-the-art on both KITTI and Nuscenes datasets.

Acknowledgements. This research is supported by National Natural Science Foundation of China (61906031, 62172070, U1903215, 6182910), and Fundamental Research Funds for Central Universities (DUT21RC(3)025).

References

1. Bian, J., et al.: Unsupervised scale-consistent depth and ego-motion learning from monocular video. Adv. Neural. Inf. Process. Syst. **32**, 35–45 (2019)
2. Bloesch, M., Czarnowski, J., Clark, R., Leutenegger, S., Davison, A.J.: Codeslam-learning a compact, optimisable representation for dense visual slam. In: Proceedings of the IEEE Conference on Computer Vision and Pattern Recognition, pp. 2560–2568 (2018)
3. Bucila, C., Caruana, R., Niculescu-Mizil, A.: Model compression. In: ACM SIGKDD International Conference on Knowledge Discovery and Data Mining (KDD 2006) (2006)
4. Caesar, H., et al.: nuscenes: a multimodal dataset for autonomous driving. In: Proceedings of the IEEE/CVF Conference on Computer Vision and Pattern Recognition, pp. 11621–11631 (2020)
5. Casser, V., Pirk, S., Mahjourian, R., Angelova, A.: Depth prediction without the sensors: leveraging structure for unsupervised learning from monocular videos. In: Proceedings of the AAAI Conference on Artificial Intelligence, vol. 33, pp. 8001–8008 (2019)
6. Chen, G., Choi, W., Yu, X., Han, T., Chandraker, M.: Learning efficient object detection models with knowledge distillation. Adv. Neural Inf. Process. Syst. **30** (2017)
7. Choi, J., Jung, D., Lee, D., Kim, C.: Safenet: self-supervised monocular depth estimation with semantic-aware feature extraction. In: Thirty-Fourth Conference on Neural Information Processing Systems, NIPS 2020. NeurIPS (2020)
8. Eigen, D., Fergus, R.: Predicting depth, surface normals and semantic labels with a common multi-scale convolutional architecture. In: Proceedings of the IEEE International Conference on Computer Vision, pp. 2650–2658 (2015)
9. Eigen, D., Puhrsch, C., Fergus, R.: Depth map prediction from a single image using a multi-scale deep network. In: 28th Annual Conference on Neural Information Processing Systems 2014, NIPS 2014, pp. 2366–2374. Neural information processing systems foundation (2014)
10. Fan, H., Hao, S., Guibas, L.: A point set generation network for 3D object reconstruction from a single image. In: 2017 IEEE Conference on Computer Vision and Pattern Recognition (CVPR) (2017)
11. Fang, J., Liu, G.: Self-supervised learning of depth and ego-motion from video by alternative training and geometric constraints from 3D to 2D. arXiv preprint arXiv:2108.01980 (2021)

12. Garg, R., B.G., V.K., Carneiro, G., Reid, I.: Unsupervised CNN for single view depth estimation: geometry to the rescue. In: Leibe, B., Matas, J., Sebe, N., Welling, M. (eds.) ECCV 2016. LNCS, vol. 9912, pp. 740–756. Springer, Cham (2016). https://doi.org/10.1007/978-3-319-46484-8_45
13. Geiger, A., Lenz, P., Urtasun, R.: Are we ready for autonomous driving? the kitti vision benchmark suite. In: 2012 IEEE Conference on Computer Vision and Pattern Recognition, pp. 3354–3361. IEEE (2012)
14. Godard, C., Mac Aodha, O., Brostow, G.J.: Unsupervised monocular depth estimation with left-right consistency. In: Proceedings of the IEEE Conference on Computer Vision and Pattern Recognition, pp. 270–279 (2017)
15. Godard, C., Mac Aodha, O., Firman, M., Brostow, G.J.: Digging into self-supervised monocular depth estimation. In: Proceedings of the IEEE/CVF International Conference on Computer Vision, pp. 3828–3838 (2019)
16. Guizilini, V., Ambrus, R., Pillai, S., Raventos, A., Gaidon, A.: 3D packing for self-supervised monocular depth estimation. In: Proceedings of the IEEE/CVF Conference on Computer Vision and Pattern Recognition, pp. 2485–2494 (2020)
17. Guizilini, V., Hou, R., Li, J., Ambrus, R., Gaidon, A.: Semantically-guided representation learning for self-supervised monocular depth. In: International Conference on Learning Representations (2019)
18. Gupta, S., Hoffman, J., Malik, J.: Cross modal distillation for supervision transfer. In: Proceedings of the IEEE Conference on Computer Vision and Pattern Recognition, pp. 2827–2836 (2016)
19. He, K., Zhang, X., Ren, S., Sun, J.: Deep residual learning for image recognition. In: Proceedings of the IEEE Conference on Computer Vision and Pattern Recognition, pp. 770–778 (2016)
20. Hinton, G., Vinyals, O., Dean, J.: Distilling the knowledge in a neural network. arXiv preprint arXiv:1503.02531 (2015)
21. Janai, J., Guney, F., Ranjan, A., Black, M., Geiger, A.: Unsupervised learning of multi-frame optical flow with occlusions. In: Proceedings of the European Conference on Computer Vision (ECCV), pp. 690–706 (2018)
22. Johnston, A., Carneiro, G.: Self-supervised monocular trained depth estimation using self-attention and discrete disparity volume. In: Proceedings of the IEEE/CVF Conference on Computer Vision and Pattern Recognition, pp. 4756–4765 (2020)
23. Kingma, D.P., Ba, J.: Adam: A method for stochastic optimization. arXiv preprint arXiv:1412.6980 (2014)
24. Lee, S., Im, S., Lin, S., Kweon, I.S.: Learning monocular depth in dynamic scenes via instance-aware projection consistency. In: Proceedings of the AAAI Conference on Artificial Intelligence, vol. 35, pp. 1863–1872 (2021)
25. Li, S., Wu, X., Cao, Y., Zha, H.: Generalizing to the open world: deep visual odometry with online adaptation. In: Proceedings of the IEEE/CVF Conference on Computer Vision and Pattern Recognition, pp. 13184–13193 (2021)
26. Li, Y., Yang, J., Song, Y., Cao, L., Li, L.J.: Learning from noisy labels with distillation. In: 2017 IEEE International Conference on Computer Vision (ICCV) (2017)
27. Lyu, X., et al.: Hr-depth: high resolution self-supervised monocular depth estimation. CoRR abs/2012.07356 (2020)
28. Peng, R., Wang, R., Lai, Y., Tang, L., Cai, Y.: Excavating the potential capacity of self-supervised monocular depth estimation. In: Proceedings of the IEEE/CVF International Conference on Computer Vision, pp. 15560–15569 (2021)

29. Pilzer, A., Lathuilière, S., Sebe, N., Ricci, E.: Refine and distill: exploiting cycle-inconsistency and knowledge distillation for unsupervised monocular depth estimation. In: 2019 IEEE/CVF Conference on Computer Vision and Pattern Recognition (CVPR) (2020)

30. Pilzer, A., Xu, D., Puscas, M., Ricci, E., Sebe, N.: Unsupervised adversarial depth estimation using cycled generative networks. In: 2018 International Conference on 3D Vision (3DV), pp. 587–595. IEEE (2018)

31. Poggi, M., Aleotti, F., Tosi, F., Mattoccia, S.: On the uncertainty of self-supervised monocular depth estimation. In: Proceedings of the IEEE/CVF Conference on Computer Vision and Pattern Recognition, pp. 3227–3237 (2020)

32. Prucksakorn, T., Jeong, S., Chong, N.Y.: A self-trainable depth perception method from eye pursuit and motion parallax. Robot. Auton. Syst. **109**, 27–37 (2018)

33. Shu, C., Yu, K., Duan, Z., Yang, K.: Feature-metric loss for self-supervised learning of depth and egomotion. In: Vedaldi, A., Bischof, H., Brox, T., Frahm, J.-M. (eds.) ECCV 2020. LNCS, vol. 12364, pp. 572–588. Springer, Cham (2020). https://doi.org/10.1007/978-3-030-58529-7_34

34. Teed, Z., Deng, J.: RAFT: recurrent all-pairs field transforms for optical flow. In: Vedaldi, A., Bischof, H., Brox, T., Frahm, J.-M. (eds.) ECCV 2020. LNCS, vol. 12347, pp. 402–419. Springer, Cham (2020). https://doi.org/10.1007/978-3-030-58536-5_24

35. Wagstaff, B., Peretroukhin, V., Kelly, J.: Self-supervised structure-from-motion through tightly-coupled depth and egomotion networks. arXiv preprint arXiv:2106.04007 (2021)

36. Wang, L., Wang, Y., Wang, L., Zhan, Y., Wang, Y., Lu, H.: Can scale-consistent monocular depth be learned in a self-supervised scale-invariant manner? In: Proceedings of the IEEE/CVF International Conference on Computer Vision, pp. 12727–12736 (2021)

37. Watson, J., Mac Aodha, O., Prisacariu, V., Brostow, G., Firman, M.: The temporal opportunist: self-supervised multi-frame monocular depth. In: Proceedings of the IEEE/CVF Conference on Computer Vision and Pattern Recognition, pp. 1164–1174 (2021)

38. Yang, J., Alvarez, J.M., Liu, M.: Self-supervised learning of depth inference for multi-view stereo. In: Proceedings of the IEEE/CVF Conference on Computer Vision and Pattern Recognition, pp. 7526–7534 (2021)

39. Yang, J., Mao, W., Alvarez, J.M., Liu, M.: Cost volume pyramid based depth inference for multi-view stereo. In: Proceedings of the IEEE/CVF Conference on Computer Vision and Pattern Recognition, pp. 4877–4886 (2020)

40. Yin, Z., Shi, J.: Geonet: Unsupervised learning of dense depth, optical flow and camera pose. In: Proceedings of the IEEE Conference on Computer Vision and Pattern Recognition, pp. 1983–1992 (2018)

41. Zhao, W., Liu, S., Shu, Y., Liu, Y.J.: Towards better generalization: joint depth-pose learning without posenet. In: 2020 IEEE/CVF Conference on Computer Vision and Pattern Recognition (CVPR) (2020)

42. Zhou, J., Wang, Y., Qin, K., Zeng, W.: Unsupervised high-resolution depth learning from videos with dual networks. In: Proceedings of the IEEE/CVF International Conference on Computer Vision, pp. 6872–6881 (2019)

43. Zhou, T., Brown, M., Snavely, N., Lowe, D.G.: Unsupervised learning of depth and ego-motion from video. In: 2017 IEEE Conference on Computer Vision and Pattern Recognition (CVPR) (2017)

Fusing Local Similarities for Retrieval-Based 3D Orientation Estimation of Unseen Objects

Chen Zhao[1]([✉]), Yinlin Hu[1,2], and Mathieu Salzmann[1,2]

[1] EPFL-CVLab, Lausanne, Switzerland
{chen.zhao,yinlin.hu,mathieu.salzmann}@epfl.ch
[2] ClearSpace SA, Renens, Switzerland

Abstract. In this paper, we tackle the task of estimating the 3D orientation of previously-unseen objects from monocular images. This task contrasts with the one considered by most existing deep learning methods which typically assume that the testing objects have been observed during training. To handle the unseen objects, we follow a retrieval-based strategy and prevent the network from learning object-specific features by computing multi-scale local similarities between the query image and synthetically-generated reference images. We then introduce an adaptive fusion module that robustly aggregates the local similarities into a global similarity score of pairwise images. Furthermore, we speed up the retrieval process by developing a fast retrieval strategy. Our experiments on the LineMOD, LineMOD-Occluded, and T-LESS datasets show that our method yields a significantly better generalization to unseen objects than previous works. Our code and pre-trained models are available at https://sailor-z.github.io/projects/Unseen_Object_Pose.html.

Keywords: Object 3D orientation estimation · Unseen objects

1 Introduction

Estimating the 3D orientation of objects from an image is pivotal to many computer vision and robotics tasks, such as robotic manipulation [7,29,41], augmented reality, and autonomous driving [6,11,22,39]. Motivated by the tremendous success of deep learning, much effort [24,32,36] has been dedicated to developing deep networks able to recognize the objects depicted in the input image and estimate their 3D orientation. To achieve this, most learning-based methods assume that the training data and testing data contain exactly the same objects [16,33] or similar objects from the same category [21,34]. However, this assumption is often violated in real-world applications, such as robotic manipulation, where one would typically like the robotic arm to be able to handle previously-unseen objects without having to re-train the network for them.

Supplementary Information The online version contains supplementary material available at https://doi.org/10.1007/978-3-031-19769-7_7.

Fig. 1. 3D orientation estimation for unseen objects. The network is trained on a limited number of objects and tested on unseen (new) objects that fundamentally differ from the training ones in shape and appearance. The goal is to predict both the category and the 3D orientation of these unseen objects.

In this paper, as illustrated in Fig. 1, we tackle the task of 3D orientation estimation for *previously-unseen* objects. Specifically, we develop a deep network that can be trained on a limited number of objects, and yet remains effective when tested on novel objects that drastically differ from the training ones in terms of both appearance and shape. To handle such previously-unseen objects, we cast the task of 3D orientation estimation as an image retrieval problem. We first create a database of synthetic images depicting objects in different orientations. Then, given a real query image of an object, we search for the most similar reference image in the database, which thus indicates both the category and 3D orientation of this object.

Intuitively, image retrieval methods [35] offer a promising potential for generalization, because they learn the relative similarity of pairwise images, which can be determined without being aware of the object category. However, most previous works [1,28,30,35,37] that follow this approach exploit a global image representation to measure image similarity, ignoring the risk that a global descriptor may integrate high-level semantic information coupled with the object category, which could affect the generalization ability to unseen objects. To address this problem, our approach relies on the similarities of local patterns, which are independent to the object category and then facilitate the generalization to new objects. Specifically, we follow a multi-scale strategy and extract feature maps of different sizes from the input image. To facilitate the image comparison process, we then introduce a similarity fusion module, adaptively aggregating multiple local similarity scores into a single one that represents the similarity between two images. To further account for the computational complexity of the resulting multi-scale local comparisons, we design a fast retrieval strategy.

We conduct experiments on three public datasets, LineMOD [13], LineMOD-Occluded (LineMOD-O) [3], and T-LESS [14], comparing our method with both hand-crafted [8] and deep learning [1,27,35,38] approaches. Our empirical results evidence the superior generalization ability of our method to previously-unseen objects. Furthermore, we perform ablation studies to shed more light on the effectiveness of each component in our method. Our contributions can be summarized as follows:

- We estimate the 3D orientation of previously-unseen objects by introducing an image retrieval framework based on multi-scale local similarities.
- We develop a similarity fusion module, robustly predicting an image similarity score from multi-scale pairwise feature maps.
- We design a fast retrieval strategy that achieves a good trade-off between the 3D orientation estimation accuracy and efficiency.

2   Related Work

Object Pose Estimation. In recent years, deep learning has been dominating the field of object pose estimation. For instance, PoseCNN [36] relies on two branches to directly predict the object orientation as a quaternion, and the 2D location of the object center, respectively. PVNet [24] estimates the 2D projections of 3D points using a voting network. The object pose is then recovered by using a PnP algorithm [10] over the predict 2D-3D correspondences. DenseFusion [32] fuses 2D and 3D features extracted from RGB-D data, from which it predicts the object pose. GDR-Net [33] predicts a dense correspondence map, acting as input to a Patch-PnP module that recovers the object pose. These deep learning methods have achieved outstanding pose estimation accuracy when the training and testing data contain the same object instances [9]. However, the patterns they learn from the input images are instance specific, and these methods cannot generalize to unseen objects [23].

Category-Level Object Pose Estimation. Some methods nonetheless loosen the constraint of observing the same object instances at training and testing time by performing category-level object pose estimation [5,18,31,34]. These methods assume that the training data contain instances belonging to a set of categories, and new instances from these categories are observed during testing. In this context, a normalized object coordinate space (NOCS) is typically used [18, 34], providing a canonical representation shared by different instances within the same category. The object pose is obtained by combining the NOCS maps, instance masks, and depth values. These methods rely on the intuition that the shapes of different instances in the same category are similar, and then the patterns learned from the training data can generalize to new instances in the testing phase. As such, these methods still struggle in the presence of testing objects from entirely new categories. Furthermore, all of these techniques require *depth* information as input. By contrast, our method relies only on RGB images, and yet can handle unseen objects from new categories at testing time.

Unseen Object Pose Estimation. A few attempts at predicting the pose of unseen objects have been made in the literature. In particular, LatentFusion [23] introduces a latent 3D representation and optimizes an object's pose by differentiable rendering. DeepIM [19] presents an iterative framework, using a matching

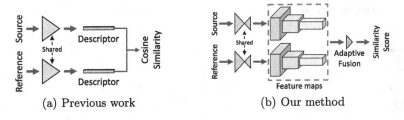

(a) Previous work   (b) Our method

Fig. 2. Difference between previous works and our method. (a) Existing works convert images into global descriptors that are used to compute a similarity score for retrieval. (b) Our method compares images using local similarities between the corresponding elements in feature maps, and adaptively fuses these local similarities into a single one that indicates the similarity of two images.

network to optimize an initial object pose. Both of these methods require an *initial* pose estimate, which is typically hard to obtain for unseen objects. Furthermore, the pose estimation step in LatentFusion leverages *depth* information. Since estimating the full 6D pose of an *unseen* object from a single RGB image is highly challenging, several works suggest simplifying this problem by focusing on estimating the 3D object orientation [2,28,35,37,38]. These methods utilize the 3D object model to generate multi-view references, which are then combined with the real image to either perform template matching [2,27,28,35,37] or directly regress the 3D orientation [38]. However, they propose to learn a global representation from an image, in which the high-level semantic information is correlated to the object and then limits the generalization to unseen objects. In this paper, we also focus on 3D orientation estimation of unseen objects, but handle this problem via a multi-scale local similarity learning network.

3   Method

3.1   Problem Formulation

Let us assume to be given a set of training objects $\mathcal{O}_{train}$ belonging to different categories $\mathcal{C}_{train}$.[1] As depicted by Fig. 1, we aim to train a model that can predict the 3D orientation of new objects $\mathcal{O}_{test}$, $\mathcal{O}_{test} \cap \mathcal{O}_{train} = \emptyset$, from entirely new categories $\mathcal{C}_{test}$, $\mathcal{C}_{test} \cap \mathcal{C}_{train} = \emptyset$. Specifically, given an RGB image $\mathbf{I}_{src}$ containing an object $O_{src} \in \mathcal{O}_{test}$, our goal is to both recognize the object category C_{src} and estimate the object's 3D orientation, expressed as a rotation matrix $\mathbf{R}_{src} \in \mathbb{R}^{3 \times 3}$. We tackle this dual problem as an image retrieval task. For each $O_{src}^i \in (\mathcal{O}_{train} \cup \mathcal{O}_{test})$, we generate references $\mathcal{I}_{ref}^i$ with different 3D orientations by rendering the corresponding 3D model $\mathbf{M}_i$. We then seek to pick $\hat{\mathbf{I}}_{ref} \in \{\mathcal{I}_{ref}^1 \cup \mathcal{I}_{ref}^2 \cdots \cup \mathcal{I}_{ref}^N\}$ that is the most similar to $\mathbf{I}_{src}$. The category label C_{src} and 3D orientation $\mathbf{R}_{src}$ of O_{src} are then taken as those of the corresponding $\hat{O}_{ref}$.

---

[1] In our scenario, and in contrast to category-level pose estimation, each object instance corresponds to its own category.

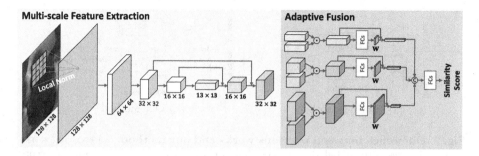

Fig. 3. Network architecture. We extract multi-scale features from a locally-normalized image. We then compute local similarities at each scale between the features of the source image and those of a reference one, and adaptively fuse them into a global similarity score.

3.2 Motivation

As illustrated in Fig. 2(a), the existing retrieval-based 3D orientation estimation methods [28,35,37] convert an image into a global descriptor. Retrieval is then performed by computing the similarity between pairs of descriptors. As a consequence, the deep network that extracts the global descriptor typically learns to encode object-specific semantic information in the descriptor, which results in a limited generalization ability to unseen objects. By contrast, we propose to compare images via local patch descriptors, in which it is harder to encode high-level semantic information thus encouraging the network to focus on local geometric attributes. As shown in Fig. 2(b), we estimate local similarities between the corresponding elements in source and reference feature maps. Furthermore, to enforce robustness to noise, such as background, we introduce an adaptive fusion module capable of robustly predicting an image similarity score from the local ones.

3.3 Multi-scale Patch-Level Image Comparison

As the source image is real but the reference ones are synthetic with discretely sampled 3D orientation, the appearance and shape variations are inevitable even for the most similar reference. Moreover, the background included in the source image, but absent from the reference ones, typically interferes with our patch-level comparisons. In practice, we have observed that small patches could be too sensitive to appearance and shape variations, while large patches tend to be affected by the background. Finding a single effective patch size balancing robustness to the domain gap and to the background therefore is challenging.

To address this issue, we introduce a multi-scale feature extraction module. As shown in Fig. 3, our network takes a grey-scale image $\mathbf{I} \in \mathbb{R}^{128 \times 128}$ as input, which shows better robustness than color images in practice. Subsequently, we employ a series of ResNet layers [12], estimating a down-sampled feature map $\mathbf{F} \in \mathbb{R}^{13 \times 13 \times C}$. We compute multi-scale feature representations by progressively up-sampling $\mathbf{F}$ using deconvolution layers [20] and bilinear interpolation. We also utilize skip connections [26] to better preserve the geometric information. The elements in the generated multi-scale feature maps then encode patches of different sizes in $\mathbf{I}$, which enables multi-scale patch-level image comparison.

To perform image retrieval, one nonetheless needs to compute a single similarity score for a pair of images. To this end, we compare the pairwise multi-scale feature maps and fuse the resulting local similarities into a single score expressed as

$$s = f(g(\mathbf{F}^1_{src}, \mathbf{F}^1_{ref}), g(\mathbf{F}^2_{src}, \mathbf{F}^2_{ref}), \cdots, g(\mathbf{F}^S_{src}, \mathbf{F}^S_{ref})), \qquad (1)$$

where $\mathbf{F}_{src}$ and $\mathbf{F}_{ref}$ represent the feature maps of $\mathbf{I}_{src}$ and $\mathbf{I}_{ref}$, respectively, and S denotes the number of scales. A straightforward solution to estimate s is to compute the per-element cosine similarity for all pairs $(\mathbf{F}^i_{src}, \mathbf{F}^i_{ref})$, with $i \in \{1, 2, \cdots, S\}$, and average the resulting local similarities. However, this strategy would not be robust to outlier patches, such as those dominated by background content. Therefore, we introduce an adaptive fusion strategy illustrated in the right part of Fig. 3. Following the same formalism as above, it computes an image similarity score as

$$s = f(\text{cat}\,[g(\mathbf{F}^*_i \odot \mathbf{w}_i)], \psi), i \in \{1, 2, \cdots, S\}, \qquad (2)$$

where cat indicates the concatenation process, $g : \mathbb{R}^{H \times W \times C} \to \mathbb{R}^C$ denotes the summation over the spatial dimensions, $\mathbf{F}^*_i$ represents the local similarities obtained by computing the cosine similarities between the corresponding elements in $\mathbf{F}^i_{src}$ and $\mathbf{F}^i_{ref}$, $\odot$ indicates the Hadamard product, and ψ represents the learnable parameters of the fully connected layers (FCs) $f(\cdot)$. The weights $\mathbf{w}_i$ encode a confidence map over $\mathbf{F}^*_i$ to account for outliers, and are computed as

$$\mathbf{w}_i = \frac{\exp(h(\mathbf{F}^*_i, \omega)) \odot \text{sigmoid}(q(\mathbf{F}^*_i, \theta))}{\sum \exp(h(\mathbf{F}^*_i, \omega)) \odot \text{sigmoid}(q(\mathbf{F}^*_i, \theta))}, \qquad (3)$$

where ω and θ are learnable parameters of the convolutional layers $h(\cdot)$ and $q(\cdot)$, respectively. This formulation accounts for both the individual confidence of each element in $\mathbf{F}^*_i$ via the sigmoid function, and the relative confidence w.r.t. all elements jointly via the softmax-liked function. As such, it models both the local and global context of $\mathbf{F}^*_i$, aiming to decrease the confidence of the outliers while increases that of the inliers. Our experimental results in Sect. 4.5 show that our adaptive fusion yields better results than the straightforward averaging process described above, even when trained in an unsupervised manner.

To further reduce the effects of object-related patterns in local regions and synthetic-to-real domain gap, we pre-process $\mathbf{I}$ via a local normalization. Each pixel $\overline{p}_{ij}$ in the normalized image is computed as

$$\overline{p}_{ij} = \frac{p_{ij} - \mu}{\sigma}, \qquad (4)$$

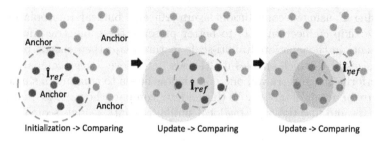

Fig. 4. Fast retrieval. The location of the source image in the reference database is initialized by comparing the source image with a set of anchors. $\hat{\mathbf{I}}_{ref}$ is dynamically updated based on the similarity scores of the references within a local region around the current $\hat{\mathbf{I}}_{ref}$ estimate.

where p_{ij} is the corresponding pixel in $\mathbf{I}$, and

$$\mu = \frac{1}{r^2} \sum_{i',j'} p_{i'j'}, \quad \sigma = \sqrt{\frac{1}{r^2} \sum_{i',j'} (p_{i'j'} - \mu)^2}, \tag{5}$$

with $i' \in [i \pm r/2]$, $j' \in [j \pm r/2]$, and r denoting the window size.

3.4 Fast Retrieval

Although the proposed patch-level image comparison integrates more local geometric information than the image-level methods [28,35], and as will be shown by our experiments thus yields better generalization to unseen objects, it suffers from a high retrieval time. Indeed, a naïve image retrieval strategy compares $\mathbf{I}_{src}$ with every reference in the database. Given N objects with R reference images each, the cost of $O(NR)$ quickly becomes unaffordable as N and R increase. This could be remedied by parallel computing, but at the cost of increasing memory consumption. Here, we therefore introduce a fast retrieval method that balances effectiveness and efficiency.

As illustrated in Fig. 4, instead of comparing $\mathbf{I}_{src}$ with all the references one-by-one, we first roughly locate $\mathbf{I}_{src}$ in the reference database and then iteratively refine this initial location. Our method is summarized in Algorithm 1. We omit some subscripts for convenience. Specifically, for each object, we first sample k_{ac} anchors from $\mathcal{I}_{ref}$ using farthest point sampling (FPS), which leads to a good coverage [25] of $\mathcal{I}_{ref}$. This is done using the geodesic distance of the corresponding 3D rotation matrices as a metric in FPS. The anchor with the largest score s computed from Eq. 2 is taken as the initial point $\hat{\mathbf{I}}_{ref}$. Subsequently, we perform retrieval using the method described in Sect. 3.3 within a local region centered at the current $\hat{\mathbf{I}}_{ref}$, and update $\hat{\mathbf{I}}_{ref}$ based on the similarity scores. Such updates are performed until convergence.

The straightforward application of this strategy would be prone to local optima. Intuitively, this can be addressed by increasing the size of the local

---

Algorithm 1: Fast Retrieval

Input: $\mathbf{I}_{src}$, $\mathcal{I}_{ref}$, $\mathcal{R}_{ref} = \{\mathbf{R}_1, \mathbf{R}_2, \cdots, \mathbf{R}_R\}$, k_{ac}, R
Output: $\hat{\mathbf{I}}_{ref}$, $\mathbf{R}_{src}$

1 Sample k_{ac} anchors from $\mathcal{I}_{ref}$ using FPS;
2 Estimate similarities using Eq. 2;
3 Initialize $\hat{\mathbf{I}}_{ref}$ as the most similar anchor;
4 $j = 1$;
5 **repeat**
6   Define a search space around $\hat{\mathbf{I}}_{ref}$ with a radius of $\lfloor R/2^j \rfloor$;
7   Compute anchors using FPS;
8   Estimate similarities using Eq. 2;
9   Update $\hat{\mathbf{I}}_{ref}$;
10      $j{+}{+}$;
11 **until** $\hat{\mathbf{I}}_{ref}$ *converges*;
12 Determine $\mathbf{R}_{src}$ as $\hat{\mathbf{R}}_{ref} \in \mathcal{R}_{ref}$.

---

region, but this would come at a higher computational cost. Therefore, we determine a search space and further make use of FPS to select anchors within the space, because FPS covers a larger region than KNN with the same number of samples. At each iteration, we decrease the radius of the search space to further improve efficiency.

3.5 Training and Testing

In the training stage, we follow the infoNCE contrastive learning formalism [4]. We associate each training sample (source image) in a mini-batch with its closest reference to form a positive pair. To better cover the entire training set, we also group each positive pair with a random sample, leading to a triplet. Note that in the standard infoNCE loss [4] for contrastive learning, all samples except for the most similar one are treated as negative. In our context, all reference images would be penalized equally, except for the one from the same category and with the closest 3D orientation. Their 3D orientation difference is thus not differentiated. To better account for the continuous nature of the 3D orientation difference and avoid over-penalizing some references, we introduce a weighted-infoNCE loss

$$\ell_{ij} = -\log \frac{\exp(s_{ij}/\tau) \cdot w_{ij}}{\sum_{k=1}^{3B} \exp(s_{ik}/\tau) \cdot w_{ik}}, \tag{6}$$

where s_{ij} is the similarity score of the positive pair $(\mathbf{I}_i, \mathbf{I}_j)$, $\tau = 0.1$ denotes a temperature parameter [4], B is the size of a mini-batch, and w_{ij} (and similarly w_{ik}) represents a weight computed as

$$w_{ij} = \begin{cases} \arccos(\frac{\operatorname{tr}(\mathbf{R}_i^{\mathrm{T}}\mathbf{R}_j)-1}{2})/\pi & \text{if } C_j = C_i \\ 1 & \text{else} \end{cases}. \tag{7}$$

In the testing phase, we store the features of the reference images and sample the initial anchors offline; we recognize the object categories and predict the 3D orientation online. More specifically, we first compare $\mathbf{I}_{src}$ with k_{ac} anchors for each $\mathcal{I}^i_{ref}$, and take C_{src} to be the category of the anchor with the largest similarity score. This process reduces the complexity of object category recognition from $O(NR)$ to $O(Nk_{ac})$. Subsequently, we restrict the search for $\hat{\mathbf{I}}_{ref}$ in our fast retrieval process to the references depicting the recognized object. $\mathbf{R}_{src}$ is finally taken as the corresponding rotation matrix $\hat{\mathbf{R}}_{ref}$.

4   Experiments

4.1   Implementation Details

In our experiments, we set the window size for local normalization and the number of anchors for fast retrieval to $r = 32$ and $k_{ac} = 1024$, respectively. We train our network for 200 epochs using the Adam [17] optimizer with a batch size of 16 and a learning rate of 10^{-4}, which is divided by 10 after 50 and 150 epochs. Training takes around 20 hours on a single NVIDIA Tesla V100.

4.2   Experimental Setup

Following the standard setting [27,28,35,38], we assume to have access to 3D object models, which provide us with canonical frames, without which the object orientation in the camera frame would be ill-defined. Note that the 3D object models are also used to generate the reference images. We generate $R = 10,000$ reference images for each object by rendering the 3D model with different 3D orientation. We randomly sample $\mathbf{R}_{ref}$ using a 6D continuous representation [40], which results in better coverage of the orientation space. Following [23,27,35], we crop the objects from the source images using the provided bounding boxes.

We compare our method with both a hand-crafted approach, i.e., HOG [8], and deep learning ones, i.e., LD [35], NetVLAD [1], PFS [38], MPE [27], and GDR-Net [33]. Note that DeepIM [19] and LatentFusion [23] are not evaluated since DeepIM requires an object pose initialization and LatentFusion needs additional depth information. We also exclude [28] because it requires training a separate autoencoder for every object, and thus cannot be used to estimate orientation for an unseen object without training a new autoencoder.

4.3   Experiments on LineMOD and LineMOD-O

We first conduct experiments on LineMOD [13] and LineMOD-O [3]. We split the cropped data into three non-overlapping groups, i.e., Split #1, Split #2, and Split #3, according to the depicted objects. Deep learning models are trained on LineMOD and tested on both LineMOD and LineMOD-O. We augment training data by random occlusions when the evaluation is performed on LineMOD-O. In the case of unseen objects, we select one of the three groups as the testing set. We remove all images belonging to this group from training data, which ensures that no testing objects are observed in the training stage. In the case of seen objects on LineMOD, we separate 10% from each group of this dataset for testing. We assume the object category to be unknown during testing, and therefore employ the evaluated methods to classify the object and then predict its 3D orientation.

We thus evaluate the tested methods in terms of both object classification accuracy and 3D orientation estimation accuracy. These are computed as

$$\text{Class. Acc.} = \begin{cases} 1 & \text{if } \hat{C}_{ref} = C_{src} \\ 0 & \text{otherwise} \end{cases} \tag{8}$$

and

$$\text{Rota. Acc.} = \begin{cases} 1 & \text{if } d(\hat{\mathbf{R}}_{ref}, \mathbf{R}_{src}) < \lambda \text{ and } \hat{C}_{ref} = C_{src} \\ 0 & \text{otherwise} \end{cases}, \tag{9}$$

respectively, with $\lambda = 30°$ a predefined threshold [38] and $d(\hat{\mathbf{R}}_{ref}, \mathbf{R}_{src})$ the geodesic distance [35] between two rotation matrices. This distance is defined as

$$d(\hat{\mathbf{R}}_{ref}, \mathbf{R}_{src}) = \arccos(\frac{\text{tr}(\mathbf{R}_{src}^{\text{T}} \hat{\mathbf{R}}_{ref}) - 1}{2})/\pi. \tag{10}$$

We provide the results for LineMOD and LineMOD-O in Table 1 and Table 2, respectively. As PFS assumes the object category is known, we only report its 3D orientation estimation accuracy. We replace the detection module in GDR-Net with the ground-truth bounding boxes in the presence of unseen objects since this detector cannot be used to detect unseen objects. Therefore, the classification accuracy of GDR-Net is not reported in this case. Being a traditional method, HOG does not differentiate seen and unseen objects, and thus achieves comparable results in both cases. However, its limited accuracy indicates that HOG suffers from other challenges, such as the appearance difference between real and synthetic images, and the presence of background and of occlusions. The previous retrieval-based methods, i.e., LD, NetVLAD, and MPE, achieve remarkable performance in the case of seen objects, but their accuracy significantly drops in the presence of unseen ones. This evidences that the global descriptors utilized in these approaches are capable of encoding 3D orientation information, but the described patterns are object specific, thus limiting the generalization ability of these methods to unseen objects. The performance of both PFS and GDR-Net also drops dramatically in the presence of unseen objects because the features extracted from 2D observations or 3D shapes remain strongly object dependent. Our method outperforms the competitors by a considerably large margin in the case of unseen objects, which demonstrates that the proposed patch-level image comparison framework generalizes better to unseen objects than previous works. This is because our use of patch-level similarities makes the network focus on local geometric attributes instead of high-level semantic information.

Table 1. Experimental results on LineMOD [13].

		Split #1		Split #2		Split #3		Mean	
		Seen	Unseen	Seen	Unseen	Seen	Unseen	Seen	Unseen
Class. acc. (%)	HOG [8]	39.57	41.26	30.24	32.65	36.19	35.19	35.33	36.37
	LD [35]	99.17	52.53	99.02	47.85	97.94	30.60	98.71	43.66
	NetVLAD [1]	100.00	68.36	100.00	52.30	100.00	47.22	100.00	55.96
	PFS [38]	–	–	–	–	–	–	–	–
	MPE [27]	98.75	57.25	83.29	69.98	97.73	87.57	93.26	71.60
	GDR-Net [33]	100.00	–	100.00	-	100.00	–	100.00	–
	Ours	100.00	**97.90**	99.44	**92.47**	98.03	**88.93**	99.16	**93.10**
Rota. acc. (%)	HOG [8]	38.89	40.17	28.21	30.74	31.02	28.48	32.71	33.13
	LD [35]	94.50	8.63	89.57	12.47	91.47	5.22	91.85	8.77
	NetVLAD [1]	100.00	36.11	98.66	20.33	99.35	23.38	99.34	26.61
	PFS [38]	100.00	6.31	99.19	6.65	**99.46**	5.54	**99.55**	6.17
	MPE [27]	91.94	38.96	66.47	41.46	87.72	61.62	82.04	47.35
	GDR-Net [33]	99.89	4.61	**99.28**	4.82	99.31	5.02	99.49	4.82
	Ours	97.49	**89.55**	94.90	**79.04**	93.67	**75.96**	95.35	**81.52**

Table 2. Experimental results on LineMOD-O [3].

		Split #1		Split #2		Split #3		Mean	
		Seen	Unseen	Seen	Unseen	Seen	Unseen	Seen	Unseen
Class. acc. (%)	HOG [8]	0.60	0.60	0.23	0.23	36.20	36.20	12.34	12.34
	LD [35]	80.61	65.72	**84.56**	58.45	66.94	46.34	77.37	56.84
	NetVLAD [1]	**85.27**	56.46	74.37	47.73	**90.33**	66.32	83.32	56.84
	PFS [38]	–	–	–	–	–	–	–	–
	MPE [27]	83.29	56.07	55.26	45.08	70.72	57.55	69.76	52.90
	GDR-Net [33]	84.48	–	83.81	–	89.19	–	**85.83**	–
	Ours	83.22	**83.99**	78.69	**74.42**	82.44	**75.06**	81.45	**77.82**
Rota. acc. (%)	HOG [8]	0.60	0.60	0.18	0.18	5.25	5.25	2.01	2.01
	LD [35]	32.21	6.25	26.56	3.26	24.57	4.57	27.78	4.69
	NetVLAD [1]	51.60	24.32	42.20	18.05	36.56	18.84	43.45	20.40
	PFS [38]	**71.40**	6.25	**60.88**	13.15	**54.67**	4.68	**62.32**	8.73
	MPE [27]	40.47	22.56	27.31	5.20	35.06	18.22	34.28	15.33
	GDR-Net [33]	63.37	3.12	55.31	2.97	49.91	2.39	56.20	2.83
	Ours	64.92	**60.75**	56.51	**52.41**	52.47	**37.85**	57.97	**50.34**

4.4 Experiments on T-LESS

To further evaluate the generalization ability to unseen objects, we conduct an experiment on T-LESS [14]. In this case, all deep learning approaches, including ours, were trained on LineMOD, and tested on the Primesense test scenes of T-LESS. As the objects' appearance and shape in T-LESS are significantly different from the ones in LineMOD, this experiment provides a challenging benchmark to evaluate generalization. As in [15,27,28], we use $err_{vsd} \leq 0.3$ as a metric on this dataset. Note that we do not use the refinement module in [27,28] since it could be applied to all evaluated methods and thus is orthogonal to our contributions. As we concentrate on 3D object orientation estimation, we only consider the error of rotation matrices when computing err_{vsd}.

Table 3. Experimental results on T-LESS [14]. All deep learning methods were trained on LineMOD and tested on the Primesense test scenes of T-LESS. We use $err_{vsd} \leq 0.3$ as a metric.

Method	HOG [8]	LD [35]	NetVLAD [1]	PFS [38]	MPE [27]	GDR-Net [33]	Ours
Rota. acc. (%)	74.22	24.19	56.46	17.92	66.88	11.89	**78.73**

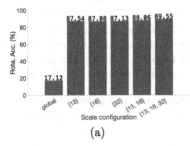
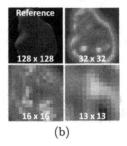

(a)   (b)

Fig. 5. Importance of local comparisons. (a) Results of our method with different scale configurations on the unseen objects of LineMOD Split #1. "Global" indicates a baseline that averages the smallest feature map into a global descriptor ($\mathbb{R}^{13 \times 13 \times C} \rightarrow \mathbb{R}^{C}$) for retrieval while maintaining the other components, except for the adaptive fusion, unchanged. (b) Feature maps used for retrieval.

The results of all the methods are provided in Table 3. For this dataset, the 3D orientation estimates of all the previous deep learning methods are less accurate than those of the traditional approach, i.e., HOG. This shows that the models pretrained on LineMOD are unreliable when tested on T-LESS. By contrast, our method outperforms both hand-crafted and deep learning competitors. It evidences that our method can still effectively estimate 3D orientation for unseen objects even when the object's appearance and shape entirely differ from those in the training data.

4.5 Ablation Studies

Local Comparisons. One of the key differences between our method and existing works [1,27,35] is our use of the local comparisons during retrieval. Figure 5(a) demonstrates the importance of local features in our framework. We start from a "global" baseline in which we average the smallest feature map along the spatial dimensions to form a global descriptor ($\mathbb{R}^{13 \times 13 \times C} \rightarrow \mathbb{R}^{C}$), which is used to compute the cosine similarity for retrieval. We keep the other components unchanged, except for the adaptive fusion. This baseline shows inferior generalization to unseen objects, while the performance significantly increases when local features are utilized. This observation indicates the importance of local comparisons for the unseen-object generalization. Moreover, the combination of multi-scale features also positively impacts the results, because it yields the ability to mix local geometric information at different scales, as illustrated by Fig. 5(b).

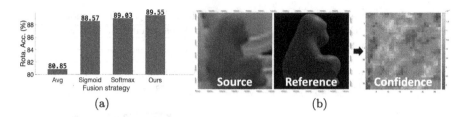

Fig. 6. (a) Our adaptive fusion significantly outperforms a simple averaging strategy, and yields a boost over the "Sigmoid" and "Softmax" alternatives. (b) Confidence map depicting the weights in Eq. 3 obtained from the feature maps of the source and reference on the left. Yellow dots indicate local similarities with high confidence.

Table 4. Comparison between the fast retrieval and greedy search. We report the 3D orientation accuracy and the test time on the unseen objects of LineMOD Split #1.

Method	Fast retrieval	Greedy search
Rota. acc. (%)	89.55	95.93
Time (s)	0.42	30.74

Adaptive Fusion. To analyze the importance of the adaptive fusion module in our method, we introduce the following three baselines. "Avg" consists of replacing the adaptive fusion with a simple averaging process, estimating the image similarity score by averaging all per-element similarities over the pairs of feature maps. Furthermore, "Sigmoid" and "Softmax" involve using only the sigmoid or softmax function in Eq. 3, respectively. As shown in Fig. 6(a), our approach yields an 8.7% increase in Rota. Acc. over "Avg", which demonstrates the superiority of our adaptive fusion strategy. The reason behind this performance improvement is that our module assigns different confidence weights to the local similarities, as illustrated in Fig. 6(b), which makes it possible to distinguish the useful information from the useless one. Moreover, the performance decreases (from 89.55% to 88.57% and 89.03%) when separately employing sigmoid and softmax functions in Eq. 3, which indicates the effectiveness of combining local and global context.

Fast Retrieval. We further conduct an experiment comparing the fast retrieval with a greedy search strategy. The greedy search compares the source image with every reference in the database during retrieval. To achieve a fair comparison, we divide all references into different groups with a group size of 1024 and perform parallel estimation over the data in each group for the greedy search. As shown in Table 4, although the greedy search achieves a better 3D orientation estimation accuracy, the time consumption (30.74 s) is not affordable in practice. Note that LineMOD contains 13 objects and we generate 10,000 reference images for each object. Therefore, the image comparison is performed 130,000 times for each source image in the greedy search, which results in an enormous

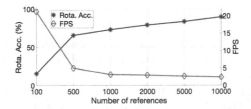

Fig. 7. Influence of the number of references. We vary the number of references for each object and report the results on the unseen objects of LineMOD Split #1.

Table 5. Influence of the individual components.

Local norm.	Weighted infoNCE	Rota. acc. (%)
✗	✗	85.91
✓	✗	89.25
✗	✓	87.14
✓	✓	**89.55**

time consumption. Note that we could not execute the comparisons w.r.t. all references in parallel because the NVIDIA Tesla V100 GPU could not store all the $130,000 \times 3$ feature maps. By contrast, our fast retrieval algorithm reduces the number of comparisons in two aspects: First, $\mathbf{I}_{src}$ is only compared with the initial anchors of each object for category recognition, reducing the complexity from $O(NR)$ to $O(Nk_{ac})$; second, the retrieval within the references that contain the recognized object for 3D orientation estimation only compares $\mathbf{I}_{src}$ with dynamically updated anchors, which reduces the complexity from $O(R)$ to $O(k_{ac})$.

Number of References. Intuitively, we expect the number of reference images to be positively correlated with the 3D object orientation estimation accuracy while negatively correlated with the retrieval speed. To shed more light on the influence of the number of reference images, we evaluate our method with a varying number of references. As shown in Fig. 7, as expected, as the number of reference images increases, Rota. Acc. increases while FPS decreases. Therefore, one can flexibly adjust the number of references in practice according to the desired accuracy and efficiency.

Effectiveness of the individual components. Finally, we conduct an ablation study to further understand the importance of the other individual components in our method, i.e., local normalization and the weighted-infoNCE loss. As shown in Table 5, each of these two components has a positive impact on the 3D orientation estimation accuracy, and the optimal performance is achieved by leveraging both of them.

5   Conclusion

In this paper, we have presented a retrieval-based 3D orientation estimation method for *previously-unseen* objects. Instead of representing an image as a global descriptor, we convert it to multiple feature maps at different resolutions, whose elements represent local patches of different sizes in the original image. We perform retrieval based on patch-level similarities, which are adaptively fused into a single similarity score for a pair of images. We have also designed a fast retrieval algorithm to speed up our method. Our experiments have demonstrated that our method outperforms both traditional and previous learning-based methods by a large margin in terms of 3D orientation estimation accuracy for unseen objects. In future work, we plan to extend our method to full 6D pose estimation of unseen objects.

Acknowledgments. This work was funded in part by the Swiss National Science Foundation and the Swiss Innovation Agency (Innosuisse) via the BRIDGE Discovery grant 40B2-0_194729.

References

1. Arandjelovic, R., Gronat, P., Torii, A., Pajdla, T., Sivic, J.: NetVLAD: CNN architecture for weakly supervised place recognition. In: Proceedings of the IEEE Conference on Computer Vision and Pattern Recognition, pp. 5297–5307 (2016)
2. Balntas, V., Doumanoglou, A., Sahin, C., Sock, J., Kouskouridas, R., Kim, T.K.: Pose guided RGBD feature learning for 3d object pose estimation. In: Proceedings of the IEEE International Conference on Computer Vision, pp. 3856–3864 (2017)
3. Brachmann, E., Krull, A., Michel, F., Gumhold, S., Shotton, J., Rother, C.: Learning 6D object pose estimation using 3D object coordinates. In: Fleet, D., Pajdla, T., Schiele, B., Tuytelaars, T. (eds.) ECCV 2014. LNCS, vol. 8690, pp. 536–551. Springer, Cham (2014). https://doi.org/10.1007/978-3-319-10605-2_35
4. Chen, T., Kornblith, S., Norouzi, M., Hinton, G.: A simple framework for contrastive learning of visual representations. In: International Conference on Machine Learning, pp. 1597–1607. PMLR (2020)
5. Chen, W., Jia, X., Chang, H.J., Duan, J., Shen, L., Leonardis, A.: FS-net: fast shape-based network for category-level 6d object pose estimation with decoupled rotation mechanism. In: Proceedings of the IEEE Conference on Computer Vision and Pattern Recognition, pp. 1581–1590 (2021)
6. Chen, X., Ma, H., Wan, J., Li, B., Xia, T.: Multi-view 3D object detection network for autonomous driving. In: Proceedings of the IEEE Conference on Computer Vision and Pattern Recognition, pp. 1907–1915 (2017)
7. Collet, A., Martinez, M., Srinivasa, S.S.: The moped framework: object recognition and pose estimation for manipulation. Int. J. Robot. Res. **30**(10), 1284–1306 (2011)
8. Dalal, N., Triggs, B.: Histograms of oriented gradients for human detection. In: Proceedings of the IEEE Conference on Computer Vision and Pattern Recognition, vol. 1, pp. 886–893. IEEE (2005)
9. Du, G., Wang, K., Lian, S., Zhao, K.: Vision-based robotic grasping from object localization, object pose estimation to grasp estimation for parallel grippers: a review. Artif. Intell. Rev. **54**(3), 1677–1734 (2020). https://doi.org/10.1007/s10462-020-09888-5

10. Fischler, M.A., Bolles, R.C.: Random sample consensus: a paradigm for model fitting with applications to image analysis and automated cartography. Commun. ACM **24**(6), 381–395 (1981)
11. Geiger, A., Lenz, P., Urtasun, R.: Are we ready for autonomous driving? The kitti vision benchmark suite. In: Proceedings of the IEEE Conference on Computer Vision and Pattern Recognition, pp. 3354–3361. IEEE (2012)
12. He, K., Zhang, X., Ren, S., Sun, J.: Deep residual learning for image recognition. In: Proceedings of the IEEE Conference on Computer Vision and Pattern Recognition, pp. 770–778 (2016)
13. Hinterstoisser, S., et al.: Model based training, detection and pose estimation of texture-less 3D objects in heavily cluttered scenes. In: Lee, K.M., Matsushita, Y., Rehg, J.M., Hu, Z. (eds.) ACCV 2012. LNCS, vol. 7724, pp. 548–562. Springer, Heidelberg (2013). https://doi.org/10.1007/978-3-642-37331-2_42
14. Hodan, T., Haluza, P., Obdržálek, Š., Matas, J., Lourakis, M., Zabulis, X.: T-less: an RGB-D dataset for 6D pose estimation of texture-less objects. In: Proceedings of the IEEE Winter Conference on Applications of Computer Vision, pp. 880–888. IEEE (2017)
15. Hodaň, T., Matas, J., Obdržálek, Š: On evaluation of 6D object pose estimation. In: Hua, G., Jégou, H. (eds.) ECCV 2016. LNCS, vol. 9915, pp. 606–619. Springer, Cham (2016). https://doi.org/10.1007/978-3-319-49409-8_52
16. Hu, Y., Fua, P., Wang, W., Salzmann, M.: Single-stage 6d object pose estimation. In: Proceedings of the IEEE Conference on Computer Vision and Pattern Recognition, pp. 2930–2939 (2020)
17. Kingma, D.P., Ba, J.: Adam: a method for stochastic optimization. arXiv preprint arXiv:1412.6980 (2014)
18. Li, X., Wang, H., Yi, L., Guibas, L.J., Abbott, A.L., Song, S.: Category-level articulated object pose estimation. In: Proceedings of the IEEE Conference on Computer Vision and Pattern Recognition, pp. 3706–3715 (2020)
19. Li, Y., Wang, G., Ji, X., Xiang, Y., Fox, D.: DeepIM: deep iterative matching for 6D pose estimation. In: Proceedings of the European Conference on Computer Vision, pp. 683–698 (2018)
20. Long, J., Shelhamer, E., Darrell, T.: Fully convolutional networks for semantic segmentation. In: Proceedings of the IEEE Conference on Computer Vision and Pattern Recognition, pp. 3431–3440 (2015)
21. Manuelli, L., Gao, W., Florence, P., Tedrake, R.: KPAM: keypoint affordances for category-level robotic manipulation. arXiv preprint arXiv:1903.06684 (2019)
22. Marchand, E., Uchiyama, H., Spindler, F.: Pose estimation for augmented reality: a hands-on survey. IEEE Trans. Visual. Comput. Graph. **22**(12), 2633–2651 (2015)
23. Park, K., Mousavian, A., Xiang, Y., Fox, D.: LatentFusion: end-to-end differentiable reconstruction and rendering for unseen object pose estimation. In: Proceedings of the IEEE Conference on Computer Vision and Pattern Recognition, pp. 10710–10719 (2020)
24. Peng, S., Liu, Y., Huang, Q., Zhou, X., Bao, H.: PVNet: pixel-wise voting network for 6DoF pose estimation. In: Proceedings of the IEEE Conference on Computer Vision and Pattern Recognition, pp. 4561–4570 (2019)
25. Qi, C.R., Yi, L., Su, H., Guibas, L.J.: PointNet++: deep hierarchical feature learning on point sets in a metric space. Adv. Neural Inf. Process. Syst. **30** (2017)
26. Ronneberger, O., Fischer, P., Brox, T.: U-Net: convolutional networks for biomedical image segmentation. In: Navab, N., Hornegger, J., Wells, W.M., Frangi, A.F. (eds.) MICCAI 2015. LNCS, vol. 9351, pp. 234–241. Springer, Cham (2015). https://doi.org/10.1007/978-3-319-24574-4_28

27. Sundermeyer, M., et al.: Multi-path learning for object pose estimation across domains. In: Proceedings of the IEEE Conference on Computer Vision and Pattern Recognition, pp. 13916–13925 (2020)
28. Sundermeyer, M., Marton, Z.C., Durner, M., Brucker, M., Triebel, R.: Implicit 3D orientation learning for 6D object detection from RGB images. In: Proceedings of the European Conference on Computer Vision, pp. 699–715 (2018)
29. Tremblay, J., To, T., Sundaralingam, B., Xiang, Y., Fox, D., Birchfield, S.: Deep object pose estimation for semantic robotic grasping of household objects. arXiv preprint arXiv:1809.10790 (2018)
30. Vaze, S., Han, K., Vedaldi, A., Zisserman, A.: Generalized category discovery. arXiv preprint arXiv:2201.02609 (2022)
31. Wang, C., et al.: 6-pack: category-level 6D pose tracker with anchor-based keypoints. In: Proceedings of the IEEE International Conference on Robotics and Automation, pp. 10059–10066. IEEE (2020)
32. Wang, C., et al.: Densefusion: 6D object pose estimation by iterative dense fusion. In: Proceedings of the IEEE Conference on computer vision and pattern recognition, pp. 3343–3352 (2019)
33. Wang, G., Manhardt, F., Tombari, F., Ji, X.: GDR-Net: geometry-guided direct regression network for monocular 6d object pose estimation. In: Proceedings of the IEEE Conference on Computer Vision and Pattern Recognition, pp. 16611–16621 (2021)
34. Wang, H., Sridhar, S., Huang, J., Valentin, J., Song, S., Guibas, L.J.: Normalized object coordinate space for category-level 6D object pose and size estimation. In: Proceedings of the IEEE/CVF Conference on Computer Vision and Pattern Recognition, pp. 2642–2651 (2019)
35. Wohlhart, P., Lepetit, V.: Learning descriptors for object recognition and 3D pose estimation. In: Proceedings of the IEEE Conference on Computer Vision and Pattern Recognition, pp. 3109–3118 (2015)
36. Xiang, Y., Schmidt, T., Narayanan, V., Fox, D.: PoseCNN: a convolutional neural network for 6D object pose estimation in cluttered scenes. arXiv preprint arXiv:1711.00199 (2017)
37. Xiao, Y., Du, Y., Marlet, R.: PoseContrast: class-agnostic object viewpoint estimation in the wild with pose-aware contrastive learning. In: 2021 International Conference on 3D Vision (3DV), pp. 74–84. IEEE (2021)
38. Xiao, Y., Qiu, X., Langlois, P.A., Aubry, M., Marlet, R.: Pose from shape: deep pose estimation for arbitrary 3D objects. arXiv preprint arXiv:1906.05105 (2019)
39. Xu, D., Anguelov, D., Jain, A.: PointFusion: deep sensor fusion for 3D bounding box estimation. In: Proceedings of the IEEE Conference on computer Vision and Pattern Recognition, pp. 244–253 (2018)
40. Zhou, Y., Barnes, C., Lu, J., Yang, J., Li, H.: On the continuity of rotation representations in neural networks. In: Proceedings of the IEEE Conference on Computer Vision and Pattern Recognition, pp. 5745–5753 (2019)
41. Zhu, M., et al.: Single image 3D object detection and pose estimation for grasping. In: Proceedings of the IEEE International Conference on Robotics and Automation, pp. 3936–3943. IEEE (2014)

Lidar Point Cloud Guided Monocular 3D Object Detection

Liang Peng[1,2], Fei Liu[3], Zhengxu Yu[1], Senbo Yan[1,2], Dan Deng[2], Zheng Yang[2], Haifeng Liu[1], and Deng Cai[1,2(✉)]

[1] State Key Lab of CAD&CG, Zhejiang University, Hangzhou, China
{pengliang,senboyan,haifengliu}@zju.edu.cn, dengcai@cad.zju.edu.cn
[2] Fabu Inc., Hangzhou, China
{dengdan,yangzheng}@fabu.ai
[3] State Key Lab of Industrial Control and Technology, Zhejiang University, Hangzhou, China
liufei21@zju.edu.cn

Abstract. Monocular 3D object detection is a challenging task in the self-driving and computer vision community. As a common practice, most previous works use manually annotated 3D box labels, where the annotating process is expensive. In this paper, we find that the precisely and carefully annotated labels may be unnecessary in monocular 3D detection, which is an interesting and counterintuitive finding. Using rough labels that are randomly disturbed, the detector can achieve very close accuracy compared to the one using the ground-truth labels. We delve into this underlying mechanism and then empirically find that: concerning the label accuracy, the 3D location part in the label is preferred compared to other parts of labels. Motivated by the conclusions above and considering the precise LiDAR 3D measurement, we propose a simple and effective framework, dubbed LiDAR point cloud guided monocular 3D object detection (LPCG). This framework is capable of either reducing the annotation costs or considerably boosting the detection accuracy without introducing extra annotation costs. Specifically, It generates pseudo labels from unlabeled LiDAR point clouds. Thanks to accurate LiDAR 3D measurements in 3D space, such pseudo labels can replace manually annotated labels in the training of monocular 3D detectors, since their 3D location information is precise. LPCG can be applied into any monocular 3D detector to fully use massive unlabeled data in a self-driving system. As a result, in KITTI benchmark, we take the first place on both monocular 3D and BEV (bird's-eye-view) detection with a significant margin. In Waymo benchmark, our method using 10% labeled data achieves comparable accuracy to the baseline detector using 100% labeled data. The codes are released at https://github.com/SPengLiang/LPCG.

Keywords: Monocular 3D detection · LiDAR point cloud · Self-driving

Supplementary Information The online version contains supplementary material available at https://doi.org/10.1007/978-3-031-19769-7_8.

S. Avidan et al. (Eds.): ECCV 2022, LNCS 13661, pp. 123–139, 2022.
https://doi.org/10.1007/978-3-031-19769-7_8

1   Introduction

3D object detection plays a critical role in many applications, such as self-driving. It gives cars the ability to perceive the world in 3D, avoiding collisions with other objects on the road. Currently, the LiDAR (Light Detection and Ranging) device is typically employed to achieve this [12,29,31,46], with the main shortcomings of the high price and limited working ranges. The single camera, as an alternative, is widely available and several orders of magnitude cheaper, consequently making monocular methods [1,6,41,45] popular in both industry and academia.

To the best of our knowledge, most previous monocular-based works [1,6,28,45] employ the precisely annotated 3D box labels. The annotation process operated on LiDAR point clouds is time-consuming and costly. In this paper, we empirically find that **the perfect manually annotated 3D box labels are not essential in monocular 3D detection**. We disturb the manually annotated labels by randomly shifting their values in a range, while the detector respectively trained by disturbed labels and perfect labels show very close performance. This is a counterintuitive finding. To explore the underlying mechanism, we divide a 3D box label into different groups according to its physical nature (including 3D locations, orientations, and dimensions of objects), and disturb each group of labels, respectively. We illustrate the experiment in Fig. 1. The results indicate that the precise location label plays the most important role and dominates the performance of monocular 3D detection, and the accuracy of other groups of labels is not as important as generally considered. The underlying reason lies in the ill-posed nature of monocular imagery. It brings difficulties in recovering the 3D location, which is the bottleneck for the performance.

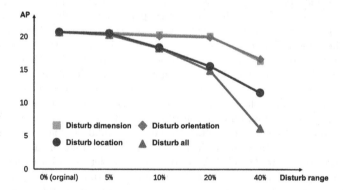

Fig. 1. We disturb the perfect manually annotated labels by randomly shifting the corresponding values within the percentage range. We can see that: 1) the disturbed labels (5%) and perfect labels lead to close accuracy; 2) the location dominates the overall accuracy (10%, 20%, 40%).

Unlike other classical computer vision tasks, manually annotating 3D boxes from monocular imagery is infeasible. It is because the depth information is lost during the camera projection process. Actually, the lost depth also is the reason why 3D location labels are the most important and difficult part for monocular 3D detection. LiDAR point clouds, which provide the crucial 3D measurements, are indispensable in the labeling procedure. As a common practice, annotators annotate 3D boxes on the LiDAR points clouds. On the other hand, concerning the data collecting process in a self-driving system, a large number of successive snippets are collected. Generally speaking, to save the high annotation costs, only some key frames in collected snippets are labeled to train networks, such as KITTI dataset [8]. Consequently, massive LiDAR point clouds holding valuable 3D information remain unlabeled.

Inspired by the 3D location label requirement and accurate LiDAR 3D measurements in 3D space, we propose a general and intuitive framework to make full use of LiDAR point clouds, dubbed **LPCG** (LiDAR point cloud guided monocular 3D object detection). Specifically, we use unlabeled LiDAR point clouds to generate pseudo labels, converting unlabeled data to training data for monocular 3D detectors. These pseudo labels are not as accurate as the manually annotated labels, but they are good enough for the training of monocular 3D detectors due to accurate LiDAR 3D measurements.

We further present two working modes in LPCG: the high accuracy mode and the low cost mode, to generate different qualities of pseudo labels according to annotation costs. The high accuracy mode requires a small amount of labeled data to train a LiDAR-based detector, and then the trained detector can produce high-quality pseudo labels on other unlabeled data. This manner can largely boost the accuracy of monocular 3D detectors. Additionally, we propose a heuristic method to produce pseudo labels without requiring any 3D box annotation. Such pseudo labels are directly obtained from the RoI LiDAR point clouds, by employing point clustering and minimum bounding box estimation. We call this manner the low cost mode. Either the high accuracy mode or the low cost mode in LPCG can be plugged into any monocular 3D detector.

Based on the above two modes, we can fully use LiDAR point clouds, allowing monocular 3D detectors to learn desired objectives on a large training set meanwhile avoiding architecture modification and removing extra annotation costs. By applying the framework (high accuracy mode), we significantly increase the 3D and BEV (bird's-eye-view) AP of prior state-of-the-art methods [1,11,13,27,45]. In summary, our contributions are two folds as below:

- **First**, we analyze requirements in terms of the label accuracy towards the training of monocular 3D detection. Based on this analysis, we introduce a general framework that can utilize massive unlabeled LiDAR point clouds, to generate new training data with valuable 3D information for monocular methods during the training.
- **Second**, experiments show that the baseline detector employing our method outperforms recent SOTA methods by a large margin, ranking 1st on KITTI [8] monocular 3D and BEV detection benchmark at the time of submission (car, March. 2022). In Waymo [36] benchmark, our method achieves close accuracy compared to the baseline detector using 100% labeled data while our method requires only 10% labeled data with 90% unlabeled data.

2   Related Work

2.1   LiDAR-based 3D Object Detection

The LiDAR device can provide accurate depth measurement of the scene, thus is employed by most state-of-the-art 3D object detection methods [12,29,31,32,42, 43]. These methods can be roughly divided into voxel-based methods and point-based methods. Voxel-based methods [47] first divide the point cloud into a voxel grid and then feed grouped points into fully connected layers, constructing unified feature representations. They then employ 2D CNNs to extract high-level voxel features to predict 3D boxes. By contrast, point-based methods [24,30] directly extract features on the raw point cloud via fully connected networks, such as PointNet [25] and PointNet++ [26]. SOTA 3D detection methods predominantly employ LiDAR point clouds both in training and inference, while we only use LiDAR point clouds in the training stage.

2.2   Image-Only-Based Monocular 3D Object Detection

As a commonly available and cheap sensor, the camera endows 3D object detection with the potential of being adopted everywhere. Thus monocular 3D object detection has become a very popular area of research and has developed quickly in recent years. Monocular works can be categorized into image-only-based methods [1,13] and depth-map-based methods [18,19,40] according to input representations. M3D-RPN [1] employs different convolution kernels in row-spaces that can explore different features in specific depth ranges and improve 3D estimates with the 2D-3D box consistency. Furthermore, RTM3D [13] predicts perspective key points and initial guesses of objects' dimensions/orientations/locations, where the key points are further utilized to refine the initial guesses by solving a constrained optimization problem. More recently, many image-only-based works utilize depth estimation embedding [45], differentiable NMS [11], and geometry properties [15,17,48], obtaining great success. There is also a related work [44] that introduces a novel autolabeling strategy of suggesting a differentiable template matching model with curriculum learning, using differentiable rendering of SDFs, while the pipeline is rather complicated.

2.3   Depth-Map-Based Monocular 3D Object Detection

Although monocular methods are developing quickly, a large performance gap still exists compared to LiDAR-based methods. Some prior works [40,41] argue that the improper choice of data representation is one of the main reasons, and propose to use transformed image-based depth maps. They first project LiDAR point clouds onto the image plane, to form depth map labels to train a depth estimator. Pseudo-LiDAR [40] converts the image into a point cloud by using an estimated depth map and then conducts 3D detection on it. They show promising results compared to previous image-only-based methods. Inspired by this, many later methods [6,18,19,41] also utilize off-the-shelf depth estimates to aim

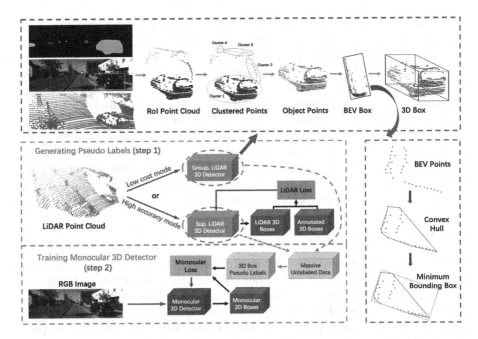

Fig. 2. Overview framework. We generate 3D box pseudo labels from unlabeled LiDAR point clouds, aiming to train the monocular 3D detector. Such 3D boxes are predicted via the well-trained LiDAR 3D detector (high accuracy mode) or obtained directly from the point cloud without training (low cost mode). "Unsup." and "Sup." in the figure denote unsupervised and supervised, respectively.

3D detection and gain performance improvements. More recently, CaDDN [28] integrates the dense depth estimation into monocular 3D detection, by using a predicted categorical depth distribution to project contextual features to the 3D space. Compared to previous depth-map-based methods, we aim to explore the potential of using LiDAR point clouds to generate pseudo labels for monocular 3D detectors.

3   LiDAR Guided Monocular 3D Detection

In this section, we detail the proposed framework, namely, LPCG (LiDAR Guided Monocular 3D Detection). First, as shown in Fig. 1, the manually annotated perfect labels are unnecessary for monocular 3D detection. The accuracy led by disturbed labels (5%) is comparable to the one led by perfect labels. When enforcing large disturbances (10% and 20%), we can see that the location dominates the performance (the AP dramatically decreases only when disturbing the location). It indicates that rough pseudo 3D box labels with precise locations may replace the perfect annotated 3D box labels.

We note that LiDAR point clouds can provide valuable 3D location information. More specifically, LiDAR point clouds provide accurate depth measurement within the scene, which is crucial for 3D object detection as precise surrounding depths indicate locations of objects. Also, LiDAR point clouds can be easily captured by the LiDAR device, allowing a large amount of LiDAR point clouds to be collected offline without manual cost. Based on the analysis above, we use LiDAR point clouds to generate 3D box pseudo labels. The newly-generated labels can be used to train monocular 3D detectors. This simple and effective way allows monocular 3D detectors to learn desired objectives meanwhile eliminating annotation costs on unlabeled data. We show the overall framework in Fig. 2, in which the method is able to work in two modes according to the reliance on 3D box annotations. If we use a small amount of 3D box annotations as prior, we call it the high accuracy mode since this manner leads to high performances. By contrast, we call it the low cost mode if we do not use any 3D box annotation.

3.1 High Accuracy Mode

To take advantage of available 3D box annotations, as shown in Fig. 2, we first train a LiDAR-based 3D detector from scratch with LiDAR point clouds and associated 3D box annotations. The pre-trained LiDAR-based 3D detector is then utilized to infer 3D boxes on other unlabeled LiDAR point clouds. Such results are treated as pseudo labels to train monocular 3D detectors. We compare the pseudo labels with manually annotated perfect labels in Sect. 5.5. Due to precise 3D location measurements, pseudo labels predicted from the LiDAR-based 3D detector are rather accurate and qualified to be used directly in the training of monocular 3D detectors. We summarize the outline in Algorithm 1.

Algorithm 1: Outline of the high accuracy mode in LPCG. Both labeled and unlabeled training data contains RGB images and associated LiDAR point clouds.

---

1 **Input:** Labeled data $A : \{A_{data}, A_{label}\}$, unlabeled data $B : \{B_{data}\}$

2 **Output:** Well-trained monocular 3D detection model M_{mono}

3 $M_{lidar} \leftarrow$ Training a supervised LiDAR-based 3D detection model on labeled data $\{A_{data}, A_{label}\}$.

4 $\{B_{pseudo-label}\} \leftarrow$ Conducting predictions from LiDAR point clouds on unlabeled data: $M_{lidar}(B_{data})$

5 $C : \{C_{data}, C_{label}\} \leftarrow$ Merging training data: $\{A_{data} \cup B_{data}, A_{label} \cup B_{pseudo-label}\}$

6 $M_{mono} \leftarrow$ Training a supervised monocular-based model on new set $\{C_{data}, C_{label}\}$.

7 Return M_{mono}

Interestingly, with different training settings for the LiDAR-based 3D detector, we empirically find that monocular 3D detectors trained by resulting pseudo labels show close performances. It indicates that monocular methods can indeed be beneficial from the guidance of the LiDAR point clouds and only a small number of 3D box annotations are sufficient to push the monocular method to achieve high performance. Thus the manual annotation cost of high accuracy mode is much lower than the one of the previous manner. Detailed experiments can be found in Sect. 5.6. Please note, the observations on label requirements and 3D locations are the core motivation of LPCG. The premise that LPCG can work well is that LiDAR points provide rich and precise 3D measurements, which offer accurate 3D locations.

3.2   Low Cost Mode

In this section, we describe the method of using LiDAR point clouds to eliminate the reliance on manual 3D box labels. First, an off-the-shelf 2D instance segmentation model [9] is adopted to perform segmentation on the RGB image, obtaining 2D box and mask estimates. These estimates are then used for building camera frustums in order to select associated LiDAR RoI points for every object, where those boxes without any LiDAR point inside are ignored. However, LiDAR points located in the same frustum consist of object points and mixed background or occluded points. To eliminate irrelevant points, we take advantage of DBSCAN [7] to divide the RoI point cloud into different groups according to the density. Points that are close in 3D spatial space will be aggregated into a cluster. We then regard the cluster containing most points as a target corresponding to the object. Finally, we seek the minimum 3D bounding box that covers all target points.

To simplify the problem of solving the 3D bounding box, we project points onto the bird's-eye-view map, reducing parameters since the height (h) and y coordinate (under camera coordinate system) of the object can be easily obtained. Therefore, we have:

$$L = \min_{B_{bev}}(Area(B_{bev})), \quad subject\ to\ p\ is\ inside\ B_{bev}\ , where\ p \in LiDAR_{RoI} \quad (1)$$

where B_{bev} refers to a bird's-eye-view (BEV) box. We solve this problem by using the convex hull of object points followed by obtaining the box by using rotating calipers [37]. Furthermore, the height h can be represented by the max spatial offset along the y-axis of points, and the center coordinate y is calculated by averaging y coordinates of points. We use a simple rule of restricting object dimensions to remove outliers. The overall training pipeline for monocular methods is summarized in Algorithm 2.

4   Applications in Real-World Self-driving System

In this section, we describe the application of LPCG to a real-world self-driving system. First, we illustrate the data collecting strategy in Fig. 3. Most self-driving systems can easily collect massive unlabeled LiDAR point cloud data and synchronized RGB images. This data is organized with many sequences, where each sequence often refers to a specific scene and contains many successive frames.

Algorithm 2: Outline of the low cost mode in LPCG. Unlabeled data contains RGB images and associated LiDAR point clouds.

---
1 **Input:** Unlabeled data $D : \{D_{data-image}, D_{data-lidar}\}$, pre-trained Mask-RCNN model: M_{mask}
2 **Output:** Well-trained monocular 3D detection model M_{mono}
3 $Mask_{2D} \leftarrow$ Conducting predictions from RGB images: $M_{mask}(D_{data-image})$.
4 $LiDAR_{RoI} \leftarrow$ Selecting and clustering LiDAR point clouds from $D_{data-lidar}$ by $Mask_{2D}$
5 $D_{pseudo-label} \leftarrow$ Gernerating pseudo labels on RoI LiDAR points $LiDAR_{RoI}$.
6 $M_{mono} \leftarrow$ Training a supervised monocular-based model on new data $\{D_{data-image}, D_{pseudo-label}\}$.
7 Return M_{mono}

Table 1. Comparisons of different modes in previous works and ours.

Approaches	Modality	3D box annotations	Unlabeled LiDAR data
Previous	Image-only based	Yes	No
	Depth-map based	Yes	Yes
Ours	High accuracy mode	Yes	Yes
	Low cost mode	No	Yes

Due to the limited time and resources in the real world, only some sequences are chosen for annotating, to train the network, such as Waymo [36]. Further, to reduce the high annotation costs, only some key frames in the selected sequences are annotated, such as KITTI [8]. Therefore, there remains massive unlabeled data in real-world applications.

Considering that LPCG can fully take advantage of the unlabeled data, it is natural to be employed in a real-world self-driving system. Specifically, the high accuracy mode only requires a small amount of labeled data. Then we can generate high-quality training data from remaining unlabeled data for monocular 3D detectors, to boost the accuracy. In experiments, we quantitatively and qualitatively show that the generated 3D box pseudo labels are good enough for monocular 3D detectors. Additionally, the low cost mode does not require any 3D box annotation, still providing accurate 3D box pseudo labels. We compare LPCG with previous methods in Table 1 in terms of the data requirements.

5   Experiments

5.1   Implementation Details

We use the image-only-based monocular 3D detector M3D-RPN [1], and adopt PV-RCNN [29] as the LiDAR 3D detector for the high accuracy mode. We filter out 3D boxes generated from LiDAR point clouds with the confidence of 0.7. Experiments on other methods are conducted by the official code that is

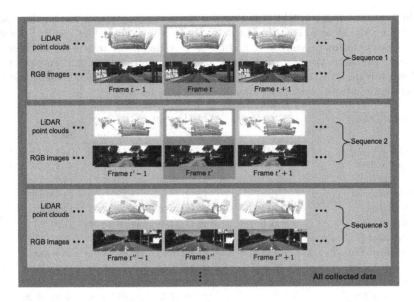

Fig. 3. Data collecting strategy in a real-world system. Only some sequences (*e.g.*, sequence 1 and 2) are chosen for annotating because of the limited time and resources in the real world, such as Waymo [36]. Further, concerning the high annotation costs, only some key frames (*e.g.*, frame t and t') in the selected sequences are annotated, such as KITTI [8].

publicly available, and all settings keep the same as the original paper. During the process of using LiDAR point clouds to train monocular 3D detectors, the learning iteration is scaled according to the number of training data. The high accuracy mode is employed by default. For the low cost mode, we use Mask-RCNN [9] pre-trained in the COCO dataset [14], and filter the final 3D bounding box by the width range of 1.2–1.8 m and the length range of 3.2–4.2 m. We filter out 2D boxes predicted from Mask-RCNN [9] with the confidence of 0.9. More details and ablations are provided in the supplementary material as the space limitation.

5.2 Dataset and Metrics

Dataset. Following prior works [1,13,18,27,35,40], experiments are conducted on the popular KITTI 3D object dataset [8], which contains 7, 481 manually annotated images for training and 7, 518 images for testing. Due to groundtruths of the test set are not available, the public training set is further split into two subsets [3]: training set (3, 712 images) and validation set (3, 769 images). Following the fashion, we report our results both on the validation set and the test set. And we use the validation set for all ablations. Also, our method and depth-map-based methods use RGB images and synchronized LiDAR point clouds from KITTI raw scenes. For depth-map-based methods, note that the original depth

Table 2. Comparisons on KITTI testing set. We use red to indicate the highest result and blue for the second-highest result and cyan for the third-highest result. ‡ denotes the baseline detector we employed, and the improvements are relative to the baseline detectors. We define the new state of the art. Please note, DD3D [22]* employs both the large private DDAD15M dataset (containing approximately 15M frames) and the KITTI depth dataset (containing approximately 26K frames).

| Approaches | Extra data | AP_{BEV}/AP_{3D} (IoU=0.7)$|_{R_{40}}$ | | |
		Easy	Moderate	Hard
ROI-10D [21]	KITTI depth	9.78/4.32	4.91/2.02	3.74/1.46
MonoGRNet [27]	None	18.19/5.74	11.17/9.61	8.73/4.25
AM3D [19]	KITTI depth	25.03/16.50	17.32/10.74	14.91/9.52
MonoPair [4]	None	19.28/13.04	14.83/9.99	12.89/8.65
D4LCN [6]	KITTI depth	22.51/16.65	16.02/11.72	12.55/9.51
RTM3D [13]	None	19.17/14.41	14.20/10.34	11.99/8.77
PatchNet [18]	KITTI depth	22.97/15.68	16.86/11.12	14.97/10.17
Neighbor-Vote [5]	KITTI depth	27.39/15.57	18.65/9.90	16.54/8.89
MonoRUn [2]	None	27.94/19.65	17.34/12.30	15.24/10.58
MonoRCNN [33]	None	25.48/18.36	18.11/12.65	14.10/10.03
Monodle [20]	None	24.79/17.23	18.89/12.26	16.00/10.29
CaDDN [28]	None	27.94/19.17	18.91/13.41	17.19/11.46
Ground-Aware [15]	None	29.81/21.65	17.98/13.25	13.08/9.91
GrooMeD-NMS [11]	None	26.19/18.10	18.27/12.32	14.05/9.65
MonoEF [48]	None	29.03/21.29	19.70/13.87	17.26/11.71
DDMP-3D [38]	KITTI depth	28.08/19.71	17.89/12.78	13.44/9.80
PCT [39]	KITTI depth	29.65/21.00	19.03/13.37	15.92/11.31
AutoShape [16]	None	30.66/22.47	20.08/14.17	15.95/11.36
GUPNet [17]	None	30.29/22.26	**21.19/15.02**	**18.20/13.12**
M3D-RPN [1] ‡	None	21.02/14.76	13.67/9.71	10.23/7.42
MonoFlex [45] ‡	None	28.23/19.94	19.75/13.89	16.89/12.07
DD3D [22] *	*DDAD15M...*	*32.35/23.19*	*23.41/16.87*	*20.42/14.36*
LPCG+M3D-RPN [1]	KITTI depth	**30.72/22.73**	20.17/14.82	16.76/12.88
Improvements (to baseline)		+9.70/+7.97	+6.50/+5.11	+6.53/+5.46
LPCG+MonoFlex [45]	KITTI depth	35.96/25.56	24.81/17.80	21.86/15.38
Improvements (to baseline)		+7.73/+5.62	+5.06/+3.91	+4.97/+3.31

training set overlaps KITTI 3D detection validation set. Therefore we exclude scenes that emerge in KITTI 3D validation set to avoid data leakage [23,34]. LiDAR point clouds in the remaining scenes are used. We call this extra dataset the KITTI depth dataset. It provides approximately 26K samples to train the depth estimator (for most depth-map-based methods) or to generate extra training samples for monocular 3D detectors (LPCG).

Additionally, to further validate the effectiveness of LPCG, we conduct experiments on the Waymo Open Dataset [36], which is a modern large dataset. It contains 798 training sequences and 202 validation sequences, and we adopt the same data processing strategy proposed in CaDDN [28]. The sampled training dataset includes approximately 50K training samples with manual annotations.

Metrics. Each manually annotated object is divided into easy, moderate, and hard levels according to the occlusion, truncation, and 2D box height [8]. Average precisions (AP) on the car class for bird's-eye-view (BEV) and 3D boxes with 0.5/0.7 IoU thresholds are commonly used metrics for monocular 3D detection. Many previous methods utilize the AP_{11} metric, which has an overrated issue [35], and AP_{40} [35] is proposed to resolve it. We report AP_{40} results to make comprehensive comparisons. For Waymo dataset, we adopt the official mAP and mAPH metrics.

Table 3. Comparisons with SDFLabel [44]. Note that here we use the same number of training samples for fair comparisons.

Approaches	Data requirements in training	AP_{3D} (IoU = 0.7)$\vert_{R_{40}}$		
		Easy	Moderate	Hard
SDFLabel [44]	2D masks+LiDAR+CAD models	1.23	0.54	–
LPCG (low cost mode)	2D masks+LiDAR	**5.36**	**3.07**	**2.32**

5.3 Results on KITTI

We evaluate LPCG on KITT test set using two base monocular detectors [1,45] with the high accuracy mode. Table 2 shows quantitative results in *test* set. Due to the space limitation, qualitative results are included in the supplementary material. We can observe that our method increases the current SOTA BEV/3D AP from **21.19/15.02** to **24.81/17.80** under the moderate setting, which is rather significant. Even using a monocular detector [1] proposed in 2019, our method still allows it to achieve new state-of-the-art compared to prior works. Note that our method still performs better, while DD3D [22] employs both the large private DDAD15M dataset (containing approximately 15M frames) and the KITTI depth dataset (containing approximately 20K frames). Also, we boost the performance on pedestrian and cyclist categories of the original method, and provide the results in Table 5. Such results prove the effectiveness of LPCG.

For the low cost mode, we note that there are few works exploring this area, namely, few works have explored monocular 3D detection without any 3D box annotation. The most related work is SDFLabel [44], which also does not require 3D box annotation. Thus we compare LPCG with the low cost mode with SDFLabel [44] in Table 3. Please note, in this experiment we use the same number of training samples, namely, the 3, 769 samples in KITTI3D training set. Our method outperforms it by a large margin, and our method is more generally usable as our pipeline is much simpler than SDFLabel [44].

5.4 Results on Waymo

To further prove the effectiveness of our method, we conduct experiments on the Waymo open dataset. Concerning its large scale, when enough perfect labels are available, in this dataset we aim to investigate the performance gap between the generated pseudo labels and the manual 3D box annotations. More specifically,

Table 4. Comparisons on Waymo. "lab." and "unlab." denote labeled and unlabeled.

Difficulty	w/ LPCG	Data requirements	Overall	0−30m	30−50m	50m−∞
Under 3D mAP metric						
LEVEL 1 (IOU = 0.5)	No	10% labeled data	4.14	14.64	1.63	0.04
	No	100% labeled data	6.42	19.50	3.04	0.17
	Yes	10% lab. + 90% unlab. data	6.23	18.39	3.44	0.19
LEVEL 2 (IOU = 0.5)	No	10% labeled data	3.88	14.59	1.58	0.04
	No	100% labeled data	6.02	19.43	2.95	0.15
	Yes	10% lab. + 90% unlab. data	5.84	18.33	3.34	0.17
Under 3D mAPH metric						
LEVEL 1 (IOU = 0.5)	No	10% labeled data	3.94	14.07	1.51	0.04
	No	100% labeled data	6.19	18.88	2.89	0.16
	Yes	10% lab. + 90% unlab. data	6.09	18.03	3.33	0.17
LEVEL 2 (IOU = 0.5)	No	10% labeled data	3.69	14.02	1.46	0.03
	No	100% labeled data	5.80	18.81	2.80	0.14
	Yes	10% lab. + 90% unlab. data	5.70	17.97	3.23	0.15

Table 5. Improvements on other categories.

Approaches	Pedestrian, AP_{3D} (IoU = 0.5)$\|_{R_{40}}$			Cyclist, AP_{3D} (IoU = 0.5)$\|_{R_{40}}$		
	Easy	Moderate	Hard	Easy	Moderate	Hard
M3D-RPN [1]	4.75	3.55	2.79	3.10	1.49	1.17
M3D-RPN+LPCG	**7.21**	**5.53**	**4.46**	**4.83**	**2.65**	**2.62**

we use the baseline detector M3D-RPN [1], training it with pseudo labels and manual annotations, respectively. We report the results in Table 4. Pseudo labels on unlabeled data are generated by the high accuracy mode in LPCG. Interestingly, we can see that the detector using 10% labeled data and 90% unlabeled data achieves comparable accuracy to the one using 100% labeled data (*e.g.*, 6.42 *vs.* 6.23 and 6.19 *vs.* 6.09). This result demonstrates the generalization ability of LPCG, indicating that LPCG can also reduce the annotation costs for the large scale dataset with slight accuracy degradation.

5.5 Comparisons on Pseudo Labels and Manually Annotated Labels

As expected, pseudo labels are not as accurate as manually annotated labels. It is interesting to quantitatively evaluate pseudo labels using manually annotated labels. We report the results in Table 6. TP, FP, FN are calculated by matching pseudo labels and annotated labels. Regarding matched objects, we average the relative error (MRE) on each group of 3D box parameters (location, dimension, and orientation). We can see that pseudo labels from the high accuracy mode can match most real objects (91.39%), and the mean relative errors are 1%-6%. Therefore pseudo labels from the high accuracy mode are good enough for monocular 3D detectors. Actually, experiments in Table 2 and 4 also verify the effectiveness. On the other hand, for the low cost mode, we can see that many real objects are missed (11834). We note that missed objects are often truncated, occluded, or faraway. The attached LiDAR points cannot indicate the full 3D outline of objects, thus they are hard to recover by geometry-based methods.

Table 6. Performance of pseudo labels on *val* set. We evaluate pseudo labels using manually annotated labels. "MRE" refers to the mean relative error (*e.g.*, the relative error of location is $\frac{Error_{Loc}}{Loc}$). "Loc., Dim., Orient." are the location (x, y, z), dimension (h, w, l), and orientation (R_y). "TP, FP, FN" are the true positive, false positive, and false negative, which are calculated by matching pseudo labels and annotated labels. Please see Sect. 5.5 for detailed analysis. Note that pseudo labels on *val* set are just for the evaluation, and they are not used in the training of monocular 3D detectors.

Pseudo label types	TP	FP	FN	Loc. MRE	Dim. MRE	Orient. MRE
Low cost mode	2551	161	11834	4%/5%/2%	8%/6%/7%	8%
High accuracy mode	13146	3299	1239	4%/4%/1%	4%/4%/6%	4%

Table 7. Ablation for annotation numbers. 3712 is the total annotations in the KITTI 3D training dataset. All results are evaluated on KITTI *val* set with metric $AP|_{R_{40}}$.

Annotations	$AP_{BEV}/AP_{3D}|$ (IoU = 0.7)$_{R_{40}}$		
	Easy	Moderate	Hard
100	30.19/20.90	21.96/15.37	19.16/13.00
200	30.39/22.55	22.44/16.17	19.60/14.32
500	32.01/23.13	23.31/17.42	20.26/14.95
1000	33.08/25.71	24.89/19.29	21.94/16.75
3712	**33.94/26.17**	**25.20/19.61**	**22.06/16.80**

5.6 Ablation Studies

We conduct the ablation studies on KITTI *val* set. Because of the space limitation, we provide extra ablation studies in the supplementary material.

Different Monocular Detectors. We plug LPCG into different monocular 3D detectors [1,11,13,27,45], to show its extension ability. Table 8 shows the results. We can see that LPCG obviously and consistently boosts original performances, *e.g.*, 7.57 → 10.06 for MonoGRNet [27], 10.06 → 19.43 for RTM3D [13], and 14.32 → 20.46 for GrooMeD-NMS [11] under the moderate setting (AP$_{3D}$ (IoU = 0.7)). Furthermore, we explore the feasibility of using a rather simple model when large data is available. We perform this experiment on RTM3D [13] with ResNet18 [10] backbone, which achieves 46.7 FPS on a NVIDIA 1080Ti GPU. [1] To the best of our knowledge, it is the simplest and fastest model for monocular 3D detection. With employing LPCG, this simple model obtains very significant improvements. LPCG endows it (46.7 FPS) with the comparable accuracy to other state-of-the-art models (*e.g.*, GrooMeD-NMS (8.3 FPS)). These results prove that LPCG is robust to the choice of monocular 3D detectors.

The Number of Annotations. We also investigate the impact of the number of annotations. We report the results in Table 7. The results indicate that a small number of annotations in LPCG can also lead to high accuracy for monocular

---

[1] From RTM3D official implementation.

Table 8. Extension on different monocular detectors. LPCG can be easily plugged into other methods. * denotes that the model is re-implemented by us. All the methods are evaluated on KITTI *val* set with metric $AP|_{R_{40}}$.

Approaches	AP_{3D} (IoU=0.5)$|_{R_{40}}$			AP_{3D} (IoU=0.7)$|_{R_{40}}$		
	Easy	Moderate	Hard	Easy	Moderate	Hard
MonoGRNet [27]	47.34	32.32	25.54	11.93	7.57	5.74
MonoGRNet+LPCG	**53.84**	**37.24**	**29.70**	**16.30**	**10.06**	**7.86**
Improvements	+6.50	+4.92	+4.16	+4.37	+2.49	+2.12
M3D-RPN [1]	48.56	35.94	28.59	14.53	11.07	8.65
M3D-RPN+LPCG	**62.92**	**47.14**	**42.03**	**26.17**	**19.61**	**16.80**
Improvements	+14.36	+11.20	+13.44	+11.64	+8.54	+8.15
RTM3D [13]	55.44	39.24	33.82	13.40	10.06	9.07
RTM3D+LPCG	**65.44**	**49.40**	**43.55**	**25.23**	**19.43**	**16.77**
Improvements	+10.00	+10.16	+9.73	+11.83	+9.37	+7.70
RTM3D (ResNet18) [13]	47.78	33.75	28.48	10.85	7.51	6.33
RTM3D (ResNet18)+LPCG	**62.98**	**45.86**	**41.63**	**22.69**	**16.78**	**14.50**
Improvements	+15.20	+12.11	+13.15	+11.84	+9.27	+8.17
GrooMeD-NMS [11]	55.62	41.07	32.89	19.67	14.32	11.27
GrooMeD-NMS +LPCG	**68.27**	**50.80**	**45.14**	**27.79**	**20.46**	**17.75**
Improvements	+12.65	+9.73	+12.25	+8.12	+6.14	+6.48
MonoFlex [45]*	56.73	42.97	37.34	20.02	15.19	12.95
MonoFlex +LPCG	**69.16**	**54.27**	**48.37**	**31.15**	**23.42**	**20.60**
Improvements	+12.43	+11.30	+11.03	+11.13	+8.23	+7.65

3D detectors. For example, the detector using 1000 annotations performs close to the full one (24.89/19.29 *vs.* 25.20/19.61 under the moderate setting).

6   Conclusion

In this paper, we first analyze the label requirements for monocular 3D detection. Experiments show that disturbed labels and perfect labels can lead to very close performance for monocular 3D detectors. With further exploration, we empirically find that the 3D location is the most important part of 3D box labels. Additionally, a self-driving system can produce massive unlabeled LiDAR point clouds, which have precise 3D measurements. Therefore, we propose a framework (LCPG), to generate pseudo 3D box labels on unlabeled LiDAR point clouds, to enlarge the training set of monocular 3D detectors. Extensive experiments on various datasets validate the effectiveness of LCPG. Furthermore, the main limitation of LCPG is more training time due to the increased training samples.

Acknowledgments. This work was supported in part by The National Key Research and Development Program of China (Grant Nos: 2018AAA0101400), in part by The National Nature Science Foundation of China (Grant Nos: 62036009, U1909203, 61936006, 61973271), in part by Innovation Capability Support Program of Shaanxi (Program No. 2021TD-05).

References

1. Brazil, G., Liu, X.: M3D-RPN: monocular 3D region proposal network for object detection. In: Proceedings of the IEEE International Conference on Computer Vision, pp. 9287–9296 (2019)
2. Chen, H., Huang, Y., Tian, W., Gao, Z., Xiong, L.: MonoRUn: monocular 3D object detection by reconstruction and uncertainty propagation. In: Proceedings of the IEEE/CVF Conference on Computer Vision and Pattern Recognition, pp. 10379–10388 (2021)
3. Chen, X., Kundu, K., Zhu, Y., Ma, H., Fidler, S., Urtasun, R.: 3D object proposals using stereo imagery for accurate object class detection. IEEE Trans. Pattern Anal. Mach. Intell. **40**(5), 1259–1272 (2017)
4. Chen, Y., Tai, L., Sun, K., Li, M.: MonoPair: Monocular 3D object detection using pairwise spatial relationships. In: Proceedings of the IEEE/CVF Conference on Computer Vision and Pattern Recognition, pp. 12093–12102 (2020)
5. Chu, X., et al.: Neighbor-vote: improving monocular 3d object detection through neighbor distance voting. arXiv preprint arXiv:2107.02493 (2021)
6. Ding, M., et al.: Learning depth-guided convolutions for monocular 3D object detection. In: Proceedings of the IEEE/CVF Conference on Computer Vision and Pattern Recognition, pp. 11672–11681 (2020)
7. Ester, M., Kriegel, H.P., Sander, J., Xu, X., et al.: A density-based algorithm for discovering clusters in large spatial databases with noise. In: Kdd, vol. 96, pp. 226–231 (1996)
8. Geiger, A., Lenz, P., Urtasun, R.: Are we ready for autonomous driving? The KITTI vision benchmark suite. In: 2012 IEEE Conference on Computer Vision and Pattern Recognition, pp. 3354–3361. IEEE (2012)
9. He, K., Gkioxari, G., Dollár, P., Girshick, R.: Mask R-CNN. In: Proceedings of the IEEE International Conference on Computer Vision, pp. 2961–2969 (2017)
10. He, K., Zhang, X., Ren, S., Sun, J.: Deep residual learning for image recognition. In: Proceedings of the IEEE Conference on Computer Vision and Pattern Recognition, pp. 770–778 (2016)
11. Kumar, A., Brazil, G., Liu, X.: GrooMeD-NMS: grouped mathematically differentiable NMS for monocular 3D object detection. In: Proceedings of the IEEE/CVF Conference on Computer Vision and Pattern Recognition, pp. 8973–8983 (2021)
12. Lang, A.H., Vora, S., Caesar, H., Zhou, L., Yang, J., Beijbom, O.: PointPillars: fast encoders for object detection from point clouds. In: Proceedings of the IEEE Conference on Computer Vision and Pattern Recognition, pp. 12697–12705 (2019)
13. Li, P., Zhao, H., Liu, P., Cao, F.: RTM3D: real-time monocular 3D detection from object keypoints for autonomous driving. arXiv preprint arXiv:2001.03343 (2020)
14. Lin, T.-Y., et al.: Microsoft COCO: common objects in context. In: Fleet, D., Pajdla, T., Schiele, B., Tuytelaars, T. (eds.) ECCV 2014. LNCS, vol. 8693, pp. 740–755. Springer, Cham (2014). https://doi.org/10.1007/978-3-319-10602-1_48
15. Liu, Y., Yixuan, Y., Liu, M.: Ground-aware monocular 3D object detection for autonomous driving. IEEE Robot. Autom. Lett. **6**(2), 919–926 (2021)
16. Liu, Z., Zhou, D., Lu, F., Fang, J., Zhang, L.: Autoshape: real-time shape-aware monocular 3D object detection. In: Proceedings of the IEEE/CVF International Conference on Computer Vision, pp. 15641–15650 (2021)
17. Lu, Y., et al.: Geometry uncertainty projection network for monocular 3D object detection. In: Proceedings of the IEEE/CVF International Conference on Computer Vision, pp. 3111–3121 (2021)

18. Ma, X., Liu, S., Xia, Z., Zhang, H., Zeng, X., Ouyang, W.: Rethinking pseudo-LiDAR representation. arXiv preprint arXiv:2008.04582 (2020)
19. Ma, X., Wang, Z., Li, H., Zhang, P., Ouyang, W., Fan, X.: Accurate monocular 3D object detection via color-embedded 3D reconstruction for autonomous driving. In: Proceedings of the IEEE International Conference on Computer Vision, pp. 6851–6860 (2019)
20. Ma, X., et al.: Delving into localization errors for monocular 3D object detection. In: Proceedings of the IEEE/CVF Conference on Computer Vision and Pattern Recognition, pp. 4721–4730 (2021)
21. Manhardt, F., Kehl, W., Gaidon, A.: ROI-10D: monocular lifting of 2D detection to 6D pose and metric shape. In: Proceedings of the IEEE Conference on Computer Vision and Pattern Recognition, pp. 2069–2078 (2019)
22. Park, D., Ambrus, R., Guizilini, V., Li, J., Gaidon, A.: Is pseudo-lidar needed for monocular 3D object detection? In: Proceedings of the IEEE/CVF International Conference on Computer Vision, pp. 3142–3152 (2021)
23. Peng, L., Liu, F., Yan, S., He, X., Cai, D.: OCM3D: object-centric monocular 3D object detection. arXiv preprint arXiv:2104.06041 (2021)
24. Qi, C.R., Liu, W., Wu, C., Su, H., Guibas, L.J.: Frustum pointnets for 3D object detection from RGB-D data. In: Proceedings of the IEEE Conference on Computer Vision and Pattern Recognition, pp. 918–927 (2018)
25. Qi, C.R., Su, H., Mo, K., Guibas, L.J.: PointNet: deep learning on point sets for 3D classification and segmentation. In: Proceedings of the IEEE Conference on Computer Vision and Pattern Recognition, pp. 652–660 (2017)
26. Qi, C.R., Yi, L., Su, H., Guibas, L.J.: PointNet++: deep hierarchical feature learning on point sets in a metric space. arXiv preprint arXiv:1706.02413 (2017)
27. Qin, Z., Wang, J., Lu, Y.: MonoGRNet: a geometric reasoning network for monocular 3D object localization. In: Proceedings of the AAAI Conference on Artificial Intelligence, vol. 33, pp. 8851–8858 (2019)
28. Reading, C., Harakeh, A., Chae, J., Waslander, S.L.: Categorical depth distribution network for monocular 3D object detection. In: Proceedings of the IEEE/CVF Conference on Computer Vision and Pattern Recognition, pp. 8555–8564 (2021)
29. Shi, S., et al.: PV-RCNN: point-voxel feature set abstraction for 3D object detection. In: Proceedings of the IEEE/CVF Conference on Computer Vision and Pattern Recognition, pp. 10529–10538 (2020)
30. Shi, S., Wang, X., Li, H.: PointRCNN: 3D object proposal generation and detection from point cloud. In: Proceedings of the IEEE/CVF Conference on Computer Vision and Pattern Recognition, pp. 770–779 (2019)
31. Shi, S., Wang, Z., Shi, J., Wang, X., Li, H.: From points to parts: 3D object detection from point cloud with part-aware and part-aggregation network. arXiv preprint arXiv:1907.03670 (2019)
32. Shi, W., Rajkumar, R.: Point-GNN: graph neural network for 3D object detection in a point cloud. In: Proceedings of the IEEE/CVF Conference on Computer Vision and Pattern Recognition, pp. 1711–1719 (2020)
33. Shi, X., Ye, Q., Chen, X., Chen, C., Chen, Z., Kim, T.K.: Geometry-based distance decomposition for monocular 3D object detection. arXiv preprint arXiv:2104.03775 (2021)
34. Simonelli, A., Bulo, S.R., Porzi, L., Kontschieder, P., Ricci, E.: Are we missing confidence in pseudo-LiDAR methods for monocular 3D object detection? In: Proceedings of the IEEE/CVF International Conference on Computer Vision, pp. 3225–3233 (2021)

35. Simonelli, A., Bulo, S.R., Porzi, L., López-Antequera, M., Kontschieder, P.: Disentangling monocular 3D object detection. In: Proceedings of the IEEE International Conference on Computer Vision, pp. 1991–1999 (2019)
36. Sun, P., et al.: Scalability in perception for autonomous driving: waymo open dataset. In: Proceedings of the IEEE/CVF Conference on Computer Vision and Pattern Recognition, pp. 2446–2454 (2020)
37. Toussaint, G.T.: Solving geometric problems with the rotating calipers. In: Proceedings of IEEE Melecon, vol. 83, p. A10 (1983)
38. Wang, L., et al.: Depth-conditioned dynamic message propagation for monocular 3D object detection. In: Proceedings of the IEEE/CVF Conference on Computer Vision and Pattern Recognition, pp. 454–463 (2021)
39. Wang, L., Zhang, L., Zhu, Y., Zhang, Z., He, T., Li, M., Xue, X.: Progressive coordinate transforms for monocular 3D object detection. In: Advances in Neural Information Processing Systems, vol. 34 (2021)
40. Wang, Y., Chao, W.L., Garg, D., Hariharan, B., Campbell, M., Weinberger, K.Q.: Pseudo-LiDAR from visual depth estimation: Bridging the gap in 3D object detection for autonomous driving. In: Proceedings of the IEEE Conference on Computer Vision and Pattern Recognition, pp. 8445–8453 (2019)
41. Weng, X., Kitani, K.: Monocular 3D object detection with pseudo-LiDAR point cloud. In: Proceedings of the IEEE International Conference on Computer Vision Workshops (2019)
42. Yang, Z., Sun, Y., Liu, S., Jia, J.: 3DSSD: point-based 3D single stage object detector. In: Proceedings of the IEEE/CVF Conference on Computer Vision and Pattern Recognition, pp. 11040–11048 (2020)
43. Ye, M., Xu, S., Cao, T.: HVNet: hybrid voxel network for lidar based 3D object detection. In: Proceedings of the IEEE/CVF Conference on Computer Vision and Pattern Recognition, pp. 1631–1640 (2020)
44. Zakharov, S., Kehl, W., Bhargava, A., Gaidon, A.: Autolabeling 3D objects with differentiable rendering of SDF shape priors. In: Proceedings of the IEEE/CVF Conference on Computer Vision and Pattern Recognition, pp. 12224–12233 (2020)
45. Zhang, Y., Lu, J., Zhou, J.: Objects are different: flexible monocular 3D object detection. In: Proceedings of the IEEE/CVF Conference on Computer Vision and Pattern Recognition, pp. 3289–3298 (2021)
46. Zheng, W., Tang, W., Jiang, L., Fu, C.W.: SE-SSD: self-ensembling single-stage object detector from point cloud. In: Proceedings of the IEEE/CVF Conference on Computer Vision and Pattern Recognition, pp. 14494–14503 (2021)
47. Zhou, Y., Tuzel, O.: VoxelNet: end-to-end learning for point cloud based 3D object detection. In: Proceedings of the IEEE Conference on Computer Vision and Pattern Recognition, pp. 4490–4499 (2018)
48. Zhou, Y., He, Y., Zhu, H., Wang, C., Li, H., Jiang, Q.: Monocular 3D object detection: an extrinsic parameter free approach. In: Proceedings of the IEEE/CVF Conference on Computer Vision and Pattern Recognition, pp. 7556–7566 (2021)

Structural Causal 3D Reconstruction

Weiyang Liu[1,2]($\boxtimes$), Zhen Liu[3], Liam Paull[3], Adrian Weller[2,4],
and Bernhard Schölkopf[1]

[1] Max Planck Institute for Intelligent Systems, Tübingen, Germany
`eulerliu@gmail.com`
[2] University of Cambridge, Cambridge, UK
[3] Mila, Université de Montréal, Montreal, Canada
[4] Alan Turing Institute, London, UK

Abstract. This paper considers the problem of unsupervised 3D object reconstruction from in-the-wild single-view images. Due to ambiguity and intrinsic ill-posedness, this problem is inherently difficult to solve and therefore requires strong regularization to achieve disentanglement of different latent factors. Unlike existing works that introduce explicit regularizations into objective functions, we look into a different space for implicit regularization – the structure of latent space. Specifically, we restrict the structure of latent space to capture a topological causal ordering of latent factors (*i.e.*, representing causal dependency as a directed acyclic graph). We first show that different causal orderings matter for 3D reconstruction, and then explore several approaches to find a task-dependent causal factor ordering. Our experiments demonstrate that the latent space structure indeed serves as an implicit regularization and introduces an inductive bias beneficial for reconstruction.

1 Introduction

Understanding the 3D structures of objects from their 2D views has been a longstanding and fundamental problem in computer vision. Due to the lack of high-quality 3D data, unsupervised single-view 3D reconstruction is typically favorable; however, it is an ill-posed problem by nature, and it typically requires a number of carefully-designed priors and regularizations to achieve good disentanglement of latent factors [6,7,15,31,33,61,75]. Distinct from these existing works that focus on introducing explicit regularizations, we aim to explore how the structure of latent space can implicitly regularize 3D reconstruction, and to answer the following question: *Can a suitable structure of latent space encode helpful implicit regularization and yield better inductive bias?*

Current single-view 3D reconstruction methods [33,39,75] typically decompose 3D objects into several semantic latent factors such as 3D shape, texture, lighting and viewpoint. These latent factors are independently extracted from single 2D images and then fed into a differentiable renderer to reconstruct the

---

Supplementary Information The online version contains supplementary material available at https://doi.org/10.1007/978-3-031-19769-7_9.

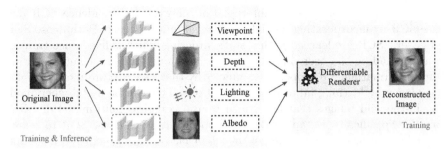

(a) Current standard unsupervised 3D reconsturction pipeline

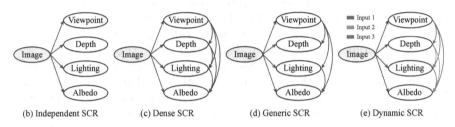

(b) Independent SCR   (c) Dense SCR   (d) Generic SCR   (e) Dynamic SCR

Fig. 1. (a) Overview of the standard 3D reconstruction pipeline. (b) Graphical model of independent SCR, which is adopted in the standard pipeline and assumes full conditional independence. (c) Graphical model of a dense SCR example. This makes no assumption on the distribution. (d) Graphical model of a generic SCR example. This assumes partial conditional independence. (e) Graphical model of a dynamic SCR example. This yields strong flexibility. All directed edges are learned in practice.

original 2D images, as illustrated in Fig. 1(a). Conditioned on the input image, these latent factors are typically assumed to be independent from each other. Such an assumption for disentanglement can be too strong and sometimes unrealistic, because it suggests that the estimated viewpoint will not affect the estimation of lighting in the image, which contradicts the formation of realistic images. This observation motivates us to explore how the dependency structure of latent factors implicitly regularizes the encoder and improves disentanglement.

Taking inspiration from structural causal models [53], we propose the **Structural Causal Reconstruction (SCR)** framework which introduces structural priors to the latent space. We consider the causal ordering of latent factors and study how different causal orderings can introduce different inductive biases.

Depending on the type of causal orderings and the corresponding flexibility, we derive three SCR variants: dense SCR which learns a chain factorization without any embedded conditional independence, generic SCR which learns a directed acyclic graph (DAG) over the latent factors, and dynamic SCR which learns a dynamic DAG that is dependent on the input image. We note that the standard 3D reconstruction pipeline can be viewed as independent SCR as shown in Fig. 1(b) (*i.e.*, viewpoint, depth, lighting and albedo are conditionally independent from each other given the input image), while dense SCR does not assume any conditional independence. Generic SCR learns a DAG over the latent factors and serves as an interpolation between independence SCR and dense SCR by incorporating partial conditional independence. Both dense SCR and generic SCR are learned with a static ordering which is fixed once trained. To accommodate the over-simplified rendering model and the complex nature of image formation, we propose dynamic SCR that can capture more complex dependency by learning input-dependent DAGs. This can be useful when modeling in-the-wild images that are drawn from a complex multi-modal distribution [47]. Specifically, we apply Bayesian optimization to dense SCR to search for the best dense causal ordering of the latent factors. For generic SCR, we first propose to directly learn a DAG with an additional regularization. Besides that, we further propose a two-phase algorithm: first running dense SCR to obtain a dense ordering and then learning the edges via masking. For dynamic SCR, we propose a self-attention approach to learn input-dependent DAGs.

From a distribution perspective, independent SCR (Fig. 1(b)) is the least expressive graphical model in the sense that it imposes strong conditional independence constraints and therefore limits potential distributions that can factorize over it. On the contrary, any conditional distribution $P(V, D, L, A|I)$ (where V, D, L, A, I denote viewpoint, depth, lighting, albedo and image, respectively) can factorize over dense SCR, making it the most expressive variant for representing distributions. Generic SCR unifies both independent SCR and dense SCR by incorporating a flexible amount of conditional independence constraints. Dynamic SCR is able to capture even more complex conditional distribution that is dynamically changing for different input images.

Intuition for why learning a latent dependency structure helps 3D reconstruction comes from the underlying entanglement among estimated viewpoint, depth, lighting and albedo. For example, conditioned on a given 2D image, a complete disentanglement between viewpoint and lighting indicates that changing the estimated viewpoint of an object will not change its estimated lighting. This makes little sense, since changing the viewpoint will inevitably affect the estimation of lighting. In contrast to existing pipelines that extract viewpoint and depth independently from the image (Fig. 2(a)), the information of viewpoint may give constraints on lighting (*e.g.*, modeled as a directed edge from V to L in Fig. 2(b)). Therefore, instead of ignoring the

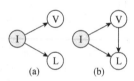

Fig. 2. (a) Viewpoint and lighting are extracted independently from the input image. (b) The extracted viewpoint gives constraints on lighting. These arrows denote encoding latent variables from the image (*i.e.*, anti-causal direction).

natural coupling among latent factors and assuming conditional independence, we argue that learning a suitable dependency structure for latent factors is crucial for intrinsic disentanglement. In general, modeling the latent dependency and causality among viewpoint, depth, lighting and albedo renders an implicit regularization for disentanglement, leading to strong generalizability. Beside the intuition from the anti-causal direction, Sect. 3.3 gives another interpretation for SCR from the causal direction. Our contributions are:

- We explicitly model the causal structure among the latent factors.
- To learn a causal ordering, we propose three SCR variants including dense SCR, generic SCR and dynamic SCR. Each one yields a different level of distribution expressiveness and modeling flexibility.
- We constrain the latent space structure to be a topological causal ordering (which can represent arbitrary DAGs), reducing the difficulty of learning.
- Our method is in parallel to most current 3D reconstruction pipelines and can be used simultaneously with different pipelines such as [15,39,62,75].
- Our empirical results show that different causal orderings of latent factors lead to significantly different 3D reconstruction performance.

2   Related Work

Multi-view 3D Reconstruction. This method usually requires multi-view images of the same target object. Classical techniques such as Structure from Motion [49] and Simultaneous Localization and Mapping [18] rely on hand-crafted geometric features and matching across different views. Owing to the availability of large 3D object datasets, modern approaches [9,32,77] can perform multi-view 3D reconstruction with neural networks that map 2D images to 3D volumes.

Shape from X. There are many alternative monocular cues that can be used for reconstructing shapes from images, such as shading [28,82], silhouettes [36], texture [74] and symmetry [16,46]. These methods are generally not applicable to in-the-wild images due to their strong assumptions. Shape-from-symmetry [16, 46,60,63] assumes the symmetry of the target object, making use of the original image and its horizontally flipped version as a stereo pair for 3D reconstruction. [60] demonstrates the possibility to detect symmetries and correspondences using descriptors. Shape-from-shading assumes a specific shading model (*e.g.*, Phong shading [54] and spherical harmonic lighting [22]), and solves an inverse rendering problem to decompose different intrinsic factors from 2D images.

Single-View 3D Reconstruction. This line of research [9,12,13,21,23,25, 29,39,64,73,75,78,86] aims to reconstruct a 3D shape from a single-view image. [50,68,71] use images and their corresponding ground truth 3D meshes as supervisory signals. This, however, requires either annotation efforts [76] or synthetic construction [5]. To avoid 3D supervision, [7,33,34,41] consider an analysis-by-synthesis approach with differentiable rendering, but they still require either multi-view images or known camera poses. To further reduce supervision, [31] learns category-specific 3D template shapes from an annotated image collection, but annotated 2D keypoints are still necessary in order to infer camera pose correctly. [26] also studies a similar category-specific 3D reconstruction from a single image. [39] estimates 3D mesh, texture and camera pose of both rigid and non-rigid objects from a single-view image using silhouette as supervision. Videos [1,48,66,72,85] are also leveraged as a form of supervision for single-view 3D reconstruction. For human bodies and faces, [4,8,14,15,19,20,30,67,72,80] reconstruct 3D shapes from single-view images with a predefined shape model such as SMPL [45], FLAME [38] or BFM [52]. Among many works in single-view 3D reconstruction, we are particularly interested in a simple and generic unsupervised framework from [75] that utilizes the symmetric object prior. This framework adopts the Shape-from-shading pipeline to extract intrinsic factors of images, including 3D shape, texture, viewpoint and illumination parameters (as shown in Fig. 1(a)). The encoders are trained to minimize the reconstruction error between the input image and the rendered image. It shows impressive results in reconstructing human faces, cat faces and synthetic cars.

For the sake of simplicity, we build the SCR pipeline based on the framework of [75] and focus on studying how the causal structure of latent factors affects the 3D reconstruction performance. We emphasize that our method is a parallel contribution to [75] and is generally applicable to any 3D reconstruction framework without the need of significant modifications.

3   Causal Ordering of Latent Factors Matters

The very first question we need to address is *"Does the causal ordering of latent factors matter for unsupervised 3D reconstruction?"*. Without an affirmative answer, it will be pointless to study how to learn a good causal ordering.

3.1   A Motivating Example from Function Approximation

We start with a motivating example to show the advantages of modeling the dependency between latent factors. We take a look at the example in Fig. 3 where the lighting factor L can be represented using either $f_L(I)$ in Fig. 3(a) or $f_L(I) + h_L(V)$ in Fig. 3(b). There are a few perspectives to compare these two representations and see their difference (also see Appendix C):

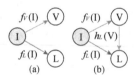

Fig. 3.   Two structures encode the lighting factor.

- We first assume the underlying data generating function for lighting is given by $L := f_L^*(I) + h_L^*(V)$ where f_L^* and h_L^* are two polynomial functions of order p. Because $V := f_V^*(I)$ where f_V^* is also a polynomial function of order p, we can then write the lighting function as $L := f_L^*(I) + h_L^* \circ f_V^*(I)$ which is a polynomial function order $2p$. The lighting function can be learned with either $L = f_L(I)$ in Fig. 3(a) or $L = f_L(I) + h_L \circ f_V(I)$ in Fig. 3(b). The previous requires the encoder $f_L(I)$ to learn a polynomial of order $2p$, while the latter requires learning that of only order p.
- From the perspective of function approximation, it is obvious that $f_L(I) + h_L \circ f_V(I)$ is always more expressive than $f_L(I)$ given that f_L, h_L, f_V are of the same representation capacity. Therefore, the structure shown in Fig. 3(b) is able to capture more complex and nonlinear lighting function.
- Making the lighting L partially dependent on the viewpoint V gives the lighting function an inherent structural prior, which may implicitly regularizes the function class and constrain its inductive bias.

3.2   Expressiveness of Representing Conditional Distributions

The flexibility of SCR can also be interpreted from a distribution perspective. Most existing 3D reconstruction pipelines can be viewed as independent SCR whose conditional distribution $P(V, D, L, A|I)$ can be factorized into

$$P(V, D, L, A|I) = P(V|I) \cdot P(D|I) \cdot P(L|I) \cdot P(A|I) \tag{1}$$

which renders the conditional independence among V, D, L, A. This is in fact a strong assumption that largely constrains the potential family of distributions that can factorize over this model, making this model less expressive in representing conditional distributions. In contrast, dense SCR does not assume any conditional independence because it yields the following factorization (this is just one of the potential orderings and we randomly choose one for demonstration):

$$P(V, D, L, A|I) = P(V|I) \cdot P(D|I, V) \cdot P(L|I, V, D) \cdot P(A|I, V, D, L) \tag{2}$$

which imposes no constraints to the factorized conditional distribution and is more expressive. Therefore, any dense ordering has this nice property of assuming no conditional independence among latent factors. However, there exists a trade-off between expressiveness and learnability. A more expressive model usually requires more data to train and is relatively sample-inefficient. Generic SCR is proposed in search of a sweet spot between expressiveness and learnability by incorporating partial conditional independence. Taking Fig. 1(c) as an example, we can observe that this model assumes $P(D \perp L|I)$ and $P(D \perp A|I)$. Going beyond generic SCR, dynamic SCR aims to tackle with the scenario where the conditional distribution $P(V, D, L, A|I)$ is dynamically changing rather than being static for all the images. This can greatly enhance the modeling flexibility.

3.3   Modeling Causality in Rendering-Based Decoding

The previous subsection shows that there is no difference for different dense orderings in representing $P(V, D, L, A|I)$. This conclusion is drawn from the perspective of modeling correlation. However, one of the most significant properties of topological ordering is its ability to model acyclic causality. In terms of causal relationships, different orderings (including both dense and generic ones) make a difference. The standard 3D reconstruction pipeline is naturally an autoencoder architecture, where the encoder and decoder can be interpreted as anti-causal and causal

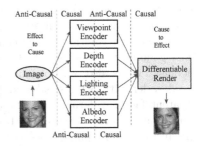

Fig. 4. Three possible partitions of anti-causal and causal mappings. (Color figure online)

mappings, respectively [3,35,37,55,56,70]. Here, the causal part is a generative mapping, and the anti-causal part is in the opposite direction, inferring causes from effects. However, how to determine which part of the pipeline should be viewed as anti-causal or causal remains unclear. Here we discuss three possible partitions of anti-causal and causal mappings, as shown in Fig. 4. The partition denoted by green dashed line uses an identity mapping as the anti-causal direction and the rest of the pipeline performs causal reconstruction. This partition does not explicitly model the causes and may not be useful. For the partition labeled by the blue dashed line, all the encoders are viewed as anti-causal, so the latent factor ordering is also part of anti-causal learning and does not necessarily benefit from the underlying causal ordering (*i.e.*, causal DAG [65], cf. [37]). The partition denoted by the red dashed views part of the encoder as anti-causal learning and the rest of the encoder along with the renderer as causal learning. This partition is particularly interesting because it puts the latent factor ordering to the causal direction and effectively connects latent factor ordering to the underlying causal

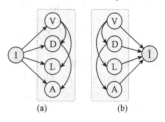

Fig. 5. Latent structure modeling from (a) anti-causal direction and (b) causal direction. Gray regions denote where the causal ordering is learned.

ordering. Our SCR framework (in Sect. 4.1) is designed based on such insight. When the underlying causal ordering is available, using it as the default ordering could be beneficial. Although the causal ordering could improve strong generalization [35], learning the causal ordering without additional knowledge (*e.g.*, interventions or manipulations such as randomized experiment) is difficult and out of our scope. *We hypothesize that the underlying causal ordering leads to fast, generalizable and disentangled 3D reconstruction, and learning causal ordering based on these criteria may help us identify crucial causal relations.* As an encouraging signal, one of the best-performing dense ordering (DAVL) well matches the conventional rendering procedures in OpenGL, which is likely to be similar to the underlying causal ordering.

In the previous examples of Fig. 2 and Fig. 3, we justify the necessity of the topological ordering from the factor estimation (*i.e.*, anti-causal) perspective. As discussed above, we can alternatively incorporate the causal ordering to the causal mapping and model the causality among latent factors in the decoding (*i.e.*, generative) process, which well matches the design of structural causal models. This is also conceptually similar to [58,79] except that SCR augments the decoder with a physics-based renderer. Figure 5 shows two interpretations of latent factor ordering from the causal and anti-causal directions. While the causal mapping encourages SCR to approximate the underlying causal ordering, the anti-causal mapping does not necessarily do so. The final learned causal ordering may be the result of a trade-off between causal and anti-causal mapping.

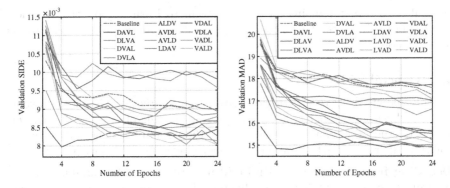

Fig. 6. The scale-invariant depth error (left) and mean angle deviation (right) on the BFM dataset [52] for different dense causal orderings. For visualization clarity, we plot the SIDE of the best three orderings, the worst three orderings and random six orderings. For MAD, we plot the same selection of orderings along with the best three and worst three orderings. We denote depth, albedo, lighting and viewpoint as D, A, L and V, respectively. For the full results, please refer to Appendix B.

3.4 Empirical Evidence on 3D Reconstruction

Most importantly, we demonstrate the empirical performance of different dense causal orderings for unsupervised 3D reconstruction. The details of our pipeline and the experimental settings are given in Sect. 4.1 and Appendix A, respectively. Here we focus on comparing different dense orderings. As can be observed from Fig. 6, different settings for dense SCR yield significantly different empirical behaviors, validating our claim that topological causal ordering of latent factors matters in unsupervised 3D reconstruction. Moreover, we discover that most of the dense orderings perform consistently for both SIDE and MAD metrics. For example, depth-albedo-viewpoint-lighting, depth-viewpoint-albedo-lighting and depth-viewpoint-lighting-albedo perform consistently better than the other dense orderings and the baseline (*i.e.*, independent SCR). This again matches our intuition in Sect. 3.3 that different dense ordering indicates different causality and leads to different disentanglement/reconstruction performance despite being equivalent in representing the conditional distribution $P(\boldsymbol{V}, \boldsymbol{D}, \boldsymbol{L}, \boldsymbol{A} | \boldsymbol{I})$.

Interestingly, the well-performing dense orderings also seem to match our knowledge about the underlying causal ordering. For example, we also tend to put viewpoint in front of lighting, because the viewpoint will cause the change of lighting effects on the object. Almost all the well-performing dense orderings have this pattern, suggesting that the well-performing orderings tend to match the intrinsic causality that is typically hard to obtain in practice.

4 Learning Causal Ordering for 3D Reconstruction

We introduce a generic framework to learn causal ordering. Our proposed pipeline and algorithms to learn different variants of SCR are by no means optimal ones and it remains an open problem to learn a good causal order-ing. We instead aim to show that a suitable causal ordering is beneficial to 3D reconstruction.

4.1 General SCR Framework

Our unsupervised 3D reconstruction pipeline is inspired by [75] but with some novel modifications to better accommodate the learning of causal ordering. Our goal is to study how causal ordering affects the disentanglement and generaliz-ability in 3D reconstruction rather than achieving state-of-the-art performance.

Decoding from a Common Embedding Space. A differentiable renderer typically takes in latent factors of different dimensions, making it less conve-nient to incorporate causal factor ordering. In order to easily combine multiple latent factors, we propose a learnable decoding method that includes additional neural networks ($f_V^2, f_D^2, f_L^2, f_A^2$ shown in Fig. 7) to the differentiable renderer. These neural networks transform the latent factors from a common d-dimensional embedding space ($\boldsymbol{u}_V, \boldsymbol{u}_D, \boldsymbol{u}_L, \boldsymbol{u}_A$) to their individual dimensions ($\boldsymbol{V}, \boldsymbol{D}, \boldsymbol{L}, \boldsymbol{A}$) such that the differentiable renderer can directly use them as inputs.

Implementing SCR in a Common Embedding Space. Since all the latent factors can be represented in a common embedding space of the same dimension, we now introduce how to implement SCR in this pipeline. We start by listing a few key desiderata: (1) all variants of SCR should have (roughly) the same number of trainable parameters as independent SCR (baseline) such that the comparison is meaningful; (2) learning SCR should be efficient, differentiable and end-to-end; (3) different structures among latent factors can be explored in a unified framework by imposing different constraints on the adjacency matrix.

We first interpret conditional probability in terms of neural networks. For example, $P(\boldsymbol{V}|\boldsymbol{I}, \boldsymbol{D})$ can be implemented as a single neural network $\boldsymbol{V} = f_V(\boldsymbol{I}, \boldsymbol{D})$ that takes both image $\boldsymbol{I}$ and depth $\boldsymbol{D}$ as input. Instead of parameter-izing the encoder f_V with one neural network, we separate f_V into two neural networks f_V^1, f_V^2 – the first one f_V^1 aims to map different factors into a common embedding space of the same dimension, and the second one f_V^2 transforms the embedding to the final factor that can be used directly for the differentiable ren-derer. Taking $P(\boldsymbol{V}|\boldsymbol{I}, \boldsymbol{D})$ as an example, we model it using $\boldsymbol{V} = f_V^2(f_V^1(\boldsymbol{I}), \boldsymbol{u}_D)$. We define the SCR adjacency matrix that characterizes the dependency struc-ture among latent factors as $\boldsymbol{M} = [\boldsymbol{M}_V, \boldsymbol{M}_D, \boldsymbol{M}_L, \boldsymbol{M}_A] \in \mathbb{R}^{4 \times 4}$ where $\boldsymbol{M}_V = [M_{VV}, M_{VD}, M_{VL}, M_{VA}]^\top \in \mathbb{R}^{4 \times 1}$ and M_{VD} denotes the weight of the directed edge from $\boldsymbol{V}$ to $\boldsymbol{D}$ (the weight can be constrained to be either binary or contin-uous). Because causal ordering is equivalent to DAG, $\boldsymbol{M}$ can be permuted into a strictly upper triangular matrix. Generally, latent factors are modeled by

$$\begin{aligned} \boldsymbol{V} = f_V^2\left(f_V^1\left(\boldsymbol{I}\right), \boldsymbol{M}_V^\top \boldsymbol{u}\right) \qquad \boldsymbol{D} = f_D^2\left(f_D^1\left(\boldsymbol{I}\right), \boldsymbol{M}_D^\top \boldsymbol{u}\right) \\ \boldsymbol{L} = f_L^2\left(f_L^1\left(\boldsymbol{I}\right), \boldsymbol{M}_L^\top \boldsymbol{u}\right) \qquad \boldsymbol{A} = f_A^2\left(f_A^1\left(\boldsymbol{I}\right), \boldsymbol{M}_A^\top \boldsymbol{u}\right) \end{aligned} \tag{3}$$

where $\boldsymbol{u} = [\boldsymbol{u}_V; \boldsymbol{u}_D; \boldsymbol{u}_L; \boldsymbol{u}_A] \in \mathbb{R}^{4 \times d}$. The input to $f_V^2, f_D^2, f_L^2, f_A^2$ can either be added element-wisely or concatenated, and we use element-wise addition in order

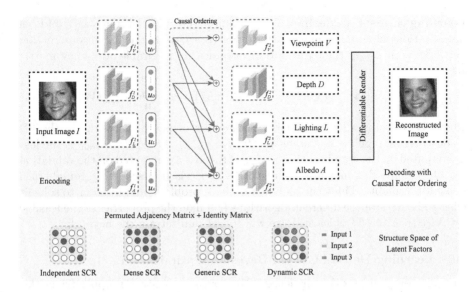

Fig. 7. Our unsupervised 3D reconstruction pipeline to explore causal ordering. The causal edges in the figure are for illustration. Actual edges are learned in practice. (Color figure online)

not to introduce additional parameters. M exactly implements causal ordering as an equivalent form of causal DAG. More generally, M characterizes the latent space structure and can also be constrained to be some other family of structures.

Interpreting SCR as a Part of Causal Mapping. After modeling the latent factors with two separate neural networks, we can view $f_V^1, f_D^1, f_L^1, f_A^1$ as the encoding process (*i.e.*, the light blue region in Fig. 7). Different from [75], we view the causal ordering, $f_V^2, f_D^2, f_L^2, f_A^2$ and differentiable renderer as the decoding process (*i.e.*, the light yellow region in Fig. 7). This can be understood as an augmented trainable physics-based renderer which performs rendering with additional neural networks and a causal ordering. More importantly, incorporating causal ordering to the decoding process makes it a part of causal mapping, which may produce more interpretable ordering due to its intrinsic connection to the underlying causality. Therefore, our novel pipeline design makes it possible to benefit from (or even estimate) the underlying causal ordering.

Loss Functions. To avoid introducing additional priors to SCR and better study the effect of causal ordering, we stick to the same loss functions as [75]. The loss function is defined as $\mathcal{L} = \mathcal{L}_{\mathrm{rec}}(\hat{I}, I) + \lambda_f \mathcal{L}_{\mathrm{rec}}(\hat{I}', I) + \lambda_p \mathcal{L}_{\mathrm{p}}(\hat{I}, I)$ where $\mathcal{L}_{\mathrm{rec}}$ is the reconstruction loss and $\mathcal{L}_{\mathrm{p}}$ is the perceptual loss. λ_f, λ_p are hyperparameters $\hat{I}$ is the reconstructed image with original depth and albedo. $\hat{I}'$ is the reconstructed image with flipped depth and albedo. Similar to [75], we also use the confidence map to compensate asymmetry. Appendix A provides the detailed formulation.

Learning Causal Ordering. We formulate the SCR learning as a bi-level optimization where the inner optimization is to train the 3D reconstruction networks with $\mathcal{L}$ and the outer optimization learns a suitable adjacency matrix M:

$$\min_{M \in \mathcal{M}_{\text{DAG}}} \mathcal{L}_{\text{val}}(W^*(M), M) \qquad \text{s.t. } W^*(M) = \arg\min_W \mathcal{L}_{\text{train}}(W, M) \qquad (4)$$

where W denotes all the trainable parameters of neural networks in the 3D reconstruction pipeline, including $f_V^1, f_V^2, f_D^1, f_D^2, f_L^1, f_L^2, f_A^1, f_A^2$. $\mathcal{L}_{\text{train}}$ is the loss $\mathcal{L}$ computed on the training set, and $\mathcal{L}_{\text{val}}$ is the loss $\mathcal{L}$ computed on the validation set. Optionally, $\mathcal{L}_{\text{val}}$ may also include other supervised losses (*e.g.*, ground truth depth) if available. This is in general a difficult problem, and in order to solve it effectively, we propose different algorithms based on the properties of the feasible set $\mathcal{M}_{\text{DAG}}$. After M is learned, we will fix M and retrain the network.

4.2   Learning Dense SCR via Bayesian Optimization

The adjacency matrix M for dense SCR is an all-one strictly upper triangular matrix after proper permutation. Therefore, we are essentially learning the ordering permutation which is a discrete and non-differentiable structure. We resort to Bayesian optimization (BO) [17] that is designed for gradient-free and "expensive to evaludate" optimization. Specifically, BO first places a Gaussian process prior on $\mathcal{L}_{\text{val}}(W^*(M), M)$ in Eq. (4) and collect all the evaluated points on M. Then BO updates posterior probability distribution on $\mathcal{L}_{\text{val}}$ using all available data and evaluates $\mathcal{L}_{\text{val}}$ on the maximizer point of the acquisition function which is computed with the current posterior distribution. Note that, evaluation on $\mathcal{L}_{\text{val}}$ requires computing $W^*(M)$. Finally, BO outputs the latest evaluated M. We use the position permutation kernel $K(\pi_1, \pi_2|\lambda) = \exp(-\lambda \cdot \sum_i |\pi_1^{-1}(i) - \pi_2^{-1}(i)|)$ where π is a permutation mapping that maps the original index to the permuted index. We use the expected improvement as the acquisition function. We note a special advantage of BO over gradient-dependent methods: the validation metric can be obtained from user study, which is often more reliable and flexible.

4.3   Learning Generic SCR via Optimization Unrolling

To solve the bi-level optimization in Eq. (4), we can unroll the inner optimization with a few gradient updates and replace $W^*(M)$ with $W - \eta \nabla_W \mathcal{L}_{\text{train}}(W, M)$. Then the optimization becomes $\min_{M \in \mathcal{M}_{\text{DAG}}} \mathcal{L}_{\text{val}}(W - \eta \nabla_W \mathcal{L}_{\text{train}}(W, M), M)$. Here we unroll 1-step gradient update as an example, but we can also unroll multiple steps for better performance in practice. In order to constrain the adjacency matrix M to be a DAG, we can turn the feasible set $M \in \mathcal{M}_{\text{DAG}}$ into a constraint [84]: $\mathcal{H}(M) = \text{tr}((I_n + \frac{c}{n} M \circ M)^n) - n = 0$ where I_n is an identity matrix of size n, c is some arbitrary positive number, n is the number of latent factors (here $n = 4$) and $\circ$ denotes the element-wise multiplication. Using Lagrangian multiplier method, we end up with the following optimization:

$$\min_M \mathcal{L}_{\text{val}}(W - \eta \nabla_W \mathcal{L}_{\text{train}}(W, M), M) + \lambda_{\text{DAG}} \mathcal{H}(M) \qquad (5)$$

where λ_{DAG} is a hyperparameter. Alternatively, we may use the augmented Lagrangian method for stronger regularization [81,84]. Although Eq. (5) is easy to optimize, it is still difficult to guarantee the learned M to be a strict DAG and the search space may also be too large. To address this, we further propose a different approach to learn M. The basic idea is to learn generic SCR based on the solution from dense SCR. We simply need to relearn/remove some edges for the given dense ordering. The final optimization is given by

$$\min_{M} \mathcal{L}_{\text{val}}(W - \eta \nabla_W \mathcal{L}_{\text{train}}(W, M \circ M^*_{\text{dense}}), M \circ M^*_{\text{dense}}) \qquad (6)$$

where M^*_{dense} is obtained from BO for dense SCR. It is a binary matrix that can be permuted to be strictly upper triangular. If we also constrain M to be binary, we will use a preset threshold to binarize the obtained M before retraining.

4.4 Learning Dynamic SCR via Masked Self-attention

In order to make the adjacency matrix M be adaptively dependent on the input, we need to turn M into the output of a function that takes the image I as input, *i.e.*, $M = \Phi(I)$. One sensible choice is to parameterize $\Phi(\cdot)$ with an additional neural network, but it will inevitably introduce significantly more parameters and increase the capacity of the framework, making it unfair to compare with the other variants. Therefore, we take a different route by utilizing self-attention to design $\Phi(\cdot)$. Specifically, we use $M = \Phi(I) = q(u) \circ M^*_{\text{dense}}$ where u is the matrix containing all the factor embeddings ($u = [u_V; u_D; u_L; u_A]$), $q(u)$ can be either the Sigmoid activation $\sigma(uu^\top)$ or cosine cross-similarity matrix among u_V, u_D, u_L, u_A (*i.e.*, $q(u)_{i,j} = \frac{\langle u_i, u_j \rangle}{\|u_i\| \|u_j\|}$, $i, j \in \{V, D, L, A\}$), and M^*_{dense} is the solution obtained from BO for dense SCR. M^*_{dense} essentially serves as a mask for the self-attention such that the resulting causal ordering is guaranteed to be a DAG. Since there is no fixed M, the entire pipeline is trained with the final objective function: $\min_W \mathcal{L}_{\text{train}}(W, q(u) \circ M^*_{\text{dense}})$ in an end-to-end fashion. We note that the function $q(u)$ has no additional parameters and meanwhile makes the causal ordering (*i.e.*, $\Phi(I)$) dynamically dependent on the input image I.

4.5 Insights and Discussion

Connection to Neural Architecture Search. We discover an intriguing connection between SCR and neural architecture search (NAS) [11,40,87]. SCR can be viewed as a special case of NAS that operates on a semantically interpretable space (*i.e.*, the dependency structure among latent factors), while standard NAS does not necessarily produce an interpretable architecture. SCR performs like a top-down NAS where a specific neural structure is derived from semantic dependency/causality and largely constrains the search space for neural networks without suffering from countless poor local minima like NAS does.

Semantic Decoupling in Common Embeddings. In order to make SCR interpretable, we require the latent embeddings u_V, u_D, u_L, u_A to be semantically decoupled. For example, u_V should contain sufficient information to decode

V. The semantic decoupling in the common embedding space can indeed be pre-served. First, the DAG constraint can naturally encourage semantic decoupling. We take an arbitrary dense ordering (*e.g.*, DAVL) as an example. u_D is the only input for f_D^2, so it contain sufficient information for D. u_D, A are the inputs for f_A^2, so the information of A will be largely encoded in u_A (u_D already encodes the information of D). The same reasoning applies to V and L. Note that, a generic DAG will have less decoupling than dense ordering due to less number of directed edges. Second, we enforce the encoders $f_V^2, f_D^2, f_L^2, f_A^2$ to be relatively simple functions (*e.g.*, shallow neural networks), such that they are unable to encode too much additional information and mostly serve as dimen-sionality transformation. They could also be constrained to be invertible. Both mechanisms ensure the semantic decoupling in the common embedding space.

5   Experiments and Results

Datasets. We evaluate our method on two human face datasets (CelebA [44] and BFM [52]), one cat face dataset that combines [83] and [51] (cropped by [75]) and one car dataset [75] rendered from ShapeNet [5] with random viewpoints and illumination. These images are split 8:1:1 into training, validation and testing.

Metrics. For fairness, we use the same metrics as [75]. The first one is Scale Invariant Depth Error (SIDE) [10] which computes the standard deviation of the difference between the estimated depth map at the input view and the ground truth depth map at the log scale. We note that this metric may not reflect the true reconstruction quality. As long as this metric is reasonably low, it may no longer be a stronger indicator for reconstruction quality, which is also verified by [27]. To make a comprehensive evaluation, we also use another metric: the mean angle deviation (MAD) [75] between normals computed from ground truth depth from the predicted depth. It measures how well the surface is reconstructed.

Implementation. For the network architecture, we fol-low [75] and only make essential changes to its setup such that the comparison is meaningful. For the detailed implementation and exper-imental settings, refer to Appendix A.

Table 1. Depth reconstruction results on BFM.

Method	SIDE ($\times 10^{-2}$) ↓	MAD (deg.) ↓
Supervised	**0.410** $\pm$ 0.103	**10.78** $\pm$ 1.01
Constant Null Depth	2.723 $\pm$ 0.371	43.34 $\pm$ 2.25
Average GT Depth	1.990 $\pm$ 0.556	23.26 $\pm$ 2.85
Wu et al. [75] (reported)	**0.793** $\pm$ 0.140	16.51 $\pm$ 1.56
Ho et al. [27] (reported)	0.834 $\pm$ 0.169	**15.49** $\pm$ 1.50
Wu et al. [75] (our run)	0.901 $\pm$ 0.190	17.53 $\pm$ 1.84
Independent SCR	0.895 $\pm$ 0.183	17.36 $\pm$ 1.78
Dense SCR (random)	1.000 $\pm$ 0.275	17.66 $\pm$ 2.09
Dense SCR (BO)	**0.830** $\pm$ 0.205	14.88 $\pm$ 1.94
Generic SCR (Eq. 5)	0.859 $\pm$ 0.215	15.17 $\pm$ 1.92
Generic SCR (Eq. 6)	**0.820** $\pm$ 0.190	14.79 $\pm$ 1.96
Dynamic SCR (Sigmoid)	0.827 $\pm$ 0.220	14.86 $\pm$ 2.02
Dynamic SCR (Cosine)	**0.815** $\pm$ 0.232	**14.80** $\pm$ 1.95

5.1   Quantitative Results

Geometry Reconstruc-tion. We train and test all the methods on the BFM dataset to evaluate the depth reconstruction quality.

The results are given in Table 1. We compare different variants of SCR with our own baseline (*i.e.*, independent SCR), two state-of-the-art methods [27,75], supervised learning upper bound, constant null depth and average ground truth depth. We note that there is a performance difference between our re-run version and the original version of [75]. This is because all our experiments are run under CUDA-10 while the original version of [75] is trained on CUDA-9. We also re-train our models on CUDA-9 and observe a similar performance boost (see Appendix D). We suspect this is because of the rendering precision on different CUDA versions. However, this will not affect the advantages of our method and our experiment settings are the same for all the other compared methods. More importantly, we build our SCR on a baseline that performs similarly to [75] (independent SCR vs. our version of [75]). SCR improves our baseline for more than 0.0065 on SIDE and 2.5 degree on MAD. Specifically, our dense SCR learns an ordering of depth-viewpoint-albedo-lighting. Generic SCR (Eq. 6) and both dynamic SCR variants are built upon this ordering. We notice that if we use a random dense ordering, then the 3D reconstruction results are even worse than our baseline, which shows that dense SCR can indeed learn crucial structures. Such a significant performance gain shows that a suitable SCR can implicitly regularize the neural networks and thus benefit the 3D reconstruction.

Effect of Dense Ordering. We also perform ablation study to see how generic SCR and dynamic SCR perform if they are fed with different dense orderings. Table 2 compares three

Table 2. MAD (degree) results on BFM.

Method	DLVA	DAVL	DVAL
Dense SCR (fixed)	15.02 ± 2.00	15.14 ± 1.91	14.88 ± 1.94
G-SCR (Eq. 6)	**14.96** ± 1.90	**14.85** ± 2.13	**14.79** ± 1.96
Dy-SCR (Sigmoid)	15.01 ± 1.99	15.03 ± 2.12	14.86 ± 2.02
Dy-SCR (Cosine)	14.99 ± 1.93	15.05 ± 2.15	14.80 ± 1.95

different dense orderings: depth-light-viewpoint-albedo (DLVA), depth-albedo-viewpoint-lighting (DAVL) and depth-viewpoint-albedo-lighting (DVAL). We show that G-SCR and D-SCR can consistently improve the 3D reconstruction results even if different dense orderings are given as the mask.

Convergence. Figure 8 plots the convergence curves of both SIDE and MAD in Fig. 8. We observe that dense SCR, generic SCR and dynamic SCR converge much faster than the baseline. Dense SCR achieves impressive performance at the very begin-

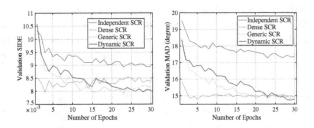

Fig. 8. Convergence curves of validation SIDE & MAD.

ning of the training. When converged, dynamic SCR and generic SCR performs better than dense SCR due to its modeling flexibility.

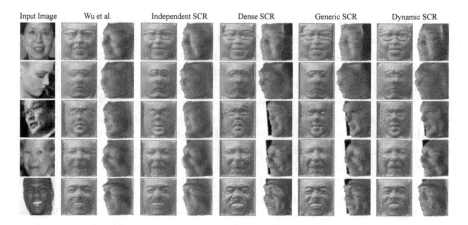

Fig. 9. One textureless view and canonical normal map on CelebA. All the methods (including Wu et al. [75]) are trained on CelebA under the same experimental settings.

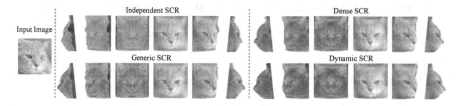

Fig. 10. Textured and textureless shapes from multiple views on cat faces.

5.2   Qualitative Results

CelebA. We show the reconstruction results for a few challenging in-the-wild face images (*e.g.*, extreme poses and expressions) in Fig. 9. We train dense SCR with BO and the other SCR variants are trained based on the best learned dense ordering. Our SCR variants including dense SCR, generic SCR and dynamic SCR are able to reconstruct fine-grained geometric details and recover more realistic shapes than both [75] and our independent baseline. This well verifies the importance of implicit regularization from latent space structure.

Cat Faces. We also train all the SCR variants on cat faces. Results in Fig. 10 show that dynamic SCR yields the best 3D reconstruction quality, while dense SCR and generic SCR can also recover reasonably good geometric details.

Cars. We train all the methods on the synthetic car dataset under the same settings, and then evaluate these

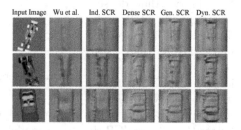

Fig. 11. Canonical normal maps on cars.

methods on car images with abundant geometric details. Figure 11 shows that our SCR variants recovers very fine-grained geometric details and produce highly realistic normal maps which are significantly better than both [75] and independent SCR.

Conclusion. We demonstrate the potential of causality in regularizing 3D reconstruction. For acknowledgements, see https://wyliu.com/papers/ECCV.txt.

References

1. Agrawal, P., Carreira, J., Malik, J.: Learning to see by moving. In: ICCV (2015)
2. Albiero, V., Chen, X., Yin, X., Pang, G., Hassner, T.: img2pose: face alignment and detection via 6DoF, face pose estimation. In: CVPR (2021)
3. Besserve, M., Sun, R., Schölkopf, B.: Intrinsic disentanglement: an invariance view for deep generative models. In: ICML 2018 Workshop on Theoretical Foundations and Applications of Deep Generative Models (2018)
4. Blanz, V., Vetter, T.: A morphable model for the synthesis of 3D faces. In: Proceedings of the 26th Annual Conference on Computer Graphics and Interactive Techniques (1999)
5. Chang, A.X., et al.: ShapeNet: an information-rich 3D model repository. arXiv preprint arXiv:1512.03012 (2015)
6. Chen, C.H., Tyagi, A., Agrawal, A., Drover, D., Stojanov, S., Rehg, J.M.: Unsupervised 3D pose estimation with geometric self-supervision. In: CVPR (2019)
7. Chen, W., et al.: Learning to predict 3D objects with an interpolation-based differentiable renderer. In: NeurIPS (2019)
8. Choutas, V., Pavlakos, G., Bolkart, T., Tzionas, D., Black, M.J.: Monocular expressive body regression through body-driven attention. In: Vedaldi, A., Bischof, H., Brox, T., Frahm, J.-M. (eds.) ECCV 2020. LNCS, vol. 12355, pp. 20–40. Springer, Cham (2020). https://doi.org/10.1007/978-3-030-58607-2_2
9. Choy, C.B., Xu, D., Gwak, J.Y., Chen, K., Savarese, S.: 3D-R2N2: a unified approach for single and multi-view 3D object reconstruction. In: Leibe, B., Matas, J., Sebe, N., Welling, M. (eds.) ECCV 2016. LNCS, vol. 9912, pp. 628–644. Springer, Cham (2016). https://doi.org/10.1007/978-3-319-46484-8_38
10. Eigen, D., Puhrsch, C., Fergus, R.: Depth map prediction from a single image using a multi-scale deep network. In: NeurIPS (2014)
11. Elsken, T., Metzen, J.H., Hutter, F.: Neural architecture search: a survey. J. Mach. Learn. Res. 20(1), 1997–2017 (2019)
12. Fahim, G., Amin, K., Zarif, S.: Single-view 3D reconstruction: a survey of deep learning methods. Comput. Graph. 94, 164–190 (2021)
13. Fan, H., Su, H., Guibas, L.J.: A point set generation network for 3D object reconstruction from a single image. In: CVPR (2017)
14. Feng, Y., Choutas, V., Bolkart, T., Tzionas, D., Black, M.J.: Collaborative regression of expressive bodies using moderation. In: 3DV (2021)
15. Feng, Y., Feng, H., Black, M.J., Bolkart, T.: Learning an animatable detailed 3D face model from in-the-wild images. ACM Trans. Graph. (TOG) 40, 1–13 (2021)
16. François, A.R., Medioni, G.G., Waupotitsch, R.: Mirror symmetry => 2-view stereo geometry. Image Vis. Comput. 21(2), 137–143 (2003)
17. Frazier, P.I.: A tutorial on Bayesian optimization. arXiv preprint arXiv:1807.02811 (2018)

18. Fuentes-Pacheco, J., Ruiz-Ascencio, J., Rendón-Mancha, J.M.: Visual simultaneous localization and mapping: a survey. Artif. Intell. Rev. **43**(1), 55–81 (2015)
19. Gecer, B., Ploumpis, S., Kotsia, I., Zafeiriou, S.: GANFit: generative adversarial network fitting for high fidelity 3D face reconstruction. In: CVPR (2019)
20. Gerig, T., et al.: Morphable face models-an open framework. In: 2018 13th IEEE International Conference on Automatic Face & Gesture Recognition (FG 2018) (2018)
21. Girdhar, R., Fouhey, D.F., Rodriguez, M., Gupta, A.: Learning a predictable and generative vector representation for objects. In: Leibe, B., Matas, J., Sebe, N., Welling, M. (eds.) ECCV 2016. LNCS, vol. 9910, pp. 484–499. Springer, Cham (2016). https://doi.org/10.1007/978-3-319-46466-4_29
22. Green, R.: Spherical harmonic lighting: the gritty details. In: Archives of the Game Developers Conference, vol. 56, p. 4 (2003)
23. Gwak, J., Choy, C.B., Chandraker, M., Garg, A., Savarese, S.: Weakly supervised 3D reconstruction with adversarial constraint. In: 3DV (2017)
24. He, K., Zhang, X., Ren, S., Sun, J.: Deep residual learning for image recognition. In: CVPR (2016)
25. Henderson, P., Ferrari, V.: Learning to generate and reconstruct 3D meshes with only 2D supervision. arXiv preprint arXiv:1807.09259 (2018)
26. Henderson, P., Ferrari, V.: Learning single-image 3D reconstruction by generative modelling of shape, pose and shading. IJCV **128**(4), 835–854 (2020)
27. Ho, L.N., Tran, A.T., Phung, Q., Hoai, M.: Toward realistic single-view 3D object reconstruction with unsupervised learning from multiple images. In: ICCV (2021)
28. Horn, B.K., Brooks, M.J.: Shape from Shading. MIT Press, Cambridge (1989)
29. Hu, T., Wang, L., Xu, X., Liu, S., Jia, J.: Self-supervised 3D mesh reconstruction from single images. In: CVPR (2021)
30. Kanazawa, A., Black, M.J., Jacobs, D.W., Malik, J.: End-to-end recovery of human shape and pose. In: CVPR (2018)
31. Kanazawa, A., Tulsiani, S., Efros, A.A., Malik, J.: Learning category-specific mesh reconstruction from image collections. In: Ferrari, V., Hebert, M., Sminchisescu, C., Weiss, Y. (eds.) ECCV 2018. LNCS, vol. 11219, pp. 386–402. Springer, Cham (2018). https://doi.org/10.1007/978-3-030-01267-0_23
32. Kar, A., Häne, C., Malik, J.: Learning a multi-view stereo machine. In: NIPS (2017)
33. Kato, H., Harada, T.: Learning view priors for single-view 3D reconstruction. In: CVPR (2019)
34. Kato, H., Ushiku, Y., Harada, T.: Neural 3D mesh renderer. In: CVPR (2018)
35. Kilbertus, N., Parascandolo, G., Schölkopf, B.: Generalization in anti-causal learning. arXiv preprint arXiv:1812.00524 (2018)
36. Koenderink, J.J.: What does the occluding contour tell us about solid shape? Perception **13**(3), 321–330 (1984)
37. Leeb, F., Lanzillotta, G., Annadani, Y., Besserve, M., Bauer, S., Schölkopf, B.: Structure by architecture: disentangled representations without regularization. arXiv preprint arXiv:2006.07796 (2020)
38. Li, T., Bolkart, T., Black, M.J., Li, H., Romero, J.: Learning a model of facial shape and expression from 4D scans. ACM Trans. Graph. **36**(6), 1–17 (2017)
39. Li, X., et al.: Self-supervised single-view 3D reconstruction via semantic consistency. In: Vedaldi, A., Bischof, H., Brox, T., Frahm, J.-M. (eds.) ECCV 2020. LNCS, vol. 12359, pp. 677–693. Springer, Cham (2020). https://doi.org/10.1007/978-3-030-58568-6_40
40. Liu, H., Simonyan, K., Yang, Y.: Darts: differentiable architecture search. In: ICLR (2019)

41. Liu, S., Li, T., Chen, W., Li, H.: Soft rasterizer: a differentiable renderer for image-based 3D reasoning. In: ICCV (2019)
42. Liu, W., Wen, Y., Raj, B., Singh, R., Weller, A.: Sphereface revived: unifying hyperspherical face recognition. TPAMI (2022)
43. Liu, W., Wen, Y., Yu, Z., Li, M., Raj, B., Song, L.: Sphereface: deep hypersphere embedding for face recognition. In: CVPR (2017)
44. Liu, Z., Luo, P., Wang, X., Tang, X.: Deep learning face attributes in the wild. In: ICCV (2015)
45. Loper, M., Mahmood, N., Romero, J., Pons-Moll, G., Black, M.J.: SMPL: a skinned multi-person linear model. ACM Trans. Graph. **34**(6), 1–16 (2015)
46. Mukherjee, D.P., Zisserman, A.P., Brady, M., Smith, F.: Shape from symmetry: detecting and exploiting symmetry in affine images. Philos. Trans. R. Soc. Lond. Series A: Phys. Eng. Sci. **351**(1695), 77–106 (1995)
47. Murphy, K.P.: Dynamic Bayesian Networks: Representation, Inference and Learning. University of California, Berkeley (2002)
48. Novotny, D., Larlus, D., Vedaldi, A.: Learning 3D object categories by looking around them. In: ICCV (2017)
49. Ozyesil, O., Voroninski, V., Basri, R., Singer, A.: A survey of structure from motion. arXiv preprint arXiv:1701.08493 (2017)
50. Pan, J., Han, X., Chen, W., Tang, J., Jia, K.: Deep mesh reconstruction from single RGB images via topology modification networks. In: ICCV (2019)
51. Parkhi, O.M., Vedaldi, A., Zisserman, A., Jawahar, C.: Cats and dogs. In: CVPR (2012)
52. Paysan, P., Knothe, R., Amberg, B., Romdhani, S., Vetter, T.: A 3D face model for pose and illumination invariant face recognition. In: 2009 Sixth IEEE International Conference on Advanced Video and Signal Based Surveillance, pp. 296–301. IEEE (2009)
53. Pearl, J.: Causality. Cambridge University Press, Cambridge (2009)
54. Phong, B.T.: Illumination for computer generated pictures. Commun. ACM **18**(6), 311–317 (1975)
55. Schölkopf, B., Janzing, D., Peters, J., Sgouritsa, E., Zhang, K., Mooij, J.: On causal and anticausal learning. In: Langford, J., Pineau, J. (eds.) Proceedings of the 29th International Conference on Machine Learning (ICML), pp. 1255–1262. Omnipress, New York (2012). http://icml.cc/2012/papers/625.pdf
56. Schölkopf, B., et al.: Toward causal representation learning. Proc. IEEE **109**(5), 612–634 (2021)
57. Schroff, F., Kalenichenko, D., Philbin, J.: FaceNet: a unified embedding for face recognition and clustering. In: CVPR (2015)
58. Shen, X., Liu, F., Dong, H., Lian, Q., Chen, Z., Zhang, T.: Disentangled generative causal representation learning. arXiv preprint arXiv:2010.02637 (2020)
59. Simonyan, K., Zisserman, A.: Very deep convolutional networks for large-scale image recognition. arXiv preprint arXiv:1409.1556 (2014)
60. Sinha, S.N., Ramnath, K., Szeliski, R.: Detecting and reconstructing 3D mirror symmetric objects. In: Fitzgibbon, A., Lazebnik, S., Perona, P., Sato, Y., Schmid, C. (eds.) ECCV 2012. LNCS, vol. 7573, pp. 586–600. Springer, Heidelberg (2012). https://doi.org/10.1007/978-3-642-33709-3_42
61. Suwajanakorn, S., Snavely, N., Tompson, J.J., Norouzi, M.: Discovery of latent 3D keypoints via end-to-end geometric reasoning. In: NeurIPS (2018)
62. Tewari, A., et al.: MoFA: model-based deep convolutional face autoencoder for unsupervised monocular reconstruction. In: ICCV (2017)

63. Thrun, S., Wegbreit, B.: Shape from symmetry. In: ICCV (2005)
64. Tulsiani, S., Zhou, T., Efros, A.A., Malik, J.: Multi-view supervision for single-view reconstruction via differentiable ray consistency. In: CVPR (2017)
65. Vowels, M.J., Camgoz, N.C., Bowden, R.: D'ya like DAGs? A survey on structure learning and causal discovery. arXiv preprint arXiv:2103.02582 (2021)
66. Wang, C., Buenaposada, J.M., Zhu, R., Lucey, S.: Learning depth from monocular videos using direct methods. In: CVPR (2018)
67. Wang, M., Shu, Z., Cheng, S., Panagakis, Y., Samaras, D., Zafeiriou, S.: An adversarial neuro-tensorial approach for learning disentangled representations. IJCV **127**(6), 743–762 (2019)
68. Wang, N., Zhang, Y., Li, Z., Fu, Y., Liu, W., Jiang, Y.-G.: Pixel2Mesh: generating 3D mesh models from single RGB images. In: Ferrari, V., Hebert, M., Sminchisescu, C., Weiss, Y. (eds.) ECCV 2018. LNCS, vol. 11215, pp. 55–71. Springer, Cham (2018). https://doi.org/10.1007/978-3-030-01252-6_4
69. Wang, Q., et al.: Exponential convergence of the deep neural network approximation for analytic functions. arXiv preprint arXiv:1807.00297 (2018)
70. Weichwald, S., Schölkopf, B., Ball, T., Grosse-Wentrup, M.: Causal and anti-causal learning in pattern recognition for neuroimaging. In: International Workshop on Pattern Recognition in Neuroimaging (2014)
71. Wen, C., Zhang, Y., Li, Z., Fu, Y.: Pixel2Mesh++: multi-view 3D mesh generation via deformation. In: ICCV (2019)
72. Wen, Y., Liu, W., Raj, B., Singh, R.: Self-supervised 3d face reconstruction via conditional estimation. In: ICCV (2021)
73. Wiles, O., Zisserman, A.: SilNet: single-and multi-view reconstruction by learning from silhouettes. In: BMVC (2017)
74. Witkin, A.P.: Recovering surface shape and orientation from texture. Artif. Intell. **17**(1–3), 17–45 (1981)
75. Wu, S., Rupprecht, C., Vedaldi, A.: Unsupervised learning of probably symmetric deformable 3D objects from images in the wild. In: CVPR (2020)
76. Xiang, Yu., et al.: ObjectNet3D: a large scale database for 3D object recognition. In: Leibe, B., Matas, J., Sebe, N., Welling, M. (eds.) ECCV 2016. LNCS, vol. 9912, pp. 160–176. Springer, Cham (2016). https://doi.org/10.1007/978-3-319-46484-8_10
77. Xie, H., Yao, H., Sun, X., Zhou, S., Zhang, S.: Pix2Vox: context-aware 3D reconstruction from single and multi-view images. In: ICCV (2019)
78. Yan, X., Yang, J., Yumer, E., Guo, Y., Lee, H.: Perspective transformer nets: learning single-view 3D object reconstruction without 3D supervision. In: NIPS (2016)
79. Yang, M., Liu, F., Chen, Z., Shen, X., Hao, J., Wang, J.: CausalVAE: disentangled representation learning via neural structural causal models. In: CVPR (2021)
80. Yi, H., et al.: MMFace: a multi-metric regression network for unconstrained face reconstruction. In: CVPR (2019)
81. Yu, Y., Chen, J., Gao, T., Yu, M.: DAG-GNN: DAG structure learning with graph neural networks. In: ICML (2019)
82. Zhang, R., Tsai, P.S., Cryer, J.E., Shah, M.: Shape-from-shading: a survey. TPAMI **21**(8), 690–706 (1999)
83. Zhang, W., Sun, J., Tang, X.: Cat head detection - how to effectively exploit shape and texture features. In: Forsyth, D., Torr, P., Zisserman, A. (eds.) ECCV 2008. LNCS, vol. 5305, pp. 802–816. Springer, Heidelberg (2008). https://doi.org/10.1007/978-3-540-88693-8_59

84. Zheng, X., Aragam, B., Ravikumar, P.K., Xing, E.P.: DAGs with no tears: continuous optimization for structure learning. In: NeurIPS (2018)
85. Zhou, T., Brown, M., Snavely, N., Lowe, D.G.: Unsupervised learning of depth and ego-motion from video. In: CVPR (2017)
86. Zhu, R., Kiani Galoogahi, H., Wang, C., Lucey, S.: Rethinking reprojection: closing the loop for pose-aware shape reconstruction from a single image. In: ICCV (2017)
87. Zoph, B., Le, Q.V.: Neural architecture search with reinforcement learning. In: ICLR (2017)

3D Human Pose Estimation Using Möbius Graph Convolutional Networks

Niloofar Azizi[1][✉], Horst Possegger[1], Emanuele Rodolà[2], and Horst Bischof[1]

[1] Graz University of Technology, Graz, Austria
{azizi,possegger,bischof}@icg.tugraz.at
[2] Sapienza University of Rome, Rome, Italy
rodola@di.uniroma1.it

Abstract. 3D human pose estimation is fundamental to understanding human behavior. Recently, promising results have been achieved by graph convolutional networks (GCNs), which achieve state-of-the-art performance and provide rather light-weight architectures. However, a major limitation of GCNs is their inability to encode all the transformations between joints explicitly. To address this issue, we propose a novel spectral GCN using the Möbius transformation (MöbiusGCN). In particular, this allows us to directly and explicitly encode the transformation between joints, resulting in a significantly more compact representation. Compared to even the lightest architectures so far, our novel approach requires 90–98% fewer parameters, *i.e.* our lightest MöbiusGCN uses only 0.042M trainable parameters. Besides the drastic parameter reduction, explicitly encoding the transformation of joints also enables us to achieve state-of-the-art results. We evaluate our approach on the two challenging pose estimation benchmarks, Human3.6M and MPI-INF-3DHP, demonstrating both state-of-the-art results and the generalization capabilities of MöbiusGCN.

1 Introduction

Estimating 3D human pose helps to analyze human motion and behavior, thus enabling high-level computer vision tasks such as action recognition [30], sports analysis [49, 64], augmented and virtual reality [15]. Although human pose estimation approaches already achieve impressive results in 2D, this is not sufficient for many analysis tasks, because several 3D poses can project to exactly the same 2D pose. Thus, knowledge of the third dimension can significantly improve the results on the high-level tasks.

Estimating 3D human joint positions, however, is challenging. On the one hand, there are only very few labeled datasets because 3D annotations are expensive. On the other hand, there are self-occlusions, complex joint inter-dependencies, small and barely visible joints, changes in appearance like clothing and lighting, and the many degrees of freedom of the human body.

To solve 3D human pose estimation, some methods utilize multi-views [50,70], synthetic datasets [46], or motion [25,55]. For improved generalization, however, we follow the most common line of work and estimate 3D poses given only the 2D estimate of a single RGB image as input, similar to [27,33,45]. First, we compute 2D pose joints given RGB images using an off-the-shelf architecture. Second, we approximate the 3D pose of the human body using the estimated 2D joints.

S. Avidan et al. (Eds.): ECCV 2022, LNCS 13661, pp. 160–178, 2022.
https://doi.org/10.1007/978-3-031-19769-7_10

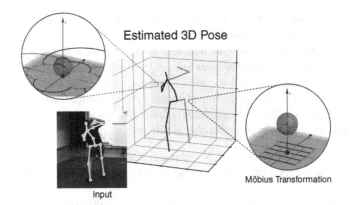

Fig. 1. Our MöbiusGCN accurately learns the transformation (particularly the rotation) between joints by leveraging the Möbius transformation given estimated 2D joint positions from a single RGB image. The spectral GCN puts a scalar-valued function on each node of the graph and compares the graph signal with the graph filter on the eigenspace of the graph Laplacian matrix and returns it to the spatial domain. We define the Möobius transformation as our scalar-valued function. Möbius transformation is mainly the composition of the rotation function and translation functions. Thus, it is capable of encoding the transformation between the human body joints.

With the advent of deep learning methods, the accuracy of 3D human pose estimation has significantly improved, *e.g.* [33,45]. Initially, these improvements were driven by CNNs (Convolutional Neural Networks). However, these assume that the input data is stationary, hierarchical, has a grid-like structure, and shares local features across the data domain. The convolution operator in the CNN assumes that the nodes have fixed neighbor positions and a fixed number of neighbor nodes. Therefore, CNNs are not applicable to graph-structured data. The input to 3D pose estimation from 2D joint positions, however, is graph-structured data. Thus, to handle this irregular nature of the data, GCNs (Graph Convolutional Networks) have been proposed [4].

GCNs are able to achieve state-of-the-art performance for 2D-to-3D human pose estimation with comparably few parameters, *e.g.* [71]. Nevertheless, to the best of our knowledge, none of the previous GCN approaches explicitly models the inter-segmental angles between joints. Learning the inter-segmental angle distribution explicitly along with the translation distribution, however, leads to encoding better feature representations. Thus, we present a novel spectral GCN architecture, *MöbiusGCN*, to accurately learn the transformation between joints and to predict 3D human poses given 2D joint positions from a single RGB image. To this end, we leverage the Möbius transformation on the eigenvalue matrix of the graph Laplacian. Previous GCNs applied for estimating the 3D pose of the human body are defined in the real domain, *e.g.* [71]. Our MöbiusGCN operates in the complex domain, which allows us to encode all the transformations (*i.e.* inter-segmental angles and translation) between nodes simultaneously (Fig. 1).

An enriched feature representation achieved by encoding the transformation distribution between joints using a Möbius transformation provides us with a compact model. A light DNN architecture makes the network independent of expensive hardware setup, enabling the use of mobile phones and embedded devices at inference time. This can be achieved by our compact MöbiusGCN architecture.

Due to a large number of weights that need to be estimated, fully-supervised state-of-the-art approaches need an enormous amount of annotated data, where data annotation is both time-consuming and requires expensive setup. Our MöbiusGCN, on the contrary, requires only a tiny fraction of the model parameters, which allows us to achieve competitive results with significantly fewer annotated data.

We summarize our main contributions as follows:

- We introduce a novel spectral GCN architecture leveraging the Möbius transformation to explicitly encode the pose, in terms of inter-segmental angles and translations between joints.
- We achieve state-of-the-art 3D human pose estimation results, despite requiring only a fraction of the model parameters (*i.e.* 2–9% of even the currently lightest approaches).
- Our light-weight architecture and the explicit encoding of transformations lead to state-of-the-art performance compared to other semi-supervised methods, by training only on a reduced dataset given estimated 2D human joint positions.

2   Related Work

Human 3D Pose Estimation. The classical approaches addressing the 3D human pose estimation task are usually based on hand-engineered features and leverage prior assumptions, *e.g.* using motion models [57] or other common heuristics [18,48]. Despite good results, their major downside is the lack of generality.

Current state-of-the-art approaches in computer vision, including 3D human pose estimation, are typically based on DNNs (Deep Neural Networks), *e.g.* [24,29,31,52]. To use these architectures, it is assumed that the statistical properties of the input data have locality, stationarity, and multi-scalability [16], which reduces the number of parameters.

Although DNNs achieve state-of-the-art in spaces governed by Euclidean geometry, a lot of the real-world problems are of a non-Euclidean nature. For these problem classes, GCNs have been introduced. There are two types of GCNs: spectral GCN and spatial GCN. Spectral GCNs rely on the Graph Fourier Transform, which analyzes the graph signals in the vector space of the graph Laplacian matrix. The second category, spatial GCN, is based on feature transformations and neighborhood aggregation on the graph. Well-known spatial GCN approaches include Message Passing Neural Networks [12] and GraphSAGE [14].

For 3D human pose estimation, GCNs achieve competitive results with comparably few parameters. Pose estimation with GCNs has been addressed, *e.g.* in [27,66, 71]. Takano [66] proposed Graph Stacked Hourglass Networks (GraphSH), in which graph-structured features are processed across different scales of human skeletal representations. Liu et al. [27] investigated different combinations of feature transformations and neighborhood aggregation of spatial features. They also showed the benefits of using separate weights to incorporate a node's self-information. Zhao et al. [71] proposed semantic GCN (SemGCN), which currently represents the lightest architecture (0.43M). The key idea is to learn the adjacency matrix, which lets the architecture encode the graph's semantic relationships between nodes. In contrast to SemGCN, we can further reduce the number of parameters by an order of magnitude (0.042M) by explicitly encoding the transformation between joints. The key ingredient to this significant reduction is the Möbius transformation.

Möbius Transformation. The Möbius transformation has been used in neural networks as an activation function [32,39], in hyperbolic neural networks [11], for data augmentation [73], and for knowledge graph embedding [37]. Our work is the first to introduce Möbius transformations for spectral graph convolutional networks. To utilize the Möbius transformation, we have to design our neural network in the complex domain. The use of complex numbers (analysis in polar coordinates) to harness phase information along with the signal amplitude is well established in signal processing [32]. By applying the Möbius transformation, we let the architecture encode the transformations (*i.e.* inter-segmental angle and translation) between joints explicitly, which leads to a very compact architecture.

Handling Rotations. Learning the rotation between joints in skeletons has been investigated previously for 3D human pose estimation [2,40,75]. Learning the rotation using Euler angles or quaternions, however, has obvious issues like discontinuities [53,77]. Continuous functions are easier to learn in neural networks [77]. Zhou et al. [77] tackle the discontinuity by lifting the problem to 5 and 6 dimensions. Another direction of research focuses on designing DNNs for inverse kinematics with the restricted assumption of putting joint angles in a specific range to avoid discontinuities. However, learning the full range of rotations is necessary for many real-world problems [77]. Our MöbiusGCN is continuous by definition and thus, allows us to elegantly encode rotations.

Data Reduction. A major benefit of light architectures is that they require smaller datasets to train. Semi-supervised methods require a small subset of annotated data and a large set of unannotated data for training the network. These methods are actively investigated in different domains, considering the difficulty of providing annotated datasets. Several semi-supervised approaches for 3D human pose estimation benefit from applying more constraints over the possible space solutions by utilizing multiview approaches using RGB images from different cameras [35,50,51,63,69]. These methods need expensive laboratory setups to collect synchronized multi-view data.

Pavllo et al. [45] phrase the loss over the back-projected estimated 3D human pose to 2D human pose space conditioned on time. Tung et al. [61] use generative adversarial networks to reduce the required annotated data for training the architecture. Iqbal et al. [19] relax the constraints using weak supervision; they introduce an end-to-end architecture that estimates 2D pose and depth independently, and uses a consistency loss to estimate the pose in 3D.

Our compact MöbiusGCN achieves competitive state-of-the-art results with only scarce training samples. MöbiusGCN does not require any multi-view setup or temporal information. Further, it does not rely on large unlabeled datasets. It just requires a small annotated dataset to train. In contrast, the previous semi-supervised methods require complicated architectures and a considerable amount of unlabeled data during the training phase.

3   Spectral Graph Convolutional Network

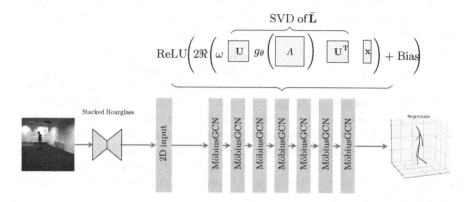

Fig. 2. The complete pipeline of the proposed MöbiusGCN architecture; The output of the off-the-shelf stacked hourglass architecture [38], *i.e.* estimated 2D joints of the human body, is the input to the MöbiusGCN architecture. The MöbiusGCN architecture locally encodes the transformation between the joints of the human body. SVD is the singular value decomposition of the normalized Laplacian matrix. Function g_θ is the Möbius transformation applied on the eigenvalues of the eigenvalue matrix independently. $\mathbf{x}$ is the graph signal and ω are the learnable parameters, both in the complex domain.

3.1   Graph Definitions

Let $\mathcal{G}(V, E)$ represent a graph consisting of a finite set of N vertices, $V = \{v_1, \ldots, v_N\}$, and a set of M edges $E = \{e_1, \ldots, e_M\}$, with $e_j = (v_i, v_k)$ where $v_i, v_k \in V$. The graph's adjacency matrix $\mathbf{A}_{N \times N}$ contains 1 in case two vertices are connected and 0 otherwise. $\mathbf{D}_{N \times N}$ is a diagonal matrix where $\mathbf{D}_{ii}$ is the degree of vertex v_i. A graph is directed if $(v_i, v_k) \neq (v_k, v_i)$, otherwise it is an undirected graph. For an undirected graph, the adjacency matrix is symmetric. The non-normalized graph Laplacian matrix is defined as $\mathbf{L} = \mathbf{D} - \mathbf{A}$, and can be normalized to $\bar{\mathbf{L}} = \mathbf{I} - \mathbf{D}^{-\frac{1}{2}} \mathbf{A} \mathbf{D}^{-\frac{1}{2}}$, where $\mathbf{I}$ is the identity matrix. $\bar{\mathbf{L}}$ is real, symmetric, and positive semi-definite. Therefore, it has N ordered, real, and non-negative eigenvalues $\{\lambda_i : i = 1, \ldots, N\}$ and corresponding orthonormal eigenvectors $\{\mathbf{u}_i : i = 1, \ldots, N\}$.

A signal $\mathbf{x}$ defined on the nodes of the graph is a vector $\mathbf{x} \in \mathbb{R}^N$, where its i-th component represents the function value at the i-th vertex in V. Similarly, $\mathbf{X} \in \mathbb{R}^{N \times d}$ is called a d-dimensional graph signal on $\mathcal{G}$ [56].

3.2   Graph Fourier Transform

Graph signals $\mathbf{x} \in \mathbb{R}^N$ admit a graph Fourier expansion $\mathbf{x} = \sum_{i=1}^{N} \langle \mathbf{u}_i, \mathbf{x} \rangle \mathbf{u}_i$, where $\mathbf{u}_i, i = 1, \ldots, N$ are the eigenvectors of the graph Laplacian [56]. Eigenvalues and eigenvectors of the graph Laplacian matrix are analogous to frequencies and sinusoidal basis functions in the classical Fourier series expansion.

3.3 Spectral Graph Convolutional Network

Spectral GCNs [5] build upon the graph Fourier transform. Let $\mathbf{x}$ be the graph signal and $\mathbf{y}$ be the graph filter on graph $\mathcal{G}$. The graph convolution $*_{\mathcal{G}}$ can be defined as

$$\mathbf{x} *_{\mathcal{G}} \mathbf{y} = \mathbf{U}(\mathbf{U}^{\top}\mathbf{x} \odot \mathbf{U}^{\top}\mathbf{y}) \tag{1}$$

where the matrix $\mathbf{U}$ contains the eigenvectors of the normalized graph Laplacian and $\odot$ is the Hadamard product. This can also be written as

$$\mathbf{x} *_{\mathcal{G}} g_{\theta} = \mathbf{U}g_{\theta}(\boldsymbol{\Lambda})\mathbf{U}^{\top}\mathbf{x}, \tag{2}$$

where $g_{\theta}(\boldsymbol{\Lambda})$ is a diagonal matrix with the parameter $\theta \in \mathbb{R}^{N}$ as a vector of Fourier coefficients.

3.4 Spectral Graph Filter

Based on the corresponding definition of g_{θ} in Eq. (2), spectral GCNs can be classified into spectral graph filters with smooth functions and spectral graph filters with rational functions.

Spectral Graph Filter with Smooth Functions. Henaff et al. [16] proposed defining $g_{\theta}(\boldsymbol{\Lambda})$ to be a smooth function (smoothness in the frequency domain corresponds to the spatial decay), to address the localization problem.

Defferrard et al. [9] proposed defining the function g_{θ} in such a way to be directly applicable over the Laplacian matrix to address the computationally costly Laplacian matrix decomposition and multiplication with the eigenvector matrix in Eq. (2).

Kipf andWelling [21] defined $g_{\theta}(\mathbf{L})$ to be the Chebychev polynomial by assuming all the eigenvalues in the range of $[-1, 1]$. Computing the polynomials of the Chebychev polynomial, however, is computationally expensive. Also, considering polynomials with higher orders causes overfitting. Therefore, Kipf andWelling [21] approximated the Chebychev polynomial with its first two orders.

Spectral Graph Filter with Rational Functions. Fractional spectral GCNs, unlike polynomial spectral GCNs, can model sharp changes in the frequency response [3]. Levie et al. [23] put the eigenvalues of the Laplacian matrix on the unit circle by applying the Cayley transform on the Laplacian matrix with a learned parameter, named spectral coefficient, that lets the network focus on the most useful frequencies.

Our proposed MöbiusGCN is also a fractional GCN which applies the Möbius transformation on the eigenvalue matrix of the normalized Laplacian matrix to encode the transformations between joints.

4 MöbiusGCN

A major drawback of previous spectral GCNs is that they do not encode the transformation distribution between nodes explicitly. We address this by applying the Möbius transformation function over the eigenvalue matrix of the decomposed Laplacian matrix. This simultaneous encoding of the rotation and translation distribution in the complex domain leads to better feature representations and fewer parameters in the network.

The input to our first block of MöbiusGCN are the joint positions in 2D Euclidean space, given as $\mathcal{J} = \{J_i \in \mathbb{R}^2 | i = 1, \ldots, \kappa\}$, which can be computed directly from the image. Our goal is then to predict the corresponding 3D Euclidean joint positions $\hat{\mathcal{Y}} = \{\hat{\mathcal{Y}}_i \in \mathbb{R}^3 | i = 1, \ldots, \kappa\}$.

We leverage the structure of the input data, which can be represented by a connected, undirected and unweighted graph. The input graphs are fixed and share the same topological structure, which means the graph structure does not change, and each training and test example differs only in having different features at the vertices. In contrast to pose estimation, tasks like protein-protein interaction [62] are not suitable for our MöbiusGCN, because there the topological structure of the input data can change across samples.

4.1 Möbius Transformation

The general form of a Möbius transformation [32] is given by $f(z) = \frac{az+b}{cz+d}$ where $a, b, c, d, z \in \mathbb{C}$ satisfy $ad - bc \neq 0$. The Möbius transformation can be expressed as the composition of simple transformations. Specifically, if $c \neq 0$, then:

- $f_1(z) = z + d/c$ defines translation by d/c,
- $f_2(z) = 1/z$ defines inversion and reflection with respect to the real axis,
- $f_3(z) = \frac{bc-ad}{c^2} z$ defines homothety and rotation,
- $f_4(z) = z + a/c$ defines the translation by a/c.

These functions can be composed to form the Möbius transformation

$$f(z) = f_4 \circ f_3 \circ f_2 \circ f_1(z) = \frac{az+b}{cz+d},$$

where $\circ$ denotes the composition of two functions f and g as $f \circ g(z) = f(g(z))$.

The Möbius transformation is analytic everywhere except at the pole $z = -\frac{d}{c}$. Since a Möbius transformation remains unchanged by scaling with a coefficient [32], we normalize it to yield the determinant 1. We observed that in our gradient-based optimization setup, the Möbius transformation in each node converges into the fixed points. In particular, the Möbius transformation can have two fixed points (loxodromic), one fixed point (parabolic or circular), or no fixed point. The fixed points can be computed by solving $\frac{az+b}{cz+d} = z$, which gives $\gamma_{1,2} = \frac{a-d+\sqrt{(a-d)^2-4bc}}{2c}$.

4.2 MöbiusGCN

To predict the 3D human pose, we explicitly encode the local transformations between joints, where each joint corresponds to a node in the graph. To do so, we define $g_\theta(\Lambda)$ in Eq. (2) to be the Möbius transformation applied to the Laplacian eigenvalues, resulting in the following fractional spectral graph convolutional network

$$\mathbf{x} *_{\mathcal{G}} g_\theta(\Lambda) = \mathbf{U} \, \text{Möbius}(\Lambda) \, \mathbf{U}^\top \mathbf{x} = \sum_{i=0}^{N-1} \text{Möbius}_i(\lambda_i) \mathbf{u}_i \mathbf{u}_i^\top \mathbf{x}, \qquad (3)$$

where

$$\text{Möbius}_i(\lambda_i) = \frac{a_i\lambda_i + b_i}{c_i\lambda_i + d_i}, \tag{4}$$

with $a_i, b_i, c_i, d_i, \lambda_i \in \mathbb{C}$.

Applying the Möbius transformation over the Laplacian matrix places the signal in the complex domain. To return back to the real domain, we sum it up with its conjugate

$$\mathbf{Z} = 2\Re\{w \, \mathbf{U} \, \text{Möbius}(\mathbf{\Lambda}) \, \mathbf{U}^\top \mathbf{x}\}, \tag{5}$$

where w is the shared complex-valued learnable weight to encode different transformation features. This causes the number of learned parameters to be reduced by a factor equal to the number of joints (nodes of the graph). The inter-segmental angles between joints are encoded by learning the rotation functions between neighboring nodes.

We can easily generalize this definition to the graph signal matrix $\mathbf{X} \in \mathbb{C}^{N \times d}$ with d input channels (*i.e.* a d-dimensional feature vector for every node) and $\mathbf{W} \in \mathbb{C}^{d \times F}$ feature maps. This defines a MöbiusGCN block

$$\mathbf{Z} = \sigma(2\Re\{\mathbf{U}\text{Möbius}(\mathbf{\Lambda})\mathbf{U}^\top \mathbf{XW}\} + \mathbf{b}), \tag{6}$$

where $\mathbf{Z} \in \mathbb{R}^{N \times F}$ is the convolved signal matrix, σ is a nonlinearity (*e.g.* ReLU [36]), and $\mathbf{b}$ is a bias term.

To encode enriched and generalized joint transformation feature representations, we make the architecture deep by stacking several blocks of MöbiusGCN. Stacking these blocks yields our complete architecture for 3D pose estimation, as shown in Fig. 2.

To apply the Möbius transformation over the matrix of eigenvalues of the Laplacian matrix, we encode the weights of the Möbius transformation for each eigenvalue in four diagonal matrices $\mathbf{A}, \mathbf{B}, \mathbf{C}, \mathbf{D}$ and compute

$$\mathbf{U}\text{Möbius}(\mathbf{\Lambda})\mathbf{U}^\top = \mathbf{U}(\mathbf{A}\mathbf{\Lambda} + \mathbf{B})(\mathbf{C}\mathbf{\Lambda} + \mathbf{D})^{-1}\mathbf{U}^\top. \tag{7}$$

4.3   Why MöbiusGCN is a Light Architecture

As a direct consequence of applying the Möbius transformation for the graph filters in polar coordinates, the filters in each block can encode the inter-segmental angle features between joints in addition to the translation features explicitly. By applying the Möbius transformation, the graph filter scales and rotates the eigenvectors of the Laplacian matrix in the graph Fourier transform simultaneously. This leads to learning better feature representations and thus, yields a more compact architecture.

For a better understanding, consider the following analogy with the classical Fourier transform: While it can be hard to construct an arbitrary signal by a linear combination of basis functions with real coefficients (i.e., the signal is built just by changing the amplitudes of the basis functions), it is significantly easier to build a signal by using *complex* coefficients, which change both the phase and amplitude.

In previous spectral GCNs, specifically [21], the Chebychev polynomials are only able to scale the eigenvectors of the Laplacian matrix, in turn requiring both more parameters and additional nonlinearities to encode the rotation distribution between joints implicitly.

4.4 Discontinuity

Our model encodes the transformation between joints in the complex domain by learning the parameters of the normalized Möbius transformation. In the definition of the Möbius transformation, if $ad - bc \neq 0$, then the Möbius transformation is an injective function and thus, continuous by the definition of continuity for neural networks given by [77]. MöbiusGCN does not suffer from discontinuities in representing intersegmental angles, in contrast to Euler angles or quaternions. Additionally, this leads to significantly fewer parameters in our architecture.

Table 1. Quantitative comparisons w.r.t. MPJPE (in mm) on Human3.6M [18] under Protocol #1. Best in bold, second-best underlined. In the upper part, all methods use stacked hourglass (HG) 2D estimates [38] as inputs, except for [66] (which uses CPN [7], indicated by ∗). In the lower part, all methods use the 2D ground truth (GT) as input.

Protocol #1	# Param.	Dir.	Disc.	Eat	Greet	Phone	Photo	Pose	Purch.	Sit	SitD.	Smoke	Wait	WalkD.	Walk	WalkT.	Average
Martinez et al. [33]	4.2M	51.8	56.2	58.1	59.0	69.5	78.4	55.2	58.1	74.0	94.6	62.3	59.1	65.1	49.5	52.4	62.9
Tekin et al. [59]	n/a	54.2	61.4	60.2	61.2	79.4	78.3	63.1	81.6	70.1	107.3	69.3	70.3	74.3	51.8	63.2	69.7
Sun et al. [58]	n/a	52.8	54.8	54.2	54.3	61.8	67.2	53.1	53.6	71.7	86.7	61.5	53.4	61.6	47.1	53.4	59.1
Yang et al. [68]	n/a	51.5	58.9	50.4	57.0	62.1	65.4	49.8	52.7	69.2	85.2	57.4	58.4	**43.6**	60.1	47.7	58.6
Hossain and Little [17]	16.9M	48.4	50.7	57.2	55.2	63.1	72.6	53.0	51.7	66.1	80.9	59.0	57.3	62.4	46.6	49.6	58.3
Fang et al. [10]	n/a	50.1	54.3	57.0	57.1	66.6	73.3	53.4	55.7	72.8	88.6	60.3	57.7	62.7	47.5	50.6	60.4
Pavlakos et al. [43]	n/a	48.5	54.4	54.5	52.0	59.4	65.3	49.9	52.9	65.8	71.1	56.6	52.9	60.9	44.7	47.8	56.2
SemGCN [71]	0.43M	48.2	60.8	51.8	64.0	64.6	**53.6**	51.1	67.4	88.7	**57.7**	73.2	65.6	48.9	64.8	51.9	60.8
Sharma et al. [54]	n/a	48.6	54.5	54.2	55.7	62.2	72.0	50.5	54.3	70.0	78.3	58.1	55.4	61.4	45.2	49.7	58.0
GraphSH [66] ∗	3.7M	**45.2**	**49.9**	47.5	50.9	**54.9**	66.1	48.5	46.3	59.7	71.5	51.4	48.6	53.9	**39.9**	**44.1**	**51.9**
Ours (HG)	0.16M	46.7	60.7	**47.3**	**50.7**	64.1	61.5	**46.2**	**45.3**	67.1	80.4	54.6	51.4	55.4	43.2	48.6	52.1
Ours (HG)	**0.04M**	52.5	61.4	47.8	53.0	66.4	65.4	48.2	46.3	71.1	84.3	57.8	52.3	57.0	45.7	50.3	54.2
Liu et al. [27] (GT)	4.2M	36.8	40.3	33.0	36.3	37.5	45.0	39.7	34.9	40.3	47.7	37.4	38.5	38.6	29.6	32.0	37.8
GraphSH [66] (GT)	3.7M	35.8	**38.1**	**31.0**	35.3	35.8	**43.2**	37.3	31.7	**38.4**	**45.5**	35.4	36.7	36.8	27.9	**30.7**	**35.8**
SemGCN [71] (GT)	0.43M	37.8	49.4	37.6	40.9	45.1	41.4	40.1	48.3	50.1	42.2	53.5	44.3	40.5	47.3	39.0	43.8
Ours (GT)	0.16M	**31.2**	46.9	**32.5**	**31.7**	41.4	44.9	**33.9**	**30.9**	49.2	55.7	35.9	36.1	37.5	29.07	33.1	36.2
Ours (GT)	**0.04M**	33.6	48.5	34.9	34.8	46.0	49.5	36.7	33.7	50.6	62.7	38.9	40.3	41.4	33.1	36.3	40.0

5   Experimental Results

5.1   Datasets and Evaluation Protocols

We use the publicly available motion capture dataset Human3.6M [18]. It contains 3.6 million images produced by 11 actors performing 15 actions. Four different calibrated RGB cameras are used to capture the subjects during training and test time. Same as previous works, *e.g.* [33,43,54,58,59,66,71], we use five subjects (S1, S5, S6, S7, S8) for training and two subjects (S9 and S11) for testing. Each sample from the different camera views is considered independently. We also use MPI-INF-3DHP dataset [34] to test the generalizability of our model. MPI-INF-3DHP contains 6 subjects for testing in three different scenarios: studio with a green screen (GS), studio without green screen (noGS), and outdoor scene (Outdoor). Note that for experiments on MPI-INF-3DHP we also only trained on Human3.6M.

Following [33,58,59,66,71], we use the MPJPE protocol, referred to as Protocol #1. MPJPE is the mean per joint position error in millimeters between predicted joint positions and ground truth joint positions after aligning the pre-defined root joints (*i.e.* the pelvis joint). Note that some works (*e.g.* [27,45]) use the P-MPJPE metric, which reports the error after a rigid transformation to align the predictions with the ground truth joints. We explicitly select the standard MPJPE metric as it is more challenging and also allows for a fair comparison to previous related works. For the MPI-INF-3DHP test set, similar to previous works [28,66], we use the percentage of correct 3D keypoints (3D PCK) within a 150 mm radius [34] as evaluation metric.

5.2 Implementation Details

2D Pose Estimation. The inputs to our architecture are the 2D joint positions estimated from the RGB images for all four cameras independently. Our method is independent of the off-the-shelf architecture used for estimating 2D joint positions. Similar to previous works [33,71], we use the stacked hourglass architecture [38] to estimate the 2D joint positions. The hourglass architecture is an autoencoder architecture that stacks the encoder-decoder with skip connections multiple times. Following [71], the stacked hourglass network is first pre-trained on the MPII [1] dataset and then fine-tuned on the Human3.6M [18] dataset. As described in [45], the input joints are scaled to image coordinates and normalized to $[-1, 1]$.

3D Pose Estimation. The ground truth 3D joint positions in the Human3.6M dataset are given in world coordinates. Following previous works [33,71], we transform the joint positions to the camera space given the camera calibration parameters. Similar to previous works [33,71], to make the architecture trainable, we chose a predefined joint (the pelvis joint) as the center of the coordinate system. We do not use any augmentations throughout all our experiments.

We trained our architecture using Adam [20] with an initial learning rate of 0.001 and used mini-batches of size 64. The learning rate is dropped with a decay rate of 0.5 when the loss on the validation set saturates. The architecture contains seven MöbiusGCN blocks, where each block, except the first and the last block with the input and the output channels 2 and 3 respectively, contains either 64 channels (leading to 0.04M parameters) or 128 channels (leading to 0.16M parameters). We initialized the weights using the Xavier method [13]. During the test phase, the scale of the outputs is calibrated by forcing the sum of the length of all 3D bones to be equal to a canonical skeleton [42,74,76]. To help the architecture differentiate between different 3D poses with the same 2D pose, similar to Poier et al. [47], we provide the center of mass of the subject to the architecture as an additional input. Same as [33,71], we predict 16 joints (*i.e.* without the 'Neck/Nose' joint).

Also, as in previous works [27,33,42,71], our network predicts the normalized locations of 3D joints. We did all our experiments on an NVIDIA GeForce RTX 2080 GPU using the PyTorch framework [41]. For the loss function, same as previous works e.g. [33,45], we use the mean squared error (MSE) between the 3D ground truth joint locations $\mathcal{Y}$ and our predictions $\hat{\mathcal{Y}}$, *i.e.*

$$\mathcal{L}(\mathcal{Y}, \hat{\mathcal{Y}}) = \sum_{i=1}^{\kappa} (\mathcal{Y}_i - \hat{\mathcal{Y}}_i)^2. \tag{8}$$

Complex-Valued MöbiusGCN. In complex-valued neural networks, the data and the weights are represented in the complex domain. A complex function is holomorphic (complex-differentiable) if not only their partial derivatives exist but also satisfy the Cauchy-Riemann equations. Complex neural networks have different applications, *e.g.* Wolter and Yao [65] proposed a complex-valued recurrent neural network which helps solving the exploding/vanishing gradients problem in RNNs. Complex-valued neural networks are easier to optimize than real-valued neural networks and have richer representational capacity [60]. Considering the Liouville theorem [32], designing a fully complex differentiable (holomorphic) neural network is hard as only constant functions are both holomorph and bounded. Nevertheless, it was shown that in practice full complex differentiability of complex neural networks is not necessary [60].

In complex-valued neural networks, the complex convolution operator is defined as

$$\mathbf{W} * \mathbf{h} = (\mathbf{A} * \mathbf{x} - \mathbf{B} * \mathbf{y}) + i(\mathbf{B} * \mathbf{x} + \mathbf{A} * \mathbf{y}),$$

where $\mathbf{W} = \mathbf{A} + i\mathbf{B}$ and $\mathbf{h} = \mathbf{x} + i\mathbf{y}$. $\mathbf{A}$ and $\mathbf{B}$ are real matrices and $\mathbf{x}$ and $\mathbf{y}$ are real vectors. We also apply the same operators on our graph signals and graph filters. The PyTorch framework [41] utilizes Wirtinger calculus [22] for backpropagation, which optimizes the real and imaginary partial derivatives independently.

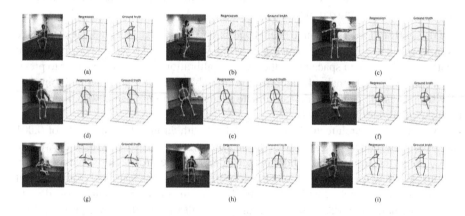

Fig. 3. Qualitative results of MöbiusGCN on Human3.6M [18].

5.3 Fully-Supervised MöbiusGCN

In the following, we compare the results of MöbiusGCN in a fully-supervised setup with the previous state-of-the-art for 3D human pose estimation on the Human3.6M and MPI-INF-3DHP datasets. For this, we use a) estimated 2D poses using the stacked hourglass architecture (HG) [38] as input and b) the 2D ground truth (GT).

Comparisons on Human3.6M. Table 1 shows the comparison of our MöbiusGCN to the state-of-the-art methods under Protocol #1 on Human3.6M dataset.

By setting the number of channels to 128 in each block of the MöbiusGCN (0.16M parameters), given estimated 2D joint positions (HG), we achieve an average MPJPE of 52.1mm over all actions and test subjects. Using the ground truth (GT) 2D joint positions as input, we achieve an MPJPE of 36.2 mm. These results are on par with the state-of-the-art, *i.e.* GraphSH [66], which achieves 51.9 mm and 35.8 mm, respectively. Note, however, that MöbiusGCN drastically reduces the number of training parameters by up to 96% (0.16M *vs.* 3.7M).

Reducing the number of channels to 64, we still achieve impressive results considering the lightness of our architecture (*i.e.* only 0.042M parameters). Compared to GraphSH [66], we reduce the number of parameters by 98.9% (0.042M vs 3.7M) and still achieve notable results, *i.e.* MPJPE of 40.0 mm *vs.* 35.8 mm (using 2D GT inputs) and 54.2 mm *vs.* 51.9 mm (using the 2D HG inputs). Note that GraphSH [66] is the only approach which uses a better 2D pose estimator as input (CPN [7] instead of HG [38]). Nevertheless, our MöbiusGCN (with 0.16M parameters) achieves competitive results. Furthermore, MöbiusGCN outperforms the previously lightest architecture SemGCN [71], *i.e.* 40.0mm vs 43.8 mm (using 2D GT inputs) and 54.2 mm vs 60.8 mm (using 2D HG input), although we require only 9.7% of their number of parameters (0.042M *vs.* 0.43M). Figure 3 shows qualitative results of our MöbiusGCN with 0.16M parameters on unseen subjects of the Human3.6M dataset given the 2D ground truth (GT) as input.

Table 2. Results on the MPI-INF-3DHP test set [34]. Best in bold, second-best underlined.

Method	# Parameters	GS	noGS	Outdoor	All (PCK)
Martinez et al. [33]	4.2M	49.8	42.5	31.2	42.5
Mehta et al. [34]	n/a	70.8	62.3	58.8	64.7
Luo et al. [28]	n/a	71.3	59.4	65.7	65.6
Yang et al. [68]	n/a	–	–	–	69.0
Zhou et al. [76]	n/a	71.1	64.7	72.7	69.2
Ci et al. [8]	n/a	74.8	70.8	77.3	74.0
Zhou et al. [72]	n/a	75.6	71.3	80.3	75.3
GraphSH [66]	3.7M	**81.5**	**81.7**	<u>75.2</u>	**80.1**
Ours	**0.16M**	<u>79.2</u>	<u>77.3</u>	**83.1**	<u>80.0</u>

Comparisons on MPI-INF-3DHP. The quantitative results on MPI-INF-3DHP [34] are shown in Table 2. Although we train MöbiusGCN only on the Human3.6M [18] dataset and our architecture is lightweight, the results indicate our strong generalization capabilities to unseen datasets, especially for the most challenging outdoor scenario. Figure 4 shows some qualitative results on unseen self-occlusion examples from the test set of MPI-INF-3DHP dataset with MöbiusGCN trained only on Human3.6M.

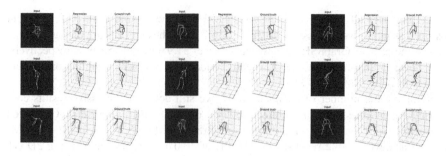

Fig. 4. Qualitative self-occlusion results of MöbiusGCN on MPI-INF-3DHP [34] (trained only on Human3.6m).

Though MöbiusGCN for 3D human pose estimation has comparably fewer parameters, it is computationally expensive, *i.e.* $\mathcal{O}(n^3)$, both in the forward and backward pass due to the decomposition of the Laplacian matrix. In practice, however, this is not a concern for human pose estimation because of the small human pose graphs, *i.e.* $\sim$20 nodes. More specifically, a single forward pass takes on average only 0.001 s.

Table 3. Supervised quantitative comparison between GCN architectures on Human3.6M [18] under Protocol #1. Best in bold, second-best underlined. All methods use 2D ground truth as input.

Method	# Parameters	MPJPE
Liu et al. [27]	4.20M	37.8
GraphSH [66]	3.70M	35.8
Liu et al. [27]	1.05M	40.1
GraphSH [66]	0.44M	39.2
SemGCN [71]	0.43M	43.8
Yan et al. [67]	0.27M	57.4
Veličković et al. [62]	<u>0.16M</u>	82.9
Chebychev-GCN	0.08M	110.6
Ours	0.66M	**33.7**
Ours	<u>0.16M</u>	<u>36.2</u>
Ours	**0.04M**	40.0

Comparison to Previous GCNs. Table 3 shows our performance in comparison to previous GCN architectures. Besides significantly reducing the number of required parameters, applying the Möbius transformation also allows us to leverage better feature representations. Thus, MöbiusGCN can outperform all other light-weight GCN architectures. It even achieves better results (36.2 mm *vs.* 39.2 mm) than the light-weight version of the state-of-the-art GraphSH [66], which requires 0.44M parameters.

We also compare our proposed spectral GCN with the vanilla spectral GCN, *i.e.* Chebychev-GCN [21]. Each block of Chebychev-GCN is the real-valued spectral GCN from [21]. We use 7 blocks, similar to our MöbiusGCN, with 128 channels each. Our complex-valued MöbiusGCN with only 0.04M clearly outperforms the Chebychev-GCN [21] with 0.08M parameters (40.0 mm *vs.* 110.6 mm). This highlights the representational power of our MöbiusGCN in contrast to vanilla spectral GCNs.

Table 4. Semi-supervised quantitative comparison on Human3.6M [18] under Protocol #1. Temp, MV, GT, and HG stand for temporal, multi-view, ground-truth, and stacked hourglass as 2D pose input respectively. Best in bold, second-best underlined.

Method	Temp	MV	Input	MPJPE
Rhodin et al. [50]	✗	✓	RGB	131.7
Pavlakos et al. [44]	✓	✓	RGB	110.7
Chen et al. [6]	✗	✓	HG	91.9
Li et al. [26]	✓	✗	RGB	<u>88.8</u>
Ours (0.16M)	✗	✗	HG	**82.3**
Iqbal et al. [19]	✗	✓	GT	<u>62.8</u>
Ours (0.16M)	✗	✗	GT	**62.3**

5.4 MöbiusGCN with Reduced Dataset

A major practical limitation with training neural network architectures is to acquire sufficiently large and accurately labeled datasets. Semi-supervised methods try to address this by combining fewer labeled samples with large amounts of unlabeled data. Another benefit of MöbiusGCN is that we require fewer training samples. Having a better feature representation in MöbiusGCN leads to a light architecture and therefore, requires less training samples.

To demonstrate this, we train MöbiusGCN with a limited number of samples. In particular, we use only one subject to train MöbiusGCN and do not need any unlabeled data. Table 4 compares the MöbiusGCN to the semi-supervised approaches [6, 19, 26, 44, 50], which were trained using both labeled and unlabeled data. As can be seen, MöbiusGCN performs favorably: we achieve an MPJPE of 82.3 mm (given 2D HG inputs) and an MPJPE of 62.3 mm (using the 2D GT input). In contrast to previous works, we neither utilize other subjects as weak supervision nor need large unlabeled datasets during training.

As shown in Table 4, MöbiusGCN also outperforms methods which rely on multi-view cues [6, 50] or leverage temporal information [26]. Additionally, we achieve better results to [19], even though, in contrast to this approach, we do not incorporate multi-view information or require extensive amounts of unlabeled data during training.

Table 5 analyzes the effect of increasing the number of training samples. As can be seen, our MöbiusGCN only needs to train on three subjects to perform on par with SemGCN [71].

Table 5. Evaluating the effects of using fewer training subjects on Human3.6M [18] under Protocol #1 (given 2D GT inputs). The first 3 experimental results are after **10** and **50** (full convergence) training epochs for MöbiusGCN and SemGCN, respectively. Best in bold, second-best underlined.

Subject	MöbiusGCN	MöbiusGCN	SemGCN [71]
S1	**56.4**$_{50}$	$\underline{62.3}_{10}$	63.9$_{50}$
S1, S5	**44.5**$_{50}$	$\underline{47.9}_{10}$	58.7$_{50}$
S1, S5, S6	**42.3**$_{50}$	$\underline{43.1}_{10}$	48.5$_{50}$
All	**36.2**	**36.2**	43.8
# Parameters	**0.16M**	**0.16M**	$\underline{0.43M}$

6 Conclusion and Discussion

We proposed a novel rational spectral GCN (MöbiusGCN) to predict 3D human pose estimation by encoding the transformation between joints of the human body, given the human body joint positions in 2D. Our proposed method achieves state-of-the-art result accuracy while preserving the compactness of the model with lower number of parameters than the most compact model existing in the literature (lower number of parameters by an order of magnitude). We verified the generalizability of our model on the MPI-INF-3DHP dataset, where we achieve state-of-the-art results on the most challenging in-the-wild (outdoor) scenario.

Our proposed simple and light-weight architecture requires less data for training. This allows us to outperform the previous lightest architecture by just training our model with three subjects on the Human3.6M dataset. We also showed promising results of our architecture in comparison to previous state-of-the-art semi-supervised architectures despite not using any temporal or multi-view information or large unlabeled datasets.

Acknowledgement. This research was funded by the Austrian Research Promotion Agency (FFG) under project no. 874065 and the ERC grant no. 802554 (SPECGEO).

References

1. Andriluka, M., Pishchulin, L., Gehler, P., Schiele, B.: 2D human pose estimation: new benchmark and state of the art analysis. In: CVPR (2014)
2. Barrón, C., Kakadiaris, I.A.: Estimating anthropometry and pose from a single uncalibrated image. Comput. Vis. Image Underst. **81**(3), 269–284 (2001)
3. Bianchi, F.M., Grattarola, D., Livi, L., Alippi, C.: Graph neural networks with convolutional Arma filters. IEEE TPAMI (2021). (Early access article)
4. Bronstein, M.M., Bruna, J., LeCun, Y., Szlam, A., Vandergheynst, P.: Geometric deep learning: going beyond euclidean data. IEEE Signal Process. Mag. **34**(4), 18–42 (2017)
5. Bruna, J., Zaremba, W., Szlam, A., LeCun, Y.: Spectral networks and locally connected networks on graphs. In: ICLR (2014)
6. Chen, X., Lin, K.-Y., Liu, W., Qian, C., Lin, L.: Weakly-supervised discovery of geometry-aware representation for 3D human pose estimation. In: CVPR (2019)

7. Chen, Y., Wang, Z., Peng, Y., Zhang, Z., Yu, G., Sun, J.: Cascaded pyramid network for multi-person pose estimation. In: CVPR (2018)
8. Ci, H., Wang, C., Ma, X., Wang, Y.: Optimizing network structure for 3D human pose estimation. In: ICCV (2019)
9. Defferrard, M., Bresson, X., Vandergheynst, P.: Convolutional neural networks on graphs with fast localized spectral filtering. In: NeurIPS (2016)
10. Fang, H.-S., Xu, Y., Wang, W., Liu, X., Zhu, S.-C.: Learning pose grammar to encode human body configuration for 3D pose estimation. In: AAAI (2018)
11. Ganea, O., Becigneul, G., Hofmann, T.: Hyperbolic Neural Networks. In: Proceedings of NeurIPS (2018)
12. Gilmer, J., Schoenholz, S.S., Riley, P.F., Vinyals, O., Dahl, G.E.: Neural message passing for quantum chemistry. In: ICML (2017)
13. Glorot, X., Bengio, Y.: Understanding the difficulty of training deep feedforward neural networks. In: AISTATS (2010)
14. Hamilton, W., Ying, Z., Leskovec, J.: Inductive representation learning on large graphs. In: NeurIPS (2017)
15. Han, X., Wu, Z., Wu, Z., Yu, R., Davis, L.s.: VITON: an image-based virtual try-on network. In: CVPR (2018)
16. Henaff,M., Bruna, J., LeCun, Y.: Deep convolutional networks on graph-structured data. arXiv preprint arXiv:1506.05163 (2015)
17. Hossain, M.R.I., Little, J.J.: Exploiting temporal information for 3D human pose estimation. In: Ferrari, V., Hebert, M., Sminchisescu, C., Weiss, Y. (eds.) ECCV 2018. LNCS, vol. 11214, pp. 69–86. Springer, Cham (2018). https://doi.org/10.1007/978-3-030-01249-6_5
18. Ionescu, C., Papava, D., Olaru, V., Sminchisescu, C.: Human3.6M: large scale datasets and predictive methods for 3D human sensing in natural environments. IEEE TPAMI **36**(7), 1325–1339 (2014)
19. Iqbal, U., Molchanov, P., Kautz, J.: Weakly-supervised 3D human pose learning via multi-view images in the wild. In: CVPR (2020)
20. Kingma, D.P., Ba, J.: Adam: a method for stochastic optimization. In ICLR (2015)
21. Kipf, T.N., Welling, M.: Semi-supervised classification with graph convolutional networks. In: ICLR (2017)
22. Kreutz-Delgado, K.: The complex gradient operator and the CR-calculus. arXiv preprint arXiv:0906.4835 (2009)
23. Levie, R., Monti, F., Bresson, X., Bronstein, M.M.: CayleyNets: graph convolutional neural networks with complex rational spectral filters. IEEE Trans. Signal Process **67**(1), 97–109 (2018)
24. Li, C., Lee, G.H.: Generating multiple hypotheses for 3D human pose estimation with mixture density network. In: CVPR (2019)
25. Li, W., Liu, H., Ding, R., Liu, M., Wang, P.: Lifting transformer for 3D human pose estimation in video. arXiv preprint arXiv:2103.14304 (2021)
26. Li, Z., Wang, X., Wang, F., Jiang, P.: On boosting single-frame 3D human pose estimation via monocular videos. In: ICCV (2019)
27. Liu, K., Ding, R., Zou, Z., Wang, L., Tang, W.: A comprehensive study of weight sharing in graph networks for 3D human pose estimation. In: Vedaldi, A., Bischof, H., Brox, T., Frahm, J.-M. (eds.) ECCV 2020. LNCS, vol. 12355, pp. 318–334. Springer, Cham (2020). https://doi.org/10.1007/978-3-030-58607-2_19
28. Luo, C., Chu, X., Yuille, A.: A fully convolutional network for 3D human pose estimation. In: BMVC (2018)
29. Luo, D., Songlin, D., Ikenaga, T.: Multi-task neural network with physical constraint for real-time multi-person 3D pose estimation from monocular camera. Multimed. Tools. Appl. **80**, 27223–27244 (2021)

30. Luvizon, D.C., Picard, D., Tabia, H.: 2D/3D pose estimation and action recognition using multitask deep learning. In: CVPR (2018)
31. Ma, X., Su, J., Wang, C., Ci, H., Wang, Y.: Context modeling in 3D human pose estimation: a unified perspective. In: CVPR (2021)
32. Mandic, D.P., Goh, V.S.L.: Complex-valued Nonlinear Adaptive Filters: Noncircularity, Widely Linear and Neural Models. Wiley, Hoboken (2009)
33. Martinez, J., Hossain, R., Romero, J., Little, J.J.: A simple yet effective baseline for 3D human pose estimation. In: ICCV (2017)
34. Mehta, D., et al.: Monocular 3D human pose estimation in the wild using improved CNN supervision. In: 3DV (2017)
35. Mitra, R., Gundavarapu, N.B., Sharma, A., Jain, A.: Multiview-consistent semi-supervised learning for 3D human pose estimation. In: CVPR (2020)
36. Nair, V., Hinton, G.E.: Rectified linear units improve restricted boltzmann machines. In: Proceedings of ICML (2010)
37. Nayyeri, M., Vahdati, S., Aykul, C., Lehmann, J.: 5* knowledge graph embeddings with projective transformations. In: AAAI (2021)
38. Newell, A., Yang, K., Deng, J.: Stacked hourglass networks for human pose estimation. In: Leibe, B., Matas, J., Sebe, N., Welling, M. (eds.) ECCV 2016. LNCS, vol. 9912, pp. 483–499. Springer, Cham (2016). https://doi.org/10.1007/978-3-319-46484-8_29
39. Özdemir, N., İskender, B.B., Özgür, N.Y.: Complex-valued neural network with Möbius activation function. Commun. Nonlinear **16**, 4698–4703 (2011)
40. Parameswaran, V., Chellappa, R.: View independent human body pose estimation from a single perspective image. In: CVPR (2004)
41. Paszke, A., et al.: PyTorch: an imperative style, high-performance deep learning library. In: NeurIPS (2019)
42. Pavlakos, G., Zhou, X., Derpanis, K.G., Daniilidis, K.: Coarse-to-fine volumetric prediction for single-image 3D human pose. In: CVPR (2017)
43. Pavlakos, G., Zhou, X., Daniilidis, K.: Ordinal depth supervision for 3D human pose estimation. In: CVPR (2018)
44. Pavlakos, G., Kolotouros, N., Daniilidis, K.: TexturePose: supervising human mesh estimation with texture consistency. In: ICCV (2019)
45. Pavllo, D., Feichtenhofer, C., Grangier, D., Auli, M.: 3D human pose estimation in video with temporal convolutions and semi-supervised training. In: CVPR (2019)
46. Peng, X., Tang, Z., Yang, F., Feris, R.S., Metaxas, D.: jointly optimize data augmentation and network training: adversarial data augmentation in human pose estimation. In: CVPR (2018)
47. Poier, G., Schinagl, D., Bischof, H.: Learning pose specific representations by predicting different views. In: CVPR (2018)
48. Ramakrishna, V., Kanade, T., Sheikh, Y.: Reconstructing 3D human pose from 2D image landmarks. In: Fitzgibbon, A., Lazebnik, S., Perona, P., Sato, Y., Schmid, C. (eds.) ECCV 2012. LNCS, vol. 7575, pp. 573–586. Springer, Heidelberg (2012). https://doi.org/10.1007/978-3-642-33765-9_41
49. Rematas, K., Kemelmacher-Shlizerman, I., Curless, B., Seitz, S.: Soccer on your tabletop. In: CVPR (2018)
50. Rhodin, H., Salzmann, M., Fua, P.: Unsupervised geometry-aware representation for 3D human pose estimation. In: Ferrari, V., Hebert, M., Sminchisescu, C., Weiss, Y. (eds.) ECCV 2018. LNCS, vol. 11214, pp. 765–782. Springer, Cham (2018). https://doi.org/10.1007/978-3-030-01249-6_46
51. Rhodin, H., et al.: Learning monocular 3D human pose estimation from multi-view images. In: CVPR (2018)

52. Sárándi, I., Linder, T., Arras, K.O., Leibe, B.: MeTRAbs: metric-scale truncation-robust heatmaps for absolute 3D human pose estimation. IEEE Trans. Biom. Behav. Identity Sci. **3**(1), 16–30 (2020)

53. Saxena, A., Driemeyer, J., Ng, A.Y.: Learning 3D object orientation from images. In: ICRA (2009)

54. Sharma, S., Varigonda, P.T., Bindal, P., Sharma, A., Jain, A.: Monocular 3D human pose estimation by generation and ordinal ranking. In: ICCV (2019)

55. Shere, M., Kim, H., Hilton, A.: Temporally consistent 3D human pose estimation using dual 360deg cameras. In: ICCV (2021)

56. Shuman, D.I., Narang, S.K., Frossard, P., Ortega, A., Vandergheynst, P.: The Emerging field of signal processing on graphs: extending high-dimensional data analysis to networks and other irregular domains. IEEE Signal Process. Mag. **30**(3), 83–98 (2013)

57. Sminchisescu, C.: 3D Human motion analysis in monocular video techniques and challenges. In: Rosenhahn, B., Klette, R., Metaxas, D. (eds.) Human Motion. Computational Imaging and Vision, vol. 36, pp. 185–211. Springer, Dordrecht (2006). https://doi.org/10.1007/978-1-4020-6693-1_8

58. Sun, X., Shang, J., Liang, S., Wei, Y.: Compositional human pose regression. In: ICCV (2017)

59. Tekin, B., Marquez-Neila, P., Salzmann, M., Fua, P.: Learning to fuse 2D and 3D image cues for monocular body pose estimation. In: ICCV (2017)

60. Trabelsi, C., et al.: Deep complex networks. In: ICLR (2018)

61. Tung, H.-Y.F., Harley, A.W., Seto, W., Fragkiadaki, K.: Adversarial inverse graphics networks: learning 2D-to-3D lifting and image-to-image translation from unpaired supervision. In: ICCV (2017)

62. Veličković, P., Cucurull, G., Casanova, A., Romero, A., Liò, P., Bengio, Y.: Graph attention networks. In: ICLR (2018)

63. Wandt, B., Rudolph, M., Zell, P., Rhodin, H., Rosenhahn, B.: CanonPose: self-supervised monocular 3D human pose estimation in the wild. In: CVPR (2021)

64. Wang, J., Qiu, K., Peng, H., Fu, J., Zhu, J.: AI coach: deep human pose estimation and analysis for personalized athletic training assistance. In: ACM-MM (2019)

65. Wolter, M., Yao, A.: Complex gated recurrent neural networks. In: NeurIPS (2018)

66. Xu, T., Takano, W.: Graph stacked hourglass networks for 3D human pose estimation. In: CVPR (2021)

67. Yan, S., Xiong, Y., Lin, D.: Spatial temporal graph convolutional networks for skeleton-based action recognition. In: AAAI (2018)

68. Yang, W., Ouyang, W., Wang, X., Ren, J., Li, H., Wang, X.: 3D human pose estimation in the wild by adversarial learning. In: CVPR (2018)

69. Yao, Y., Jafarian, Y., Park, H.S.: MONET: multiview semi-supervised keypoint detection via epipolar divergence. In ICCV (2019)

70. Zhang, Z., Wang, C., Qiu, W., Qin, W., Zeng, W.: AdaFuse: adaptive multiview fusion for accurate human pose estimation in the wild. Int. J. Comput. Vision **129**(3), 703–718 (2020). https://doi.org/10.1007/s11263-020-01398-9

71. Zhao, L., Peng, X., Tian, Y., Kapadia, M., Metaxas, D.N.: Semantic graph convolutional networks for 3D human pose regression. In: CVPR (2019)

72. Zhou, K., Han, X., Jiang, N., Jia, K., Lu, J.: HEMlets pose: learning part-centric heatmap triplets for accurate 3D human pose estimation. In: ICCV (2019)

73. Zhou, S., Zhang, J., Jiang, H., Lundh, T., Ng, A.Y.: Data augmentation with Möbius transformations. Mach. Learn. Sci. Technol. **2**(2), 025016 (2021)

74. Zhou, X., Zhu, M., Pavlakos, G., Leonardos, S., Derpanis, K.G., Daniilidis, K.: Monocap: monocular human motion capture using a CNN coupled with a geometric prior. IEEE TPAMI **41**(4), 901–914 (2018)

75. Zhou, X., Sun, X., Zhang, W., Liang, S., Wei, Y.: Deep kinematic pose regression. In: Hua, G., Jégou, H. (eds.) ECCV 2016. LNCS, vol. 9915, pp. 186–201. Springer, Cham (2016). https://doi.org/10.1007/978-3-319-49409-8_17
76. Zhou, X., Huang, Q., Sun, X., Xue, X., Wei, Y.: Towards 3D human pose estimation in the wild: a weakly-supervised approach. In: ICCV (2017)
77. Zhou, Y., Barnes, C., Lu, J., Yang, J., Li, H.: On the continuity of rotation representations in neural networks. In: CVPR (2019)

Learning to Train a Point Cloud Reconstruction Network Without Matching

Tianxin Huang[1], Xuemeng Yang[1], Jiangning Zhang[1,2], Jinhao Cui[1], Hao Zou[1],

Jun Chen[1], Xiangrui Zhao[1], and Yong Liu[1,3(✉)]

[1] APRIL Lab, Zhejiang University, Hangzhou, China
yongliu@iipc.zju.edu.cn
[2] Tencent Youtu Laboratory, Shanghai, China
[3] Huzhou Institute of Zhejiang University, Huzhou, China

Abstract. Reconstruction networks for well-ordered data such as 2D images and 1D continuous signals are easy to optimize through element-wised squared errors, while permutation-arbitrary point clouds cannot be constrained directly because their points permutations are not fixed. Though existing works design algorithms to match two point clouds and evaluate shape errors based on matched results, they are limited by pre-defined matching processes. In this work, we propose a novel framework named PCLossNet which learns to train a point cloud reconstruction network without any matching. By training through an adversarial process together with the reconstruction network, PCLossNet can better explore the differences between point clouds and create more precise reconstruction results. Experiments on multiple datasets prove the superiority of our method, where PCLossNet can help networks achieve much lower reconstruction errors and extract more representative features, with about 4 times faster training efficiency than the commonly-used EMD loss. Our codes can be found in https://github.com/Tianxinhuang/PCLossNet.

Keywords: Learning to train · No matching · Point cloud reconstruction

1 Introduction

To reconstruct a series of data, a network such as auto-encoder is usually adopted to predict an output as similar to the original data as possible, which is usually trained with reconstruction errors between original data and the network output. Reconstruction errors for 2D images or 1D signals are quite easy to calculate directly with element-wised mean squared errors (MSE) because their elements such as pixels are arranged in a certain order. However, matching algorithms are required to synchronize different data when calculating reconstruction errors for point clouds because permutations of input and output point sets in reconstruction networks may be different.

Different matching algorithms match points between point clouds according to different rules. Figure 1-(a) and (b) illustrate the matching processes adopted by two

Supplementary Information The online version contains supplementary material available at https://doi.org/10.1007/978-3-031-19769-7_11.

S. Avidan et al. (Eds.): ECCV 2022, LNCS 13661, pp. 179–194, 2022.
https://doi.org/10.1007/978-3-031-19769-7_11

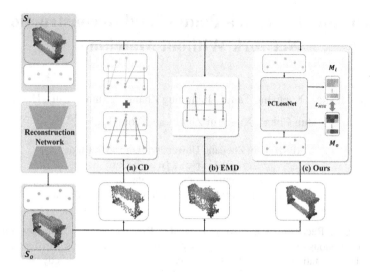

Fig. 1. (a) and (b) denote two matching processes of commonly-used CD and EMD losses, while (c) illustrates the PCLossNet framework.

commonly used structural losses: the Chamfer Distance (CD) and the Earth Mover's Distance (EMD) [4]. CD matches points in one point set with their nearest neighbors in the other set, while EMD optimizes to find a points bijection with the approximated minimum matching distance between point clouds. Many works [5,8,13,19,29,31] aims at improving the network structures to learn more representative features or construct better shapes, while all of them use the Chamfer Distance (CD) and the Earth Mover's Distance (EMD) as basic shape constraints. However, matching processes adopted by either CD or EMD are actually approximations for the shape differences. Converging well under matching may not mean the point cloud shapes are totally the same. Inevitable shape defects may exist due to biases between the pre-defined matching process and true shape differences. As illustrates in Fig. 1-(a) and (b), CD may create non-uniform surfaces because the its matching pays attention to the average neighbor distances, which allows one point to be neighbors of multiple points of another point set and lacks constraints for uniformity. Though EMD constructs bijection to create more uniform reconstructed results, the optimization processes to approximate the minimum matching distance may have biases and create distortions sometimes. Though some works introduce discriminator networks on point clouds [11,13,22] to enhance details, they still use CD or EMD to constrain basic structures of point clouds and are limited by the matching processes. In this condition, we want to get rid of the limitations of matching processes by replacing them with a differentiable network structure, in which way we can learn to train a reconstruction network without matching. In this work, we propose a novel struture named Point Cloud reconstruction Loss Network (PCLoss-Net) to train the reconstruction network without matching. As illustrated in Fig. 1-(c), PCLossNet extracts comparison matrices M_i and M_o from point clouds and evaluate their shape differences with distances between comparison matrices. To train the networks, parameters of the reconstruction network and PCLossNet are updated by turns in a generative-adversarial process. Intuitively speaking, in each iteration, PCLossNet explores the regions with larger reconstruction errors by maximizing distances between M_i and M_o, while the reconstruction network is optimized by minimizing the distances. The reconstruction network and PCLossNet promote each other to stimulate the potential of networks. Note that our work is different from existing GAN-based discriminator constraints because we design PCLossNet not only to improve reconstruction performances simply, but also to replace the pre-defined matching process. More relative discussions can be found in Sec. 3. Our contributions can be summarized as follows:

- We propose a new differentiable structure named PCLossNet to transform shape differences between point clouds to errors between extracted comparison matrices.
- By training with a generative adversarial process, PCLossNet can dynamically search for shape differences between point clouds and constrain the reconstruction network without any pre-defined matching process;
- Experiments on multiple datasets demonstrate that networks trained with PCLoss-Net can achieve better reconstruction performances and extract more representative features with higher classification accuracy.

2 Related Works

2.1 Optimization-Based Matching Losses

Point clouds are permutation-invariant, which means the data is irrelevant with the order of single points. Mean Square Error (MSE) commonly used in the reconstruction of 2D images or 1D signals is not appropriate for point clouds because the order of reconstructed point clouds may be quite different with the ground truths. Since the raise of PointNet [17] and PointNet++ [18], more and more works [9,14,19,31] have been developed to improve the reconstruction performances on point clouds. All of them are trained based on the Chamfer Distance or the Earth Mover's Distance [4]. The Chamfer Distance (CD) is defined as

$$\mathcal{L}_{CD}(S_1, S_2) = \frac{1}{2}(\frac{1}{|S_1|} \sum_{x \in S_1} \min_{y \in S_2} \|x - y\|_2 + \frac{1}{|S_2|} \sum_{x \in S_2} \min_{y \in S_1} \|x - y\|_2), \quad (1)$$

where S_1 and S_2 are two point sets. CD is actually the average distance from points in one set to their nearest neighbors in another set. A same nearest neighbor is allowed for multiple points for the calculation of CD. With the matching by nearest neighbors, CD concentrates on the differences between point clouds contours. But it usually constructs non-uniform surfaces because it lacks constraints for local uniformity.

The Earth Mover's Distance (EMD) can be presented as

$$\mathcal{L}_{EMD}(S_1, S_2) = \min_{\phi: S_1 \to S_2} \frac{1}{|S_1|} \sum_{x \in S_1} \|x - \phi(x)\|_2, \quad (2)$$

where S_1 and S_2 are two point sets. EMD aims to find an one-to-one optimal bijection ϕ from one point set to another by optimizing the minimum matching distances between point sets. An optimization process is needed to construct the bijection in each iteration. In practice, exact computation of EMD is too expensive for deep learning, even on graphics hardware. In this work, we follow [4] to conduct a $(1 + \epsilon)$ approximation scheme for EMD. The algorithm is easily parallelizable on GPU. EMD can create more uniform shapes by constructing bijection, while the optimized matching may cause distortions. Besides, EMD can only be applied to reconstructed output with the same number of points as input to solve the bijection, which limits its application. Some recent works [16,25] try different methods to improve CD or EMD, while their performances are still limited by the matching algorithms. Though DPdist [20] is a fully-network based training loss without any matching, it is proposed for registration instead of reconstruction, which is also inflexible due to the requirements of appropriate pre-training process.

2.2 Generative Adversarial Network

The Generative adversarial network [6] is an appropriate framework to train a network with another network. They are created to learn a transformation from a prior distribution such as the Gaussian distribution to the distribution of ground truths. In this way, sampling from the prior distribution can construct new data. A discriminator is adopted to judge if the generated data satisfy the same distribution as ground truths. The training strategy for a basic GAN can be presented as

$$\min_{G} \max_{D} V(G, D) = \mathbb{E}_{x \sim p_{data}(x)}[log D(x)] + \mathbb{E}_{z \sim p_z(z)}[log(1 - D(G(z)))], \quad (3)$$

where p_{data} and p_z are the distributions of real data and input noise variables, respectively. G is a generation network to transform sampled noise to fake but realistic generated data, while D is a discriminator network to distinguish generated data and real data. Many GANs [3,7,15] have been developed based on the basic generative adversarial framework. Discriminator proposed in GANs [6] is sometimes introduced to improve the detail preserved ability on reconstruction-related tasks [11,21,22] of point clouds. Nevertheless, existing works only use the discriminator as a supplement for structural losses CD or EMD to improve the reconstruction performances. They are still influenced by the pre-defined matching processes.

3 Methodology

In this section, we introduce the core ideas of our method. The whole pipeline of our algorithm is shown in Fig. 2. We propose a framework named PCLossNet to extract comparison matrices M_i and M_o from point clouds S_i and S_o and evaluate their differences by errors between corresponding comparison matrices. Details of PCLoss-Net are presented in Sect. 3.1. It is trained together with the reconstruction network in a generative-adversarial process. The training process is further demonstrated in Algorithm 1. Besides, we also conduct a simple theoretical analysis of our method in Sect. 3.3.

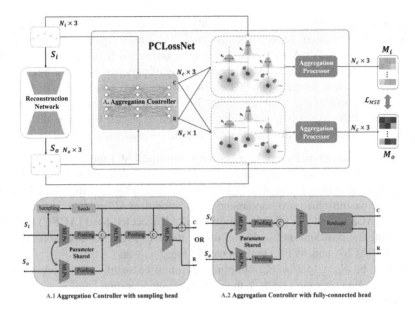

Fig. 2. The whole pipeline to train the reconstruction network with PCLossNet which transforms point clouds differences to the errors between comparison matrices M_i and M_o. N_i and N_o are the points numbers of point clouds S_i and S_o, while N_c denote the number of distributions to extract M_i and M_o, $j \in \{1, 2, ..., N_c\}$. A.Aggregation Controller and Aggregation processor are components of PCLossNet, while A.1 and A.2 are two different implements of A.Aggregation Controller. ⓒ and ⊕ denote the concatenation and element-wise addition, respectively.

3.1 The Architecture of PCLossNet

As illustrated in Fig. 2, PCLossNet plays the important role of extracting comparison matrices from point clouds. Existing point-based discriminators such as [13,22] predict scores to evaluate the similarities from point clouds with deep neural networks, which are totally non-linear structures. By running an adversarial training, the discriminators are expected to evaluate the shapes differences with the scores. However, same score may come from different outputs because the mapping from point clouds to scores is totally non-linear with unlimited searching spaces. Therefore, all existing point-based discriminators need a matching process to constrain reconstructed point clouds to similar shapes with original point clouds, which can reduce the searching spaces of discriminators and avoid the ambiguity of predicted scores as much as possible. They are actually limited by the biases between matching losses and true shapes differences.

In this condition, we decouple the non-linear discriminator structures into non-linear Aggregation Controller(AC) module to preserve the adversarial ability, and Aggregation processor (AP) module extracting comparison matrices M_i and M_o totally based on 3D Euclidean Space to naturally limit the searching spaces of comparison matrices. AP module extracts comparison codes M_i and M_o by weighting points with multiple distributions, while the centers and widths of these distributions are controlled with aggregation centers C and decay radii R predicted by AC module with MLPs from S_i and S_o. The number of weighting distributions is defined as N_c in this work. During the training process, C and R are dynamically adjusted to search for differences between S_i and S_o. The operations to aggregate points/features by weights in PCLossNet are similar as those in NetVLAD [2], while they have obvious differences on the specific network structures. NetVLAD uses a separate group of parameters for the aggregation around each clustering center, while PCLossNet uses a group of parameters in AC module to adjust aggregations around all centers. Besides, NetVLAD is trained end-to-end by loss, while PCLossNet is adversarially optimized to search for shape differences. More discussions about AC module and AP module are presented below.

Aggregation Controller. In the Aggregation Controller (AC) module, features from input S_i and reconstructed S_o are extracted with parameter-shared Multi Layer Perceptrons (MLPs) and symmetric pooling operations. To search for the differences between point clouds, we introduce two implements of AC modules: AC module with the SAmpling head (**ACSA**) and AC module with the Fully Connected head (**ACFC**) to C and R controlling the positions and widths of distributions to extract comparison matrices. Structures of ACSA and ACFC are presented in Fig. 2-A.1 and A.2.

Let input be S_i, output be S_o, sampling operation be $sam(\cdot)$, the combination of MLPs and pooling operation be $f(\cdot)$, the concatenation operation be $g(\cdot)$.

We can present the ACSA as

$$\begin{cases} seeds = sam(S_i), \\ C = seeds + MLPs(g(f(g(f(S_i),f(S_o),seeds)),seeds)), \\ R = MLPs(g(f(g(f(S_i),f(S_o),seeds)),seeds)), \end{cases} \quad (4)$$

while the ACFC can be described as

$$C, R = FCs(g(f(S_i),f(S_o))), \quad (5)$$

where C and R are the predicted aggregation centers and corresponding decay radii, respectively. $FCs(\cdot)$ and $MLPs(\cdot)$ denote the fully-connected layers and MLPs, respectively. Note that we provide two simple organized networks ACSA and ACFC as examples of AC modules in this work. More effective networks can be designed to futher improve the performance of PCLossNet.

Aggregation Processor. Aggregation processor (AP) module aggregates comparison matrices from point clouds on 3D Euclidean Space with multiple distributions controlled by aggregation centers C and decay radii R from AC module. Note that there is not any network structure in AP module. With provided centers C, decay radii R, and point cloud S, we can define AP module as follows.

(1) We build the graph $G = (V, E)$, where

$$V = C \cup S, E = \{<c, s>|\forall c \in C, \forall s \in S\}, \quad (6)$$

when $<c, s> = s - c$ exists for point clouds without direct edge information.

(2) Then we can aggregate the comparison matrices M from G by the distributions decided by aggregation centers C and decay radii R as

$$M = \{m|m = \sum_{k=1}^{N} \frac{exp(-\frac{\|e_k^j\|_2}{R_j+\sigma})}{\sum_{k=1}^{N} exp(-\frac{\|e_k^j\|_2}{R_j+\sigma}) + \delta} \cdot e_k^j, j \in \{1 \cdots N_c\}\}, \quad (7)$$

Algorithm 1. Training Process

Input: Dataset $\mathbf{S}_i$, the number of iterations $iter$, the reconstruction network $RecNet(\cdot)$

for $n = 1$ **to** $iter$ **do**

    Calculate output of the reconstruction network:

    $\mathbf{S}_o^n = RecNet(\mathbf{S}_i^n)$.

    Fix the parameter of reconstruction network and update PCLossNet by descending gradient:

    $\nabla_{\theta_L} L_{PCLossNet}(\mathbf{S}_o^n, \mathbf{S}_i^n)$.

    Fix PCLossNet and update the reconstruction network by descending gradient:

    $\nabla_{\theta_T} T_{PCLossNet}(\mathbf{S}_o^n, \mathbf{S}_i^n)$.

end for

---

where $e_k^j = <c_j, s^k>$ and $e_k^j \in E$. N is the points number of S, while N_c is the number of distributions aggregating comparison matrices, also the number of centers in C. σ is a tiny constant to ensure the equation still works when $R_j \to 0$. δ is a tiny constant to protect weights from going out of bounds when the denominator is small.

To ensures each point gets enough weight from distributions, we use the uniformity constraint to shorten distances between points and their nearest aggregation centers as

$$L_{UC} = \underset{\forall s \in S}{mean} \ \underset{<c,s>\in E}{min} \|c - s\|_2. \tag{8}$$

Besides, we restrain the decay radius R to reduce the decay radii around each center as

$$L_R = \|R\|_2^2. \tag{9}$$

3.2 Training of the Reconstruction Network

To train a point cloud reconstruction network, we update parameters of the reconstruction network and PCLossNet in a generative-adversarial process. Our training algorithm is presented in Algorithm 1. θ_L and θ_T are parameters of PCLossNet and the reconstruction network, respectively. Loss function for the updating of the reconstruction network is

$$T_{PCLossNet}(S_i, S_o) = \|M_i, M_o\|_2 + \epsilon * L_{UC}, \tag{10}$$

while loss function for the updating of PCLossNet is

$$L_{PCLossNet}(S_i, S_o) = -log(\|M_i, M_o\|_2 + \sigma_L) + L_{UC} + \epsilon_1 * L_R, \tag{11}$$

where M_i and M_o, L_{UC} and L_R are comparison matrices, uniformity constraint and decay radii constraint mentioned in Sect. 3.1. σ_L is a tiny constant to prevent the gradient of $L_{PCLossNet}$ from explosion when $\|M_i, M_o\|_2 \to 0$. ϵ and ϵ_1 are weights for components. L_{UC} is defined in Eq. 10 to ensure each point in reconstructed point clouds get enough weights to be constrained.

    Loss function for the reconstruction network is designed to reduce the error between comparison matrices, while the loss function for PCLossNet tries to explore more differences by increasing that error. We adopt $log(\cdot)$ to adjust the updating of PCLossNet dynamically, in which case the gradient would decrease as $\|M_i, M_o\|_2$ increases. In this way, PCLossNet is updated slowly when the reconstruction network is weak, fast when the reconstruction network works well and reaches a small reconstruction error.

3.3 Algorithm Analysis

To intuitively analyze our method, the training process can be modeled as a process of solving equations. As illustrated in Sect. 3.1, reconstructed output and the ground truths are abstracted into comparison matrices through Eq. 7. Let $s_i^k \in S_i$ and $s_o^k \in S_o$ be k-th point in input and output, $c \in C$ and $r \in R$ be aggregation center and decay radii. Then, for the input and reconstructed point clouds in each iteration, we have

$$\begin{cases} \|\sum_{k=1}^{N_i} w_{(s_i^k, c_1, r_1)} \cdot s_i^k - \sum_{k=1}^{N_o} w_{(s_o^k, c_1, r_1)} \cdot s_o^k\|_2 = \sigma_n^1 \\ \|\sum_{k=1}^{N_i} w_{(s_i^k, c_1, r_1)} \cdot s_i^k - \sum_{k=1}^{N_o} w_{(s_o^k, c_1, r_1)} \cdot s_o^k\|_2 = \sigma_n^2 \\ \vdots \\ \|\sum_{k=1}^{N_i} w_{(s_i^k, c_{N_c}, r_{N_c})} \cdot s_i^k - \sum_{k=1}^{N_o} w_{(s_o^k, c_{N_c}, r_{N_c})} \cdot s_o^k\|_2 = \sigma_n^{N_c}, \end{cases} \quad (12)$$

where $w_{(s_i^k, c_j, r_j)} = \dfrac{exp(-\frac{\|s_i^k - c_j\|_2}{r_j + \sigma})}{\sum_{k=1}^{N_i} exp(-\frac{\|s_i^k - c_j\|_2}{r_j + \sigma}) + \delta}$ as Eq. 7. N_c is the number of aggregation centers, while N_i and N_o are the number of input and reconstructed points, respectively. σ_n^j is the corresponding distance between comparison matrices around j-th aggregation center after n-th iteration. We can see that Eqs. 12 is undetermined in a single iteration because we usually have $N_c < N_i$ and $N_c < N_o$ to reduce the computational cost.

In each later iteration, a new group of equations are added. For $L_{PCLossNet}$ in Eq. 11, $-log(\|M_i, M_o\|_2)$ searches for equations as independent as possible from former ones during subsequent iterations, while L_{UC} and L_R improves local independence in the group of equations. L_{UC} would like to provide a near aggregation center for each point, while L_R tends to shrink the decay radii and concentrates bigger weights on fewer points. They will lead to uniform spatial positions of aggregation centers and smaller intersections between their neighbors, which will improve the local independence in each group of equations. As a result, the set of equations will approach to be determined after multiple iterations, which can constrain all points without matching.

4 Experiments

4.1 Datasets and Implementation Details

In this work, three point cloud datasets: ShapeNet [28], ModelNet10 (MN10) and ModelNet40 (MN40) [26] are adopted. Each model consists of 2048 points randomly sampled from the surfaces of original mesh models.

We train networks on train splits of ShapeNet part dataset following FoldingNet [27] and evaluate performances on both the test split of ShapeNet and MN40 to provide a robust and exhaustive evaluation. We optimize the reconstruction network by Adam Optimizer [12] with a learning rate of 0.0001, while PCLossNet is trained with a learning rate of 0.005. We compare commonly-used matching losses Chamfer Distance (CD)

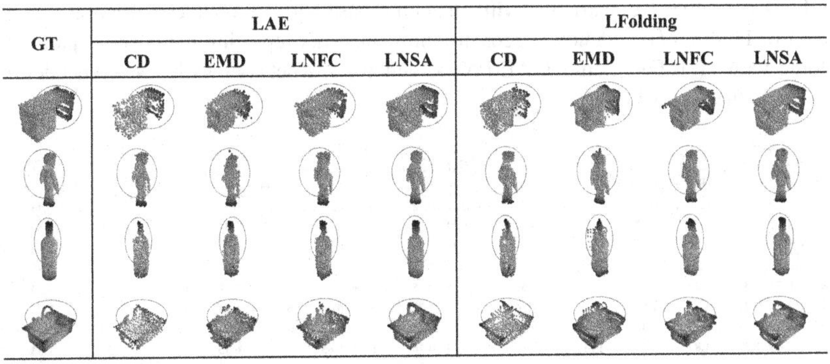

GT	LAE				LFolding			
	CD	EMD	LNFC	LNSA	CD	EMD	LNFC	LNSA

Fig. 3. Qualitative Comparison with matching losses on point clouds. We can see that our methods have great improvements on reconstruction details over CD and EMD with matching priors.

and Earth Mover's Distance (EMD) mentioned in Sect. 2.1 with two implements of PCLossNet: **PCLossNet with ACFC (LNFC)** and **PCLossNet with ACSA (LNSA)**, where ACFC and ACSA are defined in Sec. 3.1.

Metrics. To provide a clear and accurate evaluation for the reconstruction performance, we adopt multi-scale Chamfer Distance (MCD) proposed by [9] and Hausdorff distance (HD) following [24] as metrics in this work. Let input point cloud be S_i, reconstructed output point cloud be S_o, MCD can be defined as

$$MCD = \xi \cdot CD(S_i, S_o) + \frac{1}{|K|} \sum_{\forall k \in K} \frac{1}{|C|} \sum_{\forall c \in C} CD(S_i^{c,k}, S_o^{c,k}), \qquad (13)$$

where C denotes centers of evaluated local regions, which is acquired with farthest point sampling (FPS) [18] from S_i, S_o. $CD(\cdot)$ denotes the Chamfer Distance. K is a list including multiple k values to control the local region scales. $S_i^{c,k}$ means the local region on S_i with k points around center c. We can see that MCD evaluates both local and global reconstruction errors with Chamfer Distance, while ξ is a parameter to control their weights. Here, we have $K = \{4, 8, 16, 32, 64\}$ around 256 sampled C. HD can be defined as

$$HD = \frac{1}{2}(\max_{x \in S_i} \min_{y \in S_o} \|x - y\|_2^2 + \max_{x \in S_o} \min_{y \in S_i} \|x - y\|_2^2). \qquad (14)$$

We can see that HD measures the global worst reconstruction distortions. We use MCD to comprehensively consider global and local structural differences, and HD to compare the most obvious shape distortions. In this work, all metrics are multiplied with 10^2.

Reconstruction Networks. AE [1] and FoldingNet [27] are two classic and commonly used point cloud reconstruction networks, which have been used in many works [10, 13, 19, 30]. In this work, We apply PointNet [17], PointNet++ [18] and DGCNN [23] to the encoder parts of AE [1] and FoldingNet [27] to construct diverse reconstruction networks. We use 128-dim bottleneck layers in the encoder parts following AE [1].

Table 1. Quantitative Comparisons with matching-based losses on reconstruction networks. $AE(f(\cdot))$ or $Folding(f(\cdot))$ denotes reconstruction networks replacing the encoder parts of AE or Folding with $f(\cdot)$. For example, AE(PN++) denotes AE network with PN++ encoder. **Bold** values denote the best values.

Data		ShapeNet								ModelNet40							
RecNet		AE	Folding	AE (PN++)	Folding (PN++)	AE (DGCNN)	Folding (DGCNN)	LAE	LFolding	AE	Folding	AE (PN++)	Folding (PN++)	AE (DGCNN)	Folding (DGCNN)	LAE	LFolding
CD	MCD	0.32	0.42	0.37	0.34	0.30	0.52	0.31	0.28	0.75	0.83	0.88	0.79	0.76	0.75	0.44	0.39
	HD	1.87	4.20	2.50	3.37	1.88	3.84	1.02	1.20	6.08	7.35	7.38	7.55	6.40	7.03	1.69	2.16
EMD	MCD	0.25	–	0.26	–	0.21	–	0.23	0.21	0.61	–	0.66	–	0.56	–	0.33	0.32
	HD	2.23	–	2.51	–	2.09	–	2.48	2.40	6.18	–	6.47	–	5.66	–	3.82	3.88
LNFC	MCD	0.23	**0.32**	0.25	0.33	0.21	0.69	0.15	0.15	**0.58**	0.75	0.68	0.76	**0.56**	1.04	0.23	0.22
	HD	1.66	2.71	1.98	2.84	1.65	4.26	1.27	1.22	5.43	6.80	6.32	6.94	5.35	8.08	2.05	1.98
LNSA	MCD	**0.23**	0.33	**0.24**	**0.31**	**0.20**	**0.43**	**0.13**	**0.14**	0.59	0.75	**0.66**	**0.74**	0.60	**0.74**	**0.17**	**0.19**
	HD	**1.66**	**2.57**	**1.87**	**2.50**	**1.51**	**3.10**	**0.65**	**0.76**	**5.30**	**6.65**	**6.04**	**6.75**	**5.30**	**6.55**	**1.05**	**1.37**

Besides, to build stronger reconstruction networks, we divide point clouds into multiple local regions following PointNet++ [18] and apply AE and FoldingNet in each region to construct Local AE (LAE) and Local FoldingNet (LFolding) networks. To make a fair evaluation, we retrain all networks under same settings with different training losses.

4.2 Comparisons with Basic Matching-Based Losses

In this section, we conduct comparisons on multiple kinds of reconstruction networks trained with two basic matching-based losses CD and EMD. We do not introduce extra constraint in this part to make a comparison only between basic structural losses with matching and our method without matching.

Comparisons on Reconstruction Errors. The qualitative and quantitative results are presented in Fig. 3 and Table 1, respectively. We can see that our methods can achieve lowest reconstruction errors in most conditions and reconstruct models details much better than CD and EMD, which confirms that they can better constrain shape differences between point clouds. LNSA usually has better performances than LNFC because the sampling head adopted in LNSA may be easier to acquire aggregation centers around complicated shapes than the fully-connected head used in LNFC.

Besides, LNSA has greater improvements on LAE and LFolding networks, which means that it can tap into greater potential when the reconstruction network is stronger.

Comparisons on Unsupervised Classification. To further explore the effectiveness of PCLossNet, we also conduct a comparison on unsupervised classification following AE [1] and FoldingNet [27]. Specifically speaking, we train multiple auto-encoders with different loss methods and use the encoder parts to extract features from point clouds. Features extracted from train splits of MN10 and MN40 are adopted to train Supported Vector Machines (SVMs), whose classification accuracies are measured on test splits to evaluate the distinguishability of features. The results are presented in Table 2. We can see that our LNSA and LNFC can cover all best performed cases with higher classification accuracies, which means that PCLossNet can help the reconstruction network learn more representative features.

Table 2. Classification comparisons on MN10 and MN40.

RecNet		AE		Folding		AE (PN++)		Folding (PN++)		AE (DGCNN)		Folding (DGCNN)	
Dataset		MN10	MN40	MN10	MN40	MN10	MN40	MN10	MN40	MN10	MN40	MN10	MN40
Methods	CD	90.60	85.92	91.03	85.22	90.38	88.03	91.48	87.01	91.04	87.95	90.81	87.18
	EMD	89.49	85.47	–	–	90.15	88.07	–	–	90.48	87.78	–	–
	LNFC	91.15	86.08	89.93	84.82	90.71	**88.19**	91.04	**87.26**	91.81	87.91	91.26	86.12
	LNSA	**91.48**	**86.36**	**91.70**	**85.35**	**92.04**	87.54	**91.48**	86.73	**92.37**	**88.11**	**91.81**	**87.50**

Table 3. Comparison with existing GAN discriminator losses on point clouds reconstruction. The symbol * denotes training with only the discriminator and no matching loss.

Data	SP								MN40							
RecNet	AE		Folding		LAE		LFolding		AE		Folding		LAE		LFolding	
Metrics	MCD	HD	MCD	HD	MCD	HD	MCD	HD	MCD	HD	MCD	HD	MCD	HD	MCD	HD
PUGAN*	19.83	50.37	20.84	54.01	1.61	2.62	1.59	2.67	23.12	50.95	23.92	56.27	2.41	3.89	2.40	3.94
PFNet*	20.15	50.66	20.18	50.65	1.62	2.63	1.59	2.63	23.48	51.28	23.51	51.33	2.43	3.89	2.40	3.89
CRN*	14.36	50.63	20.18	50.65	1.44	2.65	3.10	5.72	17.22	49.26	23.51	51.33	2.16	3.89	3.69	6.92
PUGAN	0.32	1.88	0.36	3.83	0.32	1.02	0.27	1.11	0.73	5.85	0.77	7.29	0.45	1.71	0.38	1.97
PFNet	0.32	1.87	0.41	4.14	0.31	0.99	0.26	0.97	0.74	6.28	0.88	7.55	0.44	1.69	0.35	1.69
CRN	0.31	1.86	0.34	3.17	0.31	1.00	0.26	0.99	0.71	5.66	0.76	7.24	0.44	1.69	0.35	1.76
LNFC	0.23	1.66	**0.32**	2.71	0.15	1.27	0.15	1.22	**0.58**	5.43	0.75	6.80	0.23	2.05	0.22	1.98
LNSA	**0.23**	**1.66**	0.33	**2.57**	**0.13**	**0.65**	**0.14**	**0.76**	0.59	**5.30**	**0.75**	**6.65**	**0.17**	**1.05**	**0.19**	**1.37**

4.3 Comparisons with Discriminators-Based Losses

Some recent works such as PFNet [11], CRN [22] and PUGAN [13] also use discriminators to enhance the reconstruction performances. However, these works are quite different from PCLossNet because they still use matching-based CD or EMD loss to constrain basic structural differences. In this section, we make a comparison with discriminator-based losses from these works in Table 3. We can see that existing discriminator-based losses show inferior performances over our methods, while they cannot work without matching-based losses. Our work, especially LNSA can achieve obvious improvements over even these existing discriminator-based losses, which confirms it can break the limitation of existing matching-based reconstruction losses.

4.4 Comparisons on Training Efficiency

In this section, we evaluate the training efficiencies of different loss functions on AE networks [1], which are measured by the time consumed training a single iteration. The results are presented in Table 4. We can see that our method LNSA can achieve the best reconstruction performances while getting nearly 4 times faster than EMD. Though CD and PFNet loss are faster than LNSA, they get the worst reconstruction results.

To explore the efficiency potential of our method, we also conduct a fast implement of LNSA named FLNSA. From Algorithm 1 we can see that there are two back propagation processes in an iteration step of LNSA, which actually decreases the training efficiency. In this condition, we randomly give single back propagation process in each

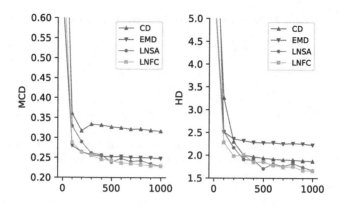

Fig. 4. Comparisons of the training processes.

iteration to improve the efficiency. In details, we assign 0.7 probability to train PCLoss-Net and 0.3 to train the reconstruction network. We can see that FLNSA would get same time efficiency with CD while getting much better reconstruction performances, which shows our potential on training efficiency.

Table 4. Training efficiency comparison on AE network. The comparisons are conducted based on an NVIDIA 2080ti with a 2.9 GHz i5-9400 CPU.

Methods	CD	EMD	PUGAN	PFNet	CRN	LNFC	LNSA	FLNSA
Time (ms)	**23**	216	77	45	97	56	57	**23**
MCD↓	0.32	0.25	0.32	0.32	0.31	0.23	**0.23**	0.24
HD ↓	1.87	2.23	1.88	1.87	1.86	1.66	**1.66**	1.80

4.5   How Is the Training Process Going?

To observe the convergence process of PCLossNet, we compare the reconstruction errors on ShapeNet and AE [1] during whole training process between different methods. The results are presented in Fig. 4. We can see that LNSA and LNFC have relatively large errors at the beginning of iterations because they need to search for the differences through training. But they will get lower errors than CD and EMD after enough iterations, which confirms that PCLossNet can help the reconstruction network converge better through the adversarial process.

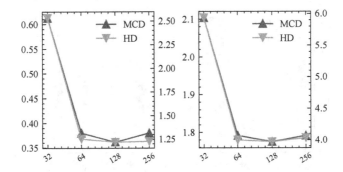

Fig. 5. Ablation study on the number of aggregation centers N_c. left and right vertical axises measure MCD and HD, while left and right pictures demonstrate results evaluates on ShapeNet and ModelNet40, respectively.

4.6 Ablation Study

Ablation for Components in PCLossNet. To clarify the function of each component in the loss, we conduct an ablation study here. $L_{log\|\cdot\|}$ denotes log operation mentioned in Sect. 3.2, while L_{UC} and L_R are components of the loss function defined in Sect. 3.1. We can see each module makes sense. Removing anyone will reduce the performances (Table 5).

Table 5. Ablation study for components in PCLossNet.

$L_{\|\cdot\|}$	$L_{log\|\cdot\|}$	L_{UC}	L_R	SP		MN40	
				MCD	HD	MCD	HD
✓	✗	✗	✗	0.26	1.91	0.66	5.95
✓	✓	✗	✗	0.25	1.90	0.62	5.80
✓	✓	✓	✗	0.23	1.72	0.59	5.52
✓	✓	✓	✓	**0.23**	**1.66**	**0.59**	**5.30**

Influence of Aggregation Centers Number. The number of aggregation centers N_c is actually a very important hyper-parameter in PCLossNet, which decides the number of distributions to extract comparison matrices. To choose an appropriate value for N_c, we conduct an related ablation study based on AE [1] and LNSA, whose results are presented in Fig. 5. We use logarithmic coordinates for vertical axises to show the differences clearer. We can see that the reconstruction network can reach the smallest reconstruction error when N_c is 128. Though larger N_c has close results, the computational cost also increase.

Influence of PCLossNet Learning Rate. The learning rate of PCLossNet is also an interesting hyper-parameter. It should be higher than the learning rate of the reconstruction network to ensure that PCLossNet can find regions with greater differences in each

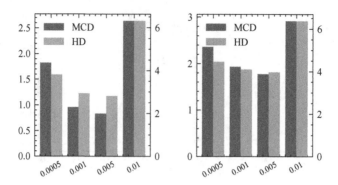

Fig. 6. Ablation study on the PCLossNet Learning rate. Left and right vertical axises measure MCD and HD, while left and right pictures demonstrate results evaluates on ShapeNet and ModelNet40, respectively.

iteration, which will help the reconstruction network converge well. However, too large learning rate may also block the convergence of PCLossNet. To determine the learning rate of PCLossNet, we conduct an ablation experiment and present it in Fig. 6. We can see that 0.005 is an proper option for PCLossNet.

5   Conclusion

In this work, we propose a novel learning-based framework named PCLossNet to help the point cloud reconstruction network to get rid of the limitation of commonly used matching processes. PCLossNet transforms differences between point clouds to the errors between extracted comparison matrices which are aggregated from point clouds with multiple distributions dynamically. By decoupling the extraction process into nonlinear Aggregation Controller module and 3D Euclidean space-based Aggregation processor, PCLossNet can get over the limitations of existing point-based discriminators and naturally supports adversarial training. By training in a generative-adversarial process together with the reconstruction network, PCLossNet can search for the main differences between reconstructed results and original point clouds and train the reconstruction network without any matching. Experiments on multiple datasets and reconstruction networks demonstrate reconstruction networks trained with PCLossNet can outperform those trained with matching-based losses, carrying smaller reconstruction errors and higher feature classification accuracy.

Acknowledgement. We thank all authors, reviewers and the chair for the excellent contributions. This work is supported by the National Science Foundation 62088101.

References

1. Achlioptas, P., Diamanti, O., Mitliagkas, I., Guibas, L.: Learning representations and generative models for 3D point clouds. In: International Conference on Machine Learning, pp. 40–49. PMLR (2018)

2. Arandjelovic, R., Gronat, P., Torii, A., Pajdla, T., Sivic, J.: NetVLAD: CNN architecture for weakly supervised place recognition. In: Proceedings of the IEEE Conference on Computer Vision and Pattern Recognition, pp. 5297–5307 (2016)

3. Arjovsky, M., Chintala, S., Bottou, L.: Wasserstein GAN. arXiv preprint arXiv:1701.07875 (2017)

4. Fan, H., Su, H., Guibas, L.J.: A point set generation network for 3D object reconstruction from a single image. In: Proceedings of the IEEE Conference on Computer Vision and Pattern Recognition, pp. 605–613 (2017)

5. Gadelha, M., Wang, R., Maji, S.: Multiresolution tree networks for 3D point cloud processing. In: Proceedings of the European Conference on Computer Vision (ECCV), pp. 103–118 (2018)

6. Goodfellow, I.J., et al.: Generative adversarial networks. arXiv preprint arXiv:1406.2661 (2014)

7. Gulrajani, I., Ahmed, F., Arjovsky, M., Dumoulin, V., Courville, A.C.: Improved training of Wasserstein GANs. In: Advances in Neural Information Processing Systems, pp. 5767–5777 (2017)

8. Han, Z., Wang, X., Liu, Y.S., Zwicker, M.: Multi-angle point cloud-VAE: unsupervised feature learning for 3D point clouds from multiple angles by joint self-reconstruction and half-to-half prediction. In: 2019 IEEE/CVF International Conference on Computer Vision (ICCV), pp. 10441–10450. IEEE (2019)

9. Huang, T., Liu, Y.: 3D point cloud geometry compression on deep learning. In: Proceedings of the 27th ACM International Conference on Multimedia, pp. 890–898 (2019)

10. Huang, T., et al.: RFNet: recurrent forward network for dense point cloud completion. In: Proceedings of the IEEE/CVF International Conference on Computer Vision, pp. 12508–12517 (2021)

11. Huang, Z., Yu, Y., Xu, J., Ni, F., Le, X.: PF-net: point fractal network for 3D point cloud completion. In: Proceedings of the IEEE/CVF Conference on Computer Vision and Pattern Recognition, pp. 7662–7670 (2020)

12. Kingma, D.P., Ba, J.: Adam: a method for stochastic optimization. arXiv preprint arXiv:1412.6980 (2014)

13. Li, R., Li, X., Fu, C.W., Cohen-Or, D., Heng, P.A.: PU-GAN: a point cloud upsampling adversarial network. In: Proceedings of the IEEE/CVF International Conference on Computer Vision, pp. 7203–7212 (2019)

14. Liu, M., Sheng, L., Yang, S., Shao, J., Hu, S.M.: Morphing and sampling network for dense point cloud completion. In: Proceedings of the AAAI Conference on Artificial Intelligence, vol. 34, pp. 11596–11603 (2020)

15. Mao, X., Li, Q., Xie, H., Lau, R.Y., Wang, Z., Paul Smolley, S.: Least squares generative adversarial networks. In: Proceedings of the IEEE International Conference on Computer Vision, pp. 2794–2802 (2017)

16. Nguyen, T., Pham, Q.H., Le, T., Pham, T., Ho, N., Hua, B.S.: Point-set distances for learning representations of 3D point clouds. In: Proceedings of the IEEE/CVF International Conference on Computer Vision, pp. 10478–10487 (2021)

17. Qi, C.R., Su, H., Mo, K., Guibas, L.J.: PointNet: deep learning on point sets for 3D classification and segmentation. In: Proceedings of the IEEE Conference on Computer Vision and Pattern Recognition, pp. 652–660 (2017)

18. Qi, C.R., Yi, L., Su, H., Guibas, L.J.: PointNet++: deep hierarchical feature learning on point sets in a metric space. In: Advances in Neural Information Processing Systems, pp. 5099–5108 (2017)

19. Rao, Y., Lu, J., Zhou, J.: Global-local bidirectional reasoning for unsupervised representation learning of 3D point clouds. In: Proceedings of the IEEE/CVF Conference on Computer Vision and Pattern Recognition, pp. 5376–5385 (2020)

20. Urbach, D., Ben-Shabat, Y., Lindenbaum, M.: DPDist: comparing point clouds using deep point cloud distance. In: Vedaldi, A., Bischof, H., Brox, T., Frahm, J.-M. (eds.) ECCV 2020. LNCS, vol. 12356, pp. 545–560. Springer, Cham (2020). https://doi.org/10.1007/978-3-030-58621-8_32

21. Wang, H., Jiang, Z., Yi, L., Mo, K., Su, H., Guibas, L.J.: Rethinking sampling in 3D point cloud generative adversarial networks. arXiv preprint arXiv:2006.07029 (2020)

22. Wang, X., Ang Jr, M.H., Lee, G.H.: Cascaded refinement network for point cloud completion. In: Proceedings of the IEEE/CVF Conference on Computer Vision and Pattern Recognition, pp. 790–799 (2020)

23. Wang, Y., Sun, Y., Liu, Z., Sarma, S.E., Bronstein, M.M., Solomon, J.M.: Dynamic graph CNN for learning on point clouds. Acm Trans. Graph. (tog) **38**(5), 1–12 (2019)

24. Wu, C.H., et al.: PCC arena: a benchmark platform for point cloud compression algorithms. In: Proceedings of the 12th ACM International Workshop on Immersive Mixed and Virtual Environment Systems, pp. 1–6 (2020)

25. Wu, T., Pan, L., Zhang, J., Wang, T., Liu, Z., Lin, D.: Density-aware chamfer distance as a comprehensive metric for point cloud completion. arXiv preprint arXiv:2111.12702 (2021)

26. Wu, Z., et al.: 3D shapenets: a deep representation for volumetric shapes. In: Proceedings of the IEEE Conference on Computer Vision and Pattern Recognition, pp. 1912–1920 (2015)

27. Yang, Y., Feng, C., Shen, Y., Tian, D.: FoldingNet: point cloud auto-encoder via deep grid deformation. In: Proceedings of the IEEE Conference on Computer Vision and Pattern Recognition, pp. 206–215 (2018)

28. Yi, L., et al.: A scalable active framework for region annotation in 3d shape collections. ACM Trans. Graph. (ToG) **35**(6), 1–12 (2016)

29. Yu, L., Li, X., Fu, C.W., Cohen-Or, D., Heng, P.A.: PU-net: point cloud upsampling network. In: Proceedings of the IEEE Conference on Computer Vision and Pattern Recognition, pp. 2790–2799 (2018)

30. Yuan, W., Khot, T., Held, D., Mertz, C., Hebert, M.: PCN: point completion network. In: 2018 International Conference on 3D Vision (3DV), pp. 728–737. IEEE (2018)

31. Zhao, Y., Birdal, T., Deng, H., Tombari, F.: 3D point capsule networks. In: Proceedings of the IEEE/CVF Conference on Computer Vision and Pattern Recognition, pp. 1009–1018 (2019)

PanoFormer: Panorama Transformer
for Indoor 360° Depth Estimation

Zhijie Shen[1,2], Chunyu Lin[1,2(✉)], Kang Liao[1,2,3], Lang Nie[1,2], Zishuo Zheng[1,2], and Yao Zhao[1,2]

[1] Institute of Information Science, Beijing Jiaotong University, Beijing, China
{zhjshen,cylin,kang_liao,nielang,zszheng,yzhao}@bjtu.edu.cn
[2] Beijing Key Laboratory of Advanced Information Science and Network Technology, Beijing, China
[3] Max Planck Institute for Informatics, Saarbrücken, Germany
https://github.com/zhijieshen-bjtu/PanoFormer

Abstract. Existing panoramic depth estimation methods based on convolutional neural networks (CNNs) focus on removing panoramic distortions, failing to perceive panoramic structures efficiently due to the fixed receptive field in CNNs. This paper proposes the panorama transformer (named *PanoFormer*) to estimate the depth in panorama images, with tangent patches from spherical domain, learnable token flows, and panorama specific metrics. In particular, we divide patches on the spherical tangent domain into tokens to reduce the negative effect of panoramic distortions. Since the geometric structures are essential for depth estimation, a self-attention module is redesigned with an additional learnable token flow. In addition, considering the characteristic of the spherical domain, we present two panorama-specific metrics to comprehensively evaluate the panoramic depth estimation models' performance. Extensive experiments demonstrate that our approach significantly outperforms the state-of-the-art (SOTA) methods. Furthermore, the proposed method can be effectively extended to solve semantic panorama segmentation, a similar pixel2pixel task.

1 Introduction

Depth information is important for computer systems to understand the real 3D world. Monocular depth estimation has attracted researchers' [6–8,15,39, 40] attention with its convenience and low cost, especially for panoramic depth estimation [34,41,43], where the depth of the whole scene can be obtained from a single 360° image.

Since estimating depth from a single image is an ill-posed and inherently ambiguous problem, current solutions almost use powerful CNNs to extract explicitly or implicitly prior geometric to realize it [3,5]. However, when applied

---

Supplementary Information The online version contains supplementary material available at https://doi.org/10.1007/978-3-031-19769-7_12.

S. Avidan et al. (Eds.): ECCV 2022, LNCS 13661, pp. 195–211, 2022.
https://doi.org/10.1007/978-3-031-19769-7_12

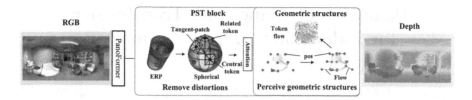

Fig. 1. We present PanoFomer to establish panoramic perception capability. The tangent-patch is proposed to remove panoramic distortions, and the token flows force the token positions to fit the structure of the sofa better. More details refer to Sec. 3

to panoramic tasks, these SOTA depth estimation solutions for perspective imagery [21] show a dramatic degradation because the 360° field-of-view (FoV) from panorama brings geometric distortions that challenge the structure perception. Specifically, distortions in panoramas (usually represented in equirectangular projection—ERP) increase from the center to both sides along the latitude direction, severely deforming objects' shapes. Due to the fixed receptive field, CNNs are inferior for dealing with distortions and perceiving geometric structures in panoramas [5]. To deal with the distortions in panoramas, some researchers [17,28,29,34] adopt the projection-fusion strategy. But this strategy needs to cover the domain gap between different projections, and the extra cross-projection fusion module increases computational burdens. Other researchers [9,10,13,16,19,22,30,30,37,38,44] employ various distortion-aware convolution filters to make CNN-based depth estimation solutions adapt to 360° images. However, the fixed sampling positions still limit their performance. Pintore *et al.* [26] focuses on the full geometric context of an indoor scene, proposing SliceNet but losing detailed information when reconstructing the depth map. We note that all the existing methods cannot perceive the distorted geometric structures with the fixed receptive field.

To address the above limitations, we propose the first panorama Transformer (PanoFormer) to enable the network's panoramic perception capability by removing distortions and perceiving geometric structures simultaneously (shown in Fig. 1). To make the Transformer suitable for panoramic dense prediction tasks (e.g., depth estimation and semantic segmentation), we redesign its structure. First, we propose a dense patches dividing method and handcrafted tokens to catch detailed features. Then, we design a relative position embedding method to reduce the negative effect of distortions, which utilizes a central token to locate the eight most relevant tokens to form a tangent patch (it differs from directly dividing patches on the ERP domain in traditional vision Transformers). To achieve this goal, we propose an efficient spherical token locating model (STLM) to guide the 'non-distortion' token sampling process on the ERP domain directly by building the Transformations among the three domains (shown in Fig. 3). Subsequently, we design a Panoramic Structure-guided Transformer (PST) block to replace the traditional block in a hierarchical architecture. Specifically, we redesign the self-attention module with additional learnable weight to push token flow, so as to flexibly capture various objects' structures.

This module encourages the PanoFormer to further perceive geometric structures effectively. In this way, we establish our network's perception capability to achieve panoramic depth estimation. Moreover, the proposed PST block can be applied to other learning frameworks as well.

Furthermore, current evaluation metrics for depth estimation are suitable for perspective imagery. However, these metrics did not consider distortions and the seamless boundary property in panoramas. To comprehensively evaluate the depth estimation for panoramic images, we design a Pole Root Mean Square Error (P-RMSE) and Left-Right Consistency Error (LRCE) to measure the accuracy on polar regions and depth consistency around the boundaries, respectively.

Extensive experiments demonstrate that our solution significantly outperforms SOTA algorithms in panoramic depth estimation. Besides, our solution achieves the best performance when applied to semantic segmentation, which is also a pixel2pixel panoramic task. The contributions of this paper are summarized as follows:

- We present *PanoFormer*, the first panorama Transformer, to establish the panoramic perception capability by reducing distortions and perceiving geometric structures for the panoramic depth estimation task.
- We propose a PST block that divides patches on the spherical tangent domain and reshapes the self-attention module with the learnable token flow. Moreover, the proposed block can be applied in other learning frameworks.
- Considering the difference between Panorama and normal images, we design two new panorama-specific metrics to evaluate the panoramic depth estimation.
- Experiments demonstrate that our method significantly outperforms the current state-of-the-art approaches on all metrics. The excellent panorama semantic segmentation results also prove the extension ability of our model.

2   Related Work

2.1   Panoramic Depth Estimation

There are two main fusion methods to reduce distortions while estimating depth on ERP maps. One is the equirectangular-cube fusion method represented by Bifuse [34], and the other is the dual-cube fusion approach described by Shen [28]. Specifically, Bifuse [34] propose a two-branch method of fusing equirectangular projection and cube projection, which improves the tolerance of the model to distortions. Moreover, UniFuse [17] also uses a dual projection fusion scheme only at the encoding stage to reduce computation cost. Noting that the single-cube projection method produces significant discontinuities at the cube boundary, Shen *et al* [28] proposed a dual-cube approach based on a 45° rotation to reduce distortions. This class of methods can attenuate the negative effect of distortions, but they need to repeatedly change the projection for fusion, increasing the model's complexity.

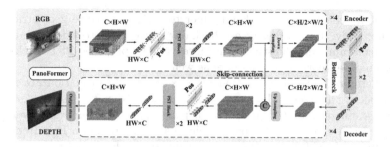

Fig. 2. Our PanoFormer takes a monocular RGB panoramic image as the input and outputs the high-quality depth map

To apply depth estimation models of normal images to panoramas, Tateno *et al.* [33] obtained exciting results by designing distortion-aware convolution filters to expand the perceptual field. Zioulis *et al.* [45] demonstrated that monocular depth estimation models trained on conventional 2D images produce low-quality results, highlighting the necessity of learning directly on the 360° domain. Jin *et al.* [18] demonstrated the effectiveness of geometric prior for panoramic depth estimation. Chen *et al.* [5] used strip pooling and deformable convolution to design a new encoding structure for accommodating different degrees of distortions. Moreover, Pintore *et al.* [26] proposed SliceNet, a network similar to HorizonNet [31], which uses a bidirectional Long Short-Term Memory (LSTM) to model long-range dependencies. However, the slicing method ignores the latitudinal distortion property and thus cannot accurately predict the depth near the poles. Besides, [2,11] proved that on large-scale datasets, Transformer-based depth estimation for normal images are superior to CNN.

2.2 Vision Transformer

Unlike CNN-based networks, the Transformer has the nature to model long-range dependencies by global self-attention [27]. Inspired by ViT [11], researchers have designed many efficient networks that have the advantages of both CNNs and Transformers. To enhance local features extraction, convolutional layers are added into muti-head self-attention (CvT [36]) and feed-forward network (FFN) (CeiT [42], LocalViT [23]) is replaced by locally-enhanced feed-forward network (LeFF) (Uformer [35]). Besides, CvT [36] demonstrates that the padding operation in CNNs implicitly encodes position, and CeiT [42] proposes the image-to-tokens embedding method. Inspired by SwinT [24], Uformer [35] proposes a shifted windows-based multi-head attention mechanism to improve the efficiency of the model. But all these solutions are developed based on normal FoV images, which cannot be applied to panoramic images directly. Based on these previous works, we further explore suitable Transformer structure for panoramic images and adapt it to the dense prediction task.

3 PanoFomer

3.1 Architecture Overview

Our primary motivation is to make the Transformer suitable for pixel-level omnidirectional vision tasks by redesigning the standard components in conventional Transformers. Specifically, we propose a pixel-level patch division strategy, a relative position embedding method, and a panoramic self-attention mechanism. The proposed pixel-level patch division strategy is to enhance local features and improve the ability of Transformers to capture detailed features. For position embedding, we renounce the conventional absolute position embedding method and get the position of other related tokens on the same patch by the central token (described in Sect. 3.3). This method not only eliminates distortions, but also provides position embedding. Furthermore, we establish a learnable flow in the panorama self-attention module to perceive panoramic structures that are essential for depth estimation.

As shown in Fig. 2, the PanoFomer is a hierarchical structure with five major parts: input stem, output stem, encoder, decoder and bottleneck. For the input stem, a 3×3 convolution layer is adopted with size $H \times W$ to form the features with dimension C. Then the features are fed into the encoder. There are four hierarchical stages in encoder and decoder, and each of them contains a position embedding, two PST blocks (sharing the same settings), and a convolution layer. Specifically, a 4×4 convolution layer is adopted for increasing dimension and down-sampling in the encoder, while a 2×2 transposed convolution layer is used in the decoder for decreasing dimension and up-sampling. Finally, the output features from the decoder share the same resolution and dimension as the features from the input stem. Furthermore, the output stem, implemented by a 3×3 convolution, is employed to recover the depth map from features. More specifically, the number of heads is sequentially set as [Encoder:1, 2, 4, 8; Bottleneck: 16; Decoder: 16, 8, 4, 2]. As for all padding operations in convolution layers, we utilize circular padding for both horizontal sides of the features.

3.2 Transformer-Customized Spherical Token

In vision Transformers, the input image is first divided into patches of the same size. For example, ViT [11] divides the input image into patches with size of 16×16 to reduce the computational burden. Then, these patches are embedded as tokens in a learning-based way via a linear layer. However, this strategy loses much detailed information, which is a fatal drawback for dense prediction tasks, such as depth estimation. To overcome this issue, we propose a pixel-level patches dividing method.

First, the input features are divided into pixel-level patches, which means each sampling position in the features corresponds to a patch centered on it. Such a dense division strategy allows the network to learn more detailed features, which is beneficial for dense prediction tasks. Furthermore, we make each patch consist of 9 features at different positions (one central position and eight

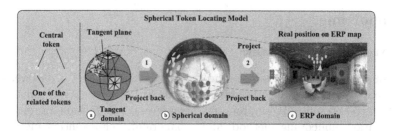

Fig. 3. Spherical Token Locating Model (STLM): locate related tokens on ERP domain. 1: tangential domain of unit sphere to spherical domain; 2: spherical domain to ERP domain

surrounding positions, illustrated in Fig. 3 left) to balance the computational burden. Unlike standard Transformers that embed patches as tokens by a linear layer, our tokens are handcrafted. We define the features at the central position as the central token and those from the other 8 surrounding positions as the related tokens. The central token can determine the position of related tokens by looking up the eight most relevant tokens among the features. To remove distortion and embed position information for the handcrafted tokens, we propose a distortion-based relative position embedding method in Sect. 3.3.

3.3 Relative Position Embedding

Inspired by the cube projection, we note that the spherical tangent projection can effectively remove the distortion (see Supplementary Materials for proof). Therefore, we propose STLM to initialize the position of related tokens. Unlike the conventional Transformers (e.g., ViT [11]), which directly adds absolute position encoding to the features, we "embed" the position information via the central token. Firstly, the central token is projected from the ERP domain to the spherical domain; then, we use the central token to look up the position of eight nearest neighbors on the tangent plane; finally, these positions are all projected back to the ERP domain (the three steps are represented by yellow arrows in Fig. 3). We call patches formed in this way as tangent patches. To facilitate locating the related tokens in the ERP domain, we further establish the relationship among the three domains (illustrated in Fig. 3).

Tangent Domain to Spherical Domain: Let the unit sphere be S^2, and $S(0,0) = (\theta_0, \phi_0) \in S^2$ is the spherical coordinate origin. $\forall S(x,y) = (\theta, \phi) \in S^2$, we can obtain other 8 points (related tokens) around it (current token) on the spherical domain.

$$S(\pm 1, 0) = (\theta \pm \Delta\theta, \phi)$$
$$S(0, \pm 1) = (\theta, \phi \pm \Delta\phi) \tag{1}$$
$$S(\pm 1, \pm 1) = (\theta \pm \Delta\theta, \phi \pm \Delta\phi)$$

where (θ, ϕ) denotes the unit spherical coordinates, and $\theta \in (-\pi, \pi)$, $\phi \in (-\frac{\pi}{2}, \frac{\pi}{2})$; $\Delta\theta, \Delta\phi$ is the sampling step size.

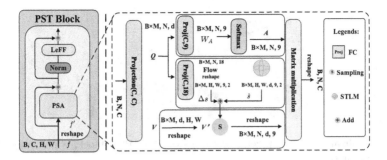

Fig. 4. The proposed PST Block can remove the negative effect of distortions and perceive geometric structures

By the geocentric projection [25], we can calculate the local coordinates $(T(x, y))$ of the sampling point in tangent domain [9] (the current token in tangent domain is represented as $T(0, 0) = T(\theta, \phi) = (0, 0)$):

$$T(\theta \pm \Delta\theta, \phi) = (\pm \tan \Delta\phi, 0)$$
$$T(\theta, \phi \pm \Delta\phi) = (0, \phi \pm \tan \Delta\theta)$$
$$T(\theta \pm \Delta\theta, \phi \pm \Delta\phi) = (\pm \tan \Delta\phi, \pm \sec \Delta\phi \tan \Delta\theta)$$

(2)

By applying the inverse projection described in [9], we can get the position of all tokens of a tangent patch in the spherical domain.

Spherical Domain to ERP Domain: Furthermore, by utilizing the projection equation [28], we can get the position of each tangent patch in the ERP domain. This whole process is named Spherical Token Locating Model (STLM).

3.4   Panorama Self-attention with Token Flow

Based on the traditional vision Transformer block, we replace the original attention mechanism with panorama self-attention. To further enhance local features interaction, we replace FFN with LeFF [42] for our pixel-level depth estimation task. Specifically, as illustrated in Fig. 4, when the features $f \in \mathbb{R}^{C \times H \times W}$ with a height of H and a width of W are fed into PST block, they are flattened and reshaped as $f' \in \mathbb{R}^{N \times C}$, where $N = H \times W$. Then a fully connected layer is applied to obtain query $Q \in \mathbb{R}^{N \times d}$ and value $V \in \mathbb{R}^{N \times d}$, where $d = C/M$, and M is the head number. The Q and V will pass through three parallel branches for computing attention score ($A \in \mathbb{R}^{N \times 9}$), token flows ($\Delta s \in \mathbb{R}^{N \times 18}$), and re-sampling features. In the top branch, a full connection layer is adopted to get attention weights $W_A \in \mathbb{R}^{N \times 9}$ from Q, and then softmax is employed to calculate the attention score A. In the middle branch, another fully connection layer is used to learn a token flow Δs and it is further reshaped to $\Delta s' \in \mathbb{R}^{d \times H \times W \times 9 \times 2}$, sharing the same dimension with $\hat{s}$ (the initialed position from the STLM). Moreover, $\Delta s'$ and $\hat{s}$ are added together to calculate the final token positions. In the bottom branch, the value V is reshaped to $V' \in \mathbb{R}^{C \times H \times W}$ and are sampled to

202   Z. Shen et al.

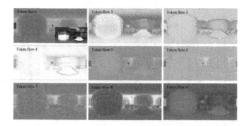

Fig. 5. Visualization of the token flows from the first PST block, which suggest the panoramic structures

form the divided patches (described in 3.2) by looking up the related tokens in the final token positions. Afterward, the PSA can be represented as follows:

$$
\text{PSA}(f, \hat{s}) = \sum_{m=1}^{M} W_m *
\left[\sum_{q=1}^{H \times W} \sum_{k=1}^{9} A_{mqk} \cdot W'_m f \left(\hat{s}_{mqk} + \Delta s_{mqk} \right) \right],
\tag{3}
$$

where $\hat{s} = \textit{\textbf{STLM}}(f)$, and $\textit{\textbf{STLM}}(\cdot)$ denotes the spherical token locating model; m indexes the head of self-attention, M is the whole heads, q index the current point (token), k indexes the tokens in a tangent patch, Δs_{mqk} is the learned flow of each token, A_{mqk} represents the attention weight of each token, and W_m and W'_m are normal learnable weights of each head.

From the above process, we can see that the final positions of the tokens are determined by two steps: position initialization from STLM and additional learnable flow. Actually, the initialized position realizes the division of the tangent patch (described in Sect. 3.3) and removes the panoramic distortion. Furthermore, the learnable flow exhibits a panoramic geometry by adjusting the spatial distribution of tokens. To verify the effectiveness of the token flow, we visualize all tokens from the first PST block in Fig. 5. It can be observed that this additional flow provides the network with clear scene geometric information, which helps the network to estimate the panorama depth with the structure as a clue.

3.5   Objective Function

For better supervision, we combine reverse Huber [14] (or Berhu [21]) loss and gradient loss [28] to design our objective function as commonly used in previous works [26,28]. In our objective function, the Berhu loss β_δ can be written as:

$$
\beta_\delta(g, p) = \begin{cases} |g - p| & \text{for } |g - p| \leq \delta \\ \frac{|(g-p)^2| + \delta^2}{2\delta} & \text{otherwise} \end{cases}
\tag{4}
$$

where g, p denote the ground truth and predicted values, respectively.

Similar to SliceNet [26], we apply gradient loss to Berhu loss. To obtain depth edges, we use two convolution kernels to obtain gradients in horizontal and vertical directions, respectively. They are represented as K_h and K_v, where

$K_h = [\text{-1 0 1, -2 0 2, -1 0 1}]$, and $K_v = (K_h)^T$. Denote the gradient function as G, the horizontal gradient I_h and vertical gradient I_v of the input image I can be expressed as $I_h = G(K_h, I)$ and $I_v = G(K_v, I)$, respectively. In this paper, $\delta = 0.2$ and the final objective function can be written as

$$\ell_{final} = \beta_{0.2}(g, p) + \beta_{0.2}(G(K_h, g), G(K_h, p)) + \beta_{0.2}(G(K_v, g), G(K_v, p)), \quad (5)$$

4 Panorama-Specific Metrics

Rethinking the spherical domain, we note that two significant properties cannot be neglected: the spherical domain is continuous and seamless everywhere; the distortion in the spherical domain is equal everywhere. For the first issue, we propose LRCE to measure the depth consistency of left-right boundaries. For the second issue, since distortions on ERP maps vary in longitude, RMSE cannot visually reflect the model's ability to adapt to distortions. Therefore, we provide P-RMSE to focus on the regions with massive distortions to verify the model's panoramic perception capability.

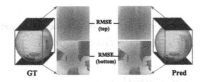

Fig. 6. P–RMSE: calculate the RMSE of the polar regions

Pole Root Mean Square Error. Cube projection is a special spherical tangent projection format that projects the sphere onto the cube's six faces. The top and bottom faces correspond to the polar regions of the spherical domain, so we select the two parts to design P-RMSE (illustrated in Fig. 6). Define the function of converting ERP to Cube as $E2C(\cdot)$, the converted polar regions of the ERP image E can be expressed as $Select(E2C(E), T, B)$, where T, B represent the top and bottom parts, respectively. The error C_e between the ground truth GT and the predicted depth map P at the polar regions can be expressed as

$$C_e = Select(E2C(GT), T, B) - Select(E2C(P), T, B) \quad (6)$$

The final P-RMSE can be written as

$$\text{P-RMSE} = \sqrt{\frac{1}{N_{C_e}} \sum_{i=1}^{N_{C_e}} |C_e^i|} \quad (7)$$

where N_{C_e} is the number of values in C_e.

Left-Right Consistency Error. We can evaluate the depth consistency of the left-right boundaries by calculating the horizontal gradient between the both sides of the depth map. Define that the horizontal gradient G_E^H of the image E can be written as $G_E^H = E_{first}^{col} - E_{last}^{col}$, where $E_{first}^{col}/E_{last}^{col}$ represent the values in the first/last columns of the image E. But consider an extreme case where if the edge of an object in the scene happens to be on the edge of the depth map, then there is ambiguity in reflecting continuity only by G_E^H. We cannot tell whether this discontinuity is real or caused by the model. Therefore, we add ground truth to our design. The horizontal gradient of ground truth and the predicted depth map are denoted as G_{GT}^H and G_P^H (where $G_{GT}^H = GT_{first}^{col} - GT_{last}^{col}$, $G_P^H = P_{first}^{col} - P_{last}^{col}$), respectively. The final expression can be as follows:

$$\text{LRCE} = \frac{1}{N_{error}} \sum_{i=1}^{N_{error}} |error_i| \tag{8}$$

where $error = G_{GT}^H - G_P^H$ and N_{error} is the number of values in $error$.

5 Experiments

In the experimental part, we compare the state-of-the-art approaches on four popular datasets and validate the effectiveness of our model.

5.1 Datasets and Implementations

Four datasets are used for our experimental validation, they are Stanford2D3D [1], Matterport3D [4], PanoSUNCG [29] and 3D60 [45].

Stanford2D3D and Matterport3D are two real-world datasets. They were rendered from a common viewpoint. Previous work used a dataset that was rendered only on the equator and its surroundings, ignoring the area near the poles, which undermined the integrity of the panorama. We strictly follow the previous works and employ the official datasets (Notice that the Stanford2D3D and Matterport3D that are contained in 3D60 have a problem that the light in the scenarios will leak the depth information). PanoSUNCG is a virtual panoramic dataset. And 3D60 is an updated version of 360D (360D is no longer available now). It consists of data from the above three datasets. There is a gap between the distributions of these three datasets, which makes the dataset more responsive to the model's generalizability. Note that we divide the dataset as the previous work and eliminate the samples that failed to render [5,34].

In the implementation, we conduct our experiments on two GTX 3090 GPUs, and the batch size is set to 4. We choose Adam [20] as the optimizer and keep the default settings. The initialized learning rate is 1×10^{-4}. The number of parameters of our model is 20.37 M.

Table 1. Quantitative comparisons on Matterport3D, Stanford2D3D, PanoSUNCG and 3D60 Datasets.

Dataset	Method	Classic metrics					
		$\delta_1 \uparrow$	$\delta_2 \uparrow$	$\delta_3 \uparrow$	**RMSE↓**	MRE↓	MAE↓
		Higher the better			Lower the better		
Matterport3D	FCRN [21]	0.7703	0.9174	0.9617	0.6704	0.2409	0.4008
	OmniDepth [45]	0.6830	0.8794	0.9429	0.7643	0.2901	0.4838
	Bifuse [34]	0.8452	0.9319	0.9632	0.6295	0.2408	0.3470
	UniFuse [17]	0.8897	0.9623	0.9831	0.4941	–	0.2814
	SliceNet [26]	0.8716	0.9483	0.9716	–	0.1764	0.3296
	Ours	**0.9184**	**0.9804**	**0.9916**	**0.3635**	**0.0571**	**0.1013**
Stanford2D3D	FCRN [21]	0.7230	0.9207	0.9731	0.5774	0.1837	0.3428
	OmniDepth [45]	0.6877	0.8891	0.9578	0.6152	0.1996	0.3743
	Bifuse [34]	0.8660	0.9580	0.9860	0.4142	0.1209	0.2343
	UniFuse [17]	0.8711	0.9664	0.9882	0.3691	–	0.2082
	SliceNet [26]	0.9031	0.9723	0.9894	–	0.0744	0.1048
	Ours	**0.9394**	**0.9838**	**0.9941**	**0.3083**	**0.0405**	**0.0619**

Dataset	Method	Classic metrics				New metrics	
		$\delta_1 \uparrow$	$\delta_2 \uparrow$	$\delta_3 \uparrow$	**RMSE↓**	P-RMSE↓	LRCE↓
PanoSUNCG	FCRN [21]	0.9532	0.9905	0.9966	0.2833	0.1094	0.1119
	OmniDepth [45]	0.9092	0.9702	0.9851	0.3171	0.0929	0.0913
	Bifuse [34]	0.9590	0.9838	0.9907	0.2596	0.0967	0.0735
	UniFuse [17]	0.9655	0.9846	0.9912	0.2802	0.0826	0.0884
	Ours	**0.9780**	**0.9961**	**0.9987**	**0.1503**	**0.0537**	**0.0442**
3D60	FCRN [21]	0.9532	0.9905	0.9966	0.2833	0.1681	0.2100
	OmniDepth [45]	0.9092	0.9702	0.9851	0.3171	0.1373	0.1941
	Bifuse [34]	0.9699	0.9927	0.9969	0.2440	0.1229	0.1357
	UniFuse [17]	0.9835	0.9965	0.9987	0.1968	0.0829	0.1021
	DAMO [5]	0.9865	0.9966	0.9987	0.1769	–	–
	SliceNet [26]	0.9788	0.9952	0.9969	–	0.1746	0.1600
	Ours	**0.9876**	**0.9975**	**0.9991**	**0.1492**	**0.0501**	**0.0898**

5.2 Comparison Results

We selected the metrics used in previous work and the two proposed metrics for the quantitative comparison, including RMSE, $\delta(1.25,\ 1.25^2,\ 1.25^3)$ and panorama-specific metrics, LRCE and P-RMSE (We cannot calculate the proposed new metrics due to limitation of the two real-world datasets). RMSE reflects the overall variability. δ exhibits the difference between ground truth and the predicted depth.

Quantitative Analysis. Table 1 shows the quantitative comparison results with the current SOTA monocular panoramic depth estimation solutions on the four popular datasets. As shown in the table, our model achieves the first place in all

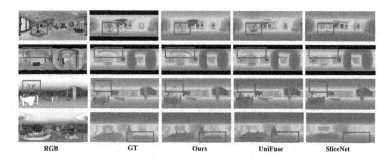

Fig. 7. Qualitative results on Matterport3D, Stanford2D3D, PanoSUNCG, and 3D60. More results can be found in Supplementary Materials

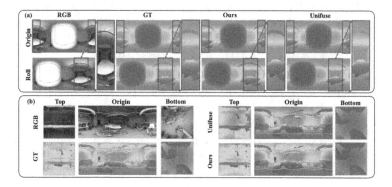

Fig. 8. Visualization of the new metrics' comparison between our method and Unifuse [17]. (a) We stitch the ERP results to observe the depth consistency. (b) We project the areas with massive distortions to cube face to compare the models' performance

metrics. In particular, the RMSE metric of our model achieves a 16% improvement on Stanford2D3D, 26% on Matterport3D. Even on the virtual dataset PanoSUNCG, there is a 42% improvement on RMSE. But there is just a 16% improvement on 3D60 dataset with RMSE. The improvement is not particularly significant compared to the other three datasets because 3D60 dataset is more extensive, the difference between the models is not obvious. The improvement on δ performance further demonstrates that our model can obtain more accurate prediction results. On the new metric P-RMSE, we achieved an average gain of about 40% on the other two virtual datasets. It indicates that our model is more resilient to the distortion in panoramas. In addition, on LRCE, our model outperforms 40% on PanoSUNCG and 12% on 3D60, showing that our model can better constrain the depth consistency of the left-right boundaries in panoramas, because our network fully considers the seamless property of the sphere.

Qualitative Analysis. Figure 7 shows the qualitative comparison with the current SOTA approaches. From the figures, we can observe that SliceNet is relatively accurate in predicting regions without distortion. However, the model performance degrades dramatically in regions with distortions or large object deformations. Although SliceNet can efficiently focus on the global panoramic structures, the depth reconstruction process cannot accurately recover the details, which affects the model's performance. UniFuse can deal with deformation effectively, but it still suffers from incorrect estimating and tends to lose detailed information. From Fig. 8, we can observe that our results are very competitive at boundary and pole areas.

Table 2. Ablation study. We trained on Stanford2D3D for 70 epochs. a is the baseline structure developed with CNNs

Index	Transformer	STLM	Token Flow	RMSE	P-RMSE	LRCE
a	✗	✗	✗	0.6704	0.2258	0.2733
b	✓	✗	✗	0.4349	0.2068	0.2155
c	✓	✓	✗	0.3739	0.1825	0.1916
d	✓	✓	✓	**0.3366**	**0.1793**	**0.1784**

5.3   Ablation Study

With the same conditions, we validated the key components of our model by ablation study on Stanford2D3D (real-world dataset, small-scale, challenging). As illustrated in Table 2, a presents the baseline structure that we use convolutional layers to replace PST blocks; Our network with the traditional attention mechanism is expressed with b; c indicates our attention module without token flow; Our entire network is shown as d.

Transformer vs. CNN. From Table 2, we can observe that the Transformer gains 35% improvements over CNNs in terms of RMSE. Furthermore, qualitative results in b are more precise than the CNNs. Essentially, CNNs are a special kind of self-attention. Since the convolutional kernel is fixed, it requires various components or structures or even deeper networks to help the model learn the data patterns. On the other hand, the attention in Transformer is more flexible, and it is relatively easier to learn the patterns.

Effectiveness of Tangent-patches for Transformer. To illustrate the effectiveness of the tangent-patch dividing method, we compared an alternative attention structure that currently performs SOTA in vision Transformers. From Table 2, our network with tangent-patches (c) outperforms the attention mechanism (b) with 21% on RMSE, 10% on P-RMSE and 12% on LRCE. It proves that tangent-patch can help networks deal with panoramic distortions.

Effectiveness of Token Flow. Since the geometric structures are essential for depth estimation, we add the additional token flows to perceive geometric structures. The results in Table 2 show that our model with the token flow can make P-RMSE more competitive. In Fig. 9, we can observe that the token flow allows the model to estimate the depth details more accurately.

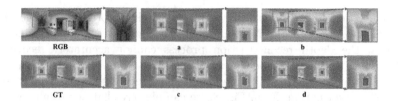

Fig. 9. Qualitative comparison of ablation study. a, b, c, d are the same as Table 2

5.4   Extensibility

We also validate the extensibility of our model by the panoramic segmentation that is also a pixel2pixel task. We did not change any structure of our network and strictly followed the experimental protocol in [32]. As listed in Table 3, the experimental results show that our model outperforms the current SOTA approaches. Due to page limitations, more qualitative comparisons and the results with a high resolution can be found in the supplementary material.

Table 3. Quantitative comparison for semantic segmentation on Stanford2D3D. Results are averaged over the official 3 folds [32]

Dataset	Method	mIoU↑	mAcc↑
Stanford2D3D	TangentImg [12]	41.8	54.9
	HoHoNet [32]	43.3	53.9
	Ours	**48.9**	**64.5**

6   Conclusion

In this paper, we propose the first panorama Transformer (PanoFormer) for indoor panoramic depth estimation. Unlike current approaches, we remove the negative effect of distortions and further model geometric structures by using learnable token flow to establish the network's panoramic perceptions. Concretely, we design a PST block, which can be effectively extended to other learning frameworks. To comprehensively measure the performance of the panoramic depth estimation models, we propose two panorama-specific metrics based on the priors of equirectangular images. Experiments demonstrate that our algorithm significantly outperforms current SOTA methods on depth estimation and other pixel2pixel panoramic tasks, such as semantic segmentation.

Acknowledgement. This work was supported by the National Key R&D Program of China (No.2021ZD0112100), and the National Natural Science Foundation of China (Nos. 62172032, U1936212, 62120106009).

References

1. Armeni, I., Sax, S., Zamir, A.R., Savarese, S.: Joint 2d–3d-semantic data for indoor scene understanding. arXiv preprint arXiv:1702.01105 (2017)
2. Bhat, S.F., Alhashim, I., Wonka, P.: Adabins: depth estimation using adaptive bins. In: Proceedings of the IEEE/CVF Conference on Computer Vision and Pattern Recognition, pp. 4009–4018 (2021)
3. Bhoi, A.: Monocular depth estimation: A survey. arXiv preprint arXiv:1901.09402 (2019)
4. Chang, A., et al.: Matterport3d: Learning from rgb-d data in indoor environments. In: 2017 International Conference on 3D Vision (3DV), pp. 667–676. IEEE Computer Society (2017)
5. Chen, H.X., Li, K., Fu, Z., Liu, M., Chen, Z., Guo, Y.: Distortion-aware monocular depth estimation for omnidirectional images. IEEE Signal Process. Lett. **28**, 334–338 (2021)
6. Cheng, H.T., Chao, C.H., Dong, J.D., Wen, H.K., Liu, T.L., Sun, M.: Cube padding for weakly-supervised saliency prediction in 360 videos. In: Proceedings of the IEEE Conference on Computer Vision and Pattern Recognition, pp. 1420–1429 (2018)
7. Cheng, X., Wang, P., Zhou, Y., Guan, C., Yang, R.: Omnidirectional depth extension networks. In: 2020 IEEE International Conference on Robotics and Automation (ICRA), pp. 589–595. IEEE (2020)
8. Cohen, T.S., Geiger, M., Köhler, J., Welling, M.: Spherical cnns. In: International Conference on Learning Representations (2018)
9. Coors, B., Condurache, A.P., Geiger, A.: Spherenet: Learning spherical representations for detection and classification in omnidirectional images. In: Proceedings of the European conference on computer vision (ECCV), pp. 518–533 (2018)
10. Dai, J., Qi, H., Xiong, Y., Li, Y., Zhang, G., Hu, H., Wei, Y.: Deformable convolutional networks. In: Proceedings of the IEEE International Conference on Computer Vision, pp. 764–773 (2017)
11. Dosovitskiy, A., et al.: An image is worth 16×16 words: Transformers for image recognition at scale. arXiv preprint arXiv:2010.11929 (2020)
12. Eder, M., Shvets, M., Lim, J., Frahm, J.M.: Tangent images for mitigating spherical distortion. In: Proceedings of the IEEE/CVF Conference on Computer Vision and Pattern Recognition, pp. 12426–12434 (2020)
13. Eigen, D., Puhrsch, C., Fergus, R.: Depth map prediction from a single image using a multi-scale deep network. Advances in neural information processing systems 27 (2014)
14. Esmaeili, A., Marvasti, F.: A novel approach to quantized matrix completion using huber loss measure. IEEE Signal Process. Lett. **26**(2), 337–341 (2019)
15. Esteves, C., Allen-Blanchette, C., Makadia, A., Daniilidis, K.: Learning so (3) equivariant representations with spherical cnns. In: Proceedings of the European Conference on Computer Vision (ECCV), pp. 52–68 (2018)
16. Jiang, C., Huang, J., Kashinath, K., Marcus, P., Niessner, M., et al.: Spherical cnns on unstructured grids. arXiv preprint arXiv:1901.02039 (2019)
17. Jiang, H., Sheng, Z., Zhu, S., Dong, Z., Huang, R.: Unifuse: unidirectional fusion for 360 panorama depth estimation. IEEE Robot. Autom. Lett. **6**(2), 1519–1526 (2021)
18. Jin, L., Xu, Y., Zheng, J., Zhang, J., Tang, R., Xu, S., Yu, J., Gao, S.: Geometric structure based and regularized depth estimation from 360 indoor imagery. In: Proceedings of the IEEE/CVF Conference on Computer Vision and Pattern Recognition. pp. 889–898 (2020)

19. Khasanova, R., Frossard, P.: Geometry aware convolutional filters for omnidirectional images representation. In: International Conference on Machine Learning, pp. 3351–3359. PMLR (2019)
20. Kingma, D.P., Ba, J.: Adam: A method for stochastic optimization. arXiv preprint arXiv:1412.6980 (2014)
21. Laina, I., Rupprecht, C., Belagiannis, V., Tombari, F., Navab, N.: Deeper depth prediction with fully convolutional residual networks. In: 2016 Fourth International Conference on 3D Vision (3DV), pp. 239–248. IEEE (2016)
22. Lee, Y., Jeong, J., Yun, J., Cho, W., Yoon, K.J.: Spherephd: applying cnns on a spherical polyhedron representation of 360deg images. In: Proceedings of the IEEE/CVF Conference on Computer Vision and Pattern Recognition, pp. 9181–9189 (2019)
23. Li, Y., Zhang, K., Cao, J., Timofte, R., Van Gool, L.: Localvit: Bringing locality to vision transformers. arXiv preprint arXiv:2104.05707 (2021)
24. Liu, Z., et al.: Swin transformer: hierarchical vision transformer using shifted windows. In: Proceedings of the IEEE/CVF International Conference on Computer Vision, pp. 10012–10022 (2021)
25. Pearson, I.F.: Map Projections: Theory and Applications (1990)
26. Pintore, G., Agus, M., Almansa, E., Schneider, J., Gobbetti, E.: Slicenet: deep dense depth estimation from a single indoor panorama using a slice-based representation. In: Proceedings of the IEEE/CVF Conference on Computer Vision and Pattern Recognition, pp. 11536–11545 (2021)
27. Ranftl, R., Bochkovskiy, A., Koltun, V.: Vision transformers for dense prediction. In: Proceedings of the IEEE/CVF International Conference on Computer Vision, pp. 12179–12188 (2021)
28. Shen, Z., Lin, C., Nie, L., Liao, K., Zhao, Y.: Distortion-tolerant monocular depth estimation on omnidirectional images using dual-cubemap. In: 2021 IEEE International Conference on Multimedia and Expo (ICME). pp. 1–6. IEEE (2021)
29. Song, S., Yu, F., Zeng, A., Chang, A.X., Savva, M., Funkhouser, T.: Semantic scene completion from a single depth image. In: Proceedings of the IEEE Conference on Computer Vision and Pattern Recognition, pp. 1746–1754 (2017)
30. Su, Y.C., Grauman, K.: Kernel transformer networks for compact spherical convolution. In: Proceedings of the IEEE/CVF Conference on Computer Vision and Pattern Recognition, pp. 9442–9451 (2019)
31. Sun, C., Hsiao, C.W., Sun, M., Chen, H.T.: Horizonnet: learning room layout with 1d representation and pano stretch data augmentation. In: Proceedings of the IEEE/CVF Conference on Computer Vision and Pattern Recognition, pp. 1047–1056 (2019)
32. Sun, C., Sun, M., Chen, H.T.: Hohonet: 360 indoor holistic understanding with latent horizontal features. In: Proceedings of the IEEE/CVF Conference on Computer Vision and Pattern Recognition, pp. 2573–2582 (2021)
33. Tateno, K., Navab, N., Tombari, F.: Distortion-aware convolutional filters for dense prediction in panoramic images. In: Proceedings of the European Conference on Computer Vision (ECCV), pp. 707–722 (2018)
34. Wang, F.E., Yeh, Y.H., Sun, M., Chiu, W.C., Tsai, Y.H.: Bifuse: Monocular 360 depth estimation via bi-projection fusion. In: Proceedings of the IEEE/CVF Conference on Computer Vision and Pattern Recognition, pp. 462–471 (2020)
35. Wang, Z., Cun, X., Bao, J., Liu, J.: Uformer: a general u-shaped transformer for image restoration. arXiv preprint arXiv:2106.03106 (2021)

36. Wu, H., Xiao, B., Codella, N., Liu, M., Dai, X., Yuan, L., Zhang, L.: CVT: Introducing convolutions to vision transformers. In: Proceedings of the IEEE/CVF International Conference on Computer Vision, pp. 22–31 (2021)

37. Xiong, B., Grauman, K.: Snap angle prediction for 360° panoramas. In: Ferrari, V., Hebert, M., Sminchisescu, C., Weiss, Y. (eds.) ECCV 2018. LNCS, vol. 11209, pp. 3–20. Springer, Cham (2018). https://doi.org/10.1007/978-3-030-01228-1_1

38. Xu, Y., Zhang, Z., Gao, S.: Spherical dnns and their applications in 360° images and videos. IEEE Trans. Pattern Anal. Mach. Intell. (2021)

39. Yan, Z., Li, X., Wang, K., Zhang, Z., Li, J., Yang, J.: Multi-modal masked pre-training for monocular panoramic depth completion. arXiv preprint arXiv:2203.09855 (2022)

40. Yan, Z., Wang, K., Li, X., Zhang, Z., Xu, B., Li, J., Yang, J.: Rignet: Repetitive image guided network for depth completion. arXiv preprint arXiv:2107.13802 (2021)

41. Yu-Chuan, S., Kristen, G.: Flat2sphere: Learning spherical convolution for fast features from 360 imagery. In: Proceedings of International Conference on Neural Information Processing Systems (NIPS) (2017)

42. Yuan, K., Guo, S., Liu, Z., Zhou, A., Yu, F., Wu, W.: Incorporating convolution designs into visual transformers. In: Proceedings of the IEEE/CVF International Conference on Computer Vision, pp. 579–588 (2021)

43. Yun, I., Lee, H.J., Rhee, C.E.: Improving 360 monocular depth estimation via non-local dense prediction transformer and joint supervised and self-supervised learning. In: Proceedings of the AAAI Conference on Artificial Intelligence, vol. 36, pp. 3224–3233 (2022)

44. Zhu, X., Su, W., Lu, L., Li, B., Wang, X., Dai, J.: Deformable DETR: deformable transformers for end-to-end object detection. In: International Conference on Learning Representations (2020)

45. Zioulis, Nikolaos, Karakottas, Antonis, Zarpalas, Dimitrios, Daras, Petros: OmniDepth: dense depth estimation for indoors spherical panoramas. In: Ferrari, Vittorio, Hebert, Martial, Sminchisescu, Cristian, Weiss, Yair (eds.) ECCV 2018. LNCS, vol. 11210, pp. 453–471. Springer, Cham (2018). https://doi.org/10.1007/978-3-030-01231-1_28

Self-supervised Human Mesh Recovery with Cross-Representation Alignment

Xuan Gong[1,2]($\boxtimes$) (ID), Meng Zheng[2] (ID), Benjamin Planche[2] (ID),
Srikrishna Karanam[2] (ID), Terrence Chen[2], David Doermann[1] (ID),
and Ziyan Wu[2] (ID)

[1] University at Buffalo, Buffalo, NY, USA
{xuangong,doermann}@buffalo.edu
[2] United Imaging Intelligence, Cambridge, MA, USA
{xuan.gong,meng.zheng,benjamin.planche,srikrishna.karanam,terrence.chen,
ziyan.wu}@uii-ai.com

Abstract. Fully supervised human mesh recovery methods are data-hungry and have poor generalizability due to the limited availability and diversity of 3D-annotated benchmark datasets. Recent progress in self-supervised human mesh recovery has been made using synthetic-data-driven training paradigms where the model is trained from synthetic paired 2D representation (*e.g.*, 2D keypoints and segmentation masks) and 3D mesh. However, on synthetic dense correspondence maps (*i.e.*, IUV) few have been explored since the domain gap between synthetic training data and real testing data is hard to address for 2D dense representation. To alleviate this domain gap on IUV, we propose cross-representation alignment utilizing the complementary information from the robust but sparse representation (2D keypoints). Specifically, the alignment errors between initial mesh estimation and both 2D representations are forwarded into regressor and dynamically corrected in the following mesh regression. This adaptive cross-representation alignment explicitly learns from the deviations and captures complementary information: robustness from sparse representation and richness from dense representation. We conduct extensive experiments on multiple standard benchmark datasets and demonstrate competitive results, helping take a step towards reducing the annotation effort needed to produce state-of-the-art models in human mesh estimation.

Keywords: Human mesh recovery · Representation alignment · Synthetic-to-real learning

Supplementary Information The online version contains supplementary material available at https://doi.org/10.1007/978-3-031-19769-7_13.

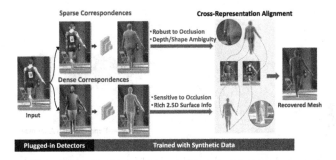

Fig. 1. Motivation: In our synthetic-data-driven pipeline, we train a model from 2D representations to 3D mesh. During test, 2D representations are inferred from off-the-shelf detectors, where sparse/dense 2D representations come with complementary advantage: 2D keypoints provide a robust but sparse representation of the skeleton, dense correspondences (IUV maps) provide rich but sensitive body information. This motivates us to explore cross-representation alignment to take advantage of both to optimize recovered human mesh.

1 Introduction

3D human analysis from images is an important task in computer vision, with a wide range of downstream applications such as healthcare [13] and computer animation [28]. We consider the problem of human mesh estimation, *i.e.*, estimating the 3D parameters of a parametric human mesh model given input data, typically RGB images. With the availability of models such as SMPL [31], there has been much recent progress in this area [2,12,19].

However, obtaining good performance with these methods requires many data samples with 3D annotations. In the SMPL model, this would be the pose and shape parameters. Generating these 3D annotations is very expensive in general and prohibitive in many specific situations, such as medical settings [62]. Developing these annotations requires expensive, and custom motion capture setups and heavily customized algorithms such as MoSh [30], which are highly impractical in many scenarios, including the aforementioned medical one. This results in a situation where there are only limited datasets with 3D pose and shape annotations, further resulting in models that tend to perform well in narrow scenarios while generalizing poorly to out of distribution data [29].

To relieve the requirement of expensive 3D labels, attempts are made to utilize more easily obtained annotations, *e.g.*, 2D landmarks and silhouettes [38,42,48,51], ordinal depth relations [37], dense correspondences [7], or 3D skeletons [26]. To get rid of weak supervision, some take a step further by exploring temporal or multi-view images [23,50] or prior knowledge such as poses with temporal consistency [55].

There has been some recent works using synthetic data for human body modeling, *e.g.*, dense correspondences estimation [65], depth estimation [49], 3D pose estimation [23,35,41,47,49], and 3D human reconstruction [63]. While these approaches show promising results, they need to render images under vari-

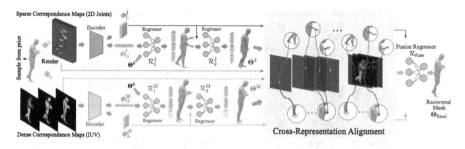

Fig. 2. Overview of the proposed pipeline with cross-representation alignment. For training, we generate paired data between 3D mesh and intermediate representations (*i.e.* 2D joints and IUV map).

ous synthetically designed conditions such as lighting and background. However, it is very challenging for such an approach to produce data (and hence the resulting trained model) that generalizes to real-world conditions. In contrast, [5,43–46,56] rely on various intermediate representations used for adjacent tasks such as keypoint, binary silhouettes, edges, and depth. Concretely, while insufficient data handicaps 3D human mesh estimation, tasks such as keypoint estimation have substantially more annotated data. This then leads to a situation where one can expect intermediate representations for these tasks (*e.g.*, 2D keypoints estimation, binary silhouettes) to generalize better than the representation learned by standard mesh estimation models such as SPIN [19]. At test time on real data, all one needs to do is to compute these representations with off-the-shelf detectors and then infer with the trained intermediate-representation-to-mesh regressor.

Although the aforementioned synthesis-based methods regress the SMPL parameters directly from intermediate representations such as 2D keypoints, binary silhouettes, and depth, none of them successfully utilize synthetic dense correspondence maps (*i.e.*, IUV), which can provide richer and complementary information to 2D joints/edge/silhouette. While adding IUV to the representations may seem incremental, [43] acknowledge it is actually challenging due to the large domain gap between real IUV and synthetic IUV.

We propose cross-representation alignment (CRA) to address the large domain gap while employing dense intermediate representation in synthetic training to handle all the above considerations. Our critical insight is that all these representations may not be wholly consistent but come with complementary advantages. For instance, while 2D keypoints provide a robust sparse representation of the skeleton, dense correspondences (via UV maps) can help further finetune/finesse the final output (shown in Fig. 1).

To this end, our proposed CRA fusion module comprises a trainable alignment scheme between the regressed mesh output and the evidential representations as part of an iterative feedback loop (shown in Fig. 2). Unlike our counterparts [43,46,56] which simply concatenate the features from each representation and regress SMPL parameters iteratively. We instead exploit the complementary information among different representations by generating feedback based on alignment error between the mesh estimation and each representation. The alignment feedback is then forwarded into the following regressor inferring the final SMPL estimation. By introducing the feedback mechanism here, our proposed method can effectively exploit the complementary knowledge between both representations and adapt to their different characteristics, not only during training but also after deployment with real data.

To summarize, our key contributions are:

- We propose a novel synthetic-training pipeline successfully utilizing both sparse and dense representation by bridging the synthetic-to-real gap in dense correspondence via adaptive representation alignment.
- We capture complementary advantages in cross-modality with a trainable cross-representation fusion module that aligns the regressed mesh output with representation evidence as part of the iterative regression.
- We conduct extensive benchmarking on standard datasets and demonstrate competitive numbers with conventional evaluation metrics and protocols.

2   Related Work

Single-Image Human 3D Pose/Mesh Estimation. The emergence of statistical body models such as SCAPE [1] and SMPL [31] makes it possible to represent the human body with low-dimensional parameters. Iterative optimization-based approaches have been leveraged to fit these parametric models to 2D observations such as keypoints [2,36] and silhouettes [25]. These model-fitting approaches are time-consuming, sensitive to initialization, and difficult to tune. Recent advances are dominated by learning-based methods which regress a parametric model (*e.g.*, pose and shape parameters for SMPL [31]) or non-parametric model (*e.g.*, mesh vertices [20]) under the supervision of 3D labels. Several works learn 3D body mesh from image through intermediate representations, *e.g.*, surface keypoints [48,51], silhouettes [38], body part segmentations [34], IUV maps [54,59,60], and 3D markers [58]. Others directly learn 3D body parameters from the input image [19]. Recent works have explored body kinematics [6,53], pose augmentation, and pose probabilistic distributions [21] to boost performance. Self-attention and graph convolutional networks have also been used to learn relationships among vertices [27], body-parts [17,66] to handle occlusions.

Weakly-Supervised Human 3D Pose/Mesh Estimation. Several works take steps to leverage a variety of easily obtained clues, such as paired 2D landmarks and silhouettes [38,42,48,51], ordinal depth relations [37], DensePose [7], 3D skeleton [26]. HMR [12] fits SMPL parameters to 2D ground-truth and utilizes adversarial learning to exploit unpaired 3D data to relieve the reliance on expensive 3D ground truth. Kundu *et al.* [22] learn human pose and shape with 2D evidence together with appearance consensus between pairs of images of the same person. Based on GHUM [52] as the parametric model, THUNDR [58] realizes weak-supervision via intermediate 3D marker representation.

Self-supervised Human 3D Pose/Mesh Estimation. Kundu *et al.* [22,23] utilize temporal and multi-view images as pairs and background/foreground disentangling for self-supervision of human pose/mesh estimation. Multi-view self-supervised 3D pose estimation methods [18,40,50] usually require additional knowledge w.r.t. the scene and camera position or multi-view images. In the absence of multi-view video sequences and other views, geometric consistency [4], kinematics knowledge [24], and temporally consistent poses [55] have been explored for auxiliary prior self-supervision. HUND [57] utilizes in-the-wild images and learns the mesh with differential rendering measures between predictions and image structures. Other synthesis-based methods generate 2D keypoints, silhouettes [43–45], and 3D skeleton [56] with existing MoCap data for training.

3   Method

3.1   Prerequisites

3D Human Mesh Parameterization: We parameterize the 3D human mesh using the Skinned Multi-Person Linear (SMPL) model. SMPL [31] is a parametric model providing independent body shape $\boldsymbol{\beta}$ and pose $\boldsymbol{\theta}$ representations with low-dimensional parameters (*i.e.*, $\boldsymbol{\beta} \in \mathbb{R}^{10}$ and $\boldsymbol{\theta} \in \mathbb{R}^{72}$). Pose parameters include global body rotation (3-DOF) and relative 3D rotations of 23 joints (23×3-DOF) in the axis-angle format. The shape parameters indicating individual heights and weights (among other parameters) are the first 10 coefficients of a PCA shape space. SMPL provides a differentiable kinematic function $\mathcal{S}$ from these pose/shape parameters to 6890 mesh vertices: $\boldsymbol{v} = \mathcal{S}(\boldsymbol{\theta}, \boldsymbol{\beta}) \in \mathbb{R}^{6890 \times 3}$. Besides, 3D joint locations for N_{J} joints of interest are obtained as $\boldsymbol{j}^{\mathrm{3D}} = \mathcal{J}\boldsymbol{v}$, where $\mathcal{J} \in \mathbb{R}^{N_{\mathrm{J}} \times 6890}$ is a learned linear regression matrix.

Dense Human Body Representation: We use DensePose [8] to establish dense correspondence between the 2D image and the mesh surface behind clothes. It semantically defines 24 body parts as $\boldsymbol{I}$ to represent Head, Torso, Lower/Upper Arms, Lower/Upper Legs, Hands and Feet, where head, torso, and lower/upper limbs are partitioned into frontal-back parts to guarantee body parts are isomorphic to a plane. For UV parametrization, each body part index has a unique UV coordinate which is geometrically consistent. In this manner, with IUV representation each pixel can be projected back to vertices on the template mesh according to a predefined bijective mapping between the 3D surface space and the IUV space. We denote the IUV map as $[\boldsymbol{I}, \boldsymbol{U}, \boldsymbol{V}] \in \mathbb{R}^{3 \times (P+1) \times H \times W}$, where $P = 24$ indicating 24 foreground body parts, H and W are the height and width of IUV map. The index channel is one-hot indicating whether it belongs to the background or specific body part: $\boldsymbol{I} \in \{0,1\}^{(P+1) \times H \times W}$. While $\boldsymbol{U}$ and $\boldsymbol{V}$ are independent channels containing the U, V values (ranging from 0 to 1) for corresponding body part [8]. IUV can be further reorganized as a more compact representation $\boldsymbol{M} = [\boldsymbol{M}^{\mathrm{I}}, \boldsymbol{M}^{\mathrm{U}}, \boldsymbol{M}^{\mathrm{V}}] \in \mathbb{R}^{3 \times H \times W}$ which is convertible with the explicit one-hot IUV version mentioned above. With

$h = 1, \ldots, H$ and $w = 1, \ldots, W$ as pixel position, we have $M_{hw}^I \in \{0, 1, \ldots, P\}$, where 0 indicates background and non-zero value indicates body part index. As at most one out of the $P+1$ channels (background and body parts) has non-zero U/V values, the simplified $\boldsymbol{M}^U$ and $\boldsymbol{M}^V$ are represented by $M_{hw}^U = U_{M_{hw}^I hw}$, $M_{hw}^V = V_{M_{hw}^I hw}$.

3.2 Training Data Synthesis

We generate paired 2D representations and 3D meshes on-the-fly with SMPL. We utilize prior poses from the existing MoCap [3,41] datasets for diverse and realistic simulation. Body shape parameters are sampled from normal distribution $\beta_n \sim \mathcal{N}(\mu_n, \sigma_n^2)(n = 1, \ldots, 10)$, where the mean and variance are empirically obtained from prior statistics [43] for generalization. We employ perspective projection with identity camera rotation $\boldsymbol{r} \in \mathbb{R}^{3 \times 3}$, dynamically sampled camera translation $\boldsymbol{t} \in \mathbb{R}^3$ as extrinsic parameters, and fixed focal length $\boldsymbol{f} \in \mathbb{R}^2$ as intrinsic parameters.

At each training step, the sampled $\boldsymbol{\theta}$ and $\boldsymbol{\beta}$ are forwarded into SMPL model to obtain mesh vertex $\boldsymbol{v}$ and 3D joints $\boldsymbol{j}^{3D}$. Then we project the 3D joints $\boldsymbol{j}^{3D}$ to 2D joints $\boldsymbol{j}^{2D}$, with sampled extrinsic and intrinsic camera parameters mentioned above: $\boldsymbol{j}^{2D} = \boldsymbol{f}\Pi\left(\boldsymbol{r}\boldsymbol{j}^{3D} + \boldsymbol{t}\right)$, where Π denotes perspective projection. We normalize the $\boldsymbol{j}^{2D}$ to be from -1 to 1, and denote normalized version as $\boldsymbol{j}^{2D}$ in the following for simplification. With these camera parameters, we render the human mesh to 2D dense IUV based on an existing rendering method [39]. Specifically, we take predefined unique IUV value for each vertex on the SMPL model as a template, project the vertex IUV into 2D and then obtain a continuous 2D IUV map via rasterization and shading.

The 2D joints $\boldsymbol{j}^{2D} \in \mathbb{R}^{N_J \times 2}$ are transformed into 2D Gaussian joint heatmaps $\boldsymbol{J} \in \mathbb{R}^{N_J \times H \times W}$ as inputs to our neural networks. The IUV map with $\boldsymbol{M} \in \mathbb{R}^{3 \times H \times W}$ is used as the other 2D representation. Note that we normalize the I channel in $\boldsymbol{M}$ to values between $[0, 1]$. For simplification, we subsequently denote the normalized version as $\boldsymbol{M}$. Finally, we have the synthesized paired data with 2D representations $\{\boldsymbol{j}^{2D}, \boldsymbol{J}, \boldsymbol{M}\}$ and 3D mesh $\{\boldsymbol{\theta}, \boldsymbol{\beta}, \boldsymbol{v}, \boldsymbol{j}^{3D}\}$.

3.3 Individual Coarse-to-Fine Regression

Given the 2D representation (either $\boldsymbol{J}$ or $\boldsymbol{M}$), we first extract features with an encoder, then forward the features into the regressor, and predict the SMPL model with pose, shape, and camera parameters $\boldsymbol{\Theta} = \{\hat{\boldsymbol{\theta}}, \hat{\boldsymbol{\beta}}, \hat{\boldsymbol{\pi}}\}$.

The encoder takes 2D representation as input and outputs features $\phi_0 \in \mathbb{R}^{C_0 \times H_0 \times W_0}$. Before forwarding the features into the following regressor, we reduce the feature dimensions spatial-wisely and channel-wisely to maintain more global and local information. For global features, we use average-pooling to reduce spatial dimension and get $\phi_G = \text{AvgPool}(\phi_0) \in \mathbb{R}^{C_0 \times 1 \times 1}$. For fine-grained features, we use a multi-layer perceptron (MLP) for channel reduction and retain the spatial dimension the same:

$$\phi_l = \begin{cases} \mathcal{P}_l(\phi_{l-1}) & \text{if } l = 1 \\ \mathcal{P}_l(\phi_{l-1} \oplus \phi_0) & \text{if } l > 1, \end{cases} \tag{1}$$

where $\oplus$ denotes concatenation, $l = 1, \ldots, L$ is the perception layer, $\mathcal{P}_l$ indicates the l-th perceptron, and $\phi_l \in \mathbb{R}^{C_l \times H_0 \times W_0}$ with channel C_l monotonically decreasing. We denote the final output after MLP as ϕ_L.

Taking the flattened feature ϕ and initialized Θ^0 as input, the regressor $\mathcal{R}$ updates $\Theta = \{\hat{\theta}, \hat{\beta}, \hat{\pi}\}$. Note here that we use continuous 6-dimensional representation [64] for optimization of joint rotation in $\hat{\theta} \in \mathbb{R}^{24 \times 6}$ which can be converted to the discontinuous Euler rotation vectors. The predicted camera parameters for the standard weak-perspective projection are represented by $\hat{\pi} = [\hat{\pi}_s, \hat{\pi}_t]$, where $\hat{\pi}_s \in \mathbb{R}$ is the scale factor and $\hat{\pi}_t \in \mathbb{R}^2$ indicates translation. Similar to the standard iterative error feedback (IEF) procedure [12], we iteratively update the prediction Θ. For each representation (J and M) stream, we have two regressors $\mathcal{R}_1$ and $\mathcal{R}_2$ estimating Θ with global feature ϕ_G and fine-grained feature ϕ_L respectively:

$$\Theta^J = \mathcal{R}_2^J(\phi_L^J; \mathcal{R}_1^J(\phi_G^J; \Theta^0)) \quad \text{and} \quad \Theta^M = \mathcal{R}_2^M(\phi_L^M; \mathcal{R}_1^M(\phi_G^M; \Theta^0)), \tag{2}$$

where Θ^J and Θ^M are the parameter predictions for the 2D joints representation J and IUV representation M respectively.

3.4 Evidential Cross-Representation Alignment

To utilize the complementary information of both representations, we design a novel fusion module $\mathcal{R}_{\text{fuse}}$ considering the misalignment between the prediction and the evidence from the intermediate representations (*i.e.*, 2D joints and IUV map). One observation is that the pose parameters are represented as relative rotations and kinematic trees where minor parameter differences can result in significant misalignment on 2D projections. Another observation is that the inferred 2D joints and IUV map are likely to be noisy and inconsistent in real scenarios. During testing, we can hardly distinguish which of the available 2D representations is more reliable, so we incorporate alignment between both pieces of evidence and both predictions.

Given $\Theta = \{\hat{\theta}, \hat{\beta}, \hat{\pi}\}$ as prediction, SMPL takes $\hat{\theta}$ and $\hat{\beta}$ to output 3D vertices $\hat{v}$ and 3D joints $\hat{j}^{3D}$. Then with predicted camera parameters $\hat{\pi}$, we have the reprojected 2D joints $\hat{j}^{2D} = \hat{\pi}_s \Pi (\hat{j}^{3D}) + \hat{\pi}_t$ with orthographic projection function Π. We denote normalized version of $\hat{j}^{2D}$ as $\hat{j}^{2D}$ in the following for simplification. We also render the IUV map $\widehat{M} \in \mathbb{R}^{3 \times H_0 \times W_0}$ with $\hat{v}$, $\hat{\pi}$ and predefined unique IUV value for each vertex on the SMPL. Note that our projections and rendering techniques are differentiable.

To evaluate the misalignment on 2D joints, we have

$$\mathcal{D}_J(\hat{j}^{2D}, j^{2D}) = \hat{j}^{2D} - j^{2D}, \tag{3}$$

where $\mathcal{D}_J(\cdot, \cdot) \in \mathbb{R}^{N_J \times 2}$ is a discrepancy vector which can also be seen as 2D joints pixel index offset between the prediction and the evidence. For misalignment between predicted IUV map $\widehat{M} = [\widehat{M}^I, \widehat{M}^U, \widehat{M}^V]$ and evidential IUV map $M \in \mathbb{R}^{3 \times H \times W}$, we downsize M to be with $\mathbb{R}^{3 \times H_0 \times W_0}$. For simplicity, we use M to represent the downsized version from now on. The discrepancy map $\mathcal{D}_M(\cdot, \cdot) \in \mathbb{R}^{H_0 \times W_0}$ can be obtained:

$$\mathcal{D}_M(\widehat{M}, M) = \frac{|\widehat{M}^I - M^I|}{|\widehat{M}^I - M^I|_d + \epsilon} + \sum_{p=1}^{P} [\mathbf{1}(\widehat{M}^I = \tfrac{p}{P}) \odot \widehat{M}^U - \mathbf{1}(M^I = \tfrac{p}{P}) \odot M^U]$$

$$+ \sum_{p=1}^{P} [\mathbf{1}(\widehat{M}^I = \tfrac{p}{P}) \odot \widehat{M}^V - \mathbf{1}(M^I = \tfrac{p}{P}) \odot M^V],$$

$$(4)$$

where the $|\cdot|$ indicates ℓ_1 norm, $(\cdot)_d$ indicates detachment from gradients, $\epsilon = 1e^{-5}$ is to prevent the denominator to be zero; thus the first term corresponds to a differentiable version of the indicator function $\mathbf{1}(\widehat{M}^I = \widetilde{M}^I)$. In the second and third terms, $P = 24$ indicates the 24 body parts, $\odot$ denotes element-wise multiplication, and the indicator function $\mathbf{1}$ judges whether $\widetilde{M}^I$ or $\widehat{M}^I$ corresponds to specific body part p, which is normalized here as $\tfrac{p}{P}$.

To simplify the notations, from this point on, we refer to $\hat{j}^{2D}$ as $\hat{j}$, we have $\{\hat{j}^J, \widehat{M}^J\}$ and $\{\hat{j}^M, \widehat{M}^M\}$ corresponding to Θ^J and Θ^M respectively. Then we have $D_J^J = \mathcal{D}_J(\hat{j}^J, j^{2D})$ and $D_J^M = \mathcal{D}_J(\hat{j}^M, j^{2D})$ as the 2D joints misalignment between the two predictions and the evidence. And $D_M^J = \mathcal{D}_M(\widehat{M}^J, M)$ and $D_M^M = \mathcal{D}_M(\widehat{M}^M, M)$ as the IUV misalignment between the two predictions and the evidences. All these misalignment representations are flattened and then taken as input of $\mathcal{R}_{fuse}$ along with the flattened features ϕ_L^J and ϕ_L^M:

$$\Theta^{final} = \mathcal{R}_{fuse}(D_J^J, D_M^J, \phi_L^J, D_J^M, D_M^M, \phi_L^M; \Theta^J, \Theta^M), \qquad (5)$$

where Θ^{final} is the final prediction initialized with both Θ^J and Θ^M. Note that each step of the fusion module is differentiable, *i.e.*, maintaining the gradients so that the following loss function is able to penalize misalignment and correct the precedent prediction from $\mathcal{R}_1^J, \mathcal{R}_2^J, \mathcal{R}_1^M, \mathcal{R}_2^M$ during training.

3.5   Loss Function

As described in Sect. 3.4, from Θ^{final} we can obtain predicted vertices $\hat{v}$, 3D joints $\hat{j}^{3D}$, and project to 2D joints $\hat{j}^{2D}$. We have prediction and supervision in terms of vertices, 2D joints, 3D joints and SMPL parameters respectively. To balance among these parts, we make the loss weights learnable using homoscedastic uncertainty as in prior works [14,43]:

$$\mathcal{L}_{\mathrm{reg}}(\hat{v}, \hat{j}^{2D}, \hat{j}^{3D}, \hat{\theta}, \hat{\beta}, v, j^{2D}, j^{3D}, \theta, \beta)$$
$$= \frac{\mathcal{L}_2(\hat{v}, v)}{\sigma_v^2} + \frac{\mathcal{L}_2(\hat{j}^{2D}, j^{2D})}{\sigma_{j2D}^2} + \frac{\mathcal{L}_2(\hat{j}^{3D}, j^{3D})}{\sigma_{j3D}^2} \tag{6}$$
$$+ \frac{\mathcal{L}_2([\hat{\theta}, \hat{\beta}], [\theta, \beta])}{\sigma_{\mathrm{SMPL}}^2} + \log(\sigma_v \sigma_{j2D} \sigma_{j3D} \sigma_{\mathrm{SMPL}}),$$

where $\mathcal{L}_2$ denotes the mean square error (MSE), and σ_v, σ_{j2D}, σ_{j3D} and σ_{SMPL} indicates weights for vertex, 2D joints, 3D joints, SMPL parameters which are adaptively adjusted during training.

Auxiliary Refinement. Our framework can naturally refine the network with available in-the-wild images. Given an image, we use an existing off-the-shelf detector to obtain IUV map M and 2D joints j^{2D}. The IUV map is downsampled and the 2D joints are processed to Gaussian heatmaps J. We take $\{M, J\}$ as input, forward through our network, and output the final prediction Θ. As described in Sect. 3.4, we obtain the reprojected $\hat{j}^{2D}$ and rendered $\widehat{M}$ in a differentiable manner. Given $\mathcal{D}_M$ defined in Eq. 4, the refinement loss function is thus computed as:

$$\mathcal{L}_{\mathrm{refine}}(\hat{j}^{2D}, \widehat{M}, j^{2D}, M) = \mathcal{L}_2(\hat{j}^{2D}, j^{2D}) + \mathcal{D}_M(\widehat{M}, M), \tag{7}$$

4 Experiments

4.1 Datasets

Training Data. To generate synthetic training data, we sample SMPL pose parameters from the training sets of UP-3D [25], 3DPW [32], and the five training subjects of Human3.6M [11] (S1, S5, S6, S7, S8). The sampling of shape parameters follows the procedure of prior work [43].

Evaluation Data. We report evaluation results on both indoor and outdoor datasets, including 3DPW [32], MPI-INF-3DHP [33], and Human3.6M [11] (Protocols 1 and 2 [12] with subjects S9, S11). For 3DPW, we report the mean per joint position error (MPJPE), mean per joint position error after rigid alignment with Procrustes analysis (PMPJPE), and after-scale correction [43] for pose estimation, and per-vertex error (PVE) for shape estimation. For MPI-INF-3DHP, we report metrics after rigid alignment, including PMPJPE, percentage of correct keypoints (PCK) thresholded at 150mm, and the area under the curve (AUC) over a range of PCK thresholds [33]. For Human3.6M, we report MPJPE and PMPJPE on protocols 1 and 2 using the H3.6M joints definition.

4.2 Implementation Details

Synthetic Data Preprocessing and Augmentation: We generate paired data on-the-fly with details described in Sect. 3.2. We follow the hyperparameters

Table 1. Comparison of our method with weakly supervised and self-supervised SOTA in terms of MPJPE and PMPJPE (both in mm) on the H3.6M Protocol #1 and Protocol #2 test sets. * indicates methods that can estimate more than 3D pose.

Method	2D Superv.	Auxiliary requirements			Protocol # 1		Protocol # 2	
		Image pairs	Multi-view imagery	Temporal prior	MPJPE↓	PMPJPE↓	MPJPE↓	PMPJPE↓
*HMR (unpaired) [12]	✓	✗	✗	✗	106.84	67.45		66.5
*SPIN (unpaired) [19]	✓	✗	✗	✗	-	-	-	62.0
*Kundu et al. [22]	✓	✓	✗	✗	**86.4**			**58.2**
*THUNDER [58]	✓	✗	✗	✗	87.0	62.2	83.4	59.7
Kundu et al. [24]	✗	✓	✗	✗	-	-	-	89.4
Kundu et al. [23]	✗	✓	✓	✗	-	-	-	85.8
*Kundu et al. [22]	✗	✓	✓	✗	102.1	-	-	74.1
CanonPose [50]	✗	✗	✓	✗	**81.9**	-	-	53
Yu et al. [55]	✗	✗	✗	✓	-	-	92.4	**52.3**
*Song et al. [46]	✗	✗	✗	✗	-	-	-	56.4
*STRAP [43]	✗	✗	✗	✗	87.0	59.3	83.1	55.4
*HUND [57]	✗	✗	✗	✗	91.8	66.0	-	-
*Skeleton2Mesh [56]	✗	✗	✗	✗	87.1	**55.4**	-	-
*Ours (synthesis only)	✗	✗	✗	✗	87.1	58.2	81.3	54.8
*Ours (w/ refinement)	✗	✗	✗	✗	**84.3**	57.8	**81.0**	**53.9**

in [43] for SMPL shape and camera translation sampling. We use $N_J = 17$ COCO joints to extract 3D joints from the SMPL model and then project to 2D joints representation. The vertices v are randomly perturbed within $[-10\,\mathrm{mm}, 10\,\mathrm{mm}]$ for augmentation. From perturbed vertices and sampled camera parameters, we render 2D IUV map M based on Pytorch3D [39]. We detect the foreground body area on 2D IUV and crop around the foreground area with a scale of 1.2 around the bounding box, which is unified for consistency between training and testing. We crop both IUV M and joints heatmaps J and then resize to the target size with $H = 256$, $W = 256$. To simulate noise and discrepancy between 2D joints and IUV prediction, we do a series of probabilistic augmentations, including randomly masking one of the six body parts (same as PartDrop in [60]), randomly masking one of the six body parts (head, torso, left/right arm, left/right leg) on IUV map, randomly occluding the IUV map with a dynamically-sized rectangle, and randomly perturbing the 2D joints position.

Architecture: We use ResNet-18 [10] as encoder and the size of the output ϕ_0 is $C_0 = 512$, $H_0 = 8$, $W_0 = 8$. Through average pooling we get ϕ_G with size $512 \times 1 \times 1$. Each perceptron $\mathcal{P}_l$ in the MLP consists of Conv1D and ReLU operations with $L = 3$ layers in total. The MLP reduce the feature channels to $[C_1, C_2, C_3] = [256, 64, 8]$ progressively, and produces the feature vector ϕ_L with size $8 \times 8 \times 8$. Each regression network for $\{\mathcal{R}_1^J, \mathcal{R}_2^J, \mathcal{R}_1^M, \mathcal{R}_2^M\}$ consists of two fully-connected layers with 512 neurons each, followed by an output layer with 157 neurons ($\Theta = \{\hat{\theta}, \hat{\beta}, \hat{\pi}\} \in \mathbb{R}^{24 \times 6 + 10 + 3}$ as explained in Sect. 3.3). Taking the input vector with dimension $2 \times (C_L \times H_0 \times W_0 + 3 \times H_0 \times W_0 + 2 \times N_J) = 1540$, the regression network for $\mathcal{R}_{\mathrm{fuse}}$ consists of two fully-connected layers with 1,540 neurons each, followed by an output layer with 157 neurons.

Table 2. A comparison with fully/weakly/self-supervised SOTA methods in terms of PVE, MPJPE, MPJPE-SC, and PMPJPE (all in mm) on the 3DPW test dataset.

Method		PVE↓	MPJPE↓	MPJPE-SC↓	PMPJPE↓
Full Superv.	HMR [12]	139.3	116.5	-	72.6
	VIBE [16]	113.4	113.4	-	56.5
	PyMAF [61]	110.1	92.8	-	58.9
Weak Superv.	HMR (unpaired) [12]	-	-	126.3	92.0
	Kundu et al. [22]	-	153.4	-	89.8
	THUNDER [58]	-	87.8	-	59.9
Self Superv.	Kundu et al. [22]	-	187.1	-	102.7
	STRAP [43]	131.4	118.3	99.0	66.8
	HUND [57]	-	_90.4_	-	63.5
	STRAP V2 [45]	-	-	90.9	61.0
	STRAP V3 [44]	-	-	84.7	59.2
	Song et al. [46]	-	-	-	**55.9**
	Ours (*synthesis only*)	117.4	91.1	_80.8_	_56.3_
	Ours (*w/ refinement*)	115.3	**89.1**	**79.0**	**55.9**

Training: With the final prediction Θ_{final}, we use Eq. 6 as a loss function to train the whole network in an end-to-end fashion. We use Adam [15] optimizer to train for 30 epochs with a learning rate of $1e^{-4}$ and a batch size of 128. On the image, we predict 2D joints and IUV maps using the off-the-shelf Keypoint-RCNN [9] and DensePose [8] models. For auxiliary refinement, we use RGB images from the corresponding training set when testing on the Human3.6M, 3DPW, and MPI-INF-3DHP. We use Adam to train for ten epochs with a learning rate of $1e^{-6}$ and a batch size of 128 for auxiliary refinement.

Testing: We infer 2D joints on the testing images with the pretrained Keypoint-RCNN [9] with ResNet-50 backbone. We obtain the IUV prediction with pretrained DensePose-RCNN [9] with ResNet-101 backbone. Since 3DPW test images may have multiple persons, we use the same protocol as [19] to get the bounding box for the target person by using the scale and center information and get the 2D representations with maximum IOU with the target bounding box. We crop both the IUV maps and 2D joints heatmaps with a scale of 1.2 before forwarding them to the network for 3D mesh inference.

4.3 Quantitative Results

Human3.6M: We evaluate our method on the Human3.6M [11] test dataset (both Protocol #1 and Protocol #2) and compare our method with SOTA weakly supervised methods and self-supervised methods in Table 1. Note that the weakly supervised methods utilized paired images and 2D ground-truth such

Table 3. Comparison with SOTA methods in terms of PCK, AUC, and PMPJPE (mm) after rigid alignment on the MPI-INF-3DHP test dataset. * indicates methods that can estimate more than 3D pose. Methods in the top half require training images paired with 2D ground-truth. Methods in the bottom half do not.

Method	Images Used	PCK↑	AUC↑	PMPJPE↓
*HMR (unpaired) [12]	H36M+3DHP	77.1	40.7	113.2
Kundu et al. [24]	H36M+3DHP	79.2	43.4	99.2
Kundu et al. [23]	H36M+YTube	**83.2**	**58.7**	97.6
CanonPose [50]	H36M+YTube	77.0	-	**70.3**
Yu et al. [55]	3DHP	86.2	51.7	-
*Skeleton2Mesh [56]	3DHP	87.0	50.8	87.4
*SPIN (unpaired) [19]	3DHP	87.0	48.5	80.4
*Ours (synthesis only)	**None**	89.4	54.0	80.2
*Ours (w/refinement)	3DHP	**89.7**	**55.0**	**79.1**

as 2D joints for supervision during training. And some self-supervised methods use auxiliary clues such as image pairs in video sequences, multi-view images, or prior knowledge of human keypoint positions on temporal sequences. Without reliance on either of these prerequisites, our method shows very competitive results compared with the prior arts with auxiliary refinement. Among the methods not requiring auxiliary clues, e.g. temporal or multi-view imagery, we achieve the best results in 3D pose estimation metrics-MPJPE Protocol #1, MPJPE, and PMPJPE on Protocol #2 of the Human3.6M test set.

3DPW: On the test set of 3DPW [32], we calculate PVE as shape evaluation metric and MPJPE, PMPJPE, MPJPE-SC [43] as pose evaluation metrics. From the comparisons in Table 2, we note that our method outperforms the prior arts, including those trained with 3D ground-truth (i.e., full supervision) and 2D ground-truth (i.e., weak supervision), on all metrics for pose evaluation. Although we do not rely on any annotated data, our method achieves results on shape estimation comparable to the prior arts trained with 3D annotation.

MPI-INF-3DHP: On the test set of MPI-INF-3DHP [33], we consider the usual metrics PCK, AUC, and PMPJPE after rigid alignment, to evaluate the 3D pose estimation. As shown in Table 3, other methods heavily rely on the related human image dataset for training, and some have additional requirements on multi-view images (i.e., Human3.6M) and continuous images in temporal sequence (i.e., YouTube videos). In contrast, our method has no such requirements and yet achieves better results on PCK than the prior arts (including weakly supervised methods). With access to the images, we can refine the network with a 0.9 mm improvement in PMPJPE. Compared with the methods relying on both temporal and multi-view images [23,50], our method achieves state-of-the-art PCK and very competitive AUC and PMPJPE without any requirements of images. Notably, we do not use any prior information of MPI-INF-3DHP during synthetic

Table 4. Ablations of one/two representations, concatenation fusion, two-stream fusion with regressor $\mathcal{R}_3$, and the evidential representation alignment on the 3DPW test dataset in terms of PVE and PMPJPE (mm). Here $^{\triangleleft}\mathcal{R}$, $^{\triangleright\triangleleft}\mathcal{R}$ denotes the regressor taking misalignment of its preceding regressor prediction in terms of $^{\triangleleft}$the other/$^{\triangleright\triangleleft}$both representation(s) as additional input. Note: no refinement applied for comparison.

Representation	Regressor	Fusion	PVE↓	PMPJPE↓
^{1}J2D	$\mathcal{R}_1\ \mathcal{R}_1\ \mathcal{R}_1$	-	181.3	75.2
^{2}IUV	$\mathcal{R}_1\ \mathcal{R}_1\ \mathcal{R}_1$	-	167.2	83.1
^{3}J2D & IUV	$\mathcal{R}_1\ \mathcal{R}_1\ \mathcal{R}_1$	input $\oplus$	121.3	60.1
^{4}J2D & IUV	$\{\mathcal{R}_1\ \mathcal{R}_1\}^{\times 2}$	$\mathcal{R}_{\text{fuse}}$	120.8	61.0
^{5}J2D & IUV	$\{\mathcal{R}_1\ \mathcal{R}_2\}^{\times 2}$	$\mathcal{R}_{\text{fuse}}$	117.7	59.6
^{6}J2D & IUV	$\{\mathcal{R}_1\ \mathcal{R}_2\}^{\times 2}$	$^{\triangleleft}\mathcal{R}_{\text{fuse}}$	118.6	58.2
^{7}J2D & IUV	$\{\mathcal{R}_1\ \mathcal{R}_2\}^{\times 2}$	$^{\triangleright\triangleleft}\mathcal{R}_{\text{fuse}}$	**117.4**	**56.3**

training but still achieve very competitive performance on MPI-INF-3DHP with model only trained with synthetic data. This demonstrates the superiority of our method's generalization ability to unseen in-the-wild data.

Table 5. Comparisons of PVE and PMPJPE (both in mm) when adding noise on IUV/2D joints representations of 3DPW test images. We study the performances when using one/two representations and using two representations with and without CRA.

	Body part occlusion prob.	0.1	0.2	0.3	0.4	0.5	0.6	0.7	0.8	0.9
PVE↓	IUV	163.4	166.1	169.0	171.7	174.6	177.0	180.1	182.7	185.5
	IUV + J2D (wo/CRA)	138.8	143.6	145.8	148.3	150.7	153.0	155.7	157.9	160.3
	IUV + J2D (w/CRA)	**118.1**	**118.7**	**119.4**	**120.1**	**120.7**	**121.4**	**122.0**	**122.7**	**123.4**
PMPJPE↓	IUV	92.9	94.9	97.0	99.0	101.1	102.9	105.1	106.9	109.1
	IUV + J2D (wo/CRA)	61.8	62.6	64.3	66.2	68.0	69.8	71.7	73.3	75.1
	IUV + J2D (w/CRA)	**56.8**	**57.3**	**57.8**	**58.2**	**58.7**	**59.2**	**59.6**	**60.1**	**60.6**
	Remove 2D joints prob	0.1	0.2	0.3	0.4	0.5	0.6	0.7	0.8	0.9
PVE↓	J2D	185.0	193.9	204.2	237.8	273.7	306.8	339.6	370.9	402.1
	J2D + IUV (wo/CRA)	150.5	164.8	181.9	201.9	222.6	246.6	270.1	294.0	318.7
	J2D + IUV (w/CRA)	**127.4**	**139.8**	**153.9**	**170.2**	**188.9**	**210.1**	**232.9**	**258.0**	**284.2**
PMPJPE↓	J2D	88.5	98.8	122.5	145.8	169.4	189.7	208.5	224.6	239.0
	J2D + IUV (wo/CRA)	68.0	78.4	90.0	102.8	114.9	127.5	138.7	148.3	156.3
	J2D + IUV (w/CRA)	**63.9**	**72.7**	**82.5**	**92.8**	**104.1**	**115.7**	**128.1**	**140.5**	**153.6**

Ablations: In Table 4, we study the efficacy of our cross-representation alignment, where $\oplus$ denotes concatenate two representations as input of the encoder for fusion. From line 1 to line 3, we note that using the complementary information of 2D joints and IUV is better than using only one. The bottom half shows the results under our two-stream fusion pipeline, demonstrating the efficacy of our alignment module. The comparison between line 4 and line 5 shows that separate regressors taking features with different scales achieve better results than iterative regression with $\mathcal{R}_1$ only taking features with size $C_0 \times 1 \times 1$. And the incorporation of our evidential representation alignment scheme (discrepancy vector/map (Eq. 3/4) between the preceding regressor's prediction and the evidence as an additional input of the regressor) achieves further improvement (line 6 over line 7). We can see that utilizing discrepancy on both representations before $\mathcal{R}_{\text{fuse}}$ achieves the best result.

To study the efficiency of our proposed cross-representation alignment, we further simulate the extremely challenging conditions by adding noise on the inferred 2D joints and IUV representations. On the IUV map, we simulate the occlusion cases by masking out one of the six coarse body parts (head, torso, left/right arm, left/right leg) with increasing probability. For 2D joints, we remove the key joints($i.e.$, left and right elbow, wrist, knee, ankle) with increasing probability. From the comparisons in Table 5, we can see that the combination of 2D joints and IUV can outperform IUV only on both shape and pose evaluations. Notably, our proposed cross-representation alignment (w/ CRA) outperforms the baseline (wo/ CRA) by a large margin, especially for the cases with severe noise.

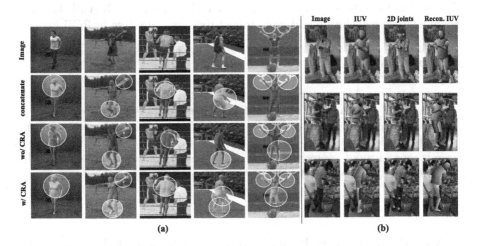

Fig. 3. (a) Comparison of qualitative results on human mesh estimation: taking 2D joints and IUV as input and processing with concatenation, two-stream fusion with and without GRU. (b) Visualization of IUV, 2D joints and reconstructed IUV from 2D joints on 3DPW test set. Note IUV (col 1) is visualized in HSV color space, which is predicted from pretrained Densepose-RCNN. 2D joints (col 2) are predicted from Keypoint-RCNN. IUV reconstructed from the 2D joints by the decoder (col 3) is trained together with CRA.

4.4   Qualitative Results

Qualitative examples are given in Fig. 3(a). We compare our proposed CRA (row 4) with typical concatentation taking 2D joints and IUV as input (row 2), and the baseline of CRA with no alignment applied (row 3). From the highlighted part we can see that our method with alignment module achieves much better shape estimation as well as pose estimation especially on joints such as wrist and knees. Notably the visualization is on images selected from SSP-3D [43] and MPI-INF-3DHP test set of which we do not utilize any prior knowledge. The results demonstrate the robustness and generalization ability of our proposed method to unseen in-the-wild data. We observe that for a small number of cases it could be difficult for CRA to recover from errors existing in all input immediate representations (*e.g.*, no detection on the lower body in both sparse and dense correspondences).

Auxiliary Reconstruction: Our two-stream pipeline enables utilization of the encoded features ϕ of one representation to reconstruct another representation (*e.g.* from 2D joints to IUV map) at the same time while recovering the human mesh. We use a symmetric version of encoder as the decoder for each representation and employ the same loss function as [61] for IUV reconstruction with the synthetic IUV as supervision during training. From Fig. 3(b), we note that the IUV prediction from off-the-shelf detector (trained with annotation) is occasionally sensitive to occlusion. While our recovered IUV trained with synthetic data can generalize to occlusion and more robust to ambiguous area in RGB image.

5   Conclusion

We propose a novel human mesh recovery framework relying only on synthetically generated intermediate representations based on pose priors. We design a Cross-Representation Alignment module to exploit complementary features from these intermediate modalities by enforcing consistency between predicted mesh parameters and input representations. Experimental results on popular benchmark datasets demonstrate the efficacy and generalizability of this framework.

References

1. Anguelov, D., Srinivasan, P., Koller, D., Thrun, S., Rodgers, J., Davis, J.: Scape: shape completion and animation of people. In: ACM SIGGRAPH 2005 Papers, pp. 408–416 (2005)
2. Bogo, F., Kanazawa, A., Lassner, C., Gehler, P., Romero, J., Black, M.J.: Keep It SMPL: automatic estimation of 3D human pose and shape from a single image. In: Leibe, B., Matas, J., Sebe, N., Welling, M. (eds.) ECCV 2016. LNCS, vol. 9909, pp. 561–578. Springer, Cham (2016). https://doi.org/10.1007/978-3-319-46454-1_34
3. C: Mocap. In: mocap. cs. cmu (2003)
4. Chen, C.H., Tyagi, A., Agrawal, A., Drover, D., Stojanov, S., Rehg, J.M.: Unsupervised 3d pose estimation with geometric self-supervision. In: Proceedings of the IEEE/CVF Conference on Computer Vision and Pattern Recognition, pp. 5714–5724 (2019)
5. Clever, H.M., Grady, P., Turk, G., Kemp, C.C.: Bodypressure-inferring body pose and contact pressure from a depth image. IEEE Transactions on Pattern Analysis and Machine Intelligence (2022)

6. Georgakis, G., Li, R., Karanam, S., Chen, T., Košecká, J., Wu, Z.: Hierarchical kinematic human mesh recovery. In: Vedaldi, A., Bischof, H., Brox, T., Frahm, J.-M. (eds.) ECCV 2020. LNCS, vol. 12362, pp. 768–784. Springer, Cham (2020). https://doi.org/10.1007/978-3-030-58520-4_45
7. Guler, R.A., Kokkinos, I.: Holopose: Holistic 3d human reconstruction in-the-wild. In: Proceedings of the IEEE/CVF Conference on Computer Vision and Pattern Recognition. pp. 10884–10894 (2019)
8. Güler, R.A., Neverova, N., Kokkinos, I.: Densepose: dense human pose estimation in the wild. In: Proceedings of the IEEE Conference on Computer Vision and Pattern Recognition, pp. 7297–7306 (2018)
9. He, K., Gkioxari, G., Dollár, P., Girshick, R.: Mask r-cnn. In: Proceedings of the IEEE International Conference on Computer Vision, pp. 2961–2969 (2017)
10. He, K., Zhang, X., Ren, S., Sun, J.: Deep residual learning for image recognition. In: Proceedings of the IEEE Conference on Computer Vision and Pattern Recognition (CVPR), June 2016
11. Ionescu, C., Papava, D., Olaru, V., Sminchisescu, C.: Human3. 6m: Large scale datasets and predictive methods for 3d human sensing in natural environments. IEEE transactions on pattern analysis and machine intelligence 36(7), 1325–1339 (2013)
12. Kanazawa, A., Black, M.J., Jacobs, D.W., Malik, J.: End-to-end recovery of human shape and pose. In: Proceedings of the IEEE Conference on Computer Vision and Pattern Recognition (CVPR), pp. 7122–7131 (2018)
13. Karanam, S., Li, R., Yang, F., Hu, W., Chen, T., Wu, Z.: Towards contactless patient positioning. IEEE Trans. Med. Imaging 39(8), 2701–2710 (2020)
14. Kendall, A., Gal, Y., Cipolla, R.: Multi-task learning using uncertainty to weigh losses for scene geometry and semantics. In: Proceedings of the IEEE Conference on Computer Vision and Pattern Recognition, pp. 7482–7491 (2018)
15. Kingma, D.P., Ba, J.: Adam: A method for stochastic optimization. arXiv preprint arXiv:1412.6980 (2014)
16. Kocabas, M., Athanasiou, N., Black, M.J.: Vibe: video inference for human body pose and shape estimation. In: Proceedings of the IEEE/CVF Conference on Computer Vision and Pattern Recognition, pp. 5253–5263 (2020)
17. Kocabas, M., Huang, C.H.P., Hilliges, O., Black, M.J.: Pare: Part attention regressor for 3d human body estimation. In: Proceedings of the IEEE/CVF International Conference on Computer Vision (ICCV). pp. 11127–11137 (October 2021)
18. Kocabas, M., Karagoz, S., Akbas, E.: Self-supervised learning of 3d human pose using multi-view geometry. In: Proceedings of the IEEE/CVF Conference on Computer Vision and Pattern Recognition, pp. 1077–1086 (2019)
19. Kolotouros, N., Pavlakos, G., Black, M.J., Daniilidis, K.: Learning to reconstruct 3d human pose and shape via model-fitting in the loop. In: Proceedings of the IEEE/CVF International Conference on Computer Vision, pp. 2252–2261 (2019)
20. Kolotouros, N., Pavlakos, G., Daniilidis, K.: Convolutional mesh regression for single-image human shape reconstruction. In: Proceedings of the IEEE/CVF Conference on Computer Vision and Pattern Recognition, pp. 4501–4510 (2019)
21. Kolotouros, N., Pavlakos, G., Jayaraman, D., Daniilidis, K.: Probabilistic modeling for human mesh recovery. In: Proceedings of the IEEE/CVF International Conference on Computer Vision (ICCV). pp. 11605–11614 (October 2021)
22. Kundu, J.N., Rakesh, M., Jampani, V., Venkatesh, R.M., Venkatesh Babu, R.: Appearance consensus driven self-supervised human mesh recovery. In: Vedaldi, A., Bischof, H., Brox, T., Frahm, J.-M. (eds.) ECCV 2020. LNCS, vol. 12346, pp. 794–812. Springer, Cham (2020). https://doi.org/10.1007/978-3-030-58452-8_46

23. Kundu, J.N., Seth, S., Jampani, V., Rakesh, M., Babu, R.V., Chakraborty, A.: Self-supervised 3d human pose estimation via part guided novel image synthesis. In: Proceedings of the IEEE/CVF Conference on Computer Vision and Pattern Recognition. pp. 6152–6162 (2020)
24. Kundu, J.N., Seth, S., Rahul, M., Rakesh, M., Radhakrishnan, V.B., Chakraborty, A.: Kinematic-structure-preserved representation for unsupervised 3d human pose estimation. In: Proceedings of the AAAI Conference on Artificial Intelligence, vol. 34, pp. 11312–11319 (2020)
25. Lassner, C., Romero, J., Kiefel, M., Bogo, F., Black, M.J., Gehler, P.V.: Unite the people: closing the loop between 3d and 2d human representations. In: Proceedings of the IEEE Conference on Computer Vision and Pattern Recognition, pp. 6050–6059 (2017)
26. Li, J., Xu, C., Chen, Z., Bian, S., Yang, L., Lu, C.: Hybrik: a hybrid analytical-neural inverse kinematics solution for 3d human pose and shape estimation. In: Proceedings of the IEEE/CVF Conference on Computer Vision and Pattern Recognition (CVPR), pp. 3383–3393, June 2021
27. Lin, K., Wang, L., Liu, Z.: End-to-end human pose and mesh reconstruction with transformers. In: Proceedings of the IEEE/CVF Conference on Computer Vision and Pattern Recognition, pp. 1954–1963 (2021)
28. Liu, S., Song, L., Xu, Y., Yuan, J.: Nech: neural clothed human model. In: 2021 International Conference on Visual Communications and Image Processing (VCIP), pp. 1–5. IEEE (2021)
29. Liu, S., Huang, X., Fu, N., Li, C., Su, Z., Ostadabbas, S.: Simultaneously-collected multimodal lying pose dataset: enabling in-bed human pose monitoring. IEEE Trans. Pattern Anal. Mach. Intell. (2022)
30. Loper, M., Mahmood, N., Black, M.J.: Mosh: Motion and shape capture from sparse markers. ACM Trans. Graph. (TOG) 33(6), 1–13 (2014)
31. Loper, M., Mahmood, N., Romero, J., Pons-Moll, G., Black, M.J.: Smpl: a skinned multi-person linear model. ACM Trans. Graph. (TOG) 34(6), 1–16 (2015)
32. von Marcard, T., Henschel, R., Black, M.J., Rosenhahn, B., Pons-Moll, G.: Recovering accurate 3d human pose in the wild using imus and a moving camera. In: Proceedings of the European Conference on Computer Vision (ECCV). pp. 601–617 (2018)
33. Mehta, D., et al.: Monocular 3d human pose estimation in the wild using improved CNN supervision. In: 2017 International Conference on 3D Vision (3DV), pp. 506–516. IEEE (2017)
34. Omran, M., Lassner, C., Pons-Moll, G., Gehler, P., Schiele, B.: Neural body fitting: Unifying deep learning and model based human pose and shape estimation. In: 2018 International Conference on 3D Vision (3DV), pp. 484–494. IEEE (2018)
35. Patel, P., Huang, C.H.P., Tesch, J., Hoffmann, D.T., Tripathi, S., Black, M.J.: Agora: Avatars in geography optimized for regression analysis. In: Proceedings of the IEEE/CVF Conference on Computer Vision and Pattern Recognition, pp. 13468–13478 (2021)
36. Pavlakos, G., et al.: Expressive body capture: 3d hands, face, and body from a single image. In: Proceedings of the IEEE/CVF Conference on Computer Vision and Pattern Recognition, pp. 10975–10985 (2019)
37. Pavlakos, G., Zhou, X., Daniilidis, K.: Ordinal depth supervision for 3d human pose estimation. In: Proceedings of the IEEE Conference on Computer Vision and Pattern Recognition, pp. 7307–7316 (2018)

38. Pavlakos, G., Zhu, L., Zhou, X., Daniilidis, K.: Learning to estimate 3d human pose and shape from a single color image. In: Proceedings of the IEEE Conference on Computer Vision and Pattern Recognition, pp. 459–468 (2018)
39. Ravi, N., et al.: Accelerating 3d deep learning with pytorch3d. arXiv:2007.08501 (2020)
40. Rhodin, H., Salzmann, M., Fua, P.: Unsupervised geometry-aware representation for 3D human pose estimation. In: Ferrari, V., Hebert, M., Sminchisescu, C., Weiss, Y. (eds.) ECCV 2018. LNCS, vol. 11214, pp. 765–782. Springer, Cham (2018). https://doi.org/10.1007/978-3-030-01249-6_46
41. Rogez, G., Schmid, C.: Mocap-guided data augmentation for 3d pose estimation in the wild. In: Advances in Neural Information Processing Systems (NeurIPS), pp. 3108–3116 (2016)
42. Rong, Y., Liu, Z., Li, C., Cao, K., Loy, C.C.: Delving deep into hybrid annotations for 3d human recovery in the wild. In: Proceedings of the IEEE/CVF International Conference on Computer Vision, pp. 5340–5348 (2019)
43. Sengupta, A., Budvytis, I., Cipolla, R.: Synthetic training for accurate 3d human pose and shape estimation in the wild. In: BMVC (2020)
44. Sengupta, A., Budvytis, I., Cipolla, R.: Hierarchical kinematic probability distributions for 3d human shape and pose estimation from images in the wild. In: Proceedings of the IEEE/CVF International Conference on Computer Vision (ICCV), pp. 11219–11229, October 2021
45. Sengupta, A., Budvytis, I., Cipolla, R.: Probabilistic 3d human shape and pose estimation from multiple unconstrained images in the wild. In: Proceedings of the IEEE/CVF Conference on Computer Vision and Pattern Recognition (CVPR), pp. 16094–16104, June 2021
46. Song, J., Chen, X., Hilliges, O.: Human body model fitting by learned gradient descent. In: Vedaldi, A., Bischof, H., Brox, T., Frahm, J.-M. (eds.) ECCV 2020. LNCS, vol. 12365, pp. 744–760. Springer, Cham (2020). https://doi.org/10.1007/978-3-030-58565-5_44
47. Song, L., Yu, G., Yuan, J., Liu, Z.: Human pose estimation and its application to action recognition: a survey. J. Vis. Commun. Image Represent. **76**, 103055 (2021)
48. Tan, J., Budvytis, I., Cipolla, R.: Indirect deep structured learning for 3d human body shape and pose prediction. In: British Machine Vision Conference 2017, BMVC 2017 (2017)
49. Varol, G., Romero, J., Martin, X., Mahmood, N., Black, M.J., Laptev, I., Schmid, C.: Learning from synthetic humans. In: Proceedings of the IEEE conference on computer vision and pattern recognition. pp. 109–117 (2017)
50. Wandt, B., Rudolph, M., Zell, P., Rhodin, H., Rosenhahn, B.: Canonpose: self-supervised monocular 3d human pose estimation in the wild. In: Proceedings of the IEEE/CVF Conference on Computer Vision and Pattern Recognition, pp. 13294–13304 (2021)
51. Wehrbein, T., Rudolph, M., Rosenhahn, B., Wandt, B.: Probabilistic monocular 3d human pose estimation with normalizing flows. In: Proceedings of the IEEE/CVF International Conference on Computer Vision (ICCV), pp. 11199–11208, October 2021
52. Xu, H., Bazavan, E.G., Zanfir, A., Freeman, W.T., Sukthankar, R., Sminchisescu, C.: Ghum & ghuml: Generative 3d human shape and articulated pose models. In: Proceedings of the IEEE/CVF Conference on Computer Vision and Pattern Recognition, pp. 6184–6193 (2020)

53. Xu, Y., Wang, W., Liu, T., Liu, X., Xie, J., Zhu, S.C.: Monocular 3d pose estimation via pose grammar and data augmentation. IEEE Trans. Pattern Anal. Mach. Intell. (2021)
54. Xu, Y., Zhu, S.C., Tung, T.: Denserac: Joint 3d pose and shape estimation by dense render-and-compare. In: Proceedings of the IEEE/CVF International Conference on Computer Vision, pp. 7760–7770 (2019)
55. Yu, Z., Ni, B., Xu, J., Wang, J., Zhao, C., Zhang, W.: Towards alleviating the modeling ambiguity of unsupervised monocular 3d human pose estimation. In: Proceedings of the IEEE/CVF International Conference on Computer Vision. pp. 8651–8660 (2021)
56. Yu, Z., Wang, J., Xu, J., Ni, B., Zhao, C., Wang, M., Zhang, W.: Skeleton2mesh: Kinematics prior injected unsupervised human mesh recovery. In: Proceedings of the IEEE/CVF International Conference on Computer Vision, pp. 8619–8629 (2021)
57. Zanfir, A., Bazavan, E.G., Zanfir, M., Freeman, W.T., Sukthankar, R., Sminchisescu, C.: Neural descent for visual 3d human pose and shape. In: Proceedings of the IEEE/CVF Conference on Computer Vision and Pattern Recognition, pp. 14484–14493 (2021)
58. Zanfir, M., Zanfir, A., Bazavan, E.G., Freeman, W.T., Sukthankar, R., Sminchisescu, C.: Thundr: transformer-based 3d human reconstruction with markers. In: Proceedings of the IEEE/CVF International Conference on Computer Vision (ICCV), pp. 12971–12980, October 2021
59. Zeng, W., Ouyang, W., Luo, P., Liu, W., Wang, X.: 3d human mesh regression with dense correspondence. In: Proceedings of the IEEE/CVF Conference on Computer Vision and Pattern Recognition, pp. 7054–7063 (2020)
60. Zhang, H., Cao, J., Lu, G., Ouyang, W., Sun, Z.: Learning 3d human shape and pose from dense body parts. IEEE Trans. Pattern Anal. Mach. Intell. (2020)
61. Zhang, H., Tian, Y., Zhou, X., Ouyang, W., Liu, Y., Wang, L., Sun, Z.: Pymaf: 3d human pose and shape regression with pyramidal mesh alignment feedback loop. In: Proceedings of the IEEE International Conference on Computer Vision (2021)
62. Zheng, M., Planche, B., Gong, X., Yang, F., Chen, T., Wu, Z.: Self-supervised 3d patient modeling with multi-modal attentive fusion. In: 25th International Conference on Medical Image Computing and Computer Assisted Intervention (MICCAI) (2022)
63. Zheng, Z., Yu, T., Wei, Y., Dai, Q., Liu, Y.: Deephuman: 3d human reconstruction from a single image. In: Proceedings of the IEEE/CVF International Conference on Computer Vision, pp. 7739–7749 (2019)
64. Zhou, Y., Barnes, C., Lu, J., Yang, J., Li, H.: On the continuity of rotation representations in neural networks. In: Proceedings of the IEEE/CVF Conference on Computer Vision and Pattern Recognition, pp. 5745–5753 (2019)
65. Zhu, T., Karlsson, P., Bregler, C.: SimPose: effectively learning densepose and surface normals of people from simulated data. In: Vedaldi, A., Bischof, H., Brox, T., Frahm, J.-M. (eds.) ECCV 2020. LNCS, vol. 12374, pp. 225–242. Springer, Cham (2020). https://doi.org/10.1007/978-3-030-58526-6_14
66. Zou, Z., Tang, W.: Modulated graph convolutional network for 3d human pose estimation. In: Proceedings of the IEEE/CVF International Conference on Computer Vision (ICCV), pp. 11477–11487, October 2021

AlignSDF: Pose-Aligned Signed Distance Fields for Hand-Object Reconstruction

Zerui Chen[1]([✉]), Yana Hasson[1,2], Cordelia Schmid[1], and Ivan Laptev[1]

[1] Inria, École Normale Supérieure, CNRS, PSL Research University,
75005 Paris, France
{Zerui.Chen,Yana.Hasson,Cordelia.Schmid,Ivan.Laptev}@inria.fr
[2] Deepmind, London, UK
https://zerchen.github.io/projects/alignsdf.html

Abstract. Recent work achieved impressive progress towards joint reconstruction of hands and manipulated objects from monocular color images. Existing methods focus on two alternative representations in terms of either parametric meshes or signed distance fields (SDFs). On one side, parametric models can benefit from prior knowledge at the cost of limited shape deformations and mesh resolutions. Mesh models, hence, may fail to precisely reconstruct details such as contact surfaces of hands and objects. SDF-based methods, on the other side, can represent arbitrary details but are lacking explicit priors. In this work we aim to improve SDF models using priors provided by parametric representations. In particular, we propose a joint learning framework that disentangles the pose and the shape. We obtain hand and object poses from parametric models and use them to align SDFs in 3D space. We show that such aligned SDFs better focus on reconstructing shape details and improve reconstruction accuracy both for hands and objects. We evaluate our method and demonstrate significant improvements over the state of the art on the challenging ObMan and DexYCB benchmarks.

Keywords: Hand-object reconstruction · Parametric mesh models · Signed distance fields (SDFs)

1 Introduction

Reconstruction of hands and objects from visual data holds a promise to unlock widespread applications in virtual reality, robotic manipulation and human-computer interaction. With the advent of deep learning, we have witnessed a large progress towards 3D reconstruction of hands [4,5,24,39,57,74] and objects [12,16,47,66]. Joint reconstruction of hands and manipulated objects, as well as detailed modeling of hand-object interactions, however, remains less explored and poses additional challenges.

Some of the previous works explore 3D cues and perform reconstruction from multi-view images [7], depth maps [1,61,73] or point clouds [9]. Here, we focus on a more challenging but also more practical setup and reconstruct hands and objects

---

Supplementary Information The online version contains supplementary material available at https://doi.org/10.1007/978-3-031-19769-7_14.

S. Avidan et al. (Eds.): ECCV 2022, LNCS 13661, pp. 231–248, 2022.
https://doi.org/10.1007/978-3-031-19769-7_14

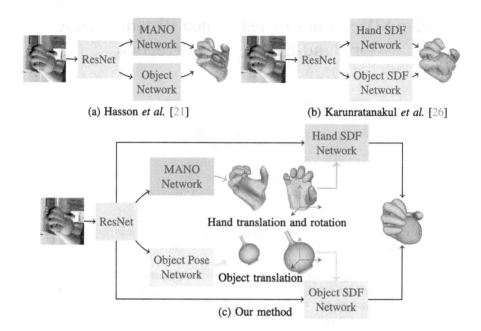

Fig. 1. Previous work on hand-object reconstruction use either (a) parametric shape models or (b) implicit 3D representations. Our proposed method (c) extends SDFs with prior knowledge on hand and object poses obtained via parametric models and can produce detailed meshes for hands and manipulated objects from monocular RGB images.

jointly from monocular RGB images. Existing methods in this setting can be generally classified as the ones using parametric mesh models [20,21,49,54,71] and methods based on implicit representations [11,26,36,45].

Methods from the first category [20,21,71] often build on MANO [54], a popular parametric hand model, see Fig. 1(a). Since MANO is derived from 3D scans of real human hands and encodes strong prior shape knowledge, such methods typically provide anthropomorphically valid hand meshes. However, the resolution of parametric meshes is limited, making them hard to recover detailed interactions. Also, reconstructing 3D objects remains a big challenge. Hasson et al. [21] propose to use Atlas-Net [16] to reconstruct 3D objects. However, their method can only reconstruct simple objects, and the reconstruction accuracy remains limited. To improve reconstruction, several methods [20,62,71] make a restricting assumption that the ground-truth 3D object model is available at test time and only predict the 6D pose of the object.

Recently, neural implicit representations have shown promising results for object reconstruction [45]. Following this direction, Karunratanakul et al. [26] propose to represent hands and objects in a unified signed distance field (SDF) and show the potential to model hand-object interactions, see Fig. 1(b). We here adopt SDF and argue that such implicit representations may benefit from explicit prior knowledge about the pose of hands and objects.

For more accurate reconstruction of hands and manipulated objects, we attempt to combine the advantages of the parametric models and SDFs. Along this direction, previous works [10, 14, 25, 56] attempt to leverage parametric models to learn SDFs from 3D poses or raw scans. In our work, we address a different and more challenging setup of reconstructing hands and objects from monocular RGB images. We hence propose a new pose-normalized SDF framework suited for our task.

Scene geometry depends both on the shape and the global pose of underlying objects. While the pose generally affects all object points with low-parametric transformations (e.g., translation and rotation), it is a common practice to separate the pose and the shape parameters of the model [13,54]. We, hence, propose to disentangle the learning of pose and shape for both hands and objects. As shown in Fig. 1(c), for the hand, we first estimate its MANO parameters and then learn hand SDF in a canonical frame normalized with respect to the rotation and trsnslation of the hand wrist. Similarly, for objects, we estimate their translation and learn object SDF in a translation-normalized canonical frame. By normalizing out the pose, we simplify the task of SDF learning which can focus on estimating the shape disregarding the global rotation and translation transformations. In our framework, the MANO network and the object pose network are responsible for solving the pose, and SDF networks focus on learning the geometry of the hand and the object under their canonical poses.

To validate the effectiveness of our approach, we conduct extensive experiments on two challenging benchmarks: ObMan [21] and DexYCB [6]. ObMan is a synthetic dataset and contains a wide range of objects and grasp types. DexYCB is currently the largest real dataset for capturing hands and manipulated objects. We experimentally demonstrate that our approach outperforms state-of-the-art methods by a significant margin on both benchmarks. Our contributions can be summarized as follows:

- We propose to combine the advantages of parametric mesh models and SDFs and present a joint learning framework for 3D reconstruction of hands and objects.
- To effectively incorporate prior knowledge into SDFs learning, we propose to disentangle the pose learning from the shape learning for this task. Within our framework, we employ parametric models to estimate poses for the hand and the object and employ SDF networks to learn hand and object shapes in pose-normalized coordinate frames.
- We show the advantage of our method by conducting comprehensive ablation experiments on ObMan. Our method produces more detailed joint reconstruction results and achieves state-of-the-art accuracy on the ObMan and DexYCB benchmarks.

2   Related Work

Our work focuses on joint reconstruction of hands and manipulated objects from monocular RGB images. In this section, we first review recent methods for object shape modeling and 3D hand reconstruction. Then, we focus on hand-and-object interaction modeling from a single color image.

3D Object Modeling. Modeling the pose and shape of 3D objects from monocular images is one of the longest standing objectives of computer vision [42,52]. Recent methods train deep neural network models to compute the object shape [11, 16, 36, 55, 70] and pose [31,32,34,68] directly from image pixels. Learned object shape reconstruction from single view images has initially focused on point-cloud [48], mesh

[16,66] and voxel [12,51] representations. In recent years, deep implicit representations [11,36,45] have gained popularity. Unlike other commonly used representations, implicit functions can theoretically model surfaces at unlimited resolution, which makes them an ideal choice to model detailed interactions. We propose to leverage the flexibility of implicit functions to reconstruct hands and arbitrary unknown objects. By conditioning the signed distance function (SDF) on predicted poses, we can leverage strong shape priors from available models. Recent work [59] also reveals that it is effective to encode structured information to improve the quality of NeRF [37] for articulated bodies.

3D Hand Reconstruction. The topic of 3D hand reconstruction has attracted wide attention since the 90s [23,50]. In the deep learning era, we have witnessed significant progress in hand reconstruction from color images. Most works focus on predicting 3D positions of sparse keypoints [24,38,40,60,69,75]. These methods can achieve high accuracy by predicting each hand joint locations independently. However sparse representations of the hand are insufficient to reason precisely about hand-object interactions, which requires millimeter level accuracy. To address this limitation, several recent works model the dense hand surface [2,4,5,8,29,30,41,44,65,74]. A popular line of work reconstructs the hand surface by estimating the parameters of MANO [54], a deformable hand mesh model. These methods can produce anthropomorphically plausible hand meshes using the strong hand prior captured by the parametric model. Such methods either learn to directly regress hand mesh parameters from RGB images [2,4,8,74] or fit them to a set of constraints as a post-processing step [41,44,65]. Unlike previous methods, we condition the hand implicit representation on MANO parameters and produce hand reconstructions of improved visual quality.

3D Hand-Object Reconstruction. Joint reconstruction of hands and objects from monocular RGB images is a very challenging task given the partial visibility and strong mutual occlusions. Methods often rely on multi-view images [3,19,43,67] or additional depth information [17,18,58,63,64] to solve this problem. Recent learning-based methods focus on reconstructing hands and objects directly from single-view RGB images. To simplify the reconstruction task, several methods [15,20,62,71] make a strong assumption that the ground-truth object model is known at test-time and predict its 6D pose. Some methods propose to model hand interactions with unseen objects at test time [21,28,53]. Most related to our approach, Hasson *et al.* [21] propose a two-branch network to reconstruct the hand and an unknown manipulated object. The object branch uses AtlasNet [16] to reconstruct the object mesh and estimate its position relative to the hand. Their method can only reconstruct simple objects which can be obtained by deforming a sphere. In contrast, SDF allows us to model arbitrary object shapes.

In order to improve the quality of hand-object reconstructions, [21] introduce heuristic interaction penalties at train time, Yang *et al.* [71] model each hand-object contact as a spring-mass system and refine the reconstruction result by an optimization process. Recent work [33] also applies an online data augmentation strategy to boost the joint reconstruction accuracy. Though these methods based on parametric mesh models can achieve relatively robust reconstruction results, the modeling accuracy is limited by the underlying parametric mesh. Closest to our approach, Karunratanakul *et al.* [26] propose to model the hand, the object and their contact areas using deep signed distance

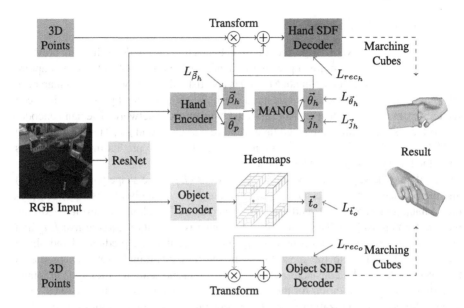

Fig. 2. Our method can reconstruct detailed hand meshes and object meshes from monocular RGB images. Two gray blocks of 3D points indicate the same set of 3D query points. The red arrows denote different loss functions applied during training. The dashed arrows denote marching cubes algorithm [35] used at test time. (Color figure online)

functions. Their method can reconstruct hand and object meshes at a high resolution and capture detailed interactions. However, their method is model-free and does not benefit from any prior knowledge about hands or objects. A concurrent work [72] uses an off-the-shelf hand pose estimator and leverages hand poses to improve hand-held object shapes, which operates in a less-challenging setting than ours. Different from previous works, our method brings together the advantages of both parametric models and deep implicit functions. By embedding prior knowledge into SDFs learning, our method can produce more robust and detailed reconstruction results.

3 Method

As illustrated in Fig. 2, our method is designed to reconstruct the hand and object meshes from a single RGB image. Our model can be generally split into two parts: the hand part and the object part. The hand part estimates MANO parameters and uses them to transform 3D points to the hand canonical coordinate frame. Then, the hand SDF decoder predicts the signed distance for each input 3D point and uses the Marching Cubes algorithm [35] to reconstruct the hand mesh at test time. Similarly, the object part estimates the object translation relative to the hand wrist and uses it to transform the same set of 3D points. The object SDF decoder takes the transformed 3D points as input and reconstructs the object mesh. In the following, we describe the three main components of our model: hand pose estimation in Sect. 3.1, object pose estimation in Sect. 3.2, and hand and object shape reconstruction in Sect. 3.3.

3.1 Hand Pose Estimation

To embed more prior knowledge about human hands into our model, following previous works [20,21,71], we employ a parametric hand mesh model, MANO [54], to capture the kinematics for the human hand. MANO is a statistical model, which could map pose (θ_p) and shape (β_h) parameters to a hand mesh. To estimate hand poses, we first feed features extracted from ResNet-18 [22] to the hand encoder network. The hand encoder network consists of fully connected layers and regresses θ_p and β_h. Then, we integrate MANO as a differentiable layer into our model and use it to predict the hand vertices (v_h), the hand joints (j_h) and hand poses (θ_h).

We define the supervision on the joint locations (L_{j_h}), the shape parameters (L_{β_h}) and the predicted hand poses (L_{θ_h}). To compute L_{j_h}, we apply L2 loss between predicted hand joints and the ground truth. However, using L_{j_h} alone can result in extreme mesh deformations [21]. Therefore, we use another two regularization terms: L_{β_h} and L_{θ_h}. The shape regularization term (L_{β_h}) constrains that the predicted hand shape $(\beta_h \in \mathbb{R}^{10})$ is close to the mean shape in the MANO training set. The predicted hand poses $(\theta_h \in \mathbb{R}^{48})$ consist of axis-angle rotation representations for sixteen joints, including one global rotation for the wrist joint and fifteen rotations for the other local joints. The pose regularization term (L_{θ_h}) constrains local joint rotations to be close to the mean pose in the MANO training set. We also apply L2 loss for the two regularization terms. For the task of hand pose estimation, the overall loss L_{hand} is the summation of all L_{j_h}, L_{β_h} and L_{θ_h} terms:

$$L_{hand} = \lambda_{j_h} L_{j_h} + \lambda_{\beta_h} L_{\beta_h} + \lambda_{\theta_h} L_{\theta_h}, \tag{1}$$

ll λ_{j_h}, λ_{β_h} and λ_{θ_h} to $5 \times 10^{-1}, 5 \times 10^{-7}$ and 5×10^{-5}, respectively.

3.2 Object Pose Estimation

In our method, we set the origin of our coordinate system as the wrist joint defined in MANO. To solve the task of object pose estimation, we usually need to predict the object rotation and its translation. However, estimating the 3D rotation for unknown objects is a challenging and ambiguous task, especially for symmetric objects. Therefore, we here only predict the 3D object translation relative to the hand wrist. To estimate the relative 3D translation (t_o), we employ volumetric heatmaps [38,46] to predict per voxel likelihood for the object centroid and use a soft-argmax operator [60] to extract the 3D coordinate from heatmaps. Then, we convert the 3D coordinate into our wrist-relative coordinate system using camera intrinsics and the wrist location.

During training, we optimize network parameters by minimizing the L2 loss between the estimated 3D object translations t_o and corresponding ground truth. For the task of object pose estimation, the resulting loss L_{obj} is the summation of L_{t_o}:

$$L_{obj} = \lambda_{t_o} L_{t_o}, \tag{2}$$

where we empirically set λ_{t_o} to 5×10^{-1}.

3.3   Hand and Object Shape Reconstruction

Following previous works [26,45], we use neural networks to approximate signed distance functions for the hand and the object. For any input 3D point x, we employ the hand SDF decoder and the object SDF decoder to predict its signed distance to the hand surface and the object surface, respectively. However, it is very challenging to directly learn neural implicit representations for this task, because SDF networks have to handle a wide range of objects and different types of grasps. As a result, Grasping Field [26] cannot achieve satisfactory results in producing detailed hand-and-object interactions.

To reduce the difficulty for this task, our method makes an attempt to disentangle the shape learning and the pose learning, which could help liberate the power of SDF networks. By estimating the hand pose, we could obtain the global rotation (θ_{hr}) and its rotation center (t_h) defined by MANO. The global rotations center (t_h) depends on the estimated MANO shape parameters (β_h). Using the estimated θ_{hr} and t_h, we transform x to the canonical hand pose (*i.e.,* the global rotation equals to zero):

$$x_{hc} = \exp(\theta_{hr})^{-1}(x - t_h) + t_h, \tag{3}$$

where $\exp(\cdot)$ denotes the transformation from the axis-angle representation to the rotation matrix using the *Rodrigues formula*. Then, we concatenate x and x_{hc} and feed them to the hand SDF decoder and predict its signed distance to the hand:

$$\text{SDF}_h(x) = f_h(I, [x, x_{hc}]), \tag{4}$$

where f_h denotes the hand SDF decoder and I denotes image features extracted from the ResNet backbone. Benefiting from this formulation, the hand SDF encoder is aware of x in the canonical hand pose and can focus on learning the hand shape. Similarly, by estimating the object pose, we obtain the object translation t_o and transform x to the canonical object pose:

$$x_{oc} = x - t_o. \tag{5}$$

Then, we concatenate x and x_{oc} and feed them to the object SDF decoder and predict its signed distance to the object:

$$\text{SDF}_o(x) = f_o(I, [x, x_{oc}]), \tag{6}$$

where f_o denotes the object SDF decoder. By feeding x_{oc} into f_o, the object SDF decoder can focus on learning the object shape in its canonical pose.

To train $\text{SDF}_h(x)$ and $\text{SDF}_o(x)$ we minimize L1 distance between predicted signed distances and corresponding ground-truth signed distances for sampled 3D points and training images. The resulting loss is the summation of L_{rec_h} and L_{rec_o}:

$$L_{rec} = \lambda_{rec_h} L_{rec_h} + \lambda_{rec_o} L_{rec_o}, \tag{7}$$

where L_{rec_h} and L_{rec_o} optimize $\text{SDF}_h(x)$ and $\text{SDF}_o(x)$, respectively. We set λ_{rec_h} and λ_{rec_o} to 5×10^{-1}. In summary, we train our model in an end-to-end fashion by minimizing the sum of losses introduced above:

$$L = L_{hand} + L_{obj} + L_{rec}. \tag{8}$$

Given the trained SDF networks, the hand and object surfaces are implicitly defined by the zero-level set of $\text{SDF}_h(x)$ and $\text{SDF}_o(x)$. We generate hand and object meshes using the Marching Cubes algorithm [35] at test time.

4 Experiments

In this section, we present a detailed evaluation of our proposed method. We introduce benchmarks in Sect. 4.1 and describe our evaluation metrics and implementation details in Sects. 4.2–4.3. We then present hand-only ablations and hand-object experiments on the ObMan benchmark in Sects. 4.5 and 4.4 respectively. Finally, we present experimental results for the DexYCB benchmark in Sect. 4.6.

4.1 Benchmarks

ObMan Benchmark [21]. ObMan contains synthetic images and corresponding 3D meshes for a wide range of hand-object interactions with varying hand poses and objects. For training, we follow [26,45] and discard meshes that contain too many double sided triangles, obtaining 87,190 samples. For each sample, we normalize the hand mesh and the object mesh so that they fit inside a unit cube and sample 40,000 points. At test time, we report results on 6285 samples following [21,26].

DexYCB Benchmark [6]. With 582 K grasping frames for 20 YCB objects, DexYCB is currently the largest real benchmark for hand-object reconstruction. Following [33], we only consider right-hand samples and use the official "S0" split. We filter out the frames for which the minimum distance between the hand mesh and the object mesh is larger than 5 mm. We also normalize the hand mesh and the object mesh to a unit cube and sample 40,000 points to generate SDF training samples for DexYCB. As a result, we obtain 148,415 training samples and 29,466 testing samples.

4.2 Evaluation Metrics

The output of our model is structured, and a single metric does not fully capture performance. Therefore, we employ different metrics to evaluate our method.

Hand Shape Error (H_{se}). We follow [26,45] and evaluate the chamfer distance between reconstructed and ground-truth hand meshes to reflect hand reconstruction accuracy. Since the scale of the hand and the translation are ambiguous in monocular images, we optimize the scale and translation to align the reconstructed mesh with the ground-truth and sample 30,000 points from both meshes to calculate the chamfer distance. H_{se} (cm^2) is the median chamfer distance over the entire test set.

Hand Validity Error (H_{ve}). Following [19,76] we perform *Procrustes analysis* by optimizing the scale, translation and global rotation with regard to the ground-truth. We report H_{ve} (cm^2), the chamfer distance after alignment.

Object Shape Error (O_{se}). We reuse the optimized hand scale and translation from the computation of H_{se} to transform the reconstructed object mesh, following [26]. We follow the same process described for H_{se} to compute O_{se} (cm^2).

Hand Joint Error (H_{je}). To measure the hand pose accuracy, we compute the mean join error (cm) relative to the hand wrist joint over 21 joints following [75].

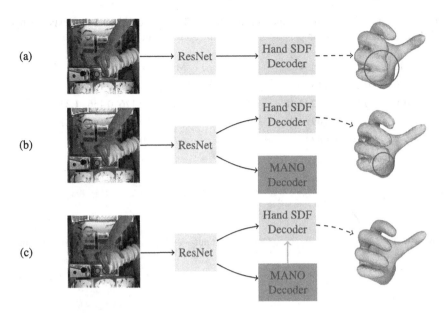

Fig. 3. Three baseline models for hand-only ablation experiments. Dashed arrows denote the marching cubes algorithm [35] used at test time.

Object Translation Error (O_{te}). As we mention in Sect. 3.2, we predict the position of the object centroid relative to the hand wrist. We compute the L2 distance (cm) between the estimated object centroid and its ground-truth to report O_{te}.

Contact Ratio (C_r). Following [26], we report the ratio of samples for which the inter-penetration depth between the hand and the object is larger than zero.

Penetration Depth. (P_d). We compute the maximum of the distances (cm) from the hand mesh vertices to the object's surface similarly to [21,26].

Intersection Volume. (I_v). Following [21], we voxelize the hand and the object using a voxel size of 0.5 cm and compute their intersection volume (cm^3).

4.3   Implementation Details

We use ResNet-18 [22] as a backbone to extract features from input images of size 256×256. To construct volumetric heatmaps, we employ three deconvolution layers to consecutively upsample feature maps from 8×8 to 64×64 and set the resolution of volumetric heatmaps to $64 \times 64 \times 64$. To train hand and object SDF decoders, we randomly sample 1,000 3D points (500 positive points outside the shape and 500 neg-ative points inside the shape) for the hand and the object, respectively. We detail our data augmentation strategies used during training in supplementary materials. We train our model with the Adam optimizer [27] with a batch size of 256. We set the initial learning rate to 1×10^{-4} and decay it by half every 600 epoch on ObMan and every

Table 1. Hand-only ablation experiments using 87K ObMan training samples.

Models	$H_{se} \downarrow$	$H_{ve} \downarrow$	$H_{je} \downarrow$
(a)	0.128	0.113	–
(b)	0.126	0.112	**1.18**
(c)	**0.124**	**0.109**	1.20
(c*)	0.101	0.087	–

Table 2. Hand-only ablation experiments using 30 K ObMan training samples.

Models	$H_{se} \downarrow$	$H_{ve} \downarrow$	$H_{je} \downarrow$
(a)	0.183	0.160	–
(b)	0.176	0.156	**1.23**
(c)	**0.168**	**0.147**	1.27
(c*)	0.142	0.126	–

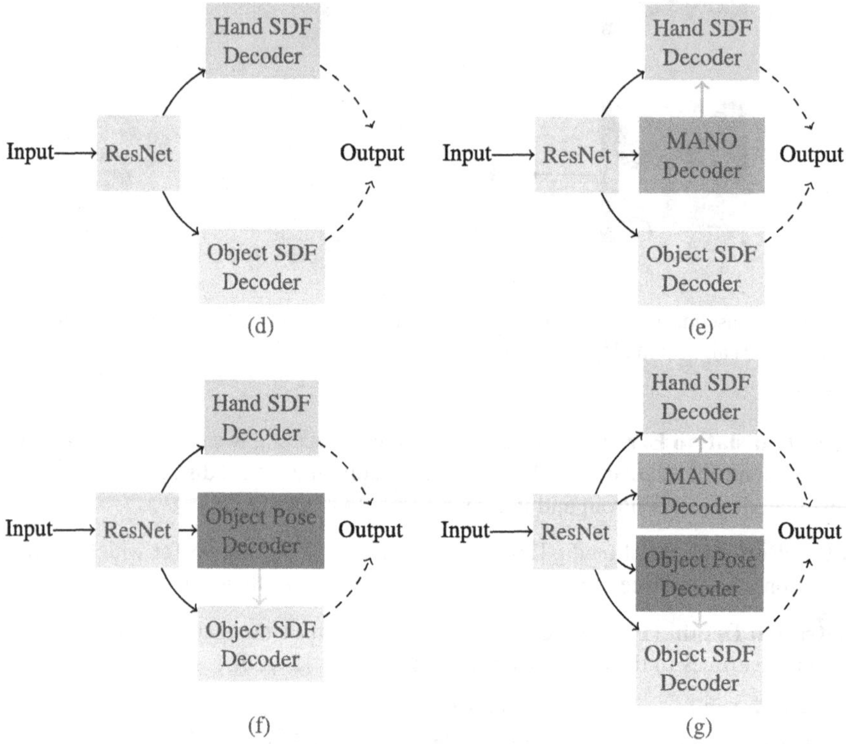

Fig. 4. Four models for hand-object ablation experiments. Dashed arrows denote the marching cubes algorithm [35] used at test time.

300 epoch on DexYCB. The total number of training epochs is 1600 for ObMan and 800 for DexYCB, which takes about 90 h on four NVIDIA 1080 Ti GPUs.

4.4 Hand-Only Experiments on ObMan

To validate the effectiveness of our method, we first conduct hand-only ablation experiments on ObMan. To this end, as shown in Fig. 3, we first build three types of baseline models. The baseline model (a) directly employs the hand SDF decoder to learn

Input   (a)   (b)   (c)   GT

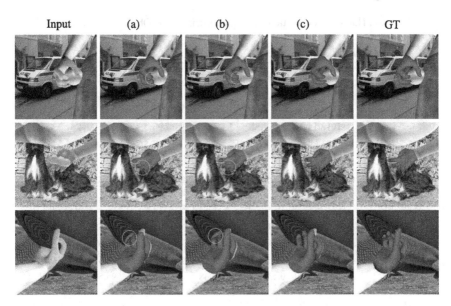

Fig. 5. Qualitative comparison of hand reconstructions between different hand-only baseline models on ObMan (87 K training samples).

$SDF_h(x)$ from backbone features, which often results in a blurred reconstructed hand. The baseline model (b) trains the hand SDF decoder and the MANO network jointly and achieves better results. However, the reconstructed hand still suffers from ill-delimited outlines, which typically result in finger merging issues, illustrated in the second and third columns of Fig. 5. Compared with the baseline model (b), the baseline model (c) further uses the estimated MANO parameters to transform sampled 3D points into the canonical hand pose, which helps disentangle the hand shape learning from the hand pose learning. As result, the hand SDF decoder can focus on learning the geometry of the hand and reconstruct a clear hand. The model (c^*) uses ground-truth hand poses, which is the upper-bound of our method. Tables 1 and 2 present quantitative results for these four models. In Table 1, we present our results using all ObMan training samples and observe that the baseline model (c) has the lowest H_{se} and H_{ve}, which indicates that it achieves the best hand reconstruction quality. The baseline model (c) can also perform hand pose estimation well and reduce the joint error to 1.2 cm. It shows that the model (c) can transform x to the hand canonical pose well with reliable θ_{hr} and t_h and benefit the learning of the hand SDFs. In Fig. 5, we also visualize results obtained from different models and observe that our method can produce more precise hands even under occlusions. To check whether our method can still function well when the training data is limited, we randomly choose 30 K samples to train these three models and summarize our results in Table 2. We observe that the advantage of the model (c) is more obvious using less training data. When compared with the model (a), our method can achieve more than 8% improvement in H_{se} and H_{ve}, which further validates the effectiveness of our approach.

Table 3. Hand-object ablation experiments using 87 K ObMan training data.

Models	$H_{se}\downarrow$	$H_{ve}\downarrow$	$O_{se}\downarrow$	$H_{je}\downarrow$	$O_{te}\downarrow$	C_r	P_d	I_v
(d)	0.140	0.124	4.09	–	–	90.3%	0.50	1.51
(e)	**0.131**	**0.114**	4.14	**1.12**	–	94.7%	0.58	2.00
(f)	0.148	0.130	**3.36**	–	3.29	92.5%	0.57	2.26
(g)	0.136	0.121	3.38	1.27	3.29	95.5%	0.66	2.81
(g*)	0.111	0.093	2.11	–	–	94.5%	0.76	3.87

Table 4. Comparison with previous state-of-the-art methods on ObMan.

Methods	$H_{se}\downarrow$	$H_{ve}\downarrow$	$O_{se}\downarrow$	$H_{je}\downarrow$	$O_{te}\downarrow$	C_r	P_d	I_v
Hasson et al. [21]	0.415	0.383	3.60	**1.13**	–	94.8%	1.20	6.25
Karunratanakul et al. [26]-1 De	0.261	0.246	6.80	–	–	5.63%	0.00	0.00
Karunratanakul et al. [26]-2 De	0.237	–	5.70	–	–	69.6%	0.23	0.20
Ours (g)	**0.136**	**0.121**	3.38	1.27	**3.29**	95.5%	0.66	2.81

4.5 Hand-Object Experiments on ObMan

Given promising results for hand-only experiments, we next validate our approach for the task of hand-object reconstruction. As shown in Fig. 4, we first build four baseline models. The baseline model (d) directly uses the hand and the object decoder to learn SDFs. Compared with the model (d), the model (e) estimates MANO parameters for the hand and uses it to improve the learning the hand SDF decoder. The model (f) estimates the object pose and uses the estimated pose to learn the object SDF decoder. The model (g) combines models (e) and (f) and uses estimated hand and object poses to improve the learning of the hand SDFs and the object SDFs, respectively. The model (g*) is trained with ground-truth hand poses and object translations, which serves as the upper-bound for our method. We summarize our experimental results for these five models in Table 3. Compared with the baseline model (d), the model (e) achieves a 6.4% and 8.8% improvement in H_{se} and H_{ve}, respectively. It shows that embedding hand prior knowledge and aligning hand poses to the canonical pose can improve learning the hand SDFs. By comparing the model (f) with the baseline model (d), we align object poses to their canonical poses using estimated object pose parameters and greatly reduce O_{se} from 4.09 cm^2 to 3.36 cm^2. Finally, our full model (g) combines the advantages of models (e) and (f) and can produce high-quality hand meshes and object meshes. In Table 4, we compare our method against previous state-of-the-art methods and show that our approach outperforms previous state-of-the-art methods [21,26] by a significant margin. When we take a closer look at metrics (C_r, P_d, I_v) that reflect hand-object interactions, we can observe that the reconstructed hand and the reconstructed object from our model are in contact with each other in more than 95.5% of test samples. Compared with the SDF method [26], our method encourages the contact between the hand mesh and the object mesh. Compared with the MANO-based method [21], the penetration depth (P_d) and intersection volume (I_v) of our model is much lower, which

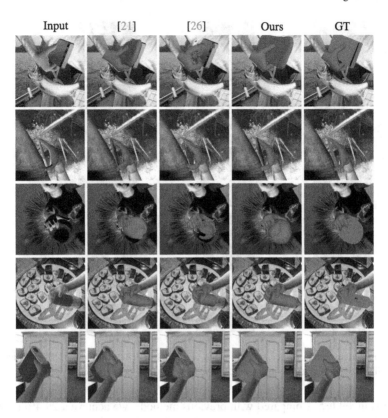

Fig. 6. Qualitative comparison between different types of methods in hand-object experiments on ObMan. Compared with recent methods [21,26], our approach produces more precise reconstructions both for the hands and objects.

Table 5. Comparison with previous state-of-the-art methods on DexYCB.

Method	H_{se} ↓	H_{ve} ↓	O_{se} ↓	H_{je} ↓	O_{te} ↓	C_r	P_d	I_v
Hasson *et al.* [21]	0.785	0.594	4.4	2.0	–	95.8%	1.32	7.67
Karunratanakul *et al.* [26]	0.741	0.532	5.8	–	–	96.7%	0.83	1.34
Ours (g)	**0.523**	**0.375**	**3.5**	**1.9**	**2.7**	96.1%	0.71	3.45

suggests that our method can produce more detailed hand-object interactions. In Fig. 6, we also visualize reconstruction results from different methods. Compared to previous methods, our model can produce more realistic joint reconstruction results even for objects with thin structures.

4.6 Hand-Object Experiments on DexYCB

To validate our method on real data, we next present experiments on the DexYCB benchmark and compare our results to the state of the art. We summarize our experimen-

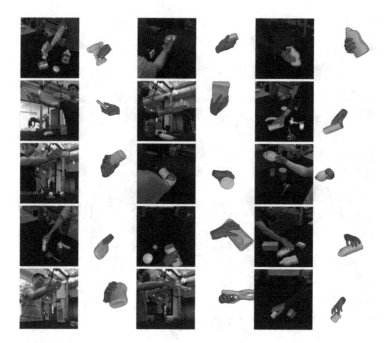

Fig. 7. Qualitative results of our model on the DexYCB benchmark. Our model produces convincing 3D hand-and-object reconstruction results in the real-world setting.

tal results in Table 5. Compared with previous methods, we achieve a 29.4% improvement in H_{se} and a 20.5% improvement in O_{se}, which shows that our method improves both the hand and object reconstruction accuracy. The hand-object interaction metrics for DexYCB also indicate that our method works well for real images. Figure 7 illustrates qualitative results of our method on the DexYCB benchmark. We can observe that our method can accurately reconstruct hand shapes under different poses and a wide range of real-world objects.

5   Conclusion

In this work, we combine advantages of parametric mesh models and SDFs for the task of a joint hand-object reconstruction. To embed prior knowledge into SDFs and to increase the learning efficiency, we propose to disentangle the shape learning and pose learning for both the hand and the object. Then, we align SDF representations with respect to estimated poses and enable learning of more accurate shape estimation. Our model outperforms previous state-of-the-art methods by a significant margin on main benchmarks. Our results also demonstrate significant improvements in visual quality.

Acknowledgements. This work was granted access to the HPC resources of IDRIS under the allocation AD011013147 made by GENCI. This work was funded in part by the French government under management of Agence Nationale de la Recherche as part of the "Investissements d'avenir" program, reference ANR19-P3IA-0001 (PRAIRIE 3IA Institute) and by Louis Vuitton ENS Chair on Artificial Intelligence.

References

1. Baek, S., Kim, K.I., Kim, T.K.: Augmented skeleton space transfer for depth-based hand pose estimation. In: CVPR (2018)
2. Baek, S., Kim, K.I., Kim, T.K.: Pushing the envelope for RGB-based dense 3D hand pose estimation via neural rendering. In: CVPR (2019)
3. Ballan, L., Taneja, A., Gall, J., Gool, L.V., Pollefeys, M.: Motion capture of hands in action using discriminative salient points. In: ECCV (2012). https://doi.org/10.1007/978-3-642-33783-3_46
4. Boukhayma, A., Bem, R.D., Torr, P.H.: 3D hand shape and pose from images in the wild. In: CVPR (2019)
5. Cai, Y., Ge, L., Cai, J., Thalmann, N.M., Yuan, J.: 3D hand pose estimation using synthetic data and weakly labeled RGB images. In: TPAMI (2020)
6. Chao, Y.W., et al.: DexYCB: a benchmark for capturing hand grasping of objects. In: CVPR (2021)
7. Chen, L., Lin, S.Y., Xie, Y., Lin, Y.Y., Xie, X.: MVHM: a large-scale multi-view hand mesh benchmark for accurate 3D hand pose estimation. In: WACV (2021)
8. Chen, X., et al.: Camera-space hand mesh recovery via semantic aggregation and adaptive 2D–1D registration. In: CVPR (2021)
9. Chen, X., Wang, G., Zhang, C., Kim, T.K., Ji, X.: SHPR-Net: deep semantic hand pose regression from point clouds. IEEE Access **6**, 43425–43439 (2018)
10. Chen, X., Zheng, Y., Black, M.J., Hilliges, O., Geiger, A.: SNARF: differentiable forward skinning for animating non-rigid neural implicit shapes. In: ICCV (2021)
11. Chen, Z., Zhang, H.: Learning implicit fields for generative shape modeling. In: CVPR (2019)
12. Choy, C.B., Xu, D., Gwak, J.Y., Chen, K., Savarese, S.: 3D-R2N2: a unified approach for single and multi-view 3D object reconstruction. In: Leibe, B., Matas, J., Sebe, N., Welling, M. (eds.) ECCV 2016. LNCS, vol. 9912, pp. 628–644. Springer, Cham (2016). https://doi.org/10.1007/978-3-319-46484-8_38
13. Cootes, T.F., Taylor, C.J., Cooper, D.H., Graham, J.: Active shape models-their training and application. CVIU **61**(1), 38–59 (1995)
14. Deng, B., Lewis, J.P., Jeruzalski, T., Pons-Moll, G., Hinton, G., Norouzi, M., Tagliasacchi, A.: NASA neural articulated shape approximation. In: Vedaldi, A., Bischof, H., Brox, T., Frahm, J.-M. (eds.) ECCV 2020. LNCS, vol. 12352, pp. 612–628. Springer, Cham (2020). https://doi.org/10.1007/978-3-030-58571-6_36
15. Doosti, B., Naha, S., Mirbagheri, M., Crandall, D.: HOPE-Net: a graph-based model for hand-object pose estimation. In: CVPR (2020)
16. Groueix, T., Fisher, M., Kim, V.G., Russell, B.C., Aubry, M.: A papier-mâché approach to learning 3D surface generation. In: CVPR (2018)
17. Hamer, H., Gall, J., Weise, T., Van Gool, L.: An object-dependent hand pose prior from sparse training data. In: CVPR (2010)
18. Hamer, H., Schindler, K., Koller-Meier, E., Van Gool, L.: Tracking a hand manipulating an object. In: ICCV (2009)
19. Hampali, S., Rad, M., Oberweger, M., Lepetit, V.: HOnnotate: a method for 3D annotation of hand and object poses. In: CVPR (2020)
20. Hasson, Y., Tekin, B., Bogo, F., Laptev, I., Pollefeys, M., Schmid, C.: Leveraging photometric consistency over time for sparsely supervised hand-object reconstruction. In: CVPR (2020)
21. Hasson, Y., et al.: Learning joint reconstruction of hands and manipulated objects. In: CVPR (2019)

22. He, K., Zhang, X., Ren, S., Sun, J.: Deep residual learning for image recognition. In: CVPR (2016)

23. Heap, T., Hogg, D.: Towards 3D hand tracking using a deformable model. In: FG (1996)

24. Iqbal, U., Molchanov, P., Gall, T.B.J., Kautz, J.: Hand pose estimation via latent 2.5D heatmap regression. In: ECCV (2018)

25. Karunratanakul, K., Spurr, A., Fan, Z., Hilliges, O., Tang, S.: A skeleton-driven neural occupancy representation for articulated hands. In: 3DV (2021)

26. Karunratanakul, K., Yang, J., Zhang, Y., Black, M.J., Muandet, K., Tang, S.: Grasping field: learning implicit representations for human grasps. In: 3DV (2020)

27. Kingma, D.P., Ba, J.: Adam: a method for stochastic optimization. arXiv preprint arXiv:1412.6980 (2014)

28. Kokic, M., Kragic, D., Bohg, J.: Learning to estimate pose and shape of hand-held objects from RGB images. In: IROS (2019)

29. Kulon, D., Güler, R.A., Kokkinos, I., Bronstein, M., Zafeiriou, S.: Weakly-supervised mesh-convolutional hand reconstruction in the wild. In: CVPR (2020)

30. Kulon, D., Wang, H., Güler, R.A., Bronstein, M.M., Zafeiriou, S.: Single image 3D hand reconstruction with mesh convolutions. In: BMVC (2019)

31. Labbé, Y., Carpentier, J., Aubry, M., Sivic, J.: CosyPose: consistent multi-view multi-object 6D pose estimation. In: Vedaldi, A., Bischof, H., Brox, T., Frahm, J.-M. (eds.) ECCV 2020. LNCS, vol. 12362, pp. 574–591. Springer, Cham (2020). https://doi.org/10.1007/978-3-030-58520-4_34

32. Labbé, Y., Carpentier, J., Aubry, M., Sivic, J.: Single-view robot pose and joint angle estimation via render & compare. In: CVPR (2021)

33. Li, K., et al.: ArtiBoost: boosting articulated 3D hand-object pose estimation via online exploration and synthesis. In: CVPR (2022)

34. Li, Y., Wang, G., Ji, X., Xiang, Y., Fox, D.: DeepIM: deep iterative matching for 6D pose estimation. In: ECCV (2018)

35. Lorensen, W.E., Cline, H.E.: Marching cubes: a high resolution 3D surface construction algorithm. In: TOG (1987)

36. Mescheder, L., Oechsle, M., Niemeyer, M., Nowozin, S., Geiger, A.: Occupancy networks: learning 3D reconstruction in function space. In: CVPR (2019)

37. Mildenhall, B., Srinivasan, P.P., Tancik, M., Barron, J.T., Ramamoorthi, R., Ng, R.: NeRF: representing scenes as neural radiance fields for view synthesis. In: ECCV (2020)

38. Moon, G., Chang, J.Y., Lee, K.M.: V2V-PoseNet: voxel-to-voxel prediction network for accurate 3D hand and human pose estimation from a single depth map. In: CVPR (2018)

39. Moon, G., Shiratori, T., Lee, K.M.: DeepHandMesh: a weakly-supervised deep encoder-decoder framework for high-fidelity hand mesh modeling. In: Vedaldi, A., Bischof, H., Brox, T., Frahm, J.-M. (eds.) ECCV 2020. LNCS, vol. 12347, pp. 440–455. Springer, Cham (2020). https://doi.org/10.1007/978-3-030-58536-5_26

40. Mueller, F., et al.: GANerated hands for real-time 3D hand tracking from monocular RGB. In: CVPR (2018)

41. Mueller, F., et al.: Real-time pose and shape reconstruction of two interacting hands with a single depth camera. In: TOG (2019)

42. Mundy, J.L.: Object recognition in the geometric era: a retrospective. In: Ponce, J., Hebert, M., Schmid, C., Zisserman, A. (eds.) Toward Category-Level Object Recognition. LNCS, vol. 4170, pp. 3–28. Springer, Heidelberg (2006). https://doi.org/10.1007/11957959_1

43. Oikonomidis, I., Kyriazis, N., Argyros, A.A.: Full DOF tracking of a hand interacting with an object by modeling occlusions and physical constraints. In: ICCV (2011)

44. Panteleris, P., Oikonomidis, I., Argyros, A.: Using a single RGB frame for real time 3D hand pose estimation in the wild. In: WACV (2018)

45. Park, J.J., Florence, P., Straub, J., Newcombe, R., Lovegrove, S.: DeepSDF: learning continuous signed distance functions for shape representation. In: CVPR (2019)
46. Pavlakos, G., Zhou, X., Derpanis, K.G., Daniilidis, K.: Coarse-to-fine volumetric prediction for single-image 3D human pose. In: CVPR (2017)
47. Peng, S., Jiang, C., Liao, Y., Niemeyer, M., Pollefeys, M., Geiger, A.: Shape as points: a differentiable poisson solver. In: NeurIPS (2021)
48. Qi, C.R., Su, H., Mo, K., Guibas, L.J.: PointNet: deep learning on point sets for 3D classification and segmentation. In: CVPR (2017)
49. Qian, N., Wang, J., Mueller, F., Bernard, F., Golyanik, V., Theobalt, C.: HTML: a parametric hand texture model for 3D hand reconstruction and personalization. In: Vedaldi, A., Bischof, H., Brox, T., Frahm, J.-M. (eds.) ECCV 2020. LNCS, vol. 12356, pp. 54–71. Springer, Cham (2020). https://doi.org/10.1007/978-3-030-58621-8_4
50. Rehg, J.M., Kanade, T.: Visual tracking of high DOF articulated structures: an application to human hand tracking. In: Eklundh, J.-O. (ed.) ECCV 1994. LNCS, vol. 801, pp. 35–46. Springer, Heidelberg (1994). https://doi.org/10.1007/BFb0028333
51. Riegler, G., Ulusoy, A.O., Geiger, A.: OctNet: learning deep 3D representations at high resolutions. In: CVPR (2017)
52. Roberts, L.G.: Machine perception of three-dimensional solids. Ph.D. thesis, Massachusetts Institute of Technology (1963)
53. Romero, J., Kjellström, H., Kragic, D.: Hands in action: real-time 3D reconstruction of hands in interaction with objects. In: ICRA (2010)
54. Romero, J., Tzionas, D., Black, M.J.: Embodied hands: modeling and capturing hands and bodies together. In: TOG (2017)
55. Saito, S., Huang, Z., Natsume, R., Morishima, S., Kanazawa, A., Li, H.: PIFu: pixel-aligned implicit function for high-resolution clothed human digitization. In: ICCV (2019)
56. Saito, S., Yang, J., Ma, Q., Black, M.J.: SCANimate: weakly supervised learning of skinned clothed avatar networks. In: CVPR (2021)
57. Spurr, A., Dahiya, A., Wang, X., Zhang, X., Hilliges, O.: Self-supervised 3D hand pose estimation from monocular RGB via contrastive learning. In: CVPR (2021)
58. Sridhar, S., Mueller, F., Zollhöfer, M., Casas, D., Oulasvirta, A., Theobalt, C.: Real-time joint tracking of a hand manipulating an object from RGB-D input. In: Leibe, B., Matas, J., Sebe, N., Welling, M. (eds.) ECCV 2016. LNCS, vol. 9906, pp. 294–310. Springer, Cham (2016). https://doi.org/10.1007/978-3-319-46475-6_19
59. Su, S.Y., Yu, F., Zollhöfer, M., Rhodin, H.: A-NeRF: articulated neural radiance fields for learning human shape, appearance, and pose. In: NeurIPS (2021)
60. Sun, X., Xiao, B., Wei, F., Liang, S., Wei, Y.: Integral human pose regression. In: ECCV (2018)
61. Supančič, J.S., Rogez, G., Yang, Y., Shotton, J., Ramanan, D.: Depth-based hand pose estimation: methods, data, and challenges. Int. J. Comput. Vis. **126**(11), 1180–1198 (2018). https://doi.org/10.1007/s11263-018-1081-7
62. Tekin, B., Bogo, F., Pollefeys, M.: H+O: unified egocentric recognition of 3D hand-object poses and interactions. In: CVPR (2019)
63. Tsoli, A., Argyros, A.A.: Joint 3D tracking of a deformable object in interaction with a hand. In: ECCV (2018)
64. Tzionas, D., Gall, J.: 3D object reconstruction from hand-object interactions. In: ICCV (2015)
65. Wang, J., et al.: RGB2Hands: real-time tracking of 3D hand interactions from monocular RGB video. TOG **39**(6), 1–16 (2020)
66. Wang, N., Zhang, Y., Li, Z., Fu, Y., Liu, W., Jiang, Y.G.: Pixel2Mesh: generating 3D mesh models from single RGB images. In: ECCV (2018)

67. Wang, Y., et al.: Video-based hand manipulation capture through composite motion control. TOG **34**(4), 1–14 (2013)
68. Xiang, Y., Schmidt, T., Narayanan, V., Fox, D.: PoseCNN: a convolutional neural network for 6D object pose estimation in cluttered scenes. In: RSS (2018)
69. Xiong, F., et al.: A2J: anchor-to-joint regression network for 3D articulated pose estimation from a single depth image. In: ICCV (2019)
70. Xu, Q., Wang, W., Ceylan, D., Mech, R., Neumann, U.: DISN: deep implicit surface network for high-quality single-view 3D reconstruction. In: NeurIPS (2019)
71. Yang, L., Zhan, X., Li, K., Xu, W., Li, J., Lu, C.: CPF: learning a contact potential field to model the hand-object interaction. In: ICCV (2021)
72. Ye, Y., Gupta, A., Tulsiani, S.: What's in your hands? 3D reconstruction of generic objects in hands. In: CVPR (2022)
73. Zhang, H., Zhou, Y., Tian, Y., Yong, J.H., Xu, F.: Single depth view based real-time reconstruction of hand-object interactions. TOG **40**(3), 1–12 (2021)
74. Zhou, Y., Habermann, M., Xu, W., Habibie, I., Theobalt, C., Xu, F.: Monocular real-time hand shape and motion capture using multi-modal data. In: CVPR (2020)
75. Zimmermann, C., Brox, T.: Learning to estimate 3D hand pose from single RGB images. In: ICCV (2017)
76. Zimmermann, C., Ceylan, D., Yang, J., Russell, B., Argus, M., Brox, T.: FreiHAND: a dataset for markerless capture of hand pose and shape from single RGB images. In: ICCV (2019)

A Reliable Online Method for Joint Estimation of Focal Length and Camera Rotation

Yiming Qian[1] and James H. Elder[2]([✉])

[1] A*star, Institute of High Performance Computing, Singapore, Singapore
qian_yiming@ihpc.a-star.edu.sg
[2] Centre for Vision Research, York University, Toronto, Canada
jelder@yorku.ca

Abstract. Linear perspective cues deriving from regularities of the built environment can be used to recalibrate both intrinsic and extrinsic camera parameters online, but these estimates can be unreliable due to irregularities in the scene, uncertainties in line segment estimation and background clutter. Here we address this challenge through four initiatives. First, we use the PanoContext panoramic image dataset [27] to curate a novel and realistic dataset of planar projections over a broad range of scenes, focal lengths and camera poses. Second, we use this novel dataset and the YorkUrbanDB [4] to systematically evaluate the linear perspective deviation measures frequently found in the literature and show that the choice of deviation measure and likelihood model has a huge impact on reliability. Third, we use these findings to create a novel system for online camera calibration we call $f\mathbf{R}$, and show that it outperforms the prior state of the art, substantially reducing error in estimated camera rotation and focal length. Our fourth contribution is a novel and efficient approach to estimating uncertainty that can dramatically improve online reliability for performance-critical applications by strategically selecting which frames to use for recalibration.

1 Introduction

Online camera calibration is an essential task for applications such as traffic analytics, mobile robotics, architectural metrology and sports videography. While intrinsic parameters can be estimated in the lab, these parameters drift due to mechanical fluctuations. Online estimation of extrinsic camera parameters is crucial for translating observations made in the image to quantitative inferences about the 3D scene. If a fixed camera is employed, extrinsics can potentially be estimated manually at deployment, but again there will be drift due to mechanical and thermal variations, vibrations and wind, for outdoor applications. With

Supplementary Information The online version contains supplementary material available at https://doi.org/10.1007/978-3-031-19769-7_15.

longer viewing distances, rotational drift can lead to major errors that may be devastating for performance-critical tasks, such as judging the 3D distance between a pedestrian and a car. For PTZ cameras and mobile applications, camera rotation varies over time. PTZ encoder readings are not always available and become less reliable over time [22], and for mobile applications IMU data are subject to drift. For all of these reasons, reliable visual methods for online geometric recalibration of camera intrinsics and extrinsics are important.

Coughlan and Yuille [3] introduced an approach to online estimation of 3D camera rotation based on a "Manhattan World" (3-point perspective) assumption, i.e., that a substantial portion of the linear structure in the image projects from three mutually orthogonal directions: one vertical and two horizontal. This assumption can apply quite generally in the built environment and is relevant for diverse application domains including traffic analytics, mobile robotics, architecture and sports videography (Fig. 1).

Fig. 1. Example application domains of online camera calibration. We show here labelling of line segments according to Manhattan direction, as determined by our novel $f\mathbf{R}$ system that simultaneously estimates focal length and camera orientation.

Subsequent work, reviewed below, has built on this idea to improve accuracy and incorporate simultaneous estimation of focal length. However, we have found that these methods vary a lot in accuracy depending upon the scene and whether focal length is known or estimated. This motivates our attempt to better understand how these methods generalize to diverse scenes and knowledge of intrinsics, and to develop methods to estimate uncertainty of individual estimates.

We address these limitations through four contributions: **1)** From the PanoContext dataset [27] we curate a novel and realistic PanoContext-$f\mathbf{R}$ dataset of planar projections over a broad range of scenes, focal lengths and camera poses that can be used to evaluate systems for simultaneous estimation of focal length and camera rotation (tilt/roll). **2)** We use this novel dataset and the YorkUr-banDB [4] to evaluate the linear perspective deviation measures found in the literature and show that a) the choice of deviation measure has a huge impact on reliability, and b) it is critical to also employ the correct likelihood models. **3)** We use these findings to create a novel system for online camera calibration we call $f\mathbf{R}$, and show that it outperforms the state of the art. **4)** We develop a novel approach to estimating uncertainty in parameter estimates. This is important for two reasons. First, knowledge of uncertainty can be factored into risk models employed in engineering applications, co-determining actions designed to mitigate this risk. Second, especially for cases where parameters are expected to be slowly varying, systems can sparsely select frames that are more likely to generate trustable estimates of camera parameters. The source code and PanoContext-$f\mathbf{R}$ dataset are available on GitHub[1].

2   Prior Work

2.1   Image Features and Deviation Measures

To apply linear perspective cues to camera calibration, two key design choices are 1) what features of the image to use and 2) how to measure the agreement of these features with hypothesized vanishing points. Coughlan and Yuille's original approach [3] employed a likelihood measure on the angular deviation between the line connecting a pixel to the hypothesized vanishing point (which we will refer to in the following as the *vanishing line* and the ray passing through the pixel in the direction orthogonal to the local luminance gradient.

There have since been diverse attempts to improve on this approach, in large part by using different features and different measures of agreement. Deutscher et al. [5] and Košeckà and Zhang [9] first generalized this approach to allow simultaneous estimation of focal length. While Deutscher et al. retained the gradient field representation of Coughlan and Yuille, in their Video Compass system, Košeckà and Zhang switched to using sparser line segments, and proposed a Gaussian likelihood model based upon the projection distance from the line passing through a line segment to its associated vanishing point (Fig. 2(a)).

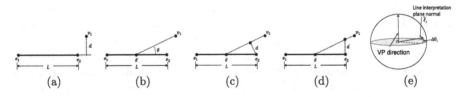

(a)   (b)   (c)   (d)   (e)

Fig. 2. Five deviation measures evaluated (see also [25]): (a) distance from vanishing point to line passing through line segment. (b) angular deviation between line segment and the vanishing line through the line segment midpoint. (c) distance from line segment endpoint to vanishing line (d) distance from line segment endpoint to vanishing line, measured orthogonal to line segment. (e) angle between the interpretation plane normal and the plane orthogonal to the vanishing direction [19].

---

[1] https://github.com/ElderLab-York-University/OnlinefR.

Rother [16] used line segments as well, employing a heuristic version of the deviation measure similar to Coughlan and Yuille's (Fig. 2(b)). Lee et al. [10] followed a similar approach to simultaneous estimation of focal length and camera rotation, and have open-sourced their code, allowing us to evaluate their system below. Denis et al. [4] employed the same angular deviation measure for edges, but returned to Coughlan and Yuille's probabilistic framework, using a learned generalized exponential distribution to model these deviations.

Tardif [20] continued the practice of using line segments, but proposed an alternative deviation measure based on the squared distance between segment endpoints and the vanishing line (Fig. 2(c)). Wildenauer [21] later adopted the same measure but, similar to Denis et al. [4], returned to Coughlan and Yuille's probabilistic framework, using a Cauchy distribution to model the signed distance of segment endpoints to the vanishing line.

These three deviation measures were reviewed by Xu et al. [25], who used intuitive arguments to suggest that each of these measures is flawed. Based upon their analysis, they proposed a novel deviation measure that also uses the distance of the segment endpoints to the vanishing line, but measures this distance orthogonal to the *segment*, instead of orthogonal to the vanishing line (Fig. 2(d)). Their full probabilistic model also favours vanishing points that lie far from the segment midpoint, along the segment direction.

Hough maps can be used to accurately detect lines in an image. Motivated by the early work of Collins & Weiss [2], Tal & Elder [19] proposed a probabilistic Hough method to estimate lines and then estimated camera rotation based on a probabilistic model of deviation in the Gauss sphere, specifically, the angular deviation between the interpretation plane normal and the plane orthogonal to the vanishing direction (Fig. 2(e)).

2.2   Benchmarks

The most prevalent dataset used for Manhattan scene analysis is the YorkUrbanDB dataset [4], which does sample a broad range of scenes but is limited in that focal length is fixed and roll and tilt vary over a relatively modest range. Also, the 101 images in the dataset are too few to train a deep network. More recently there have been efforts [8,12] to create datasets large enough to train deep networks by curating random planar projections from existing panoramic image datasets such as SUN360 [24] and Google StreetView. This makes it easy to generate enough images to train a deep network, and also ensures that we have exact ground truth for camera focal length, tilt and roll angle. Unfortunately, there is no ground truth for camera pan, and so networks trained on these datasets are unable to recover the full camera rotation $\mathbf{R}$.

2.3   State-of-the-Art Systems

In this paper we are focused on the problem of online estimation of focal length f and camera rotation $\mathbf{R}$. Unfortunately, many of the systems reviewed above assumed known focal length, and/or did not open-source their code to allow comparison. The exception, as mentioned, is the line segment method of Lee et al. [10] (Fig. 2(b)) which we compare against below.

More recently, Simon et al. [18] use the angular deviation measure in Fig. 2(b) but introduce a preprocessing step to estimate the horizon as a guide to estimating Manhattan vanishing points. They have open-sourced their code.

Hold-Geoffroy et al. [8] kicked off the use of deep learning for online camera calibration, using the Sun360 panoramic image dataset [24] to generate planar projections based on random samplings of focal length, horizon, roll and camera aspect ratio. They used this dataset to train, validate and test a network based on a denseNet backbone and three separate heads to estimate the horizon height and angle and the vertical FOV, from which camera focal length, roll and tilt can be derived. While the Hold-Geoffroy et al. [8] dataset has not been released and the method has not been open-sourced, they do provide a web interface for testing their method on user-provided images, allowing us to evaluate and compare their approach on public datasets.

Lee et al. [12] followed a similar approach to generating large datasets (see above) to train a *neural geometric parser*. Their system takes line segments detected by LSD [7] as input, first estimating the zenith vanishing point, selecting segments that are close to vertical, generating vanishing point hypotheses from pairs of these segments, and then feeding the segments and the hypotheses into a deep network to score the hypotheses. High-scoring zenith candidates are then used to generate camera tilt/roll and focal length hypotheses that are combined with the image and Manhattan line maps as input to a second network to generate final camera tilt/roll and focal length estimates. In a more recent paper, Lee et al. [11] have introduced an updated system called CTRL-C that continues to employ LSD line segments but shifts to a transformer architecture for estimating the parameters. While the neural geometric parser has not been open-sourced, CTRL-C has, and we compare with their approach below.

There are other recent deep learning approaches to estimation of vanishing points and/or camera rotation [13,14,23], but these do not estimate focal length.

3 $f\mathbf{R}$ Method

To develop our $f\mathbf{R}$ method for online camera calibration, we will assume a camera-centred world frame aligned with the Manhattan structure of the scene and employ a standard projection model $\tilde{\mathbf{x}} = \mathbf{KR}\bar{\mathbf{X}}$ where $\tilde{\mathbf{x}}$ is an image point in homogenous coordinates, $\mathbf{K}$ is the camera's intrinsic matrix, $\mathbf{R}$ is the camera rotation matrix and $\bar{\mathbf{X}}$ is a 3D world point in augmented coordinates. Unless otherwise noted, we will assume that any nonlinear distortions in the camera have been calibrated out in the lab, and that after laboratory calibration the camera has a central principal point, square pixels and zero skew, leaving a single intrinsic unknown, the focal length f. While these assumptions will not be met exactly by real cameras in the field, they are the standard assumptions employed by all of the methods reviewed above and the methods we compare against below, and in our experience are reasonable approximations for most higher-quality modern cameras.

The goal of $f\mathbf{R}$ then is to estimate the unknown focal length f, together with the 3D rotation $\mathbf{R}$ of the camera relative to the Manhattan frame of the scene. Again, we note that this goes beyond the capacity of recent deep learning approaches, which are only able to estimate two of the three rotational degrees of freedom (roll and tilt). Standard formulae can be used to convert between camera Euler angles (roll, tilt, pan) and the camera rotation matrix $\mathbf{R}$ [6].

We collect these unknowns into the parameter set $\Psi = \{f, \mathbf{R}\}$. Note that Ψ completely determines the locations of the three Manhattan vanishing points: The ith vanishing point is given by $\tilde{\mathbf{x}}_i = \mathbf{K}\mathbf{R}_i$, where $\mathbf{R}_i$ is the ith column of $\mathbf{R}$.

In order to better understand the relative advantages of the various deviation measures proposed in the literature, and the role of probabilistic modeling, we will adopt the original mixture model framework first proposed by Coughlan & Yuille [3] for luminance gradients and later extended to line segments [9], edges [4] and lines [19]. We will focus on the use of line segments as features, as there are numerous high-quality open-source line segment detectors now available [1,7,15] and line segments have been shown to be effective features for Manhattan scene analysis. We will use the MCMLSD line segment detector [1] for most of the experiments below, but will also evaluate the advantages and disadvantages of alternative detectors.

3.1   Probabilistic Model

$f\mathbf{R}$ will assume that each line segment $\boldsymbol{l}_i$ in the image is generated by a model $m_i \in M$ corresponding to one of four possible processes: the three Manhattan families of parallel lines or a background process assumed to generate a uniform distribution of lines. Given camera parameters Ψ, the probability of observing the segment $\boldsymbol{l}_i$ is then given by:

$$p(\boldsymbol{l}_i|\Psi) = \sum_{m_i \in M} p(\boldsymbol{l}_i|\Psi, m_i)p(m_i) \tag{1}$$

where $p(m_i)$ is the prior probability that a segment is generated by process m_i and $p(\boldsymbol{l}_i|\Psi, m_i)$ is the likelihood of an observed line generated by m_i under camera parameters Ψ.

Since, as noted above, the parameters Ψ completely determine the vanishing points, the likelihoods $p(\boldsymbol{l}_i|\Psi, m_i)$ for Manhattan processes m_i can be parameterized by one of the non-negative measures of deviation between the line $\boldsymbol{l}_i$ and the corresponding Manhattan vanishing point (deviation measures **a-e** reviewed above and summarized in Fig. 2). We will generally use the YorkUrbanDB training partition to fit exponential models $p(x) = \frac{1}{\lambda} \exp(-x/\lambda)$ for these deviations, and assume that the background process is uniform on this measure - see supplemental material for the fits. The one exception is for deviation measure d, where we employ the central Gaussian model proposed by Xu et al. [25], which also favours vanishing points that lie far from the segment midpoint, along the segment direction. Here the dispersion parameter σ does not have a clear interpretation in terms of the YorkUrbanDB so we use the value of $\sigma = 1$ pixel recommended by Xu et al. Table 1 summarizes these parameters.

Table 1. Dispersion parameters for likelihood models.

Deviation measure	Model	Param	Horiz	Vert
a	Exponential	λ	94.46 pix	17.26 pix
b	Exponential	λ	1.46 deg	0.57 deg
c	Exponential	λ	0.39 pix	0.53 pix
d	Gaussian	σ	1.00 pix	1.00 pix
e	Exponential	λ	0.80 deg	0.57 deg

The priors $p(m_i)$ are also estimated from the YorkUrbanDB training set: 45% of lines are expected to arise from the vertical process, 26% from each of the horizontal processes and 3% from the background process.

Assuming conditional independence over n observed line segments l_i and a flat prior over the camera parameters Ψ, the optimal solution should be found by maximizing the sum of log likelihoods, however we find empirically that we obtain slightly better results if we weight by line segment length $|l_i|$, solving for

$$\Psi^* = \arg\max_{\Psi} \sum_{i=1,\dots,n} |l_i| \log p\left(l_i | \Psi\right) \tag{2}$$

We conjecture that the small improvement achieved by weighting by line length may derive from a statistical dependence between line length and the likelihood model: Longer lines may be more accurate.

3.2 Parameter Search

The $f\mathbf{R}$ objective function (Eq. 2) is generally non-convex. We opt for a simple search method that can be used to compare the deviation measures a–e summarized in Fig. 2 and to assess the role of probabilistic modeling. The method proceeds in two stages. In Stage 1, we do a coarse $k \times k \times k \times k$ grid search over the four-dimensional parameter space of Ψ, evaluating Eq. 2 at each of the k^4 parameter proposals. Specifically, we sample uniformly over the range $[-45, +45]$ deg for pan, $[-15, +15]$ deg for roll, $[-35, +35]$ deg for tilt and $[50, 130]$ deg for horizontal FOV. (We also evaluated a RANSAC approach [21] for Stage 1 but found it be less accurate - see supplementary material for results.) We then deploy a nonlinear iterative search (MATLAB fmincon) initialized at each of the top l of these k^4 proposals and constrained to solutions within the ranges above. In the following we will use $k = 8, l = 4$.

Note that focal length f and FOV are monotonically related through FOV $= 2 \arctan\left(\frac{w}{2f}\right)$, where w is the image width. While we search over FOV we will generally convert to focal length when reporting results.

3.3   Error Prediction

One advantage of a probabilistic approach to online camera calibration is the opportunity to predict when estimates can be trusted. Since we explicitly model the distribution of deviations between line segments and hypothesized vanishing points, we might hope to estimate uncertainty in camera parameters using standard error propagation methods. Unfortunately, we find empirically that this local method does not lead to very accurate estimates of uncertainty, perhaps because the derivative of the camera parameters with respect to the segment orientations at the estimated parameters $\hat{\psi}$ is not very predictive of the derivative at the true camera parameters ψ^* (see supplementary material for details). We therefore propose instead three easily computable global predictors to inform the level of trust that should be invested in estimated camera parameters:

1) **Number of Line Segments.** We assign each detected segment l_i to the most likely generating process m_i under our mixture model (Eq. 1), count those assigned to each Manhattan direction, and take the minimum, to reflect the intuition that this will be the weakest link in the global parameter estimation process. 2) **Entropy.** This cue takes advantage of the first stage of our parameter search. We conjecture that more reliable parameter estimation will be reflected in a more peaked distribution of likelihoods over the k^4 parameter proposals evaluated. To capture this intuition, we normalize the likelihoods into a discrete probability distribution and compute its entropy. Low entropy distributions are expected to be more reliable. 3) **Likelihood.** We compute the mean log likelihood of the final estimate output from the second nonlinear iterative stage of our search, normalized by the number of line segments.

To predict camera parameter MAE from these cues we fit a KNN regression model on the PanoContext-fR training partition - see Sect. 5.4 for details.

4   Datasets

We evaluate fR and competing methods on two datasets: the YorkUrbanDB [4] test partition, and our novel PanoContext-fR dataset, curated from [27].

The YorkUrbanDB test partition consists of 51 indoor and outdoor images. Camera focal length f was fixed and estimated in the lab. Manhattan line segments were hand-labelled to allow estimation of the camera rotation $\mathbf{R}$ [4]. While the camera was handheld, the variation in camera roll and tilt is fairly modest: standard deviation of 0.83 deg for roll and 4.96 deg for tilt.

To allow evaluation over a wider range of camera parameters and to assess how well recent deep networks generalize compared to geometry-driven approaches, we introduce PanoContext-fR, a curation of planar projections from the PanoContext dataset [27], which contains 706 indoor panoramic scenes. We randomly divide this dataset into two equal training and test partitions of 353 scenes each. We will use the test partition to evaluate and compare deviation measures, probabilistic models and state-of-the-art systems for camera parameter estimation. While our fR system is not a machine learning method and does not require the training partition, we will explore methods for learning to predict estimation error in Sect. 3.3 and will use the training dataset there.

For each of these scenes we generated 15 planar projections, using a standard 640 × 480 pixel resolution. Three images were sampled for each of 5 fixed horizontal FOVs from 60 to 120 deg in steps of 15 deg. Pan, roll and tilt were randomly and uniformly sampled over $[-180, +180]$, $[-10, +10]$ and $[-30, +30]$ ranges respectively (Fig. 3). This generated 5,295 images for each of the training and test partitions. This dataset is not intended for network training, but rather for evaluating generalization of pre-trained and geometry-driven models.

We sampled FOVs discretely for PanoContext-fR so we could clearly see whether algorithms are able to estimate focal length (see below). However, we have also created and will make publicly available an alternative PanoContext-ufR that samples horizontal FOV randomly and uniformly over $[60, 120]$ deg. (See supplementary material.) Since the YorkUrbanDB provides the ground truth rotation matrix, we can evaluate the frame angle error, i.e., the magnitude of the rotation required to align the estimated camera frame with the ground truth frame. Since for PanoContext-fR we do not have ground truth pan, we report mean absolute roll and tilt errors.

Fig. 3. Three example planar projections from our new PanoContext-fR dataset.

5   Experiments

5.1   Evaluating Deviation Measures

We begin by applying the fR search method (Sect. 3.2) to identify the parameters Ψ maximizing agreement between line segments and vanishing points, based on the five deviation measures **a-e** summarized in Fig. 2. Without probabilistic modeling, we employed the objective function $\sum_i \log \delta_i$, where $\delta_i = d_i, \theta_i, \Delta\theta_i$ depending on which of the five deviation measures is employed. (We found empirically that logging the deviations before summing improved results.). We found that none of the deviation measures yielded good results: Mean focal length error remained above 11% on the York Urban DB and above 34% on the PanoContext-fR dataset (See Supplementary Material for detailed results.)

Probabilistic modeling of the deviation measure greatly improves results (see Table 2 for results on the PanoContext-fR dataset and supplementary material for results on the YorkUrbanDB). We note that this represents the learning of only 4 parameters (horizontal and vertical dispersions and priors). Moreover this learning seems to generalize well, as the parameters used to evaluate on the PanoContext-*f*R dataset were learned on the YorkUrbanDB training set.

Table 2. Evaluation of deviation measures on the PanoContext-*f*R training set. Numbers are mean ± standard error.

Dev. measure	Roll MAE (deg)	Tilt MAE (deg)	Focal length MAE (%)
a	4.53 ± 0.045	14.2 ± 0.14	35.6 ± 0.35
b (*f*R)	$\mathbf{0.78 \pm 0.02}$	$\mathbf{1.59 \pm 0.06}$	$\mathbf{8.4 \pm 0.26}$
c	0.90 ± 0.009	2.26 ± 0.022	14.1 ± 0.14
d	0.81 ± 0.008	1.67 ± 0.015	10.1 ± 0.08
e	0.89 ± 0.009	2.13 ± 0.021	19.5 ± 0.19

Our second observation is that performance also depends strongly on the deviation measure, even when carefully modeling the likelihood. The large errors produced by measure **a** likely reflect the fact that vanishing points are often far from the principal point, resulting generally in very unpredictable deviations from the line passing through an associated line segment. The other measures yield better performance, but we note that deviation measure **e** tends to have higher errors for focal length. We believe this is because the Gauss Sphere measure of deviation suffers from a degeneracy: As focal length tends to infinity, the interpretation plane normals collapse to a great circle parallel to the image plane. Thus a hypothesized vanishing point near the principal point will always generate small deviations, leading to a large likelihood.

This leaves deviation measures **b-d**. Method **d** performs less well on the YorkUrbanDB dataset and method **c** has higher errors on the PanoContext-*f*R dataset. Overall, method **b**, the method originally used by Coughlan & Yuille for isophotes [3], Rother for line segments [16] and Denis et al. for edges [4], appears to generate the most consistent performance, and so we adopt this method as our standard *f*R deviation measure for the remainder of the paper.

5.2 Evaluating Line Segment Detectors

All of the experiments above employed the MCMLSD line segment detector [1]. Table 3 assesses the sensitivity of our system to this choice by substituting two alternative detectors: The well-known LSD detector [7] and a more recent deep learning detector called HT-LCNN [15]. Performance on the PanoContext-*f*R dataset is significantly better using MCMLSD than the more recent HT-LCNN, despite the fact that HT-LCNN is reported [15] to have much better precision-recall performance on the YorkUrban DB and Wireframe [26] datasets.

We believe this discrepancy stems largely from the incompleteness of these datasets, which do not claim to label *all* Manhattan line segments. Methods like MCMLSD, which attempt to detect all line segments (Manhattan and non-Manhattan), thus tend to achieve lower precision scores on these datasets, while deep learning methods like HT-LCNN learn to detect only the labelled segments and thus achieve higher precision scores. Our experiment reveals that this higher precision does not translate into better performance but rather *worse* performance, presumably because HT-LCNN fails to generate Manhattan segments that may be unlabelled but are nevertheless useful for estimating the camera parameters. We proceed with the MCMLSD line segment detector but will return to this choice when considering run-time efficiency (Sect. 5.5).

Table 3. Evaluating the choice of line segment detector on the PanoContext-$f\mathbf{R}$ training set. Numbers are mean ± standard error.

Methods	Run time (sec)	Roll MAE (deg)	Tilt MAE (deg)	Focal length MAE (%)
MCMLSD [1]	4.23	**0.78 ± 0.02**	**1.59 ± 0.06**	**8.4 ± 0.26**
LSD [7]	0.40	1.23 ± 0.012	3.35 ± 0.03	8.6 ± 0.11
HT-LCNN [15]	2.91	0.93 ± 0.009	1.91 ± 0.02	10.1 ± 0.14

5.3   Comparison with State of the Art

We compare our $f\mathbf{R}$ system against the four state-of-the-art systems [8,10,11,18] reviewed in Sect. 2. Three of these [10,11,18] are open-sourced and we downloaded the code. The fourth (Hold-Geoffroy) [8] provides a web interface. We test two versions of the CTRL-C [11] network: CTRL-C S360 was trained on a dataset curated from the SUN360 dataset [24]. CTRL-C GSV was trained on a dataset curated from the Google Street View dataset [12].

Table 4(top) shows results of this comparison on the York UrbanDB test set. We report roll and tilt error instead of frame error, since the deep learning methods (Hold-Geoffroy [8] and CTRL-C [11]) do not estimate pan angle. (See supplemental material for comparison of pan angle error with the methods of Lee et al. [10] and Simon et al. [18]). (The Simon et al. method returned a valid result for only 69% of the York UrbanDB test images and 46% of the PanoContext-$f\mathbf{R}$ test images; we therefore report their average performance only for these.)

All methods have fairly low error in roll, but remember that in the York UrbanDB, variation in camera roll is limited (standard deviation of 0.83 deg). Our $f\mathbf{R}$ system improves on the next best method (Lee et al. [10]) by 38%. For tilt estimation, variability in performance across systems is more pronounced, but three systems (our $f\mathbf{R}$ system, Lee et al. [10] and CTRL-C S360 [11] all perform well. Focal length estimation is the big differentiator. Our $f\mathbf{R}$ system performs much better than all of the other systems, beating the next best system (Lee et al. [10]) by 57% (reducing MAE from 11.5% to 4.9%). Notice that the deep learning methods (Hold-Geoffroy [8] and CTRL-C [11]) do not perform well.

Table 4. Performance comparison with SOA on the York UrbanDB (top) and PanoContext-*f*R (bottom) test sets. Numbers are mean ± standard error.

YorkUrbanDB	Run time (sec)	Roll (MAE deg)	Tilt MAE (deg)	FOV MAE (%)	Focal len. MAE (%)
Lee[10]	1.13	0.80±0.2	1.33±0.2	9.2 ± 1.15	11.5 ± 1.62
Simon[18]	0.44	2.63±0.6	11.0±1	11.4 ± 2.56	15.1 ± 2.08
Hold-Geoffroy[8]	n/a	1.47±0.2	7.57±0.6	48.7 ± 2.11	37.7 ± 7.02
CTRL-C[11] S360	**0.32**	1.38±0.2	**1.09±0.1**	29.0 ± 2.50	24.7 ± 1.80
CTRL-C [11] GSV	**0.32**	1.37±0.2	3.90±0.3	69.4 ± 1.66	48.8 ± 0.77
*f*R	6.40	**0.50±0.1**	1.16±0.1	**3.8 ± 0.61**	**4.6 ± 0.78**
Fast-*f*R	0.89	1.13±0.4	1.62±0.3	5.1 ± 1.04	5.7 ± 1.21

PanoContext-*f*R	Roll MAE (deg)	Tilt MAE (deg)	FOV MAE (%)	Focal len. (MAE %)
Lee[10]	2.02± 0.02	3.35±0.03	13.4 ± 0.22	22.3 ± 0.22
Simon[18]	1.50±0.02	3.18± 0.05	32.8 ± 0.46	54.0 ± 0.53
Hold-Geoffroy[8]	1.58±0.02	3.69±0.04	16.1 ± 0.12	20.5 ± 0.20
CTRL-C[11] S360	1.03±0.01	2.19±0.02	8.1 ± 0.07	13.1 ± 0.13
CTRL-C [11] GSV	2.24±0.02	8.98± 0.09	17.3 ± 0.11	21.9 ± 0.31
*f*R	**0.78±0.02**	**1.59±0.06**	**5.3±0.15**	8.4±0.26
Fast-*f*R	0.89±0.02	1.90±0.07	5.4 ± 0.16	**7.6 ± 0.23**

Table 4(bottom) compares these systems on our novel PanoContext-*f*R dataset. (While FOV and focal length are monotonically related, we show MAE for both for the reader's convenience.) We find that our *f*R system performs significantly better that all of the other systems on all three benchmarks (roll, tilt and focal length), improving on the next best system (CTRL-C S360 [11]) by 23%, 25% and 33%, respectively. (The relatively strong performance of the CTRL-C S360 system here may derive from the fact that the PanoContext dataset was original drawn from the SUN360 dataset on which CTRL-C S360 is trained.)

Figure 4 compares distributions of estimated and ground truth parameters for the PanoContext-*f*R dataset. The distributions for camera rotation are generally reasonable, although the method of Simon et al. [18] exhibits odd preferences for specific roll and tilt angles, the Hold-Geoffroy system [8] has an overly strong bias to zero roll, and when trained on GSV, the CTRL-C system [11] develops a bias against small upward tilts.

The sampling of five discrete FOVs in the PanoContext-*f*R dataset allows us to visualize the sensitivity of the algorithms to focal length. While the geometry-based algorithms all show modulation with the sampled FOVs (albeit weak for Simon et al. [18]), the deep learning methods do not, each forming a single broad peak. Our *f*R system appears to be best able to pick up the FOV signal from the data without a strong prior bias.

The ability of the deep learning methods to estimate FOV/focal length might in part be cue to limitations in the range of FOVs in their training datasets.

However, in the supplementary material we show that this does not fully account for the superiority of our *f*R system, which we believe derives from more direct geometry and probabilistic modeling.

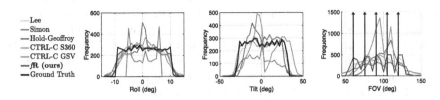

Fig. 4. Distribution of ground truth and estimated camera parameters for the PanoContext-*f*R test sets.

5.4 Predicting Reliability

We employ three global cues to predict camera parameter estimation error: 1) The minimum number of segments over the three Manhattan directions, 2) Entropy over our parameter grid search and 3) mean log likelihood of the final parameter estimate: See supplementary material for a visualization of how parameter error varies as a function of these cues. To estimate the reliability of *f*R estimates at inference, we use the PanoContext-*f*R training set to fit a regression model that uses these three cues jointly to predict the absolute error in each of the camera parameters: We collect these cues into a 3-vector, use the training data to compute a whitening transform and then use KNN regression in the 3D whitened space to estimate error, again selecting K by 5-fold cross-validation.

Table 5 assesses the ability of this model to predict error on the held-out PanoContext-*f*R test set. The table shows the mean error obtained if we accept only the top 25%, 50%, 75% or 100% of estimates predicted to have least error. The model is remarkably effective. For example, by accepting only the 25% of estimates predicted to be most reliable, mean error declines by 24% for roll, 52% for tilt and 66% for focal length. This ability to predict estimation error can be extremely useful in applications where the camera can be recalibrated on sparse frames, or when conservative actions can be taken to mitigate risks.

Table 5. Parameter MAE for 25%, 50%, 75% and 100% of held-out PanoContext-*f*R test images predicted to be most reliable.

% of images	Roll MAE (deg)	Tilt MAE (deg)	Focal length MAE (%)
25%	0.56	0.68	2.8%
50%	0.59	0.75	3.9%
75%	0.63	0.86	5.2%
100%	0.74	1.42	8.3%

5.5 Run Time

We ran our experiments on a 2.3 GHz 8-Core Intel i9 CPU with 32 GB RAM and an NVIDIA Tesla V100-32GB GPU. Our *f*R system as implemented is slower than competing systems (Table 4), primarily due to our line segment detection system MCMLSD, which consumes on average 4.23 s per image (Table 3).

To address this issue, we have created a more efficient version of our system we call Fast-*f*R, through two innovations. First, we replace MCMLSD with LSD, which consumes only 40 ms per image on average. Second, we limit the number of iterations in our second search stage to 10. This reduces average run time from 6.4 s to 0.83 s with only a modest decline in accuracy (Table 4).

6 Limitations

*f*R relies on the Manhattan World assumption, i.e., that images contain aligned rectilinear structure. The superior performance of *f*R on the York Urban DB and PanoContext-*f*R datasets, which are representative of common outdoor and indoor built environments, attests to the real-world applicability of this assumption. Nevertheless, the method will typically fail on scenes that do not conform to the Manhattan assumption. This may include images with insufficient structure (see Supplementary Material for examples), nature scenes and non-Manhattan built environments (e.g., the Guggenheim museums in New York or Bilbao). Generalizing the *f*R method to exploit additional scene regularities, e.g., Atlanta World [17] or quadrics, is an interesting direction for future research.

Like all SOA methods reviewed and evaluated here, our *f*R system assumes no non-linear distortions, a central principal point, square pixels and zero skew. These are reasonable approximations, since deviations from these assumptions can be calibrated out in the factory or lab, and subsequent drift of these parameters is typically minor relative to changes in focal length and rotation.

While the Manhattan constraint can also be used to estimate the principal point, we find that more accurate estimates of focal length and rotation are obtained by assuming a central principal point. Simultaneous online estimation of focal length, principal point and camera rotation, as well as radial distortion, thus remains an interesting and challenging direction for future research.

7 Conclusions

We have employed existing and novel datasets to assess the reliability of online systems for joint estimation of focal length and camera rotation. We show that the reliability of geometry-driven systems depends profoundly on the deviation measure employed and accurate probabilistic modeling. Based on these findings, we have proposed a novel probabilistic geometry-driven approach called *f*R that outperforms four state-of-the-art competitors, including two geometry-based systems and two deep learning systems. The *f*R advantage is most pronounced for estimation of focal length, where the deep learning systems seem to struggle.

We note that the deep learning systems also do not predict pan angle, and so do not fully solve for the camera rotation. We further demonstrate an ability to estimate the reliability of specific $f\mathbf{R}$ predictions, leading to substantial gains in accuracy when systems can choose to reject parameter estimates judged to be unreliable. We believe this will be useful in many real-world applications, particularly in mobile robotics and autonomous vehicles.

Acknowledgement. This work was supported by the University of Toronto, NSERC, the Ontario Research Fund, the York University VISTA and Research Chair programs (Canada), and the Agency for Science, Technology and Research (A*STAR) under its RIE2020 Health and Biomedical Sciences (HBMS) Industry Alignment Fund Pre-Positioning (IAF-PP) Grant No. H20c6a0031 and AI3 HTPO Seed Fund (C211118014) (Singapore).

References

1. Almazan, E.J., Tal, R., Qian, Y., Elder, J.H.: MCMLSD: a dynamic programming approach to line segment detection. In: IEEE Conference on Computer Vision and Pattern Recognition, pp. 2031–2039 (2017)
2. Collins, R.T., Weiss, R.S.: Vanishing point calculation as a statistical inference on the unit sphere. In: International Conference on Computer Vision, vol. 90, pp. 400–403. Citeseer (1990)
3. Coughlan, J.M., Yuille, A.L.: Manhattan world: compass direction from a single image by Bayesian inference. In: International Conference on Computer Vision, vol. 2, pp. 941–947. IEEE (1999)
4. Denis, P., Elder, J.H., Estrada, F.J.: Efficient edge-based methods for estimating manhattan frames in urban imagery. In: Forsyth, D., Torr, P., Zisserman, A. (eds.) ECCV 2008. LNCS, vol. 5303, pp. 197–210. Springer, Heidelberg (2008). https://doi.org/10.1007/978-3-540-88688-4_15
5. Deutscher, J., Isard, M., MacCormick, J.: Automatic camera calibration from a single manhattan image. In: Heyden, A., Sparr, G., Nielsen, M., Johansen, P. (eds.) ECCV 2002. LNCS, vol. 2353, pp. 175–188. Springer, Heidelberg (2002). https://doi.org/10.1007/3-540-47979-1_12
6. Faugeras, O.: Three-Dimensional Computer Vision: A Geometric Viewpoint. MIT Press, Cambridge (1993)
7. von Gioi, R.G., Jakubowicz, J., Morel, J.M., Randall, G.: Lsd: a fast line segment detector with a false detection control. IEEE Trans. Pattern Anal. Mach. Intell. **4**, 722–732 (2008)
8. Hold-Geoffroy, Y., et al.: A perceptual measure for deep single image camera calibration. In: IEEE Conference on Computer Vision and Pattern Recognition, pp. 2354–2363 (2018)
9. Košecká, J., Zhang, W.: Video compass. In: Heyden, A., Sparr, G., Nielsen, M., Johansen, P. (eds.) European Conference on Computer Vision, pp. 476–490. Springer, Berlin Heidelberg, Berlin, Heidelberg (2002)
10. Lee, D.C., Hebert, M., Kanade, T.: Geometric reasoning for single image structure recovery. In: IEEE Conference on Computer Vision and Pattern Recognition, pp. 2136–2143. IEEE (2009)

11. Lee, J., Go, H., Lee, H., Cho, S., Sung, M., Kim, J.: CTRL-C: camera calibration transformer with line-classification. In: International Conference on Computer Vision, pp. 16228–16237 (2021)

12. Lee, J., Sung, M., Lee, H., Kim, J.: Neural geometric parser for single image camera calibration. In: Vedaldi, A., Bischof, H., Brox, T., Frahm, J.-M. (eds.) ECCV 2020. LNCS, vol. 12357, pp. 541–557. Springer, Cham (2020). https://doi.org/10.1007/978-3-030-58610-2_32

13. Li, H., Chen, K., Kim, P., Yoon, K.J., Liu, Z., Joo, K., Liu, Y.H.: Learning icosahedral spherical probability map based on bingham mixture model for vanishing point estimation. In: International Conference on Computer Vision, pp. 5661–5670 (2021)

14. Li, H., Zhao, J., Bazin, J.C., Chen, W., Liu, Z., Liu, Y.H.: Quasi-globally optimal and efficient vanishing point estimation in Manhattan world. In: International Conference on Computer Vision, pp. 1646–1654 (2019)

15. Lin, Y., Pintea, S.L., van Gemert, J.C.: Deep hough-transform line priors. In: Vedaldi, A., Bischof, H., Brox, T., Frahm, J.-M. (eds.) ECCV 2020. LNCS, vol. 12367, pp. 323–340. Springer, Cham (2020). https://doi.org/10.1007/978-3-030-58542-6_20

16. Rother, C.: A new approach to vanishing point detection in architectural environments. Image Vis. Comput. **20**(9–10), 647–655 (2002)

17. Schindler, G., Dellaert, F.: Atlanta world: An expectation maximization framework for simultaneous low-level edge grouping and camera calibration in complex man-made environments. In: Proceedings of the 2004 IEEE Computer Society Conference on Computer Vision and Pattern Recognition, 2004. CVPR 2004. vol. 1, pp. I-I. IEEE (2004)

18. Simon, G., Fond, A., Berger, M.O.: A Simple and Effective Method to Detect Orthogonal Vanishing Points in Uncalibrated Images of Man-Made Environments. In: Bashford-Rogers, T., Santos, L.P. (eds.) EuroGraphics - Short Papers. The Eurographics Association (2016). https://doi.org/10.2312/egsh.20161008

19. Tal, R., Elder, J.H.: An accurate method for line detection and manhattan frame estimation. In: Park, J.-I., Kim, J. (eds.) ACCV 2012. LNCS, vol. 7729, pp. 580–593. Springer, Heidelberg (2013). https://doi.org/10.1007/978-3-642-37484-5_47

20. Tardif, J.P.: Non-iterative approach for fast and accurate vanishing point detection. In: IEEE International Conference on Computer Vision, pp. 1250–1257. IEEE (2009)

21. Wildenauer, H., Hanbury, A.: Robust camera self-calibration from monocular images of Manhattan worlds. In: IEEE Conference on Computer Vision and Pattern Recognition, pp. 2831–2838. IEEE (2012)

22. Wu, Z., Radke, R.: Keeping a pan-tilt-zoom camera calibrated. Pattern Analysis and Machine Intelligence, IEEE Transactions on **35**(8), 1994–2007 (2013). https://doi.org/10.1109/TPAMI.2012.250

23. Xian, W., Li, Z., Fisher, M., Eisenmann, J., Shechtman, E., Snavely, N.: UprightNet: geometry-aware camera orientation estimation from single images. In: International Conference on Computer Vision, pp. 9974–9983 (2019)

24. Xiao, J., Ehinger, K.A., Oliva, A., Torralba, A.: Recognizing scene viewpoint using panoramic place representation. In: IEEE Conference on Computer Vision and Pattern Recognition, pp. 2695–2702. IEEE (2012)

25. Xu, Y., Oh, S., Hoogs, A.: A minimum error vanishing point detection approach for uncalibrated monocular images of man-made environments. In: IEEE Conference on Computer Vision and Pattern Recognition, pp. 1376–1383 (2013)

26. Xu, Z., Shin, B., Klette, R.: A statistical method for line segment detection. Comput. Vis. Image Underst. **138**, 61–73 (2015)
27. Zhang, Y., Song, S., Tan, P., Xiao, J.: PanoContext: a whole-room 3D context model for panoramic scene, pp. 668–686 (2014)

PS-NeRF: Neural Inverse Rendering for Multi-view Photometric Stereo

Wenqi Yang[1], Guanying Chen[2(✉)], Chaofeng Chen[3], Zhenfang Chen[4], and Kwan-Yee K. Wong[1]

[1] The University of Hong Kong, Pok Fu Lam, Hong Kong
[2] FNii and SSE, CUHK-Shenzhen, Shenzhen, China
chenguanying@cuhk.edu.cn
[3] Nanyang Technological University, Singapore, Singapore
[4] MIT-IBM Watson AI Lab, Cambridge, USA

Abstract. Traditional multi-view photometric stereo (MVPS) methods are often composed of multiple disjoint stages, resulting in noticeable accumulated errors. In this paper, we present a neural inverse rendering method for MVPS based on implicit representation. Given multi-view images of a non-Lambertian object illuminated by multiple unknown directional lights, our method jointly estimates the geometry, materials, and lights. Our method first employs multi-light images to estimate per-view surface normal maps, which are used to regularize the normals derived from the neural radiance field. It then jointly optimizes the surface normals, spatially-varying BRDFs, and lights based on a shadow-aware differentiable rendering layer. After optimization, the reconstructed object can be used for novel-view rendering, relighting, and material editing. Experiments on both synthetic and real datasets demonstrate that our method achieves far more accurate shape reconstruction than existing MVPS and neural rendering methods. Our code and model can be found at https://ywq.github.io/psnerf.

Keywords: Multi-view photometric stereo · Inverse rendering · Neural rendering

1 Introduction

Multi-view stereo (MVS) is a technique for automated 3D scene reconstruction from a set of images captured from different viewpoints. As MVS methods rely on feature matching across different images, they generally assume the scene to be composed of textured Lambertian surfaces [1,12,45] and their reconstructions often lack fine details. Photometric stereo (PS), on the other hand, can recover per-pixel surface normals of a scene from single-view images captured under varying light directions. By utilizing shading information, PS method can recover

Supplementary Information The online version contains supplementary material available at https://doi.org/10.1007/978-3-031-19769-7_16.

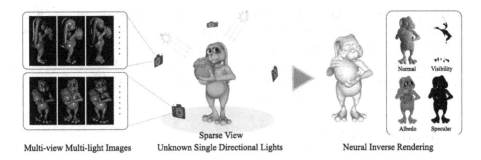

Fig. 1. Our method takes multi-view multi-light images as input, and is able to reconstruct accurate surface and faithful BRDFs based on our shadow-aware renderer. Specifically, we only take images under **sparse** views with each view illuminated by multiple **unknown** single directional lights.

fine surface details for both non-Lambertian and textureless objects [14,46,56]. However, single-view PS methods are not capable of recovering a full 3D shape.

To combine the merits of both techniques, multi-view photometric stereo (MVPS) methods are proposed to recover high-quality full 3D shapes for non-Lambertian and textureless objects [11,26,41]. Traditional MVPS methods are often composed of multiple disjoint stages [26,41], leading to noticeable accumulated errors.

Neural rendering methods have recently been introduced to tackle the problems of multi-view reconstruction and novel-view synthesis [36,39,60]. These methods work on multi-view images captured under fixed illuminations and show spectacular results. In addition to viewing directions, Kaya *et al.* [23] use surface normals estimated from an observation map-based PS method to condition the neural radiance field (NeRF). Although improved rendering results have been reported, this method has 4 fundamental limitations, namely 1. it requires calibrated lights as input to estimate per-view normal maps; 2. it takes surface normals as inputs for NeRF, making novel-view rendering difficult (if not impossible); 3. it does not recover BRDFs for the surface, making it not suitable for relighting; and 4. the PS network and NeRF are disjointed, and normal estimation errors from PS network will propagate to NeRF and cannot be eliminated.

To solve these challenges, we propose a neural inverse rendering method for multi-view photometric stereo (See Fig. 1). Our method does not require calibrated lights. We first estimate per-view normal maps to constrain the gradient of density in NeRF. Surface normals, BRDFs, and lights are then jointly optimized based on a shadow-aware differentiable rendering layer. By taking advantage of both multi-view and multi-light images, our method achieves far more accurate shape reconstruction results. Moreover, as our method explicitly models BRDFs and lights, it allows novel-view rendering, relighting, and material editing.

In summary, the key contributions of this paper are as follows:

- We introduce a neural inverse rendering method for multi-view photometric stereo, which jointly optimizes shape, BRDFs, and lights based on a shadow-aware differentiable rendering layer.

- We propose to regularize the surface normals derived from the radiance field with normals estimated from multi-light images, which significantly improves surface reconstruction, especially for sparse input views (e.g., 5 views).
- Our method achieves state-of-the-art results in MVPS, and demonstrates that incorporating multi-light information appropriately can produce a far more accurate shape reconstruction.

2   Related Work

Single-View Photometric Stereo (PS). Traditional PS methods rely on outlier rejection [37,58,59], reflectance model fitting [10,18,51], or exemplars [15,16] to deal with non-Lambertian surfaces. Deep learning based PS methods solve this problem by learning the surface reflectance prior from a dataset [7,8, 17,31,44,49]. These methods typically train a deep network to learn the mapping from image measurements to surface normals. Recently, many efforts have been devoted to reduce the number of images required [25,65] and to estimate light directions in an uncalibrated setup [6,22,33].

Multi-view Photometric Stereo (MVPS). MVPS combines the advantages of MVS and PS methods, leading to more accurate surface reconstructions. Traditional MVPS methods assume a simplified surface reflectance [11,29,32,41,57]. They first apply MVS to reconstruct a coarse shape from multi-view images and adopt PS to obtain per-view surface normals from multi-light images. They then refine the coarse shape using the obtained normals. Zhou *et al.* [26,66] propose a method to deal with isotropic materials. Their method first reconstructs sparse 3D points from multi-view images and identifies per-view iso-depth contours using multi-light images. It then recovers a complete 3D shape by propagating the sparse 3D points along the iso-depth contours. Kaya *et al.* [21] propose an uncertainty-aware deep learning based method to integrate MVS depth maps with PS normal maps to produce a full shape. Another branch of methods adopts a co-located camera-light setup for joint reflectance and shape recovery [9,27,38,53].

Neural Rendering. Neural rendering methods have achieved great successes in novel-view synthesis and multi-view reconstruction [39,47,50,60]. In particular, neural radiance field (NeRF) [36] achieves photo-realistic view synthesis by representing a continuous space with an MLP which maps 5D coordinates (*i.e.*, 3D point and view direction) to density and color. Many follow-up works are proposed to improve the reconstructed shape [40,52], rendering speed [13,30,42], and robustness [2,34,62]. However, these methods essentially treat surface points as light sources, and thus cannot disentangle materials and lights [55].

Some methods have been proposed to jointly recover shape, materials, and lights [4,61]. PhySG [61] and NeRD [4] adopt Spherical Gaussian (SG) representation for BRDFs and environment illumination to enable fast rendering. Neural-PIL [5] replaces the costly illumination integral operation with a simple network query. These methods [4,5,61], however, ignore cast shadows during optimization. NeRV [48] explicitly models shadow and indirect illumination, but it requires a

Table 1. Comparisons among different neural inverse rendering methods (MVI stands for multi-view images).

Method	Input	Shape	BRDF	Lighting	Shadow
NeRV [48]	MVI (Fixed lighting)	Density	Microfacet model	Known Envmap	Yes
PhySG [61]	MVI (Fixed lighting)	SDF	Microfacet (SGs)	Unknown Envmap (SGs)	No
NeRFactor [64]	MVI (Fixed lighting)	Density	Learned BRDF	Unknown Envmap	Yes
NeRD [5]	MVI (Varying lighting)	Density	Microfacet (SGs)	Unknown Envmap (SGs)	No
NRF [3]	MVI (Co-located light)	Density	Microfacet model	Known Co-located light	Yes
KB22 [23]	MVI (Multi-light)	Density	No	No	No
Ours	MVI (Multi-light)	Density	Mixture of SGs	Unknown Multi-light	Yes

known environment map. NeRFactor [64] proposes a learned auto-encoder to represent BRDFs and pre-extracts a light visibility buffer using the obtained mesh. NRF [3] optimizes a neural reflectance field assuming a co-located camera-light setup.

Similar to our proposed method, Kaya *et al.* [23] introduce a NeRF-based method for MVPS. Their method [23] assumes calibrated light directions and takes the estimated normals as input to condition their NeRF. It thus cannot disentangle surface materials and light directions. Table 1 summarizes the differences between our method and existing neural inverse rendering methods. Our method is the only one that explicitly models surface reflectances and lights under a multi-view photometric stereo setup.

3 Methodology

3.1 Overview

Given multi-view and multi-light images[1] of an object taken from M sparse views, our goal is to simultaneously reconstruct its shape, materials, and lights. We denote the MVPS image set as $\mathcal{I}$, and multi-light images for each view m as $\mathcal{I}^m = \{I_1^m, I_2^m, \ldots, I_{L_m}^m\}$. Note that the number of lights for each view can vary. Figure 2 illustrates the overall pipeline of our method.

Inspired by the recent success of neural radiance field [36] for 3D scene representation, we propose to represent the object shape with a density field. Our method consists of two stages to make full use of multi-view multi-light images.

In the first stage, we estimate a guidance normal map $\mathcal{N}_m$ for each view, which is used to supervise the normals derived from the density field. This direct normal supervision is expected to provide a strong regularization on the density field, leading to an accurate surface. In the second stage, based on the learned density field as the shape prior, we jointly optimize the surface normals, materials, and lights using a shadow-aware rendering layer.

---

[1] *i.e.*, multiple images are captured for each view, where each image is illuminated by a single unknown directional light.

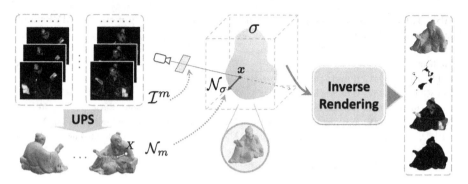

Fig. 2. Given multi-view multi-light images, we first obtain guidance normal maps via uncalibrated photometric stereo (UPS) to regularize the neural density field, which encourages accurate surface reconstruction. We then perform neural inverse rendering to jointly optimize surface normals, BRDFs and lights based on the initial shape.

3.2 Stage I: Initial Shape Modeling

In the first stage, we optimize a neural radiance field with surface normal regularizations to represent the object shape.

Neural Radiance Field. NeRF employs an MLP to map a 3D point x_k in space and view direction d to density σ_k and color c_k, *i.e.*,

$$(\sigma_k, c_k) = \text{MLP}_{\text{NeRF}}(x_k, d). \tag{1}$$

Given a pixel in an image, its color can be computed by integrating the color of the points sampled along its visual ray r through volume rendering, *i.e.*,

$$\tilde{C}(r) = \sum_{k=1}^{K} T_k(1 - \exp(-\sigma_k \delta_k))c_k, \quad T_k = \exp\left(-\sum_{j=1}^{k-1} \sigma_j \delta_j\right), \tag{2}$$

where $\delta_k = t_{k+1} - t_k$ is the distance between adjacent sampled points.

Typically, a NeRF is fitted to a scene by minimizing the reconstruction error between the rendered image C and the input image I, *i.e.*,

$$\mathcal{L}'_c = \sum \|C - I\|_2^2. \tag{3}$$

However, as the surface geometry has no direct supervision, the density field is generally very noisy. Recently, UNISURF [40] improves the surface quality of NeRF by gradually shrinking the sampling range of a ray for volume rendering, leading to a smoother surface. However, as shown in our experiments, the shape recovered by UNISURF [40] is still not satisfactory.

Surface Normal Regularization. The above observation motivates us to introduce regularizations for the surface geometry. Notably, state-of-the-art uncalibrated photometric stereo (UPS) method, such as [6], can estimate a good

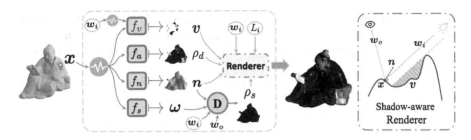

Fig. 3. In Stage II, we model normals, BRDFs, and light visibility of the scene with MLPs. The weights of the MLPs and lights are jointly optimized to fit the input images.

normal map from single-view multi-light images. We therefore use the normal map $\mathcal{N}_m$ estimated by [6] to regularize the density field by minimizing

$$\mathcal{L}'_n = \sum \|\mathcal{N}_\sigma - \mathcal{T}_{m2w}(\mathcal{N}_m)\|_2^2, \quad \mathcal{N}_\sigma(\boldsymbol{x}) = \frac{\nabla \sigma(\boldsymbol{x})}{\|\nabla \sigma(\boldsymbol{x})\|_2}, \tag{4}$$

where $\mathcal{N}_\sigma$ is the surface normal derived from the gradient of the density filed σ based on the expected depth location, and $\mathcal{T}_{m2w}$ is transformation that transforms the view-centric normals from the camera coordinate system to the world coordinate system. We also include the normal smoothness regularization [40] with $\epsilon \sim \mathcal{N}(0, 0.01)$, *i.e.*,

$$\mathcal{L}'_{ns} = \sum \|\mathcal{N}_\sigma(\boldsymbol{x}) - \mathcal{N}_\sigma(\boldsymbol{x} + \epsilon)\|_2^2. \tag{5}$$

We adopt UNISURF [40] as our radiance field method for initial shape modeling, and the overall loss function for Stage I is given by

$$\mathcal{L}' = \alpha_1 \mathcal{L}'_c + \alpha_2 \mathcal{L}'_n + \alpha_3 \mathcal{L}'_{ns}, \tag{6}$$

where α_* denotes the loss weights.

3.3 Stage II: Joint Optimization with Inverse Rendering

With the initial shape (*i.e.*, σ and $\mathcal{N}_\sigma$) from Stage I, we are able to jointly optimize the surface normals, spatially-varying BRDFs, and lights based on a shadow-aware rendering layer (see Fig. 3). Specifically, we first extract the surface from the density field via root-finding [35,39] similar to [40]. We then model surface normals, BRDFs, and light visibility of the scene with MLPs. The weights of the MLPs and the lights are then jointly optimized to fit the input multi-view and multi-light images. In the following subsections, we will describe the formulation of our rendering layer and each component in detail.

Rendering Equation. The rendering equation for a non-Lambertian surface point $\boldsymbol{x}$ can be written as [20]

$$\hat{I}(\boldsymbol{w}_o; \boldsymbol{x}) = \int_\Omega L_i(\boldsymbol{w}_i) f_r(\boldsymbol{w}_o, \boldsymbol{w}_i; \boldsymbol{x})(\boldsymbol{w}_i \cdot \boldsymbol{n}) \, \mathrm{d}\boldsymbol{w}_i, \tag{7}$$

where w_i and w_o are the incident light direction and view direction respectively, $f_r(w_o, w_i; x)$ represents a general BRDF at location x. $L_i(w_i)$ is the light intensity along w_i and $\hat{I}(w_o; x)$ is the integrated radiance over the upper-hemisphere Ω.

By assuming a directional light and considering light visibility, the rendering equation can be rewritten as

$$\hat{I}(w_o, w_i; x) = f_v(w_i; x) L_i(w_i) f_r(w_o, w_i; x)(w_i \cdot n), \tag{8}$$

where $f_v(w_i; x)$ indicates the visibility of light along w_i at x, and models cast-shadow in the rendered image.

Shape Modeling. In Stage I, we optimize a radiance field with surface normal regularizations to produce an initial density field σ. Note the normals used to regularize $\mathcal{N}_\sigma$ are estimated by a PS method [6] which inevitably contain estimation errors. As a result, the derived normals $\mathcal{N}_\sigma$ might not be accurate in some regions. To refine the normals, we use an MLP $f_n(x) \mapsto n$ to model the distribution of surface normals. This MLP will be optimized by an image fitting loss. To encourage the refined normals to not deviate too much from the derived normals, we use the derived normals to regularize the output of $f_n(x)$ by minimizing

$$\mathcal{L}_n'' = \sum \|f_n(x) - \mathcal{N}_\sigma(x)\|_2^2. \tag{9}$$

Visibility Modeling. Given the density field σ, a surface point x, and a light direction w_i, the light visibility $\mathcal{V}_\sigma(x, w_i)$ can be computed by applying volume rendering to calculate the accumulated density along the ray directed from x to the light source [64].

Since we perform ray-marching to calculate visibility for each point and each query light direction, it will be time-consuming to compute light visibility for an environment map lighting. Besides, the computed visibility might also be noisy. We therefore model the distribution of light visibility using another MLP $f_v(x, w_i) \mapsto v$, which is regularized by the computed visibility by minimizing

$$\mathcal{L}_v'' = \|f_v(x, w_i) - \mathcal{V}_\sigma(x, w_i)\|_2^2. \tag{10}$$

Material Modeling. As in previous works [4,61,64], we assume the BRDF model can be decomposed into diffuse color ρ_d and specular reflectance ρ_s, i.e., $f_r(x, w_i, w_o) = \rho_d + \rho_s(x, w_i, w_o)$. For diffuse color, we use an MLP $f_a(x) \mapsto \rho_d$ to model the albedo for a surface point x.

For specular component, one may adopt a reflectance model (e.g., Microfacet) to model the specular reflectance and estimate its parameters (e.g., roughness) [28,48]. However, we found it difficult to model the specular effects of real-world objects by directly estimating the roughness parameter. Instead, we propose to fit the specular reflectance with a weighted combination of specular basis following [16,24].

We assume isotropic materials and simplify the input to a half-vector h and normal n according to [43], and define a set of Sphere Gaussian (SG) basis as

$$D(h, n) = G(h, n; \lambda) = \left[e^{\lambda_1(h^T n - 1)}, \cdots, e^{\lambda_k(h^T n - 1)} \right]^T, \tag{11}$$

where $\lambda_* \in \mathcal{R}_+$ denotes specular sharpness. We introduce an MLP $f_s(\boldsymbol{x}) \mapsto \boldsymbol{\omega}$ to model the spatially-varying SG weights, and specular reflectance can then be recovered as

$$\rho_s = \boldsymbol{\omega}^T D(\boldsymbol{h}, \boldsymbol{n}). \tag{12}$$

To encourage a smooth albedo and specular reflectance distribution, we impose smoothness losses $\mathcal{L}_{as}^R$ and $\mathcal{L}_{ss}^R$ (defined similarity as in Eq. (5)) for $f_a(\boldsymbol{x})$ and $f_s(\boldsymbol{x})$ respectively.

Light Modeling. Each image is illuminated by a directional light, which is parameterized by a 3-vector light direction and a scalar light intensity. We set the light directions and intensities as learnable parameters, and initialize them by the lights estimated by the UPS method [6]. The light parameters will be refined after the joint optimization.

Joint Optimization. Based on our scene representation, we can rerender the input images using the differentiable rendering equation. Given multi-view and multi-light images, we optimize the normal, visibility, and BRDF MLPs together with the light parameters to minimize the image reconstruction loss, given by

$$\mathcal{L}_c'' = \sum \|\hat{I} - I\|_2^2, \tag{13}$$

where $\hat{I}$ is the rerendered image and I is the corresponding input image.

The overall loss function used for our neural inverse rendering stage is

$$\mathcal{L}'' = \beta_1 \mathcal{L}_c'' + \beta_2 \mathcal{L}_n'' + \beta_3 \mathcal{L}_v'' + \beta_4 \mathcal{L}_{as}'' + \beta_5 \mathcal{L}_{ss}'', \tag{14}$$

where β_* denotes the corresponding loss weight.

4 Experiments

4.1 Implementation Details

Please refer to supplementary materials for implementation details. We first use all 96 lights for training and evaluation, and then demonstrate that our framework can support arbitrary sparse light sources in Sect. 4.5.

Evaluation Metrics. We adopt commonly used quantitative metrics for different outputs. Specifically, we use mean angular error (MAE) in degree for surface normal evaluation under test views, and Chamfer distance for mesh evaluation[2]. Following [40], we extract meshes using the MISE algorithm [35]. PSNR, SSIM [54], and LPIPS [63] are used to evaluate the reconstructed images.

---

[2] We rescale the meshes into the range of $[-1, 1]$ for all the objects, and show Chamfer distance in the unit of mm).

Table 2. Results of different MVPS methods on DiLiGenT-MV benchmark.

Method	Chamfer Dist↓						Normal MAE↓					
	BEAR	BUDDHA	COW	POT2	READING	Average	BEAR	BUDDHA	COW	POT2	READING	Average
PJ16 [41]	19.58	11.77	**9.25**	24.82	22.62	17.61	12.78	14.68	13.21	15.53	12.92	13.83
LZ20 [26]	8.91	13.29	14.01	7.40	24.78	13.68	4.39	11.45	**4.14**	6.70	8.73	7.08
Ours	**8.65**	**8.61**	10.21	**6.11**	**12.35**	**9.19**	**3.54**	10.87	4.42	**5.93**	8.42	**6.64**

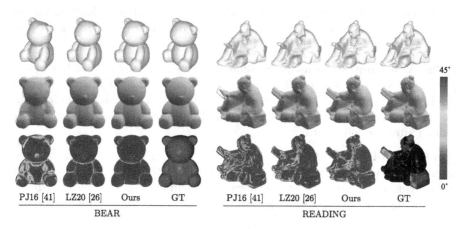

PJ16 [41]   LZ20 [26]   Ours   GT   PJ16 [41]   LZ20 [26]   Ours   GT

BEAR   READING

Fig. 4. Qualitative results of shape and normal on DiLiGenT-MV benchmark.

4.2 Dataset

Real Data. We adopted the widely used DiLiGenT-MV benchmark [26] for evaluation. It consists of 5 objects with diverse shapes and materials. Each object contains images captured from 20 views. For each view, 96 images are captured under varying light directions and intensities. Ground-truth meshes are provided. In our experiments, we sample 5 testing views with equal interval, and take the rest 15 views for training. Note that our method assumes unknown light direction and light intensity in evaluation.

Synthetic Data. To enable more comprehensive analysis, we rendered a synthetic dataset with 2 objects (*i.e.*, *BUNNY, ARMADILLO*) with Mitsuba [19]. We rendered objects under two sets of lightings, one with directional lights, denoted as Synth^{PS} dataset, and the other with environment map, denoted as Synth^{Env} dataset. We randomly sampled 20 camera views on the upper hemisphere, where 15 views for training and 5 views for testing. We used directional lights in the same distribution as DiLiGenT-MV benchmark for each view. For the synthetic dataset, we set the same light intensity for each light source.

4.3 Comparison with MVPS Methods

We compared our method with state-of-the-art MVPS methods [26,41] on DiLiGenT-MV benchmark, where all 20 views are used for optimization. It should be noted that our method does not use calibrated lights as [26,41].

Table 3. Comparison of novel view rendering on the DiLiGenT-MV benchmark.

Method	BEAR PSNR↑	SSIM↑	LPIPS↓	BUDDHA PSNR↑	SSIM↑	LPIPS↓	COW PSNR↑	SSIM↑	LPIPS↓	POT2 PSNR↑	SSIM↑	LPIPS↓	READING PSNR↑	SSIM↑	LPIPS↓
NeRF [36]	30.98	0.9887	1.09	29.31	0.9664	2.80	33.03	0.9907	0.54	32.52	0.9842	1.41	26.58	0.9664	1.84
KB22 [23]	39.63	**0.9960**	**0.24**	33.62	**0.9844**	0.56	31.38	0.9800	0.87	33.39	0.9767	1.02	22.45	0.9560	2.11
UNISURF [40]	40.13	0.9954	0.40	30.98	0.9707	1.98	40.17	0.9953	0.39	43.06	0.9954	0.50	22.19	0.9579	2.29
NeRFactor [61]	29.28	0.9791	2.38	26.34	0.9385	6.33	27.60	0.9630	1.67	32.32	0.9738	1.84	25.62	0.9468	2.99
NeRD [5]	26.24	0.9661	3.56	20.94	0.8701	6.25	23.98	0.8914	1.97	26.34	0.8422	2.19	20.13	0.9023	3.90
PhySG [61]	34.01	0.9841	1.60	29.64	0.9594	2.65	34.38	0.9856	1.02	35.92	0.9814	1.23	24.19	0.9531	2.88
Ours	**41.58**	0.9959	0.31	**33.73**	0.9829	**0.54**	**42.39**	**0.9962**	**0.22**	**45.44**	**0.9960**	**0.15**	**30.47**	**0.9808**	**0.75**

Table 4. Shape reconstruction results of neural rendering methods on both real and synthetic datasets. We use both $Synth^{Env}$ dataset and $Synth^{PS}$ dataset for evaluation.

Method	BEAR MAE↓	BEAR CD↓	BUDDHA MAE↓	BUDDHA CD↓	COW MAE↓	COW CD↓	POT2 MAE↓	POT2 CD↓	READING MAE↓	READING CD↓	BUNNY MAE↓ PS	Env	BUNNY CD↓ PS	Env	ARMADILLO MAE↓ PS	Env	ARMADILLO CD↓ PS	Env
NeRF [36]	73.90	66.68	59.89	29.28	55.14	70.07	69.71	42.28	55.75	48.26	49.42	48.22	19.67	20.09	44.27	41.54	20.95	24.34
KB22 [23]	53.19	66.18	39.72	17.92	85.11	82.43	87.30	63.82	70.13	86.79	31.64	37.00	9.61	18.32	35.88	51.08	14.50	7.58
UNISURF [40]	6.48	9.24	17.11	9.83	8.25	13.25	13.05	10.21	19.72	62.89	10.00	11.46	6.89	8.74	8.12	10.12	3.76	3.96
NeRFactor [61]	12.68	26.21	25.71	26.97	17.87	50.65	15.46	29.00	21.24	47.36	21.97	21.50	17.29	18.29	36.07	19.27	7.86	18.44
NeRD [5]	19.49	13.90	30.41	18.54	33.18	38.62	28.16	9.00	30.83	30.05	17.51	19.12	11.08	13.44	19.36	18.43	7.02	9.34
PhySG [61]	11.22	19.07	26.31	21.66	11.53	22.23	13.74	32.29	25.74	46.90	23.53	25.66	22.23	21.87	14.46	19.53	8.27	12.08
Ours	**3.21**	**7.24**	**10.10**	**8.93**	**4.08**	**11.33**	**5.67**	**5.76**	**8.83**	**12.83**	**5.14**	–	**5.32**	–	**5.18**	–	**3.61**	–

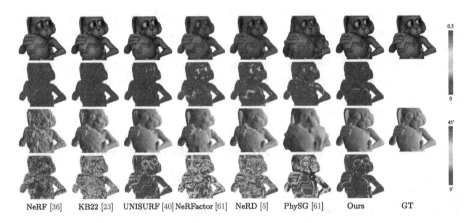

NeRF [36]   KB22 [23]   UNISURF [40]   NeRFactor [61]   NeRD [5]   PhySG [61]   Ours   GT

Fig. 5. Novel view rendering and normal estimation results on the synthetic dataset.

Table 2 summarizes the quantitative comparison. The results of Chamfer distance are calculated with ground truth shape. The normal MAE results of DiLiGenT test views are calculated inside the intersection of input mask and our predicted mask. We can see that our method outperforms previous methods in 4 out of 5 objects by a large margin in terms of both two metrics. In particular, our method significantly reduces the Chamfer distance of the most challenging *READING* object from 22.62mm to 12.35mm. On average, our method improves Chamfer distance by 33%, and normal MAE by 6%. Such remarkable performance proves that our method is much better in shape reconstruction than existing approaches. We notice our results of the *COW* object are slightly worse. Detailed discussions are in the supplementary material.

Table 5. Analysis on normal supervision.

Method	Normal MAE↓						Chamfer Dist↓					
	BEAR	BUDDHA	COW	POT2	READING	BUNNY	BEAR	BUDDHA	COW	POT2	READING	BUNNY
NeRF	73.45	59.62	55.10	69.41	55.55	49.24	66.68	29.28	70.07	42.28	48.26	19.67
NeRF^{+N}	7.03	13.50	8.26	7.93	14.01	12.04	16.02	9.65	12.04	7.25	13.31	5.44
UNISURF	6.51	17.13	8.26	13.04	19.68	10.04	9.24	9.83	13.25	10.21	62.89	6.89
UNISURF^{+N}	**4.26**	**11.29**	**5.05**	**6.37**	**9.58**	**7.76**	**7.24**	**8.93**	**11.33**	**5.76**	**12.83**	**5.32**

NeRF   NeRF^{+N} UNISURF   UNISURF^{+N} GT

Fig. 6. Qualitative results of estimated surface normal on w/ or w/o normal supervision.

To show the advantages of our method intuitively, we present two visual example comparisons in Fig. 4. We show the easiest *BEAR* with a smooth surface and the most challenging *READING* with many wrinkles. It can be observed that our results have fewer noises than LZ20 and PJ16 on the smooth surface of *BEAR*, and our reconstruction of the detailed wrinkles of *READING* is much better. It demonstrates that our method is superior in reconstructing both smooth and rugged surfaces.

4.4 Comparison with Neural Rendering Based Methods

We also compared our method with existing neural rendering methods, including NeRF [36], KB22 [23], UNISURF [40], PhySG [61], NeRFactor [61], and NeRD [5]. Among them, the first three methods can only support novel-view rendering and cannot perform scene decomposition.

For KB22 [23], we re-implemented it by adding the normals estimated by PS method as input to NeRF following their paper. For other methods, we used their released codes for experiments.

Evaluation on DiLiGenT-MV Benchmark. Following the training of our radiance field stage, we use the multi-light averaged image for each view (equivalent to image lit by frontal hemisphere lights) for training and testing the baselines. For our method, we rendered directional light images and computed the light-averaged image for each view in testing phases. As the light intensity can only be estimated up to an unknown scale, we reported the scale-invariant PSNR, SSIM and LPIPS for all methods by first finding a scalar to rescale the rendered image using least-square.

Table 3 and Table 4 show that our method achieves the best rendering and normal estimation results, and significantly outperforms the existing best

Table 6. Normal improvement.

Method	BEAR	BUDDHA	COW	POT2	READING	BUNNY
SDPS-Net	7.52	11.47	9.57	7.98	15.94	10.65
Stage I	4.26	11.29	5.05	6.39	9.58	7.79
Ours	3.25	10.20	4.12	5.73	8.87	5.24

Table 7. Light improvement.

Method	BEAR	BUDDHA	COW	POT2	READING	BUNNY
SDPS-Net	4.90	7.17	8.55	4.73	9.09	8.96
Ours	2.27	2.75	2.59	2.89	4.26	1.53

Table 8. Ablation study on Stage II.

Method	BEAR			BUDDHA			COW			POT2			READING			BUNNY		
	PSNR↑	SSIM↑	MAE↓	PSNR↑	SSIM↑	MAE↓	PSNR↑	SSIM↑	MAE↓	PSNR↑	SSIM↑	MAE↓	PSNR↑	SSIM↑	MAE↓	PSNR↑	SSIM↑	MAE↓
fixed-light	34.51	0.9812	3.62	29.48	0.9661	10.51	35.58	0.9872	4.45	39.47	0.9845	5.84	24.58	0.9708	9.18	24.04	0.9833	6.94
w/o vis	33.44	0.9794	3.57	28.06	0.9571	10.29	36.94	0.9890	4.09	39.38	0.9837	5.75	24.49	0.9699	8.91	21.29	0.9761	5.34
Ours	35.68	0.9837	3.25	29.58	0.9670	10.20	37.06	0.9890	4.12	40.01	0.9860	5.73	24.89	0.9725	8.87	25.88	0.9871	5.24

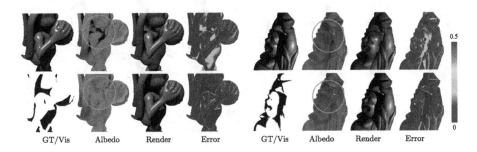

GT/Vis   Albedo   Render   Error   GT/Vis   Albedo   Render   Error

Fig. 7. Effectiveness of visibility modeling. (left: BUNNY, right: BUDDHA)

method UNISURF [40] for shape reconstruction. We attribute the success in view rendering to the faithful decomposition of shape, material, and light components and our shadow-aware design. The multi-light images provide abundant high-frequency information related to the shape and material, which ensures high rendered quality under diverse illuminations.

Evaluation on Synthetic Data. Since existing neural rendering methods mainly assume a fixed environment lighting, we additionally evaluate on the synthetic dataset that is rendered with both PS lights (Synth^{PS} dataset) and environment light (Synth^{Env} dataset) to investigate the shape reconstruction performance under different light conditions.

The results in Table 4 reports similar geometry reconstruction performance for all baselines methods in both PS and environment lighting. They failed to reconstruct accurate surface in both light conditions. This may be due to the sparse input views, and the ambiguity in shape and material joint estimation, especially for non-textured regions, which exists in both PS and Env lightings. In contrast, our method successfully estimates more faithful geometry. We also show the qualitative comparison in Fig. 5. Our method achieves the best performance on both re-rendered quality and reconstructed surface normal, especially for details such as eye regions.

Table 9. Quantitative results of different methods when trained with different number of views.

Method	BUNNY						ARMADILLO					
	Normal MAE ↓			Chamfer Dist ↓			Normal MAE ↓			Chamfer Dist ↓		
	5	15	30	5	15	30	5	15	30	5	15	30
PhySG [61]	34.53	25.69	27.25	36.75	21.87	42.25	30.81	19.53	20.19	49.38	12.08	16.60
UNISURF [40]	20.31	11.46	9.11	18.09	8.74	6.85	14.43	10.11	8.63	11.80	3.96	3.65
Ours	**6.96**	**5.17**	**5.71**	**7.71**	**5.32**	**4.07**	**6.51**	**5.15**	**5.02**	**5.96**	**3.61**	**3.31**

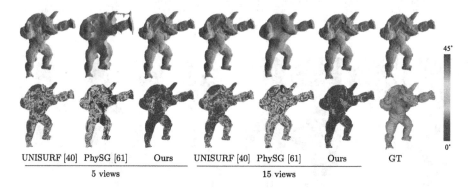

UNISURF [40]   PhySG [61]   Ours   UNISURF [40]   PhySG [61]   Ours   GT

5 views   15 views

Fig. 8. Qualitative comparison on training with different number of views.

Table 10. Quantitative results of our method with different number of light directions.

N.light	Render			Normal MAE↓			Shape	Light Dir MAE↓	
	PSNR↑	SSIM↑	LPIPS↓	SDPS	Stage I	Ours	Chamfer Dist↓	SDPS	Ours
1	20.21	0.9686	2.08	48.52	17.81	14.21	11.94	47.22	3.82
2	20.69	0.9713	1.80	34.96	16.53	13.40	12.86	18.74	2.55
4	24.78	0.9855	0.84	14.76	9.20	6.75	7.37	10.32	1.59
8	25.84	0.9873	0.70	11.05	8.29	5.39	5.27	10.24	1.59
16	26.00	0.9876	0.69	10.86	8.07	5.30	5.43	9.71	**1.31**
32	**26.03**	**0.9880**	**0.67**	10.69	7.84	5.21	**4.92**	9.33	2.08
96	25.90	0.9873	0.68	**10.65**	**7.74**	**5.18**	5.32	**8.96**	1.53

4.5   Method Analysis

Effectiveness of Normal Regularization. Our method exploits multi-light images to infer surface normals to regularize the surface geometry in radiance field. It eliminates the ambiguity in density estimation especially for concave-shaped objects. We show the reconstruction results of before and after adding the normal constraint for both our method (*i.e.*, UNISURF [40] as our backbone of Stage I) and NeRF [36] on all objects in DiLiGenT-MV benchmark and *BUNNY* in Synth^{PS} dataset. As shown in Table 5 and Fig. 6, the introduced normal regularization greatly improves the shape accuracy and recovered surface details for all objects. For example, on the *READING* object, NeRF decreases the normal MAE from 55.55 to 14.01 and UNISURF from 19.68 to 9.58, which verifies the effectiveness of the normal regularization.

Effectiveness of the Joint Optimization. Benefiting from the rich shading information in multi-light images and our shadow-aware renderer, we are able to reconstruct faithful surface through joint optimizing normal, BRDFs and lights. Table 6 shows that joint optimization consistently improves the surface normal accuracy. Besides, the light directions are also refined through joint optimization as Table 7 shows.

We also investigated the design of our stage II by ablating the light direction optimization and visibility modeling. For ablation study on our own method, we calculate metrics of re-rendered images quality for each light sources instead of light-averaged image. Table 8 indicates that either fixing the initialized light direction or removing the visibility modeling will decrease the rendering quality and shape reconstruction accuracy. In particular, when light visibility is not being considered, the estimated albedo will be entangled with the cast-shadow, leading to inaccurate material estimation (see Fig. 7).

Effect of Training View Numbers. Since our method can make full use of multi-light image to resolve the depth ambiguity in plain RGB image, it is able to reconstruct high-quality shape just from sparse views. To justify it, we compared our method with UNISURF [40] and PhySG [61] for surface reconstruction using 5, 15, and 30 training views. Note that the baseline methods were trained on Synth[Env] dataset as they assume a fixed environment lighting. Table 9 and Fig. 8 show the reconstruction results using different number of views. Our method achieves satisfying reconstruction results even when only 5 views are given, whereas other methods all fail when view numbers are insufficient.

Effect of Light Direction Numbers. While most previous PS methods requires calibrated lights, our method assumes uncalibrated lights and can handle an arbitrary number of lights. We also conduct experiments to explore how many light numbers are needed for reconstructing high-quality shapes, where 15 views are used. Similar to ablation study, we estimate per-light error for the re-rendered images. For light direction error, we take the mean values of their own used lights. Table 10 shows that increasing the number of lights generally increases the reconstruction accuracy. Our full method consistently outperforms SDPS-Net and Stage I under different number of lights. Given only 4 images illuminated under directional light for each view, our method achieves results comparable to that uses 96 images (*e.g.*, 6.75 vs. 5.18 for MAE), demonstrating the our method is quite robust to numbers of light directions.

5 Conclusions

In this paper, we have introduced a neural inverse rendering method for multi-view photometric stereo under uncalibrated lights. We first represent an object with neural radiance field whose surface geometry is regularized by the normals estimated from multi-light images. We then jointly optimize the surface normals, BRDFs, visibility, and lights to minimize the image reconstruction loss based on a shadow-aware rendering layer. Experiments on both synthetic and real dataset show that our method outperforms existing MVPS methods and neural rendering methods. Notably, our method is able to recover high-quality surface using as few as 5 input views.

Limitation. Although our method has been successfully applied to recover high-quality shape reconstruction for complex real-world objects, it still has the following limitations. First, we ignore surface inter-reflections in the rendering equation. Second, we assume a solid object to locate its surface locations, thus cannot handle non-solid objects (*e.g.*, fog). Last, similar to most of the neural rendering methods, we assume the camera poses are given. In the future, we are interested in extending our method to solve the above limitations.

Acknowledgements. This work was partially supported by the National Key R&D Program of China (No. 2018YFB1800800), the Basic Research Project No. HZQB-KCZYZ-2021067 of Hetao Shenzhen-HK S&T Cooperation Zone, NSFC-62202409, and Hong Kong RGC GRF grant (project# 17203119).

References

1. Agarwal, S., et al.: Building Rome in a day. Communications of the ACM (2011)
2. Barron, J.T., Mildenhall, B., Tancik, M., Hedman, P., Martin-Brualla, R., Srinivasan, P.P.: Mip-NeRF: a multiscale representation for anti-aliasing neural radiance fields. In: Proceedings of the IEEE/CVF International Conference on Computer Vision (ICCV) (2021)
3. Bi, S., et al.: Neural reflectance fields for appearance acquisition. arXiv preprint arXiv:2008.03824 (2020)
4. Boss, M., Braun, R., Jampani, V., Barron, J.T., Liu, C., Lensch, H.: NeRD: Neural reflectance decomposition from image collections. In: Proceedings of the IEEE/CVF International Conference on Computer Vision (ICCV), pp. 12684–12694 (2021)
5. Boss, M., Jampani, V., Braun, R., Liu, C., Barron, J., Lensch, H.: Neural-PIL: Neural pre-integrated lighting for reflectance decomposition. In: Proceedings of the Advances in Neural Information Processing Systems (NeurIPS), vol. 34 (2021)
6. Chen, G., Han, K., Shi, B., Matsushita, Y., Wong, K.Y.K.: Self-calibrating deep photometric stereo networks. In: Proceedings of the IEEE/CVF Conference on Computer Vision and Pattern Recognition (CVPR), pp. 8739–8747 (2019)
7. Chen, G., Han, K., Shi, B., Matsushita, Y., Wong, K.Y.K.: Deep photometric stereo for non-Lambertian surfaces. IEEE Trans. Pattern Anal. Mach. Intell. (T-PAMI) **44**(1), 129–142 (2020)
8. Chen, G., Han, K., Wong, K.Y.K.: PS-FCN: A flexible learning framework for photometric stereo. In: Proceedings of the European Conference on Computer Vision (ECCV). pp. 3–18 (2018)

9. Cheng, Z., Li, H., Asano, Y., Zheng, Y., Sato, I.: Multi-view 3d reconstruction of a texture-less smooth surface of unknown generic reflectance. In: Proceedings of the IEEE/CVF Conference on Computer Vision and Pattern Recognition (CVPR), pp. 16226–16235 (2021)

10. Chung, H.S., Jia, J.: Efficient photometric stereo on glossy surfaces with wide specular lobes. In: Proceedings of the IEEE/CVF Conference on Computer Vision and Pattern Recognition (CVPR) (2008)

11. Esteban, C.H., Vogiatzis, G., Cipolla, R.: Multiview photometric stereo. IEEE Transactions on Pattern Analysis and Machine Intelligence (T-PAMI) (2008)

12. Furukawa, Y., Ponce, J.: Accurate, dense, and robust multiview stereopsis. IEEE Trans. Pattern Anal. Mach. Intell. (T-PAMI) **32**(8), 1362–1376 (2009)

13. Garbin, S.J., Kowalski, M., Johnson, M., Shotton, J., Valentin, J.: FastNeRF: High-fidelity neural rendering at 200fps. In: Proceedings of the IEEE/CVF International Conference on Computer Vision (ICCV), pp. 14346–14355 (2021)

14. Hayakawa, H.: Photometric stereo under a light source with arbitrary motion. JOSA A (1994)

15. Hertzmann, A., Seitz, S.M.: Example-based photometric stereo: Shape reconstruction with general, varying BRDFs. IEEE Trans. Pattern Anal. Mach. Intell. (T-PAMI) (2005)

16. Hui, Z., Sankaranarayanan, A.C.: Shape and spatially-varying reflectance estimation from virtual exemplars. IEEE Trans. Pattern Anal. Mach. Intell. (T-PAMI) (2017)

17. Ikehata, S.: CNN-PS: CNN-based photometric stereo for general non-convex surfaces. In: Proceedings of the European Conference on Computer Vision (ECCV) (2018)

18. Ikehata, S., Aizawa, K.: Photometric stereo using constrained bivariate regression for general isotropic surfaces. In: Proceedings of the IEEE/CVF Conference on Computer Vision and Pattern Recognition (CVPR) (2014)

19. Jakob, W.: Mitsuba renderer (2010)

20. Kajiya, J.T.: The rendering equation. In: Proceedings of the 13th Annual Conference on Computer Graphics and Interactive Techniques, pp. 143–150 (1986)

21. Kaya, B., Kumar, S., Oliveira, C., Ferrari, V., Van Gool, L.: Uncertainty-aware deep multi-view photometric stereo. arXiv preprint arXiv:2010.07492 (2020)

22. Kaya, B., Kumar, S., Oliveira, C., Ferrari, V., Van Gool, L.: Uncalibrated neural inverse rendering for photometric stereo of general surfaces. In: Proceedings of the IEEE/CVF Conference on Computer Vision and Pattern Recognition, pp. 3804–3814 (2021)

23. Kaya, B., Kumar, S., Sarno, F., Ferrari, V., Van Gool, L.: Neural radiance fields approach to deep multi-view photometric stereo. In: Proceedings of the IEEE/CVF Winter Conference on Applications of Computer Vision (WACV), pp. 1965–1977 (2022)

24. Li, J., Li, H.: Neural reflectance for shape recovery with shadow handling. In: Proceedings of the IEEE/CVF Conference on Computer Vision and Pattern Recognition (CVPR) (2022)

25. Li, J., Robles-Kelly, A., You, S., Matsushita, Y.: Learning to minify photometric stereo. In: Proceedings of the IEEE/CVF Conference on Computer Vision and Pattern Recognition (CVPR) (2019)

26. Li, M., Zhou, Z., Wu, Z., Shi, B., Diao, C., Tan, P.: Multi-view photometric stereo: a robust solution and benchmark dataset for spatially varying isotropic materials. IEEE Trans. Image Process. (TIP) (2020)

27. Li, Z., Sunkavalli, K., Chandraker, M.: Materials for masses: Svbrdf acquisition with a single mobile phone image. In: Proceedings of the European Conference on Computer Vision (ECCV). pp. 72–87 (2018)
28. Li, Z., Xu, Z., Ramamoorthi, R., Sunkavalli, K., Chandraker, M.: Learning to reconstruct shape and spatially-varying reflectance from a single image. ACM Trans. Graph. (TOG) **37**(6), 1–11 (2018)
29. Lim, J., Ho, J., Yang, M.H., Kriegman, D.: Passive photometric stereo from motion. In: Proceedings of the IEEE/CVF International Conference on Computer Vision (ICCV), pp. 1635–1642 (2005)
30. Liu, L., Gu, J., Lin, K.Z., Chua, T.S., Theobalt, C.: Neural sparse voxel fields. In: Proceedings of the Advances in Neural Information Processing Systems (NeurIPS), vol. 33, pp. 15651–15663 (2020)
31. Logothetis, F., Budvytis, I., Mecca, R., Cipolla, R.: Px-net: simple and efficient pixel-wise training of photometric stereo networks. In: Proceedings of the IEEE/CVF International Conference on Computer Vision, pp. 12757–12766 (2021)
32. Logothetis, F., Mecca, R., Cipolla, R.: A differential volumetric approach to multiview photometric stereo. In: Proceedings of the IEEE/CVF International Conference on Computer Vision (ICCV), pp. 1052–1061 (2019)
33. Lu, F., Chen, X., Sato, I., Sato, Y.: SymPS: BRDF symmetry guided photometric stereo for shape and light source estimation. IEEE Trans. Pattern Anal. Mach. Intell. (T-PAMI) 40, 221–234 (2018)
34. Martin-Brualla, R., Radwan, N., Sajjadi, M.S., Barron, J.T., Dosovitskiy, A., Duckworth, D.: NeRF in the wild: Neural radiance fields for unconstrained photo collections. In: Proceedings of the IEEE/CVF Conference on Computer Vision and Pattern Recognition (CVPR), pp. 7210–7219 (2021)
35. Mescheder, L., Oechsle, M., Niemeyer, M., Nowozin, S., Geiger, A.: Occupancy networks: learning 3d reconstruction in function space. In: Proceedings of the IEEE/CVF Conference on Computer Vision and Pattern Recognition (CVPR), pp. 4460–4470 (2019)
36. Mildenhall, B., Srinivasan, P.P., Tancik, M., Barron, J.T., Ramamoorthi, R., Ng, R.: NeRF: representing scenes as neural radiance fields for view synthesis. In: Proceedings of the European Conference on Computer Vision (ECCV), pp. 405–421 (2020)
37. Mukaigawa, Y., Ishii, Y., Shakunaga, T.: Analysis of photometric factors based on photometric linearization. JOSA A (2007)
38. Nam, G., Lee, J.H., Gutierrez, D., Kim, M.H.: Practical svbrdf acquisition of 3d objects with unstructured flash photography. ACM Trans. Graph. (TOG) **37**(6), 1–12 (2018)
39. Niemeyer, M., Mescheder, L., Oechsle, M., Geiger, A.: Differentiable volumetric rendering: learning implicit 3d representations without 3d supervision. In: Proceedings of the IEEE/CVF Conference on Computer Vision and Pattern Recognition (CVPR), pp. 3504–3515 (2020)
40. Oechsle, M., Peng, S., Geiger, A.: UNISURF: Unifying neural implicit surfaces and radiance fields for multi-view reconstruction. In: Proceedings of the IEEE/CVF International Conference on Computer Vision (ICCV), pp. 5589–5599 (2021)
41. Park, J., Sinha, S.N., Matsushita, Y., Tai, Y.W., Kweon, I.S.: Robust multiview photometric stereo using planar mesh parameterization. IEEE Trans. Pattern Anal. Mach. Intell. (T-PAMI) **39**, 1591–1604 (2016)
42. Reiser, C., Peng, S., Liao, Y., Geiger, A.: KiloNeRF: Speeding up neural radiance fields with thousands of tiny mlps. In: Proceedings of the IEEE/CVF International Conference on Computer Vision (ICCV). pp. 14335–14345 (2021)

43. Rusinkiewicz, S.M.: A new change of variables for efficient BRDF representation. In: Drettakis, G., Max, N. (eds.) EGSR 1998. E, pp. 11–22. Springer, Vienna (1998). https://doi.org/10.1007/978-3-7091-6453-2_2
44. Santo, H., Samejima, M., Sugano, Y., Shi, B., Matsushita, Y.: Deep photometric stereo network. In: Proceedings of the IEEE International Conference on Computer Vision Workshops (ICCVW) (2017)
45. Seitz, S.M., Curless, B., Diebel, J., Scharstein, D., Szeliski, R.: A comparison and evaluation of multi-view stereo reconstruction algorithms. In: Proceedings of the IEEE/CVF Conference on Computer Vision and Pattern Recognition (CVPR) (2006)
46. Shi, B., Mo, Z., Wu, Z., Duan, D., Yeung, S.K., Tan, P.: A benchmark dataset and evaluation for non-Lambertian and uncalibrated photometric stereo. IEEE Trans. Pattern Anal. Mach. Intell. (T-PAMI) (2019)
47. Sitzmann, V., Zollhöfer, M., Wetzstein, G.: Scene representation networks: continuous 3d-structure-aware neural scene representations. In: Proceedings of the Advances in Neural Information Processing Systems (NeurIPS) 32 (2019)
48. Srinivasan, P.P., Deng, B., Zhang, X., Tancik, M., Mildenhall, B., Barron, J.T.: NeRV: Neural reflectance and visibility fields for relighting and view synthesis. In: Proceedings of the IEEE/CVF Conference on Computer Vision and Pattern Recognition (CVPR), pp. 7495–7504 (2021)
49. Taniai, T., Maehara, T.: Neural inverse rendering for general reflectance photometric stereo. In: Proceedings of the ACM International Conference on Machine Learning (ICML) (2018)
50. Tewari, A., et al.: Advances in neural rendering. arXiv preprint arXiv:2111.05849 (2021)
51. Tozza, S., Mecca, R., Duocastella, M., Del Bue, A.: Direct differential photometric stereo shape recovery of diffuse and specular surfaces. Journal of Mathematical Imaging and Vision (2016)
52. Wang, P., Liu, L., Liu, Y., Theobalt, C., Komura, T., Wang, W.: NeuS: learning neural implicit surfaces by volume rendering for multi-view reconstruction. In: Proceedings of the Advances in Neural Information Processing Systems (NeurIPS), vol. 34 (2021)
53. Wang, X., Jian, Z., Ren, M.: Non-Lambertian photometric stereo network based on inverse reflectance model with collocated light. IEEE Trans. Image Process. (TIP) **29**, 6032–6042 (2020)
54. Wang, Z., Bovik, A.C., Sheikh, H.R., Simoncelli, E.P.: Image quality assessment: From error visibility to structural similarity. IEEE Trans. Image Process. (TIP) (2004)
55. Wood, D.N., et al.: Surface light fields for 3d photography. In: Proceedings of the ACM SIGGRAPH Conference and Exhibition on Computer Graphics and Interactive Techniques in Asia (SIGGRAPH Aisa), pp. 287–296 (2000)
56. Woodham, R.J.: Photometric method for determining surface orientation from multiple images. Optical Engineering (1980)
57. Wu, C., Liu, Y., Dai, Q., Wilburn, B.: Fusing multiview and photometric stereo for 3d reconstruction under uncalibrated illumination. IEEE Transactions on Visualization and Computer Graphics (TVCG) **17**(8), 1082–1095 (2010)
58. Wu, L., Ganesh, A., Shi, B., Matsushita, Y., Wang, Y., Ma, Y.: Robust photometric stereo via low-rank matrix completion and recovery. In: Proceedings of the Asian Conference on Computer Vision (ACCV) (2010)
59. Wu, T.P., Tang, C.K.: Photometric stereo via expectation maximization. IEEE Transactions on Pattern Analysis and Machine Intelligence (T-PAMI) (2010)

60. Yariv, L., Kasten, Y., Moran, D., Galun, M., Atzmon, M., Basri, R., Lipman, Y.: Multiview neural surface reconstruction by disentangling geometry and appearance. In: Proceedings of the Advances in Neural Information Processing Systems (NeurIPS), vol. 33, pp. 2492–2502 (2020)

61. Zhang, K., Luan, F., Wang, Q., Bala, K., Snavely, N.: PhySG: Inverse rendering with spherical gaussians for physics-based material editing and relighting. In: Proceedings of the IEEE/CVF Conference on Computer Vision and Pattern Recognition (CVPR), pp. 5453–5462 (2021)

62. Zhang, K., Riegler, G., Snavely, N., Koltun, V.: NeRF++: Analyzing and improving neural radiance fields. arXiv preprint arXiv:2010.07492 (2020)

63. Zhang, R., Isola, P., Efros, A.A., Shechtman, E., Wang, O.: The unreasonable effectiveness of deep features as a perceptual metric. In: Proceedings of the IEEE/CVF Conference on Computer Vision and Pattern Recognition (CVPR), pp. 586–595 (2018)

64. Zhang, X., Srinivasan, P.P., Deng, B., Debevec, P., Freeman, W.T., Barron, J.T.: NeRFactor: Neural factorization of shape and reflectance under an unknown illumination. ACM Trans. Graph. (TOG) 40(6) (2021)

65. Zheng, Q., Jia, Y., Shi, B., Jiang, X., Duan, L.Y., Kot, A.C.: SPLINE-Net: Sparse photometric stereo through lighting interpolation and normal estimation networks. In: Proceedings of the IEEE/CVF International Conference on Computer Vision (ICCV) (2019)

66. Zhou, Z., Wu, Z., Tan, P.: Multi-view photometric stereo with spatially varying isotropic materials. In: Proceedings of the IEEE/CVF Conference on Computer Vision and Pattern Recognition (CVPR), pp. 1482–1489 (2013)

Share with Thy Neighbors: Single-View Reconstruction by Cross-Instance Consistency

Tom Monnier[1]([⊠]), Matthew Fisher[2], Alexei A. Efros[3], and Mathieu Aubry[1]

[1] LIGM, Ecole des Ponts, Univ Gustave Eiffel, Marne-la-vallée, France
`tom.monnier@enpc.fr`
[2] Adobe Research, Stanford, USA
[3] UC Berkeley, Berkeley, USA

Abstract. Approaches for single-view reconstruction typically rely on viewpoint annotations, silhouettes, the absence of background, multiple views of the same instance, a template shape, or symmetry. We avoid all such supervision and assumptions by explicitly leveraging the consistency between images of different object instances. As a result, our method can learn from large collections of unlabelled images depicting the same object category. Our main contributions are two ways for leveraging cross-instance consistency: (i) *progressive conditioning*, a training strategy to gradually specialize the model from category to instances in a curriculum learning fashion; and (ii) *neighbor reconstruction*, a loss enforcing consistency between instances having similar shape or texture. Also critical to the success of our method are: our structured autoencoding architecture decomposing an image into explicit shape, texture, pose, and background; an adapted formulation of differential rendering; and a new optimization scheme alternating between 3D and pose learning. We compare our approach, UNICORN, both on the diverse synthetic ShapeNet dataset—the classical benchmark for methods requiring multiple views as supervision—and on standard real-image benchmarks (Pascal3D+ Car, CUB) for which most methods require known templates and silhouette annotations. We also showcase applicability to more challenging real-world collections (CompCars, LSUN), where silhouettes are not available and images are not cropped around the object.

Keywords: Single-view reconstruction · Unsupervised learning

1 Introduction

One of the most magical human perceptual abilities is being able to see the 3D world behind a 2D image – a mathematically impossible task! Indeed, the ancient

Supplementary Information The online version contains supplementary material available at https://doi.org/10.1007/978-3-031-19769-7_17.

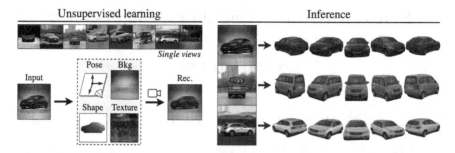

Fig. 1. Single-View Reconstruction by Cross-Instance Consistency. (left) Given a collection of single-view images from an object category, we learn without additional supervision an autoencoder that explicitly generates shape, texture, pose and background. **(right)** At inference time, our approach reconstructs high-quality textured meshes from raw single-view images.

Greeks were so incredulous at the possibility that humans could be "hallucinating" the third dimension, that they proposed the utterly implausible Emission Theory of Vision [10] (eye emitting light to "sense" the world) to explain it to themselves. In the history of computer vision, single-view reconstruction (SVR) has had an almost cult status as one of the holy grail problems [19,20,46]. Recent advancements in deep learning methods have dramatically improved results in this area [6,36]. However, the best methods still require costly supervision at training time, such as multiple views [34,43]. Despite efforts to remove such requirements, the works with the least supervision still rely on two signals limiting their applicability: (i) silhouettes and (ii) strong assumptions such as symmetries [21,25], known template shapes [12,50], or the absence of background [56]. Although crucial to achieve reasonable results, priors like silhouettes and symmetry can also harm the reconstruction quality: silhouette annotations are often coarse [5] and small symmetry prediction errors can yield unrealistic reconstructions [12,56].

In this paper, we propose the most unsupervised approach to single-view reconstruction to date, which we demonstrate to be competitive for diverse datasets. Table 1 summarizes the differences between our approach and representative prior works. More precisely, we learn in an analysis-by-synthesis fashion a network that predicts for each input image: 1) a 3D shape parametrized as a deformation of an ellipsoid, 2) a texture map, 3) a camera viewpoint, and 4) a background image (Fig. 1). Our main insight to remove the supervision and assumptions required by other methods is to leverage the consistency across different instances. First, we design a training procedure, *progressive conditioning*, which encourages the model to share elements between images by strongly constraining the variability of shape, texture and background at the beginning of training and progressively allowing for more diversity (Fig. 2a). Second, we introduce a *neighbor reconstruction* loss, which explicitly enforces neighboring instances from different viewpoints to share the same shape or texture

Table 1. Comparison with selected works. For each method, we outline the supervision and priors used (3D, Multi-Views, Camera or Keypoints, Silhouettes, Assumptions like $\Diamond$ template shape, † symmetry, ‡ solid of revolution, $\leftrightarrow$ semantic consistency, $\boxtimes$ no/limited background, $<$ frontal view, $\varnothing$ no texture), which data it has been applied to and the model output (3D, Texture, Pose, Depth).

Method	Supervision	Data	Output
Pix2Mesh [53], AtlasNet [14], OccNet [36]	3D	ShapeNet	3D
PTN [59], NMR [30]	MV, CK, S	ShapeNet	3D
DRC [51], SoftRas [34], DVR [43]	MV, CK, S	ShapeNet	3D, T
GANverse3D [66]	MV, CK, S	Bird, Car, Horse	3D, T
DPC [23], MVC [49]	MV, S	ShapeNet	3D, P
Vicente *et al.* [52], CSDM [26], DRC [51]	CK, S	Pascal3D	3D
CMR [25]	CK, S, A(†)	Bird, Car, Plane	3D, T
SDF-SRN [33], TARS [8]	CK, S	ShapeNet, Bird, Car, Plane	3D, T
TexturedMeshGen [17]	CK, A(†)	ShapeNet, Bird, Car	3D, T
UCMR [12], IMR [50]	S, A($\Diamond$, †)	Animal, Car, Moto	3D, T, P
UMR [32]	S, A($\leftrightarrow$, †)	Animal, Car, Moto	3D, T, P
RADAR [55]	S, A(‡)	Vase	3D, T, P
SMR [21]	S, A(†)	ShapeNet, Animal, Moto	3D, T, P
Unsup3D [56]	A($\boxtimes$, $<$, †)	Face	D, T, P
Henderson & Ferrari [16]	A($\boxtimes$, $\varnothing$)	ShapeNet	3D, P
Ours	None	ShapeNet, Animal, Car, Moto	3D, T, P

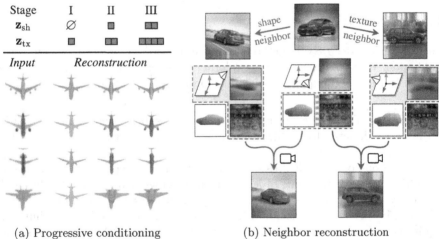

(a) Progressive conditioning

(b) Neighbor reconstruction

Fig. 2. Leveraging cross-instance consistency. (a) Progressive conditioning amounts to gradually increasing, in a multi-stage fashion, the size of the conditioning latent spaces, here associated to shape $\mathbf{z}_{sh}$ and texture $\mathbf{z}_{tx}$. **(b)** We explicitly share the shape and texture models across neighboring instances by swapping their characteristics and applying a loss to associated neighbor reconstructions.

model (Fig. 2b). Note that these simple yet effective techniques are data-driven and not specific to any dataset. Our only remaining assumption is the knowledge of the semantic class of the depicted object.

We also provide two technical insights that we found critical to learn our model without viewpoint and silhouette annotations: (i) a differentiable rendering formulation inspired by layered image models [24,38] which we found to perform better than the classical SoftRasterizer [34], and (ii) a new optimization strategy which alternates between learning a set of pose candidates with associated probabilities and learning all other components using the most likely candidate.

We validate our approach on the standard ShapeNet [4] benchmark, real image SVR benchmarks (Pascal3D+ Car [57], CUB [54]) as well as more complex real-world datasets (CompCars [60], LSUN Motorbike and Horse [63]). In all scenarios, we demonstrate high-quality textured 3D reconstructions.

Summary. We present UNICORN, a framework leveraging **UN**supervised cross-**I**nstance **CO**nsistency for 3D **R**econstructio**N**. Our main contributions are: 1) the first fully unsupervised SVR system, demonstrating state-of-the-art textured 3D reconstructions for both generic shapes and real images, and not requiring supervision or restrictive assumptions beyond a categorical image collection; 2) two data-driven techniques to enforce cross-instance consistency, namely *progressive conditioning* and *neighbor reconstruction*. Code and video results are available at imagine.enpc.fr/ monniert/UNICORN.

2   Related Work

We first review deep SVR methods and mesh-based differential renderers we build upon. We then discuss works exploring cross-instance consistency and curriculum learning techniques, to which our progressive conditioning is related.

Deep SVR. There is a clear trend to remove supervision from deep SVR pipelines to directly learn 3D from raw 2D images, which we summarize in Table 1.

A first group of methods uses strong supervision, either paired 3D and images or multiple views of the same object. Direct 3D supervision is successfully used to learn voxels [6], meshes [53], parametrized surfaces [14] and implicit functions [36,58]. The first methods using silhouettes and multiple views initially require camera poses and are also developed for diverse 3D shape representations: [51,59] opt for voxels, [5,30,34] introduce mesh renderers, and [43] adapts implicit representations. Works like [23,49] then introduce techniques to remove the assumption of known poses. Except for [66] which leverages GAN-generated images [13,27], these works are typically limited to synthetic datasets.

A second group of methods aims at removing the need for 3D and multi-view supervision. This is very challenging and they hence typically focus on learning 3D from images of a single category. Early works [26,51,52] estimate camera poses with keypoints and minimize the silhouette reprojection error. The ability to predict textures is first incorporated by CMR [25] which, in addition to keypoints and silhouettes, uses symmetry priors. Recent works [8,33] replace the mesh representation of CMR with implicit functions that do not require symmetry priors, yet the predicted texture quality is strongly deteriorated. [17] improves upon CMR and develops a framework for images with camera annotations that does not rely on silhouettes. Two works managed to further avoid the need for camera estimates but at the cost of additional hypothesis: [16] shows results with textureless synthetic objects, [56] models 2.5D objects like faces with limited background and viewpoint variation. Finally, recent works only require object silhouettes but they also make additional assumptions: [12,50] use known template shapes, [32] assumes access to an off-the-shelf system predicting part semantics, and [55] targets solids of revolution. Other related works [11,18,21,29,44,62] leverage in addition generative adversarial techniques to improve the learning.

In this work, we *do not* use camera estimates, keypoints, silhouettes, nor strong dataset-specific assumptions, and demonstrate results for both diverse shapes and real images. To the best of our knowledge, we present the first generic SVR system learned from raw image collections.

Mesh-Based Differentiable Rendering. We represent 3D models as meshes with parametrized surfaces, as introduced in AtlasNet [14] and advocated by [50]. We optimize the mesh geometry, texture and camera parameters associated to an image using differentiable rendering. Loper and Black [35] introduce the first generic differentiable renderer by approximating derivatives with local filters, and [30] proposes an alternative approximation more suitable to learning neural networks. Another set of methods instead approximates the rendering function to allow differentiability, including SoftRasterizer [34,45] and DIB-R [5]. We refer the reader to [28] for a comprehensive study. We build upon SoftRasterizer [34], but modify the rendering function to learn without silhouette information.

Cross-Instance Consistency. Although all methods learned on categorical image collections implicitly leverage the consistency across instances, few recent works explicitly explore such a signal. Inspired by [31], the SVR system of [21] is learned by enforcing consistency between the interpolated 3D attributes of two instances and attributes predicted for the associated reconstruction. [61] discovers 3D parts using the inconsistency of parts across instances. Closer to our approach, [39] introduces a loss enforcing cross-silhouette consistency. Yet it differs from our work in two ways: (i) the loss operates on silhouettes, whereas our loss is adapted to image reconstruction by modeling background and separating two terms related to shape and texture, and (ii) the loss is used as a refinement on top of two cycle consistency losses for poses and 3D reconstructions, whereas we demonstrate results without additional self-supervised losses.

Curriculum Learning. The idea of learning networks by "starting small" dates back to Elman [9] where two curriculum learning schemes are studied: (i) increasing the difficulty of samples, and (ii) increasing the model complexity. We respectively coin them *curriculum sampling* and *curriculum modeling* for differentiation. Known to drastically improve the convergence speed [2], curriculum sampling is widely adopted across various applications [1,22,47]. On the contrary, curriculum modeling is typically less studied although crucial to various methods. For example, [53] performs SVR in a coarse-to-fine manner by increasing the number of mesh vertices, and [37] clusters images by aligning them with transformations that increase in complexity. We propose a new form of curriculum modeling dubbed *progressive conditioning* which enables us to avoid bad minima.

3   Approach

Our goal is to learn a neural network that reconstructs a textured 3D object from a single input image. We assume we have access to a raw collection of images depicting objects from the same category, without any further annotation. We propose to learn in an analysis-by-synthesis fashion by autoencoding images in a structured way (Fig. 3). We first introduce our structured autoencoder (Sect. 3.1). We then present how we learn models consistent across instances (Sect. 3.2). Finally, we discuss one more technical contribution necessary to our system: an alternate optimization strategy for joint 3D and pose estimation (Sect. 3.3).

Notations. We use bold lowercase for vectors (e.g., $\mathbf{a}$), bold uppercase for images (e.g., $\mathbf{A}$), double-struck uppercase for meshes (e.g., $\mathbb{A}$), calligraphic uppercase for the main modules of our system (e.g., $\mathcal{A}$), lowercase indexed with generic parameters θ for networks (e.g., a_θ), and write $a_{1:N}$ the ordered set $\{a_1, \ldots, a_n\}$.

3.1   Structured Autoencoding

Overview. Our approach can be seen as a structured autoencoder: it takes an image as input, computes parameters with an encoder, and decodes them into explicit and interpretable factors that are composed to generate an image. We model images as the rendering of textured meshes on top of background images. For a given image $\mathbf{I}$, our model thus predicts a shape, a texture, a pose and a background which are composed to get the reconstruction $\hat{\mathbf{I}}$, as shown in Fig. 3. More specifically, the image $\mathbf{I}$ is fed to convolutional encoder networks e_θ which output parameters $e_\theta(\mathbf{I}) = \{\mathbf{z}_{\text{sh}}, \mathbf{z}_{\text{tx}}, \mathbf{a}, \mathbf{z}_{\text{bg}}\}$ used for the decoding part. $\mathbf{a}$ is a 9D vector including the object pose, while the dimension of the latent codes $\mathbf{z}_{\text{sh}}, \mathbf{z}_{\text{tx}}$ and $\mathbf{z}_{\text{bg}}$ will vary during training (see Sect. 3.2). In the following, we describe the decoding modules using these parameters to build the final image by generating a shape, adding texture, positioning it and rendering it over a background.

Shape Deformation. We follow [50] and use the parametrization of AtlasNet [14] where different shapes are represented as deformation fields applied to the unit sphere. We apply the deformation to an icosphere slightly stretched into an ellipsoid mesh $\mathbb{E}$ using a fixed anisotropic scaling. More specifically, given a

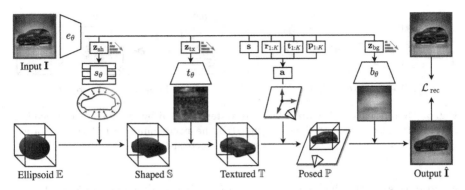

Fig. 3. Structured autoencoding. Given an input, we predict parameters that are decoded into 4 factors (shape, texture, pose, background) and composed to generate the output. Progressive conditioning is represented with ◥.

3D vertex $\mathbf{x}$ of the ellipsoid, our shape deformation module $\mathcal{S}_{\mathbf{z}_{sh}}$ is defined by $\mathcal{S}_{\mathbf{z}_{sh}}(\mathbf{x}) = \mathbf{x} + s_\theta(\mathbf{x}, \mathbf{z}_{sh})$, where s_θ is a Multi-Layer Perceptron taking as input the concatenation of a 3D point $\mathbf{x}$ and a shape code $\mathbf{z}_{sh}$. Applying this displacement to all the ellipsoid vertices enables us to generate a shaped mesh $\mathbb{S} = \mathcal{S}_{\mathbf{z}_{sh}}(\mathbb{E})$. We found that using an ellipsoid instead of a raw icosphere was very effective in encouraging the learning of objects aligned w.r.t. the canonical axes. Learning surface deformations is often preferred to vertex-wise displacements as it enables mapping surfaces, and thus meshes, at any resolution. For us, the mesh resolution is kept fixed and such a representation is a way to regularize the deformations.

Texturing. Following the idea of CMR [25], we model textures as an image UV-mapped onto the mesh through the reference ellipsoid. More specifically, given a texture code $\mathbf{z}_{tx}$, a convolutional network t_θ is used to produce an image $t_\theta(\mathbf{z}_{tx})$, which is UV-mapped onto the sphere using spherical coordinates to associate a 2D point to every vertex of the ellipsoid, and thus to each vertex of the shaped mesh. We write $\mathcal{T}_{\mathbf{z}_{tx}}$ this module generating a textured mesh $\mathbb{T} = \mathcal{T}_{\mathbf{z}_{tx}}(\mathbb{S})$.

Affine Transformation. To render the textured mesh $\mathbb{T}$, we define its position w.r.t. the camera. In addition, we found it beneficial to explicitly model an anisotropic scaling of the objects. Because predicting poses is difficult, we predict K poses candidates, defined by rotations $\mathbf{r}_{1:K}$ and translations $\mathbf{t}_{1:K}$, and associated probabilities $\mathbf{p}_{1:K}$. This involves learning challenges we tackle with a specific optimization procedure described in Sect. 3.3. At inference, we select the pose with highest probability. We combine the scaling and the most likely 6D pose in a single affine transformation module $\mathcal{A}_\mathbf{a}$. More precisely, $\mathcal{A}_\mathbf{a}$ is parametrized by $\mathbf{a} = \{\mathbf{s}, \mathbf{r}, \mathbf{t}\}$, where $\mathbf{s}, \mathbf{r}, \mathbf{t} \in \mathbb{R}^3$ respectively correspond to the three scales of an anisotropic scaling, the three Euler angles of a rotation and the three coordinates of a translation. A 3D point $\mathbf{x}$ on the mesh is then transformed by $\mathcal{A}_\mathbf{a}(\mathbf{x}) = \mathrm{rot}(\mathbf{r})\mathrm{diag}(\mathbf{s})\mathbf{x} + \mathbf{t}$ where $\mathrm{rot}(\mathbf{r})$ is the rotation matrix associated to

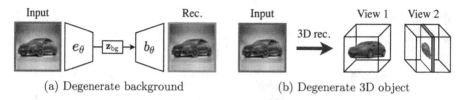

(a) Degenerate background   (b) Degenerate 3D object

Fig. 4. Degenerate solutions. An SVR system learned by raw image autoencoding is prone to degenerate solutions through **(a)** the background or **(b)** the 3D object model. We alleviate the issue with cross-instance consistency.

$\mathbf{r}$ and diag($\mathbf{s}$) is the diagonal matrix associated to $\mathbf{s}$. Our module is applied to all points of the textured mesh $\mathbb{T}$ resulting in a posed mesh $\mathbb{P} = \mathcal{A}_{\mathbf{a}}(\mathbb{T})$.

Rendering with Background. The final step of our process is to render the mesh over a background image. The background image is generated from a background code $\mathbf{z}_{\mathrm{bg}}$ by a convolutional network b_{θ}. A differentiable module $\mathcal{B}_{\mathbf{z}_{\mathrm{bg}}}$ renders the posed mesh $\mathbb{P}$ over this background image $b_{\theta}(\mathbf{z}_{\mathrm{bg}})$ resulting in a reconstructed image $\hat{\mathbf{I}} = \mathcal{B}_{\mathbf{z}_{\mathrm{bg}}}(\mathbb{P})$. We perform rendering through soft rasterization of the mesh. Because we observed divergence results when learning geometry from raw photometric comparison with the standard SoftRasterizer [34,45], we propose two key changes: a layered aggregation of the projected faces and an alternative occupancy function. We provide details in our supplementary material.

3.2 Unsupervised Learning with Cross-Instance Consistency

We propose to learn our structured autoencoder without any supervision, by synthesizing 2D images and minimizing a reconstruction loss. Due to the unconstrained nature of the problem, such an approach typically yields degenerate solutions (Fig. 4a and Fig. 4b). While previous works leverage silhouettes and dataset-specific priors to mitigate this issue, we instead propose two unsupervised data-driven techniques, namely *progressive conditioning* (a training strategy) and *neighbor reconstruction* (a training loss). We thus optimize the shape, texture and background by minimizing for each image $\mathbf{I}$ reconstructed as $\hat{\mathbf{I}}$:

$$\mathcal{L}_{\mathrm{3D}} = \mathcal{L}_{\mathrm{rec}}(\mathbf{I}, \hat{\mathbf{I}}) + \lambda_{\mathrm{nbr}}\mathcal{L}_{\mathrm{nbr}} + \lambda_{\mathrm{reg}}\mathcal{L}_{\mathrm{reg}}, \tag{1}$$

where λ_{nbr} and λ_{reg} are scalar hyperparameters, and $\mathcal{L}_{\mathrm{rec}}$, $\mathcal{L}_{\mathrm{nbr}}$ and $\mathcal{L}_{\mathrm{reg}}$ are respectively the reconstruction, neighbor reconstruction, and regularization losses, described below. In all experiments, we use $\lambda_{\mathrm{nbr}} = 1$ and $\lambda_{\mathrm{reg}} = 0.01$. Note that we optimize pose prediction using a slightly different loss in an alternate optimization scheme described in Sec. 3.3.

Reconstruction and Regularization Losses. Our reconstruction loss has two terms, a pixel-wise squared L_2 loss $\mathcal{L}_{\mathrm{pix}}$ and a perceptual loss [65] $\mathcal{L}_{\mathrm{perc}}$

defined as an L_2 loss on the `relu3_3` layer of a pre-trained VGG16 [48], similar to [56]. While pixel-wise losses are common for autoencoders, we found it crucial to add a perceptual loss to learn textures that are discriminative for the pose estimation. Our full reconstruction loss can be written $\mathcal{L}_{rec}(\mathbf{I},\hat{\mathbf{I}}) = \mathcal{L}_{pix}(\mathbf{I},\hat{\mathbf{I}}) + \lambda_{perc}\mathcal{L}_{perc}(\mathbf{I},\hat{\mathbf{I}})$ and we use $\lambda_{perc} = 10$ in all experiments. While our deformation-based surface parametrization naturally regularizes the shape, we sometimes observe bad minima where the surface has folds. Following prior works [5,12,34, 64], we thus add a small regularization term $\mathcal{L}_{reg} = \mathcal{L}_{norm} + \mathcal{L}_{lap}$ consisting of a normal consistency loss [7] $\mathcal{L}_{norm}$ and a Laplacian smoothing loss [40] $\mathcal{L}_{lap}$.

Progressive Conditioning. The goal of *progressive conditioning* is to encourage the model to share elements (*e.g.*, shape, texture, background) across instances to prevent degenerate solutions. Inspired by the curriculum learning philosophy [9,37,53], we propose to do so by gradually increasing the latent space representing the shape, texture and background. Intuitively, restricting the latent space implicitly encourages maximizing the information shared across instances. For example, a latent space of dimension 0 (*i.e.*, no conditioning) amounts to learning a global representation that is the same for all instances, while a latent space of dimension 1 restricts all the generated shapes, textures or backgrounds to lie on a 1-dimensional manifold. Progressively increasing the size of the latent code during training can be interpreted as gradually specializing from category-level to instance-level knowledge. Figure 2a illustrates the procedure with example results where we can observe the progressive specialization to particular instances: reactors gradually appear/disappear, textures get more accurate. Because common neural network implementations have fixed-size inputs, we implement progressive conditioning by masking, stage-by-stage, a decreasing number of values of the latent code. All our experiments share the same 4-stage training strategy where the latent code dimension is increased at the beginning of each stage and the network is then trained until convergence. We use dimensions 0/2/8/64 for the shape code, 2/8/64/512 for the texture code and 4/8/64/256 for the background code. We provide real-image results for each stage in our supplementary material.

Neighbor Reconstruction. The idea behind *neighbor reconstruction* is to explicitly enforce consistency between different instances. Our key assumption is that neighboring instances with similar shape or texture exist in the dataset. If such neighbors are correctly identified, switching their shape or texture in our generation model should give similar reconstruction results. For a given input image, we hence propose to use its shape or texture attribute in the image formation process of neighboring instances and apply our reconstruction loss on associated renderings. Intuitively, this process can be seen as mimicking a multi-view supervision without actually having access to multi-view images by finding neighboring instances in well-designed latent spaces. Figure 2b illustrates the procedure with an example.

More specifically, let $\{\mathbf{z}_{sh}, \mathbf{z}_{tx}, \mathbf{a}, \mathbf{z}_{bg}\}$ be the parameters predicted by our encoder for a given input training image $\mathbf{I}$, let Ω be a memory bank storing the images and parameters of the last M instances processed by the network. We write $\Omega^{(m)} = \{\mathbf{I}^{(m)}, \mathbf{z}_{sh}^{(m)}, \mathbf{z}_{tx}^{(m)}, \mathbf{a}^{(m)}, \mathbf{z}_{bg}^{(m)}\}$ each of these M instances and associated parameters. We first select the closest instance from the memory bank Ω in the texture (respectively shape) code space using the L_2 distance, $m_t = \mathrm{argmin}_m \|\mathbf{z}_{tx} - \mathbf{z}_{tx}^{(m)}\|_2$ (respectively $m_s = \mathrm{argmin}_m \|\mathbf{z}_{sh} - \mathbf{z}_{sh}^{(m)}\|_2$). We then swap the codes and generate the reconstruction $\hat{\mathbf{I}}_{tx}^{(m_t)}$ (respectively $\hat{\mathbf{I}}_{sh}^{(m_s)}$) using the parameters $\{\mathbf{z}_{sh}^{(m_t)}, \mathbf{z}_{tx}, \mathbf{a}^{(m_t)}, \mathbf{z}_{bg}^{(m_t)}\}$ (respectively $\{\mathbf{z}_{sh}, \mathbf{z}_{tx}^{(m_s)}, \mathbf{a}^{(m_s)}, \mathbf{z}_{bg}^{(m_s)}\}$). Finally, we compute the reconstruction loss between the images $\mathbf{I}^{(m_t)}$ and $\hat{\mathbf{I}}_{tx}^{(m_t)}$ (respectively $\mathbf{I}^{(m_s)}$ and $\hat{\mathbf{I}}_{sh}^{(m_s)}$). Our full loss can thus be written:

$$\mathcal{L}_{nbr} = \mathcal{L}_{rec}(\mathbf{I}^{(m_t)}, \hat{\mathbf{I}}_{tx}^{(m_t)}) + \mathcal{L}_{rec}(\mathbf{I}^{(m_s)}, \hat{\mathbf{I}}_{sh}^{(m_s)}). \qquad (2)$$

Note that we recompute the parameters of the selected instances with the current network state, to avoid uncontrolled effects of changes in the network state. Also note that, for computational reasons, we do not use this loss in the first stage where codes are almost the same for all instances.

To prevent latent codes from specializing by viewpoint, we split the viewpoints into V bins w.r.t. the rotation angle, uniformly sample a target bin for each input and look for the nearest instances only in the subset of instances within the target viewpoint range (see the supplementary for details). In all experiments, we use $V = 5$ and a memory bank of size $M = 1024$.

3.3 Alternate 3D and Pose Learning

Because predicting 6D poses is hard due to self-occlusions and local minima, we follow prior works [12,16,23,50] and predict multiple pose candidates and their likelihood. However, we identify failure modes in their optimization framework (detailed in our supplementary material) and instead propose a new optimization that alternates between 3D and pose learning. More specifically, given an input image $\mathbf{I}$, we predict K pose candidates $\{(\mathbf{r}_1, \mathbf{t}_1), \ldots, (\mathbf{r}_K, \mathbf{t}_K)\}$, and their associated probabilities $\mathbf{p}_{1:K}$. We render the model from the different poses, yielding K reconstructions $\hat{\mathbf{I}}_{1:K}$. We then alternate the learning between two steps: (i) the *3D-step* where shape, texture and background branches of the network are updated by minimizing $\mathcal{L}_{3D}$ using the pose associated to the highest probability, and (ii) the *P-step* where the branches of the network predicting candidate poses and their associated probabilities are updated by minimizing:

$$\mathcal{L}_P = \sum_k \mathbf{p}_k \mathcal{L}_{rec}(\mathbf{I}, \hat{\mathbf{I}}_k) + \lambda_{uni} \mathcal{L}_{uni}, \qquad (3)$$

where $\mathcal{L}_{rec}$ is the reconstruction loss described in Sect. 3.2, $\mathcal{L}_{uni}$ is a regularization loss on the predicted poses and λ_{uni} is a scalar hyperparameter. More precisely, we use $\mathcal{L}_{uni} = \sum_k |\bar{\mathbf{p}}_k - 1/K|$ where $\bar{\mathbf{p}}_k$ is the averaged probabilities for candidate k in a particular training batch. Similar to [16], we found it crucial to introduce this regularization term to encourage the use of all pose candidates. In particular, this prevents a collapse mode where only few pose candidates are used. Note that we do not use the neighbor reconstruction loss nor the mesh regularization loss which are not relevant for viewpoints. In all experiments, we use $\lambda_{uni} = 0.02$.

Inspired by the camera multiplex of [12], we parametrize rotations with the classical Euler angles (azimuth, elevation and roll) and rotation candidates correspond to offset angles w.r.t. reference viewpoints. Since in practice elevation has limited variations, our reference viewpoints are uniformly sampled along the azimuth dimension. Note that compared to [12], we do not directly optimize a set of pose candidates per training image, but instead learn a set of K predictors for the entire dataset. We use $K = 6$ in all experiments.

4 Experiments

We validate our approach in two standard setups. It is first quantitatively evaluated on ShapeNet where state-of-the-art methods use multiple views as supervision. Then, we compare it on standard real-image benchmarks and demonstrate its applicability to more complex datasets. Finally, we present an ablation study.

4.1 Evaluation on the ShapeNet Benchmark

We compare our approach to state-of-the-art methods using multi-views, viewpoints and silhouettes as supervision. Our method is instead learned without supervision. For all compared methods, one model is trained per class. We adhere to community standards [30, 34, 43] and use the renderings and splits from [30] of the ShapeNet dataset [4]. It corresponds to a subset of 13 classes of 3D objects, each object being rendered into a 64×64 image from 24 viewpoints uniformly spaced along the azimuth dimension. We evaluate all methods using the standard Chamfer-L_1 distance [36, 43], where predicted shapes are pre-aligned using our gradient-based version of the Iterative Closest Point (ICP) [3] with anisotropic scaling. Indeed, compared to competing methods having access to the ground-truth viewpoint during training, we need to predict it for each input image in addition to the 3D shape. This yields to both shape/pose ambiguities (*e.g.*, a small nearby object or a bigger one far from the camera) and small misalignment errors that dramatically degrade the performances. We provide evaluation details as well as results without ICP in our supplementary.

We report quantitative results and compare to the state of the art in Table 2, where methods using multi-views are visually separated. We evaluate the pretrained weights for SDF-SRN [33] and train the models from scratch using the official implementation for DVR [43]. We tried evaluating SMR [21] but could not reproduce the results. We do not compare to TARS [8] which is based on SDF-SRN and share the same performances. Our approach achieves results that are on average better than the state-of-the-art methods supervised with silhouette and viewpoint annotations. This is a strong result: while silhouettes are trivial in this benchmark, learning without viewpoint annotations is extremely challenging as it involves solving the pose estimation and shape reconstruction

Table 2. ShapeNet comparison. We report Chamfer-L_1 ↓ obtained after ICP, **best** results are highlighted. Supervisions are: Multi-Views, Camera or Keypoints, Silhouettes.

Method	Ours	SDF-SRN [33]	DVR [43]	DVR [43]
MV				✓
CK		✓	✓	✓
S		✓	✓	✓
Airplane	0.110	0.128	0.114	0.111
Bench	0.159	–	0.255	0.176
Cabinet	0.137	–	0.254	0.158
Car	0.168	0.150	0.203	0.153
Chair	0.253	0.262	0.371	0.205
Display	0.220	–	0.257	0.163
Lamp	0.523	–	0.363	0.281
Phone	0.127	–	0.191	0.076
Rifle	0.097	–	0.130	0.083
Sofa	0.192	–	0.321	0.160
Speaker	0.224	–	0.312	0.215
Table	0.243	–	0.303	0.230
Vessel	0.155	–	0.180	0.151
Mean	0.201	–	0.250	0.166

Fig. 5. Visual comparisons. We compare to DVR [43] and Soft-Ras [34] learned with full supervision (MV, CK, S).

problems simultaneously. For some categories, our performances are even better than DVR supervised with multiple views. This shows that our system learned on raw images generates 3D reconstructions comparable to the ones obtained from methods using geometry cues like silhouettes and multiple views. Note that for the lamp category, our method predicts degenerate 3D shapes; we hypothesize this is due to their rotation invariance which makes the viewpoint estimation ambiguous.

We visualize and compare the quality of our 3D reconstructions in Fig. 5. The first three examples correspond to examples advertised in DVR [43], the last two corresponds to examples we selected. Our method generates textured meshes of high-quality across all these categories. The geometry obtained is sharp and accurate, and the predicted texture mostly corresponds to the input.

4.2   Results on Real Images

Pascal3D+ Car and CUB Benchmarks. We compare our approach to state-of-the-art SVR methods on real images, where multiple views are not available. All competing methods use silhouette supervision and output meshes that are symmetric. CMR [25] additionally use keypoints, UCMR [12] and IMR [50] starts learning from a given template shape; we *do not* use any of these and directly learn from raw images. We strictly follow the community standards [12,25,50] and use the train/test splits of Pascal3D+ Car [57] (5000/220 images) and CUB-200-2011 [54] (5964/2874 images). Images are square-cropped around the object using bounding box annotations and resized to 64 × 64.

Table 3. Real-image quantitative comparisons. Supervision corresponds to Camera or Keypoints, Silhouettes, Assumptions (see Table 1 for details).

| Method | Supervision | | | Pascal3D+ Car | | | CUB-200-2011 | |
	CK	S	A	3D IoU ↑	Ch-L_1 ↓	Mask IoU ↑	PCK@0.1 ↑	Mask IoU ↑
CMR [25]	✓	✓	✓	64	–	–	48.3	70.6
IMR [50]		✓	✓	–	–	–	53.5	–
UMR [32]		✓	✓	62	–	–	58.2	73.4
UCMR [12]		✓	✓	**67.3**	0.172	73.7	–	63.7
SMR [21]		✓	✓	–	–	–	**62.2**	**80.6**
Ours				65.9	**0.163**	**83.9**	49.0	71.4

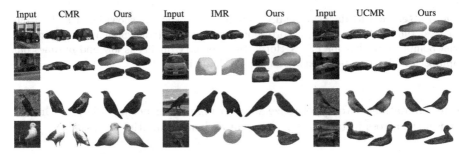

Fig. 6. Real-image comparisons. We show reconstructions on Pascal3D+ Cars (top) and CUB (bottom) and compare to CMR [25], IMR [50], UCMR [12].

A quantitative comparison is summarized in Table 3, where we report 3D IoU, Chamfer-L_1 (with ICP alignment), Mask IoU for Pascal3D+ Car, and Percentage of Correct Keypoints thresholded at $\alpha = 0.1$ (PCK@0.1) [31], Mask IoU for CUB. Our approach yields competitive results across all metrics although it does not rely on any supervision used by other works. On Pascal3D+ Car, we achieve significantly better results than UCMR for Chamfer-L_1 and Mask IoU, which we argue are less biased metrics than the standard 3D IoU [25,51] computed on unaligned shapes (see supplementary). On CUB, our approach achieves reasonable results that are however slightly worse than the state of the art. We hypothesize this is linked to our pose regularization term encouraging the use of all viewpoints whereas these bird images clearly lack back views.

We qualitatively compare our approach to the state of the art in Fig. 6. For each input, we show the mesh rendered from two viewpoints. For our car results, we additionally show meshes with synthetic textures to emphasize correspondences. Qualitatively, our approach yields results on par with prior works both in terms of geometric accuracy and overall realism. Although the textures obtained in [50] look more accurate, they are modeled as pixel flows, which has a clear limitation when synthesizing unseen texture parts. Note that we do not recover details like the bird legs which are missed by prior works due to coarse silhou-

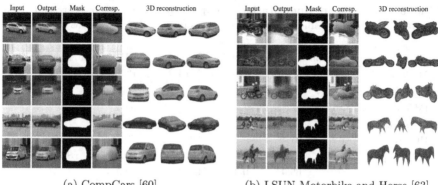

(a) CompCars [60]   (b) LSUN Motorbike and Horse [63]

Fig. 7. Real-world dataset results. From left to right, we show: input and output images, the predicted mask, a correspondence map and the mesh rendered from 3 viewpoints. Note that for LSUN Horse, the geometry quality is low and outlines our approach limitations (see text). Best viewed digitally.

ette annotations. We hypothesize we also miss them because they are hardly consistent across instances, *e.g.*, legs can be bent in multiple ways.

Real-Word Datasets. Motivated by 3D-aware image synthesis methods learned in-the-wild [41,42], we investigate whether our approach can be applied to real-world datasets where silhouettes are not available and images are not methodically cropped around the object. We adhere to standards from the 3D-aware image synthesis community [41,42] and apply our approach to 64×64 images of CompCars [60]. In addition, we provide results for the more difficult scenario of LSUN images [63] for motorbikes and horses. Because many LSUN images are noise, we filter the datasets as follows: we manually select 16 reference images with different poses, find the nearest neighbors from the first 200k images in a pre-trained ResNet-18 [15] feature space, and keep the top 2k for each reference image. We repeat the procedure with flipped reference images yielding 25k images.

Our results are shown in Fig. 7. For each input image, we show from left to right: the output image, the predicted mask, a correspondence map, and the 3D reconstruction rendered from the predicted viewpoint and two other viewpoints. Although our approach is trained to synthesize images, these are all natural by-products. While the quality of our 3D car reconstructions is high, the reconstructions obtained for LSUN images lack some realism and accuracy (especially for horses), thus outlining limitations of our approach. However, our segmentation and correspondence maps emphasize our system ability to accurately localize the object and find correspondences, even when the geometry is coarse.

Table 4. Ablation results on ShapeNet [4].

Model	Full	w/o $\mathcal{L}_{nbr}$	w/o PC
Airplane	0.110	0.124	**0.107**
Bench	**0.159**	0.188	0.206
Car	**0.168**	0.179	0.173
Chair	**0.253**	0.319	0.527
Table	**0.243**	0.246	0.598
Mean	**0.187**	0.211	0.322

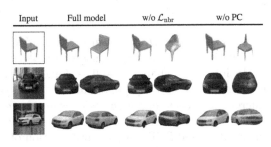

Fig. 8. Ablation visual results. For each input, we show the mesh rendered from two viewpoints.

Limitations. Even if our approach is a strong step towards generic unsupervised SVR, we can outline three main limitations. First, the lack of different views in the data harms the results (*e.g.*, most CUB birds have concave backs); this can be linked to our uniform pose regularization term which is not adequate in these cases. Second, complex textures are not predicted correctly (*e.g.*, motorbikes in LSUN). Although we argue it could be improved by more advanced autoencoders, the neighbor reconstruction term may prevent unique textures to be generated. Finally, despite its applicability to multiple object categories and diverse datasets, our multi-stage progressive training is cumbersome and an automatic way of progressively specializing to instances is much more desirable.

4.3 Ablation Study

We analyze the influence of our neighbor reconstruction loss $\mathcal{L}_{nbr}$ and progressive conditioning (PC) by running experiments without each component.

First, we provide quantitative results on ShapeNet in Table 4. When $\mathcal{L}_{nbr}$ is removed, the results are worse for almost all categories, outlining that it is important to the predicted geometry accuracy. When PC is removed, results are comparable to the full model for airplane and car but much worse for chair and table. Indeed, they involve more complex shapes and our system falls into a bad minimum with degenerate solutions, a scenario that is avoided with PC.

Second, we perform a visual comparison on ShapeNet and CompCars examples in Fig. 8. For each input, we show the mesh rendered from the predicted viewpoint and a different viewpoint. When $\mathcal{L}_{nbr}$ is removed, we observe that the reconstruction seen from the predicted viewpoint is correct but it is either wrong for chairs and degraded for cars when seen from the other viewpoint. Indeed, the neighbor reconstruction explicitly enforces the unseen reconstructed parts to be consistent with other instances. When PC is removed, we observe degenerate reconstructions where the object seen from a different viewpoint is not realistic.

5   Conclusion

We presented UNICORN, an unsupervised SVR method which, in contrast to all prior works, learns from raw images only. We demonstrated it yields high-quality results for diverse shapes as well as challenging real-world image collections. This was enabled by two key contributions aiming at leveraging consistency across different instances: our *progressive conditioning* training strategy and *neighbor reconstruction* loss. We believe our work includes both an important step forward for unsupervised SVR and the introduction of a valuable conceptual insight.

Acknowledgement. We thank François Darmon for inspiring discussions; Robin Champenois, Romain Loiseau, Elliot Vincent for feedback on the manuscript; and Michael Niemeyer, Shubham Goel for details on the evaluation. This work was supported in part by ANR project EnHerit ANR-17-CE23-0008, project Rapid Tabasco, gifts from Adobe and HPC resources from GENCI-IDRIS (2021-AD011011697R1, 2022-AD011013538).

References

1. Bengio, S., Vinyals, O., Jaitly, N., Shazeer, N.: Scheduled sampling for sequence prediction with recurrent neural networks. In: NIPS (2015)
2. Bengio, Y., Louradour, J., Collobert, R., Weston, J.: Curriculum learning. In: ICML (2009)
3. Besl, P., McKay, N.D.: A method for registration of 3-D shapes. TPAMI 14(2) (1992)
4. Chang, A.X., et al.: ShapeNet: an information-rich 3d model repository. arXiv:1512.03012 [cs] (2015)
5. Chen, W., et al.: Learning to Predict 3D Objects with an Interpolation-based Differentiable Renderer. In: NeurIPS (2019)
6. Choy, C.B., Xu, D., Gwak, J.Y., Chen, K., Savarese, S.: 3D-R2N2: a unified approach for single and multi-view 3d object reconstruction. In: Leibe, B., Matas, J., Sebe, N., Welling, M. (eds.) ECCV 2016. LNCS, vol. 9912, pp. 628–644. Springer, Cham (2016). https://doi.org/10.1007/978-3-319-46484-8_38
7. Desbrun, M., Meyer, M., Schröder, P., Barr, A.H.: Implicit fairing of irregular meshes using diffusion and curvature flow. In: SIGGRAPH (1999)
8. Duggal, S., Pathak, D.: Topologically-aware deformation fields for single-view 3D reconstruction. In: CVPR (2022)
9. Elman, J.L.: Learning and development in neural networks: The importance of starting small. Cognition (1993)
10. Finger, S.: Origins of neuroscience: a history of explorations into brain function. Oxford University Press (1994)
11. Gadelha, M., Maji, S., Wang, R.: 3D shape induction from 2D views of multiple objects. In: 3DV (2017)
12. Goel, S., Kanazawa, A., Malik, J.: Shape and viewpoint without keypoints. In: Vedaldi, A., Bischof, H., Brox, T., Frahm, J.-M. (eds.) ECCV 2020. LNCS, vol. 12360, pp. 88–104. Springer, Cham (2020). https://doi.org/10.1007/978-3-030-58555-6_6
13. Goodfellow, I., et al.: Generative adversarial nets. In: NIPS (2014)

14. Groueix, T., Fisher, M., Kim, V.G., Russell, B.C., Aubry, M.: AtlasNet: a papier-Mâché approach to learning 3D surface generation. In: CVPR (2018)
15. He, K., Zhang, X., Ren, S., Sun, J.: Deep residual learning for image recognition. In: CVPR (2016)
16. Henderson, P., Ferrari, V.: Learning single-image 3D reconstruction by generative modelling of shape, pose and shading. IJCV (2019)
17. Henderson, P., Tsiminaki, V., Lampert, C.H.: Leveraging 2D data to learn textured 3D mesh generation. In: CVPR (2020)
18. Henzler, P., Mitra, N., Ritschel, T.: Escaping Plato's Cave: 3D shape from adversarial rendering. In: ICCV (2019)
19. Hoiem, D., Efros, A.A., Hebert, M.: Geometric context from a single image. In: ICCV (2005)
20. Hoiem, D., Efros, A.A., Hebert, M.: Putting objects in perspective. IJCV (2008)
21. Hu, T., Wang, L., Xu, X., Liu, S., Jia, J.: Self-supervised 3D mesh reconstruction from single images. In: CVPR (2021)
22. Ilg, E., Mayer, N., Saikia, T., Keuper, M., Dosovitskiy, A., Brox, T.: FlowNet 2.0: evolution of optical flow estimation with deep networks. In: CVPR (2017)
23. Insafutdinov, E., Dosovitskiy, A.: Unsupervised learning of shape and pose with differentiable point clouds. In: NIPS (2018)
24. Jojic, N., Frey, B.J.: Learning Flexible Sprites in Video Layers. In: CVPR (2001)
25. Kanazawa, A., Tulsiani, S., Efros, A.A., Malik, J.: Learning category-specific mesh reconstruction from image collections. In: Ferrari, V., Hebert, M., Sminchisescu, C., Weiss, Y. (eds.) ECCV 2018. LNCS, vol. 11219, pp. 386–402. Springer, Cham (2018). https://doi.org/10.1007/978-3-030-01267-0_23
26. Kar, A., Tulsiani, S., Carreira, J., Malik, J.: Category-specific object reconstruction from a single image. In: CVPR (2015)
27. Karras, T., Laine, S., Aila, T.: A style-based generator architecture for generative adversarial networks. In: CVPR (2019)
28. Kato, H., et al.: Differentiable rendering: a survey. arXiv:2006.12057 [cs] (2020)
29. Kato, H., Harada, T.: Learning view priors for single-view 3D reconstruction. In: CVPR (2019)
30. Kato, H., Ushiku, Y., Harada, T.: Neural 3D mesh renderer. In: CVPR (2018)
31. Kulkarni, N., Gupta, A., Tulsiani, S.: Canonical surface mapping via geometric cycle consistency. In: ICCV (2019)
32. Li, X., Liu, S., Kim, K., De Mello, S., Jampani, V., Yang, M.-H., Kautz, J.: Self-supervised single-view 3D reconstruction via semantic consistency. In: Vedaldi, A., Bischof, H., Brox, T., Frahm, J.-M. (eds.) ECCV 2020. LNCS, vol. 12359, pp. 677–693. Springer, Cham (2020). https://doi.org/10.1007/978-3-030-58568-6_40
33. Lin, C.H., Wang, C., Lucey, S.: SDF-SRN: learning signed distance 3D object reconstruction from static images. In: NeurIPS (2020)
34. Liu, S., Li, T., Chen, W., Li, H.: Soft rasterizer: a differentiable renderer for image-based 3D reasoning. In: ICCV (2019)
35. Loper, M.M., Black, M.J.: OpenDR: An Approximate Differentiable Renderer. In: ECCV 2014, vol. 8695 (2014)
36. Mescheder, L., Oechsle, M., Niemeyer, M., Nowozin, S., Geiger, A.: Occupancy networks: learning 3D reconstruction in function space. In: CVPR (2019)
37. Monnier, T., Groueix, T., Aubry, M.: Deep transformation-invariant clustering. In: NeurIPS (2020)
38. Monnier, T., Vincent, E., Ponce, J., Aubry, M.: Unsupervised layered image decomposition into object prototypes. In: ICCV (2021)

39. Navaneet, K.L., Mathew, A., Kashyap, S., Hung, W.C., Jampani, V., Babu, R.V.: From image collections to point clouds with self-supervised shape and pose networks. In: CVPR (2020)
40. Nealen, A., Igarashi, T., Sorkine, O., Alexa, M.: Laplacian mesh optimization. In: GRAPHITE (2006)
41. Nguyen-Phuoc, T., Li, C., Theis, L., Richardt, C., Yang, Y.L.: HoloGAN: unsupervised learning of 3D representations from natural images. In: ICCV (2019)
42. Niemeyer, M., Geiger, A.: GIRAFFE: Representing Scenes as Compositional Generative Neural Feature Fields. In: CVPR (2021)
43. Niemeyer, M., Mescheder, L., Oechsle, M., Geiger, A.: Differentiable volumetric rendering: learning implicit 3D representations without 3D supervision. In: CVPR (2020)
44. Pavllo, D., Spinks, G., Hofmann, T., Moens, M.F., Lucchi, A.: Convolutional generation of textured 3D meshes. In: NeurIPS (2020)
45. Ravi, N., et al.: Accelerating 3D deep learning with PyTorch3D. arXiv:2007.08501 [cs] (2020)
46. Saxena, A., Min Sun, Ng, A.: Make3D: learning 3D scene structure from a single still image. TPAMI (2009)
47. Schroff, F., Kalenichenko, D., Philbin, J.: FaceNet: a unified embedding for face recognition and clustering. In: CVPR (2015)
48. Simonyan, K., Zisserman, A.: Very Deep Convolutional Networks for Large-Scale Image Recognition. In: ICLR (2015)
49. Tulsiani, S., Efros, A.A., Malik, J.: Multi-view consistency as supervisory signal for learning shape and pose prediction. In: CVPR (2018)
50. Tulsiani, S., Kulkarni, N., Gupta, A.: Implicit mesh reconstruction from unannotated image collections. arXiv:2007.08504 [cs] (2020)
51. Tulsiani, S., Zhou, T., Efros, A.A., Malik, J.: Multi-view Supervision for Single-view Reconstruction via Differentiable Ray Consistency. In: CVPR (2017)
52. Vicente, S., Carreira, J., Agapito, L., Batista, J.: Reconstructing PASCAL VOC. In: CVPR (2014)
53. Wang, N., Zhang, Y., Li, Z., Fu, Y., Liu, W., Jiang, Y.-G.: Pixel2Mesh: generating 3D mesh models from single RGB images. In: Ferrari, V., Hebert, M., Sminchisescu, C., Weiss, Y. (eds.) ECCV 2018. LNCS, vol. 11215, pp. 55–71. Springer, Cham (2018). https://doi.org/10.1007/978-3-030-01252-6_4
54. Welinder, P., Branson, S., Mita, T., Wah, C., Schroff, F., Belongie, S., Perona, P.: Caltech-UCSD Birds 200. Technical report, California Institute of Technology (2010)
55. Wu, S., Makadia, A., Wu, J., Snavely, N., Tucker, R., Kanazawa, A.: De-rendering the world's revolutionary artefacts. In: CVPR (2021)
56. Wu, S., Rupprecht, C., Vedaldi, A.: Unsupervised learning of probably symmetric deformable 3D objects from images in the wild. In: CVPR (2020)
57. Xiang, Y., Mottaghi, R., Savarese, S.: Beyond PASCAL: a benchmark for 3D object detection in the wild. In: WACV (2014)
58. Xu, Q., Wang, W., Ceylan, D., Mech, R., Neumann, U.: DISN: Deep Implicit Surface Network for High-quality Single-view 3D Reconstruction. In: NeurIPS (2019)
59. Yan, X., Yang, J., Yumer, E., Guo, Y., Lee, H.: Perspective transformer nets: learning single-view 3D object reconstruction without 3D supervision. In: NeurIPS (2016)
60. Yang, L., Luo, P., Loy, C.C., Tang, X.: A large-scale car dataset for fine-grained categorization and verification. In: CVPR (2015)

61. Yao, C.H., Hung, W.C., Jampani, V., Yang, M.H.: Discovering 3D parts from image collections. In: ICCV (2021)
62. Ye, Y., Tulsiani, S., Gupta, A.: Shelf-supervised mesh prediction in the wild. arXiv:2102.06195 [cs] (2021)
63. Yu, F., Seff, A., Zhang, Y., Song, S., Funkhouser, T., Xiao, J.: LSUN: construction of a large-scale image dataset using deep learning with humans in the loop. arXiv:1506.03365 [cs] (2016)
64. Zhang, J.Y., Yang, G., Tulsiani, S., Ramanan, D.: NeRS: neural reflectance surfaces for sparse-view 3D reconstruction in the wild. In: NeurIPS (2021)
65. Zhang, R., Isola, P., Efros, A.A., Shechtman, E., Wang, O.: The unreasonable effectiveness of deep features as a perceptual metric. In: CVPR (2018)
66. Zhang, Y., et al.: Image GANs meet differentiable rendering for inverse graphics and interpretable 3D neural rendering. In: ICLR (2021)

Towards Comprehensive Representation Enhancement in Semantics-Guided Self-supervised Monocular Depth Estimation

Jingyuan Ma, Xiangyu Lei, Nan Liu, Xian Zhao, and Shiliang Pu[⊠]

Hikvision Research Institute, Hangzhou, China
{majingyuan,leixiangyu,liunan6,zhaoxian,pushiliang.hri}@hikvision.com

Abstract. Semantics-guided self-supervised monocular depth estimation has been widely researched, owing to the strong cross-task correlation of depth and semantics. However, since depth estimation and semantic segmentation are fundamentally two types of tasks: one is regression while the other is classification, the distribution of depth feature and semantic feature are naturally different. Previous works that leverage semantic information in depth estimation mostly neglect such representational discrimination, which leads to insufficient representation enhancement of depth feature. In this work, we propose an attention-based module to enhance task-specific feature by addressing their feature uniqueness within instances. Additionally, we propose a metric learning based approach to accomplish comprehensive enhancement on depth feature by creating a separation between instances in feature space. Extensive experiments and analysis demonstrate the effectiveness of our proposed method. In the end, our method achieves the state-of-the-art performance on KITTI dataset.

Keywords: Monocular depth estimation · Self-supervised learning · Feature metric learning · Representation enhancement

1 Introduction

Depth estimation is one of the fundamentals in many computer vision applications such as robotics, augmented reality and autonomous driving. A depth map reflects the distance between image plane and corresponding objects in real world. Such depth map can be acquired from various sensor setups. Owing to the low cost of single camera setup, monocular depth estimation has been

---

J. Ma and X. Lei—First two authors contribute equally to this work.

Supplementary Information The online version contains supplementary material available at https://doi.org/10.1007/978-3-031-19769-7_18.

actively researched. Although conventional methods, using SfM or SLAM algorithm [11,36,39], fail to produce satisfying results, deep-learning based methods [1,14,17,29,63] have achieved significant improvement. Still, estimating depth from monocular image remains challenging.

Deep-learning based monocular depth estimation can be generally categorized into supervised and self/un-supervised learning. Currently, supervised methods [12,14,29,41] have yielded satisfying advancement in monocular depth estimation. However, since acquiring accurate pixel-level annotations is expensive and limited, self-supervised learning has gained attention because of its independence from annotations and better scalability in data. Under self-supervised settings, monocular depth and egomotion are jointly estimated from separate networks [1,16,17,42,57,59,63], whose training process is self-supervised by photometric loss [51]. Recently, since depth and semantics are spatially aligned, some approaches attempt to leverage semantic information in depth estimation via direct feature fusion [18,20,27] or representational enhancement [6,25]. However, depth estimation and semantic segmentation are two different tasks. Thus, their feature distributions are significantly different as shown in Fig. 1. Such cross-task feature inconsistency exists between instances and within an instance. Thus, enhancing depth feature solely from cross-task spatial consistency is not sufficient. Task-specific representational uniqueness should be identified.

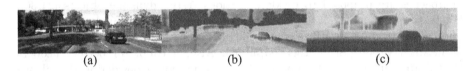

(a)   (b)   (c)

Fig. 1. The visualizations of feature heatmap for depth and semantic segmentation. (a) Colored image. (b) Semantic feature heatmap from Deeplabv3 [3]. (c) Feature heatmap of depth feature heatmap from Monodepth2 [16].

To address task-specific feature uniqueness within an instance, we design a novel and efficient intra/cross-task multi-head attention module (IC-MHA) that adequately fuses task-specific representational uniqueness with cross-task spatial consistency. Inspired by the recent success of vision transformers [9,21,33,58], task-specific representational uniqueness is addressed as window-based self-attention mechanism on task-specific feature. Additionally, the similarity between depth feature and semantic feature is computed using cross-attention mechanism [25], which represents cross-task spatial consistency. A simple fusion layer is implemented to incorporate the generated task-specific self attention and cross attention with input task-specific feature.

To further enhance depth feature by addressing its representational uniqueness between instances, we propose a hardest non-boundary triplet loss whose anchors, positives and negatives are sampled with minimum-distance based candidate mining strategy. Such triplet loss achieves comprehensive enhancement on depth feature over all regions of an image. Extensive experiments and analysis prove the effectiveness of our proposed method, which achieves the state-of-the-art self-supervised monocular depth prediction on KITTI Eigen split [10]. Here, we summarize our contribution in three-fold.

– A novel and efficient intra/cross-task multi-head attention module is proposed to enhance task-specific features by emphasizing their representational uniqueness within instances while preserving their cross-task spatial consistency.
– An effective hardest non-boundary triplet loss using minimum-distance based candidate mining strategy is proposed to further enhance depth feature by addressing its representational uniqueness between instances.
– Our proposed method outperforms previous state-of-the-art self-supervised monocular depth estimation works on KITTI Eigen split.

2   Related Work

2.1   Self-supervised Monocular Depth Estimation

As a pioneering work in self-supervised monocular depth estimation, SfMLearner [63] jointly estimates pose and depth information using two networks, which are self-supervised by photometric loss [51]. Later, under this framework, many approaches are proposed to tackle occlusions [1,16], dynamic objects [24,30], low-texture regions [42,57] and scale-inconsistency [49,59]. Furthermore, some approaches attempt to utilize consistency between consecutive frames [1,38,52] or between various SfM tasks [55,59,66]. Also, a couple of better encoders [19,62] are implemented to improve depth estimation. Considering that depth and semantics of an image are spatially aligned, some recent works propose to improve depth prediction by targeting dynamic objects [27] or exploiting semantics-aware depth feature [6,20,25,32]. In our work, since task-specific feature has its unique distribution, we propose to further refine depth feature by efficiently fusing task-specific representational uniqueness with cross-task spatial consistency.

2.2   Vision Transformer

Inspired by [46], various vision transformers [21] have been proposed and demonstrated superior performance on many tasks such as image recognition [4,9], object detection [2,45,60] and semantic segmentation [50,61]. Amongst them, some works propose to perform local attention inside of image patches [9] or windows [33]. Inspired by window-based vision transformer [33], we propose to address the uniqueness of task-specific feature as multi-head self-attention within locally partitioned windows, which is then efficiently fused with cross-task spatial consistency.

2.3 Deep Metric Learning

Deep metric learning [28,44,48] aims to cluster samples with similar characteristics closer in feature space using proper candidate mining strategy. It has
proven its success in various fields like face recognition [23], image retrieval
[44,48], keypoint detection [7,43,54] and depth estimation [25]. To further refine
depth feature, we propose a hardest non-boundary triplet loss whose positives
and negatives are sampled based on the distance between their anchor sample
and semantic boundary.

3 Methods

3.1 Proposed Model

To properly emphasize task-specific representational uniqueness and cross-task
spatial consistency, we propose an intra-/cross-task multi-head attention (IC-
MHA) module to enhance features for two task: depth estimation and semantic
segmentation. Our overall pipeline is shown in Fig. 2. Following [16,25,42,63],
a 6-DoF $T \in \mathbb{SE}(3)$ is estimated from PoseNet, whose input is a concatenated
consecutive image pair. Taking the target image of size $[H, W]$ as input, our
DepthSegNet consists of a shared encoder, task-specific decoders and our proposed IC-MHA modules. The IC-MHA module, whose input features' spatial
dimension is $[\frac{H}{2^s}, \frac{W}{2^s}]$, is inserted between task-specific decoders at multiple levels s. Inside of IC-MHA module, task-specific representational uniqueness is
addressed as intra-task local attention, and cross-task spatial consistency is represented by cross-task attention [25].

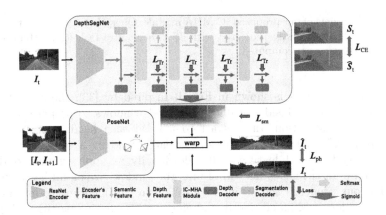

Fig. 2. An overview of our pipeline. DepthSegNet and PoseNet are implemented separately. Proposed IC-MHA module is inserted between task-specific decoders at multiple
levels. $\mathcal{L}_{T_r}$ is the metric learning loss, consisting of boundary triplet loss $\mathcal{L}_{BT}$ in [25]
and proposed hardest non-boundary triplet loss $\mathcal{L}_{NBT}$ in Eq. 8.

Intra-/Cross-Task Multi-Head Attention(IC-MHA) Module. The architecture of IC-MHA module is shown in Fig. 3. At each level $s < 4$, the upsampled task-specific features from level $s + 1$ are fed into IC-MHA module. Within each IC-MHA module, linear projections with expansion ratio r, $\{\Psi_t^j : \mathbb{R}^{H \times W \times C} \to \mathbb{R}^{H \times W \times (r \times C)} | t \in \{D, S\}, j \in \{Q, K, V\}\}$, are applied on task-specific features, $\{F_t^s | t \in \{D, S\}, s < 4\}$, to generate a query-key-value triplet for each task, $\{(Q_t, K_t, V_t) | t \in \{D, S\}\}$. Following [46], we implement these linear projections on multiple heads H. In each head, the intra-task local attention and cross-task attention are computed in parallel. Here, to save computational cost, only one query-key-value triplet is projected for each feature to compute intra-task local attention and cross-task attention. For simplicity, we illustrate such computation on depth feature as an example. Then, the same process is symmetrically identical on semantic feature.

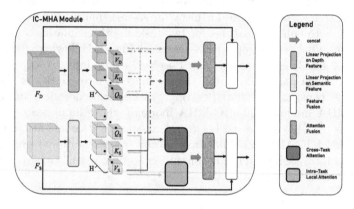

Fig. 3. An overview of our proposed IC-MHA module. Query-key-value triplet of depth and semantic feature firstly projected with Ψ_t^j. Then, intra-task local attention and cross-task attention are computed.

The overall process is shown in Fig. 4. Inspired by [33], the self-attention mechanism is applied inside locally partitioned windows to properly compute intra-task local attention. In contrast to [33], in each head, query-key-value triplet, (Q_D^h, K_D^h, V_D^h), of depth feature is partitioned by a square window of size w_h, instead of depth feature itself. Additionally, we apply windows of different sizes on different heads instead of uniform window size on all heads in [33], such that our proposed self-attention mechanism can incorporate information from various local region efficiently and effectively. Denoting partitioned query, key, value as $(\hat{Q}_D^h, \hat{K}_D^h, \hat{V}_D^h)$, the intra-task local attention is computed as:

$$F_{S_D}^h(i) = \frac{e^{(\hat{Q}_D^h(i)(\hat{K}_D^h(i))^{\mathsf{T}}/\sqrt{C'})}}{\sum_{i' < w_h^2} e^{(\hat{Q}_D^h(i')(\hat{K}_D^h(i'))^{\mathsf{T}}/\sqrt{C'})}} \cdot \hat{V}_D^h(i) \qquad (1)$$

where $i, i^{'} \in \mathbb{N}$ is the local index of feature map within one window and $C^{'} = r \cdot C$. Then, the local attention of each head is reversed back to the spatial dimension of the inputs and concatenated along channel dimension. The concatenated local attention map is projected back to the feature dimension of the inputs, C, to form the final intra-task local attention map F_{S_D}.

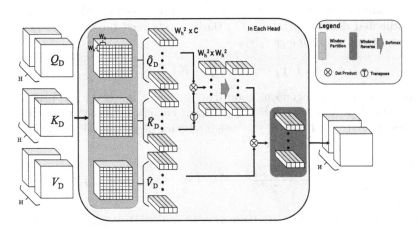

Fig. 4. The computation process of our proposed intra-task local attention. Here, we draw one head as an example. Query-key-value triplet of depth feature is firstly partitioned by a window of size w_h. Then, intra-task local attention is computed as in Eq. 1. In the end, local attention of each head is reversed back to the spatial resolution of inputs, $[H^{'}, W^{'}]$.

To represent cross-task spatial consistency, we compute cross-task attention [25] from the key-value pair of depth feature and the query of semantic feature. Here, we do not apply window partition on the input query, key and value because the purpose of addressing such consistency is to align depth boundaries with semantic boundaries. Thus, computing cross-task attention directly from query-key-value triplets of depth feature and semantic feature is more optimal. Then, for each head, such attention is computed as:

$$F_{C_D}^h(j) = \frac{e^{(Q_S^h(j)(K_D^h(j))^{\mathrm{T}}/\sqrt{C^{'}})}}{\sum_{h^{'} < H} e^{(Q_S^{h^{'}}(j)(K_D^{h^{'}}(j))^{\mathrm{T}}/\sqrt{C^{'}})}} \cdot V_D^h(j) \qquad (2)$$

where j is the spatial index of feature map and $h \in \mathbb{N}$ and $C^{'} = r \cdot C$. Then, the cross attention is summed over head h and projected back to the feature dimension of the inputs, C. The process for computing cross-task attention is visualized is in Fig 2 of Supplementary Material.

Later, a linear projection is applied on the concatenated feature $[F_{S_D}, F_{C_D}]$ to generate the final attention map, F_{A_D}. In the end, a fusion layer, consisting of two convolution layers, is implemented to incorporate attention map, F_{A_D}, with input depth feature of IC-MHA module. The output of IC-MHA module is fed into depth decoder to generate depth estimation of this level. Detailed schematic of IC-MHA module is shown in Sec 3 of Supplementary Material.

3.2   Photometric Loss and Edge-Aware Smoothness Loss

Given a pair of consecutive color images, I_s and I_t, estimated pose $T \in \mathbb{SE}(3)$ and estimated dense depth map D_t, the reconstructed target image $\hat{I}_t$ can be generated from source image I_t via:

$$\hat{I}_t(p) = I_s(\hat{p}), \quad \hat{p} = KTD_t K^{-1}p \tag{3}$$

where p is pixel's homogeneous coordinate in target image I_t, $\hat{p}$ is transformed coordinate of p, and $K \in \mathbb{R}^{3\times3}$ is a known camera intrinsic. Then, the photometric loss [16,25,42,59] is the weighted sum of structural similarity index measure(SSIM) [51] and L1-loss [17]:

$$L_{ph} = \sum_{p \in I_t}(\alpha\frac{1 - SSIM(I_s(\hat{p}), I_t(p))}{2} + (1 - \alpha)|I_s(\hat{p}) - I_t(p)|) \cdot M(p) \tag{4}$$

where $\alpha = 0.85$. Following [16], two pairs of consecutive images, $[I_{t_0}, I_{t_{-1}}]$ and $[I_{t_0}, I_{t_1}]$, are used, and minimum reprojection with auto-masking is applied, which is M in Eq. 4. To further encourage depth prediction aligned with edges of objects in an image, an edge-aware smoothness loss [16,63] is computed as:

$$L_{sm} = \sum_{p \in I_t} \sum_{i \in \{x,y\}} |\partial_i d_t^*| e^{-|\partial_i I_t|} \tag{5}$$

where $d_t^* = d_t / \bar{d}_t$ is the mean-normalized inverse depth from [47].

3.3   Hardest Non-boundary Triplet Loss with Minimum-Distance Based Candidate Mining Strategy

Although IC-MHA module identifies task-specific representational uniqueness and cross-task spatial consistency, it can only identifies task-specific representational uniqueness within instances because of its local windowed attention mechanism. Thus, further enhancement on depth feature can be achieved by creating separation in feature space between various instances using deep metric learning techniques. In [25], Jung et al. propose semantics-guided triplet loss using pseudo semantic labels. Here, we name such triplet loss as boundary triplet loss, $\mathcal{L}_{BT}$. Such boundary triplet loss is effective but not sufficient since the depth feature at non-boundary region remains unrefined. This leads to higher prediction error for pixels that are away from the boundary, shown in Fig. 8. Therefore, we propose a triplet loss that aims to fine-grain the depth feature at non-boundary region. However, where to sample an anchor in a non-boundary triplet requires careful design. Within an image, objects from the same semantic class might have various depth value, and one object can be separated or occluded, e.g., a turning car is sliced into parts by traffic signs. Moreover, how to sample a non-boundary triplet's positives and negatives should be properly handled. Considering that instances in a scene are either static or rigidly moving, depth value or feature of distant pixels in the same objects could be different or dissimilar, e.g., centers and edges of road. Thus, incorrect sampling strategy could result in overly similarity between depth feature within the same object. To overcome these two issues, we propose a *minimum-distance based candidate mining strategy* to properly sample anchors, positives and negatives for non-boundary-triplet loss.

Minimum-Distance Based Candidate Mining Strategy. To correctly mine non-boundary triplet, anchors and their positives should be sampled from the same instance, which is defined as a group of connected pixels with the same semantic labels. Such instances, denoted as $\mathcal{I}$, can be generated by applying labeling algorithm [13,53] on pixels with the same semantic label, shown as (a) in Fig. 5, and then over all semantic classes in an image. Concurrently, a boundary mask B is generated using patch-based sampling strategy[1] [25] to identify non-boundary region, shown as (b) in Fig. 5, along with boundary anchors $\mathcal{P}_B$. For each boundary anchor, $b \in \mathcal{P}_B$, we denote its positives as $\mathcal{P}_b^+$ and its negatives as $\mathcal{P}_b^-$. With these information, our sampling strategy within each instance $\mathcal{I}_i \in \mathcal{I}$ is described as follow.

1. Mask out non-boundary pixels in each instance $\mathcal{I}_i$ and randomly sample non-boundary anchors $\mathcal{P}_{NB}^i$ with $|\mathcal{P}_{NB}^i| = N_i^s$ from them.
2. Mask out boundary pixels in the same instance $\mathcal{I}_i$, denoted as $\mathcal{P}_B^i$. Here, $\mathcal{P}_B^i$ is a subset of boundary anchors $\mathcal{P}_B$.
3. For every sampled non-boundary anchor $j \in \mathcal{P}_{NB}^i$ in Step 1, find its spatially nearest boundary anchor $b_j \in \mathcal{P}_B^i$ from Step 2.
4. Then, for each non-boundary anchor $j \in \mathcal{P}_{NB}^i$, its set of positives $\mathcal{P}_j^+$ and its set of negatives $\mathcal{P}_j^-$ are sampled from the positives, $\mathcal{P}_{b_j}^+$, and negatives, $\mathcal{P}_{b_j}^-$, of its nearest boundary anchor b_j from Step 3.

Visualizations of above process are shown as (c) and (d) in Fig. 5.

In above mining process, we do not randomly sample positives from the same instance and negatives from other semantic classes because depth value within an object might change greatly. Such sampling process will lead to overly similarity between depth features within the same object. Therefore, it is more optimal to sample positives and negatives of each non-boundary anchor from its closet boundary region. The intuition behind this is that we would like to encourage non-boundary anchors' features more similar to spatially nearest boundary pixels' features and to decouple the features of non-boundary anchors from that of boundary anchors concurrently. For computational efficiency, the non-boundary anchors is sampled randomly. For each instance $\mathcal{I}_i$ at each level s, the total number of non-boundary anchors sampled is N_i^s, with $N_i^s = \frac{N_0}{4^s} \frac{|\mathcal{I}_i|}{H'W'}$. Here, N_0 is

---

[1] Detailed patch-based sampling process and $\mathcal{L}_{BT}$ [25] is in Sec 1 of Supplementary Material.

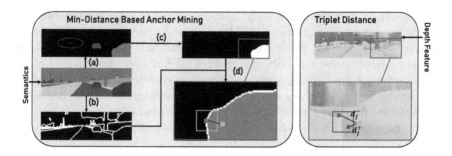

Fig. 5. An Overview of Mining strategy and triplet distance. Inside the left gray box is the minimum-distance based candidate sampling process. Here we visualize such process on one non-boundary anchor as an example: (a)generate instances; (b)generate boundary mask B; (c)Mask out one instance $\mathcal{I}_i$; (d)Randomly sample one non-boundary anchor $j \in \mathcal{P}_{NB}^i$(yellow dot), and find its closest boundary anchor b_j (green dot). The black region in blue box are the set of negatives $\mathcal{P}_j^-$, and the dark gray region in blue box are the set of positives set $\mathcal{P}_j^+$. The white curve inside of the red box is the boundary region of this instance. Inside the right gray box is the positive and negative distance in Eq. 6: hardest positive(red dot) and hardest negative (brown dot) for the non-boundary anchor(yellow dot) (Color figure online)

the total number of non-boundary anchors sampled at level $s = 0$ and $[H', W']$ is the spatial resolution of depth feature at level s.

Hardest Non-boundary Triplet Loss. For each non-boundary anchor $j \in \mathcal{P}_{NB}^i$ in instance $\mathcal{I}_i$, a set of positives $\mathcal{P}_j^+$ and a set of negatives $\mathcal{P}_j^-$ are sampled with the process described above. Inspired by [7], we select the hardest positives and negatives to compute positive distance d_j^+ and negative distance d_j^-, *i.e.* the most dissimilar positive feature in $\mathcal{P}_j^+$ and most similar negative feature in $\mathcal{P}_j^-$, shown in the right gray box in Fig. 5:

$$d_j^+ = \max_{j^+ \in P_j^+}(\|F_D(j) - F_D(j^+)\|_2), \quad d_j^- = \min_{j^- \in P_j^-}(\|F_D(j) - F_D(j^-)\|_2) \quad (6)$$

where F_D is the normalized depth feature, j^+ and j^- are the positives or negatives of non-boundary anchor i. Thus, the triplet margin loss for all non-boundary anchors $j \in \mathcal{P}_{NB}^i$ in one instance $\mathcal{I}_i$ is

$$\mathcal{L}(\mathcal{P}_{NB}^i) = \frac{\sum_{j \in \mathcal{P}_{NB}^i} \max(0, d_j^+ + m - d_j^-)}{|\mathcal{P}_{NB}^i|} \quad (7)$$

where $m = 0.3$ is the margin for feature separation. In practice, instances generated by labeling algorithm could be false because of misclassification in pseudo-labels, shown as red circle in instance mask after process (a) in Fig. 5. Thus, our final hardest non-boundary triplet loss is the mean of $L(\mathcal{P}_{NB}^i)$ over all instances,

$\mathcal{I}$, whose number of non-boundary pixels is larger than a threshold δ.

$$\mathcal{L}_{NBT} = \frac{\sum_{\mathcal{I}_i \in \mathcal{I}} \mathbb{I}\{|\mathcal{I}_i| > \delta\} \cdot L(\mathcal{P}_{NB}^i)}{\sum_{\mathcal{I}_i \in \mathcal{I}} \mathbb{I}\{|\mathcal{I}_i| > \delta\}} \tag{8}$$

where $\mathbb{I}$ is the indicator function. Our final training loss is the weighted sum of photometric loss $\mathcal{L}_{ph}$, edge-aware smoothness loss $\mathcal{L}_{sm}$, boundary triplet loss $\mathcal{L}_{BT}$, hardest non-boundary triplet loss $\mathcal{L}_{NBT}$ and semantic cross-entropy loss $\mathcal{L}_{CE}$:

$$\mathcal{L} = \sum_{s \in S}(\mathcal{L}_{ph} + \beta \cdot \mathcal{L}_{sm} + \gamma \cdot \mathcal{L}_{CE}) + \sum_{s \in S_{BT}} \eta \cdot \mathcal{L}_{BT} + \sum_{s \in S_{NBT}} \kappa \cdot \mathcal{L}_{NBT} \tag{9}$$

where β, γ, η, κ are control parameters and s represents the output level.

4   Experiments

4.1   Datasets

To ensure fair comparison with previous state-of-the-art works, we conduct experiments on widely-used KITTI dataset [15]. Following [1,16,63], we use the Eigen split [10] for depth training and evaluation, which consists of 39,810 images for training, 4,424 images for validation and 697 images for evaluation.

For the supervision of semantic segmentation, following [25], pseudo-labels for the training and validation set of Eigen split are generated using a well-trained segmentation network [65]. To evaluate semantic segmentation, the training set of the KITTI 2015 [35] is used, which contains 200 images with fine-annotated semantic labels.

4.2   Implementation Details

The encoders of DepthSegNet and PoseNet are implemented with ResNet-18 [22] with pretrained weight from ImageNet [8] loaded at initialization. For both PoseNet and DepthSegNet, input image is resized to 192×640. In addition, for fair comparison with previous state-of-the-art works, we also implement our DepthSegNet with ResNet-50 [22] whose input image is of various resolution: 192×640 and 320×1024.

For our IC-MHA module, the number of heads is set to be four($H = 4$) at each level with size of the window for each head $w_h = [2, 2, 4, 4]$. Within each head, the expansion ratio is set as: $r = 2$. We implement of our IC-MHA module at four levels, *i.e.* $s = [0, 1, 2, 3]$.

For hyperparameters of our final training loss $\mathcal{L}$ in Eq. 9, we set them as $\beta = 0.001$, $\gamma = 0.3$, $\eta = 0.1$, $\kappa = 0.1$. S and S_{BT} is set to be $\{3, 2, 1, 0\}$, and S_{NBT} is $\{1, 0\}$. Additionally, we set the threshold for pixel number in Eq. 8 as: $\delta = 80$. And the total number of non-boundary anchors at $s = 0$, *i.e.* N_0, is 8000. The non-boundary mining process is only employed only during training process, not at inference time.

During the training of our network, the data preprocessing in [16] is applied. We implement our proposed method on PyTorch [37], and the Adam optimizer [26] is used with initial learning rate as 1.5×10^{-4} for 20 epoches. At epoch 10 and 15, the learning rate is decayed to 1.5×10^{-5} and 1.5×10^{-6} respectively. The batch size for training is 12.

4.3   Evaluation Metrics

For depth prediction, we firstly set its maximum to be 80 m and conduct median-scaling using ground-truth as that in [16]. Then, the depth is evaluated by seven standard metrics [16,25,42,63], which are AbsRel, SqRel, RMSE, RMSElog, δ_1, δ_2, δ_3. For semantic segmentation, we evaluate it with the mean intersection over union(mIoU), which is the standard evaluation metric for this task.

Table 1. Comparison with recent state-of-the-art works in self-supervised monocular depth estimation. All methods are trained with monocular video sequences. Methods with (*) utilize semantic information

Methods	Input Res.	BackBone	Lower is better				Higher is better		
			AbsRel	SqRel	RMSE	RMSElog	δ_1	δ_2	δ_3
SfMLearner [63]	128 × 416	R18	0.208	1.768	6.958	0.283	0.678	0.885	0.957
SC-SfMLearner [1]	128 × 416	R18	0.137	1.089	5.439	0.217	0.830	0.942	0.975
(*)SceneNet [5]	256 × 512	DRN [56]	0.118	0.905	5.096	0.211	0.839	0.945	0.977
MonoDepth2 [16]	192 × 640	R18	0.115	0.903	4.863	0.193	0.877	0.959	0.981
(*)Guizilini *et al.* [20]	192 × 640	R18	0.117	0.854	4.714	0.191	0.873	0.963	0.981
(*)SGDepth [27]	192 × 640	R18	0.113	0.835	4.693	0.191	0.873	0.963	0.981
R-MSFM [64]	192 × 640	R18	0.112	0.806	4.704	0.191	0.878	0.960	0.981
(*)Lee *et al.* [30]	256 × 832	R18	0.112	0.777	4.772	0.191	0.872	0.959	0.982
Poggi *et al.* [40]	192 × 640	R18	0.111	0.863	4.756	0.188	0.881	0.961	0.982
Patil *et al.* [38]	192 × 640	R18	0.111	0.821	4.650	0.187	0.883	0.961	0.982
(*)SAFENet [6]	192 × 640	R18	0.112	0.788	4.582	0.187	0.878	0.963	0.983
Zhao *et al.* [59]	256 × 832	R18	0.113	0.704	4.581	0.184	0.871	0.961	0.984
HRDepth [34]	192 × 640	R18	0.109	0.792	4.632	0.185	0.884	0.962	0.983
Wang *et al.* [49]	192 × 640	R18	0.109	0.779	4.641	0.186	0.883	0.962	0.982
(*)FSRE [25][†]	192 × 640	R18	0.107	0.730	4.530	0.182	**0.886**	0.964	**0.984**
(*)FSRE [25]	192 × 640	R18	0.105	0.722	4.547	0.182	**0.886**	0.964	**0.984**
(*)**Ours**	192 × 640	R18	**0.104**	**0.690**	**4.473**	**0.179**	**0.886**	**0.965**	**0.984**
(*)SGDepth [27]	192 × 640	R50	0.112	0.833	4.688	0.190	0.884	0.961	0.981
(*)Guizilini *et al.* [20]	192 × 640	R50	0.113	0.831	4.663	0.189	0.878	**0.971**	0.983
MonoDepth2 [16]	192 × 640	R50	0.110	0.831	4.642	0.187	0.883	0.962	0.982
(*)Li *et al.* [31]	192 × 640	R50	0.103	0.709	4.471	0.180	0.892	0.966	0.984
(*)FSRE [25]	192 × 640	R50	**0.102**	0.675	4.393	0.178	**0.893**	0.966	0.984
(*)**Ours**	192 × 640	R50	**0.102**	**0.656**	**4.339**	**0.175**	0.892	**0.967**	**0.985**
PackNet [19]	375 × 1224	PackNet	0.104	0.758	4.386	0.182	0.895	0.964	0.982
FeatDepth [42]	320 × 1024	R50	0.104	0.729	4.481	0.179	0.893	0.965	0.984
(*)Guizilini *et al.* [20]	375 × 1224	PackNet	0.100	0.761	4.270	0.175	**0.902**	0.965	0.982
(*)**Ours**	320 × 1024	R50	**0.099**	**0.624**	**4.165**	**0.171**	**0.902**	**0.969**	**0.986**

[†] We re-trained [25] with its official implementation, since no pretrained model is available.

4.4 Experiment Results and Ablation Study

Comparison with Previous State-of-the-art Methods. Comparison with recent state-of-the-art results is shown in Table 1. The table shows that our method achieves the state-of-the-art performance on KITTI Eigen test split. Specifically, our proposed method outperforms previous state-of-the-art work significantly on SqRel, RMSE, RMSElog. For AbsRel, δ_1, δ_2, δ_3, our proposed method yields comparable or better performance than previous state-of-the-art methods. In addition, our method with low input resolution and lighter backbone outperforms some previous state-of-the-art approaches with higher resolution [30,59] or heavier backbone [19,20,27]. See Table 4 for detailed timing and parameter number of ours and previous methods. The testing device is NVidia V100 GPU. Furthermore, our proposed method with high resolution input and deeper network achieves significant improvement over previous state-of-the-art methods, which indicates that our proposed approach gains performance boost with better backbone network(ResNet-50).

Table 2. Ablations on proposed IC-MHA module and non-boundary triplet loss $\mathcal{L}_{NBT}$. IC-MHA* represents IC-MHA module with intra-task attention only.

CMA	$\mathcal{L}_{BT}$	IC-MHA*	IC-MHA(All)	$\mathcal{L}_{NBT}$	AbsRel	SqRel	RMSE	RMSElog	δ_1	δ_2	δ_3
✓	✓				0.107	0.730	4.530	0.182	0.886	0.964	**0.984**
✓	✓	✓			0.106	0.731	4.527	0.181	0.886	0.964	**0.984**
✓			✓		0.105	0.734	4.516	0.180	**0.887**	**0.965**	**0.984**
✓			✓	✓	**0.104**	**0.690**	**4.473**	**0.179**	0.886	**0.965**	**0.984**

Table 3. The segmentation result of proposed methods against baseline [25]

Methods	mIoU
FSRE† [25]	55.8
Ours	56.3

Table 4. Inference time and parameter number of ours against previous methods

	Time(ms)	Param.#(M)
FSRE [25](R18)	10	28.6M
Ours(R18)	12	30.3M
Ours(R50)	31	45.5M
[20] (PackNet)	60	70M

Table 5. Ablations on Different Loss term on proposed IC-MHA module.

$\mathcal{L}_{ph} + \mathcal{L}_{sm}$	$\mathcal{L}_{CE}$	$\mathcal{L}_{BT}$	$\mathcal{L}_{NBT}$	AbsRel	SqRel	RMSE	RMSElog	δ_1	δ_2	δ_3
✓	✓			0.110	0.794	4.610	0.187	0.879	0.962	0.982
✓	✓	✓		0.105	0.734	4.516	0.180	**0.887**	**0.965**	**0.984**
✓	✓		✓	0.105	0.727	4.498	0.180	0.885	**0.965**	**0.984**
✓	✓	✓	✓	**0.104**	**0.690**	**4.473**	**0.179**	0.886	**0.965**	**0.984**

Ablation Study. Ablations on our proposed IC-MHA module and our non-boundary triplet loss $\mathcal{L}_{NBT}$ is shown in Table 2. We compare our proposed methods with our baseline [25], which consists of CMA module and boundary triplet loss $\mathcal{L}_{BT}$ to emphasize cross-task correlation. Since no pretrained model is available on the official implementation of [25], we re-train the model using its official implementation for multiple times and take the best result. The experiment result verifies the effectiveness of our proposed IC-MHA module and non-boundary triplet loss $\mathcal{L}_{NBT}$. Also, the result in Table 2 shows effectiveness of our proposed intra-task local attention. To further demonstrate the effectiveness of our proposed method, the visualization of feature heatmap of IC-MHA module and CMA module is shown in Fig. 6. Such visualization is generated via PCA decomposition by normalizing the summation of top principle channels, who contributes 90% totally to feature map. It shows that our refined depth feature is much more smoothing within instances, and it is more aligned with the actual depth distribution. The heatmap of cross-task attention and intra-task attention is shown in Fig. 7. It suggests that cross-task attention is consistent with semantic feature, while intra-task attention is consistent with depth feature. The ablation study on different loss term on IC-MHA module is shown in Table 5. It shows that IC-MHA module gains performance boost from $\mathcal{L}_{BT}$ or $\mathcal{L}_{NBT}$. These two loss terms together achieve best result.

(a)   (b)   (c)

Fig. 6. Depth feature heatmap visualization: (a) Colored image. (b) Depth feature heatmap of IC-MHA module. (c) Depth feature heatmap of CMA module [25].

Additionally, in Table 3, we compare the segmentation result of our methods with that of baseline. The mean intersection-of-union (mIoU) of proposed method is better than that of baseline by 0.5. This verifies that addressing representational uniqueness of task-specific feature in IC-MHA module can improve prediction of both tasks: depth and semantic segmentation. Extra ablation study on hyperparameters of IC-MHA module and $\mathcal{L}_{NBT}$ is included in Sec 5 of Supplementary Material.

Fig. 7. Heatmap of cross-task attention feature (middle) and intra-task attention feature (right)

4.5   Qualitative Results

Qualitative comparison between our proposed method and previous state-of-the-art work, FSRE [25] is shown in Fig. 8. Error distribution[2] uses absolute error between reference depth and predicted depth. The maximum of the absolute error map is set to be 10, and then it is rescaled to $[0, 1]$. The figure proves that our proposed method not only preserves object boundaries but improves estimation at non-boundary region of each object as well. More qualitative examples are shown in Supplementary Material.

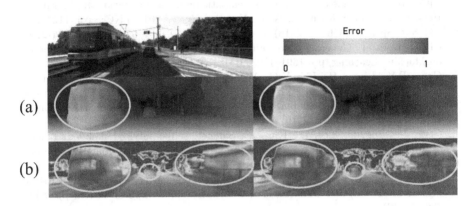

Fig. 8. Qualitative results of depth estimation. (a) Depth output of [25] (left) and ours (right). (b) Error map of [25] (left) and ours (right).

5   Conclusions

In this work, we propose a novel method for self-supervised monocular depth estimation by emphasizing task-specific uniqueness in feature space. Specifically, our proposed IC-MHA module exploits more fine-grained features by addressing representational uniqueness of task-specific feature within instances in parallel with cross-task spatial consistency. Additionally, the proposed hardest non-boundary triplet loss further enhances depth feature by addressing its uniqueness between instances, which is a full refinement on depth feature of all regions in an image. Our whole method is end-to-end trainable and achieves state-of-the-art performance on KITTI Eigen test split.

Acknowledgments. This work is supported by National Key R&D Program of China (Grant No. 2020AAA010400X).

---

[2] Considering ground-truth depth is sparse, we use estimation of top-performance supervised depth network [29] as reference.

References

1. Bian, J., Li, Z., Wang, N., Zhan, H., Shen, C., Cheng, M.M., Reid, I.: Unsupervised scale-consistent depth and ego-motion learning from monocular video. In: Advances in Neural Information Processing Systems, vol. 32. Curran Associates, Inc. (2019)

2. Carion, N., Massa, F., Synnaeve, G., Usunier, N., Kirillov, A., Zagoruyko, S.: End-to-end object detection with transformers. In: Vedaldi, A., Bischof, H., Brox, T., Frahm, J.-M. (eds.) ECCV 2020. LNCS, vol. 12346, pp. 213–229. Springer, Cham (2020). https://doi.org/10.1007/978-3-030-58452-8_13

3. Chen, L.-C., Zhu, Y., Papandreou, G., Schroff, F., Adam, H.: Encoder-decoder with atrous separable convolution for semantic image segmentation. In: Ferrari, V., Hebert, M., Sminchisescu, C., Weiss, Y. (eds.) ECCV 2018. LNCS, vol. 11211, pp. 833–851. Springer, Cham (2018). https://doi.org/10.1007/978-3-030-01234-2_49

4. Chen, M., et al.: Generative pretraining from pixels. In: International Conference on Machine Learning, pp. 1691–1703. PMLR (2020)

5. Chen, P.Y., Liu, A.H., Liu, Y.C., Wang, Y.C.F.: Towards scene understanding: unsupervised monocular depth estimation with semantic-aware representation. In: 2019 IEEE/CVF Conference on Computer Vision and Pattern Recognition (CVPR), pp. 2619–2627 (2019)

6. Choi, J., Jung, D., Lee, D., Kim, C.: Safenet: self-supervised monocular depth estimation with semantic-aware feature extraction. In: Thirty-fourth Conference on Neural Information Processing Systems, NIPS 2020. NeurIPS (2020)

7. Choy, C., Park, J., Koltun, V.: Fully convolutional geometric features. In: Proceedings of the IEEE/CVF International Conference on Computer Vision, pp. 8958–8966 (2019)

8. Deng, J., Dong, W., Socher, R., Li, L.J., Li, K., Fei-Fei, L.: Imagenet: a large-scale hierarchical image database. In: 2009 IEEE Conference on Computer Vision and Pattern Recognition, pp. 248–255. IEEE (2009)

9. Dosovitskiy, A., et al.: An image is worth 16x16 words: transformers for image recognition at scale. In: ICLR (2021)

10. Eigen, D., Puhrsch, C., Fergus, R.: Depth map prediction from a single image using a multi-scale deep network. In: Advances in Neural Information Processing Systems, vol. 27 (2014)

11. Engel, J., Koltun, V., Cremers, D.: Direct sparse odometry. IEEE Trans. Pattern Anal. Mach. Intell. **40**(3), 611–625 (2017)

12. Farooq Bhat, S., Alhashim, I., Wonka, P.: Adabins: depth estimation using adaptive bins. In: 2021 IEEE/CVF Conference on Computer Vision and Pattern Recognition (CVPR), pp. 4008–4017 (2021)

13. Fiorio, C., Gustedt, J.: Two linear time union-find strategies for image processing. Theoret. Comput. Sci. **154**(2), 165–181 (1996)

14. Fu, H., Gong, M., Wang, C., Batmanghelich, K., Tao, D.: Deep ordinal regression network for monocular depth estimation. In: IEEE Conference on Computer Vision and Pattern Recognition (CVPR) (2018)

15. Geiger, A., Lenz, P., Urtasun, R.: Are we ready for autonomous driving? The Kitti vision benchmark suite. In: Conference on Computer Vision and Pattern Recognition (CVPR) (2012)

16. Godard, C., Aodha, O.M., Firman, M., Brostow, G.: Digging into self-supervised monocular depth estimation. In: 2019 IEEE/CVF International Conference on Computer Vision (ICCV), pp. 3827–3837 (2019)

17. Godard, C., Mac Aodha, O., Brostow, G.J.: Unsupervised monocular depth estimation with left-right consistency. In: CVPR (2017)
18. Goel, K., Srinivasan, P., Tariq, S., Philbin, J.: Quadronet: Multi-task learning for real-time semantic depth aware instance segmentation. In: Proceedings of the IEEE/CVF Winter Conference on Applications of Computer Vision (WACV), pp. 315–324, January 2021
19. Guizilini, V., Ambrus, R., Pillai, S., Raventos, A., Gaidon, A.: 3D packing for self-supervised monocular depth estimation. In: IEEE Conference on Computer Vision and Pattern Recognition (CVPR) (2020)
20. Guizilini, V., Hou, R., Li, J., Ambrus, R., Gaidon, A.: Semantically-guided representation learning for self-supervised monocular depth. In: International Conference on Learning Representations (ICLR), April 2020
21. Han, K., et al.: A survey on vision transformer. IEEE Trans. Pattern Anal. Mach. Intell. 1–1 (2022)
22. He, K., Zhang, X., Ren, S., Sun, J.: Deep residual learning for image recognition. In: Proceedings of the IEEE Conference on Computer Vision and Pattern Recognition, pp. 770–778 (2016)
23. Hu, J., Lu, J., Tan, Y.P.: Discriminative deep metric learning for face verification in the wild. In: 2014 IEEE Conference on Computer Vision and Pattern Recognition, pp. 1875–1882 (2014)
24. Jiang, H., Ding, L., Sun, Z., Huang, R.: Unsupervised monocular depth perception: focusing on moving objects. IEEE Sens. J. **21**(24), 27225–27237 (2021)
25. Jung, H., Park, E., Yoo, S.: Fine-grained semantics-aware representation enhancement for self-supervised monocular depth estimation. In: Proceedings of the IEEE/CVF International Conference on Computer Vision (ICCV), pp. 12642–12652, October 2021
26. Kingma, D.P., Ba, J.: Adam: a method for stochastic optimization. arXiv preprint arXiv:1412.6980 (2014)
27. Klingner, M., Termöhlen, J.-A., Mikolajczyk, J., Fingscheidt, T.: Self-supervised monocular depth estimation: solving the dynamic object problem by semantic guidance. In: Vedaldi, A., Bischof, H., Brox, T., Frahm, J.-M. (eds.) ECCV 2020. LNCS, vol. 12365, pp. 582–600. Springer, Cham (2020). https://doi.org/10.1007/978-3-030-58565-5_35
28. Kulis, B.: Metric learning: a survey. Found. TrendsR Mach. Learn. **5**(4), 287–364 (2012). https://doi.org/10.1561/2200000019
29. Lee, J.H., Han, M.K., Ko, D.W., Suh, I.H.: From big to small: multi-scale local planar guidance for monocular depth estimation. arXiv preprint arXiv:1907.10326 (2019)
30. Lee, S., Im, S., Lin, S., Kweon, I.S.: Learning monocular depth in dynamic scenes via instance-aware projection consistency. In: Proceedings of the AAAI Conference on Artificial Intelligence (AAAI) (2021)
31. Li, R., et al.: Learning depth via leveraging semantics: self-supervised monocular depth estimation with both implicit and explicit semantic guidance (2021)
32. Li, R., et al.: Semantic-guided representation enhancement for self-supervised monocular trained depth estimation (2020)
33. Liu, Z., et al.: Swin transformer: hierarchical vision transformer using shifted windows (2021)
34. Lyu, X., et al.: HR-depth: High resolution self-supervised monocular depth estimation. arXiv preprint arXiv:2012.07356 6 (2020)
35. Menze, M., Geiger, A.: Object scene flow for autonomous vehicles. In: Conference on Computer Vision and Pattern Recognition (CVPR) (2015)

36. Mur-Artal, R., Montiel, J.M.M., Tardós, J.D.: Orb-slam: a versatile and accurate monocular slam system. IEEE Trans. Rob. **31**(5), 1147–1163 (2015)
37. Paszke, A., et al.: Automatic differentiation in pytorch (2017)
38. Patil, V., Van Gansbeke, W., Dai, D., Van Gool, L.: Don't forget the past: recurrent depth estimation from monocular video. IEEE Robot. Autom. Lett. **5**(4), 6813–6820 (2020)
39. Pire, T., Fischer, T., Castro, G., De Cristóforis, P., Civera, J., Jacobo Berlles, J.: S-PTAM: stereo parallel tracking and mapping. Robot. Auton. Syst. (RAS) **93**, 27–42 (2017)
40. Poggi, M., Aleotti, F., Tosi, F., Mattoccia, S.: On the uncertainty of self-supervised monocular depth estimation. In: 2020 IEEE/CVF Conference on Computer Vision and Pattern Recognition (CVPR), pp. 3224–3234 (2020)
41. Ranftl, R., Lasinger, K., Hafner, D., Schindler, K., Koltun, V.: Towards robust monocular depth estimation: mixing datasets for zero-shot cross-dataset transfer. IEEE Trans. Pattern Anal. Mach. Intelligence (TPAMI) **44**, 1623–1637 (2020)
42. Shu, C., Yu, K., Duan, Z., Yang, K.: Feature-metric loss for self-supervised learning of depth and egomotion. In: Vedaldi, A., Bischof, H., Brox, T., Frahm, J.-M. (eds.) ECCV 2020. LNCS, vol. 12364, pp. 572–588. Springer, Cham (2020). https://doi.org/10.1007/978-3-030-58529-7_34
43. Simo-Serra, E., Trulls, E., Ferraz, L., Kokkinos, I., Fua, P., Moreno-Noguer, F.: Discriminative learning of deep convolutional feature point descriptors. In: 2015 IEEE International Conference on Computer Vision (ICCV), pp. 118–126 (2015)
44. Song, H.O., Jegelka, S., Rathod, V., Murphy, K.: Deep metric learning via facility location. In: 2017 IEEE Conference on Computer Vision and Pattern Recognition (CVPR), pp. 2206–2214 (2017)
45. Sun, Z., Cao, S., Yang, Y., Kitani, K.M.: Rethinking transformer-based set prediction for object detection. In: Proceedings of the IEEE/CVF International Conference on Computer Vision, pp. 3611–3620 (2021)
46. Vaswani, A., et al.: Attention is all you need. In: Proceedings of the 31st International Conference on Neural Information Processing Systems. NIPS 2017, Red Hook, NY, USA, pp. 6000–6010. Curran Associates Inc. (2017)
47. Wang, C., Buenaposada, J.M., Zhu, R., Lucey, S.: Learning depth from monocular videos using direct methods. In: Proceedings of the IEEE Conference on Computer Vision and Pattern Recognition, pp. 2022–2030 (2018)
48. Wang, J., Zhou, F., Wen, S., Liu, X., Lin, Y.: Deep metric learning with angular loss. In: 2017 IEEE International Conference on Computer Vision (ICCV), pp. 2612–2620 (2017)
49. Wang, L., Wang, Y., Wang, L., Zhan, Y., Wang, Y., Lu, H.: Can scale-consistent monocular depth be learned in a self-supervised scale-invariant manner? In: Proceedings of the IEEE/CVF International Conference on Computer Vision (ICCV). pp. 12727–12736 (October 2021)
50. Wang, Y., et al.: End-to-end video instance segmentation with transformers. In: Proceedings of the IEEE/CVF Conference on Computer Vision and Pattern Recognition, pp. 8741–8750 (2021)
51. Wang, Z., Bovik, A., Sheikh, H., Simoncelli, E.: Image quality assessment: from error visibility to structural similarity. IEEE Trans. Image Process. **13**(4), 600–612 (2004)
52. Watson, J., Mac Aodha, O., Prisacariu, V., Brostow, G., Firman, M.: The temporal opportunist: self-supervised multi-frame monocular depth. In: 2021 IEEE/CVF Conference on Computer Vision and Pattern Recognition (CVPR), pp. 1164–1174 (2021)

53. Wu, K., Otoo, E., Shoshani, A.: Optimizing connected component labeling algorithms. In: Medical Imaging 2005: Image Processing, vol. 5747, pp. 1965–1976. SPIE (2005)
54. Yi, K.M., Trulls, E., Lepetit, V., Fua, P.: LIFT: learned invariant feature transform. In: Leibe, B., Matas, J., Sebe, N., Welling, M. (eds.) ECCV 2016. LNCS, vol. 9910, pp. 467–483. Springer, Cham (2016). https://doi.org/10.1007/978-3-319-46466-4_28
55. Yin, Z., Shi, J.: Geonet: Unsupervised learning of dense depth, optical flow and camera pose. In: CVPR (2018)
56. Yu, F., Koltun, V., Funkhouser, T.: Dilated residual networks. In: 2017 IEEE Conference on Computer Vision and Pattern Recognition (CVPR), pp. 636–644 (2017)
57. Zhan, H., Garg, R., Weerasekera, C.S., Li, K., Agarwal, H., Reid, I.M.: Unsupervised learning of monocular depth estimation and visual odometry with deep feature reconstruction. In: 2018 IEEE/CVF Conference on Computer Vision and Pattern Recognition, pp. 340–349 (2018)
58. Zhao, H., Jia, J., Koltun, V.: Exploring self-attention for image recognition. In: CVPR (2020)
59. Zhao, W., Liu, S., Shu, Y., Liu, Y.J.: Towards better generalization: joint depth-pose learning without posenet. In: Proceedings of IEEE Conference on Computer Vision and Pattern Recognition (CVPR) (2020)
60. Zheng, M., et al.: End-to-end object detection with adaptive clustering transformer. arXiv preprint arXiv:2011.09315 (2020)
61. Zheng, S., et al.: Rethinking semantic segmentation from a sequence-to-sequence perspective with transformers. In: Proceedings of the IEEE/CVF Conference on Computer Vision and Pattern Recognition, pp. 6881–6890 (2021)
62. Zhou, H., Greenwood, D., Taylor, S.: Self-supervised monocular depth estimation with internal feature fusion. In: British Machine Vision Conference (BMVC) (2021)
63. Zhou, T., Brown, M., Snavely, N., Lowe, D.G.: Unsupervised learning of depth and ego-motion from video. In: 2017 IEEE Conference on Computer Vision and Pattern Recognition (CVPR), pp. 6612–6619 (2017)
64. Zhou, Z., Fan, X., Shi, P., Xin, Y.: R-MSFM: Recurrent multi-scale feature modulation for monocular depth estimating. In: Proceedings of the IEEE/CVF International Conference on Computer Vision (ICCV). pp. 12777–12786 (October 2021)
65. Zhu, Y., et al.: Improving semantic segmentation via video propagation and label relaxation. In: 2019 IEEE/CVF Conference on Computer Vision and Pattern Recognition (CVPR), pp. 8848–8857 (2019)
66. Zou, Y., Luo, Z., Huang, J.-B.: DF-Net: unsupervised joint learning of depth and flow using cross-task consistency. In: Ferrari, V., Hebert, M., Sminchisescu, C., Weiss, Y. (eds.) ECCV 2018. LNCS, vol. 11209, pp. 38–55. Springer, Cham (2018). https://doi.org/10.1007/978-3-030-01228-1_3

AvatarCap: Animatable Avatar Conditioned Monocular Human Volumetric Capture

Zhe Li$^{(\boxtimes)}$, Zerong Zheng, Hongwen Zhang, Chaonan Ji, and Yebin Liu

Department of Automation, Tsinghua University, Beijing, China
`liz19@mails.tsinghua.edu.cn`

Abstract. To address the ill-posed problem caused by partial observations in monocular human volumetric capture, we present AvatarCap, a novel framework that introduces animatable avatars into the capture pipeline for high-fidelity reconstruction in both visible and invisible regions. Our method firstly creates an animatable avatar for the subject from a small number ($\sim$20) of 3D scans as a prior. Then given a monocular RGB video of this subject, our method integrates information from both the image observation and the avatar prior, and accordingly reconstructs high-fidelity 3D textured models with dynamic details regardless of the visibility. To learn an effective avatar for volumetric capture from only few samples, we propose GeoTexAvatar, which leverages both geometry and texture supervisions to constrain the pose-dependent dynamics in a decomposed implicit manner. An avatar-conditioned volumetric capture method that involves a canonical normal fusion and a reconstruction network is further proposed to integrate both image observations and avatar dynamics for high-fidelity reconstruction in both observed and invisible regions. Overall, our method enables monocular human volumetric capture with detailed and pose-dependent dynamics, and the experiments show that our method outperforms state of the art.

1 Introduction

Human volumetric capture has been a popular research topic in computer vision for decades due to its potential value in Metaverse, holographic communication, video games, etc. Multi-view systems [4,5,10,12,24,29,37,38,47,51,55,62,71,73, 77] can reconstruct high-resolution 3D human models using multiple RGB(D) sensors, but the sophisticated setup restricts their deployment in practice. To overcome this limitation, researchers have developed various technologies for monocular human reconstruction based on template tracking [16,19,20,84], volumetric fusion [49,60,74] or single-image reconstruction [22,25,33,52,53,81].

Supplementary Information The online version contains supplementary material available at https://doi.org/10.1007/978-3-031-19769-7_19.

Fig. 1. Overview of AvatarCap. We present AvatarCap that leverages an animatable avatar learned from only a small number (~20) of scans for monocular human
volumetric capture to realize high-fidelity reconstruction regardless of the visibility.

Despite the rapid development in monocular volumetric capture, most of the
existing methods mainly focus on reconstructing visible surfaces according to
direct observations and fail to recover the dynamic details in invisible regions.
POSEFusion [35] addressed this limitation via integrating keyframes of similar
poses from the whole RGBD sequence for invisible region reconstruction. However, it requires the subject to perform similar motions for multiple times facing
different directions. What's worse, the fused invisible details are copied unaltered
from other depth frames, thus suffering from poor pose generalization.

How to recover temporally coherent and pose-dependent details on invisible surfaces is an urgent and essential problem in monocular human voluemtric
capture. Recently, many works on pose-driven human avatars have arisen in the
community. They create animatable avatars from various inputs, including scans
[7,8,41,43,54], multi-view RGB videos [36,50] and monocular depth measurements [6,65]. In this paper, our key insight is that the pose-driven dynamics of
person-specific avatars are exactly what is missing in monocular human volumetric capture. With this in mind, we propose *AvatarCap*, the first pipeline
that combines person-specific animatable avatars with monocular human volumetric capture. Intuitively, the avatar encodes a data-driven prior about the
pose-dependent dynamic details, which can compensate for the lack of complete observation in monocular inputs, enabling high-quality reconstruction of
3D models with dynamic details regardless of visibility.

Although introducing person-specific avatars into volumetric capture adds
overhead in pipeline preparation, we believe that a data-driven prior of pose-
dependent dynamics is indispensable for the future dynamic monocular human
capture. In this paper, to make a trade-off between the ease of data acquisition
and reconstruction quality, we choose to use only a small number (~20) of textured scans as the database. Note that it is challenging to learn a generalized
avatar from only few scans, and state-of-the-art methods typically require hundreds of scans for creating one avatar [43,54]. If only twenty scans are available,
their results tend to be overfit and lack geometric details because they condition
all the surface details (including pose-dependent and pose-agnostic ones) on the
pose input. To address this challenge, we propose *GeoTexAvatar*, a decom-

posed representation that guarantees detail representation power and generalization capability. To be more specific, our representation distills pose-agnostic details as much as possible into a common implicit template [78], and models the remaining pose-driven dynamics with a pose-conditioned warping field. Such a disentanglement promotes better generalization since a large portion of geometric details are factored out as the common template and consequently the pose-dependent warping field is much easier to learn. On the other hand, previous methods rely on solely geometric cues to learn the conditional warping fields [78], but we find that it is not enough because many types of cloth dynamics (e.g., cloth sliding) cannot be supervised by only geometry due to the ambiguity when establishing geometric correspondences. Therefore, we introduce an extra texture template represented by NeRF [46] to jointly constrain the pose-dependent warping field using both geometry and texture supervisions, which makes it possible to learn an accurate pose-conditioned warping field. As a result, the proposed GeoTexAvatar can not only preserve more details but also produce more reasonable pose-dependent dynamics for animation.

However, it is still not trivial to leverage the animatable avatar in the monocular capture pipeline. The main reason is the huge domain gap between the avatar prior and the monocular color input without any explicit 3D information. Fortunately, a 2D normal map with plentiful details can be extracted from the monocular color image [53], and we can use it to bridge the 3D avatar and the 2D RGB input. However, directly optimizing the avatar geometry using extremely dense non-rigid deformation [61] to fit the 2D normal map is difficult, if not infeasible, because it is ill-posed to force the surface to be consistent with the normal map without explicit 3D correspondences. To overcome this challenge, we propose *Avatar-conditioned Volumetric Capture* that splits the integration between the avatar and the normal maps into two steps, i.e., canonical normal fusion and model reconstruction. Specifically, the canonical normal fusion integrates the avatar normal and the image-observed one on the unified 2D canonical image plane. In this procedure, we formulate the fusion as an optimization on both the rotation grids and the normal maps to correct low-frequency normal orientation errors caused by inaccurate SMPL [39] fitting while maintaining high-frequency details. After that, a reconstruction network pretrained on a large-scale 3D human dataset [73] is used as a strong prior for producing a high-fidelity 3D human with full-body details from the fused normal maps.

This paper proposes the following contributions: 1) AvatarCap, a new framework that introduces animatable avatars into the monocular human volumetric capture pipeline to achieve detailed and dynamic capture regardless of the visibility (Sect. 3). 2) GeoTexAvatar, a new decomposed avatar representation that contains a pose-agnostic Geo-Tex implicit template and a pose-dependent warping field to jointly constrain the pose-dependent dynamics using both geometry and texture supervisions for more detailed and well-generalized animation (Sect. 4). 3) Avatar-conditioned volumetric capture that contains a canonical normal fusion method and a reconstruction network to overcome the domain gap between the avatar prior and the monocular input for full-body high-fidelity reconstruction (Sect. 5). Code is available at https://github.com/lizhe00/AvatarCap.

2 Related Work

Template Tracking. Given a monocular RGB(D) video, many works utilize a template to fit each frame using skeletal motion [44] or non-rigid deformation [61]. Specifically, [16,31,84] solved the non-rigid warp field to track the input depth stream, while [15,23,70,82] tracked the skeletal motion of the template to fit the monocular input. LiveCap [19] and DeepCap [20] jointly solved or inferred both skeletal and non-rigid motions from a monocular RGB video. MonoClothCap [68] built a statistical deformation model based on SMPL to capture visible cloth dynamics. However, these methods only focus on fitting the template to explain the image observation while neglecting the dynamics in invisible regions.

Volumetric Fusion. Meanwhile, to realize real-time reconstruction from a single depth sensor, Newcombe *et al.* [49] pioneered to propose DynamicFusion that tracks and completes a canonical model in an incremental manner. It inspired a lot of following works [17,26,30,56,57,60,72,74,79] to incorporate different body priors or other cues to improve the performance. However, similar to methods based on template tracking, these works do not take into account the dynamic deformations in invisible regions. SimulCap [75] introduced cloth simulation into the volumetric fusion pipeline but its reconstruction quality is limited by a simple cloth simulator. POSEFusion [35] proposed to integrate multiple keyframes of similar poses to recover the dynamic details for the whole body, but this scheme leads to poor pose generalization, i.e., only those poses that are seen in different frames can be faithfully reconstructed.

Single-Image Reconstruction. Recently, researchers have paid more and more attention to recovering 3D humans from single RGB(D) images by volume regression [27,63,81], visual hull [48], depth maps [11,58], template deformation [1,83] and implicit functions [21,22,25,33,52,69]. For the implicit function representation, PIFuHD [53] introduced normal estimation to produce detailed geometry. PaMIR [80] and IPNet [3] combined a parametric body model (e.g., SMPL [39]) into the implicit function to handle challenging poses. However, without direct observation, these methods only recover over-smoothed invisible geometry without details. NormalGAN [64] inferred the back-view RGBD image from the input RGBD using a GAN [13] and then seamed them together. Unfortunately, the inferred details may be inconsistent with the pose or cloth type due to the limited variation in training data.

Animatable Human Avatar. To create animatable human avatars, previous methods usually reconstruct a template and then model the pose-dependent dynamics of the character by physical simulation [14,59] or deep learning [2, 18,67]. Recent works proposed to directly learn an animatable avatar from the

database, including scans [7,41–43,54], multi-view RGB videos [36,50] and depth
frames [6,9,65]. These works usually require a large amount of data to train a
person-specific avatar; when only a small number of scans are available, they
suffer from overfitting and struggle with pose generalization. Wang *et al.* [65]
learned a meta prior to overcome this issue, but it remains difficult to apply
their method for texture modeling.

3 Overview

As shown in Fig. 1, the whole framework of AvatarCap contains two main steps:

1. **Avatar Creation.** Before performing monocular volumetric capture, we collect a small number (∼20) of textured scans for the subject as the database to
   construct his/her animatable avatar, which will be used to facilitate dynamic
   detail capture. To create an avatar with realistic details and generalization
   capability, we propose GeoTexAvatar, a representation that decomposes the
   dynamic level set function [54] into an implicit template (including occupancy [45] and radiance [46] fields) and a pose-dependent warping field, as
   shown in Fig. 2. We train the GeoTexAvatar network by supervising both the
   geometry and the texture using the textured scans.
2. **Avatar-conditioned Volumetric Capture.** With the avatar prior, we perform volumetric capture given the monocular RGB video input, as illustrated
   in Fig. 3. To address the domain gap between the avatar and the RGB input,
   we propose to use the surface normals as the intermediate to bridge each other.
   Specifically, we firstly estimate the visible normals from each RGB image,
   which are then mapped to the canonical space using the estimated SMPL
   pose [28,76]. Then we generate the canonical avatar with pose-dependent
   dynamics given the pose and render the canonical normal maps from both
   the front and back views. The next step is to integrate the rendered normal
   maps with their image-based counterparts. To do so, we propose canonical
   normal fusion, which aims to correct low-frequency local normal orientations
   while maintaining high-frequency details from image observations. Finally, a
   pretrained reconstruction network is used to produce a high-fidelity human
   model conditioned on the integrated normal maps.

4 Avatar Creation

In this section, our goal is to learn an animatable avatar for volumetric capture.
Following the practice of SCANimate [54], we fit SMPL to the raw 3D scans and
transform them to a canonical pose via inverse skinning. We aim to construct
an animatable avatar, represented as a pose-conditioned implicit function, from
these canonicalized scans. Since only a small number (∼20) of textured scans are
available, we propose a decomposed implicit function to guarantee representation
power and generalization capability (Sect. 4.1), which allows us to better leverage
the geometry and texture information of training data (Sect. 4.2).

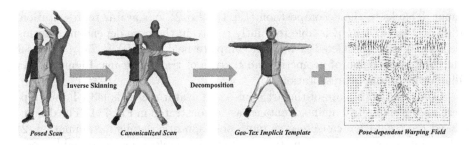

Posed Scan   Canonicalized Scan   Geo-Tex Implicit Template   Pose-dependent Warping Field

Fig. 2. Illustration of the GeoTexAvatar representation. We decompose the canonical scans into a pose-agnostic Geo-Tex implicit template and a pose-dependent warping field to enable joint supervisions by both geometry and texture for more detailed and well-generalized animation.

4.1 GeoTexAvatar Representation

Our representation is built upon the pose-conditioned implicit function in SCANimate [54], which is defined as $f(\mathbf{x}_c, \boldsymbol{\theta}) = s$, where $s \in [0, 1]$ is an occupancy value, $\mathbf{x}_c$ is a 3D point in the canonical space and $\boldsymbol{\theta}$ is the SMPL pose parameters. The pose-dependent surface is represented by the level set of this implicit function: $f(\mathbf{x}_c, \boldsymbol{\theta}) = 0.5$. However, such an entangled representation conditions all the surface dynamics, including the pose-dependent deformations and the pose-agnostic details, on the pose input. Consequently, the animation results tend to lack pose-agnostic details when given an unseen pose.

In order to not only model the pose-dependent deformations but also preserve the pose-agnostic details among different training samples, we propose a decomposed representation based on [78]:

$$T_{\mathrm{Geo}}(W(\mathbf{x}_c, \boldsymbol{\theta})) = s, \qquad (1)$$

where $W(\mathbf{x}_c, \boldsymbol{\theta}) = \mathbf{x}_c + \Delta W(\mathbf{x}, \boldsymbol{\theta})$ represents the pose-dependent warping field that takes the pose parameters and a point as input and returns its template position, and $T_{\mathrm{Geo}}(\cdot)$ is the pose-agnostic occupancy template.

Note that previous avatars learned from scans [41,43,54] ignore the texture information even though their databases contain texture. However, we find the texture is essential to constrain the pose-dependent cloth deformations, because only geometrically closest constraints cannot establish correct correspondences, especially for common tangential cloth motions (e.g., cloth sliding). Therefore, we further introduce an extra texture template using the neural radiance field [46] (NeRF) in the same decomposed manner, i.e.,

$$T_{\mathrm{Tex}}(W(\mathbf{x}_c, \boldsymbol{\theta})) = (\sigma, \mathbf{c}), \qquad (2)$$

where $T_{\mathrm{Tex}}(\cdot)$ is a template radiance field that maps a template point to its density σ and color $\mathbf{c}$. Note that we utilize the template NeRF to represent the scan texture without view-dependent variation, so we discard the view direction

Z. Li et al.

input. Thanks to the decomposition (Eq. 1 & Eq. 2), our avatar representation, dubbed *GeoTexAvatar*, is able to jointly constrain the pose-dependent warping field $W(\cdot)$ with the Geo-Tex implicit template field ($T_{\text{Geo}}(\cdot)$ & $T_{\text{Tex}}(\cdot)$) under the joint supervision of geometry and texture of training scans. Figure 2 is an illustration of our representation.

Compared with state-of-the-art scan-based avatar methods [43,54], our representation shows two main advantages as demonstrated in Fig. 7. 1) The decomposed representation can preserve more pose-agnostic details for animation. 2) The joint supervision of geometry and texture enables more reasonable pose-dependent deformations. What's more, the decomposed representation allows us to finetune the texture template for high-quality rendering, which is also an advantage over other entangled methods as shown in Fig. 8.

4.2 GeoTexAvatar Training

The training loss for our GeoTexAvatar network contains a geometry loss, a texture loss and a regularization loss for the warping field, i.e., $\mathcal{L} = \lambda_{\text{geo}}\mathcal{L}_{\text{geo}} + \lambda_{\text{tex}}\mathcal{L}_{\text{tex}} + \lambda_{\text{reg}}\mathcal{L}_{\text{reg}}$, where λ_{geo}, λ_{tex} and λ_{reg} are the loss weights.

Geometry Loss. $\mathcal{L}_{\text{geo}}$ penalizes the difference between the inferred occupancy and the ground truth:

$$\mathcal{L}_{\text{geo}} = \frac{1}{|\mathcal{P}|} \sum_{\mathbf{x}_p \in \mathcal{P}} \text{BCE}\left(s(\mathbf{x}_p), s^*(\mathbf{x}_p)\right), \tag{3}$$

where $\mathcal{P}$ is the sampled point set, $s(\mathbf{x}_p)$ and $s^*(\mathbf{x}_p)$ are inferred and ground-truth occupancy, respectively, and $\text{BCE}(\cdot)$ measures the binary cross entropy.

Texture Loss. To jointly train the NeRF template, we render the textured scans to different views for the supervision. $\mathcal{L}_{\text{tex}}$ measures the error between the color rendered by the network and the real one:

$$\mathcal{L}_{\text{tex}} = \frac{1}{|\mathcal{R}|} \sum_{\mathbf{r} \in \mathcal{R}} \left\| \hat{C}(\mathbf{r}) - C^*(\mathbf{r}) \right\|^2, \tag{4}$$

where $\mathcal{R}$ is the set of ray samples in the image view frustum, $\hat{C}(\cdot)$ is the volume rendering function as in [46], and $C^*(\mathbf{r})$ is the ground-truth color.

Regularization Loss. $\mathcal{L}_{\text{reg}}$ constrains the warped points by $W(\cdot)$ to be close with the input because the canonical pose-dependent dynamics are usually small:

$$\mathcal{L}_{\text{reg}} = \frac{1}{|\mathcal{P} \cup \mathcal{P}_{\mathcal{R}}|} \sum_{\mathbf{x}_c \in \mathcal{P} \cup \mathcal{P}_{\mathcal{R}}} \left\| \Delta W(\mathbf{x}_c, \boldsymbol{\theta}) \right\|^2, \tag{5}$$

where $\mathcal{P}_{\mathcal{R}}$ is the sampled points along each ray in $\mathcal{R}$ during volume rendering.

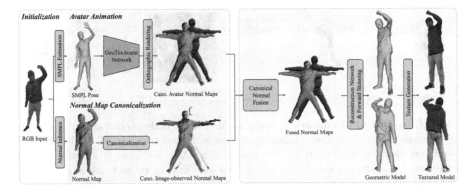

Fig. 3. Avatar-conditioned volumetric capture pipeline. Given a RGB image from the monocular video, we firstly infer the SMPL pose and normal map. Then the pose-driven GeoTexAvatar generates and renders canonical avatar normal maps, while the image-observed normal map is warped into the canonical space. The canonical normal fusion integrates both avatar and observed normals together and feeds the fused normal maps into the reconstruction network to output a high-fidelity 3D human model. Finally, a high-resolution texture is generated using the GeoTexAvatar network.

5   Avatar-Conditioned Volumetric Capture

Next, we move forward to the avatar-conditioned volumetric capture. The main difficulty lies in the enormous domain gap between the avatar representation and the input image, i.e., the image provides no 3D measurement to associate with the avatar geometry. As illustrated in Fig. 3, to overcome this challenge, we propose to employ the normal maps as the intermediate representation to bridge the gap between the image inputs and the avatar prior. Specifically, we conduct the integration between the two modals on a unified canonical image plane, and then split the integration as canonical normal fusion and model reconstruction.

Initialization. Given a RGB image, our approach firstly prepares both the canonical avatar and image normal maps as illustrated in Fig. 3. Specifically, 1) Avatar Animation: The GeoTexAvatar network outputs the animated canonical avatar using the SMPL pose, then renders the front & back canonical normal maps denoted as $\mathbf{F}_{avatar}$ & $\mathbf{B}_{avatar}$, respectively. 2) Normal Map Canonicalization: In a parallel branch, the input RGB image is fed into a 2D convolutional network [66] to infer the normal map $\mathbf{N}$ that represents the visible details. Then it is mapped to the canonical space, with the results denoted as $\mathbf{F}_{image}$ and $\mathbf{B}_{image}$. Implementation details of the two steps can be found in the Supp. Mat.

5.1   Canonical Normal Fusion

Given the prepared avatar & image-observed normal maps, we integrate them on the 2D canonical image plane. However, directly replacing the avatar normals

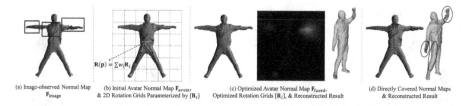

(a) Image-observed Normal Map $\mathbf{F}_{\text{image}}$

(b) Initial Avatar Normal Map $\mathbf{F}_{\text{avatar}}$ & 2D Rotation Grids Parameterized by $\{\mathbf{R}_i\}$

(c) Optimized Avatar Normal Map $\mathbf{F}_{\text{fused}}$, Optimized Rotation Grids $\{\mathbf{R}_i\}$, & Reconstructed Result

(d) Directly Covered Normal Maps & Reconstructed Result

Fig. 4. Illustration of canonical normal fusion. Directly replacing visible regions using image-observed normals causes severe reconstructed artifacts (d), while the proposed canonical normal fusion corrects the low-frequency local batch orientations and preserves high-frequency details for robust and high-fidelity reconstruction (c).

with the corresponding visible image-based ones is not feasible, because the canonicalized normal orientations may be incorrect due to the inaccurate SMPL estimation (e.g., rotation of the forearm) as shown in Fig. 4(a), leading to severe artifacts in the reconstruction (Fig. 4(d)). Therefore, we propose a new canonical normal fusion method to not only preserve high-frequency image-observed normals but also correct low-frequency local batch orientations.

Without loss of generality, we take the front avatar normal map and the image-based map ($\mathbf{F}_{\text{avatar}}$ and $\mathbf{F}_{\text{image}}$) as the example to introduce the formulation. As illustrated in Fig. 4(a), $\mathbf{F}_{\text{image}}$ contains plentiful observed details estimated from the input color, but the orientations of normals are possibly incorrect due to SMPL estimation error. On the other hand, even though the visible region of $\mathbf{F}_{\text{avatar}}$ does not completely follow the image observation, the low-frequency normal orientations are accurate in the canonical space as shown in Fig. 4(b). To this end, we propose to optimize the avatar normal map $\mathbf{F}_{\text{avatar}}$ to integrate high-frequency details from the image-observed one $\mathbf{F}_{\text{image}}$ while maintaining its initial correct low-frequency orientations as shown in Fig. 4(d). To do so, we introduce 2D rotation grids to factor out the low-frequency orientation differences between $\mathbf{F}_{\text{avatar}}$ and $\mathbf{F}_{\text{image}}$, so that the remaining high-frequency details on $\mathbf{F}_{\text{image}}$ can be rotated back to $\mathbf{F}_{\text{avatar}}$ with correct orientations. As illustrated in Fig. 4(b), each grid is assigned a rotation matrix $\mathbf{R}_i \in SO(3)$, and the rotation of a 2D point $\mathbf{p} = (x, y)$ on the map is defined as $\mathbf{R}(\mathbf{p}) = \sum_i w_i(\mathbf{p})\mathbf{R}_i$, a linear combination of $\{\mathbf{R}_i\}$ using bilinear interpolation, where $w_i(\mathbf{p})$ is the interpolation weight. With such a parameterization, we optimize the rotation grids $\{\mathbf{R}_i\}$ and avatar normal map $\mathbf{F}_{\text{avatar}}$ by minimizing

$$E(\mathbf{R}_i, \mathbf{F}_{\text{avatar}}) = \lambda_{\text{fitting}} E_{\text{fitting}}(\mathbf{R}_i, \mathbf{F}_{\text{avatar}}) + \lambda_{\text{smooth}} E_{\text{smooth}}(\mathbf{R}_i), \quad (6)$$

where E_{fitting} and E_{smooth} are energies of misalignment between rotated avatar normals and observed ones and smooth regularization of grids, respectively.

Fitting Term. The fitting term measures the residuals between the avatar normal rotated by its transformation matrix and the target image-observed one:

$$E_{\text{fitting}}(\mathbf{R}_i, \mathbf{F}_{\text{avatar}}) = \sum_{\mathbf{p} \in \mathcal{D}} \|\mathbf{R}(\mathbf{p})\mathbf{F}_{\text{avatar}}(\mathbf{p}) - \mathbf{F}_{\text{image}}(\mathbf{p})\|^2, \quad (7)$$

where $\mathcal{D}$ is the valid intersection region of $\mathbf{F}_{\text{avatar}}$ and $\mathbf{F}_{\text{image}}$.

Smooth Term. The smooth term regularizes the rotation grids to be low-frequency by constrain the rotation similarity between adjacent grids:

$$E_{\text{smooth}}(\mathbf{R}_i) = \sum_i \sum_{j \in \mathcal{N}(i)} \|\text{Rod}(\mathbf{R}_i) - \text{Rod}(\mathbf{R}_j)\|^2, \tag{8}$$

where $\mathcal{N}(i)$ is the neighbors of the i-th grid, and Rod : $SO(3) \rightarrow so(3)$ maps the rotation matrix to the axis-angle vector.

Delayed Optimization of $\mathbf{F}_{\text{avatar}}$. We firstly initialize $\{\mathbf{R}_i\}$ as identity matrices. Note that both the avatar normal map $\mathbf{F}_{\text{avatar}}$ and rotation grids $\{\mathbf{R}_i\}$ are optimizable variables, so that the solutions are not unique. If we jointly optimizes both variables, $\mathbf{F}_{\text{avatar}}$ tends to be equal with $\mathbf{F}_{\text{image}}$ which is not desired. To this end, we firstly solve the low-frequency rotation grids $\{\mathbf{R}_i\}$, then optimize $\mathbf{F}_{\text{avatar}}$ to integrate high-frequency details from $\mathbf{F}_{\text{image}}$ with $\{\mathbf{R}_i\}$ fixed. As a result, we obtain the optimized $\mathbf{F}_{\text{avatar}}$ as the fused normal map $\mathbf{F}_{\text{fused}}$ with high-frequency details and correct low-frequency orientations as shown in Fig. 4(c).

5.2   Model Reconstruction

Geometric Reconstruction. To reconstruct 3D geometry from the fused canonical normal maps $\mathbf{F}_{\text{fused}}$ & $\mathbf{B}_{\text{fused}}$, we pretrain a reconstruction network on a large-scale 3D human dataset [73]. With such a strong data prior, we can efficiently and robustly recover the 3D geometry with high-fidelity full body details from the complete normal maps. The reconstruction network is formulated as an image-conditioned implicit function $g(h(\pi(\mathbf{x}); \mathbf{F}_{\text{fused}}, \mathbf{B}_{\text{fused}}), \mathbf{x}_z)$, where $\mathbf{x}$ is a 3D point in the canonical space, $h(\cdot)$ is a function to sample convoluted image features, $\pi(\cdot)$ is the orthographic projection, $\mathbf{x}_z$ is the z-axis value, and $g(\cdot)$ is an implicit function that maps the image feature and $\mathbf{x}_z$ to an occupancy value. We perform Marching Cubes [40] on this implicit function to reconstruct the canonical model, then deform it to the posed space by forward skinning.

Texture Generation. Based on the GeoTexAvatar representation, we can generate the texture of the reconstructed geometry by mapping the radiance field to it. Specifically, given a vertex $\mathbf{v}$ of the canonical model and its normal $\mathbf{n_v}$, based on Eq. 2, we can calculate its color using volume rendering in NeRF [46] with the camera ray $\mathbf{r}(t) = \mathbf{v} - t\mathbf{n_v}$ and near and far bounds $-\delta$ and δ ($\delta > 0$).

6   Results

The volumetric captured results of our method are demonstrated in Fig. 5. For the experiments, we collect textured scans of 10 subjects and their monocular videos, and partial scans are utilized as the evaluation dataset. More details about data preprocessing and implementation can be found in the Supp. Mat.

Fig. 5. Example volumetric captured results of our method. From top to bottom are the monocular RGB input, geometric and textured results, respectively.

6.1   Comparison

Volumetric Capture. As shown in Fig. 6, we compare AvatarCap, our whole volumetric capture framework, against state-of-the-art fusion and single-RGB(D)-image reconstruction methods, including POSEFusion [35], PIFuHD [53] and NormalGAN [64]. We conduct this comparison on the sequences captured by one Kinect Azure to also compare against RGBD-based methods [35,64], and all the learning-based methods are finetuned on the person-specific scans used in our avatar creation for fairness. Figure 6 shows that our method can achieve high-fidelity reconstruction with detailed observations and reasonable pose-dependent invisible dynamics. Though POSEFusion [35] can integrate invisible surfaces from other frames, it entirely relies on each time captured sequence without pose generalization. PIFuHD [53] only considers to recover visible details from the normal map inferred by the color input without pose-conditioned person-specific dynamics, so the invisible regions are usually over-smoothed. Though NormalGAN [64] can infer a plausible back RGBD map from the RGBD input, the inferred invisible appearance may be inconsistent with the person-specific dynamics. We also conduct quantitative comparison on the testing dataset with ground-truth scans, and report the averaged errors in Tab. 1. Note that POSEFusion is a sequence-based method, but the testing scans are under discrete poses, so we only compare with the other methods. Overall, our method achieves state-of-the-art capture on both quality and accuracy.

Animatable Avatar. As shown in Fig. 7, we compare our avatar module, Geo-TexAvatar, against state-of-the-art avatar works based on person-specific scans, SCANimate [54], SCALE [41] and POP [43]. Note that POP is a multi-subject-outfit representation, in this comparison we train it from scratch using the same few (∼20) scans as other methods. Our method outperforms these methods on

Fig. 6. Qualitative comparison against monocular volumetric capture methods. We show reconstructed results of our method (AvatarCap), POSEFusion [35], PIFuHD [53] and NormalGAN [64]. And our method outperforms others on the capture of pose-dependent dynamics in the invisible regions (red circles). (Color figure online)

Table 1. Quantitative comparision of AvatarCap with PIFuHD [53] and NormalGAN [64]. We report the averaged Chamfer and Scan-to-Model distance errors ($\times 10^{-2}$ m) of differnt methods on the whole testing dataset.

Metric/Method	AvatarCap (ours)	PIFuHD [53]	NormalGAN [64]
Chamfer Distance	**1.097**	3.400	2.852
Scan-to-Mesh Distance	**1.096**	3.092	2.855

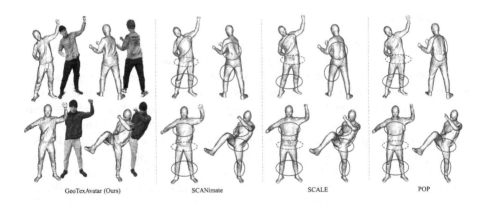

Fig. 7. Qualitative comparison against animatable avatar methods. We show animated results of our method (also with high-quality texture), SCANimate [54], SCALE [41] and POP [43]. And our method shows the superiority on the modeling of wrinkles (solid circles) and pose-dependent cloth tangential motions (dotted circles).

Table 2. Quantitative comparison of GeoTexAvatar with SCANimate [54], **SCALE** [41] **and POP** [43]. We report the averaged Chamfer distance errors ($\times 10^{-3}$ m) between the animated results of different methods and the ground-truth scans.

Case/Method	GeoTexAvatar (Ours)	SCANimate [54]	SCALE [41]	POP [43]
HOODY_1	**6.29**	7.38	8.19	6.83
SHIRT_1	**2.80**	5.72	4.72	3.08

the recovery of dynamic details as well as tangential cloth motion benefiting from the proposed decomposed representation and joint supervisions of both geometry and texture, respectively. We further quantitatively evaluate the animation accuracy of GeoTexAvatar and other works on the testing dataset in Table 2, and our method achieves more accurate animated results.

Fig. 8. Evaluation of the decomposed representation. (a), (b) and (c) are the animated geometric and textured results of the entangled representation [54] and decomposed representations with pose-vector and positional-map encoding, respectively.

6.2   Evaluation

Decomposed Representation of GeoTexAvatar. We evaluate the proposed decomposed representation compared with the entangled one [54] in Fig. 8. Firstly, similar to SCANimate [54], we choose the local pose vector as the pose encoding of the warping field in our representation. Compared with the entangled representation (Fig. 8 (a)), the decomposed one (Fig. 8 (b)) produces more detailed animation results, e.g., the zippers, facial and leg details, thanks to the decomposition of pose-dependent dynamics and pose-agnostic details. Besides, the decomposition allows us to finetune the texture template on a single scan to recovery high-quality texture, while the texture is totally blurred in the entangled learning. Furthermore, we empirically find that a SMPL positional map defined in the canonical space shows more powerful expression for pose-dependent dynamics than the local pose vector as shown in Fig. 8 (b) and (c).

Texture Supervision in GeoTexAvatar. We evaluate the effectiveness of texture supervision to the pose-dependent warping field by visualizing the correspondences during animation in Fig. 9. We firstly train the avatar network

Fig. 9. Evaluation of the effectiveness of texture supervision in GeoTexA-vatar. We visualize the correspondences among different frames by the vertex color which indicates whether the vertex belongs to the upper or lower body.

with and without texture template individually. To visualize the correspondences among animated results by different poses, we firstly generate the geometric template using [40] on $T_{\text{Geo}}(\cdot)$, then manually segment the template mesh as upper and lower body parts. Given a new pose, the avatar network outputs a canonical avatar model, then each vertex on this model can be warped to the template using the pose-dependent warping field. Finally, we determine whether the warped vertex belongs to the upper or lower part by its closest point on the template. Figure 9 demonstrates that the texture supervision can implicitly constrain the warping field by jointly learning an extra texture template $T_{\text{Tex}}(\cdot)$, thus enabling more reasonable pose generalization for animation. However, training only with geometry supervision results in overfitted animation due to the ambiguity when establishing correspondences under only the geometric closest constraint.

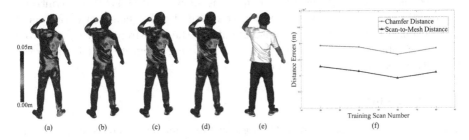

Fig. 10. Evaluation of the effect of the training scan number on the animation accuracy of the GeoTexAvatar. From (a) to (d) are visualized vertex-to-surface error of animated results trained by 20, 40, 60 and 80 scans, respectively, (e) is the ground-truth scan, and (f) is the chart of the averaged Chamfer and Scan-to-Mesh distance errors on the whole testing dataset.

Training Scan Number. We quantitatively evaluate the effect of the training scan number on the animation accuracy of GeoTexAvatar. We choose one subset ("SHORT_SLEEVE_1") of our dataset that contains 100 scans, and randomly choose 80 items as the training dataset, and the rest for evaluation. Figure 10 shows the visualized and numerical animation errors using different numbers of scans. More training samples does not always lead to more accurate results because the mapping from poses to cloth details may be one-to-many in the training dataset.

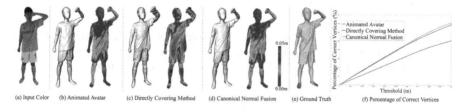

(a) Input Color   (b) Animated Avatar   (c) Directly Covering Method   (d) Canonical Normal Fusion   (e) Ground Truth   (f) Percentage of Correct Vertices

Fig. 11. Evaluation of canonical normal fusion. We visualize the per-vertex point-to-surface errors between the reconstructed models and the ground truth, (f) is the percentage of correct vertices under different thresholds

Canonical Normal Fusion. We evaluate the proposed canonical normal fusion compared with the directly covering method both qualitatively and quantitatively. Figure 11 (c) and (d) show the reconstructed results using directly covering and canonical normal fusion, respectively, as well as their per-vertex point-to-surface errors to the ground-truth scans. Due to the inaccurate SMPL estimation and the camera view difference with the orthographic hypothesis in normal map inference, the canonicalized normal tends to be fallacious. The directly covering method maintains the wrong image-observed normal, thus leading to inaccurate reconstruction and ghosting artifacts as shown in Fig. 11(c). By contrary, the canonical normal fusion not only corrects the low-frequency orientations of canonicalized image normal, but also maintains the high-frequency details from the image observation, thus enabling the following accurate and high-fidelity reconstruction as shown in Fig. 11 (d) and (f).

7   Discussion

Conclusion. We present AvatarCap, a novel monocular human volumetric capture framework, that leverages an animatable avatar learned from only few scans to capture body dynamics regardless of the visibility. Based on the proposed GeoTexAvatar and avatar-conditioned volumetric capture, our method effectively integrates the information from image observations and the avatar prior. Overall, our method outperforms other state-of-the-art capture approaches, and we believe that the avatar-conditioned volumetric capture will make progress towards dynamic and realistic 3D human with the advance of animatable avatars.

Limitation. The main limitation of our method is the 3D scan collection, a possible solution is to capture scans using 3D self-portrait methods [32, 34] with an RGBD camera. Moreover, our method may fail for loose clothes, e.g., long skirts, because the SMPL skeletons cannot correctly deform such garments.

Acknowledgement. This paper is supported by National Key R&D Program of China (2021ZD0113501) and the NSFC project No. 62125107.

References

1. Alldieck, T., Pons-Moll, G., Theobalt, C., Magnor, M.: Tex2shape: Detailed full human body geometry from a single image. In: ICCV. pp. 2293–2303 (2019)
2. Bagautdinov, T., Wu, C., Simon, T., Prada, F., Shiratori, T., Wei, S.E., Xu, W., Sheikh, Y., Saragih, J.: Driving-signal aware full-body avatars. TOG **40**(4), 1–17 (2021)
3. Bhatnagar, B.L., Sminchisescu, C., Theobalt, C., Pons-Moll, G.: Combining implicit function learning and parametric models for 3D human reconstruction. In: Vedaldi, A., Bischof, H., Brox, T., Frahm, J.-M. (eds.) ECCV 2020. LNCS, vol. 12347, pp. 311–329. Springer, Cham (2020). https://doi.org/10.1007/978-3-030-58536-5_19
4. Bradley, D., Popa, T., Sheffer, A., Heidrich, W., Boubekeur, T.: Markerless garment capture. TOG **27**(3), 1–9 (2008)
5. Brox, T., Rosenhahn, B., Gall, J., Cremers, D.: Combined region and motion-based 3d tracking of rigid and articulated objects. IEEE T-PAMI **32**(3), 402–415 (2009)
6. Burov, A., Nießner, M., Thies, J.: Dynamic surface function networks for clothed human bodies. In: ICCV, pp. 10754–10764 (2021)
7. Chen, X., Zheng, Y., Black, M.J., Hilliges, O., Geiger, A.: Snarf: differentiable forward skinning for animating non-rigid neural implicit shapes. In: ICCV, pp. 11594–11604 (2021)
8. Deng, B., et al.: NASA neural articulated shape approximation. In: Vedaldi, A., Bischof, H., Brox, T., Frahm, J.-M. (eds.) ECCV 2020. LNCS, vol. 12352, pp. 612–628. Springer, Cham (2020). https://doi.org/10.1007/978-3-030-58571-6_36
9. Dong, Z., Guo, C., Song, J., Chen, X., Geiger, A., Hilliges, O.: Pina: learning a personalized implicit neural avatar from a single RGB-D video sequence. In: CVPR (2022)
10. Dou, M., et al.: Fusion4d: real-time performance capture of challenging scenes. TOG **35**(4), 1–13 (2016)
11. Gabeur, V., Franco, J.S., Martin, X., Schmid, C., Rogez, G.: Moulding humans: non-parametric 3d human shape estimation from single images. In: ICCV, pp. 2232–2241 (2019)
12. Gall, J., Stoll, C., De Aguiar, E., Theobalt, C., Rosenhahn, B., Seidel, H.P.: Motion capture using joint skeleton tracking and surface estimation. In: CVPR, pp. 1746–1753. IEEE (2009)
13. Goodfellow, I., et al.: Generative adversarial nets. NeurIPS 27 (2014)
14. Guan, P., Reiss, L., Hirshberg, D.A., Weiss, A., Black, M.J.: Drape: dressing any person. TOG **31**(4), 1–10 (2012)
15. Guo, C., Chen, X., Song, J., Hilliges, O.: Human performance capture from monocular video in the wild. In: 3DV, pp. 889–898. IEEE (2021)
16. Guo, K., Xu, F., Wang, Y., Liu, Y., Dai, Q.: Robust non-rigid motion tracking and surface reconstruction using l0 regularization. In: ICCV, pp. 3083–3091 (2015)
17. Guo, K., Xu, F., Yu, T., Liu, X., Dai, Q., Liu, Y.: Real-time geometry, albedo and motion reconstruction using a single RGBD camera. TOG **36**(3), 32:1-32:13 (2017)
18. Habermann, M., Liu, L., Xu, W., Zollhoefer, M., Pons-Moll, G., Theobalt, C.: Real-time deep dynamic characters. TOG **40**(4), 1–16 (2021)

19. Habermann, M., Xu, W., Zollhoefer, M., Pons-Moll, G., Theobalt, C.: Livecap: real-time human performance capture from monocular video. TOG **38**(2), 1–17 (2019)
20. Habermann, M., Xu, W., Zollhofer, M., Pons-Moll, G., Theobalt, C.: Deepcap: monocular human performance capture using weak supervision. In: CVPR, pp. 5052–5063 (2020)
21. He, T., Collomosse, J., Jin, H., Soatto, S.: Geo-PIFU: geometry and pixel aligned implicit functions for single-view human reconstruction. NeurIPS **33**, 9276–9287 (2020)
22. He, T., Xu, Y., Saito, S., Soatto, S., Tung, T.: Arch++: animation-ready clothed human reconstruction revisited. In: ICCV, pp. 11046–11056 (2021)
23. He, Y., et al.: Challencap: Monocular 3d capture of challenging human performances using multi-modal references. In: CVPR, pp. 11400–11411 (2021)
24. Hong, Y., Zhang, J., Jiang, B., Guo, Y., Liu, L., Bao, H.: Stereopifu: depth aware clothed human digitization via stereo vision. In: CVPR, pp. 535–545 (2021)
25. Huang, Z., Xu, Y., Lassner, C., Li, H., Tung, T.: Arch: animatable reconstruction of clothed humans. In: CVPR, pp. 3093–3102 (2020)
26. Innmann, M., Zollhöfer, M., Nießner, M., Theobalt, C., Stamminger, M.: VolumeDeform: real-time volumetric non-rigid reconstruction. In: Leibe, B., Matas, J., Sebe, N., Welling, M. (eds.) ECCV 2016. LNCS, vol. 9912, pp. 362–379. Springer, Cham (2016). https://doi.org/10.1007/978-3-319-46484-8_22
27. Jackson, A.S., Manafas, C., Tzimiropoulos, G.: 3D human body reconstruction from a single image via volumetric regression. In: Leal-Taixé, L., Roth, S. (eds.) 3d human body reconstruction from a single image via volumetric regression. LNCS, vol. 11132, pp. 64–77. Springer, Cham (2019). https://doi.org/10.1007/978-3-030-11018-5_6
28. Kolotouros, N., Pavlakos, G., Black, M.J., Daniilidis, K.: Learning to reconstruct 3D human pose and shape via model-fitting in the loop. In: ICCV, pp. 2252–2261 (2019)
29. Leroy, V., Franco, J.S., Boyer, E.: Multi-view dynamic shape refinement using local temporal integration. In: ICCV, pp. 3094–3103 (2017)
30. Li, C., Zhao, Z., Guo, X.: ArticulatedFusion: real-time reconstruction of motion, geometry and segmentation using a single depth camera. In: Ferrari, V., Hebert, M., Sminchisescu, C., Weiss, Y. (eds.) ECCV 2018. LNCS, vol. 11212, pp. 324–340. Springer, Cham (2018). https://doi.org/10.1007/978-3-030-01237-3_20
31. Li, H., Adams, B., Guibas, L.J., Pauly, M.: Robust single-view geometry and motion reconstruction. TOG **28**(5), 1–10 (2009)
32. Li, H., Vouga, E., Gudym, A., Luo, L., Barron, J.T., Gusev, G.: 3D self-portraits. TOG **32**(6), 1–9 (2013)
33. Li, R., Xiu, Y., Saito, S., Huang, Z., Olszewski, K., Li, H.: Monocular real-time volumetric performance capture. In: Vedaldi, A., Bischof, H., Brox, T., Frahm, J.-M. (eds.) ECCV 2020. LNCS, vol. 12368, pp. 49–67. Springer, Cham (2020). https://doi.org/10.1007/978-3-030-58592-1_4
34. Li, Z., Yu, T., Pan, C., Zheng, Z., Liu, Y.: Robust 3d self-portraits in seconds. In: CVPR, pp. 1344–1353 (2020)
35. Li, Z., Yu, T., Zheng, Z., Guo, K., Liu, Y.: Posefusion: pose-guided selective fusion for single-view human volumetric capture. In: CVPR. pp. 14162–14172 (2021)
36. Liu, L., Habermann, M., Rudnev, V., Sarkar, K., Gu, J., Theobalt, C.: Neural actor: neural free-view synthesis of human actors with pose control. TOG **40**(6), 1–16 (2021)

37. Liu, Y., Dai, Q., Xu, W.: A point-cloud-based multiview stereo algorithm for free-viewpoint video. TVCG **16**(3), 407–418 (2009)
38. Liu, Y., Stoll, C., Gall, J., Seidel, H.P., Theobalt, C.: Markerless motion capture of interacting characters using multi-view image segmentation. In: CVPR, pp. 1249–1256. IEEE (2011)
39. Loper, M., Mahmood, N., Romero, J., Pons-Moll, G., Black, M.J.: SMPL: a skinned multi-person linear model. TOG **34**(6), 1–16 (2015)
40. Lorensen, W.E., Cline, H.E.: Marching cubes: a high resolution 3D surface construction algorithm. TOG **21**(4), 163–169 (1987)
41. Ma, Q., Saito, S., Yang, J., Tang, S., Black, M.J.: Scale: modeling clothed humans with a surface codec of articulated local elements. In: CVPR, pp. 16082–16093 (2021)
42. Ma, Q., Yang, J., Ranjan, A., Pujades, S., Pons-Moll, G., Tang, S., Black, M.J.: Learning to dress 3d people in generative clothing. In: CVPR. pp. 6469–6478 (2020)
43. Ma, Q., Yang, J., Tang, S., Black, M.J.: The power of points for modeling humans in clothing. In: ICCV, pp. 10974–10984 (2021)
44. Magnenat-Thalmann, N., Laperrire, R., Thalmann, D.: Joint-dependent local deformations for hand animation and object grasping. In: In Proceedings on Graphics Interface. Citeseer (1988)
45. Mescheder, L., Oechsle, M., Niemeyer, M., Nowozin, S., Geiger, A.: Occupancy networks: learning 3D reconstruction in function space. In: CVPR, pp. 4460–4470 (2019)
46. Mildenhall, B., Srinivasan, P.P., Tancik, M., Barron, J.T., Ramamoorthi, R., Ng, R.: NeRF: representing scenes as neural radiance fields for view synthesis. In: Vedaldi, A., Bischof, H., Brox, T., Frahm, J.-M. (eds.) ECCV 2020. LNCS, vol. 12346, pp. 405–421. Springer, Cham (2020). https://doi.org/10.1007/978-3-030-58452-8_24
47. Mustafa, A., Kim, H., Guillemaut, J.Y., Hilton, A.: General dynamic scene reconstruction from multiple view video. In: ICCV, pp. 900–908 (2015)
48. Natsume, R., et al.: Siclope: Silhouette-based clothed people. In: CVPR, pp. 4480–4490 (2019)
49. Newcombe, R.A., Fox, D., Seitz, S.M.: Dynamicfusion: reconstruction and tracking of non-rigid scenes in real-time. In: CVPR, pp. 343–352 (2015)
50. Peng, S., Dong, J., Wang, Q., Zhang, S., Shuai, Q., Zhou, X., Bao, H.: Animatable neural radiance fields for modeling dynamic human bodies. In: ICCV, pp. 14314–14323 (2021)
51. Pons-Moll, G., Pujades, S., Hu, S., Black, M.J.: Clothcap: seamless 4D clothing capture and retargeting. TOG **36**(4), 1–15 (2017)
52. Saito, S., Huang, Z., Natsume, R., Morishima, S., Kanazawa, A., Li, H.: PIFU: pixel-aligned implicit function for high-resolution clothed human digitization. In: ICCV, pp. 2304–2314 (2019)
53. Saito, S., Simon, T., Saragih, J., Joo, H.: Pifuhd: multi-level pixel-aligned implicit function for high-resolution 3d human digitization. In: CVPR, June 2020
54. Saito, S., Yang, J., Ma, Q., Black, M.J.: Scanimate: weakly supervised learning of skinned clothed avatar networks. In: CVPR, pp. 2886–2897 (2021)
55. Shao, R., et al.: Doublefield: Bridging the neural surface and radiance fields for high-fidelity human reconstruction and rendering. In: CVPR (2022)
56. Slavcheva, M., Baust, M., Cremers, D., Ilic, S.: Killingfusion: non-rigid 3d reconstruction without correspondences. In: CVPR, pp. 1386–1395 (2017)

57. Slavcheva, M., Baust, M., Ilic, S.: Sobolevfusion: 3D reconstruction of scenes under-going free non-rigid motion. In: CVPR, pp. 2646–2655. IEEE, Salt Lake City, June 2018
58. Smith, D., Loper, M., Hu, X., Mavroidis, P., Romero, J.: Facsimile: fast and accu-rate scans from an image in less than a second. In: ICCV, pp. 5330–5339 (2019)
59. Stoll, C., Gall, J., De Aguiar, E., Thrun, S., Theobalt, C.: Video-based reconstruc-tion of animatable human characters. TOG **29**(6), 1–10 (2010)
60. Su, Z., Xu, L., Zheng, Z., Yu, T., Liu, Y., Fang, L.: RobustFusion: human volu-metric capture with data-driven visual cues using a RGBD camera. In: Vedaldi, A., Bischof, H., Brox, T., Frahm, J.-M. (eds.) ECCV 2020. LNCS, vol. 12349, pp. 246–264. Springer, Cham (2020). https://doi.org/10.1007/978-3-030-58548-8_15
61. Sumner, R.W., Schmid, J., Pauly, M.: Embedded deformation for shape manipu-lation. TOG **26**(3), 80-es (2007)
62. Suo, X., et al.: Neuralhumanfvv: real-time neural volumetric human performance rendering using RGB cameras. In: CVPR, pp. 6226–6237 (2021)
63. Varol, G., et al.: BodyNet: volumetric inference of 3D human body shapes. In: Ferrari, V., Hebert, M., Sminchisescu, C., Weiss, Y. (eds.) ECCV 2018. LNCS, vol. 11211, pp. 20–38. Springer, Cham (2018). https://doi.org/10.1007/978-3-030-01234-2_2
64. Wang, L., Zhao, X., Yu, T., Wang, S., Liu, Y.: NormalGAN: learning detailed 3D human from a single RGB-D image. In: Vedaldi, A., Bischof, H., Brox, T., Frahm, J.-M. (eds.) ECCV 2020. LNCS, vol. 12365, pp. 430–446. Springer, Cham (2020). https://doi.org/10.1007/978-3-030-58565-5_26
65. Wang, S., Mihajlovic, M., Ma, Q., Geiger, A., Tang, S.: Metaavatar: learning ani-matable clothed human models from few depth images. NeurIPS 34 (2021)
66. Wang, T.C., Liu, M.Y., Zhu, J.Y., Tao, A., Kautz, J., Catanzaro, B.: High-resolution image synthesis and semantic manipulation with conditional GANs. In: CVPR, pp. 8798–8807 (2018)
67. Xiang, D., et al.: Modeling clothing as a separate layer for an animatable human avatar. TOG **40**(6), 1–15 (2021)
68. Xiang, D., Prada, F., Wu, C., Hodgins, J.: Monoclothcap: towards temporally coherent clothing capture from monocular RGB video. In: 3DV, pp. 322–332. IEEE (2020)
69. Xiu, Y., Yang, J., Tzionas, D., Black, M.J.: Icon: implicit clothed humans obtained from normals. In: CVPR (2022)
70. Xu, W., et al.: Monoperfcap: human performance capture from monocular video. TOG **37**(2), 1–15 (2018)
71. Ye, G., Liu, Y., Hasler, N., Ji, X., Dai, Q., Theobalt, C.: Performance capture of interacting characters with handheld kinects. In: Fitzgibbon, A., Lazebnik, S., Perona, P., Sato, Y., Schmid, C. (eds.) ECCV 2012. LNCS, vol. 7573, pp. 828–841. Springer, Heidelberg (2012). https://doi.org/10.1007/978-3-642-33709-3_59
72. Yu, T., et al.: Bodyfusion: real-time capture of human motion and surface geometry using a single depth camera. In: ICCV, Venice, pp. 910–919. IEEE (2017)
73. Yu, T., et al.: Function4d: real-time human volumetric capture from very sparse consumer RGBD sensors. In: CVPR, pp. 5746–5756 (2021)
74. Yu, T., et al.: Doublefusion: real-time capture of human performances with inner body shapes from a single depth sensor. In: CVPR, Salt Lake City, pp. 7287–7296. IEEE, June 2018
75. Yu, T., et al.: Simulcap: single-view human performance capture with cloth simu-lation. In: CVPR, pp. 5499–5509. IEEE (2019)

76. Zhang, H., et al.: Pymaf: 3D human pose and shape regression with pyramidal mesh alignment feedback loop. In: ICCV, pp. 11446–11456 (2021)
77. Zheng, Y., et al.: Deepmulticap: performance capture of multiple characters using sparse multiview cameras. In: ICCV (2021)
78. Zheng, Z., Yu, T., Dai, Q., Liu, Y.: Deep implicit templates for 3D shape representation. In: CVPR, pp. 1429–1439 (2021)
79. Zheng, Z., et al.: HybridFusion: real-time performance capture using a single depth sensor and sparse IMUs. In: Ferrari, V., Hebert, M., Sminchisescu, C., Weiss, Y. (eds.) Hybridfusion: real-time performance capture using a single depth sensor and sparse imus. LNCS, vol. 11213, pp. 389–406. Springer, Cham (2018). https://doi.org/10.1007/978-3-030-01240-3_24
80. Zheng, Z., Yu, T., Liu, Y., Dai, Q.: Pamir: parametric model-conditioned implicit representation for image-based human reconstruction. IEEE T-PAMI (2021)
81. Zheng, Z., Yu, T., Wei, Y., Dai, Q., Liu, Y.: Deephuman: 3D human reconstruction from a single image. In: ICCV, pp. 7739–7749 (2019)
82. Zhi, T., Lassner, C., Tung, T., Stoll, C., Narasimhan, S.G., Vo, M.: TexMesh: reconstructing detailed human texture and geometry from RGB-D Video. In: Vedaldi, A., Bischof, H., Brox, T., Frahm, J.-M. (eds.) ECCV 2020. LNCS, vol. 12355, pp. 492–509. Springer, Cham (2020). https://doi.org/10.1007/978-3-030-58607-2_29
83. Zhu, H., Zuo, X., Wang, S., Cao, X., Yang, R.: Detailed human shape estimation from a single image by hierarchical mesh deformation. In: CVPR, pp. 4491–4500 (2019)
84. Zollhöfer, M., et al.: Real-time non-rigid reconstruction using an RGB-D camera. TOG **33**(4), 1–12 (2014)

Cross-Attention of Disentangled Modalities for 3D Human Mesh Recovery with Transformers

Junhyeong Cho[1], Kim Youwang[2], and Tae-Hyun Oh[2,3]([✉])

[1] Department of CSE, POSTECH, Pohang, Korea
junhyeong99@postech.ac.kr
[2] Department of EE, POSTECH, Pohang, Korea
{youwang.kim,taehyun}@postech.ac.kr
[3] Graduate School of AI, POSTECH, Pohang, Korea
https://github.com/postech-ami/FastMETRO

Abstract. Transformer encoder architectures have recently achieved state-of-the-art results on monocular 3D human mesh reconstruction, but they require a substantial number of parameters and expensive computations. Due to the large memory overhead and slow inference speed, it is difficult to deploy such models for practical use. In this paper, we propose a novel transformer encoder-decoder architecture for 3D human mesh reconstruction from a single image, called *FastMETRO*. We identify the performance bottleneck in the encoder-based transformers is caused by the token design which introduces high complexity interactions among input tokens. We disentangle the interactions via an encoder-decoder architecture, which allows our model to demand much fewer parameters and shorter inference time. In addition, we impose the prior knowledge of human body's morphological relationship via attention masking and mesh upsampling operations, which leads to faster convergence with higher accuracy. Our FastMETRO improves the Pareto-front of accuracy and efficiency, and clearly outperforms image-based methods on Human3.6M and 3DPW. Furthermore, we validate its generalizability on FreiHAND.

1 Introduction

3D human pose and shape estimation models aim to estimate 3D coordinates of human body joints and mesh vertices. These models can be deployed in a wide range of applications that require human behavior understanding, *e.g.*, human motion analysis and human-computer interaction. To utilize such models for practical use, monocular methods [2, 8, 15, 16, 20–22, 24, 25, 34, 36, 40, 44] estimate

---

T. H. Oh—Joint affiliated with Yonsei University, Korea.

Supplementary Information The online version contains supplementary material available at https://doi.org/10.1007/978-3-031-19769-7_20.

the 3D joints and vertices without using 3D scanners or stereo cameras. This task is essentially challenging due to complex human body articulation, and becomes more difficult by occlusions and depth ambiguity in monocular settings.

To deal with such challenges, state-of-the-art methods [24,25] exploit non-local relations among human body joints and mesh vertices via transformer encoder architectures. This leads to impressive improvements in accuracy by consuming a substantial number of parameters and expensive computations as trade-offs; efficiency is less taken into account, although it is crucial in practice.

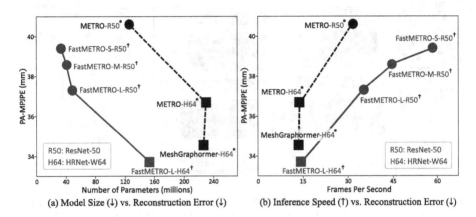

Fig. 1. Comparison with encoder-based transformers [24,25] and our models on Human3.6M [14]. Our FastMETRO substantially improves the Pareto-front of accuracy and efficiency. † indicates training for 60 epochs, and * denotes training for 200 epochs.

In this paper, we propose **FastMETRO** which employs a novel transformer encoder-decoder architecture for 3D human pose and shape estimation from an input image. Compared with the transformer encoders [24,25], FastMETRO is more practical because it achieves competitive results with much fewer parameters and faster inference speed, as shown in Fig. 1. Our architecture is motivated by the observation that the encoder-based methods overlook the importance of the token design which is a key-factor in accuracy and efficiency.

The encoder-based transformers [24,25] share similar transformer encoder architectures. They take K joint and N vertex tokens as input for the estimation of 3D human body joints and mesh vertices, where K and N denote the number of joints and vertices in a 3D human mesh, respectively. Each token is constructed by the concatenation of a global image feature vector $\mathbf{x} \in \mathbb{R}^C$ and 3D coordinates of a joint or vertex in the human mesh. This results in the input tokens of dimension $\mathbb{R}^{(K+N)\times(C+3)}$ which are fed as input to the transformer encoders.[1] This token design introduces the same sources of the performance bottleneck: 1) spatial information is lost in the global image feature $\mathbf{x}$, and 2) the same image feature $\mathbf{x}$ is used in an overly-duplicated way. The former is caused by the

---

[1] For simplicity, we discuss the input tokens mainly based on METRO [24]. Mesh Graphormer [25] has subtle differences, but the essence of the bottleneck is shared.

average pooling operation to obtain the global image feature $\mathbf{x}$. The latter leads to considerable inefficiency, since expensive computations are required to process mostly duplicated information, where distinctively informative signals are only in 0.15% of the input tokens.[2] Furthermore, the computational complexity of each transformer layer is quadratic as $O(L^2C + LC^2)$, where $L \geq K + N$. Once either L or C is dominantly larger, it results in unfavorable efficiency. Both methods [24,25] are such undesirable cases.

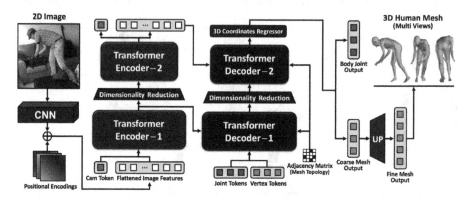

Fig. 2. Overall architecture of FastMETRO. Our model estimates 3D coordinates of human body joints and mesh vertices from a single image. We extract image features via a CNN backbone, which are fed as input to our transformer encoder. In addition to image features produced by the encoder, our transformer decoder takes learnable joint and vertex tokens as input. To effectively learn non-local joint-vertex relations and local vertex-vertex relations, we mask self-attentions of non-adjacent vertices according to the topology of human triangle mesh. Following [24,25], we progressively reduce the hidden dimension sizes via linear projections in our transformer.

In contrast, our FastMETRO does not concatenate an image feature vector for the construction of input tokens. As illustrated in Fig. 2, we disentangle the image encoding part and mesh estimation part via an encoder-decoder architecture. Our joint and vertex tokens focus on certain image regions through cross-attention modules in the transformer decoder. In this way, the proposed method efficiently estimates the 3D coordinates of human body joints and mesh vertices from a 2D image. To effectively capture non-local joint-vertex relations and local vertex-vertex relations, we mask self-attentions of non-adjacent vertices according to the topology of human triangle mesh. To avoid the redundancy caused by the spatial locality of human mesh vertices, we perform coarse-to-fine mesh upsampling as in [22,24,25]. By leveraging the prior knowledge of human body's morphological relationship, we substantially reduce optimization difficulty. This leads to faster convergence with higher accuracy.

We present the proposed method with model-size variants by changing the number of transformer layers: FastMETRO-S, FastMETRO-M, FastMETRO-L. Compared with the encoder-based transformers [24,25], FastMETRO-S requires

---

[2] 3-dimensional coordinates out of $(C+3)$-dimensional input tokens, where $C = 2048$.

only about 9% of the parameters in the transformer architecture, but shows competitive results with much faster inference speed. In addition, the large variant (FastMETRO-L) achieves the state of the art on the Human3.6M [14] and 3DPW [32] datasets among image-based methods, which also demands fewer parameters and shorter inference time compared with the encoder-based methods. We demonstrate the effectiveness of the proposed method by conducting extensive experiments, and validate its generalizability by showing 3D hand mesh reconstruction results on the FreiHAND [47] dataset.

Our contributions are summarized as follows:

- We propose FastMETRO which employs a novel transformer encoder-decoder architecture for 3D human mesh recovery from a single image. Our method resolves the performance bottleneck in the encoder-based transformers, and improves the Pareto-front of accuracy and efficiency.
- The proposed model converges much faster by reducing optimization difficulty. Our FastMETRO leverages the prior knowledge of human body's morphological relationship, *e.g.*, masking attentions according to the human mesh topology.
- We present model-size variants of our FastMETRO. The small variant shows competitive results with much fewer parameters and faster inference speed. The large variant clearly outperforms existing image-based methods on the Human3.6M and 3DPW datasets, which is also more lightweight and faster.

2   Related Work

Our proposed method aims to estimate the 3D coordinates of human mesh vertices from an input image by leveraging the attention mechanism in the transformer architecture. We briefly review relevant methods in this section.

Human Mesh Reconstruction. The reconstruction methods belong to one of the two categories: parametric approach and non-parametric approach. The parametric approach learns to estimate the parameters of a human body model such as SMPL [29]. On the other hand, the non-parametric approach learns to directly regress the 3D coordinates of human mesh vertices. They obtain the 3D coordinates of human body joints via linear regression from the estimated mesh.

The reconstruction methods in the parametric approach [2,8,11,15,16,20,21, 36,40,44] have shown stable performance in monocular 3D human mesh recovery. They have achieved the robustness to environment variations by exploiting the human body prior encoded in a human body model such as SMPL [29]. However, their regression targets are difficult for deep neural networks to learn; the pose space in the human body model is expressed by the 3D rotations of human body joints, where the regression of the 3D rotations is challenging [31].

Recent advances in deep neural networks have enabled the non-parametric approach with promising performance [6,22,24,25,34]. Kolotouros *et al.* [22] propose a graph convolutional neural network (GCNN) [18] to effectively learn local vertex-vertex relations, where the graph structure is based on the topology of SMPL human triangle mesh [29]. They extract a global image feature vector through a CNN backbone, then construct vertex embeddings by concatenating the image feature vector with the 3D coordinates of vertices in the human mesh. After iterative updates via graph convolutional layers, they estimate the 3D locations of human mesh vertices. To improve the robustness to partial occlusions, Lin *et al.* [24,25] propose transformer encoder architectures which effectively learn the non-local relations among human body joints and mesh vertices via the attention mechanism in the transformer. Their models, METRO [24] and Mesh Graphormer [25], follow the similar framework with the GCNN-based method [22]. They construct vertex tokens by attaching a global image feature vector to the 3D coordinates of vertices in the human mesh. After several updates via transformer encoder layers, they regress the 3D coordinates of human mesh vertices.

Among the reconstruction methods, METRO [24] and Mesh Graphormer [25] are the most relevant work to our FastMETRO. We found that the token design in those methods leads to a substantial number of unnecessary parameters and computations. In their architectures, transformer encoders take all the burdens to learn complex relations among mesh vertices, along with the highly non-linear mapping between 2D space and 3D space. To resolve this issue, we disentangle the image-encoding and mesh-estimation parts via an encoder-decoder architecture. This makes FastMETRO more lightweight and faster, and allows our model to learn the complex relations more effectively.

Transformers. Vaswani *et al.* [41] introduce a transformer architecture which effectively learns long-range relations through the attention mechanism in the transformer. This architecture has achieved impressive improvements in diverse computer vision tasks [3–5,7,12,17,24,25,27,28,30,38,43,45]. Dosovitskiy *et al.* [7] present a transformer encoder architecture, where a learnable token aggregates image features via self-attentions for image classification. Carion *et al.* [3] propose a transformer encoder-decoder architecture, where learnable tokens focus on certain image regions via cross-attentions for object detection. Those transformers have the most relevant architectures to our model.

Our FastMETRO employs a transformer encoder-decoder architecture, whose decoupled structure is favorable to learn the complex relations between the heterogeneous modalities of 2D image and 3D mesh. Compared with the existing transformers [3–5,7,12,17,27,30,38,43,45], we progressively reduce hidden dimension sizes in the transformer architecture as in [24,25]. Our separate decoder design enables FastMETRO to easily impose the human body prior by masking self-attentions of decoder input tokens, which leads to stable optimization and higher accuracy. This is novel in transformer architectures.

3   Method

We propose a novel method, called **Fast MEsh TRansfOrmer** (FastMETRO). FastMETRO has a transformer encoder-decoder architecture for 3D human mesh recovery from an input image. The overview of our method is shown in Fig. 2. The details of our transformer encoder and decoder are illustrated in Fig. 3.

3.1 Feature Extractor

Given a single RGB image, our model extracts image features $\mathbf{X}_I \in \mathbb{R}^{H \times W \times C}$ through a CNN backbone, where $H \times W$ denotes the spatial dimension size and C denotes the channel dimension size. A 1×1 convolution layer takes the image features $\mathbf{X}_I$ as input, and reduces the channel dimension size to D. Then, a flatten operation produces flattened image features $\mathbf{X}_F \in \mathbb{R}^{HW \times D}$. Note that we employ positional encodings for retaining spatial information in our transformer, as illustrated in Fig. 3.

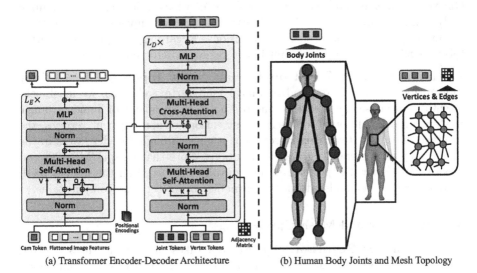

(a) Transformer Encoder-Decoder Architecture   (b) Human Body Joints and Mesh Topology

Fig. 3. Details of our transformer architecture and 3D human body mesh. For simplicity, we illustrate the transformer without progressive dimensionality reduction. Note that the camera feature is not fed as input to the decoder. We mask attentions using the adjacency matrix obtained from the human triangle mesh of SMPL [29].

3.2 Transformer with Progressive Dimensionality Reduction

Following the encoder-based transformers [24,25], FastMETRO progressively reduces the hidden dimension sizes in the transformer architecture via linear projections, as illustrated in Fig. 2.

Transformer Encoder. Our transformer encoder (Fig. 3a) takes a learnable camera token and the flattened image features $\mathbf{X}_F$ as input. The camera token captures essential features to predict weak-perspective camera parameters through the attention mechanism in the transformer; the camera parameters are used for fitting the 3D estimated human mesh to the 2D input image. Given the camera token and image features, the transformer encoder produces a camera feature and aggregated image features $\mathbf{X}_A \in \mathbb{R}^{HW \times D}$.

Transformer Decoder. In addition to the image features $\mathbf{X}_A$ obtained from the encoder, our transformer decoder (Fig. 3a) takes the set of learnable joint tokens and the set of learnable vertex tokens as input. Each token in the set of joint tokens $\mathcal{T}_J = \{\mathbf{t}_1^J, \mathbf{t}_2^J, \ldots, \mathbf{t}_K^J\}$ is used to estimate 3D coordinates of a human body joint, where $\mathbf{t}_i^J \in \mathbb{R}^D$. The joint tokens correspond to the body joints in Fig. 3b. Each token in the set of vertex tokens $\mathcal{T}_V = \{\mathbf{t}_1^V, \mathbf{t}_2^V, \ldots, \mathbf{t}_N^V\}$ is used to estimate 3D coordinates of a human mesh vertex, where $\mathbf{t}_j^V \in \mathbb{R}^D$. The vertex tokens correspond to the mesh vertices in Fig. 3b. Given the image features and tokens, the transformer decoder produces joint features $\mathbf{X}_J \in \mathbb{R}^{K \times D}$ and vertex features $\mathbf{X}_V \in \mathbb{R}^{N \times D}$ through self-attention and cross-attention modules. Our transformer decoder effectively captures non-local relations among human body joints and mesh vertices via self-attentions, which improves the robustness to environment variations such as occlusions. Regarding the joint and vertex tokens, each focuses on its relevant image region via cross-attentions.

Attention Masking Based on Human Mesh Topology. To effectively capture local vertex-vertex and non-local joint-vertex relations, we mask self-attentions of non-adjacent vertices according to the topology of human triangle mesh in Fig. 3b. Although we mask the attentions of non-adjacent vertices, the coverage of each vertex token increases as it goes through decoder layers in the similar way with iterative graph convolutions. Note that GraphCMR [22] and Mesh Graphormer [25] perform graph convolutions based on the human mesh topology, which demands additional learnable parameters and computations.

3.3 Regressor and Mesh Upsampling

3D Coordinates Regressor. Our regressor takes the joint features $\mathbf{X}_J$ and vertex features $\mathbf{X}_V$ as input, and estimates the 3D coordinates of human body joints and mesh vertices. As a result, 3D joint coordinates $\hat{\mathbf{J}}_{3D} \in \mathbb{R}^{K \times 3}$ and 3D vertex coordinates $\hat{\mathbf{V}}_{3D} \in \mathbb{R}^{N \times 3}$ are predicted.

Coarse-to-Fine Mesh Upsampling. Following [22,24,25], our FastMETRO estimates a coarse mesh, then upsample the mesh. In this way, we avoid the redundancy caused by the spatial locality of human mesh vertices. As in [22], FastMETRO obtains the fine mesh output $\hat{\mathbf{V}}_{3D}' \in \mathbb{R}^{M \times 3}$ from the coarse mesh output $\hat{\mathbf{V}}_{3D}$ by performing matrix multiplication with the upsampling matrix $\mathbf{U} \in \mathbb{R}^{M \times N}$, i.e., $\hat{\mathbf{V}}_{3D}' = \mathbf{U}\hat{\mathbf{V}}_{3D}$, where the upsampling matrix $\mathbf{U}$ is pre-computed by the sampling algorithm in [37].

3.4 Training FastMETRO

3D Vertex Regression Loss. To train our model for the regression of 3D mesh vertices, we use $L1$ loss function. This regression loss L_{3D}^V is computed by

$$L_{3D}^V = \frac{1}{M} \|\hat{\mathbf{V}}_{3D}' - \bar{\mathbf{V}}_{3D}\|_1, \tag{1}$$

where $\bar{\mathbf{V}}_{3D} \in \mathbb{R}^{M \times 3}$ denotes the ground-truth 3D vertex coordinates.

3D Joint Regression Loss. In addition to the estimated 3D joints $\hat{\mathbf{J}}_{3D}$, we also obtain 3D joints $\hat{\mathbf{J}}'_{3D} \in \mathbb{R}^{K \times 3}$ regressed from the fine mesh $\hat{\mathbf{V}}'_{3D}$, which is the common practice in the literature [6,8,16,20–22,24,25,34,40,44]. The regressed joints $\hat{\mathbf{J}}'_{3D}$ are computed by the matrix multiplication of the joint regression matrix $\mathbf{R} \in \mathbb{R}^{K \times M}$ and the fine mesh $\hat{\mathbf{V}}'_{3D}$, *i.e.*, $\hat{\mathbf{J}}'_{3D} = \mathbf{R}\hat{\mathbf{V}}'_{3D}$, where the regression matrix $\mathbf{R}$ is pre-defined in SMPL [29]. To train our model for the regression of 3D body joints, we use $L1$ loss function. This regression loss L^J_{3D} is computed by

$$L^J_{3D} = \frac{1}{K}(\|\hat{\mathbf{J}}_{3D} - \bar{\mathbf{J}}_{3D}\|_1 + \|\hat{\mathbf{J}}'_{3D} - \bar{\mathbf{J}}_{3D}\|_1), \tag{2}$$

where $\bar{\mathbf{J}}_{3D} \in \mathbb{R}^{K \times 3}$ denotes the ground-truth 3D joint coordinates.

2D Joint Projection Loss. Following the literature [8,16,20–22,24,25,40,44], for the alignment between the 2D input image and the 3D reconstructed human mesh, we train our model to estimate weak-perspective camera parameters $\{s, \mathbf{t}\}$; a scaling factor $s \in \mathbb{R}$ and a 2D translation vector $\mathbf{t} \in \mathbb{R}^2$. The weak-perspective camera parameters are estimated from the camera feature obtained by the transformer encoder. Using the camera parameters, we get 2D body joints via an orthographic projection of the estimated 3D body joints. The projected 2D body joints are computed by

$$\hat{\mathbf{J}}_{2D} = s\Pi(\hat{\mathbf{J}}_{3D}) + \mathbf{t}, \tag{3}$$

$$\hat{\mathbf{J}}'_{2D} = s\Pi(\hat{\mathbf{J}}'_{3D}) + \mathbf{t}, \tag{4}$$

where $\Pi(\cdot)$ denotes the orthographic projection; $\left[\begin{smallmatrix} 1 & 0 & 0 \\ 0 & 1 & 0 \end{smallmatrix}\right]^{\mathsf{T}} \in \mathbb{R}^{3 \times 2}$ is used for this projection in FastMETRO. To train our model with the projection of 3D body joints onto the 2D image, we use $L1$ loss function. This projection loss L^J_{2D} is computed by

$$L^J_{2D} = \frac{1}{K}(\|\hat{\mathbf{J}}_{2D} - \bar{\mathbf{J}}_{2D}\|_1 + \|\hat{\mathbf{J}}'_{2D} - \bar{\mathbf{J}}_{2D}\|_1), \tag{5}$$

where $\bar{\mathbf{J}}_{2D} \in \mathbb{R}^{K \times 2}$ denotes the ground-truth 2D joint coordinates.

Total Loss. Following the literature [6,8,16,19–22,24,25,34,40,44], we train our model with multiple 3D and 2D training datasets to improve its accuracy and robustness. This total loss L_{total} is computed by

$$L_{\text{total}} = \alpha(\lambda^V_{3D}L^V_{3D} + \lambda^J_{3D}L^J_{3D}) + \beta\lambda^J_{2D}L^J_{2D}, \tag{6}$$

where $\lambda^V_{3D}, \lambda^J_{3D}, \lambda^J_{2D} > 0$ are loss coefficients and $\alpha, \beta \in \{0, 1\}$ are binary flags which denote the availability of ground-truth 3D and 2D coordinates.

Table 1. Configurations for the variants of FastMETRO. Each has the same transformer architecture with a different number of layers. Only transformer parts are described.

Model	#Params	Time (ms)	Enc–1 & Dec–1		Enc–2 & Dec–2	
			#Layers	Dimension	#Layers	Dimension
FastMETRO–S	9.2M	9.6	1	512	1	128
FastMETRO–M	17.1M	15.0	2	512	2	128
FastMETRO–L	24.9M	20.8	3	512	3	128

4   Implementation Details

We implement our proposed method with three variants: FastMETRO-S, FastMETRO-M, FastMETRO-L. They have the same architecture with a differ-

Table 2. Comparison with transformers for monocular 3D human mesh recovery on Human3.6M [14]. † and * indicate training for 60 epochs and 200 epochs, respectively.

Model	CNN Backbone		Transformer		Overall		
	#Params	Time (ms)	#Params	Time (ms)	#Params	FPS	PA-MPJPE ↓
METRO–R50* [24]	23.5M	7.5	102.3M	24.2	125.8M	31.5	40.6
METRO–H64† [24]	128.1M	49.0	102.3M	24.2	230.4M	13.7	38.0
METRO–H64* [24]	128.1M	49.0	102.3M	24.2	230.4M	13.7	36.7
MeshGraphormer–H64† [25]	128.1M	49.0	98.4M	24.5	226.5M	13.6	35.8
MeshGraphormer–H64* [25]	128.1M	49.0	98.4M	24.5	226.5M	13.6	34.5
FastMETRO–S–R50†	23.5M	7.5	**9.2M**	**9.6**	**32.7M**	**58.5**	39.4
FastMETRO–M–R50†	23.5M	7.5	17.1M	15.0	40.6M	44.4	38.6
FastMETRO–L–R50†	23.5M	7.5	24.9M	20.8	48.4M	35.3	37.3
FastMETRO–L–H64†	128.1M	49.0	24.9M	20.8	153.0M	14.3	**33.7**

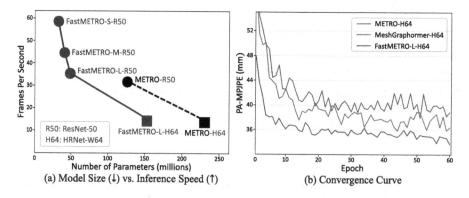

(a) Model Size (↓) vs. Inference Speed (↑)   (b) Convergence Curve

Fig. 4. Comparison with encoder-based transformers [24,25] and our proposed models on Human3.6M [14]. The small variant of our FastMETRO shows much faster inference speed, and its large variant converges faster than the transformer encoders.

ent number of layers in the transformer encoder and decoder. Table 1 shows the configuration for each variant. Our transformer encoder and decoder are initialized with Xavier Initialization [9]. Please refer to the supplementary material for complete implementation details.

5 Experiments

5.1 Datasets

Following the encoder-based transformers [24, 25], we train our FastMETRO with **Human3.6M** [14], **UP-3D** [23], **MuCo-3DHP** [33], **COCO** [26] and **MPII** [1] training datasets, and evaluate the model on P2 protocol in Human3.6M. Then, we fine-tune our model with **3DPW** [32] training dataset, and evaluate the model on its test dataset.

Following the common practice [6, 24, 25, 34], we employ the pseudo 3D human mesh obtained by SMPLify-X [35] to train our model with Human3.6M [14]; there is no available ground-truth 3D human mesh in the Human3.6M training dataset due to the license issue. For fair comparison, we employ the ground-truth 3D human body joints in Human3.6M during the evaluation of our model. Regarding the experiments on 3DPW [32], we use its training dataset for fine-tuning our model as in the encoder-based transformers [24, 25].

Table 3. Comparison with the state-of-the-art monocular 3D human pose and mesh recovery methods on 3DPW [32] and Human3.6M [14] among image-based methods.

Model	3DPW			Human3.6M	
	MPVPE ↓	MPJPE ↓	PA-MPJPE ↓	MPJPE ↓	PA-MPJPE ↓
HMR–R50 [16]	–	130.0	76.7	88.0	56.8
GraphCMR–R50 [22]	–	–	70.2	–	50.1
SPIN–R50 [21]	116.4	96.9	59.2	62.5	41.1
I2LMeshNet–R50 [34]	–	93.2	57.7	55.7	41.1
PyMAF–R50 [44]	110.1	92.8	58.9	57.7	40.5
ROMP–R50 [40]	105.6	89.3	53.5	–	–
ROMP–H32 [40]	103.1	85.5	53.3	–	–
PARE–R50 [20]	99.7	82.9	52.3	–	–
METRO–R50 [24]	–	–	–	56.5	40.6
DSR–R50 [8]	99.5	85.7	51.7	60.9	40.3
METRO–H64 [24]	88.2	77.1	47.9	54.0	36.7
PARE–H32 [20]	88.6	74.5	46.5	–	–
MeshGraphormer–H64 [25]	87.7	74.7	45.6	**51.2**	34.5
FastMETRO–S–R50	91.9	79.6	49.3	55.7	39.4
FastMETRO–M–R50	91.2	78.5	48.4	55.1	38.6
FastMETRO–L–R50	90.6	77.9	48.3	53.9	37.3
FastMETRO–L–H64	**84.1**	**73.5**	**44.6**	52.2	**33.7**

5.2   Evaluation Metrics

We evaluate our FastMETRO using three evaluation metrics: MPJPE [14], PA-MPJPE [46], MPVPE [36]. The unit of each metric is millimeter.

MPJPE. This metric denotes Mean-Per-Joint-Position-Error. It measures the Euclidean distances between the predicted and ground-truth joint coordinates.

PA-MPJPE. This metric is often called *Reconstruction Error*. It measures MPJPE after 3D alignment using Procrustes Analysis (PA) [10].

MPVPE. This metric denotes Mean-Per-Vertex-Position-Error. It measures the Euclidean distances between the predicted and ground-truth vertex coordinates.

5.3   Experimental Results

We evaluate the model-size variants of our FastMETRO on the 3DPW [32] and Human3.6M [14] datasets. In this paper, the inference time is measured using a single NVIDIA V100 GPU with a batch size of 1.

Comparison with Encoder-Based Transformers. In Table 2, we compare our models with METRO [24] and Mesh Graphormer [25] on the Human3.6M [14] dataset. Note that encoder-based transformers [24,25] are implemented with ResNet-50 [13] (**R50**) or HRNet-W64 [42] (**H64**). FastMETRO-S outperforms

Fig. 5. Qualitative results of our FastMETRO on Human3.6M [14] and 3DPW [32]. We visualize the 3D human mesh estimated by FastMETRO-L-H64. By leveraging the attention mechanism in the transformer, our model is robust to partial occlusions.

METRO when both models employ the same CNN backbone (R50), although our model demands only 8.99% of the parameters in the transformer architecture. Regarding the overall inference speed, our model is 1.86× faster. It is worth noting that FastMETRO-L-R50 achieves similar results with METRO-H64, but our model is 2.58× faster. FastMETRO-L outperforms Mesh Graphormer when both models employ the same CNN backbone (H64), while our model demands only 25.30% of the parameters in the transformer architecture. Also, our model converges much faster than the encoder-based methods as shown in Fig. 4.

Comparison with Image-Based Methods. In Table 3, we compare our FastMETRO with the image-based methods for 3D human mesh reconstruction on 3DPW [32] and Human3.6M [14]. Note that existing methods are implemented with ResNet-50 [13] (**R50**) or HRNet-W32 [42] (**H32**) or HRNet-W64 [42] (**H64**). When all models employ R50 as their CNN backbones, FastMETRO-S achieves the best results without iterative fitting procedures or test-time optimizations. FastMETRO-L-H64 achieves the state of the art in every evaluation metric on the 3DPW dataset and PA-MPJPE metric on the Human3.6M dataset.

Visualization of Self-Attentions. In Fig. 6, the first and second rows show the visualization of the attention scores in self-attentions between a specified body joint and mesh vertices. We obtain the scores by averaging attention scores from

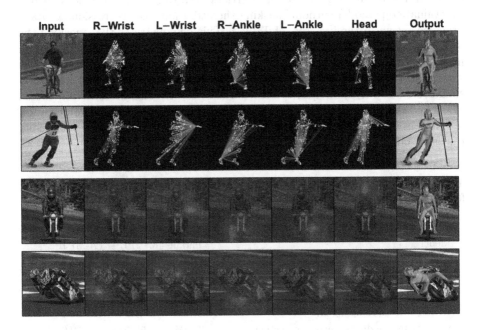

Fig. 6. Qualitative results of FastMETRO-L-H64 on COCO [26]. We visualize the attentions scores in self-attentions (top two rows) and cross-attentions (bottom two rows). The brighter lines or regions indicate higher attention scores.

all attention heads of all multi-head self-attention modules in our transformer decoder. As shown in Fig. 6, our FastMETRO effectively captures the non-local relations among joints and vertices via self-attentions in the transformer. This improves the robustness to environment variations such as occlusions.

Visualization of Cross-Attentions. In Fig. 6, the third and fourth rows show the visualization of the attention scores in cross-attentions between a specified body joint and image regions. We obtain the scores by averaging attention scores from all attention heads of all multi-head cross-attention modules in our transformer decoder. As shown in Fig. 6, the input tokens used in our transformer decoder focus on their relevant image regions. By leveraging the cross-attentions between disentangled modalities, our FastMETRO effectively learns to regress the 3D coordinates of joints and vertices from a 2D image.

5.4   Ablation Study

We analyze the effects of different components in our FastMETRO as shown in Table 4. Please refer to the supplementary material for more experiments.

Attention Masking. To effectively learn the local relations among mesh vertices, GraphCMR [22] and Mesh Graphormer [25] perform graph convolutions based on the topology of SMPL human triangle mesh [29]. For the same goal, we mask self-attentions of non-adjacent vertices according to the topology. When we evaluate our model without masking the attentions, the regression accuracy drops as shown in the first row of Table 4. This demonstrates that masking the attentions of non-adjacent vertices is effective. To compare the effects of attention masking with graph convolutions, we train our model using graph convolutions without masking the attentions. As shown in the second row of Table 4, we obtain similar results but this requires more parameters. We also evaluate our model when we mask the attentions in half attention heads, *i.e.*, there is no attention masking in other half attention heads. In this case, we get similar results using the same number of parameters as shown in the third row of Table 4.

Table 4. Ablation study of our FastMETRO on Human3.6M [14]. The effects of different components are evaluated. The default model is FastMETRO-S-R50.

Model	#Params	Human3.6M	
		MPJPE ↓	PA-MPJPE ↓
w/o attention masking	**32.7M**	58.0	40.7
w/o attention masking + w/ graph convolutions	33.1M	56.6	**39.4**
w/ attention masking in half attention heads	**32.7M**	55.8	**39.4**
w/ learnable upsampling layers	45.4M	58.1	41.1
w/o progressive dimensionality reduction	39.5M	**55.5**	39.6
FastMETRO–S–R50	**32.7M**	55.7	**39.4**

Coarse-to-Fine Mesh Upsampling. The existing transformers [24, 25] also first estimate a coarse mesh, then upsample the mesh to obtain a fine mesh. They employ two learnable linear layers for the upsampling. In our FastMETRO, we use the pre-computed upsampling matrix $\mathbf{U}$ to reduce optimization difficulty as in [22]; this upsampling matrix is a sparse matrix which has only about $25K$ non-zero elements. When we perform the mesh upsampling using learnable linear layers instead of the matrix $\mathbf{U}$, the regression accuracy drops as shown in the fourth row of Table 4, although it demands much more parameters.

Progressive Dimensionality Reduction. Following the existing transformer encoders [24, 25], we also progressively reduce the hidden dimension sizes in our transformer via linear projections. To evaluate its effectiveness, we train our model using the same number of transformer layers but without progressive dimensionality reduction, *i.e.*, hidden dimension sizes in all transformer layers are the same. As shown in the fifth row of Table 4, we obtain similar results but this requires much more parameters. This demonstrates that the dimensionality reduction is helpful for our model to achieve decent results using fewer parameters.

Generalizability. Our model can reconstruct any arbitrary 3D objects by changing the number of input tokens used in the transformer decoder. Note that we can employ learnable layers for coarse-to-fine mesh upsampling without masking attentions. For 3D hand mesh reconstruction, there is a pre-computed upsampling matrix and a human hand model such as MANO [39]. Thus, we can leverage the matrix for mesh upsampling and mask self-attentions of non-adjacent vertices in the same way with 3D human mesh recovery. As illustrated in Fig. 7, we can obtain an adjacency matrix and construct joint and vertex tokens from the human hand mesh topology. To validate the generalizability of our method, we train FastMETRO-L-H64 on the FreiHAND [47] training dataset and evaluate the model. As shown in Table 5, our proposed model achieves competitive results on FreiHAND.

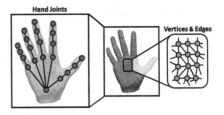

Fig. 7. Hand Joints and Mesh Topology.

Table 5. Comparison with transformers for monocular 3D hand mesh recovery on FreiHAND [14]. Test-time augmentation is not applied to these transformers.

Model	Transformer #Params	Overall #Params	FreiHAND PA-MPJPE ↓	F@15mm ↑
METRO–H64 [24]	102.3M	230.4M	6.8	0.981
FastMETRO–L–H64	**24.9M**	**153.0M**	**6.5**	**0.982**

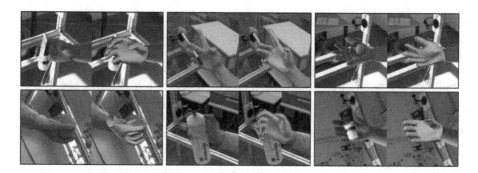

Fig. 8. Qualitative results of our FastMETRO on FreiHAND [47]. We visualize the 3D hand mesh estimated by FastMETRO-L-H64. By leveraging the attention mechanism in the transformer, our model is robust to partial occlusions.

6   Conclusion

We identify the performance bottleneck in the encoder-based transformers is due to the design of input tokens, and resolve this issue via an encoder-decoder architecture. This allows our model to demand much fewer parameters and shorter inference time, which is more appropriate for practical use. The proposed method leverages the human body prior encoded in SMPL human mesh, which reduces optimization difficulty and leads to faster convergence with higher accuracy. To be specific, we mask self-attentions of non-adjacent vertices and perform coarse-to-fine mesh upsampling. We demonstrate that our method improves the Pareto-front of accuracy and efficiency. Our FastMETRO achieves the robustness to occlusions by capturing non-local relations among body joints and mesh vertices, which outperforms image-based methods on the Human3.6M and 3DPW datasets. A limitation is that a substantial number of samples are required to train our model as in the encoder-based transformers.

Acknowledgments. This work was supported by Institute of Information & communications Technology Planning & Evaluation (IITP) grant funded by the Korea government (MSIT) (No. 2022-0-00290, Visual Intelligence for Space-Time Understanding and Generation based on Multi-layered Visual Common Sense; and No. 2019-0-01906, Artificial Intelligence Graduate School Program (POSTECH)).

References

1. Andriluka, M., Pishchulin, L., Gehler, P., Schiele, B.: 2D human pose estimation: new benchmark and state of the art analysis. In: Proceedings of the IEEE Conference on Computer Vision and Pattern Recognition (CVPR) (2014)
2. Bogo, F., Kanazawa, A., Lassner, C., Gehler, P., Romero, J., Black, M.J.: Keep it SMPL: automatic estimation of 3D human pose and shape from a single image. In: Leibe, B., Matas, J., Sebe, N., Welling, M. (eds.) ECCV 2016. LNCS, vol. 9909, pp. 561–578. Springer, Cham (2016). https://doi.org/10.1007/978-3-319-46454-1_34

3. Carion, N., Massa, F., Synnaeve, G., Usunier, N., Kirillov, A., Zagoruyko, S.: End-to-end object detection with transformers. In: Vedaldi, A., Bischof, H., Brox, T., Frahm, J.-M. (eds.) ECCV 2020. LNCS, vol. 12346, pp. 213–229. Springer, Cham (2020). https://doi.org/10.1007/978-3-030-58452-8_13

4. Cho, J., Yoon, Y., Kwak, S.: Collaborative transformers for grounded situation recognition. In: Proceedings of the IEEE/CVF Conference on Computer Vision and Pattern Recognition (CVPR) (2022)

5. Cho, J., Yoon, Y., Lee, H., Kwak, S.: Grounded situation recognition with transformers. In: Proceedings of the British Machine Vision Conference (BMVC) (2021)

6. Choi, H., Moon, G., Lee, K.M.: Pose2Mesh: graph convolutional network for 3D human pose and mesh recovery from a 2D human pose. In: Vedaldi, A., Bischof, H., Brox, T., Frahm, J.-M. (eds.) ECCV 2020. LNCS, vol. 12352, pp. 769–787. Springer, Cham (2020). https://doi.org/10.1007/978-3-030-58571-6_45

7. Dosovitskiy, A., et al.: An image is worth 16x16 words: transformers for image recognition at scale. In: International Conference on Learning Representations (ICLR) (2021)

8. Dwivedi, S.K., Athanasiou, N., Kocabas, M., Black, M.J.: Learning to regress bodies from images using differentiable semantic rendering. In: Proceedings of the IEEE/CVF International Conference on Computer Vision (ICCV) (2021)

9. Glorot, X., Bengio, Y.: Understanding the difficulty of training deep feedforward neural networks. In: Proceedings of the Thirteenth International Conference on Artificial Intelligence and Statistics, pp. 249–256 (2010)

10. Gower, J.C.: Generalized procrustes analysis. Psychometrika **40**, 33–51 (1975)

11. Guan, P., Weiss, A., Balan, A.O., Black, M.J.: Estimating human shape and pose from a single image. In: Proceedings of the IEEE International Conference on Computer Vision (ICCV) (2009)

12. Guo, L., Liu, J., Zhu, X., Yao, P., Lu, S., Lu, H.: Normalized and geometry-aware self-attention network for image captioning. In: Proceedings of the IEEE/CVF Conference on Computer Vision and Pattern Recognition (CVPR) (2020)

13. He, K., Zhang, X., Ren, S., Sun, J.: Deep residual learning for image recognition. In: Proceedings of the IEEE Conference on Computer Vision and Pattern Recognition (CVPR), pp. 770–778 (2016)

14. Ionescu, C., Papava, D., Olaru, V., Sminchisescu, C.: Human3.6M: large scale datasets and predictive methods for 3D human sensing in natural environments. IEEE Trans. Pattern Anal. Mach. Intell. TPAMI **36**(7), 1325–1339 (2014)

15. Jiang, W., Kolotouros, N., Pavlakos, G., Zhou, X., Daniilidis, K.: Coherent reconstruction of multiple humans from a single image. In: Proceedings of the IEEE/CVF Conference on Computer Vision and Pattern Recognition (CVPR) (2020)

16. Kanazawa, A., Black, M.J., Jacobs, D.W., Malik, J.: End-to-end recovery of human shape and pose. In: Proceedings of the IEEE/CVF Conference on Computer Vision and Pattern Recognition (CVPR) (2018)

17. Kim, B., Lee, J., Kang, J., Kim, E.S., Kim, H.J.: HOTR: end-to-end human-object interaction detection with transformers. In: Proceedings of the IEEE/CVF Conference on Computer Vision and Pattern Recognition (CVPR), pp. 74–83 (2021)

18. Kipf, T.N., Welling, M.: Semi-supervised classification with graph convolutional networks. In: International Conference on Learning Representations (ICLR) (2017)

19. Kocabas, M., Athanasiou, N., Black, M.J.: VIBE: video inference for human body pose and shape estimation. In: Proceedings of the IEEE/CVF Conference on Computer Vision and Pattern Recognition (CVPR) (2020)

20. Kocabas, M., Huang, C.H.P., Hilliges, O., Black, M.J.: PARE: part attention regressor for 3D human body estimation. In: Proceedings of the IEEE/CVF International Conference on Computer Vision (ICCV) (2021)
21. Kolotouros, N., Pavlakos, G., Black, M.J., Daniilidis, K.: Learning to reconstruct 3D human pose and shape via model-fitting in the loop. In: Proceedings of the IEEE/CVF International Conference on Computer Vision (ICCV) (2019)
22. Kolotouros, N., Pavlakos, G., Daniilidis, K.: Convolutional mesh regression for single-image human shape reconstruction. In: Proceedings of the IEEE/CVF Conference on Computer Vision and Pattern Recognition (CVPR) (2019)
23. Lassner, C., Romero, J., Kiefel, M., Bogo, F., Black, M.J., Gehler, P.V.: Unite the people: closing the loop between 3D and 2D human representations. In: Proceedings of the IEEE Conference on Computer Vision and Pattern Recognition (CVPR) (2017)
24. Lin, K., Wang, L., Liu, Z.: End-to-end human pose and mesh reconstruction with transformers. In: Proceedings of the IEEE/CVF Conference on Computer Vision and Pattern Recognition (CVPR), pp. 1954–1963 (2021)
25. Lin, K., Wang, L., Liu, Z.: Mesh graphormer. In: Proceedings of the IEEE/CVF International Conference on Computer Vision (ICCV), pp. 12939–12948 (2021)
26. Lin, T.-Y., et al.: Microsoft COCO: common objects in context. In: Fleet, D., Pajdla, T., Schiele, B., Tuytelaars, T. (eds.) ECCV 2014. LNCS, vol. 8693, pp. 740–755. Springer, Cham (2014). https://doi.org/10.1007/978-3-319-10602-1_48
27. Liu, S., et al.: Paint transformer: feed forward neural painting with stroke prediction. In: Proceedings of the IEEE/CVF International Conference on Computer Vision (ICCV), pp. 6598–6607 (2021)
28. Liu, Z., et al.: Swin transformer: hierarchical vision transformer using shifted windows. In: Proceedings of the IEEE/CVF International Conference on Computer Vision (ICCV) (2021)
29. Loper, M., Mahmood, N., Romero, J., Pons-Moll, G., Black, M.J.: SMPL: a skinned multi-person linear model. ACM Trans. Graphics (SIGGRAPH Asia) **34**(6), 248 (2015)
30. Lu, Y., et al.: Context-aware scene graph generation with Seq2Seq transformers. In: Proceedings of the IEEE/CVF International Conference on Computer Vision (ICCV), pp. 15931–15941 (2021)
31. Mahendran, S., Ali, H., Vidal, R.: A mixed classification-regression framework for 3D pose estimation from 2D images. In: Proceedings of the British Machine Vision Conference (BMVC) (2018)
32. von Marcard, T., Henschel, R., Black, M.J., Rosenhahn, B., Pons-Moll, G.: Recovering accurate 3D human pose in the wild using IMUs and a moving camera. In: Ferrari, V., Hebert, M., Sminchisescu, C., Weiss, Y. (eds.) ECCV 2018. LNCS, vol. 11214, pp. 614–631. Springer, Cham (2018). https://doi.org/10.1007/978-3-030-01249-6_37
33. Mehta, D., et al.: Single-shot multi-person 3D pose estimation from monocular RGB. In: International Conference on 3D Vision (3DV) (2018)
34. Moon, G., Lee, K.M.: I2L-MeshNet: image-to-lixel prediction network for accurate 3D Human pose and mesh estimation from a single RGB image. In: Vedaldi, A., Bischof, H., Brox, T., Frahm, J.-M. (eds.) ECCV 2020. LNCS, vol. 12352, pp. 752–768. Springer, Cham (2020). https://doi.org/10.1007/978-3-030-58571-6_44
35. Pavlakos, G., et al.: Expressive body capture: 3D hands, face, and body from a single image. In: Proceedings of the IEEE/CVF Conference on Computer Vision and Pattern Recognition (CVPR) (2019)

36. Pavlakos, G., Zhu, L., Zhou, X., Daniilidis, K.: Learning to estimate 3D human pose and shape from a single color image. In: Proceedings of the IEEE/CVF Conference on Computer Vision and Pattern Recognition (CVPR) (2018)
37. Ranjan, A., Bolkart, T., Sanyal, S., Black, M.J.: Generating 3D faces using convolutional mesh autoencoders. In: Ferrari, V., Hebert, M., Sminchisescu, C., Weiss, Y. (eds.) ECCV 2018. LNCS, vol. 11207, pp. 725–741. Springer, Cham (2018). https://doi.org/10.1007/978-3-030-01219-9_43
38. Rao, Y., Zhao, W., Liu, B., Lu, J., Zhou, J., Hsieh, C.J.: DynamicViT: efficient vision transformers with dynamic token sparsification. In: Advances in Neural Information Processing Systems (NIPS) (2021)
39. Romero, J., Tzionas, D., Black, M.J.: Embodied hands: modeling and capturing hands and bodies together. ACM Trans. Graphics (Proc. SIGGRAPH Asia) **36**(6) (2017)
40. Sun, Y., Bao, Q., Liu, W., Fu, Y., Black, M.J., Mei, T.: Monocular, one-stage, regression of multiple 3D people. In: Proceedings of the IEEE/CVF International Conference on Computer Vision (ICCV) (2021)
41. Vaswani, A., et al.: Attention is all you need. In: Advances in Neural Information Processing Systems (NIPS) (2017)
42. Wang, J., et al.: Deep high-resolution representation learning for visual recognition. In: IEEE Transactions on Pattern Analysis and Machine Intelligence (TPAMI) (2019)
43. Wang, Y., et al.: End-to-end video instance segmentation with transformers. In: Proceedings of the IEEE/CVF Conference on Computer Vision and Pattern Recognition (CVPR), pp. 8741–8750 (2021)
44. Zhang, H., et al.: PyMAF: 3D human pose and shape regression with pyramidal mesh alignment feedback loop. In: Proceedings of the IEEE/CVF International Conference on Computer Vision (ICCV) (2021)
45. Zheng, C., Zhu, S., Mendieta, M., Yang, T., Chen, C., Ding, Z.: 3D human pose estimation with spatial and temporal transformers. In: Proceedings of the IEEE/CVF International Conference on Computer Vision (ICCV) (2021)
46. Zhou, X., Zhu, M., Pavlakos, G., Leonardos, S., Derpanis, K.G., Daniilidis, K.: MonoCap: monocular human motion capture using a CNN coupled with a geometric prior. In: IEEE Transactions on Pattern Analysis and Machine Intelligence (TPAMI) (2018)
47. Zimmermann, C., Ceylan, D., Yang, J., Russell, B., Argus, M., Brox, T.: FreiHAND: a dataset for markerless capture of hand pose and shape from single RGB images. In: Proceedings of the IEEE/CVF International Conference on Computer Vision (ICCV) (2019)

GeoRefine: Self-supervised Online Depth Refinement for Accurate Dense Mapping

Pan Ji[✉], Qingan Yan, Yuxin Ma, and Yi Xu

OPPO US Research Center, InnoPeak Technology, Inc., Palo Alto, USA
peterji530@gmail.com

Abstract. We present a robust and accurate depth refinement system, named GeoRefine, for geometrically-consistent dense mapping from monocular sequences. GeoRefine consists of three modules: a hybrid SLAM module using learning-based priors, an online depth refinement module leveraging self-supervision, and a global mapping module via TSDF fusion. The proposed system is online by design and achieves great robustness and accuracy via: (i) a robustified hybrid SLAM that incorporates learning-based optical flow and/or depth; (ii) self-supervised losses that leverage SLAM outputs and enforce long-term geometric consistency; (iii) careful system design that avoids degenerate cases in online depth refinement. We extensively evaluate GeoRefine on multiple public datasets and reach as low as 5% absolute relative depth errors.

1 Introduction

Over the years, monocular geometric methods have been continuously improved and become very accurate in recovering 3D map points. Representative open-source systems along this line include COLMAP [43] – an offline SfM system, and ORB-SLAM [5,35,36] – an online SLAM system.

Recently, deep-learning-based methods [9,14,16] have achieved impressive results in predicting a *dense* depth map from a single image. Those models are either trained in a supervised manner [9,39,40] using ground-truth depths, or through a self-supervised framework [14,16] leveraging photometric consistency between stereo and/or monocular image pairs. During inference, with the prior knowledge learned from data, the depth models can generate *dense* depth images even in textureless regions. However, the errors in the predicted depths are still relatively high.

A few methods [33,54] aim to get the best out of geometric systems and deep methods. Tiwari *et al.* [54] let monocular SLAM and learning-based depth prediction form a self-improving loop to improve the performance of each module. Luo *et al.* [33] adopt a test-time fine tuning strategy to enforce geometric consistency using outputs from COLMAP. Nonetheless, both methods pre-compute

---

P. Ji and Q. Yan—Joint first authorship.

Supplementary Information The online version contains supplementary material available at https://doi.org/10.1007/978-3-031-19769-7_21.

S. Avidan et al. (Eds.): ECCV 2022, LNCS 13661, pp. 360–377, 2022.
https://doi.org/10.1007/978-3-031-19769-7_21

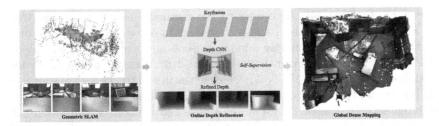

Fig. 1. We present an online depth refinement system for geometrically-consistent dense mapping from monocular data. Our system starts with geometric SLAM that is made robust by incorporating learning-based priors. Together with map points and camera poses from SLAM, a depth CNN is continuously updated using self-supervised losses. A globally consistent map is finally reconstructed from refined depth maps via an off-the-shell TSDF fusion method.

and store sparse map points and camera poses from SfM or SLAM in an offline manner, which is not applicable to many applications where data pre-processing is not possible. For example, after we deploy an agent to an environment, we want it to automatically improve its 3D perception capability as it moves around. In such a scenario, an online learning method is more desirable.

In this paper, we propose to combine geometric SLAM and a single-image depth model within an **online** learning scheme (see Fig. 1). The depth model can be any model that has been pretrained either with a self-supervised method [16] or a supervised one [39,40]. Our goal is then to incrementally refine this depth model on the test sequences in an online manner to achieve geometrically consistent depth predictions over the entire image sequence. Note that SLAM in itself is an online system that perfectly fits our online learning framework, but on the other hand, front-end tracking of SLAM often fails under challenging conditions (*e.g.*, with fast motion and large rotation). We propose to enhance the robustness of geometric SLAM with learning-based priors, *e.g.*, RAFT-flow [52], which has been shown to be both robust and accurate in a wide range of *unseen* scenes [25,53,64]. We then design a parallel depth refinement module that optimizes the neural weights of depth CNN with self-supervised losses. We perform a careful analysis of failure cases of self-supervised refinement and propose a simple yet effective keyframe mechanism to make sure that no refinement step worsens depth results. We further propose a novel occlusion-aware depth consistency loss to promote long-term consistency over temporally distant keyframes. We perform detailed ablation study to verify the effectiveness of each new component of our proposed *GeoRefine*, and conduct extensive experiments on several public datasets [3,7,15,49], demonstrating state-of-the-art performance in terms of dense mapping from monocular images.

2   Related Work

Geometric Visual SLAM. SLAM is an online geometric system that reconstructs a 3D map consisting of 3D points and simultaneously localizes camera poses w.r.t. the map [4]. According to the methods used in front-end tracking, SLAM systems can be roughly classified into two categories: (i) direct SLAM [10,11,44], which directly minimizes photometric error between adjacent frames and optimizes the geometry using semi-dense measurements; (ii) feature-based (indirect) SLAM [5,35,36,47], which extracts and tracks a set of sparse feature points and then computes the geometry in the back-end using these sparse measurements. Geometric SLAM systems have become accurate and robust due to a number of techniques developed over the years, including robust motion estimation [12], keyframe mechanism [23], bundle adjustment [55], and pose-graph optimization [26]. In our work, we build our system upon one of the state-of-the-art feature-based systems, *i.e.*, ORB-SLAM [5]. We use ORB-SLAM because it is open-source, delivers accurate 3D reconstructions, and supports multiple sensor modes.

Learning-Based SLAM. CNN-SLAM [50] is a hybrid SLAM system that uses CNN depth to bootstrap back-end optimization of sparse geometric SLAM and helps recover metric scale 3D reconstruction. In contrast, DROID-SLAM [53] builds SLAM from scratch with a deep learning framework and achieves unprecedented accuracy in camera poses, but does not have the functionality of dense mapping. TANDEM [24] presents a monocular tracking and dense mapping framework that relies on photometric bundle adjustment and a supervised multiview stereo CNN model. CodeSLAM [2] is a real-time learning-based SLAM system that optimizes a compact depth code over a conditional variational auto-encoder (VAE) and simultaneously performs dense mapping. DeepFactor [6] extends CodeSLAM by using fully-differentiable factor-graph optimization. CodeMapping [34] further improves over CodeSLAM via introducing a separate dense mapping thread to ORB-SLAM3 [5] and additionally conditioning VAE on sparse map points and reprojection errors. Our system bears the most similarity with CodeMapping in terms of overall functionalities, but is significantly different in system design and far more accurate in dense mapping.

Supervised Depth Estimation. Supervised depth estimation methods dominate the early trials [9,31,32,38,51,56,66] of learning-based depth estimation. Eigen *et al.* [9] propose the first deep learning based method to predict depth maps via a convolutional neural network and introduce a set of depth evaluation metrics that are still widely used today. Liu *et al.* [31] formulate depth estimation as a continuous conditional random field (CRF) learning problem. Fu *et al.* [13] leverage a deep ordinal regression loss to train the depth network. A few other methods combine depth estimation with additional tasks, *e.g.*, pose estimation [51,56,66] and surface normal regression [38].

Self-supervised Depth Estimation. Self-supervised depth estimation has recently become popular [14,16,20,27,28,42,45,60,67]. Garg *et al.* [14] are the first to apply the photometric loss between left-right stereo image pairs to train a monocular depth model in an unsupervised/self-supervised way. Zhou *et al.* [67] further introduce a pose network to facilitate using a photometric loss across neighboring temporal images. Later self-supervised methods are proposed to

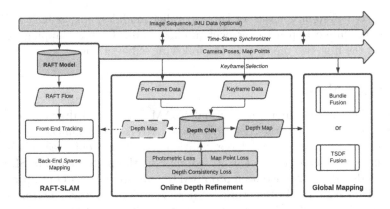

Fig. 2. The system workflow of our **GeoRefine**. Our system consists of three main modules, *i.e.*, a RAFT-SLAM module, an Online Depth Refinement module, and a Global Mapping module. Note that keyframe selection in *Online Dense Refinement* uses a **different** strategy than in SLAM.

improve the photometric self-supervision. Some methods [41,63,69] leverage an extra flow network to enforce cross-task consistency, while a few others [1,17,57] employ new loss terms during training. A notable recent method Monodepth2 by Godard *et al.* [16] achieves great improvements via a few thoughtful designs, including a per-pixel minimum photometric loss, an auto-masking strategy, and a multi-scale framework. New network architectures are also introduced to boost the performance. Along this line, Wang *et al.* [58] and Zou *et al.* [68] exploit recurrent networks for the pose and/or depth networks. Ji *et al.* [21] propose a depth factorization module and an iterative residual pose module to improve depth prediction for indoor environments. Our system is theoretically compatible with all those methods in the pretraining stage.

Instead of ground-truth depths, some methods [29,30,33,40,54,65] obtain the training depth data from the off-the-shell SfM or SLAM. Li and Snavely [30] perform 3D reconstruction of Internet photos via geometric SfM [43] and then use the reconstructed depths to train a depth network. Li *et al.* [29] learn the depths of moving people by watching and reconstructing static people. Ranftl *et al.* [39, 40] improve generalization performance of the depth model by training the depth network with various sources, including ground-truth depths and geometrically reconstructed ones. Zhang *et al.* [64] extend the work of [33] to handling moving objects by unrolling scene flow prediction. Kopf *et al.* [25] further bypass the need of running COLMAP via the use of deformation splines to estimate camera poses. Most of those methods require a pre-processing step to compute and store 3D reconstructions. In contrast, our system runs in an online manner without the need of performing offline 3D reconstruction.

3 Method – GeoRefine

In this section, we present *GeoRefine*, a self-supervised depth refinement system for geometrically consistent dense mapping from monocular sequences. As shown in Fig. 2, our system consists of three parallel modules, *i.e.*, a RAFT-SLAM module, an Online Depth Refinement module, and a Global Mapping module. We detail the first two modules in the following sub-sections.

3.1 RAFT-SLAM

It is well-known that monocular visual SLAM has several drawbacks: (i) its front-end often fails to track features under adverse environments, *e.g.*, with low-texture, fast motion, and large rotation; (ii) it can only reconstruct the scene up to an *unknown* global scale. To improve the performance of SLAM, a few methods [50,61,62] have been proposed to improve back-end optimization of *direct* LSD-SLAM [11]. In this work, we instead seek to improve the front-end of *feature-based* SLAM based on the observation that front-end tracking lose is one of the most common causes for failures and accuracy decrease. We thus present RAFT-SLAM, a hybrid SLAM system that runs a learning-based flow front-end and a traditional back-end optimizer.

3.1.1 RAFT-Flow Tracking

RAFT [52] is one of the state-of-the-art optical flow methods that has shown strong cross-dataset generalization performance. It constructs a correlation volume for all pairs of pixels and uses a gated recurrent unit (GRU) to iteratively update the flow. In our system, we replace the front-end feature matching in ORB-SLAM [5] with RAFT-flow, but still sample sparse points for robust estimation in the back-end. This simple strategy allows us to have the advantages of both learning-based flow and traditional robust estimator in one system.

More specifically, for each feature from last frame I_{i-1}, once it is associated with a map point, we locate its correspondence in incoming frame I_i by adding the flow $F_{(i-1) \to i}$. If there are multiple candidates within a predefined radius around a target pixel in I_i, we choose the one with the smallest descriptor residual; or if there is none, we create a new feature instead, with the descriptor being copied from I_{i-1}. In all our experiments, we set the radius to 1 pixel. For the sake of robustness, we only keep $N_f = 0.1 \cdot N_t$ matched correspondences for initial pose calculation, where N_t is the total ORB features within the current frame. We note that, compared to leveraging the entire flow, sampling a subset of pixels is more beneficial to the accuracy. We do a forward-backward consistency check on predicted flows to obtain a valid flow mask by using a stringent threshold of 1 pixel. Similar to [5], we then perform a local map point tracking step to densify potential associations from other views and further optimize the camera pose. The reason for combining ORB features with flow is that traditional features can help us keep the structural information, mitigating drifting caused by flow mapping in long sequential tracking.

3.1.2 Multiple Sensor Modes

Our RAFT-SLAM inherits the good properties of ORB-SLAM3 [5] in supporting multiple sensor modes. In our system, we consider a minimum sensor setup, *i.e.*, using a monocular camera with (or without) IMU sensor. Thus, two SLAM modes are supported, *i.e.*, the monocular and Visual-Inertial (VI) modes. As we have a CNN depth model to infer the depth map for every image, we additionally form a pseudo-RGBD (pRGBD) mode as in [54].

Monocular Mode. Under the monocular mode, RAFT-SLAM reconstructs camera poses and 3D map points in an arbitrary scale. Since we have a pretrained depth model available in our system, we then leverage the CNN predicted depth maps to adapt the scale of map points and camera poses for SLAM. This scale alignment step is necessary in our system because SLAM outputs will be used in the downstream task of refining the depth model. If the scales between these two modules differ too much, depth refinement will be sub-optimal or even totally fail. After initial map points are constructed in our system, we continuously align the scale for a few steps by solving the following least-squares problem:

$$\min_s \sum_{\mathbf{x}} \left(d(\mathbf{x}) - s \cdot \hat{d}(\mathbf{x})\right)^2 , \tag{1}$$

where s is the scale alignment factor to be estimated, and $d(\mathbf{x}), \hat{d}(\mathbf{x})$ are the depth values from a pretrained depth model and SLAM map points respectively. However, if the scales of two modules are already in the same order, *e.g.*, when SLAM runs in the VI or pRGBD mode, such an alignment step is not necessary.

VI Mode. VI SLAM is usually more robust than monocular SLAM under challenging environments with low-texture, motion blur, and occlusions [5]. Since the inertial sensors provide scale information, camera poses and 3D map points from VI RAFT-SLAM are recovered in metric scale. In this mode, given a scale-aware depth model (*i.e.*, a model that predicts depth in metric scale), we can run the online depth refinement module without taking special care of the scale discrepancy between the two modules.

pRGBD Mode. The pRGBD mode provides a convenient way to incorporate deep depth priors into geometric SLAM. However, we observe that it results in sub-optimal SLAM performance if we naïvely treat depth predictions as the groundtruth to run the RGBD mode (as done in [54]) due to noisy predictions. In the RGBD mode of ORB-SLAM3 [5], the depth is mainly used in two stages, *i.e.*, system initialization and bundle adjustment. By using the input depth, the system can initialize instantly from the first frame, without the need of waiting for enough temporal baselines. For each detected feature point, employing the depth and camera parameters, the system creates a *virtual right correspondence*, which leads to an extra reprojection error term in bundle adjustment [5]. To mitigate the negative impact of the noise in depth predictions, we make two simple yet effective changes in the pRGBD mode as compared to the original RGBD mode: i) we take as input the refined depth maps from the online refinement module (as described in the next subsection) to ensure that the input depth maps are more accurate and temporally consistent; ii) we remove the reprojection error term for the virtual right points in bundle adjustment. Note that the input CNN depth is still used in the map point initialization and new keypoint insertion, benefiting the robustness of the SLAM system.

3.2   Online Depth Refinement

The depth refinement module receives map points and camera poses from RAFT-SLAM. The depth model is then incrementally refined with self-supervised losses, including a photometric loss, an edge-aware depth smoothness loss, a map-point loss, and a depth consistency loss.

Similar to [67], the photometric loss is defined as the difference between a target frame $\mathbf{I}_i$ and a synthesized frame $\mathbf{I}_{j\to i}$ warped from a source frame $\mathbf{I}_j$ using the depth image $\mathbf{D}_i$ and the relative pose $\mathbf{T}_{j\to i}$, $i.e.$,

$$L_p = \sum_j pe(\mathbf{I}_i, \mathbf{I}_{j\to i}) \,, \tag{2}$$

where $pe()$ is the photometric loss function computed with the ℓ_1 norm and the SSIM [59]. Instead of only using 3 neighboring frames to construct the photo-consistency as in [16,67], we employ a wider baseline photometric loss, $e.g.$, by using a 5-keyframe snippet with $j \in \mathcal{A}_i = \{i-9, i-6, i-3, i+1\}$. Another important difference is that the relative pose $\mathbf{T}_{j\to i}$ comes from our RAFT-SLAM, which is more accurate than the one predicted by a pose network.

Following [16], we use an edge-aware normalized smoothness loss, $i.e.$,

$$L_s = |\partial_x d_i^*|e^{-|\partial_x I_i|} + |\partial_y d_i^*|e^{-|\partial_y I_i|} \,, \tag{3}$$

where $d_i^* = d_i/\bar{d}_i$ is the mean-normalized inverse depth to prevent depth scale diminishing [57].

The map points from RAFT-SLAM have undergone extensive optimization through bundle adjustment [55], so the depths of these map points are usually more accurate than the pretrained CNN depths. As in [54,62], we also leverage the map-point depths to build a map-point loss as a supervision signal to the depth model. The map-point loss is simply the difference between SLAM map points and the corresponding CNN depths as follows,

$$L_m = \frac{1}{N_i} \sum_{n=1}^{N_i} \left| \mathbf{D}_{i,n} - D_{i,n}^{slam} \right| \,, \tag{4}$$

where we have N_i 3D map points from RAFT-SLAM after filtering with a stringent criterion (see Sect. 4.1) to ensure that only accurate map points are used as supervision. In addition to the above loss terms, we propose an occlusion-aware depth consistency loss and a keyframe strategy to build our online depth refinement pipeline.

3.2.1   Occlusion-Aware Depth Consistency

Given the depth images of two adjacent images, $i.e.$, $\mathbf{D}_i$ and $\mathbf{D}_j$, and their relative pose $\mathbf{T} = [\mathbf{R}|\mathbf{t}]$, we aim to build a robust consistency loss between $\mathbf{D}_i$ and $\mathbf{D}_j$ to make the depth predictions consistent with each other. Note that the depth values are not necessarily equal at corresponding positions of frame i and

j as the camera can move over time. With camera pose $\mathbf{T}$, the depth map $\mathbf{D}_j$ can be warped and then transformed to a depth map $\tilde{\mathbf{D}}_i$ of frame i, via image warping and coordinate system transformation [1,21]. We then define our initial depth consistency loss as,

$$L_c(\mathbf{D}_i, \mathbf{D}_j) = \left| 1 - \tilde{\mathbf{D}}_i / \mathbf{D}_i \right| . \qquad (5)$$

However, the loss in Eq. (5) will inevitably include pixels in occluded regions, which hamper model refinement. To effectively handle occlusions, following the per-pixel photometric loss in [16], we devise a per-pixel depth consistency loss by taking the minimum instead of the average over a set of neighboring frames:

$$L_c = \min_{j \in \mathcal{A}_i} L_c(\mathbf{D}_i, \mathbf{D}_j) . \qquad (6)$$

3.2.2 Degenerate Cases and Keyframe Selection

Self-supervised photometric losses are not without degenerate cases. If they are not carefully considered, self-supervised training or finetuning will deteriorate, leading to worse depth predictions. A first degenerate case happens when the camera stays *static*. This degeneracy has been well considered in the literature. For example, Zhou *et al.* [67] remove static frames in an image sequence by computing and thresholding the average optical flow of consecutive frames. Godard *et al.* [16] propose an auto-masking strategy to automatically mask out static pixels when calculating the photometric loss.

A second degenerate case is when the camera undergoes *purely rotational* motion. This degeneracy is well-known in traditional computer vision geometry [19], but has not been considered in self-supervised depth estimation. Under pure rotation, motion recovery using the fundamental matrix suffers from ambiguity, so homography-based methods are preferred [19]. In the context of the photometric loss, if the camera motion is pure rotation, *i.e.*, the translation $\mathbf{t} = 0$, the view synthesis (or reprojection) step does not depend on depth anymore (*i.e.*, depth cancels out after applying the projection function). This is no surprise as their 2D correspondences are directly related by a homography matrix. So in this case, as long as the camera motion is accurately given, any arbitrary depth can minimize the photometric loss, which is undesirable when we train or finetune the depth network (as depth will be arbitrarily wrong).

To circumvent the degenerate cases described above, we propose a simple yet effective keyframe mechanism to facilitate online depth refinement without deterioration. After we receive camera poses from RAFT-SLAM, we can simply select keyframes for depth refinement according to the magnitude of camera *translations*. Only if the norm of the camera translation is over a certain threshold (see Sect. 4.1), we set its corresponding frame as a keyframe, *i.e.*, the candidate for applying self-supervised losses. This ensures that we have enough baselines for the photometric loss to be effective.

368   P. Ji et al.

Algorithm 1. *GeoRefine*: self-supervised online depth refinement for geometrically consistent dense mapping.

1: **Pretrain** the depth model.   ▷ supervised or self-supervised
2: Run RAFT-SLAM.   ▷ on separate threads
3: **Data preparation:** buffer time-synchronized keyframe data into a fixed-sized queue $\mathcal{Q}^*$; (optionally) form another data queue $\mathcal{Q}$ for per-frame data.
4: **while** True **do**
5:      Check stop condition.   ▷ stop-signal from SLAM
6:      Check SLAM failure signal.   ▷ clear data queue if received
7:      **for** $k \leftarrow 1$ to K^* **do**   ▷ *keyframe refinement*
8:          Load data in $\mathcal{Q}^*$ to GPU,   ▷ batch size as 1
9:          Compute losses as in Eq. (7),
10:         Update depth model via one gradient descent step.   ▷ ADAM optimizer
11:     **end for**
12:     Run inference and save refined depth for current keyframe.
13:     **for** $k \leftarrow 1$ to K **do**   ▷ *Per-frame refinement*
14:         Check camera translation from last frame,   ▷ skip if too small
15:         Load data in $\mathcal{Q}^*$ and $\mathcal{Q}$ to GPU,   ▷ batch size as 1
16:         Compute losses as in Eq. (7),
17:         Update depth model via one gradient descent step.   ▷ ADAM optimizer
18:     **end for**
19:     Run inference and save refined depth for current frame.
20: **end while**
21: Run global mapping.   ▷ TSDF or bundle fusion
22: **Output:** refined depth maps and global TSDF meshes.

3.2.3   Overall Refinement Strategy

Our overall refinement loss writes as

$$L = L_p + \lambda_s L_s + \lambda_m L_m + \lambda_c L_c \,, \tag{7}$$

where $\lambda_s, \lambda_m, \lambda_c$ are the weights balancing the contribution of each loss term.

GeoRefine aims to refine any pretrained depth models to achieve geometrically-consistent depth prediction for each frame of an image sequence. As RAFT-SLAM runs on separate threads, we buffer the keyframe data, including images, map points, and camera poses, into a time-synchronized data queue of a fixed size. If depth refinement is demanded for every frame, we additionally maintain a small data queue for per-frame data and construct the 5-frame snippet by taking 3 recent keyframes and 2 current consecutive frames. We conduct online refinement for the current keyframe (or frame) by minimizing the loss term in Eq. (7) and performing gradient descent for K^* (or K) steps. After depth refinement steps, we run depth inference using the refined depth model and save the depth map for the current keyframe (or frame). Global maps can be finally reconstructed by performing TSDF or bundle fusion [8,37]. The whole *GeoRefine* algorithm is summarized in Algorithm 1.

4   Experiments

We mainly conduct experiments on three public datasets: EuRoC [3], TUM-RGBD [49], and ScanNet [7]. For quantitative depth evaluation, we employ the standard error and accuracy metrics, including the Mean Absolute Error (MAE), Absolute Relative (*Abs Rel*) error, *RMSE*, $\delta < 1.25$ (namely δ_1), $\delta < 1.25^2$ (namely δ_2), and $\delta < 1.25^3$ (namely δ_3) as defined in [9]. All of our experiments are conducted on an Intel i7 CPU machine with 16-GB memory and one 11-GB NVIDIA GTX 1080.

4.1   Implementation Details

Our GeoRefine includes a RAFT-SLAM module and an online depth refinement module. RAFT-SLAM is implemented based on ORB-SLAM3 [5] (other SLAM systems are also applicable) which support monocular, visual-inertial, and RGBD modes. In our experiments, we test the three modes and show that GeoRefine achieves consistent improvements over pretrained models. The pose data queue is maintained and updated in the SLAM side, where a frame pose is stored relative to its reference keyframe which is continuously optimized by BA and pose graph. The online learning module refines a pretrained depth model with customized data loader and training losses. In our experiments, we choose a supervised model, *i.e.*, DPT [39], to showcase the effectiveness of our system. The initial DPT model is trained on a variety of public datasets and then finetuned on NYUv2 [46]. We utilize Robot Operating System (ROS) [48] to exchange data between modules for cross-language compatibility. We use ADAM [22] as the optimizer and set the learning rate to $1.0e^{-5}$. The weighting parameters λ_s, λ_m, and λ_c are set to $1.0e^{-4}$, $5.0e^{-2}$, and $1.0e^{-1}$ respectively. We freeze its decoder layers of DPT for the sake of speed and stability. We filter map points with stringent criterion to ensure good supervision signal for online depth refinement. To this end, we discard map points observed in fewer than 5 keyframes or with reprojection errors greater than 1 pixel. We maintain a keyframe data queue of length 11 and a per-frame data queue of length 2. The translation threshold for keyframe (or per-frame) refinement is set to 0.05 m (or 0.01 m). The number of refinement steps for keyframes (or per-frame) is set to 3 (or 1). All system hyper-parameters are tuned on a validation sequence (EuRoC V2_01).

4.2   EuRoC Indoor MAV Dataset

The EuRoC MAV dataset [3] is an indoor dataset which provides stereo image sequences, IMU data, and camera parameters. An MAV mounted with global shutter stereo cameras is used to capture the data in a large machine hall and a VICON room. Five sequences are recorded in the machine hall and six are in the VICON room. The ground-truth camera poses and depths are obtained with a VICON device and a Leica MS50 laser scanner, so we use all Vicon sequences as the test set. We rectify the images with the provided intrinsics to remove image distortion. To generate ground-truth depths, we project the laser point cloud

Table 1. Quantitative depth evaluation on EuRoC under different SLAM modes.

Method	Monocular				Visual-Inertial				pRGBD			
	MAE ↓	AbsRel ↓	RMSE ↓	δ_1 ↑	MAE ↓	AbsRel ↓	RMSE ↓	δ_1 ↑	MAE ↓	AbsRel ↓	RMSE ↓	δ_1 ↑
V1_01												
DPT [39]	0.387	0.140	0.484	0.832	0.501	0.174	0.598	0.709	0.387	0.140	0.484	0.832
CodeMapping [34]	–	–	–	–	0.192	–	0.381	–	–	–	–	–
Ours-DPT	0.153	0.050	0.241	0.980	0.147	0.048	0.241	0.980	0.151	0.049	0.239	0.982
V1_02												
DPT [39]	0.320	0.119	0.412	0.882	0.496	0.182	0.586	0.712	0.320	0.119	0.412	0.882
CodeMapping [34]	–	–	–	–	0.259	–	0.369	–	–	–	–	–
Ours-DPT	0.171	0.058	0.255	0.967	0.166	0.058	0.251	0.972	0.160	0.056	0.240	0.973
V1_03												
Monodepth2 [16]	0.305	0.111	0.413	0.886	0.360	0.132	0.464	0.815	0.305	0.111	0.413	0.886
DPT [39]	0.305	0.112	0.396	0.890	0.499	0.185	0.581	0.700	0.305	0.112	0.396	0.890
CodeMapping [34]	–	–	–	–	0.283	–	0.407	–	–	–	–	–
Ours-DPT	0.202	0.074	0.297	0.949	0.188	0.067	0.278	0.956	0.190	0.068	0.286	0.949
V2_01												
Monodepth2 [16]	0.423	0.153	0.581	0.800	0.490	0.181	0.648	0.730	0.423	0.153	0.581	0.800
DPT [39]	0.325	0.128	0.436	0.854	0.482	0.205	0.571	0.703	0.325	0.128	0.436	0.854
CodeMapping [34]	–	–	–	–	0.290	–	0.428	–	–	–	–	–
MonoIndoor [21]	–	0.125	0.466	0.840	–	–	–	–	–	–	–	–
Ours-DPT	0.170	0.054	0.258	0.973	0.162	0.052	0.258	0.970	0.181	0.057	0.0269	0.970
V2_02												
Monodepth2 [16]	0.597	0.191	0.803	0.723	0.769	0.233	0.963	0.562	0.597	0.191	0.803	0.723
DPT [39]	0.404	0.134	0.540	0.838	0.601	0.191	0.727	0.699	0.404	0.134	0.540	0.838
CodeMapping [34]	–	–	–	–	0.415	–	0.655	–	–	–	–	–
Ours-DPT	0.177	0.053	0.208	0.976	0.193	0.063	0.312	0.966	0.167	0.053	0.267	0.976
V2_03												
Monodepth2 [16]	0.601	0.211	0.784	0.673	0.764	0.258	0.912	0.498	0.601	0.211	0.784	0.673
DPT [39]	0.283	0.099	0.366	0.905	0.480	0.154	0.564	0.746	0.283	0.099	0.366	0.905
CodeMapping [34]	–	–	–	–	0.686	–	0.952	–	–	–	–	–
Ours-DPT	0.163	0.053	0.231	0.970	0.159	0.055	0.220	0.973	0.152	0.051	0.214	0.975

onto the image plane of the left camera using the code by [18]. The original images have a size of 480 × 754 and are resized to 384 × 384 for DPT.

Quantitative Depth Results in the Monocular Mode. We conduct quantitative evaluation by running GeoRefine under *monocular* RAFT-SLAM on the EuRoC VICON sequences, and present the depth evaluation results in the left columns of Table 1. Following [16], we perform per-frame scale alignment between the depth prediction and the groundtruth. From Table 1, we can observe consistent and significant improvements by our method over the baseline model on all test sequences. In particular, on V1_01, "Ours-DPT" reduces *Abs Rel* from 14.0% (by DPT) to 5.0%, achieving over two-times reduction in depth errors.

Quantitative Depth Results in the Visual-Inertial Mode. When IMU data are available, we can also run GeoRefine under *visual-inertial* (VI) RAFT-SLAM to get camera poses and map points directly in metric scale. Note that, in the visual-inertial mode, no scale alignment is needed. We present the quantitative depth results in Table 1, from which we can see that our system under the VI mode performs on par with the monocular mode even without scale alignment. Compared to a similar dense mapping, *i.e.*, CodeMapping [34], our GeoRefine is significantly more accurate with similar runtime (*i.e.*, around 1 sec. per keyframe; see the supplementary), demonstrating the superiority of our system design.

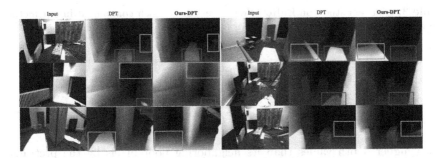

Fig. 3. Visual comparison of depth maps by the pretrained DPT and our system. Regions with salient improvements are highlighted with green/blue boxes. (Color figure online)

Fig. 4. Global reconstruction on EuRoC (left) and TUM-RGBD (right) using the refined depth maps by GeoRefine.

Table 2. Monocular SLAM results on EuRoC (RMSE ATE in meters).

Method	MH_01	MH_02	MH_03	MH_04	MH_05	V1_01	V1_02	V1_03	V2_01	V2_02	V2_03	Mean
DeepFactor [6]	1.587	1.479	3.139	5.331	4.002	1.520	0.679	0.900	0.876	1.905	1.021	2.040
DeepV2D [51]	0.739	1.144	0.752	1.492	1.567	0.981	0.801	1.570	0.290	2.202	2.743	1.298
D3VO [61]	–	–	0.080	–	0.090	–	–	0.110	–	0.050	0.019	–
DROID-SLAM [53]	0.013	**0.014**	0.022	**0.043**	0.043	0.037	0.012	**0.020**	**0.017**	0.013	**0.014**	**0.022**
ORB-SLAM [35]	0.071	0.067	0.071	0.082	0.060	**0.015**	0.020	x	0.021	0.018	x	–
DSO [10]	0.046	0.046	0.172	3.810	0.110	0.089	0.107	0.903	0.044	0.132	1.152	0.601
ORB-SLAM3 [5]	0.016	0.027	0.028	0.138	0.072	0.033	0.015	0.033	0.023	0.029	x	–
RAFT-SLAM (Ours)	**0.012**	0.018	0.023	0.045	**0.041**	0.032	**0.010**	0.022	0.019	**0.011**	0.025	0.023

Quantitative Depth Results in the pRGBD Mode. We present the quantitative depth evaluation under the *pRGBD* mode in the right columns of Table 1. We can see that the pRGBD mode performs slightly better than the other two modes in terms of depth results. This may be attributed to the fact that under this mode, the SLAM and depth refinement modules form a loosely-coupled loop so that each module benefits from the other.

Qualitative Depth Results. We show some visual comparisons in Fig. 3, from which we can clearly observe the qualitative improvements brought by our online depth refinement method. In particular, our system can correct the inaccurate geometry that is commonly present in the pretrained model. For example, in the first row of Fig. 3, a piece of thin paper lying on the floor is predicted to have

Table 3. Ablation study on EuRoC Sequence V2_03. Each component in our method improves the depth results.

Method	Monocular						Method	pRGBD					
	MAE ↓	Abs Rel ↓	RMSE ↓	δ_1 ↑	δ_2 ↑	δ_3 ↑		MAE ↓	Abs Rel ↓	RMSE ↓	δ_1 ↑	δ_2 ↑	δ_3 ↑
DPT [39]	0.283	0.099	0.366	0.905	0.979	0.994	DPT [39]	0.283	0.099	0.366	0.905	0.979	0.994
Our BaseSystem	0.269	0.090	0.347	0.905	0.983	0.997	Our BaseSystem	0.216	0.076	0.288	0.933	0.989	0.998
+ RAFT-flow	0.248	0.083	0.331	0.915	0.985	0.997	+ Refined Depth	0.199	0.065	0.268	0.958	0.995	0.999
+ Scale Alignment	0.199	0.064	0.274	0.952	0.991	0.998	+ RAFT-flow	0.171	0.056	0.237	0.972	0.995	0.998
+ Depth Consistency	0.163	0.053	0.231	0.970	0.995	0.999	+ Remove BA Term	0.152	0.051	0.214	0.975	0.997	0.999

much higher depth values than its neighboring floor pixels by the pretrained models (DPT); our GeoRefine is able to rectify its depth to be consistent with the floor. A global map of the EuRoC VICON room is shown in Fig. 1 and 4, where we can reach geometrically consistent reconstruction.

Odometry Results. Table 2 shows the odometry comparisons of our proposed RAFT-SLAM with current state-of-the-art methods on the EuRoC dataset in the monocular mode. For fairness, we adopt the same parameter settings with ORB-SLAM3 [5] in all our experiments. Note that, although our system is not elaborately designed for SLAM, it significantly outperforms other monocular baselines both in terms of accuracy and robustness, and achieves comparable results against DROID-SLAM [53] (with 19 steps of global bundle adjustment).

Ablation Study. Without loss of generality, we perform an ablation study on Seq. V2_03 to gauge the contribution of each component to our method under both monocular and pRGBD modes. Specifically, we first construct a base system by running a vanilla online refinement algorithm with the photometric loss as in Eq. (2), the depth smoothness loss as in Eq. (3), and the map-point loss as in Eq. (4). Note that the photometric loss uses camera poses from RAFT-SLAM instead of a pose network. Under the monocular mode, we denote this base model as "Our BaseSystem". We then gradually add new components to this base model, including the RAFT-flow in SLAM front-end ("+RAFT-flow"), the scale alignment strategy in RAFT-SLAM ("+Scale Alignment"), and the occlusion-aware depth consistency loss ("+Depth Consistency"). Under the pRGBD mode, "Our BaseSystem" takes the pretrained depth as input without using our proposed changes, and this base system uses the depth consistency loss. We then gradually add new components to the base system, *i.e.*, using refined depth from the online depth refinement module ("+Refined Depth"), using the RAFT-flow in SLAM front-end ("+RAFT-flow"), and removing the reprojection error term in bundle adjustment ("+Remove BA Term").

We show a complete set of ablation results in Table 3. Under the monocular mode, "Our BaseSystem" reduces the absolute relative depth error from 9.9% (by the pretrained DPT model) to 9.0%, which verifies the effectiveness of the basic self-supervised refinement method. However, the improvement brought by our base model is not significant and the SLAM module fails. Using RAFT-flow in SLAM front-end makes SLAM more robust, generating more accurate pose estimation, which in turn improves the depth refinement module. Adding our scale self-alignment in RAFT-SLAM ("+Scale Alignment") improves the depth

Table 4. Quantitative depth evaluation on TUM-RGBD.

Method	Monocular						pRGBD					
	MAE $\downarrow$	AbsRel $\downarrow$	RMSE $\downarrow$	$\delta_1 \uparrow$	$\delta_2 \uparrow$	$\delta_3 \uparrow$	MAE $\downarrow$	AbsRel $\downarrow$	RMSE $\downarrow$	$\delta_1 \uparrow$	$\delta_2 \uparrow$	$\delta_3 \uparrow$
freiburg3_structure_texture_near												
DPT [39]	0.280	0.140	0.529	0.794	0.924	0.968	0.280	0.140	0.529	0.794	0.924	0.968
Ours-DPT	**0.138**	**0.057**	**0.314**	**0.943**	**0.977**	**0.990**	**0.140**	**0.056**	**0.317**	**0.941**	**0.974**	**0.992**
freiburg3_structure_texture_far												
DPT [39]	0.372	0.134	0.694	0.810	0.939	0.968	0.372	0.134	0.694	0.810	0.939	0.968
Ours-DPT	**0.108**	**0.035**	**0.317**	**0.974**	**0.985**	**0.997**	**0.105**	**0.036**	**0.290**	**0.975**	**0.985**	**0.996**

quality significantly in all metrics, *e.g.*, *Abs Rel* decreases from 8.3% to 6.4% and δ_1 increases from 91.5% to 95.2%. Our occlusion-aware depth consistency loss ("Depth Consistency") further achieves an improvement of 1.1% in terms of *Abs Rel* and 1.8% in terms of δ_1. From this ablation study, it is evident that each component of our method makes non-trivial contributions in improving depth results. We can draw a similar conclusion under the pRGBD mode.

4.3   TUM-RGBD Dataset

TUM-RGBD is a well-known dataset mainly for benchmarking performance of RGB-D SLAM or odometry [49]. This dataset was created using a Microsoft Kinect sensor and eight high-speed tracking cameras to capture monocular images, their corresponding depth images, and camera poses. This dataset is particularly difficult for monocular systems as it contains a large amount of motion blur and rolling-shutter distortion caused by fast camera motion. We take two monocular sequences from this dataset, *i.e.*, "freiburg3_structure_texture_near" and "freiburg3_structure_texture_far", to test our system, as they satisfy our system's requirement of sufficient camera translations. The quantitative depth results are presented in Table 4. As before, under both SLAM modes, our GeoRe-fine improves upon the pretrained DPT model by a significant margin, achieving 2–4 times' reduction in terms of *Abs Rel*. A global reconstruction is visualized in Fig. 4, where the scene geometry is faithfully recovered.

5   Conclusions

In this paper, we have introduced *GeoRefine*, an online depth refinement system that combines geometry and deep learning. The core contribution of this work lies in the system design itself, where we show that accurate dense mapping from monocular sequences is possible via a robust hybrid SLAM, an online learning paradigm, and a careful consideration of degenerate cases. The self-supervised nature of the proposed system also suggests that it can be deployed in any unseen environments by virtue of its self-adaptation capability. We have demonstrated the state-of-the-art performance on several challenging public datasets.

Limitations. Our system does not have a robust mechanism for handling moving objects which are outliers both for SLAM and self-supervised losses. Hence, datasets with plenty of foreground moving objects such as KITTI [15] would not be the best test-bed for GeoRefine. Another limitation is that GeoRefine cannot deal with scenarios where camera translations are small over the entire sequence. This constraint is intrinsic to our system design, but it is worth exploring how to relax it while maintaining robustness.

References

1. Bian, J.W., et al.: Unsupervised scale-consistent depth and ego-motion learning from monocular video. In: NeurIPS (2019)
2. Bloesch, M., Czarnowski, J., Clark, R., Leutenegger, S., Davison, A.J.: Codeslam-learning a compact, optimisable representation for dense visual slam. In: CVPR, pp. 2560–2568 (2018)
3. Burri, M., et al.: The EuRoC micro aerial vehicle datasets. Int. J. Robot. Res. **35**, 1157–1163 (2016)
4. Cadena, C., et al.: Past, present, and future of simultaneous localization and mapping: toward the robust-perception age. IEEE Trans. Rob. **32**(6), 1309–1332 (2016)
5. Campos, C., Elvira, R., Rodríguez, J.J.G., Montiel, J.M., Tardós, J.D.: Orb-slam3: An accurate open-source library for visual, visual-inertial and multi-map slam. arXiv preprint arXiv:2007.11898 (2020)
6. Czarnowski, J., Laidlow, T., Clark, R., Davison, A.J.: Deepfactors: real-time probabilistic dense monocular slam. IEEE Robot. Autom. Let. **5**(2), 721–728 (2020)
7. Dai, A., Chang, A.X., Savva, M., Halber, M., Funkhouser, T., Nießner, M.: Scannet: richly-annotated 3D reconstructions of indoor scenes. In: CVPR, pp. 5828–5839 (2017)
8. Dai, A., Nießner, M., Zollhöfer, M., Izadi, S., Theobalt, C.: Bundlefusion: real-time globally consistent 3D reconstruction using on-the-fly surface reintegration. ToG **36**(4), 1 (2017)
9. Eigen, D., Puhrsch, C., Fergus, R.: Depth map prediction from a single image using a multi-scale deep network. arXiv preprint arXiv:1406.2283 (2014)
10. Engel, J., Koltun, V., Cremers, D.: Direct sparse odometry. TPAMI **40**(3), 611–625 (2017)
11. Engel, J., Schöps, T., Cremers, D.: LSD-SLAM: large-scale direct monocular SLAM. In: Fleet, D., Pajdla, T., Schiele, B., Tuytelaars, T. (eds.) ECCV 2014. LNCS, vol. 8690, pp. 834–849. Springer, Cham (2014). https://doi.org/10.1007/978-3-319-10605-2_54
12. Fischler, M.A., Bolles, R.C.: Random sample consensus: a paradigm for model fitting with applications to image analysis and automated cartography. Commun. ACM **24**(6), 381–395 (1981)
13. Fu, H., Gong, M., Wang, C., Batmanghelich, K., Tao, D.: Deep ordinal regression network for monocular depth estimation. In: CVPR (2018)
14. Garg, R., B.G., V.K., Carneiro, G., Reid, I.: Unsupervised CNN for single view depth estimation: geometry to the rescue. In: Leibe, B., Matas, J., Sebe, N., Welling, M. (eds.) ECCV 2016. LNCS, vol. 9912, pp. 740–756. Springer, Cham (2016). https://doi.org/10.1007/978-3-319-46484-8_45
15. Geiger, A., Lenz, P., Stiller, C., Urtasun, R.: Vision meets robotics: the kitti dataset. Int. J. Robot. Res. **32**, 1231–1237 (2013)
16. Godard, C., Aodha, O.M., Firman, M., Brostow, G.J.: Digging into self-supervised monocular depth estimation. In: ICCV (2019)

17. Godard, C., Mac Aodha, O., Brostow, G.J.: Unsupervised monocular depth estimation with left-right consistency. In: CVPR, pp. 270–279 (2017)
18. Gordon, A., Li, H., Jonschkowski, R., Angelova, A.: Depth from videos in the wild: unsupervised monocular depth learning from unknown cameras. In: CVPR (2019)
19. Hartley, R., Zisserman, A.: Multiple View Geometry in Computer Vision. Cambridge University Press, Cambridge (2003)
20. Hermann, M., Ruf, B., Weinmann, M., Hinz, S.: Self-supervised learning for monocular depth estimation from aerial imagery. arXiv preprint arXiv:2008.07246 (2020)
21. Ji, P., Li, R., Bhanu, B., Xu, Y.: Monoindoor: towards good practice of self-supervised monocular depth estimation for indoor environments. In: ICCV, pp. 12787–12796 (2021)
22. Kingma, D.P., Ba, J.L.: Adam: a method for stochastic gradient descent. In: ICLR, pp. 1–15 (2015)
23. Klein, G., Murray, D.: Parallel tracking and mapping for small AR workspaces. In: ISMAR (2007)
24. Koestler, L., Yang, N., Zeller, N., Cremers, D.: Tandem: tracking and dense mapping in real-time using deep multi-view stereo. In: CoLR, pp. 34–45 (2022)
25. Kopf, J., Rong, X., Huang, J.B.: Robust consistent video depth estimation. In: CVPR, pp. 1611–1621 (2021)
26. Kümmerle, R., Grisetti, G., Strasdat, H., Konolige, K., Burgard, W.: g 2 o: A general framework for graph optimization. In: ICRA (2011)
27. Li, Q., et al.: Deep learning based monocular depth prediction: datasets, methods and applications. arXiv preprint arXiv:2011.04123 (2020)
28. Li, S., Wu, X., Cao, Y., Zha, H.: Generalizing to the open world: deep visual odometry with online adaptation. In: CVPR, pp. 13184–13193 (2021)
29. Li, Z., et al.: Learning the depths of moving people by watching frozen people. In: CVPR, pp. 4521–4530 (2019)
30. Li, Z., Snavely, N.: Megadepth: learning single-view depth prediction from internet photos. In: CVPR, pp. 2041–2050 (2018)
31. Liu, F., Shen, C., Lin, G., Reid, I.: Learning depth from single monocular images using deep convolutional neural fields. TPAMI 38(10), 2024–2039 (2015)
32. Liu, J., et al.: Planemvs: 3d plane reconstruction from multi-view stereo. In: CVPR (2022)
33. Luo, X., Huang, J.B., Szeliski, R., Matzen, K., Kopf, J.: Consistent video depth estimation. TOG 39(4), 71–1 (2020)
34. Matsuki, H., Scona, R., Czarnowski, J., Davison, A.J.: Codemapping: real-time dense mapping for sparse slam using compact scene representations. IEEE Robot. Autom. Lett. 6(4), 7105–7112 (2021)
35. Mur-Artal, R., Montiel, J.M.M., Tardos, J.D.: ORB-SLAM: a versatile and accurate monocular slam system. IEEE Trans. Rob. 31(5), 1147–1163 (2015)
36. Mur-Artal, R., Tardós, J.D.: ORB-SLAM2: an open-source slam system for monocular, stereo, and RGB-D cameras. IEEE Trans. Rob. 33(5), 1255–1262 (2017)
37. Nießner, M., Zollhöfer, M., Izadi, S., Stamminger, M.: Real-time 3D reconstruction at scale using voxel hashing. ToG 32(6), 1–11 (2013)
38. Qi, X., Liao, R., Liu, Z., Urtasun, R., Jia, J.: Geonet: geometric neural network for joint depth and surface normal estimation. In: CVPR, pp. 283–291 (2018)
39. Ranftl, R., Bochkovskiy, A., Koltun, V.: Vision transformers for dense prediction. In: ICCV, pp. 12179–12188 (2021)
40. Ranftl, R., Lasinger, K., Hafner, D., Schindler, K., Koltun, V.: Towards robust monocular depth estimation: mixing datasets for zero-shot cross-dataset transfer. arXiv preprint arXiv:1907.01341 (2019)

41. Ranjan, A., et al.: Competitive collaboration: joint unsupervised learning of depth, camera motion, optical flow and motion segmentation. In: CVPR, pp. 12240–12249 (2019)

42. Ruhkamp, P., Gao, D., Chen, H., Navab, N., Busam, B.: Attention meets geometry: geometry guided spatial-temporal attention for consistent self-supervised monocular depth estimation. In: 3DV, pp. 837–847 (2021)

43. Schönberger, J.L., Frahm, J.M.: Structure-from-motion revisited. In: CVPR, pp. 4104–4113 (2016)

44. Schubert, D., Demmel, N., Usenko, V., Stückler, J., Cremers, D.: Direct sparse odometry with rolling shutter. In: Ferrari, V., Hebert, M., Sminchisescu, C., Weiss, Y. (eds.) ECCV 2018. LNCS, vol. 11212, pp. 699–714. Springer, Cham (2018). https://doi.org/10.1007/978-3-030-01237-3_42

45. Shu, C., Yu, K., Duan, Z., Yang, K.: Feature-metric loss for self-supervised learning of depth and egomotion. In: Vedaldi, A., Bischof, H., Brox, T., Frahm, J.-M. (eds.) ECCV 2020. LNCS, vol. 12364, pp. 572–588. Springer, Cham (2020). https://doi.org/10.1007/978-3-030-58529-7_34

46. Silberman, N., Hoiem, D., Kohli, P., Fergus, R.: Indoor segmentation and support inference from RGBD images. In: Fitzgibbon, A., Lazebnik, S., Perona, P., Sato, Y., Schmid, C. (eds.) ECCV 2012. LNCS, vol. 7576, pp. 746–760. Springer, Heidelberg (2012). https://doi.org/10.1007/978-3-642-33715-4_54

47. Song, S., Chandraker, M., Guest, C.C.: Parallel, real-time monocular visual odometry. In: ICRA (2013)

48. Stanford Artificial Intelligence Laboratory et al.: Robotic operating system. http://www.ros.org

49. Sturm, J., Engelhard, N., Endres, F., Burgard, W., Cremers, D.: A benchmark for the evaluation of RGB-D slam systems. In: IROS, pp. 573–580 (2012)

50. Tateno, K., Tombari, F., Laina, I., Navab, N.: CNN-SLAM: real-time dense monocular slam with learned depth prediction. In: CVPR (2017)

51. Teed, Z., Deng, J.: DeepV2D: video to depth with differentiable structure from motion. arXiv preprint arXiv:1812.04605 (2018)

52. Teed, Z., Deng, J.: RAFT: recurrent all-pairs field transforms for optical flow. In: Vedaldi, A., Bischof, H., Brox, T., Frahm, J.-M. (eds.) ECCV 2020. LNCS, vol. 12347, pp. 402–419. Springer, Cham (2020). https://doi.org/10.1007/978-3-030-58536-5_24

53. Teed, Z., Deng, J.: Droid-slam: deep visual slam for monocular, stereo, and RGB-D cameras. arXiv preprint arXiv:2108.10869 (2021)

54. Tiwari, L., Ji, P., Tran, Q.-H., Zhuang, B., Anand, S., Chandraker, M.: Pseudo RGB-D for self-improving monocular SLAM and depth prediction. In: Vedaldi, A., Bischof, H., Brox, T., Frahm, J.-M. (eds.) ECCV 2020. LNCS, vol. 12356, pp. 437–455. Springer, Cham (2020). https://doi.org/10.1007/978-3-030-58621-8_26

55. Triggs, B., McLauchlan, P., Hartley, R., Fitzgibbon, A.: Bundle adjustment-a modern synthesis. Vision Algorithms: Theory and Practice, pp. 153–177 (2000)

56. Ummenhofer, B., et al.: DEMON: depth and motion network for learning monocular stereo. In: CVPR (2017)

57. Wang, C., Buenaposada, J.M., Zhu, R., Lucey, S.: Learning depth from monocular videos using direct methods. In: CVPR, pp. 2022–2030 (2018)

58. Wang, R., Pizer, S.M., Frahm, J.M.: Recurrent neural network for (un-) supervised learning of monocular video visual odometry and depth. In: CVPR (2019)

59. Wang, Z., Bovik, A.C., Sheikh, H.R., Simoncelli, E.P.: Image quality assessment: from error visibility to structural similarity. TIP **13**(4), 600–612 (2004)

60. Xiong, M., Zhang, Z., Zhong, W., Ji, J., Liu, J., Xiong, H.: Self-supervised monocular depth and visual odometry learning with scale-consistent geometric constraints. In: IJCAI, pp. 963–969 (2021)
61. Yang, N., Stumberg, L.v., Wang, R., Cremers, D.: D3VO: deep depth, deep pose and deep uncertainty for monocular visual odometry. In: CVPR (2020)
62. Yang, N., Wang, R., Stückler, J., Cremers, D.: Deep virtual stereo odometry: leveraging deep depth prediction for monocular direct sparse odometry. In: Ferrari, V., Hebert, M., Sminchisescu, C., Weiss, Y. (eds.) ECCV 2018. LNCS, vol. 11212, pp. 835–852. Springer, Cham (2018). https://doi.org/10.1007/978-3-030-01237-3_50
63. Yin, Z., Shi, J.: GeoNet: unsupervised learning of dense depth, optical flow and camera pose. In: CVPR (2018)
64. Zhang, Z., Cole, F., Tucker, R., Freeman, W.T., Dekel, T.: Consistent depth of moving objects in video. ACM TOG **40**(4), 1–12 (2021)
65. Zhao, W., Liu, S., Shu, Y., Liu, Y.J.: Towards better generalization: joint depth-pose learning without posenet. In: CVPR, pp. 9151–9161 (2020)
66. Zhou, H., Ummenhofer, B., Brox, T.: DeepTAM: deep tracking and mapping. In: Ferrari, V., Hebert, M., Sminchisescu, C., Weiss, Y. (eds.) ECCV 2018. LNCS, vol. 11220, pp. 851–868. Springer, Cham (2018). https://doi.org/10.1007/978-3-030-01270-0_50
67. Zhou, T., Brown, M., Snavely, N., Lowe, D.G.: Unsupervised learning of depth and ego-motion from video. In: CVPR, pp. 1851–1858 (2017)
68. Zou, Y., Ji, P., Tran, Q.-H., Huang, J.-B., Chandraker, M.: Learning monocular visual odometry via self-supervised long-term modeling. In: Vedaldi, A., Bischof, H., Brox, T., Frahm, J.-M. (eds.) ECCV 2020. LNCS, vol. 12359, pp. 710–727. Springer, Cham (2020). https://doi.org/10.1007/978-3-030-58568-6_42
69. Zou, Y., Luo, Z., Huang, J.-B.: DF-Net: unsupervised joint learning of depth and flow using cross-task consistency. In: Ferrari, V., Hebert, M., Sminchisescu, C., Weiss, Y. (eds.) ECCV 2018. LNCS, vol. 11209, pp. 38–55. Springer, Cham (2018). https://doi.org/10.1007/978-3-030-01228-1_3

Multi-modal Masked Pre-training for Monocular Panoramic Depth Completion

Zhiqiang Yan, Xiang Li, Kun Wang, Zhenyu Zhang, Jun Li$^{(\boxtimes)}$, and Jian Yang$^{(\boxtimes)}$

PCA Lab, Nanjing University of Science and Technology, Nanjing, China
{Yanzq,xiang.li.implus,kunwang,junli,csjyang}@njust.edu.cn,
zhangjesse@foxmail.com

Abstract. In this paper, we formulate a potentially valuable panoramic depth completion (PDC) task as panoramic 3D cameras often produce 360° depth with missing data in complex scenes. Its goal is to recover dense panoramic depths from raw sparse ones and panoramic RGB images. To deal with the PDC task, we train a deep network that takes both depth and image as inputs for the dense panoramic depth recovery. However, it needs to face a challenging optimization problem of the network parameters due to its non-convex objective function. To address this problem, we propose a simple yet effective approach termed M³PT: multi-modal masked pre-training. Specifically, during pre-training, we simultaneously cover up patches of the panoramic RGB image and sparse depth by shared random mask, then reconstruct the sparse depth in the masked regions. To our best knowledge, it is the first time that we show the effectiveness of masked pre-training in a multi-modal vision task, instead of the single-modal task resolved by masked autoencoders (MAE). Different from MAE where fine-tuning completely discards the decoder part of pre-training, there is no architectural difference between the pre-training and fine-tuning stages in our M³PT as they only differ in the prediction density, which potentially makes the transfer learning more convenient and effective. Extensive experiments verify the effectiveness of M³PT on three panoramic datasets. Notably, we improve the state-of-the-art baselines by averagely 29.2% in RMSE, 51.7% in MRE, 49.7% in MAE, and 37.5% in RMSElog on three benchmark datasets.

Keywords: 360° depth completion · Multi-modal masked pre-training · Network optimization · Shared random mask · 3D reconstruction

1 Introduction

Panoramic depth perception (see Table 1) has received increasing attention in both academic and industrial communities due to its crucial role in a wide variety

---

Z. Yan and X. Li—Equal contribution.

Supplementary Information The online version contains supplementary material available at https://doi.org/10.1007/978-3-031-19769-7_22.

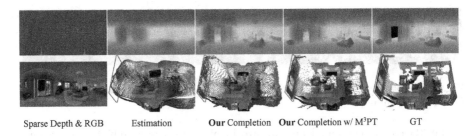

Sparse Depth & RGB   Estimation   **Our** Completion   **Our** Completion w/ M³PT   GT

Fig. 1. Comparisons of the predicted depth and 3D reconstruction results between panoramic depth estimation (with RGB) and completion (with RGB and sparse depth).

of downstream applications, such as virtual reality [1], scene understanding [46], and autonomous navigation [17]. With the development of hardware devices, panoramic 3D cameras become easier and cheaper to capture both RGB and depth (RGB-D) data with 360° field of view (FoV). Depending on the captured RGB images, all recent perception technologies [3,23,31,37,41,42,46,63,67], to the best of our knowledge, concentrate on panoramic depth estimation (PDE) that predicts dense depth from a single 360° RGB image. In this paper, we focus on exploring the 360° RGB-D pairs for the panorama perception with an effective pre-training strategy. We show our motivations as follows:

Two Motivations for Panoramic Depth Completion (PDC). *One is the 360° depth maps with missing areas.* During the collection process, the popular panoramic 3D cameras (*e.g.*, Matterport Pro2[1] and FARO Focus[2]) still produce 360° depth with missing areas when facing bright, glossy, thin or far surfaces, especially indoor rooms in Fig. 2. These depth maps will result in a poor panorama perception. To overcome this problem, we consider a new panoramic depth completion task, completing the depth channel of a single 360° RGB-D pair captured from a panoramic 3D camera. *Another is an experimental investigation that in contrast with PDE, PDC is much fitter for the panoramic depth perception.* For simplicity and fairness, we directly employ the same network architectures (*e.g.*, UniFuse [23]) to estimate or complete the depth map. Figure 7 reports that the PDC has much lower root mean square error than PDE. Furthermore, Fig. 1 shows that the PDC can recover more precise 360° depth values, leading to better 3D reconstruction. This observation reveals that PDC is more important than PDE for 3D scene understanding.

One Motivation for Pre-training. When using deep networks to perceive the depth information, there is a challenging problem: *how to optimize the parameters of the deep networks?* It is well-known that the objective function is highly non-convex, resulting in many distinct local minima in the network parameter space. Although the completion can lead to better network parameters and higher accuracy on the depth perception, it is still not to satisfy practical needs

---

[1] https://matterport.com/cameras/pro2-3D-camera.
[2] https://www.faro.com/en/Products/Hardware/Focus-Laser-Scanners.

Fig. 2. Panoramic depth maps with large missing areas shown in darkest blue color. (Color figure online)

of the 3D reconstruction. Here, we are inspired by the greedy layer-wise pre-training technology [14,30] that stacks two-layer unsupervised autoencoders to initialize the networks to a point in parameter space, and then fine-tunes them in a supervised setting. This technology drives the optimization process more effective, achieving a 'good' local minimum. Recently, the single-modal masked autoencoders [19,58] are also applied into object detection and semantic segmentation, achieving amazing improvements on their benchmarks. These interpretations and improvements motivate us to explore a new pre-training technology for the multi-modal panoramic depth completion.

In this paper, we propose a Multi-Modal Masked Pre-Training (M^{3}PT) technology to directly initialize all parameters of deep completion networks. Specifically, the key idea of M^{3}PT is to employ a shared random mask to simultaneously corrupt the RGB image and sparse depth, and then use the invisible pixels of the sparse depth as supervised signal to reconstruct the original sparse depth. After this pre-training, no-masked RGB-D pairs are fed into the pre-trained network supervised by dense ground-truth depths. Different from the layer-wise pre-training [14], M^{3}PT is to pre-train all layers of the deep network. Compared to MAE [19], M^{3}PT has no architectural difference between the pre-training and fine-tuning stages, where they differ in only the prediction density of target depth. This characteristic probably makes it convenient and effective for the transfer learning, including but not limited to the multi-modal depth completion, denoising, and super-resolution guided by RGB images. In summary, our contributions are as follows:

- We introduce a new panoramic depth completion (PDC) task that aims to complete the depth channel of a single 360° RGB-D pair captured from a panoramic 3D camera. To the best of our knowledge, we are the first to study the PDC task to facilitate 360° depth perception.
- We propose the multi-modal masked pre-training (M^{3}PT) for the multi-modal vision task. Different from the layer-wise pre-training [14] and MAE [19], M^{3}PT is to pre-train all layers of the deep network, and does not change the network architecture in the pre-training and fine-tuning stages.
- On three benchmarks, *i.e.*, Matterport3D [1], Stanford2D3D [2], and 3D60 [68], extensive experiments demonstrate that (i) PDC achieves higher accuracy of panoramic depth perception than PDE, and (ii) our M^{3}PT technology achieves the state-of-the-art performance.

Table 1. Comparisons of different depth related tasks. FoV denotes the field of vision.

Task	Depth estimation	Depth completion	Panoramic depth estimation	Panoramic depth completion
New task	No	No	No	Yes
Data FoV	<180°	<180°	360°	360°
Data modal	RGB	RGB-D	RGB	RGB-D
Camera	Perspective	Perspective	Panoramic	panoramic
Target	depth	Depth	Depth & 3D reconstruction	depth & 3D reconstruction

2   Related Work

Since this paper aims to learn the new task of monocular panoramic depth completion, we report three related but different topics whose detailed differences are listed in Table 1. First, we review depth completion approaches that input single RGB-D pair with limited FoV. Second, we elaborate on panoramic depth estimation works which predict 360° depths from panoramic color images. At last, we introduce the masked image encoding technology.

2.1   Monocular Depth Completion with Limited FoV

Existing monocular depth completion methods primarily focus on sparse depths and color images with a narrow FoV less than 180°. Up to now, based on the commonly used KITTI [50] and NYUv2 [43] datasets, a great deal of methods have been proposed to tackle the task, which can be broadly divided into depth-only [10,13,35,50] and multi-sensor fusion based [8,9,18,32,40,61,62] categories. For example, the literatures [22,51] take sparse depths as the only input to recover dense ones without using color images. Further, Lu *et al.* [34] use color images as auxiliary supervision during training and is discarded when testing. Recently, as technology quickly develops, multi-modal information can be captured by sensors, which is beneficial for depth completion. For example, S2D [35] directly concatenate RGB-D pairs and feed them into networks, contributing to promising improvement. Li *et al.* [29] propose multi-scale guided cascade hourglass network to handle diverse patterns. PENet [21] proposes to refine depth recovery at three stages. FCFRNet [33] designs channel-shuffle technology to enhance RGB-D feature fusion. GuideNet [47] proposes dynamic convolution to adaptively generate convolution kernels according to color image contents. ACMNet [64] conducts graph propagation to extract multi-modal representations. Furthermore, DeepLiDAR [38] and PwP [59] jointly utilize color images, surface normals, and sparse depths to recover dense depth. FusionNet [51] and Zhu *et al.* [66] present to estimate uncertainty for robust recovery. NLSPN [36] and DSPN [60] introduce recurrent non-local and dynamic spatial propagation networks, which significantly improve depth accuracy nearby object boundaries.

In addition, several unsupervised depth completion works [25,49,54–56] also contribute to the development of this domain. For example, KBNet [57] proposes the fantastic calibrated backprojection network which achieves very superior performances. However, as mentioned above these methods are designed for dense depth recovery from FoV-limited sparse depth, whilst we aim to learn 360° depth completion and 3D reconstruction from panoramic RGB-D input.

2.2   Monocular Depth Estimation with Full FoV

Given panoramic color images, current monocular panoramic depth estimation works mainly devote into predicting 360° depths and 3D reconstructions. This topic springs up as soon as the large indoor panoramic datasets Matterport3D [5] and Stanford2D3D [2] are constructed in 2017. For this domain in the last five years, supervised methods play a primary role while unsupervised approaches develop slowly. Next we will introduce each of them.

Supervised category: In 2018, OmniDepth [69] synthesizes 360° data with high-quality ground-truth depth annotations by rendering existing datasets. DistConv [48] proposes distortion-aware convolutional filters to address the inherent distortion of equirectangular projection (EPR) panoramic data. In 2019, Eder et al. [12] utilize surface normal and plane boundaries to train a plane-aware network to benefit depth estimation. SpherePHD [27,28] explores a new data representation via spherical polyhedron, which resolves the shape distortion of spherical panoramas. In 2020, Jin et al. [24] and Feng et al. [15] use geometric priors to help with depth estimation. Wang et al. [53] adopt a two-branch network leveraging EPR and cubemap projections, which are the two most common data forms. In 2021, PanoDepth [31] develops a two-stage framework containing view synthesis and stereo matching. UniFuse [23] further improves [53] with better accuracy and fewer parameters. SliceNet [37] transforms the EPR data into slice-based representation, which can tackle the inherent distortion. Sun et al. [45,46] focus on horizontal and vertical contents of a scene for 3D reconstruction. 360MonoDepth [39] projects the high-resolution spherical image into tangent image for efficient training. In 2022, SegFuse [16] utilizes geometric and temporal consistency to constraint depth recovery. GLPanoDepth [3] employs vision transformer and CNNs to encode cubemap and spherical images respectively, obtaining global-to-local representation of panoramas. ACDNet [67] designs adaptively combined dilated convolution to extend receptive field in the EPR and achieves state-of-the-art performances.

Unsupervised category: In 2019, Nikolaos et al. [68] explore spherical view synthesis for monocular 360° depth estimation in a self-supervised manner. In 2021, OlaNet [26] adopts the distortion-aware view synthesis, atrous spatial pyramid pooling, and L1-norm regularized smooth term to effectively and robustly deal with self-supervised panoramic depth estimation. Zhou et al. [65] combine supervised and unsupervised learning methods to facilitate network training. In 2022, Yun et al. [63] propose a self-supervised method based on gravity-aligned videos. Similarly, they also utilize the complementarity of supervised and self-supervised learning to improve their models' robustness.

Different from them that only utilize 360° color image, our goal is to recover dense depth and 3D reconstruction from the aligned 360° color image and sparse depth, which could help improve the accuracy with large margins.

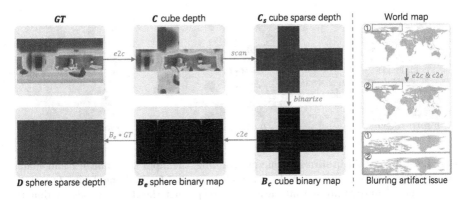

Fig. 3. A flowchart of data synthesis (left) and an example of the blurring artifact issue (right) that has negative effects on data processing. Best color of view.

2.3 Masked Image Encoding for Vision Tasks

Recently, several Transformer [52] based approaches [4,7,11,19,58] have proved it effective to learn representations from masked images. Specifically, iGPT [7] trains a sequence Transformer to auto-regressively predict unknown pixels. ViT [11] conducts masked patch prediction to learn mean color. BEiT [4] presents to predict tokenization. SimMIM [58] and MAE [19] propose to recover raw pixels of randomly masked patches by a lightweight one-layer head and an asymmetric decoder, respectively. In contrast to them, our M³PT is technically designed for multi-modal vision tasks instead of the single-modal image-based recognition.

3 Method

In this section, we first introduce how to synthesize sparse depth data and then elaborate on the multi-modal masked pre-training strategy.

3.1 Data Synthesis

All existing panoramic datasets do not provide sparse depth for 360° depth completion task. However, the sparse depth data can be possibly captured by some actual products such as Matterport Pro2 and FARO Focus 3D cameras. Limited by the lack of these hardware devices, in this paper, we imitate the principle of laser scanning to produce 360° sparse depth sampled from the dense ground-truth depth annotation, aiming at synthesizing the sparse depth data that matches the actual products as much as possible. The sampling principle is similar to that of KITTI benchmark [50] which provides depth with about 7% density captured by 64-line LiDAR scanning.

As illustrated in the left of Fig. 3, the ground-truth depth GT is stored in spherical view by equirectangular projection, which brings inherent distortion. Hence, it's inaccurate to produce sparse depth directly based on GT in scanning

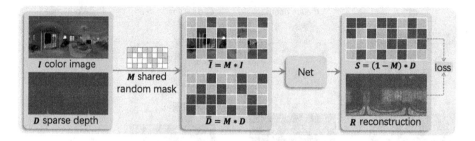

Fig. 4. Our M³PT pipeline. During pre-training, the input color image I and sparse depth D are masked out by the shared random mask M, obtaining $\bar{I}$ and $\bar{D}$ respectively. Then $\bar{I}$ and $\bar{D}$ are fed into a network to predict the depth reconstruction R, supervised by the signal S which is the complementary set of $\bar{D}$ in D. After pre-training, with I and D as input, the network with learned initial weights is applied to recover target depths supervised by dense ground-truth depth annotations.

mode. As an alternative, we first project the equirectangular GT into cubical map C by *e2c* function, ignoring the inherent distortion. Next, we generate cube sparse depth C_s via imitating the laser scanning, *e.g.*, taking one pixel for every eight pixels horizontally and one pixel for every two pixels vertically. Then C_s is binarized to obtain cube binary map B_c. B_c is thus converted into B_e by *c2e* function. Finally, we acquire the desired sparse depth D multiply B_e by GT. The process can be simply defined as:

$$D = B_e * GT, \tag{1}$$

where $B_e = f\left(B_c | C_s, C, GT\right)$, $f\left(\cdot\right)$ refers to the combination of *e2c*, *scan*, *binarize*, and *c2e*. The details of *e2c* and *c2e* functions refer to this project[3].

Note that, it is theoretically possible to use *c2e* to directly project the cubical C_s into the equirectangular D. However, this would lead to blurring artifacts in the polar region, as evidenced in the right part of Fig. 3. Instead, we choose to project a binary map and then use it to accurately sample valid points from GT. In this way, our method can reduce error pixels as much as possible.

3.2   Multi-modal Masked Pre-training

As shown in Fig. 4, our multi-modal masked pre-training (M³PT) for 360° depth completion is a simple strategy that reconstructs the sparse depth signal given partial observations of the RGB-D pair under shared random mask. Here we introduce the key components of M³PT and explicitly analyze the differences between recent visual masked pre-training approaches (*e.g.*, MAE [19], Sim-MIM [58]) and M³PT.

Shared Random Mask. Different from MAE where masking is performed on single RGB data, we propose to mask both RGB image and sparse depth with

---

[3] https://github.com/sunset1995/py360convert.

the *shared random mask* to produce incomplete RGB-D pair as input for pre-training. In fact, there are other options of masking strategies for PDC task, including (i) only mask the RGB image, (ii) only mask the sparse depth, and (iii) mask both RGB image and sparse depth but with different random masks. Unfortunately, all of these strategies have a risk of leaking information from another modality, preventing the pre-training task from learning robust semantics based on the multi-modal context. We will show the comparisons between different masking strategies later in the experimental part.

Backbone. Our method is flexible and can be applied to any existing approach which receives the RGB-D pair as input. In this paper, we mainly choose GuideNet [47] as backbone for the majority of our experiments. In addition, we also test the effectiveness of M³PT using UniFuse [23] and HoHoNet [46]. Note that there is no need to design extra modules (e.g., a decoder) for the architectures of these existing approaches, even when they have an additional pre-training stage in M³PT. It is because that the regression targets are physically similar between pre-training and fine-tuning stages. See more details in the 'Reconstruction target' part as follows.

Reconstruction Target. The reconstruction target of M³PT is the sparse depth on the masked regions. It is quite different from the popular masked pre-training methods [19,58] in vision where the missing image pixels are predicted. Compared to the vision pre-training counterparts [19,58], this design has two obvious advantages: (i) it closes the gap between pre-training and fine-tuning tasks, as they differ only in the prediction density; (ii) it leads to *no architectural modification* between pre-training and fine-tuning stages, which can potentially make the transfer learning more smooth and effective.

4   Experiments

Here, we first report datasets and metrics. Next, extensive ablation studies are conducted to verify the effectiveness of the proposed M³PT. Then, we compare against other state-of-the-art (SoTA) works on three datasets. At last, we validate the generalization capability of M³PT on KITTI benchmark [50].

4.1   Datasets

We conduct our experiments on three commonly used benchmark datasets of real world, *i.e.*, Matterport3D[4] [1], Stanford2D3D[5] [2], and 3D60[6] [68]. Matterport3D is a scanned dataset collected by Matterport's Pro 3D camera. The latest Matterport3D (512 × 256) consists of 7,907 panoramic RGB-D pairs, of which 5636 for training, 744 for validating, and 1527 for testing. Stanford2D3D is composed of 1,413 panoramic color images and corresponding depth maps,

---

[4] https://vcl3d.github.io/Pano3D/download/.

[5] http://buildingparser.stanford.edu/dataset.html.

[6] https://vcl3d.github.io/3D60/.

Table 2. Ablation on different masked input data on Stanford2D3D dataset, where the metric is RMSE (mm). The error of baseline without pre-training is **196.7**.

Masked data	Shared mask	Mask ratio	Mask size			
			4	8	16	32
RGB	–	0.75	195.7	198.1	194.0	196.2
D	–		190.7	177.6	186.1	203.3
RGB-D	No		183.4	178.5	182.2	193.0
RGB-D	Yes		**166.8**	**168.9**	**167.6**	**169.9**

Table 3. Ablation on different mask sizes and mask ratios on Stanford2D3D dataset.

Mask size	4			8			16						32		
Mask ratio	0.45	0.6	0.75	0.45	0.6	0.75	0.15	0.3	0.45	0.6	0.75	0.9	0.45	0.6	0.75
RMSE	172.4	170.3	168.8	171.2	169.3	168.9	173.5	169.4	168.9	169.7	**167.6**	169.3	173.7	172.4	169.9

whose training and testing splits contain 1,040 and 373 RGB-D pairs, respectively. We resize them to 512×256. 3D60 is initially made up of Matterport3D, Stanford2D3D, and SunCG [44]. But now it skips the entire SunCG dataset considering legal matters. As a result, the latest 3D60 (512×256) consists of 6,669 RGB-D pairs for training, 906 for validating, and 1831 for testing, 9,406 in total.

4.2 Metrics

Following previous works [37,46,63,67], we use five common and standard metrics to evaluate our methods, including MRE, MAE, RMSE, RMSElog, and δ_i ($i = 1.25, 1.25^2, 1.25^3$). Please refer to our appendix for more details.

4.3 Ablation Studies

Settings: We employ GuideNet [47] as the default backbone. The model is pre-trained for 300 epochs and fine-tuned for 100 epochs on every dataset. The mask is randomly generated following [19,58] with different sizes and ratios.

(1) Masking strategy

(i) We explore how to corrupt RGB-D data during pre-training in Table 2. We can find that shielding only RGB, only Depth, or RGB-D without shared mask, all of which lead to worse performances because these operations destroy the model's learning of unknown areas. In contrast, the model achieves the best results when employing the proposed shared random mask, indicating that corrupting the same areas can contribute to improvement for multi-modal vision tasks. The following experiments are based on the random shared mask.

(ii) We study the effect of different mask sizes and ratios on the model's representation learning in Table 3. First, the model performs better when the mask size is changed from 4 to 16. We hold the opinion that the larger mask urges the model to learn long-range dependency between invisible and visible pixels. However, when setting the mask size to 32, the model has a degraded

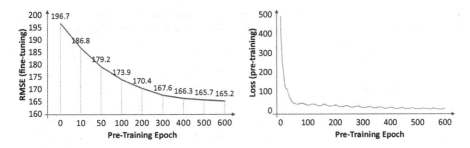

Fig. 5. Ablation on different pre-training epochs with mask size 16 and mask ratio 0.75 on Stanford2D3D dataset. The loss value is magnified by 10^4 for clear visualization.

Table 4. Ablation on different data amounts used during pre-training (Pret.). The mask size and ratio are 16 and 0.75, respectively. "300+100 epochs" denotes 300 pre-training epochs and 100 fine-tuning epochs.

Dataset	Data amount	w/o Pret (100 epochs)	w/o Pret (400 epochs)	Pret. on Self Data (300+100 epochs)	Pret. on All Data (300+100 epochs)
Matterport3D	5.6k	168.1	168.5	146.2	**138.9**
Stanford2D3D	1k	196.7	196.3	167.6	**149.0**
3D60	6.7k	159.9	160.2	142.3	**127.2**

performance as it is too large to establish remote dependency. Second, when the mask size is set to 16, the model tends to perform better from 15% to 90% ratios, which could enforce the model to predict more unseen areas and acquires representation that is closer to the real domain.

(2) Number of pre-training epochs and data amounts

(i) The left of Fig. 5 demonstrates the influence of different pre-training epochs on fine-tuning, and the right shows the loss of pre-training. We can find that the model's error gradually decreases with the increase of pre-training epochs. This is because the model can learn better representation with more epochs, which is also reflected in the lower loss in the right of Fig. 5. For the trade-off between speed and accuracy, unless otherwise stated, we pre-train the model for 300 epochs by default.

(ii) Table 4 explores the effect of different data amounts for pre-training on fine-tuning. For a fair comparison, we report the results of the 400th epoch without pre-training, whose cost roughly aligns with the setting of pre-training for 300 epochs and fine-tuning for 100 epochs. It can be found that without pre-training, the performance of 400 epochs has no improvements over 100 epochs. However, when conducting M³PT just on single dataset, it leads to 12.9% improvements averagely on three datasets, demonstrating the significant effectiveness of M³PT. Further, when pre-training on all the data of these three datasets, the performances are always superior to that only using a single dataset. Therefore, it is concluded that more data involved in M³PT can consistently prevents the overfitting risks during fine-tuning.

Table 5. Quantitative comparisons of panoramic depth completion on three datasets. The best and the second best results are highlighted in bold and underline, respectively.

Dataset	Method	Error Metric ↓				Accuracy Metric ↑		
		MRE	MAE	RMSE	RMSElog	$\delta_{1.25}$	$\delta_{1.25^2}$	$\delta_{1.25^3}$
Matterport3D	UniFuse [23]	0.0475	95.2	229.1	0.0381	0.9710	0.9924	0.9970
	HoHo-R [46]	0.0355	75.0	199.2	0.0311	0.9806	0.9945	0.9977
	HoHo-H [46]	0.0406	85.7	215.5	0.0337	0.9772	0.9938	0.9975
	PENet [21]	0.0493	91.5	248.0	0.0350	0.9728	0.9935	0.9970
	GuideNet [47]	0.0438	87.2	192.9	0.0327	0.9806	0.9948	0.9981
	M³PT	**0.0164**	**36.2**	**138.9**	**0.0193**	**0.9927**	**0.9976**	**0.9990**
Stanford2D3D	UniFuse [23]	0.0489	93.4	216.2	0.0392	0.9661	0.9919	0.9973
	HoHo-R [46]	0.0677	123.9	242.5	0.0478	0.9463	0.9862	0.9959
	HoHo-H [46]	0.0695	127.9	254.8	0.0497	0.9434	0.9852	0.9957
	PENet [21]	0.0530	95.9	200.6	0.0404	0.9694	0.9934	0.9981
	GuideNet [21]	0.0506	92.1	196.7	0.0380	0.9689	0.9926	0.9978
	M³PT	**0.0274**	**52.9**	**149.0**	**0.0263**	**0.9859**	**0.9963**	**0.9988**
3D60	UniFuse [23]	0.0446	94.1	215.6	0.0342	0.9749	0.9947	0.9984
	HoHo-R [46]	0.0338	75.6	196.9	0.0294	0.9818	0.9954	0.9983
	HoHo-H [46]	0.0376	81.9	205.8	0.0317	0.9788	0.9947	0.9981
	PENet [21]	0.0680	120.3	233.9	0.0321	0.9743	0.9926	0.9980
	GuideNet [21]	0.0689	144.2	239.3	0.0418	0.9711	0.9954	0.9987
	M³PT	**0.0144**	**34.1**	**127.2**	**0.0165**	**0.9944**	**0.9985**	**0.9995**

4.4 Comparisons with SoTA Methods

In this subsection, we compare with recent SoTA works, including UniFuse [23], HoHoNet [46], PENet [21], and GuideNet [47]. HoHo-R and HoHo-H severally refer to using ResNet [20] and HardNet [6] as its backbone. Table 5 and Fig. 6 demonstrate the quantitative and qualitative results, respectively. Based on different baselines, Fig. 7 further shows the influence of additional sparse depth information and proposed M³PT on the recovery of panoramic depth.

(1) Quantitative Results

Overall, as illustrated in Table 5, the proposed M³PT is consistently superior to other methods in all metrics on three datasets.

(i) On Matterport3D dataset, M³PT greatly exceeds the second-best HoHo-R by 53.8%, 51.7%, and 37.9% in MRE, MAE, and RMSElog, severally. Compared with the suboptimal GuideNet in RMSE, the error is reduced from 192.9 mm to 138.9 mm, improving the performance nearly by 28.0%. Besides, M³PT achieves the highest accuracies in δ_i with different thresholds, outperforming the second-best method by 1.21, 0.29, and 0.09% point in δ_1, δ_2, and δ_3, respectively.

(ii) On Stanford2D3D dataset, M³PT is superior to the suboptimal Uni-Fuse with 44.0% improvement in MRE. Also, the MAE, RMSE, RMSElog is severally reduced by 42.6%, 24.3%, and 30.8% when comparing M³PT with the

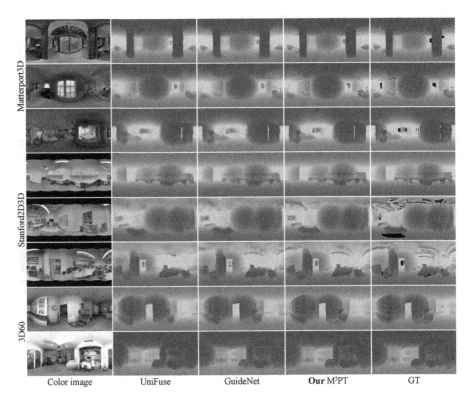

Fig. 6. Qualitative comparison of different methods, including UniFuse [23], GuideNet [47], and our M³PT. More visualizations can be found in our supplementary material.

second-best GuideNet. In addition, the accuracy metric verifies the effectiveness of M³PT again, which plays a prominent role in all approaches.

(iii) On 3D60 dataset, M³PT surpasses the second-best HoHo-R with large margins, improving it by 57.4% in MRE, 54.9% in MAE, 35.4% in RMSE, and 43.9% in RMSElog, severally. Furthermore, M³PT is more accurate than other approaches and prevail over the suboptimal methods with 1.26, 0.31, and 0.08% point in δ_1, δ_2, and δ_3, respectively.

(iv) Last but not least, apart from GuideNet that has been reported in Table 5, we further employ UniFuse, HoHo-R, and HoHo-H as baselines to see the influence of the additional sparse depth data and proposed mask strategy on the recovery of panoramic depth. As shown in Fig. 7, gray bar: only using RGB, orange bar: only using sparse depth, light orange bar: using both RGB and its sparse depth, and blue bar: using RGB and sparse depth with the proposed mask strategy M³PT. We can find that the error of only using sparse depth is much lower than that of only using RGB. Also, adding sparse depth data can benefit models with very large margins. Specifically, comparing light orange bar with gray bar, the errors of UniFuse, HoHo-R, and HoHo-H are severally reduced by 55.8%, 41.5%, and 50.5% on average on three datasets. What's more, adopting

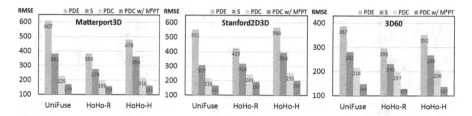

Fig. 7. Comparisons of different baselines with different-modal input data and M³PT. PDE: panoramic depth estimation only from color images. PDC: panoramic depth completion from not only color images but the corresponding sparse depths.

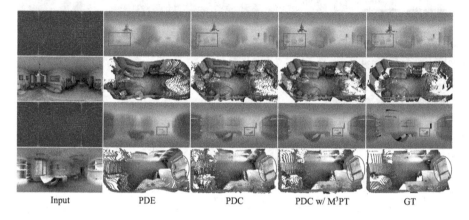

Fig. 8. Visual results of UniFuse [23] with different-modal input data and M³PT.

M³PT contributes to their significant improvements of 27.0%, 25.4%, and 26.3% on average compared with the orange bars on three datasets. These facts indicate sparse depth information has great reference value for depth recovery, and also prove that the panoramic depth completion is a potentially valuable task.

(2) Qualitative Results
(i) As shown in Fig. 6, our M³PT can recover more detailed objects and precise depth with reasonable visual effect. For example, for one thing, as illustrated in the first, fourth, and seventh rows, M³PT succeeds in predicting clearer edges of doors, tables, chairs, windows, *etc.*. For another thing, as shown in the fifth and sixth rows, although the color images of Stanford2D3D do not have pixels at both the top and bottom, M³PT can still predict more accurate depth values in the invisible areas. It strongly demonstrates the effectiveness and generalization of the proposed masked pre-training strategy via corrupting RGB-D data. In addition, M³PT is also good at distinguishing from background and foreground, *e.g.*, the furniture can be clearly discriminated from the wall.

(ii) As illustrated in Fig. 8, based on only color images (PDE), the depth predicted by UniFuse is extremely blurry that the corresponding 3D reconstruction introduces plenty of wrong location information, which causes negative defor-

Table 6. Performances of M³PT with different baselines on KITTI validation split.

Method	RMSE	MAE	iRMSE	iMAE
S2D [35]	858.02	311.47	3.07	1.67
+M³PT	**844.16**	**267.64**	**3.01**	**1.51**
GuideNet [47]	777.78	221.59	2.39	**1.00**
+M³PT	**761.57**	**217.68**	**2.26**	**1.00**
ACMNet [64]	789.72	216.65	2.32	0.96
+M³PT	**774.63**	**209.31**	**2.25**	**0.93**

mation, especially nearby walls. By contrast, adding sparse depth data (PDC) vastly improves the visual effect of both depth recovery and 3D reconstruction. Furthermore, when deploying the proposed M³PT with RGB-D data as input, both objects' structures and details tend to be more clear and abundant.

4.5 Generalization Capability

In this subsection, we further verify the generalization capability of M³PT on KITTI depth completion benchmark, whose sparse depth data is obtained by a 64-line LiDAR, and the RGB-D pairs have limited field of vision. As reported in Table 6, M³PT consistently improves the performances of S2D, GuideNet, and ACMNet. For example, M³PT reduces RMSE/MAE by 15.05mm/18.36mm averagely, indicating that our M³PT possesses robust generalization capability.

5 Conclusion

In this paper, we introduced a potentially valuable task, *i.e.*, panoramic depth completion, to help with dense panoramic depth recovery and 3D reconstruction from monocular 360° RGB-D data. Furthermore, we proposed the multi-modal masked pre-training (M³PT) framework to handle this task. It was the first time we showed that the masked pre-training could be very effective in modeling multi-modal tasks for vision, instead of the single-modal image recognition which was popularized by the masked autoencoders (MAE). As a result, comprehensive evaluations demonstrated the superiority of M³PT on three benchmark datasets. At last, we hope our exploration in this paper can facilitate future studies concerned with multi-modal vision tasks. In the future, we are going to extend M³PT to related topics such as depth denoising and super-resolution.

Acknowledgement. The authors would like to thank reviewers for their detailed comments and instructive suggestions. This work was supported by the National Science Fund of China under Grant Nos. U1713208, 62072242 and Postdoctoral Innovative Talent Support Program of China under Grant BX20200168, 2020M681608. Note that the PCA Lab is associated with, Key Lab of Intelligent Perception and Systems for High-Dimensional Information of Ministry of Education, and Jiangsu Key Lab of Image and Video Understanding for Social Security, Nanjing University of Science and Technology.

References

1. Albanis, G., et al.: Pano3d: A holistic benchmark and a solid baseline for 360° depth estimation. In: CVPRW, pp. 3722–3732. IEEE (2021)
2. Armeni, I., Sax, S., Zamir, A.R., Savarese, S.: Joint 2D–3D-semantic data for indoor scene understanding. arXiv preprint arXiv:1702.01105 (2017)
3. Bai, J., Lai, S., Qin, H., Guo, J., Guo, Y.: Glpanodepth: global-to-local panoramic depth estimation. arXiv preprint arXiv:2202.02796 (2022)
4. Bao, H., Dong, L., Wei, F.: Beit: Bert pre-training of image transformers. arXiv preprint arXiv:2106.08254 (2021)
5. Chang, A., et al.: Matterport3d: Learning from RGB-D data in indoor environments. In: 3DV (2017)
6. Chao, P., Kao, C.Y., Ruan, Y.S., Huang, C.H., Lin, Y.L.: Hardnet: a low memory traffic network. In: ICCV. pp. 3552–3561 (2019)
7. Chen, M., et al.: Generative pretraining from pixels. In: ICML, pp. 1691–1703. PMLR (2020)
8. Cheng, X., Wang, P., Guan, C., Yang, R.: Cspn++: learning context and resource aware convolutional spatial propagation networks for depth completion. In: AAAI, pp. 10615–10622 (2020)
9. Cheng, X., Wang, P., Yang, R.: Learning depth with convolutional spatial propagation network. In: ECCV, pp. 103–119 (2018)
10. Chodosh, N., Wang, C., Lucey, S.: Deep convolutional compressed sensing for LiDAR depth completion. In: Jawahar, C.V., Li, H., Mori, G., Schindler, K. (eds.) ACCV 2018. LNCS, vol. 11361, pp. 499–513. Springer, Cham (2019). https://doi.org/10.1007/978-3-030-20887-5_31
11. Dosovitskiy, A., et al.: An image is worth 16×16 words: transformers for image recognition at scale. In: ICLR (2021)
12. Eder, M., Moulon, P., Guan, L.: Pano popups: indoor 3D reconstruction with a plane-aware network. In: 3DV, pp. 76–84. IEEE (2019)
13. Eldesokey, A., Felsberg, M., Khan, F.S.: Confidence propagation through CNNs for guided sparse depth regression. IEEE Trans. Pattern Anal. Mach. Intell. 42(10), 2423–2436 (2019)
14. Erhan, D., Bengio, Y., Courville, A., Manzagol, P.A., Vincent, P., Bengio, S.: Why does unsupervised pre-training help deep learning? J. Mach. Learn. Res. 11, 625–660 (2010)
15. Feng, B.Y., Yao, W., Liu, Z., Varshney, A.: Deep depth estimation on 360 images with a double quaternion loss. In: 3DV, pp. 524–533. IEEE (2020)
16. Feng, Q., Shum, H.P., Morishima, S.: 360 depth estimation in the wild-the depth360 dataset and the segfuse network. In: VR. IEEE (2022)
17. Gordon, A., Li, H., Jonschkowski, R., Angelova, A.: Depth from videos in the wild: Unsupervised monocular depth learning from unknown cameras. In: ICCV. pp. 8977–8986 (2019)
18. Gu, J., Xiang, Z., Ye, Y., Wang, L.: Denselidar: a real-time pseudo dense depth guided depth completion network. IEEE Robot. Autom. Lett. 6(2), 1808–1815 (2021)
19. He, K., Chen, X., Xie, S., Li, Y., Dollár, P., Girshick, R.: Masked autoencoders are scalable vision learners. arXiv preprint arXiv:2111.06377 (2021)
20. He, K., Zhang, X., Ren, S., Sun, J.: Deep residual learning for image recognition. In: CVPR, pp. 770–778 (2016)

21. Hu, M., Wang, S., Li, B., Ning, S., Fan, L., Gong, X.: PENet: towards precise and efficient image guided depth completion. In: ICRA (2021)
22. Jaritz, M., De Charette, R., Wirbel, E., Perrotton, X., Nashashibi, F.: Sparse and dense data with CNNs: Depth completion and semantic segmentation. In: 3DV, pp. 52–60 (2018)
23. Jiang, H., Sheng, Z., Zhu, S., Dong, Z., Huang, R.: Unifuse: unidirectional fusion for 360 panorama depth estimation. IEEE Robot. Autom. Lett. **6**(2), 1519–1526 (2021)
24. Jin, L., : Geometric structure based and regularized depth estimation from 360 indoor imagery. In: CVPR, pp. 889–898 (2020)
25. Krauss, B., Schroeder, G., Gustke, M., Hussein, A.: Deterministic guided lidar depth map completion. arXiv preprint arXiv:2106.07256 (2021)
26. Lai, Z., Chen, D., Su, K.: Olanet: self-supervised 360° depth estimation with effective distortion-aware view synthesis and l1 smooth regularization. In: ICME, pp. 1–6. IEEE (2021)
27. Lee, Y., Jeong, J., Yun, J., Cho, W., Yoon, K.J.: SpherePHD: applying CNNs on a spherical polyhedron representation of 360deg images. In: CVPR, pp. 9181–9189 (2019)
28. Lee, Y., Jeong, J., Yun, J., Cho, W., Yoon, K.J.: SpherePHD: applying CNNs on 360° images with non-euclidean spherical polyhedron representation. IEEE Trans. Pattern Anal. Mach. Intell. (2020)
29. Li, A., Yuan, Z., Ling, Y., Chi, W., Zhang, C., et al.: A multi-scale guided cascade hourglass network for depth completion. In: WACV, pp. 32–40 (2020)
30. Li, J., Zhang, T., Luo, W., Yang, J., Yuan, X.T., Zhang, J.: Sparseness analysis in the pretraining of deep neural networks. IEEE Trans. Neural Networks Learn. Syst. **28**(6), 1425–1438 (2016)
31. Li, Y., Yan, Z., Duan, Y., Ren, L.: Panodepth: a two-stage approach for monocular omnidirectional depth estimation. In: 3DV, pp. 648–658. IEEE (2021)
32. Lin, Y., Cheng, T., Zhong, Q., Zhou, W., Yang, H.: Dynamic spatial propagation network for depth completion. In: AAAI (2022)
33. Liu, L., et al.: FCFR-net: feature fusion based coarse-to-fine residual learning for depth completion. In: AAAI, vol. 35, pp. 2136–2144 (2021)
34. Lu, K., Barnes, N., Anwar, S., Zheng, L.: From depth what can you see? Depth completion via auxiliary image reconstruction. In: CVPR, pp. 11306–11315 (2020)
35. Ma, F., Cavalheiro, G.V., Karaman, S.: Self-supervised sparse-to-dense: self-supervised depth completion from lidar and monocular camera. In: ICRA (2019)
36. Park, J., Joo, K., Hu, Z., Liu, C.-K., So Kweon, I.: Non-local spatial propagation network for depth completion. In: Vedaldi, A., Bischof, H., Brox, T., Frahm, J.-M. (eds.) ECCV 2020. LNCS, vol. 12358, pp. 120–136. Springer, Cham (2020). https://doi.org/10.1007/978-3-030-58601-0_8
37. Pintore, G., Agus, M., Almansa, E., Schneider, J., Gobbetti, E.: Slicenet: deep dense depth estimation from a single indoor panorama using a slice-based representation. In: CVPR, pp. 11536–11545 (2021)
38. Qiu, J., et al.: DeepLiDAR: deep surface normal guided depth prediction for outdoor scene from sparse lidar data and single color image. In: CVPR, pp. 3313–3322 (2019)
39. Rey-Area, M., Yuan, M., Richardt, C.: 360monodepth: high-resolution 360° monocular depth estimation. arXiv e-prints pp. arXiv-2111 (2021)
40. Schuster, R., Wasenmuller, O., Unger, C., Stricker, D.: SSGP: sparse spatial guided propagation for robust and generic interpolation. In: WACV, pp. 197–206 (2021)

41. Shen, Z., Lin, C., Liao, K., Nie, L., Zheng, Z., Zhao, Y.: Panoformer: panorama transformer for indoor 360 depth estimation. arXiv e-prints pp. arXiv-2203 (2022)
42. Shen, Z., Lin, C., Nie, L., Liao, K., Zhao, Y.: Distortion-tolerant monocular depth estimation on omnidirectional images using dual-cubemap. In: ICME, pp. 1–6. IEEE (2021)
43. Silberman, N., Hoiem, D., Kohli, P., Fergus, R.: Indoor segmentation and support inference from RGBD images. In: Fitzgibbon, A., Lazebnik, S., Perona, P., Sato, Y., Schmid, C. (eds.) ECCV 2012. LNCS, vol. 7576, pp. 746–760. Springer, Heidelberg (2012). https://doi.org/10.1007/978-3-642-33715-4_54
44. Song, S., Yu, F., Zeng, A., Chang, A.X., Savva, M., Funkhouser, T.: Semantic scene completion from a single depth image. In: CVPR, pp. 1746–1754 (2017)
45. Sun, C., Hsiao, C.W., Wang, N.H., Sun, M., Chen, H.T.: Indoor panorama planar 3D reconstruction via divide and conquer. In: CVPR, pp. 11338–11347 (2021)
46. Sun, C., Sun, M., Chen, H.T.: Hohonet: 360 indoor holistic understanding with latent horizontal features. In: CVPR, pp. 2573–2582 (2021)
47. Tang, J., Tian, F.P., Feng, W., Li, J., Tan, P.: Learning guided convolutional network for depth completion. IEEE Trans. Image Process. **30**, 1116–1129 (2020)
48. Tateno, K., Navab, N., Tombari, F.: Distortion-aware convolutional filters for dense prediction in panoramic images. In: Ferrari, V., Hebert, M., Sminchisescu, C., Weiss, Y. (eds.) ECCV 2018. LNCS, vol. 11220, pp. 732–750. Springer, Cham (2018). https://doi.org/10.1007/978-3-030-01270-0_43
49. Teutscher, D., Mangat, P., Wasenmüller, O.: PDC: piecewise depth completion utilizing superpixels. In: ITSC, pp. 2752–2758. IEEE (2021)
50. Uhrig, J., Schneider, N., Schneider, L., Franke, U., Brox, T., Geiger, A.: Sparsity invariant CNNs. In: 3DV, pp. 11–20 (2017)
51. Van Gansbeke, W., Neven, D., De Brabandere, B., Van Gool, L.: Sparse and noisy lidar completion with RGB guidance and uncertainty. In: MVA, pp. 1–6 (2019)
52. Vaswani, A., et al.: Attention is all you need. In: NeurIPS, vol. 30 (2017)
53. Wang, F.E., Yeh, Y.H., Sun, M., Chiu, W.C., Tsai, Y.H.: Bifuse: monocular 360 depth estimation via bi-projection fusion. In: CVPR, pp. 462–471 (2020)
54. Wong, A., Cicek, S., Soatto, S.: Learning topology from synthetic data for unsupervised depth completion. IEEE Robo. Autom. Lett. **6**(2), 1495–1502 (2021)
55. Wong, A., Fei, X., Hong, B.W., Soatto, S.: An adaptive framework for learning unsupervised depth completion. IEEE Robot. Autom. Lett. **6**(2), 3120–3127 (2021)
56. Wong, A., Fei, X., Tsuei, S., Soatto, S.: Unsupervised depth completion from visual inertial odometry. IEEE Robot. Autom. Lett. **5**(2), 1899–1906 (2020)
57. Wong, A., Soatto, S.: Unsupervised depth completion with calibrated backprojection layers. In: ICCV (2021)
58. Xie, Z., et al.: SimMIM: a simple framework for masked image modeling. arXiv preprint arXiv:2111.09886 (2021)
59. Xu, Y., Zhu, X., Shi, J., Zhang, G., Bao, H., Li, H.: Depth completion from sparse lidar data with depth-normal constraints. In: ICCV, pp. 2811–2820 (2019)
60. Xu, Z., Yin, H., Yao, J.: Deformable spatial propagation networks for depth completion. In: ICIP, pp. 913–917. IEEE (2020)
61. Yan, L., Liu, K., Gao, L.: Dan-conv: depth aware non-local convolution for lidar depth completion. Electron. Lett. **57**(20), 754–757 (2021)
62. Yan, Z., et al.: Rignet: repetitive image guided network for depth completion. arXiv preprint arXiv:2107.13802 (2021)
63. Yun, I., Lee, H.J., Rhee, C.E.: Improving 360 monocular depth estimation via non-local dense prediction transformer and joint supervised and self-supervised learning. arXiv preprint arXiv:2109.10563 (2021)

64. Zhao, S., Gong, M., Fu, H., Tao, D.: Adaptive context-aware multi-modal network for depth completion. IEEE Trans. Image Process. **30**, 5264–5276 (2021)
65. Zhou, K., Yang, K., Wang, K.: Panoramic depth estimation via supervised and unsupervised learning in indoor scenes. Appl. Opt. **60**(26), 8188–8197 (2021)
66. Zhu, Y., Dong, W., Li, L., Wu, J., Li, X., Shi, G.: Robust depth completion with uncertainty-driven loss functions. arXiv preprint arXiv:2112.07895 (2021)
67. Zhuang, C., Lu, Z., Wang, Y., Xiao, J., Wang, Y.: ACDNet: adaptively combined dilated convolution for monocular panorama depth estimation. In: AAAI (2022)
68. Zioulis, N., Karakottas, A., Zarpalas, D., Alvarez, F., Daras, P.: Spherical view synthesis for self-supervised 360 depth estimation. In: 3DV, pp. 690–699. IEEE (2019)
69. Zioulis, N., Karakottas, A., Zarpalas, D., Daras, P.: OmniDepth: dense depth estimation for indoors spherical panoramas. In: Ferrari, V., Hebert, M., Sminchisescu, C., Weiss, Y. (eds.) ECCV 2018. LNCS, vol. 11210, pp. 453–471. Springer, Cham (2018). https://doi.org/10.1007/978-3-030-01231-1_28

GitNet: Geometric Prior-Based Transformation for Birds-Eye-View Segmentation

Shi Gong[1], Xiaoqing Ye[2], Xiao Tan[2], Jingdong Wang[2], Errui Ding[2], Yu Zhou[1(✉)], and Xiang Bai[1]

[1] Huazhong University of Science and Technology, Wuhan, China
{gongshi,yuzhou}@hust.edu.cn
[2] Baidu Inc., Beijing, China

Abstract. Birds-eye-view (BEV) semantic segmentation is critical for autonomous driving for its powerful spatial representation ability. It is challenging to estimate the BEV semantic maps from monocular images due to the spatial gap, since it is implicitly required to realize both the perspective-to-BEV transformation and segmentation. We present a novel two-stage **G**eometry Pr**I**or-based **T**ransformation framework named GitNet, consisting of (i) the geometry-guided pre-alignment and (ii) ray-based transformer. In the first stage, we decouple the BEV segmentation into the perspective image segmentation and geometric prior-based mapping, with explicit supervision by projecting the BEV semantic labels onto the image plane to learn visibility-aware features and learnable geometry to translate into BEV space. Second, the pre-aligned coarse BEV features are further deformed by ray-based transformers to take visibility knowledge into account. GitNet achieves the leading performance on the challenging nuScenes and Argoverse Datasets.

Keywords: Birds-eye-view · Segmentation · Geometric prior-based

1 Introduction

The birds-eye-view (BEV) semantic map is a compact representation of the surrounding environment for autonomous driving, which provides both the layout of road elements and the occupancy of objects. Such representations are useful for downstream tasks such as path planning, collision avoidance. In this work, we focus on BEV map estimation from monocular images.

The BEV semantic segmentation is particularly challenging for two reasons. First, the BEV segmentation implicitly involves two coupled tasks: mapping from perspective view to the birds-eye-view, and pixel-wise classification. Most

---

S. Gong and X. Ye—Contribute equally.

Supplementary Information The online version contains supplementary material available at https://doi.org/10.1007/978-3-031-19769-7_23.

existing methods [7–10,13,16] learn to convert the image features from the perspective view to the BEV and then perform segmentation. The training process is supervised by the loss function defined in the BEV space alone, and thus the learning procedure of mapping and pixel-wise classification is coupled in these approaches. How to explicitly incorporate the geometry prior knowledge to decouple the feature for mapping and classification remains unexplored. Secondly, a fundamental difference between monocular image segmentation and BEV segmentation lies in that the latter requires inferring the labels of occluded objects behind foreground objects, which places a tremendous difficulty for the network to learn effective feature representation to differentiate the invisible from the visible. In the previous IPM-based methods [12,18,22], the features of foreground visible objects occupy the invisible regions in the BEV space. Since the visibility of pixels is not encoded in the features, it is tough for a convolutional neural network to recover the missing information in the invisible regions.

To address the aforementioned concerns, we derive a novel two-stage transformation from the perspective space to the BEV space. In the first stage, we leverage the proposed Geometry-guided Pre-Alignment (GPA) to obtain coarse pre-aligned BEV features. In the GPA, we decouple the BEV segmentation into the perspective image segmentation and geometric prior-based mapping, with explicit supervision by projecting the BEV semantic labels onto the image plane. As the projected labels reflect all ground regions including visible and invisible ones in the perspective view, while the perspective image appearance features only reflect the visible regions, we obtain the visibility-aware image features by fusing the information of projected labels and appearance features. We warp the visibility-aware features into BEV space via the learnable geometry.

In the second stage, the pre-aligned BEV features are further enhanced by the proposed Ray-based Transformer (RT), which adopts the efficient ray-based attention mechanism that we compute the attention map in a single column so as to keep the high-resolution of feature maps. The pre-aligned BEV features conveying *appearance* and *visibility* information, along with BEV positional encoding, work as *Queries*, and the augmented perspective features serve as *Keys* and *Values*. Cooperating with the projected labels, the novel Depth-Aware Dice loss is proposed to alleviate the dominant effect by closer instances in perspective view. Besides, since those pixels that have easily-classified appearances or follow a simple perspective-to-BEV mapping, such as most road regions, comprise the majority of the loss, we present a Self-Weighted Dice loss to balance the easy-hard samples among categories. To sum up, the main contributions of our work are as follows:

- We propose a novel two-stage transformation from perspective view to birds-eye-view. In the first stage, we decouple the BEV segmentation into the perspective image segmentation and geometric prior-based mapping, and provide visibility-aware and pre-aligned BEV features. In the second stage, the warped features are deformed by aggregating appearance information.
- We introduce a Depth-aware Dice loss that removes the perspective effect on the perspective image segmentation and a Self-weighted Dice loss to re-weight the easy-hard samples.

– Our framework presents new state-of-the-art performance on two large-scale datasets including nuScenes and Argoverse.

2   Related Work

Most BEV segmentation works follow the similar pipeline to first extract features from the monocular image, and then convert the features from the perspective view (PV) to the birds-eye-view (BEV). Based on different PV-BEV transformation strategies, the methods can be grouped into four categories as follows:

IPM-Based Methods: An early work [18] performs semantic segmentation in the image plane and then transforms the semantic results into the BEV space via a homography. This approach works well for predicting flat road layout but fails for objects such as cars that stand above the ground plane. [22] alleviates this problem by training a generative adversarial network to refine the predictions from the IPM. More recently [12] transforms the image features into BEV, which is then fed into a deep segmentation network for further refinement.

Depth-Based Methods: The depth-based methods are one of the main streams in this field. [5] adopts RGB-D images to learn an implicit representation for 3D localization. [17] leverages an in-painting CNN to infer the semantic labels and depths of the scene to obtain the BEV map by projecting the produced semantic point cloud onto the ground plane. EPOSH [4] first performs monocular depth estimation and then exploits depth maps to transform 2D image features to the BEV space. [6,10,11,16] learn a depth distribution within pixels to lift 2D images to 3D point clouds, and then project the point clouds onto BEV space.

Bottleneck-Based Methods: VED [7] uses the fully-connected bottleneck to realize the transformation, which loses the spatial information. Therefore the output is fairly coarse and fails to segment small objects. VPN [9] predicts the semantic BEV map from a stack of surround-view images, via a fully-connected view-transformer module. PON [13] proposes a column-wise fully-connected layer to realize the transformation of features from image space to BEV space.

Attention Based Methods: The attention-based methods are attracting increasing attention. NEAT [3] proposes a novel representation termed neural attention fields, which compresses 2D image features into the BEV representation based on the attention map. TIM [15] transforms image columns to BEV polar rays via cross-attention. Though similar to our work, the BEV features are initialized to constant, and the geometric prior is not exploited in their method, which limits the capacity of reasoning in 3D space.

3   Method

In this section, we first briefly present our GitNet approach, which learns the birds-eye-view (BEV) segmentation map from a monocular image $I \in \mathbb{R}^{H \times W \times 3}$.

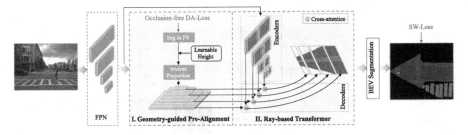

Fig. 1. The overview of the GitNet framework to predict the BEV semantic map from the perspective image. The multi-level pyramid image features extracted by the FPN are transformed to the BEV features by our *two-stage* transformation pipeline, which includes the *Geometry-guided Pre-Alignment (GPA)* and *Ray-based Transformer (RT)*. The explicit supervision is enforced to the GPA Stage guided by the learnable camera height to learn visibility-aware features, which are then converted to pre-aligned BEV features. The *RT* column-wisely refines the PV features and pre-aligned BEV features with the mechanism of attention. The refined BEV features are fed into the BEV segmentation layers, which output C pixel-wise binary classification.

The predicted BEV semantic map $S \in \mathbb{R}^{Z \times X \times C}$ is in the ego camera coordinate, with Z and X are the spatial dimensions of the regular lattice grid in BEV space and C is the number of semantic categories including road layout and objects.

3.1 Overview

The goal of our network is to predict the semantic map of the scene on the birds-eye-view space from a monocular perspective image. The challenge of predicting the BEV semantic map lies in that the input and output representations exist in different spaces and thus the network is acquired to learn the transformation from perspective image view to orthographic BEV space. As depicted in Fig. 1, our framework is a two-stage pipeline that transforms the perspective view (PV) to the birds-eye-view. It mainly consists of four modules, (i) the feature pyramid network (FPN) for multi-scale perspective feature representation, (ii) Geometry-guided Pre-Alignment that transfers features into BEV space based on the learnable camera height, (iii) the ray-based transformer module for attention-based feature enhancement before BEV segmentation, and (iv) the specially designed loss functions for re-weighting different pixels.

In our network, the core design is the two-stage transformation from the perspective space to the BEV space. Firstly we leverage the geometric guidance to provide *appearance* and *visibility* for initializing the transformed BEV features. To solve the ambiguity caused by the mounting height of the camera, we specially learn the height for better alignment between the perspective space and the birds-eye-view space. After obtaining the pre-aligned BEV features, we further adopt the ray-based transformer module based on the column-wise attention for further enhancing the feature deformation in BEV space for conducting semantic segmentation. The explicit supervision is enforced to the GPA Stage guided

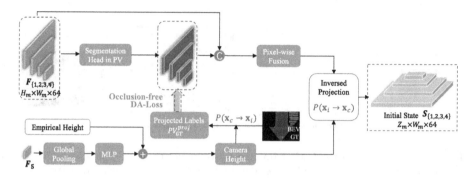

Fig. 2. Geometry-guided Pre-Alignment Module. The pyramid image features are first fed into the segmentation head to predict the BEV-consistent probability maps enforced by the *Occlusion-free DA loss* with *projected labels*. The BEV-consistent probability maps and the perspective features are further encoded by pixel-wise fusion to extract visibility-aware features. In the other branch, the smallest-scale features F_5 are used to predict the offset w.r.t. to the empirical predefined height. Then the learned camera height is applied to inversely project the visibility-aware perspective features to BEV features, which serve as initial queries of the follow-up transformer stage.

by the learnable camera height to learn visibility-aware features, which are then converted to pre-aligned BEV features. In addition, to alleviate the perspective effect caused by the imaging, we organize the projected supervision loss in a depth-aware manner and further propose the Self-Weighted Dice (SW-Dice) loss to re-weight the easy-hard samples. We will introduce the detailed design of each component in the following parts.

3.2   Geometry-Guided Pre-alignment

In this section, we introduce the first stage, *i.e.*, the Geometry-guided Pre-Alignment module. We first present the geometric relation between the perspective view and BEV. Then we detail the consistency between image features and projected BEV labels, and describe our visibility-aware feature learning method. Finally, we describe geometry-based warping to obtain the pre-aligned BEV features. The detailed pipeline of this module is depicted in Fig. 2.

Learnable Geometric Relation. The transformation from perspective view (PV) to BEV can be given by a projection matrix P. We first introduce the coordinate systems: a certain point in the camera coordinate system is represented by $\mathbf{x}_c = [x_c, y_c, z_c]^T \in \mathbb{R}^3$. The ground space is by setting the y-coordinate of the camera coordinate system to h and a certain point lying on the ground plane turns out to be $\mathbf{x}_c = [x_c, h, z_c]^T$, where h denotes the height of the mounted camera from the ground. The BEV coordinates simply remove the y-dim and can be denoted as $\mathbf{x}^B = [x_c, z_c]^T \in \mathbb{R}^2$. In the following, we do not particularly distinguish the BEV space from the ground space in the camera coordinate system. The homogeneous image coordinates $\mathbf{x}_i = [x_i, y_i, 1]^T$ have a one-to-one correspondence with the ground coordinates, which can be expressed by:

$$P(\mathbf{x}_c \to \mathbf{x}_i): \quad \mathbf{x}_i = K\mathbf{x}_c/z_c = K[x_c/z_c, h/z_c, 1]^T \tag{1}$$

where K is the camera intrinsic matrix: $K = [[f_x, 0, c_x], [0, f_y, c_y], [0, 0, 1]]^T$, and the inverse transformation from image to ground coordinates is formulated as:

$$P(\mathbf{x}_i \to \mathbf{x}_c): \left\{ \begin{array}{l} x_c = \frac{(x_i - c_x)z_c}{f_x} \\ z_c = \frac{f_y h}{y_i - c_y} \end{array} \right. \tag{2}$$

Based on the geometric correspondences illustrated in the Eq. (1) and (2), we are able to transform from the perspective space to the BEV. In this way, we are able to recover the coarse ground coordinates given the image coordinates and the camera height. However, as is acknowledged in [13], the camera height h is unavailable for a real monocular perception system. Alternatively, we enforce the network to learn the camera height parameters. The image features $\boldsymbol{F_5}$ with the scale of $\times 1/128$ are compressed into a vector by global average pooling and followed by an MLP to leverage the global context for predicting the offset of the camera height to the empirically predefined height.

Visibility-Aware Perspective Feature Learning. The BEV semantic segmentation is an implicit mapping-segmentation coupling task. Here we decouple the BEV segmentation into the geometric prior-based mapping and perspective segmentation. The latter is supervised by an explicit segmentation loss with projecting the BEV GT labels onto the image plane following the transformation $P(\boldsymbol{x}_c \to \boldsymbol{x}_i)$ in the Eq. (1) to generate the projected labels PV_{gt}^{proj}. PV_{gt}^{proj} reflects the whole perspective-view ground including visible or invisible regions. However, the perspective features extracted from images only reflect the visible foreground regions. Therefore, the projected labels PV_{gt}^{proj} can be used to obtain the visibility-aware image features by fusing the information of projected labels and image features. In specific, the pyramid features $\boldsymbol{F_{\{1,2,3,4\}}} \in \mathbb{R}^{H_i \times W_i \times 64}$ are separately fed into the weight-shared segmentation head to generate the corresponding probability maps $\boldsymbol{P_{\{1,2,3,4\}}} \in \mathbb{R}^{H_i \times W_i \times C}$ under the supervision of our depth-aware Dice (DA-Dice) segmentation loss with projected labels. We concatenate the feature maps and the corresponding probability maps, and learn the visibility-aware features $\boldsymbol{A}_i$ with pixel-wise fusion (MLP) by:

$$\boldsymbol{A}_m = \mathrm{MLP}(\boldsymbol{F}_m, \boldsymbol{P}_m) \tag{3}$$

Geometry-Based Warping. From the Eq. (2), we can derive that $z_c/x_c = f_x/(x_i - c_x)$, which indicates that pixels of perspective view lying on *the same column* (*i.e.*, with the same x-coordinate x_i) map onto *the same polar ray* in BEV space with a slope of $f_x/(x_i - c_x)$. Following the transformation $P(y_i \to z_c)$ in Eq. (2), the $\boldsymbol{j}$-th column of augmented image features $\boldsymbol{A}_m^j$ are warped into the BEV space with the learned camera height h by inverse projection. That is:

$$\boldsymbol{S}_m^j = \mathrm{Warp}(\boldsymbol{A}_m^j; P(y_i \to z_c)) \tag{4}$$

402   S. Gong et al.

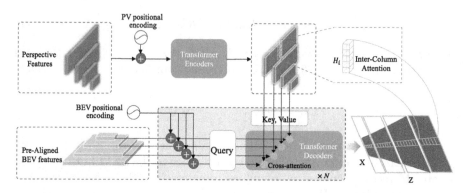

Fig. 3. Ray-based Transformer. The pyramid perspective features, along with the positional encoding in perspective view, are fed into the transformer encoders to integrate the knowledge within the same column. In the next decoder module where the inter-column cross-attention is conducted, the output features of encoder serve as *Key* and *Value*, and the pre-aligned BEV features, along with the BEV positional encoding, work as *Query*. The initial queries $\widetilde{S}_m$ are refined by $N\times$ stacked decoder layers. Finally, the refined BEV features $\widetilde{S}_m \in \mathbb{R}^{Z_m \times W_m \times 64}$ are warped to $M_m \in \mathbb{R}^{Z_m \times X \times 64}$ in the X-Z coordinate system, all of which are concatenated along with Z axis to output the BEV feature map M.

where $\{S_m^1, S_m^2, ..., S_m^{W_m}\}$ are computed in parallel in our implementation, and are concatenated to output the tensor $S_m \in \mathbb{R}^{Z_m \times W_m \times 64}$. The warped BEV features take advantage of both the appearance from multi-scale perspective-view features and the visibility with BEV projected-to-PV labels guidance. The geometry-guided transformation module provides initial queries for the follow-up transformer stage for further tuning the features for the BEV segmentation task.

3.3 Ray-Based Transformer

The second step of our two-stage transformation pipeline is the ray-based transformer, which is depicted in the Fig. 3. In this stage, we extend the common multi-head attention [20] into our Ray-based Transformer (RT). The multi-head attention needs three inputs of queries (Q), keys (K), and values (V), which is denoted as MultiHead(Q, K, V). We refer the reader to the literature [20] and see the appendix for more detailed descriptions. Since our BEV semantic segmentation task requires high-resolution feature maps, computing the attention of a full image, like most existing works, will bring high computation complexity and GPU memory. As derived in Sect. 3.2, pixels of perspective view lying on the *same column* correspond to the *same polar ray* of birds-eye-view. This motivates us to compute attention in *a single column or ray*, which greatly reduces the complexity of attention.

Differences with the Original Transformer. Our method draws on the core idea of Transformer, *i.e.*, employing the multi-head attention mechanism. But we have two new designs for our BEV segmentation task. Firstly, We use column-wise attention so that we can perform on high-resolution feature maps, which is indispensable in our pixel-wise recognition. Secondly, we introduce the pre-aligned features encoding the appearance and visibility, along with BEV positional encoding, as queries in the cross attention. The ablation study in the Sect. 4.3 validates the superiority of our designs. In the following part, we detail two attention mechanisms in our transformer, and omit other components such as normalization for the sake of simplicity. The complete structure can be seen in our supplementary material.

Column Context Augment (CCA) in Encoder. As illustrated in Fig. 3, the inputs for the transformer encoders are perspective features $\{F_1, F_2, F_3, F_4\}$ extracted from the FPN, where F_m has a spatial resolution of $H_m \times W_m$. In the CCA, each pixel adaptively integrates the information from other pixels of the same column, by using multi-head self-attention. We further introduce spatial positional encodings P_m to the input F_m to distinguish the positions of the input features. We use a sine function to generate spatial positional encoding. Let $F_m^j, P_m^j \in \mathbb{R}^{H_m \times 64}$ denote the j-th column of F_m and P_m, respectively. The mechanism of CCA can be summarized as

$$
\begin{aligned}
\widetilde{F}_m^j &= F_m^j + \text{MultiHead}(F_m^j + P_m^j, F_m^j + P_m^j, F_m^j), \\
\widetilde{F}_m &= \text{CCA}(F_m) = \text{Concat}(\widetilde{F}_m^1, \widetilde{F}_m^2, ..., \widetilde{F}_m^{W_m})
\end{aligned}
\tag{5}
$$

Ray-Based Cross-Attention (RCA) in Decoder. RCA in the transformer decoder aims to refine the output of pre-alignment block, $\{S_1, S_2, S_3, S_4\}$, based on the augmented image features $\{\widetilde{F}_1, \widetilde{F}_2, \widetilde{F}_3, \widetilde{F}_4\}$. As depicted in Fig. 3, the RCA receives the pre-aligned BEV feature as *Query*, the augmented features built from the encoder as *Key* and *Value*. Similar to CCA, spatial positional encoding P_m' is also adopted in RCA. The difference is that P_m' represents the position in the BEV, while P_m is on the image plane. The mechanism of RCA can be summarized as

$$
\begin{aligned}
\widetilde{S}_m^j &= S_m^j + \text{MultiHead}\left(S_m^j + P_m'^j, \widetilde{F}_m^j + P_m^j, \widetilde{F}_m^j\right), \\
\widetilde{S}_m &= \text{RCA}(S_m, F_m) = \text{Concat}(\widetilde{S}_m^1, \widetilde{S}_m^2, ..., \widetilde{S}_m^{W_m})
\end{aligned}
\tag{6}
$$

Since the columns of $\widetilde{S}_m$ are still in the image coordinate, we warp them to rays in the camera coordinate following the transformation $P(x_i \to x_c)$ in the Eq. (2) to obtain $\{M_1, M_2, M_3, M_4\}$, which are responsible for different depth ranges. We concatenate all features along the depth axis to obtain the feature maps of the whole scene:

$$
\begin{aligned}
M_m &= \text{Warp}(\widetilde{S}_m; P(x_i \to x_c)), \\
M &= \text{Concat}(M_1, M_2, M_3, M_4)
\end{aligned}
\tag{7}
$$

The final BEV feature maps M are fed to the downstream convolutional segmentation network. Thanks to the CCA and RCA-based transformer, the network take appearance and visibility knowledge into account, and further tunes the invisible regions of the pre-aligned BEV features, based on the context information from the perspective features.

3.4   Loss Functions

The Dice loss is commonly adopted in segmentation for alleviating the data imbalance problem. The GT semantic label of i-th pixel in the BEV map is $[y_i^1, y_i^2,..., y_i^C]$, and the predicted probability is $[p_i^1, p_i^2,..., p_i^C]$, where $y_i^k \in \{0,1\}$ and $p_i^k \in [0,1]$, and C is the number of classes. The dice loss can be formulated as:

$$L_{dice} = 1 - \frac{1}{C} \sum_{k=1}^{C} \frac{2 \sum_i^N y_i^k p_i^k}{\sum_i^N y_i^k + p_i^k + \epsilon} \tag{8}$$

where N is the number of pixels in a mini-batch and ϵ is a constant used to prevent division by zero.

For the BEV semantic segmentation task which is actually an implicit multi-task problem involving 3D location and segmentation, there are two problems that can affect the performance. For one thing, due to the perspective projection from the real world to the image plane, distant objects appear to be smaller than nearer objects. In other words, the closer instances occupy much more pixels than farther ones, which dominate the overall segmentation loss in the perspective view. For another, those pixels that have easily-classified appearances or follow a simple perspective-to-BEV mapping, such as most road regions, comprise the majority of the loss.

Occlusion-Free Depth-Aware Dice Loss: As discussed in Sect. 3.2, we project the BEV labels onto the image plane to generate the projected labels PV_{gt}^{proj}, which supervises the segmentation on the perspective view. To tackle the first problem caused by the domination of nearer objects in perspective images, we propose the novel Depth-aware Dice loss by re-weighting the loss in a depth-aware manner. In specific, the Jacobian determinant gives the ratio of the area ratio between *image ground plane* (ΔA_i) and the *BEV ground plane* (ΔA_c) as:

$$R_{A_c \to A_i} = \frac{\partial A_i}{\partial A_c} = |J| = \begin{vmatrix} \frac{\partial x_i}{\partial z_c} & \frac{\partial x_i}{\partial x_c} \\ \frac{\partial y_i}{\partial z_c} & \frac{\partial y_i}{\partial x_c} \end{vmatrix} = \begin{vmatrix} \frac{-f_x x_c}{z_c^2} & \frac{f_x}{z_c} \\ \frac{-f_y h}{z_c^2} & 0 \end{vmatrix} = \frac{f_x f_y h}{z_c^3} \tag{9}$$

We find that the area ratio is proportional to $(1/z_c)^3$, thus we re-weight the pixels with the weight z_c^3 to solve the imbalance. The depth-aware dice loss is:

$$L_{DA\_dice} = 1 - \frac{1}{C} \sum_{k=1}^{C} \frac{2 \sum_i^N z_{ci}^3 y_i^k p_i^k}{\sum_i^N z_{ci}^3 (y_i^k + p_i^k) + \epsilon} \tag{10}$$

Self-weighted Dice Loss: To further alleviate the second problem, *i.e.*, the dominating influence from easy samples in training, we propose to associate training samples with dynamically adjusted weights to emphasize hard examples. We first propose a weighting function I_i^k to adjust the hard-mining strength by a parameter α in Eq. (11) and then utilize I_i^k to reweight the Dice loss and obtain the self-weighted dice loss in Eq. (12).

$$I_i^k = 1 + \alpha[y_i^k(1 - p_i^k) + (1 - y_i^k)p_i^k]_{\text{stop\_grad}} \tag{11}$$

$$L_{\text{sw\_dice}} = 1 - \frac{1}{C}\sum_{k=1}^{C}\frac{2\sum_i^N I_i^k y_i^k p_i^k}{\sum_i^N I_i^k(y_i^k + p_i^k) + \epsilon} \tag{12}$$

Note that we detach the weighting function I_i^k to stop the backward propagation of the gradient in Eq. (12). Otherwise, the term $p_i^k(1 - p_i^k)$ within $I_i^k y_i^k p_i^k$ will be maximized to make p_i^k fall around the undesired value 0.5.

4 Experiments

4.1 Experimental Setup

Dataset. We conduct extensive experiments on two large-scale datasets: The nuScenes [1] and Argoverse [2] road-scene datasets. Since the two datasets are predominantly collected for 3D object detection task rather than BEV semantic segmentation task, we follow the data generation method in [13] to convert the ground truth 3D bounding box annotations and the vectorized road maps into GT semantic maps in BEV. In addition, for fair comparisons, we also follow the same training and validation splits with other methods. The nuScenes includes 4 road layout categories and 7 object categories, and the Argoverse includes 7 object categories as well as drivable road. For both datasets, the ground truth of birds-eye-view expands from $1\,\text{m}$ to $50\,\text{m}$ in front of the camera *i.e.*, along the z-direction and $25\,\text{m}$ to either side (*i.e.*, along the x-direction). Due to the greater diversity of nuScenes, we choose this dataset for all ablation studies.

Implementation Details. For fair comparisons, we adopt a pretrained ResNet-50 with a feature pyramid on top as the backbone. We adopt a simplified HRNet [21] as the BEV segmentation head. In our implementation, we use a simplified HRNet32 by halving the number of blocks in each stage. We use two encoder layers and four decoder layers in the Ray-based transformer. The hyperparameter α in Eq. 11 for the SW-Dice loss is set as 0.5. We adopt a similar depth-interval assignment strategy with [13], but we only use the former four scales of the FPN. The concatenated BEV feature maps from different depth intervals are of 98×100 pixels, with each pixel covering $0.5\,\text{m}$. We obtain the final output map with a resolution of 196×200 pixels by upsampling, which is consistent with other methods. The model is trained using four Tesla V100 cards, each with 32G memory. We optimize the network with Adam policy for gradients accumulated over every 8 iterations and train for 40 epochs. The initial learning rate is set to 0.0002, with a weight decay of 0.99 and batch size 12.

Evaluation Metric. Our evaluation metric is the Intersection over Union (IoU) score, which we compute by binarizing the output probability maps with the threshold of 0.5. Invisible regions are ignored during evaluation following [13].

Table 1. Results of IoU (%) on nuScenes validation set. "Mean" refers to the average IoU over all classes. "Crossing": Pedestrian Crossing, "C.V.": Construction Vehicle, "Motor.": Motorcycle, "Ped.": Pedestrian, "Cone": Traffic Cone.

Method	Layout				Object										Mean
	Drivable	Crossing	Walkway	Carpark	Bus	Bike	Car	C.V.	Motor.	Trailer	Truck	Ped.	Cone	Barrier	
IPM [13]	40.1	–	14.0	–	3.0	0.2	4.9	–	0.8	–	–	0.6	–	–	–
Depth Unpr. [13]	27.1	–	14.1	–	6.7	1.3	11.3	–	2.8	–	–	2.2	–	–	–
VED [7]	54.7	12.0	20.7	13.5	0.0	0.0	8.8	0.0	0.0	7.4	0.2	0.0	0.0	4.0	8.7
VPN [9]	58.0	27.3	29.4	12.3	20.0	4.4	25.5	4.9	5.6	16.6	17.4	7.1	4.6	10.8	17.5
Sim2real [12]	60.5	27.1	19.2	18.3	6.9	3.8	7.1	0.3	4.5	3.2	4.7	1.8	4.2	12.1	12.4
OFT [14]	62.4	30.9	34.5	23.5	23.2	4.6	34.7	3.7	6.6	18.2	17.3	1.2	1.1	12.9	19.6
PON [13]	60.4	28.0	31.0	18.4	20.8	9.4	24.7	12.3	7.0	16.6	16.3	8.2	5.7	8.1	19.1
STA-S [16]	71.1	31.5	32.0	28.0	22.8	14.6	34.6	10.0	7.1	11.4	18.1	7.4	5.8	10.8	21.8
EPOSH [4]	61.1	33.5	37.8	25.4	31.8	6.7	37.8	2.7	10.5	14.2	20.4	5.9	7.6	13.4	22.1
Ours	65.1	**41.6**	**42.1**	**31.9**	**35.4**	13.8	**43.4**	9.7	**15.0**	**22.5**	**25.5**	**14.1**	**11.6**	**18.6**	**27.9**

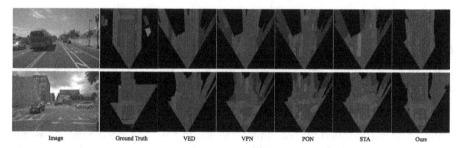

| Image | Ground Truth | VED | VPN | PON | STA | Ours |

Fig. 4. Qualitative results on the nuScenes validation set. We compare with the published works and follow their colour scheme.

4.2 Main Results

We evaluate our method on nuScenes and Argoverse, and compare against the recently published works which belong to different branches: **(i)** IPM-based methods: IPM [13], Sim2real [12]; **(ii)** Bottleneck-based methods: VED [7], VPN [9] and PON [13]; **(iii)** Depth-based methods: Depth Unprojection-based (Depth-Unpr.) [13], OFT [14], EPOSH [4] and STA-S [16]. All these works report the results on nuScenes as shown in Table 1, or provide the results on Argoverse in Table 2. Among all these methods, our method achieves the best performance for most categories and our method surpasses the previous approaches with a significant margin of mean IoU, 6.1% and 3.2% on nuScenes and Argoverse, respectively. Figure 4 further shows the visual comparisons against other methods on the nuScenes Dataset. The two fully-connected bottleneck-based works, VPN and VED, achieve a comparable IoU on the road drivable area, but they fail to recognize the smaller objects such as vehicles due to the image features are compressed into a vector. In contrast, our method leverages multi-scale spatial information for different depth intervals to keep the fine details. For example, as shown in Fig. 4, our approach accurately predicts the vehicles within all depth ranges. Compared with other relatively better methods like PON and STA-S, we exploit the geometric prior which helps to accurately locate and identify the road elements, like walkway and pedestrian crossing, which is supported by the qualitative results in Fig. 4.

Table 2. Results of IoU (%) on the Argoverse validation set.

Method	Drivable	Vehicle	Ped.	Large veh.	Bicycle	Bus	Trailer	Motorcy.	Mean
IPM [13]	43.7	7.5	1.5	–	0.4	7.4	–	0.8	–
Depth Unpr. [13]	33.0	12.7	3.3	–	1.1	20.6	–	1.6	–
VED [7]	62.9	14.0	1.0	3.9	0.0	12.3	1.3	0.0	11.9
VPN [9]	64.9	23.9	6.2	9.7	0.9	3.0	0.4	1.9	13.9
PON [13]	65.4	31.4	7.4	11.1	3.6	11.0	0.7	5.7	17.0
Ours	**67.1**	**35.9**	**9.8**	**15.7**	**4.9**	**31.7**	**11.3**	**6.2**	**20.2**

Table 3. Effects of different key components. GPA and RT denote the Geometry-guided Pre-Alignment and Ray-based Transformer, respectively. SW and DA refer to the Self-Weight Dice loss and Depth-aware Dice loss.

Group	Network		Loss		mIoU (%)		
	GPA	RT	SW	DA	Layout	Object	Total
(a)					31.2	4.9	12.4
(b)	✓				38.7	15.4	22.1
(c)		✓			40.6	16.8	23.6
(d)	✓	✓			43.2	19.1	26.0
(e)	✓	✓	✓		43.8	19.9	26.7
(f)	✓	✓		✓	44.1	20.5	27.2
(g)	✓	✓	✓	✓	**45.2**	**21.0**	**27.9**

4.3 Ablation Study

We conduct ablation studies to evaluate the key designs in our method. Unless otherwise specified, we evaluate on the nuScenes validation set. GPA denotes *Geometry-guided Pre-Alignment*, and RT is the *Ray-based Transformer* for short.

Effects of Different Components. To analyze the effects of the key designs, we try different combinations and summarize the ablation results in Table 3.

- **Baseline.** Group (a) is the baseline that is similar to [12]. We transform the image features onto the ground plane via a homography matrix. The difference between it and our GPA is that it adopts a fixed camera height and is not supervised by the projected semantic maps. The transformed features are further processed by a segmentation network that is the same as our best model for fair comparisons. From Row 1 in Table 3, we can see that the baseline achieves reasonable results in road layout areas, but fails to distinguish the objects that standing above the road.
- **Network.** In Group (b), the GPA provides a reliable prior for feature transformation and relieves the effects of occlusion by supervision of projection, improving the mIoU by +9.7% in total. The RT (c) transforms the image features to the BEV space by multi-scale column-based attention, which improves the mIoU by +11.2% in total. If we combine the GPA and RT as

disscussed in Sec. 3.1, their joint effect (d) further enhances the performance by +13.6% in total. It shows the geometric prior provides complementary information for the RT.

- **Loss.** Groups (e)(f)(g) show the improvements in our proposed loss function. The SW-Dice loss (e) automatically puts higher weights on these pixels that are hard to classify in birds-eye-view, improving the mIoU by +0.7% in total. The DA-Dice loss (f) balances the pixels of different depth ranges in the perspective view by reweighting the Dice loss under the guidance of the cubic depth when learning the geometric prior-based pre-alignment module, which improves the mIoU by +1.2% in total. The joint of both losses (g) brings a further mIoU gain of +1.9% in total.

Table 4. Effects of components of GPA, where "learnable h" denotes learning the jitter of camera height; "proj. sup." denotes supervising the image features with projected labels from BEV to image space; "pixel. fusion" denotes pixel-wise fusion between image features and probability maps of segmentation.

Group	Learnable h	proj. sup.	pixel. fusion	Layout	Object	Total
I				40.2	16.7	23.4
II	✓			41.1	17.3	24.1
III	✓	✓		42.6	18.3	25.2
IV	✓	✓	✓	**43.2**	**19.1**	**26.0**

Effects of Components of GPA: Three key designs are presented in Geometry-guided Pre-Alignment to better convert the perspective image features to BEV features, including the learnable camera height, the projection supervision, and pixel-wise fusion between probability maps and image features. Comparing Group I and II, where we enforce the network to learn the offset of the camera height, we observe a 0.7% mIoU gain. Group III leverages the projected labels from BEV space to image space to supervise the feature learning procedure, which further improves the performance by 1.1% mIoU. The further pixel-wise fusion between perspective features and segmentation probability maps in Group IV further lifts the performance, resulting in a total of 2.6% gain.

Effects of α in SW-Dice Loss: The SW-Dice loss introduces the hyperparameter α to control the strength of the modulating term with respect to the predicted probability. As is shown in Table 5a, $\alpha = 0$ means our loss is equivalent to the plain Dice loss. As α increases, the predicted probability gets dominant in the weighting function. Under all settings of α, the proposed SW-Dice loss stably outperforms the baseline ($\alpha = 0$). With the best setting, the SW-Dice loss yields a 0.7% improvement over the plain Dice loss.

Effects of Decoder Layers in RT: Table 5b shows the performance with various number of decoder layers within the ray-based transformer. Our model

Table 5. Effects of α in the proposed SW-Dice loss and the number of decoder layers within the ray-based transformer module.

(a) Effects of hyperparameter α

α	0	0.25	0.5	1.0	2.0
Layout	44.1	44.2	45.2	**45.3**	45.1
Object	20.5	20.9	**21.0**	20.9	20.6
Total	27.2	27.5	**27.9**	**27.9**	27.6

(b) Effects of decoder layers in RT

Layers	0	1	2	3	4
Layout	38.7	43.7	44.1	44.8	**45.2**
Object	15.4	19.2	20.8	**21.2**	21.0
Total	22.1	26.2	27.5	**27.9**	**27.9**

Fig. 5. An example of late-fusion of six surrounding birds-eye-view semantic maps, which predict consistent full 360° BEV semantic maps.

can yield 4.1% improvement even using one layer. The gain reflects that the pre-aligned BEV features can provide a good initialization for the decoder. With the decoder layers increasing, higher performance can be achieved. We observe that it becomes saturated when adopting more than three layers.

4.4 Multiple Views Fusion

Due to the limited field of view (FOV) of a single camera, it is essential to make full use of all surrounding cameras from multi-view to perceive the integrated scope of the scene. For this purpose, we introduce a late-fusion technique based on Bayesian filtering [13,19]. Suppose that $R_i \in \mathbb{R}^{2 \times 2}$ and $t_i \in \mathbb{R}^{2 \times 1}$ are the BEV rotation and translation matrix of i-th camera with respect to the ego car coordinates. Let O_i denote the predicted logits (before the sigmoid activation) in i-th view. O_i is warped to the car coordinate system, and we sum over all warped logits maps. The sum of logits are normalized by the sigmoid function σ to output the fused probability map P_{fuse}. In Fig. 5, we give an example of the fused 360° BEV semantic maps from six surround-view cameras. It validates that our approach can be applied seamlessly to predict consistent maps across views.

5 Conclusion

In this paper, we proposed a novel method GitNet for predicting semantic birds-eye-view maps from monocular images. The GitNet leverages a two-stage pipeline to transform the perspective view into the birds-eye-view, which first performs geometry-guided pre-alignment and then further enhances the BEV features based on ray-based transformers. Our approach can also be easily adapted to multi-view scenarios to build a full-scene BEV map.

Acknowledgments. This research was supported by the National Key Research and Development Program of China under Grant No. 2018AAA0100400, the National Natural Science Foundation of China (62176098, 61703049) and the Natural Science Foundation of Hubei Province of China under Grant 2019CFA022.

References

1. Caesar, H., et al.: nuScenes: a multimodal dataset for autonomous driving. In: Proceedings of the IEEE/CVF Conference on Computer Vision and Pattern Recognition (CVPR), pp. 11621–11631 (2020)
2. Chang, M.F., et al.: Argoverse: 3D tracking and forecasting with rich maps. In: Proceedings of the IEEE/CVF Conference on Computer Vision and Pattern Recognition (CVPR), pp. 8748–8757 (2019)
3. Chitta, K., Prakash, A., Geiger, A.: NEAT: neural attention fields for end-to-end autonomous driving. In: Proceedings of the IEEE/CVF International Conference on Computer Vision (ICCV), pp. 15793–15803 (2021)
4. Dwivedi, I., Malla, S., Chen, Y., Dariush, B.: Bird's eye view segmentation using lifted 2D semantic features. In: 32nd British Machine Vision Conference (BMVC), p. 383 (2021)
5. Henriques, J.F., Vedaldi, A.: MapNet: an allocentric spatial memory for mapping environments. In: Proceedings of the IEEE Conference on Computer Vision and Pattern Recognition (CVPR), pp. 8476–8484 (2018)
6. Hu, A., et al.: FIERY: future instance prediction in bird's-eye view from surround monocular cameras. In: Proceedings of the IEEE/CVF International Conference on Computer Vision (ICCV), pp. 15273–15282 (2021)
7. Lu, C., van de Molengraft, M.J.G., Dubbelman, G.: Monocular semantic occupancy grid mapping with convolutional variational encoder-decoder networks. IEEE Robot. Autom. Lett. 4(2), 445–452 (2019)
8. Mani, K., Daga, S., Garg, S., Narasimhan, S.S., Krishna, M., Jatavallabhula, K.M.: MonoLayout: Amodal scene layout from a single image. In: Proceedings of the IEEE/CVF Winter Conference on Applications of Computer Vision (WACV), pp. 1689–1697 (2020)
9. Pan, B., Sun, J., Leung, H.Y.T., Andonian, A., Zhou, B.: Cross-view semantic segmentation for sensing surroundings. IEEE Robot. Autom. Lett. 5(3), 4867–4873 (2020)
10. Philion, J., Fidler, S.: Lift, splat, shoot: encoding images from arbitrary camera rigs by implicitly Unprojecting to 3D. In: Vedaldi, A., Bischof, H., Brox, T., Frahm, J.-M. (eds.) ECCV 2020. LNCS, vol. 12359, pp. 194–210. Springer, Cham (2020). https://doi.org/10.1007/978-3-030-58568-6_12

11. Reading, C., Harakeh, A., Chae, J., Waslander, S.L.: Categorical depth distribution network for monocular 3D object detection. In: Proceedings of the IEEE/CVF Conference on Computer Vision and Pattern Recognition (CVPR), pp. 8555–8564 (2021)
12. Reiher, L., Lampe, B., Eckstein, L.: A Sim2Real deep learning approach for the transformation of images from multiple vehicle-mounted cameras to a semantically segmented image in bird's eye view. In: 2020 IEEE 23rd International Conference on Intelligent Transportation Systems (ITSC), pp. 1–7 (2020)
13. Roddick, T., Cipolla, R.: Predicting semantic map representations from images using pyramid occupancy networks. In: Proceedings of the IEEE/CVF Conference on Computer Vision and Pattern Recognition (CVPR), pp. 11138–11147 (2020)
14. Roddick, T., Kendall, A., Cipolla, R.: Orthographic feature transform for monocular 3D object detection. In: 30th British Machine Vision Conference (BMVC), p. 285 (2019)
15. Saha, A., Maldonado, O.M., Russell, C., Bowden, R.: Translating images into maps. arXiv preprint arXiv:2110.00966 (2021)
16. Saha, A., Mendez, O., Russell, C., Bowden, R.: Enabling spatio-temporal aggregation in birds-eye-view vehicle estimation. In: 2021 IEEE International Conference on Robotics and Automation (ICRA), pp. 5133–5139 (2021)
17. Schulter, S., Zhai, M., Jacobs, N., Chandraker, M.: Learning to look around objects for top-view representations of outdoor scenes. In: Proceedings of the European Conference on Computer Vision (ECCV), pp. 787–802 (2018)
18. Sengupta, S., Sturgess, P., Ladický, L., Torr, P.H.: Automatic dense visual semantic mapping from street-level imagery. In: 2012 IEEE/RSJ International Conference on Intelligent Robots and Systems, pp. 857–862 (2012)
19. Thrun, S.: Probabilistic robotics. Commun. ACM **45**(3), 52–57 (2002)
20. Vaswani, A., et al.: Attention is all you need. Adv. Neural Inf. Process. Syst. (NIPS) **30** (2017)
21. Wang, J., et al.: Deep high-resolution representation learning for visual recognition. IEEE Trans. Patt. Anal. Mach. Intell. (TPAMI) **43**(10), 3349–3364 (2020)
22. Zhu, X., Yin, Z., Shi, J., Li, H., Lin, D.: Generative adversarial frontal view to bird view synthesis. In: 2018 International conference on 3D Vision (3DV), pp. 454–463 (2018)

Learning Visibility for Robust Dense Human Body Estimation

Chun-Han Yao[1]([✉]), Jimei Yang[2], Duygu Ceylan[2], Yi Zhou[2], Yang Zhou[2],
and Ming-Hsuan Yang[1,3,4]

[1] UC Merced, Merced, USA
cyao6@ucmerced.edu
[2] Adobe, San Jose, USA
[3] Google, Mountain View, USA
[4] Yonsei University, Seoul, South Korea

Abstract. Estimating 3D human pose and shape from 2D images is a crucial yet challenging task. While prior methods with model-based representations can perform reasonably well on whole-body images, they often fail when parts of the body are occluded or outside the frame. Moreover, these results usually do not faithfully capture the human silhouettes due to their limited representation power of deformable models (e.g., representing only the naked body). An alternative approach is to estimate dense vertices of a predefined template body in the image space. Such representations are effective in localizing vertices within an image but cannot handle out-of-frame body parts. In this work, we learn dense human body estimation that is robust to partial observations. We explicitly model the visibility of human joints and vertices in the x, y, and z axes separately. The visibility in x and y axes help distinguishing out-of-frame cases, and the visibility in depth axis corresponds to occlusions (either self-occlusions or occlusions by other objects). We obtain pseudo ground-truths of visibility labels from dense UV correspondences and train a neural network to predict visibility along with 3D coordinates. We show that visibility can serve as 1) an additional signal to resolve depth ordering ambiguities of self-occluded vertices and 2) a regularization term when fitting a human body model to the predictions. Extensive experiments on multiple 3D human datasets demonstrate that visibility modeling significantly improves the accuracy of human body estimation, especially for partial-body cases. Our project page with code is at: https://github.com/chhankyao/visdb.

1 Introduction

Estimating 3D human pose and shape from monocular images is a crucial task for various applications such as performance retargeting, virtual avatars, and

Supplementary Information The online version contains supplementary material available at https://doi.org/10.1007/978-3-031-19769-7_24.

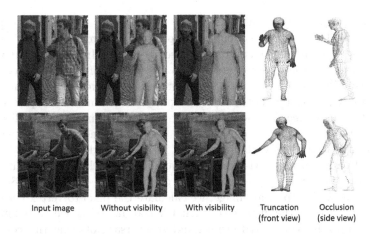

Input image   Without visibility   With visibility   Truncation   Occlusion
                                                        (front view)   (side view)

Fig. 1. Dense human body estimation with/without visibility modeling. We propose to learn dense visibility to improve human body estimation in terms of faithfulness to the input image and robustness to truncation (top) or occlusions (bottom). We show the estimated meshes without/with visibility modeling in columns 2–3 and the vertex visibility labels in columns 4–5 (purple:visible, orange:invisible). (Color figure online)

human action recognition. It is a fundamentally challenging problem due to the depth ambiguity and the complex nature of human appearances that vary with articulation, clothing, lighting, viewpoint, and occlusions. To represent the complicated 3D human bodies via compact parameters, model-based methods like SMPL [24] have been widely used in the community. However, SMPL parameters represent human bodies in a holistic manner, causing their limited flexibility to fit real-world images faithfully via direct regression. More importantly, the regression-based methods tend to fail when a human body is not fully visible in the image, *e.g.*, occluded or out of frame [16]. In this work, we aim to learn human body estimation that is faithful to the input images and robust to partial-body cases.

Instead of directly regressing SMPL parameters, we train a neural network to predict the coordinate heatmaps in three dimensions for each human joint and mesh vertex. The dense heatmap-based representation can preserve the spatial relationship in the image domain and model the uncertainty of predictions. It is shown to be effective in localizing visible joints/vertices and flexible to fit an input image faithfully [27–29,39]. Nonetheless, the x and y-axis heatmaps are defined in the image coordinates, which cannot represent the out-of-frame (*i.e.*, truncated by image boundaries) body parts. In addition, occlusions by objects or the human body itself could cause ambiguity for depth-axis predictions. Without knowing which joints/vertices are visible, the network tends to produce erroneous outputs on partial-body images. To address this, we propose *Visibility-aware Dense Body (VisDB)*, a heatmap-based dense representation augmented by visibility. Specifically, we train a network to predict binary truncation and occlusion labels along with the heatmaps for each human joint and vertex. With the visibility modeling, the proposed network can learn to make more accurate predictions based on the observable cues. In addition, the vertex-level occlusion predictions can serve as a depth ordering signal to constrain depth predictions. Finally, by using visibility as the confidence of 3D mesh prediction, we demonstrate that VisDB is a powerful intermediate representation which allows us to regress and/or optimize SMPL parameters more effectively. In Fig. 1, we show examples of truncation and occlusions as well as the dense human body estimations with and without visibility modeling.

Considering that most existing 3D human datasets lack dense visibility annotations, we obtain pseudo ground-truths from dense UV estimations [8]. Given the estimated UV map of an image, we calculate the pixel-to-vertex correspondence by minimizing the distance of their UV coordinates. Each vertex mapped to a human pixel is considered visible, and vice versa. Note that this covers the cases of truncation, self-occlusions, and occlusions by other objects. We further show that the dense vertex-to-pixel correspondence provides a good supervisory signal to localize vertices in the image space. Since dense UV estimations are based on part-wise segmentation masks which are robust to partial-body images, the dense correspondence loss can mitigate the inaccurate pseudo ground-truth meshes and better align the outputs with human silhouettes. To demonstrate the effectiveness of our method, we conduct extensive experiments on multiple human datasets used by prior arts. Both qualitative and quantitative results on the Human3.6M [12], 3DPW [25], 3DPW-OCC [25,45], and 3DOH [45] datasets show that learning visibility significantly improves the accuracy of dense human body estimation, especially on images with truncated or occluded human bodies.

The main contributions of our work are:

- We propose VisDB, a heatmap-based human body representation augmented with dense visibility. We train a neural network to predict the 3D coordinates of human joints and vertices as well as their truncation and occlusion labels. We obtain pseudo ground-truths of visibility labels from image-based dense UV estimates, which are also used as additional supervision signal to better align our predictions with the input image.
- We show how the dense visibility predictions can be used for robust human body estimation. First, we exploit occlusion labels to supervise vertex depth predictions. Second, we regress and optimize SMPL parameters to fit VisDB (partial-body) outputs by using visibility as confidence weighting.

2   Related Work

Model-Based Human Body Estimation. Most existing methods on human body estimation adopt a model-based representation. For instance, SMPL [24] is a widely-used statistical human body model that maps a set of pose $\theta \in \mathbb{R}^{72}$ and shape $\beta \in \mathbb{R}^{10}$ parameters to a 3D human mesh $V \in \mathbb{R}^{6890 \times 3}$. In SMPL, θ represents the axis-angle 3D rotations of 24 joints, and β is the top-10 PCA coefficients of a statistical human shape space. Early methods iteratively optimize the SMPL parameters to fit the estimated 2D keypoints [2] or silhouettes [20]. Several recent works [13,17,19,31,34,35] train a deep neural network to directly regress SMPL parameters from an input image. However, the SMPL representation is not always informative enough for a network to learn as it embeds the articulated body shapes in a low dimensional space. The regression-based methods often fail on truncation and occlusion cases since the networks tend to make holistic predictions based on certain body parts only [16]. Instead, we show that localizing 3D vertices is a more suitable task to learn for such scenarios. The network needs to learn the relationship between the parameters and the shape as well in order to estimate accurate SMPL parameters.

Dense Human Body Representations. To fit the complicated shapes more faithfully, dense human body representations have been proposed, including volumetric space [41], occupancy field [37,38], dense UV correspondence [1,44], and 3D mesh [4,18,21,22,29]. Among these methods, I2L-MeshNet [29] proposes an efficient heatmap representation to estimate human joints and vertices in the image space and root-relative depth axis. It can fit the input images accurately since heatmaps preserve the spatial relationship in image features extracted by a convolutional neural network (CNN). Nonetheless, even when certain body parts are not visible in the image, this model is designed to localize all the joints and vertices within the image frame. We show that it can negatively affect the model performance and emphasize the importance of additional visibility information.

Occlusion-Aware Methods. Several methods have been proposed to deal with the challenging scenarios where human bodies are partially truncated or occluded. Muller *et al.* [30] and Hassan *et al.* [9] introduce explicit modeling of human body self-contact and human-scene interactions, respectively. These methods require ground-truth annotations which are hard to obtain. Other methods leverage human-centric heatmaps, part segmentation masks, or dense UV estimations [8], to increase the model robustness on truncated images [36], crowded scenes (occluded by other people) [40] or general occlusions [7,16,43,45]. Although effective in particular scenarios, most of them directly regress SMPL parameters which still suffer from the limited representation strength. To the best our our knowledge, the proposed VisDB representation is the first to explicitly model dense human body visibility (including truncation and all occlusion scenarios), which is trained with pseudo ground-truth visibility labels from dense UV estimates.

3   Approach

We illustrate our overall framework in Fig. 2. In Sect. 3.1, we describe a heatmap-based representation which we build our method upon. Then, we introduce the proposed Visibility-aware Dense Body (VisDB) in Sect. 3.2. Each human joint and mesh vertex is represented by 1) three 1D heatmaps (x, y, z dimensions) which define its 3D coordinate and 2) three binary labels indicating its visibility in three dimensions. We train a network model to predict the dense heatmaps

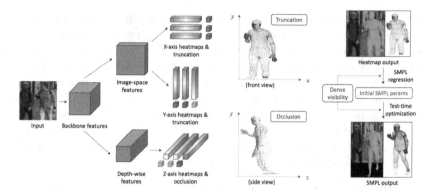

Fig. 2. VisDB framework overview (best viewed in color). Given an input image, the network extracts features in the image and depth coordinates, from where we predict the x, y, z heatmaps for each human joint and vertex. In addition, we predict a binary visibility label (purple:visible, orange:invisible) of each axis, *i.e.*, x-truncation, y-truncation, To obtain a more regularized and complete human body, we train a regression network to estimate SMPL parameters based on the dense 3D coordinates and visibility. At test time, we can further optimize the regressed SMPL parameters to fit the partial-body predictions from heatmaps. (Color figure online)

and visibility, which represents a partial body faithful to the input image. The visibility estimations can be interpreted as depth ordering signals or prediction confidence. In Sect. 3.3, we design a visibility-guided depth ordering loss to self-supervise depth estimation. In Sect. 3.4, we show that VisDB outputs can be used to fit SMPL models accurately and efficiently. We train a regression network to estimate SMPL parameters based on the joint and vertex coordinates as well as their visibility labels. During inference, we initialize the SMPL parameters by the regressor and further optimize them to align with the VisDB predictions. Finally, in Sect. 3.5, we exploit dense UV correspondence to obtain robust pseudo labels of visibility and weakly supervise vertex localization in the image space.

3.1   Preliminaries: Heatmap-Based Representation

Given an input image, a prior heatmap-based method [29] estimates three 1D heatmaps $H = \{H^x, H^y, H^z\}$ for each human joint and mesh vertex. The x and y-axis heatmaps H^x, H^y are defined in the image space, and the z-axis heatmaps H^z are defined in the depth space relative to root joint. We denote the joint heatmaps as $H_J \in \mathbb{R}^{N_J \times D \times 3}$ and vertex heatmaps as $H_V \in \mathbb{R}^{N_V \times D \times 3}$, where N_J is the number of joints, N_V is the number of vertices, and D is the heatmap resolution. The heatmaps are predicted based on image features $F \in \mathbb{R}^{c \times h \times w}$ extracted by a backbone network as follows:

$$
\begin{aligned}
H^x &= f^{\,1\text{D},x}(\text{avg}^y(f^{\,\text{up}}(F))), \\
H^y &= f^{\,1\text{D},y}(\text{avg}^x(f^{\,\text{up}}(F))), \\
H^z &= f^{\,1\text{D},z}(\psi(\text{avg}^{x,y}(F))),
\end{aligned}
\tag{1}
$$

where $f^{1\mathrm{D},i}$ is 1-by-1 1D convolution for the i-th axis heatmaps, avg^i is i-axis marginalization by averaging, f^{up} denotes up-sampling by deconvolution, and ψ is a 1D convolution layer followed by reshaping operation. Finally, the continuous 3D coordinates of joints $J \in \mathbb{R}^{N_J \times 3}$ and vertices $V \in \mathbb{R}^{N_V \times 3}$ can be obtained by applying soft-argmax on the discrete heatmaps H_J and H_V, respectively. More details can be found in [29] and supplementary material.

3.2   Visibility-Aware Dense Body

Heatmap-based representations are shown effective in estimating human bodies in the image space. However, they often fail when the human bodies are occluded or truncated since the predictions are based on spatial image features and limited by the image boundaries. Without knowing which joints/vertices are invisible, fitting a SMPL model on the entire body tends to generate erroneous outputs. To deal with more practical scenarios where only partial bodies are visible, we make the following adaptations to a heatmap-based representation: 1) To augment the x and y-axis heatmaps, we predict binary truncation labels S^x, S^y, indicating whether a joint or vertex is within the image frame, 2) For the z-axis heatmaps, we predict a binary occlusion label S^z which specifies the depth-wise visibility. The visibility labels are predicted in a similar fashion as the heatmaps in Eq. (1):

$$S^x = \sigma(\mathrm{avg}^x(g^{1\mathrm{D},x}(\mathrm{avg}^y(f^{\mathrm{up}}(F))))),$$
$$S^y = \sigma(\mathrm{avg}^y(g^{1\mathrm{D},y}(\mathrm{avg}^x(f^{\mathrm{up}}(F))))), \qquad (2)$$
$$S^z = \sigma(\mathrm{avg}^z(g^{1\mathrm{D},z}(\psi(\mathrm{avg}^{x,y}(F))))),$$

where $g^{1\mathrm{D}}$ is a 1-by-1 1D convolutional layer similar to $f^{1\mathrm{D}}$ and σ is a sigmoid operator. We then concatenate the $\{S^x, S^y, S^z\}$ predictions to obtain joint visibility $S_J \in \mathbb{R}^{N_J \times 3}$ and vertex visibility $S_V \in \mathbb{R}^{N_V \times 3}$. By applying the soft-argmax operators to the predicted 1D heatmaps, the final output of our network becomes $\{J, V, S_J, S_V\}$, referred to as Visibility-aware Dense Body (VisDB). With the visibility information, the network model can learn to focus on the visible body parts and push the invisible parts towards the image boundaries. In our experiments (Table 3), we demonstrate that visibility modeling significantly reduces the errors of visible vertices. Moreover, the visibility labels can be seen as the confidence of coordinate predictions, which are essential to mesh regularization and completion via SMPL model fitting as described in Sect. 3.4.

We denote the ground-truth VisDB as $\{J^*, V^*, S_J{}^*, S_V{}^*\}$ and train the network by using the following losses. The joint coordinate loss $\mathcal{L}_{joint}$ is defined as:

$$\mathcal{L}_{joint} = \|J - J^*\|_1. \qquad (3)$$

The vertex coordinate loss $\mathcal{L}_{vert}$ is defined as:

$$\mathcal{L}_{vert} = \|V - V^*\|_1. \qquad (4)$$

We also regress the joints from vertices using a pre-defined regressor $W \in \mathbb{R}^{N_V \times N_J}$ and calculate a regressed-joint loss $\mathcal{L}_{r-joint}$:

$$\mathcal{L}_{r-joint} = \|WV - J^*\|_1. \qquad (5)$$

Similar to [29], we apply losses on the mesh surface normal and edge length as shape regularization. The normal loss $\mathcal{L}_{norm}$ and edge loss $\mathcal{L}_{edge}$ are:

$$\mathcal{L}_{norm} = \sum_f \sum_{\{v_i, v_j\} \subset f} \left| \left\langle \frac{v_i - v_j}{\|v_i - v_j\|_2}, n_f^* \right\rangle \right|, \tag{6}$$

$$\mathcal{L}_{edge} = \sum_f \sum_{\{v_i, v_j\} \subset f} \left| \|v_i - v_j\|_2 - \|v_i^* - v_j^*\|_2 \right|, \tag{7}$$

where f is a mesh surface, n_f is the unit normal vector of f, and v_i, v_j are the coordinates of vertex i and j, respectively. Finally, we define the joint and vertex visibility loss $\mathcal{L}_{vis}$ with binary cross entropy (BCE):

$$\mathcal{L}_{vis} = \text{BCE}(S_J, S_J^*) + \text{BCE}(S_V, S_V^*). \tag{8}$$

The VisDB prediction is illustrated in Fig. 2 (left).

3.3   Resolving Depth Ambiguity via Visibility

Vertex-level visibility can not only be seen as model confidence for SMPL fitting but also provide depth ordering information. Intuitively, visible vertices should have lower depth value compared to the invisible vertices projected to the same pixel. We observe that VisDB network generally predicts accurate 2D coordinates and visibility, but sometimes fails at depth predictions when the human body occludes itself and the pose is less common in the training datasets. To resolve the depth ambiguity in self-occlusion cases, we propose a depth ordering loss $\mathcal{L}_{depth}$ based on vertex visibility as follows:

$$\mathcal{L}_{depth} = \sum_x \sum_y \text{ReLU}\left(\max_{v \in Q(x,y)} v^z - \min_{\overline{v} \in \overline{Q}(x,y)} \overline{v}^z \right), \tag{9}$$

where $Q(x, y)$ is the set of vertices projected to a discretized image coordinate (x, y) which belong to the front (occluding) part, and $\overline{Q}$ contains the vertices of the back (occluded) part(s). The definition can be written as:

$$Q(x,y) = \left\{ v \middle| v \mapsto (x,y) \wedge P(v) = p^*(x,y) \right\}$$
$$\overline{Q}(x,y) = \left\{ v \middle| v \mapsto (x,y) \wedge P(v) \neq p^*(x,y) \right\}, \tag{10}$$

where $\mapsto$ denotes the discrete projection and $P(v)$ is the part label of vertex v defined in DensePose [8]. We define the front part $p^*(x, y)$ by finding the vertex with highest z-axis visibility score s^z as:

$$p^*(x,y) = P\left(\arg\max_{v \mapsto (x,y)} s_v^z \right). \tag{11}$$

$\mathcal{L}_{depth}$ is designed to push the self-occluded part(s) $\overline{Q}$ to the back and non-occluded part Q to the front, where the occlusion information is given by the z-axis visibility. Note that we compare the maximum depth (back side) of Q and the minimum depth (front side) of $\overline{Q}$, and thus $\mathcal{L}_{depth}$ will be nonzero if the depth ordering disagrees with occlusion prediction and zero if the parts do not overlap anymore. Since this loss depends on accurate visibility estimations, we only apply it during the fine-tuning stage.

3.4   SMPL Fitting from Visible Dense Body

From the VisDB predictions, we can obtain the 3D coordinates and visibility of human joints and vertices. While the partial-body outputs are faithful to the input image from the front view, they sometimes look abnormal from a side view or contain rough surfaces. To regularize the body shape and complete the truncated parts, we perform model fitting on the visible dense body predictions. Given the coordinates and visibility of joints and vertices, we train a regression network to estimate SMPL pose $\theta \in \mathbb{R}^{72}$ and shape $\beta \in \mathbb{R}^{10}$ parameters. The regressed parameters are then forwarded to the SMPL model to generate the mesh coordinates denoted as $\text{SMPL}(\theta, \beta) \in \mathbb{R}^{N_V \times 3}$. Unlike prior art [29] which regresses a SMPL model from all the joints regardless of their visibility, our VisDB representation allows us to fit the visible partial body only. The training objectives of the SMPL regressor include SMPL parameter error, vertex error, joint error, and the negative log-likelihood of a pose prior distribution. The SMPL parameter loss $\mathcal{L}_{smpl}$ is defined as:

$$\mathcal{L}_{smpl} = \|\theta - \theta^*\|_1 + \|\beta - \beta^*\|_1, \tag{12}$$

where θ^* and β^* are the ground-truth pose and shape parameters. The SMPL vertex loss $\mathcal{L}_{smpl-vert}$ and joint loss $\mathcal{L}_{smpl-joint}$ are defined similarly as in Eq. (4) and (5) but weighted by visibility S_V, S_J as:

$$\mathcal{L}_{smpl-vert} = S_V \odot \|\text{SMPL}(\theta, \beta) - V_c^*\|_1, \tag{13}$$

$$\mathcal{L}_{smpl-joint} = S_J \odot \|W\text{SMPL}(\theta, \beta) - J_c^*\|_1, \tag{14}$$

where $\odot$ denotes element-wise multiplication, and (V_c^*, J_c^*) are the ground-truth root-relative coordinates of vertices and joints in the camera space. Ideally, the VisDB network makes more confident predictions on the clearly visible joints and vertices. Hence, we see the visibility labels as prediction confidence and use them to weight the coordinate losses. In addition, we apply a pose prior loss $\mathcal{L}_{prior}$ using a fitted Gaussian Mixture Model (GMM) provided by [33]:

$$\mathcal{L}_{prior} = -\log\left(\sum_i G_i(\theta)\right), \tag{15}$$

where G_i is the i-th component of GMM.

We observe that the regressed SMPL meshes roughly capture the human pose and shape but do not always align with the VisDB predictions in details. Therefore, we use the regressed parameters as initialization and propose efficient

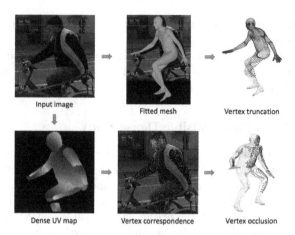

Fig. 3. Dense UV correspondence and visibility labels. Given an input image, we obtain a fitted SMPL mesh and dense UV estimation from off-the-shelf algorithms. To acquire the dense visibility labels for training, we identify the truncated vertices from the fitted mesh. From the dense UV map, we calculate the pixel-to-vertex correspondence to obtain pseudo ground-truths of vertex occlusions as well as image-space coordinates for weak supervision.

test-time optimization to further optimize the SMPL parameters against VisDB predictions. For this optimization, we apply similar losses as in Eqs. (13)–(15), except that the ground-truths $\{V_c{}^*, J_c{}^*\}$ are replaced by the VisDB predictions converted into root-relative coordinates in the camera space. Please refer to the supplemental material for details on estimating the root joint coordinate in the camera space. Since we initialize the SMPL parameters by the regression network and the use strong supervisory signal, *i.e.*, 3D joint and vertex coordinates, the test-time optimization only takes around 100 iterations to converge using an Adam optimizer [14]. We illustrate the process of SMPL regression and optimization in Fig. 2 (right).

3.5   Exploiting Dense UV Correspondence

Most existing 3D human datasets do not provide joint visibility labels, and none annotates vertex visibility. To train our VisDB network, we obtain pseudo ground-truths from the fitted SMPL meshes and dense UV estimations. For x and y-axis truncation, we can simply identify the truncated joints/vertices by projecting the fitted mesh onto the image plane. Occlusion, however, cannot be easily inferred from the input image or fitted mesh alone. One can estimate self-occlusion by rendering a fitted mesh, but this does not capture occlusions by other objects. More importantly, the fitting algorithm used to get the pseudo ground-truth meshes is not robust to partial-body cases. To address this, we propose to exploit dense UV correspondence between the input image and a SMPL mesh. Dense UV estimation provides the part-based segmentation mask of a human body as well as continuous UV coordinates of each human pixel, which are robust to truncation and occlusions. We calculate the UV coordinate of each pixel by applying an off-the-shelf dense UV estimation method [8]. For each human pixel p, we then find the corresponding mesh vertex v whose UV coordinate is closest to the pixel. The pixel-to-vertex M_P and vertex-to-pixel M_V mappings can be expressed as:

$$M_P = \{p \rightarrow v | v = \mathrm{argmin}_{v'} \|\mathrm{UV}(v') - \mathrm{UV}(p)\|_2 \forall p\}$$
$$M_V = \{v \rightarrow \{p'\} | M_P(p') = v \forall v\}.$$

(16)

A vertex mapped to at least one pixel is labeled as visible or occluded otherwise.

Similar to [7,43,44], we also utilize the dense vertex-pixel correspondence as weak supervision for better alignment with the human silhouettes. For each vertex v, we calculate the center of its corresponding pixels $M_V(v)$ and define a UV correspondence loss $\mathcal{L}_{uv}$ as:

$$\mathcal{L}_{uv} = \sum_v s_v{}^z \left\| v^{x,y} - \sum_{p \in M_V(v)} \frac{p}{|M_V(v)|} \right\|_1,$$

(17)

where $v^{x,y}$ is the 2D projection of vertex v and $s_v{}^z$ is the binary occlusion label with $s_v{}^z = 1$ indicating that the vertex v is visible. The UV correspondence loss can not only mitigate the inaccurate pseudo ground-truth meshes, but improve the faithfulness to human silhouettes since it is based on segmentation mask predictions. We empirically discover that this direct vertex-level supervision is more efficient and effective for VisDB training compared to rendering-based losses [6,43]. The proposed vertex-pixel correspondence and visibility labeling are illustrated in Fig. 3.

3.6   Model Training and Inference

We first train the VisDB network on 3D data with mesh annotations, then fine-tune it on all training data by adding the depth ordering and UV correspondence losses. The regressor network is trained to estimate the SMPL parameters based on the estimated coordinates and visibility of joints and vertices. During inference, we apply optional optimization on the regressed SMPL parameters to best align with the VisDB predicted mesh. For the VisDB network backbone, we use a ResNet50 [11] model pre-trained on the ImageNet dataset [5]. The weights are updated by the Adam optimizer [14] with a mini-batch size of 64. We represent a human body by $N_J = 30$ joints and $N_V = 6890$ vertices, and the heatmap resolution $D = 64$. In addition, we use the ground-truth bounding boxes to crop the human region from an input image and resize it to 256×256. The bounding boxes of testing data are estimated by a pre-trained Mask R-CNN [10] model if not available in the dataset. We apply common data augmentations such as random scaling ($\pm 25\%$), rotation ($\pm 45°$), horizontal flip, and color jittering ($\pm 20\%$) during training. Considering that truncation and occlusion examples are rare in most 3D human datasets, we include random occlusion masks and bounding box

Table 1. Quantitative evaluations on Human3.6M [12] **and 3DPW** [25]. To align the settings, we train our baseline, I2L-MeshNet [29], on the same datasets, and denote it by I2L-MeshNet†. Both our mesh and SMPL parameter outputs perform favorably against the prior state-of-the-arts.

Method	Output	Human3.6M		3DPW		
		MPJPE ↓	PA-MPJPE ↓	MPJPE ↓	PA-MPJPE ↓	MPVE ↓
GraphCMR [18]	Mesh	-	50.1	-	70.2	-
Pose2Mesh [4]	Mesh	64.9	47.0	89.2	58.9	109.3
I2L-MeshNet [29]	Mesh	55.7	41.1	93.2	57.7	109.2
I2L-MeshNet† [29]	Mesh	-	-	84.5	51.1	98.2
METRO [21]	Mesh	54.0	36.7	77.1	47.9	88.2
Mesh Graphormer [22]	Mesh	51.2	**34.5**	74.7	45.6	87.7
VisDB (mesh)	Mesh	**51.0**	**34.5**	**73.5**	**44.9**	**85.5**
NBF [31]	Param	-	59.9	-	-	-
HMR [13]	Param	88.0	56.8	-	81.3	-
DenseRaC [43]	Param	76.8	48.0	-	-	-
I2L-MeshNet [29]	Param	-	-	100.0	60.0	121.5
OOH [45]	Param	-	41.7	-	-	-
SPIN [17]	Param	-	41.1	-	59.2	116.4
I2L-MeshNet† [29]	Param	-	-	88.0	55.5	102.3
DSR [6]	Param	60.9	40.3	85.7	51.7	99.5
VIBE [15]	Param	65.6	41.4	82.0	51.9	99.1
TCMR [3]	Param	62.3	41.1	-	-	-
DecoMR [44]	Param	60.6	39.3	-	-	-
PARE [16]	Param	-	-	79.1	46.4	94.2
VisDB (param)	Param	**50.0**	**33.8**	**72.1**	**44.1**	**83.5**

shifting (±25%) as additional augmentations to increase the partial-body/whole-body ratio. Our models are implemented with PyTorch [32] and trained with NVIDIA Tesla V100 GPUs. More implementation details are presented in the supplemental material.

4   Experiments

4.1   Datasets and Metrics

Following most prior arts, we adopt mixed 2D-3D training on the MSCOCO [23], Human3.6M [12], MuCo-3DHP [26], and 3DPW [25] datasets. The pseudo ground-truth meshes of Human3.6M and MSCOCO are obtained by applying SMPLify-X [33] to fit the joint annotations. We evaluate our models on the Human3.6M, 3DPW, 3DPW-OCC [25,45], and 3DOH [45] testing sets. Note that 3DOH is composed of images with object occlusions and 3DPW-OCC contains a subset of 3DPW sequences where the human bodies are partially occluded. For quantitative evaluation, we calculate the common joint and vertex error metrics in the camera space and report them in millimeters (mm), including MPJPE (mean per-joint position error) [12], PA-MPJPE (Procrustes-aligned mean per-joint position error) [46], and MPVE (mean per-vertex error) [35].

Table 2. Quantitative evaluations on **3DOH** [45] and **3DPW-OCC** [25,45]. We compare VisDB with prior occlusion-aware methods to demonstrate its robustness on partial-body cases. For VisDB and I2L-MeshNet† [29], We report both the mesh and SMPL parameter (mesh/param) results.

Method	3DOH		3DPW-OCC		
	MPJPE ↓	PA-MPJPE ↓	MPJPE ↓	PA-MPJPE ↓	MPVE ↓
OOH [45]	-	58.5	-	72.2	-
I2L-MeshNet† [29]	67.0/69.3	46.3/47.9	96.5/98.0	61.0/62.6	120.2/127.0
PARE [16]	63.3	44.3	91.4	57.4	115.3
VisDB	**62.1/60.9**	**43.2/42.7**	**90.3/87.3**	**57.1/56.0**	**114.0/110.5**

4.2   Quantitative Comparisons

Human3.6M and 3DPW. In Table 1, we compare the performance of our method and prior arts on the Human3.6M [12] and 3DPW [25] datasets. For VisDB and I2L-MeshNet [29], we report the results of both heatmap-based mesh outputs (mesh) and SMPL parameters (param). Our SMPL parameters are obtained from regression and test-time optimization. Note that each method uses different network backbone, human body representation, training datasets, and inference strategy. For instance, METRO [21] and Mesh Graphormer [22] adopt a transformer-based [42] network while the others use CNN backbones. VIBE [15] and TCMR [3] are video-based approaches whereas the others only take images as input. Despite these differences, VisDB performs favorably against prior methods in term of most evaluation metrics. Particularly, our method achieves larger performance gains on the 3DPW dataset since it contains more truncation and occlusion cases. The VisDB performance is most directly comparable with I2L-MeshNet [29] as we adopt similar training settings. For fair comparisons, we retrain its model on the same datasets and denote it as I2L-MeshNet†. The results demonstrate that our visibility learning improves both the mesh and SMPL outputs significantly. In prior literature, SMPL parameters generally lead to higher errors compared to dense mesh outputs, which we conjecture is caused by the difficulty to directly regress low-dimensional parameters. On the contrary, VisDB is a powerful intermediate representation that provides dense 3D information of visible partial body, allowing us to regress and optimize SMPL parameters more accurately. In our experiments, we observe that VisDB (mesh) captures the human silhouettes better but VisDB (param) produces lower errors since the ground-truth meshes are also regularized by SMPL representation.

3DPW-OCC and 3DOH. To emphasize the robustness on partial-body images, we further evaluate on two occlusion datasets: 3DPW-OCC [25,45] and 3DOH [45]. As shown in Table 2, VisDB produces lower errors on both datasets compared to prior occlusion-aware methods. While I2L-MeshNet† performs considerably worse on these images, the errors by our model remain relatively low.

C.-H. Yao et al.

Table 3. Ablation studies of VisDB. We compare the joint/vertex errors of VisDB mesh outputs on 3DPW [25] with/without individual components. The results show that truncation modeling ($\mathcal{L}_{vis}^{x,y}$), occlusion modeling ($\mathcal{L}_{vis}^{z}$), depth ordering loss $\mathcal{L}_{depth}$, and UV correspondence loss $\mathcal{L}_{uv}$ each reduces the errors by a clear margin.

$\mathcal{L}_{vis}^{x,y}$	$\mathcal{L}_{vis}^{z}$	$\mathcal{L}_{depth}$	$\mathcal{L}_{uv}$	MPJPE	PA-MPJPE	MPVE
				84.5	51.1	98.2
	✓	✓	✓	79.4	47.8	91.1
✓		✓	✓	75.8	45.5	88.0
✓	✓		✓	77.3	46.3	88.9
✓	✓	✓		74.9	45.6	87.1
✓	✓	✓	✓	**73.5**	**44.9**	**85.5**

Table 4. Ablation studies of SMPL models. We report the performance of SMPL outputs on the 3DPW dataset [25], which shows the effectiveness of our optimization and the importance of visibility in both regression and optimization work flows.

Regression	Optimization	MPJPE	PA-MPJPE	MPVE
-	-	73.5	44.9	85.5
w/o vis	-	79.0	48.8	96.2
w/o vis	w/o vis	77.6	47.0	93.9
w/ vis	-	74.9	45.3	87.3
w/ vis	w/ vis	**72.1**	**44.1**	**83.5**

4.3 Ablation Studies

VisDB Network Training. To evaluate the contribution of individual components in our method, we perform ablation studies on the 3DPW dataset [25]. Table 3 shows the performance of VisDB mesh outputs with/without truncation modeling $\mathcal{L}_{vis}^{x,y}$, occlusion modeling $\mathcal{L}_{vis}^{z}$, depth ordering loss $\mathcal{L}_{depth}$, and dense UV correspondence loss $\mathcal{L}_{uv}$. Without $\mathcal{L}_{vis}^{x,y}$, $\mathcal{L}_{vis}^{z}$, $\mathcal{L}_{depth}$, and $\mathcal{L}_{uv}$, the vertex error increases by 6.3mm, 3.1mm, 3.9mm, and 1.9mm, respectively. These results show that both visibility modeling and depth ordering loss play a crucial role in VisDB training.

SMPL Parameter Fitting. In Table 4, we quantitatively compare the SMPL models obtained from different methods. Given an estimated VisDB mesh, we can regress the SMPL parameters and/or optimize them during inference, and each process can be done with/without dense visibility weighting (Eqs. (13) and (14)). By using visibility, the mean vertex error of regressed SMPL models drops by 8.7 mm. With the proposed test-time optimization, we can further reduce the error by 3.8 mm.

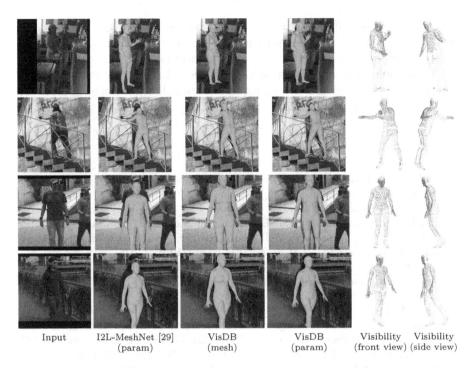

| Input | I2L-MeshNet [29]
(param) | VisDB
(mesh) | VisDB
(param) | Visibility
(front view) | Visibility
(side view) |

Fig. 4. Qualitative results on the 3DPW dataset [25]. For each example, we show the results of I2L-MeshNet [29] SMPL model, our VisDB mesh, our optimized SMPL model, as well as visibility predictions in the front and side views (purple:visible, orange:invisible). When the human body is occluded (top two rows) or truncated (bottom two rows), both our VisDB output and optimized SMPL mesh capture the human silhouettes faithfully (*e.g.*, the left hand in row 1 and the head region in rows 2,3,4). (Color figure online)

4.4 Qualitative Results

Figure 4 shows sample results by VisDB and I2L-MeshNet [29] on the 3DPW dataset [25]. I2L-MeshNet [29] regresses SMPL parameters from the entire heatmap-based mesh output, which leads to erroneous output meshes on truncated or occluded examples. VisDB predicts accurate vertex visibility labels, improving both the image-space dense body estimation and SMPL parameter optimization. The results show that VisDB (mesh) outputs can fit the human silhouettes faithfully, and VisDB (params) further regularizes and smooths the mesh surfaces. More qualitative results are shown in the supplemental material.

5 Conclusions

In this work, we address the problem of dense human body estimation from monocular images. Particularly, we identify the limitations of existing model-based and heatmap-based representations on truncated or occluded bodies. As such, we propose a visibility-aware dense body representation, VisDB. We obtain visibility pseudo ground-truths from dense UV correspondences and train a network to predict 3D coordinates as well as truncation and occlusion labels for each human joint and vertex. Extensive experimental results show that visibility modeling can facilitate human body estimation and allow accurate SMPL fitting from partial-body predictions.

References

1. Alldieck, T., Pons-Moll, G., Theobalt, C., Magnor, M.: Tex2Shape: detailed full human body geometry from a single image. In: ICCV (2019)
2. Bogo, F., Kanazawa, A., Lassner, C., Gehler, P., Romero, J., Black, M.J.: Keep it SMPL: automatic estimation of 3D human pose and shape from a single image. In: Leibe, B., Matas, J., Sebe, N., Welling, M. (eds.) ECCV 2016. LNCS, vol. 9909, pp. 561–578. Springer, Cham (2016). https://doi.org/10.1007/978-3-319-46454-1_34
3. Choi, H., Moon, G., Chang, J.Y., Lee, K.M.: Beyond static features for temporally consistent 3D human pose and shape from a video. In: CVPR, pp. 1964–1973 (2021)
4. Choi, H., Moon, G., Lee, K.M.: Pose2Mesh: graph convolutional network for 3D human pose and mesh recovery from a 2D human pose. In: Vedaldi, A., Bischof, H., Brox, T., Frahm, J.-M. (eds.) ECCV 2020. LNCS, vol. 12352, pp. 769–787. Springer, Cham (2020). https://doi.org/10.1007/978-3-030-58571-6_45
5. Deng, J., Dong, W., Socher, R., Li, L.J., Li, K., Fei-Fei, L.: ImageNet: a large-scale hierarchical image database. In: CVPR, pp. 248–255 (2009)
6. Dwivedi, S.K., Athanasiou, N., Kocabas, M., Black, M.J.: Learning to regress bodies from images using differentiable semantic rendering. In: ICCV, pp. 11250–11259 (2021)
7. Guler, R.A., Kokkinos, I.: Holopose: holistic 3D human reconstruction in-the-wild. In: CVPR, pp. 10884–10894 (2019)
8. Güler, R.A., Neverova, N., Kokkinos, I.: Densepose: dense human pose estimation in the wild. In: CVPR, pp. 7297–7306 (2018)
9. Hassan, M., Choutas, V., Tzionas, D., Black, M.J.: Resolving 3D human pose ambiguities with 3D scene constraints. In: ICCV, pp. 2282–2292 (2019)
10. He, K., Gkioxari, G., Dollár, P., Girshick, R.: Mask R-CNN. In: ICCV, pp. 2961–2969 (2017)
11. He, K., Zhang, X., Ren, S., Sun, J.: Deep residual learning for image recognition. In: CVPR, pp. 770–778 (2016)
12. Ionescu, C., Papava, D., Olaru, V., Sminchisescu, C.: Human3. 6m: large scale datasets and predictive methods for 3D human sensing in natural environments. PAMI **36**(7), 1325–1339 (2013)
13. Kanazawa, A., Black, M.J., Jacobs, D.W., Malik, J.: End-to-end recovery of human shape and pose. In: CVPR, pp. 7122–7131 (2018)
14. Kingma, D.P., Ba, J.: Adam: a method for stochastic optimization. arXiv preprint arXiv:1412.6980 (2014)
15. Kocabas, M., Athanasiou, N., Black, M.J.: Vibe: video inference for human body pose and shape estimation. In: CVPR, pp. 5253–5263 (2020)

16. Kocabas, M., Huang, C.H.P., Hilliges, O., Black, M.J.: PARE: part attention regressor for 3D human body estimation. In: ICCV, pp. 11127–11137 (2021)
17. Kolotouros, N., Pavlakos, G., Black, M.J., Daniilidis, K.: Learning to reconstruct 3D human pose and shape via model-fitting in the loop. In: ICCV, pp. 2252–2261 (2019)
18. Kolotouros, N., Pavlakos, G., Daniilidis, K.: Convolutional mesh regression for single-image human shape reconstruction. In: CVPR, pp. 4501–4510 (2019)
19. Kolotouros, N., Pavlakos, G., Jayaraman, D., Daniilidis, K.: Probabilistic modeling for human mesh recovery. In: ICCV (2021)
20. Lassner, C., Romero, J., Kiefel, M., Bogo, F., Black, M.J., Gehler, P.V.: Unite the people: closing the loop between 3D and 2D human representations. In: CVPR, pp. 6050–6059 (2017)
21. Lin, K., Wang, L., Liu, Z.: End-to-end human pose and mesh reconstruction with transformers. In: CVPR, pp. 1954–1963 (2021)
22. Lin, K., Wang, L., Liu, Z.: Mesh graphormer. In: ICCV (2021)
23. Lin, T.-Y., et al.: Microsoft COCO: common objects in context. In: Fleet, D., Pajdla, T., Schiele, B., Tuytelaars, T. (eds.) ECCV 2014. LNCS, vol. 8693, pp. 740–755. Springer, Cham (2014). https://doi.org/10.1007/978-3-319-10602-1_48
24. Loper, M., Mahmood, N., Romero, J., Pons-Moll, G., Black, M.J.: SMPL: a skinned multi-person linear model. TOG **34**(6), 1–16 (2015)
25. von Marcard, T., Henschel, R., Black, M.J., Rosenhahn, B., Pons-Moll, G.: Recovering accurate 3D human pose in the wild using IMUs and a moving camera. In: Ferrari, V., Hebert, M., Sminchisescu, C., Weiss, Y. (eds.) ECCV 2018. LNCS, vol. 11214, pp. 614–631. Springer, Cham (2018). https://doi.org/10.1007/978-3-030-01249-6_37
26. Mehta, D., et al.: Single-shot multi-person 3D pose estimation from monocular RGB. In: 3DV, pp. 120–130 (2018)
27. Moon, G., Chang, J.Y., Lee, K.M.: V2V-posenet: voxel-to-voxel prediction network for accurate 3D hand and human pose estimation from a single depth map. In: CVPR, pp. 5079–5088 (2018)
28. Moon, G., Chang, J.Y., Lee, K.M.: Camera distance-aware top-down approach for 3D multi-person pose estimation from a single RGB image. In: ICCV, pp. 10133–10142 (2019)
29. Moon, G., Lee, K.M.: I2L-MeshNet: image-to-Lixel prediction network for accurate 3D human pose and mesh estimation from a single RGB image. In: Vedaldi, A., Bischof, H., Brox, T., Frahm, J.-M. (eds.) ECCV 2020. LNCS, vol. 12352, pp. 752–768. Springer, Cham (2020). https://doi.org/10.1007/978-3-030-58571-6_44
30. Muller, L., Osman, A.A., Tang, S., Huang, C.H.P., Black, M.J.: On self-contact and human pose. In: CVPR, pp. 9990–9999 (2021)
31. Omran, M., Lassner, C., Pons-Moll, G., Gehler, P., Schiele, B.: Neural body fitting: unifying deep learning and model based human pose and shape estimation. In: 3DV, pp. 484–494 (2018)
32. Paszke, A., et al.: Pytorch: an imperative style, high-performance deep learning library. NeurIPS **32**, 8026–8037 (2019)
33. Pavlakos, G., et al.: Expressive body capture: 3D hands, face, and body from a single image. In: CVPR, pp. 10975–10985 (2019)
34. Pavlakos, G., Kolotouros, N., Daniilidis, K.: Texturepose: supervising human mesh estimation with texture consistency. In: ICCV, pp. 803–812 (2019)
35. Pavlakos, G., Zhu, L., Zhou, X., Daniilidis, K.: Learning to estimate 3D human pose and shape from a single color image. In: CVPR, pp. 459–468 (2018)

36. Rockwell, C., Fouhey, D.F.: Full-body awareness from partial observations. In: Vedaldi, A., Bischof, H., Brox, T., Frahm, J.-M. (eds.) ECCV 2020. LNCS, vol. 12362, pp. 522–539. Springer, Cham (2020). https://doi.org/10.1007/978-3-030-58520-4_31

37. Saito, S., Huang, Z., Natsume, R., Morishima, S., Kanazawa, A., Li, H.: Pifu: pixel-aligned implicit function for high-resolution clothed human digitization. In: ICCV, pp. 2304–2314 (2019)

38. Saito, S., Simon, T., Saragih, J., Joo, H.: PifuHD: multi-level pixel-aligned implicit function for high-resolution 3d human digitization. In: CVPR, pp. 84–93 (2020)

39. Sun, X., Xiao, B., Wei, F., Liang, S., Wei, Y.: Integral human pose regression. In: Ferrari, V., Hebert, M., Sminchisescu, C., Weiss, Y. (eds.) ECCV 2018. LNCS, vol. 11210, pp. 536–553. Springer, Cham (2018). https://doi.org/10.1007/978-3-030-01231-1_33

40. Sun, Y., Bao, Q., Liu, W., Fu, Y., Black, M.J., Mei, T.: Monocular, one-stage, regression of multiple 3D people. In: ICCV, pp. 11179–11188 (2021)

41. Varol, G., et al.: BodyNet: volumetric inference of 3D human body shapes. In: Ferrari, V., Hebert, M., Sminchisescu, C., Weiss, Y. (eds.) ECCV 2018. LNCS, vol. 11211, pp. 20–38. Springer, Cham (2018). https://doi.org/10.1007/978-3-030-01234-2_2

42. Vaswani, A., et al.: Attention is all you need. In: NeurIPS, pp. 5998–6008 (2017)

43. Xu, Y., Zhu, S.C., Tung, T.: DenseRAC: joint 3D pose and shape estimation by dense render-and-compare. In: ICCV, pp. 7760–7770 (2019)

44. Zeng, W., Ouyang, W., Luo, P., Liu, W., Wang, X.: 3D human mesh regression with dense correspondence. In: CVPR, pp. 7054–7063 (2020)

45. Zhang, T., Huang, B., Wang, Y.: Object-occluded human shape and pose estimation from a single color image. In: CVPR, pp. 7376–7385 (2020)

46. Zhou, X., Zhu, M., Pavlakos, G., Leonardos, S., Derpanis, K.G., Daniilidis, K.: MonoCap: monocular human motion capture using a CNN coupled with a geometric prior. PAMI **41**(4), 901–914 (2018)

Towards High-Fidelity Single-View Holistic Reconstruction of Indoor Scenes

Haolin Liu[1,2], Yujian Zheng[1,2], Guanying Chen[1,2], Shuguang Cui[1,2], and Xiaoguang Han[1,2(✉)]

[1] School of Science and Engineering, CUHK-Shenzhen, Shenzhen, China
hanxiaoguang@cuhk.edu.cn
[2] The Future Network of Intelligence Institute, CUHK-Shenzhen, Shenzhen, China

Abstract. We present a new framework to reconstruct holistic 3D indoor scenes including both room background and indoor objects from single-view images. Existing methods can only produce 3D shapes of indoor objects with limited geometry quality because of the heavy occlusion of indoor scenes. To solve this, we propose an instance-aligned implicit function (InstPIFu) for detailed object reconstruction. Combining with instance-aligned attention module, our method is empowered to decouple mixed local features toward the occluded instances. Additionally, unlike previous methods that simply represents the room background as a 3D bounding box, depth map or a set of planes, we recover the fine geometry of the background via implicit representation. Extensive experiments on the SUN RGB-D, Pix3D, 3D-FUTURE, and 3D-FRONT datasets demonstrate that our method outperforms existing approaches in both background and foreground object reconstruction. Our code and model will be made publicly available.

1 Introduction

With the development of virtual reality (VR) and augmented reality (AR), the requirements for understanding and digitizing real-world 3D scenes are getting higher, especially for the indoor environment. If reconstructing the holistic indoor scene can be as simple as taking a picture using a mobile phone, we can efficiently generate a large scale of high-quality 3D content and further promote the development of VR and AR. Also, robots can better understand the real-world with the advance of single-view scene reconstruction. Hence, the problem of holistic indoor scene reconstruction from a single image has attracted considerable attention in recent years.

Early methods simplify this problem as estimating the room layout [5,15,24, 28,40] and indoor objects [2,7,17] as 3D bounding boxes. However, such a coarse

---

H. Liu and Y. Zheng—Contributed equally to this paper.

Supplementary Information The online version contains supplementary material available at https://doi.org/10.1007/978-3-031-19769-7_25.

(a) Input image   (b) Reconstructed scene   (c) Total3D   (d) Im3D   (e) Ours

Fig. 1. Given a single indoor scene image, we reconstruct the holistic scene with detailed geometry, including the room background and indoor objects. From left to right: input image, the scene reconstructed by our method, results of Total3D [34], Im3D [56] and our method in a different camera pose.

representation can only provide scene context information but cannot provide shape-level reconstruction. Mesh retrieval based approaches [18–20] improve the object shapes by substituting the 3D object boxes with meshes searched from a database. Due to the various categories and appearances of indoor objects, the size and diversity of the database directly influence the accuracy and time efficiency of these methods.

Inspired by learning-based shape reconstruction methods, voxel representation [22,25,49] is first applied to recover the 3D geometry of indoor scenes, but the shape quality is far from satisfactory due to the limited resolution. Mesh R-CNN [12] can reconstruct meshes for multiple instances from a single-view image, but lacks of scene understanding. Recently, Total3D [34] and Im3D [56] are proposed to reconstruct the 3D indoor scene from a single image, where the instance-level objects are represented in the form of explicit mesh and implicit surface, respectively. Although they have achieved state-of-the-art results on this task, they still have the following limitation. First, they often output shapes lacking details, due to the issue of limited training data and the use of global image feature for shape reconstruction. Second, the room layout in their methods is expressed as a simplified representation (*i.e.*, the 3D bounding box) which cannot recover backgrounds with complex geometries, like non-planar surfaces.

Recently, pixel-aligned implicit function (PIFu) has achieved promising results for detailed and generalizable 3D human reconstruction from a single image [42]. Motivated by the success of PIFu, we address the limitations of existing methods by introducing an *instance-aligned implicit function* (InstPIFu) for holistic and detailed indoor scene reconstruction from a single image. Note that pixel-aligned feature cannot be straightforwardly applied to indoor scene reconstruction, as objects (*e.g.*, sofa, chair, bed, and other furniture) are often occluded in a cluttered scene (see Fig. 1), such that the extracted local feature might contain mixed information of multiple objects. It is sub-optimal to directly use such a contaminated local feature for implicit surface reconstruction. To tackle this problem, we introduce an *instance-aligned attention module*, consisting of *attentional channel filtering*, and *spatial-guided supervision* strategies, to decouple the mixed local features for different instances in the overlapping regions.

Unlike previous methods that simply recover the room layout as a 3D bounding box [5, 15, 24, 28, 34, 40, 56], sparse depth [49] or room layout structure [44, 55, 57] without non-planar geometry, our implicit surface representation allows the detailed shape reconstruction of the room background (*e.g.*, floor, wall, and ceiling). Compared with existing approaches that encode the latent shape code with global image features [34, 56], the instance-aligned local features utilized in our encoder help alleviate the over-fitting problem and recover more detailed geometry of indoor objects. Extensive experiments on the SUN RGB-D, Pix3D, 3D-FUTURE, and 3D-FRONT datasets demonstrate the superiority of our method.

The key contributions of this paper are summarized as follows:

- We introduce a new pipeline to reconstruct the holistic and detailed 3D indoor scene from a single RGB image using implicit representation. To our best knowledge, this is the first system that uses pixel-aligned feature to recover the 3D indoor scene from a single view.
- We are the first to attempt to reconstruct the room background via implicit representation. Compared to previous methods that represent room layout as a 3D box, depth map or a set of planes, our method is capable to recover background with more complex geometries, like non-planar surfaces.
- We propose a new method, called InstPIFu, to use the instance-aligned feature, extracted by a novel instance-aligned attention module, for detailed indoor object reconstruction. Our method is more robust to object occlusion and has a better generalization ability on real-world datasets.
- Our method achieves state-of-the-art performance on both the synthetic and real-world indoor scene datasets.

2   Related Work

Single-View Indoor Scene Reconstruction. The long-standing problem of indoor scene reconstruction from a single image aims to construct the holistic 3D scene, which entails room layout estimation, object detection and pose estimation, as well as 3D shape reconstruction. Early works first recover the room layout as a 3D room bounding box [5, 15, 24, 28, 40]. Follow-up works make rapid progress toward object pose recovery [2, 7, 17], but still represent objects as 3D boxes without shape details.

To recover object shapes, some methods search for models with a similar appearance from a database [18–20]. However, the mismatch between objects in images and the database often leads to unsatisfactory results. Other methods [22, 25, 49] try to reconstruct the voxel representation for each object instance, but they are subjected to the problem of limited resolution. Mesh R-CNN [12] is capable to reconstruct meshes for multiple objects from a single-view image, but ignores scene understanding. To overcome the above limitations of previous solutions, Total3D [34] proposes an end-to-end system to jointly reconstruct room layout, object bounding boxes, and meshes from a single image. But its mesh generation network can only produce non-watertight mesh when handling shapes with complex topology. The following Im3D [56] represents each object with the implicit surface function that can be converted to a watertight mesh via marching cube algorithm while preserving geometry details in the meantime. However, the state-of-the-art solution of Im3D [56] still suffers from shape over-fitting due to the problem of limited training data and the use of global image features for shape reconstruction.

Room Background Representation. Early methods [5,15,24,28,40] simply recover the room background as a 3D bounding box, but room is usually not a cuboid. The state-of-the-art single-view indoor scene reconstruction methods [34,56] are still using this representation for room background. [49] predicts the background via depth estimation, which recovers more details for background. However, the accuracy of background depth estimation is far from satisfactory because of the occlusion of foreground, *i.e.*, indoor objects. Recent works try to reconstruct the room layout structure [44,55,57] with the assumption that the background of the room (*e.g.*, floor, wall, and ceiling) is mainly composed of planes. Hence, only planar geometry can be recovered and nonplanar information is missed by these methods.

Learning-Based 3D Shape Reconstruction. Recent learning-based methods have adopted different surface representations for 3D shape reconstruction, such as voxel, mesh, point cloud, patches, primitives, and implicit surface.

Voxel-based methods [4,26,41,47,51,53] benefit from 2D CNNs because of the regularity of the voxel representation, but suffer from the balance between resolution and efficiency. Mesh-based methods reconstruct the mesh of an object through deforming a template (*e.g.*, a unit sphere), but the topology of the obtained mesh is restricted [13,21,36,52]. To modify the topology, some approaches learn to remove extra edges and vertices [34,36,46], which results in non-watertight meshes. Methods based on point cloud [9,23,29,32], patches [13,53], and primitives [6,38,48,50] are adaptable to complex topology, but require post-processing to convert to structural representations. However, the post-processing is difficult to preserve the detailed geometry of the shape. Recently, implicit surface function [3,30,31,37,54] has been widely adopted as it can achieve detailed reconstruction for shape with an arbitrary topology and is easy to be converted to fine mesh.

Pixel-Aligned Image Features. Single-view implicit surface reconstruction methods often adopt an encoder-decoder pipeline and learn a latent code from the input image for shape recovery. For time and memory efficiency, global image feature [3,8,30,35,37] is often adopted, but it cannot recover the local detailed information existed in the input image. As a result, coarse results often occur in these approaches. Recently, pixel-aligned local image features have been demonstrated to recover complex geometries from a single view [42,54].

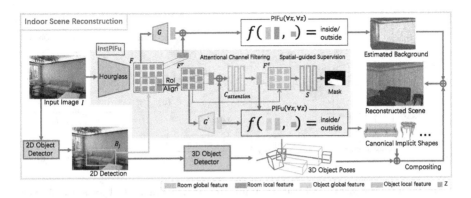

Fig. 2. Overview of the proposed InstPIFu. Given a single indoor scene image as input, our method simultaneously performs room background estimation, object detection and camera pose estimation, as well as detailed 3D object reconstruction.

3 Instance-Aligned Implicit Representation

In this section, we first review the pixel-aligned implicit function (PIFu) and point out its limitation in dealing with the occluded object in the indoor scene. We then introduce our instance-aligned implicit function to perform better indoor object reconstruction where objects are often occluded in the cluttered scene.

3.1 Review of Pixel-Aligned Implicit Modeling

Single-view scene reconstruction benefits from implicit representation [56], but the usage of global image features often causes coarse results. PIFu with pixel-aligned local features has been witnessed to recover detailed shapes in 3D human reconstruction [42].

A 3D surface can be defined by an implicit function as a level set of function f, $e.g. f(X) = 0$, where X is a 3D point. Similarly, a pixel-aligned implicit function f, represented by multi-layer perceptrons (MLPs), defines the surface as a level set of

$$f(F(x), z(X)) = s : s \in \mathbb{R}, \tag{1}$$

where $x = \pi(X)$ gives the 2D image projection point of X, $F(x) = g(I(x))$ is the local image feature at x extracted by a fully convolutional image encoder g, and $z(X)$ is the depth value in weak-perspective camera coordinate. We observe that adding the global image feature as an extra input helps in shape reconstruction. The adapted PIFu used in this work is defined as

$$f(F(x), F^G(I), z(X)) = s : s \in \mathbb{R}, \tag{2}$$

where $F^G(I)$ represents the global features of image I encoded by G.

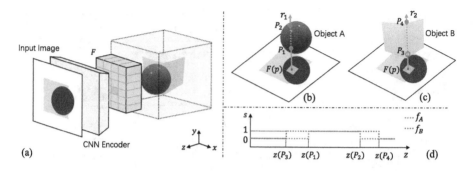

Fig. 3. Occlusion causes local feature ambiguity among different objects. (a) A scene contains two objects, and F is the extracted local feature from the image. (b)–(c) Object reconstruction in canonical coordinate system, where points along the rays are projected at p to sample local feature $F(p)$. (d) Variations of occupancy s with depth z along the ray r_1 and r_2 for f_A and f_B.

3.2 Limitation of Pixel-Aligned Feature

Although PIFu demonstrates detailed reconstruction results in single human reconstruction, applying PIFu for indoor object reconstruction straightforwardly is not good, as it suffers a lot from the object occlusion that leads to feature ambiguity. Multiple 3D points belonging to different objects can be projected into similar 2D image location and get the same local image feature, such that the local feature will contain mixed information from different instances, which is not desirable for shape reconstruction.

As an example in Fig. 3(a), a scene consists of a sphere A and a cube B, where A occludes B in the captured image. Figure 3 (b)–(c) show that when sampling pixel-aligned features for 3D points, points along the ray r_1 and r_2 (*e.g.*, P) are all projected at the point p in the overlapping region. This means that the same local feature $F(p)$ will be used to compute the occupancy value s for implicit function of A (f_A) and B (f_B), *i.e.*, $s = f_A(F(p), z(P))$ and $s = f_B(F(p), z(P))$. As PIFu implemented f_A and f_B using the same MLPs, adopting the same local feature $F(p)$ raises feature ambiguity in occupancy estimation for A and B. This is illustrated in Fig. 3(d), where variations of s with z for f_A and f_B are apparently different. Note that here we simply represent the PIFu f as an ideal occupancy field where levels of points inside the object are 1, otherwise 0.

One possible solution might be adding the global feature of the instance as extra inputs to the shape decoder. But only using global features to tackle the ambiguity in occlusion region is not enough (see our ablation study). Because the local features still contain mixed information from different instances.

3.3 Instance-Aligned Feature Concentration

To address the above limitation of PIFu, we propose InstPIFu, which adopts an instance-aligned attention module to disentangle the mixed feature information caused by object occlusion, for indoor object reconstruction. The proposed instance-aligned attention module reduces the ambiguity of the local image feature by three sequential steps, *i.e.*, *RoI alignment*, *attentional channel filtering*, and *spatial-guided supervision* (see Fig. 2).

RoI Alignment. The first step is to extract instance-related features for each instance. A straight-forward solution is to extract features independently from the cropped image patch of each target instance. However, it is inefficient when there are multiple objects in a cluttered scene, and the useful scene contextual information will be ignored. Instead, we follow Mask R-CNN [14] to use RoI alignment for instance-related feature extraction. Given an image I and the 2D bounding box B_j of an instance j, we first crop out the corresponding local features of the region of interest (RoI) from the whole pixel-aligned feature map F and align them to F^r as [14]

$$F^r = \text{RoIAlign}(F, B_j). \tag{3}$$

Note that F^r has a fixed size of $W_r \times H_r$ for input feature maps with different shapes and the 2D bounding box B_j of object j is obtained by a Faster R-CNN detector [39]. We then extract a global instance feature for instance j as $G'(F^r)$, where G' is a global instance image encoder. The global instance feature will be used to compute the channel-wise attention for local feature filtering.

Attentional Channel Filtering. Each local feature in the aligned RoI feature map F^r will be concatenated with the global instance feature $G'(F^r)$ as input for a channel-wise attention layer, similar to the Squeeze-and-Excitation block in [16] structurally, to generate an attention map with the same channel number of L_c as the local feature. This attention map will multiply with the local feature to filter out irrelevant feature by channel filtering to allow the updated local feature to concentrate on the target instance. This operation can be expressed as

$$F^c(x) = C_{\text{attention}}(F^r(x), G'(F^r)) \times F^r(x), \tag{4}$$

where $F^c(x)$ is the filtered local image feature at the 2D projection x of a 3D point X for instance j. Note that $F^r(x)$ in Eq. (4) adopts bilinear interpolation to access the features, and x in $F^r(x)$ should be shifted and scaled as well.

Spatial-Guided Supervision. To better guide the learning of the channel filtering, we need a module that can encourage the filtered feature to focus more on the target instance. Thus, we exploit a spatial-guided supervision on the filtered local feature map F^c that is the output of the channel-wise attention layer with the same shape as F^r. The Feature map F^c will be fed into a fully convolutional layer S to estimate a complete mask M for the target instance, *i.e.*, $M = S(F^c)$. This spatial-guided supervision can filter out irrelevant information out of the mask.

Instance-Aligned Implicit Function. Given the instance-aligned feature, we define InstPIFu f_o as

$$f_o(F^c(x), G'(F^r), z(X)) = s : s \in \mathbb{R}. \tag{5}$$

By applying the proposed instance-aligned attention module for decoupling the mixed local feature, compared with PIFu, the local feature used in our Inst-PIFu provides more discriminative information for accurate and detailed shape reconstruction. And this can be demonstrated by our ablation study.

4   Holistic Indoor Scene Reconstruction

Given a single image of an indoor scene, we aim to recover the holistic and detailed 3D scene in implicit representation (see Fig. 2). This problem is normally divided into several sub-tasks, including room background estimation, 3D object detection (pose estimation), as well as instance-level object reconstruction [34,56]. We first process these three tasks individually and then perform scene compositing for holistic scene reconstruction. Note that our method recovers the room background with geometry details instead of just a simplified 3D bounding box.

4.1   Room Background Estimation

Room is usually not a cuboid. Thus, it is inappropriate to represent the room background as a 3D bounding box like [34,56]. Depth map [49] is also not an ideal representation, because the accuracy of background depth estimation is heavily influenced by the occlusions of indoor objects in front of the background. Also, methods [44,55] based on plane detection cannot recover small planes and non-planar background geometries. To address the above issues, we explore to use the implicit representation for room background reconstruction in this work.

The ground-truth room surface is represented as a 0.5 level set and then discretized to a 3D occupancy field:

$$f_r^*(X) = \begin{cases} 1, & \text{if } X \text{ is inside the room} \\ 0, & \text{otherwise} \end{cases}. \tag{6}$$

Compared with indoor objects that have various styles and complicated geometries, the shape of the room background is much simpler. We find that applying the adapted PIFu (see Eq. (2)) which takes pixel-aligned features and global features for room background reconstruction already achieves good results. We train our room estimation PIFu f_r by minimizing the average of mean squared error (MSE):

$$\mathcal{L}_r = \frac{1}{n} \sum_{i=1}^{n} |f_r(F(x_i), G(F), z(X_i)) - f_r^*(X_i)|^2, \tag{7}$$

where n is the number of sample points, $X_i \in \mathbb{R}^3$ is a point in the camera coordinate system, $F(x) = g(I(x))$ is the local image feature located at x, $G(F)$ is the global image feature of the room background and F is the whole feature map produced by Hourglass network. The local and global image features are both from a stacked hourglass network [33], but an extra global encoder G is needed to encode the whole feature map F to the global feature. The obtained implicit room background can be easily converted to an explicit mesh via marching cube algorithm.

4.2 Indoor Object Reconstruction

As discussed in Sect. 3, due to the heavy occlusions between indoor objects, directly applying PIFu for instance reconstruction suffers from the problem of ambiguous local features. We adopt the proposed InstPIFu, which applies instance-aligned attention module for feature filtering, to reconstruct the indoor objects. We define the ground-truth surface of an indoor object as the room background (see Eq. (6)). The InstPIFu f_o is also trained by minimizing the average of MSE:

$$\mathcal{L}_o = \frac{1}{n} \sum_{i=1}^{n} |f_o\left(F^c\left(x_i\right), G'(F^r), z\left(X_i\right)\right) - f_o^*\left(X_i\right)|^2, \tag{8}$$

where $X_i \in \mathbb{R}^3$ is a point in the canonical coordinate system. Note that the projection from X_i to x_i is different from the original PIFu. Because X_i is in object coordinate system, extra camera and object poses are needed when projecting. We follow [56] to predict these parameters for projecting. The channel-wise attention layer is implemented as MLPs. During training, we add an extra instance mask loss for the instance-aligned attention module to enforce the feature to be constrained on the corresponding instance mask. The mask loss is simply implemented by the MSE between the predicted mask and the ground truth.

4.3 Scene Compositing

The room background is obtained in the camera coordinate system, while the objects are recovered in their canonical coordinate system to ease the learning of reconstructing indoor objects with various poses and scales. To embed objects into the scene together with the room background, the camera pose $\mathbf{R}(\beta, \gamma)$ and object bounding box parameters (δ, d, s, θ) are required. We use similar camera estimator and 3D object detector to predict above parameters as [34, 56]. Additionally, the Scene Graph Convolutional Network proposed in [56] is also used in our work to improve the performance of camera and object pose estimation. Note that we use perspective camera model.

5 Experiment

5.1 Experiment Setup

Datasets. We conduct experiments on both synthetic and real datasets. The proposed pipeline is trained on 3D-FRONT [10] which is a large-scale repository of synthetic indoor scenes, consisting of professionally designed rooms populated by 3D furniture models with high-quality geometry and texture in various styles. The furniture models come from 3D-FUTURE [11]. We use about 20K scene images for training and 6K for testing, where more than 16K objects from 3D-FUTURE are included. Following [34,56], we also evaluate our method on real-world datasets: SUN RGB-D [43] and Pix3D [45].

Metrics. We adopt the commonly used Chamfer distance (CD) to evaluate the background reconstruction, as it is difficult to compare our background results with layout Intersection over Union (IoU) [34,44,55,56] (detailed reasons in Sect. 5.2). The reconstructed indoor objects are evaluated with CD and F-Score [34,52,56].

5.2 Evaluation on Room Background Estimation

We first evaluate the effectiveness of our room background estimation module. Layout IoU is a commonly used metric when comparing the room background. It is computed using the layout structure of the whole room. However, our method only reconstructs partial room background within the camera view. Hence, to compare our room background results with existing methods quantitatively, we firstly sample 10K points within the camera frustum from the reconstructed background in representations of bounding box [56], depth map [1,49], plane sets [27] and our implicit surface, then compute CD with points on ground truth background. We choose to compare with PlaneRCNN [27] since it is popular and has decent performance in plane estimation. Because Factored3D [49] is based on depth estimation, we also compare it with Adabins [1] that is the state-of-the-art depth estimation approach. Quantitative comparisons in Table 1 shows the superiority of our method in detailed background recovery. Visual results of background reconstruction on 3D-FRONT and SUN RGB-D show that our method can recover the geometry details of the room background (see Fig. 5). More visual comparisons are given in the Supplementary Material.

Table 1. Quantitative comparisons of room background estimation on 3D-FRONT.

Method	Factored3D [49]	Adabins [1]	Im3D [56]	PlaneRCNN [27]	Ours
CD on 3D-FRONT ↓	0.697	0.573	1.974	0.717	**0.481**

5.3 Evaluation on Indoor Object Reconstruction

We compare our InstPIFu against the MGN of Total3D [34] and the LIEN of Im3D [56] on indoor object reconstruction. Quantitative and qualitative comparisons are shown on both 3D-FUTURE and Pix3D. Furthermore, we also train and test these object reconstruction networks on Pix3D with a non-overlapped split to evaluate their generalization ability. CD is used to evaluate on the 10K

Fig. 4. Qualitative comparisons of indoor object reconstruction. From left to right of every quintuplet: (1) Input images and results from (2) MGN [34], (3) LDIF [56], (4) Ours, (5) Ground truth. The first two rows are compared on 3D-FUTURE, and the last two rows are on Pix3D. Note that results of the last row are generated by models trained and tested on non-overlapped split.

points sampled from the reconstructed mesh after being aligned with the ground-truth using ICP. Note that results generated by InstPIFu and LIEN are in implicit representation which are converted to mesh using marching cube algorithm with a resolution of 256.

Evaluation on 3D-FUTURE. Table 2 summarizes the quantitative results on 3D-FUTURE evaluated on 2000 indoor objects in 8 different categories. We use scene images in 3D-FRONT as the input for our InstPIFu, and cropped patches by ground-truth 2D bounding boxes (following [34,56]) from every scene image as the input for MGN and LIEN. In these input images, object occlusions often occur. And thanks to the use of the instance-aligned feature, our method achieves the best on F-Score and shows decent results on CD (see Table 2). Although explicit methods like MGN achieve better CD loss as they directly optimize the CD loss during training, the reconstructed meshes lack details [30,36,54]. Also, MGN can not generate watertight mesh which is desired in object reconstruction. Figure 4 (first two rows) shows that the results of our method have the most similar appearances to objects in the input images.

Comparison on Pix3D. Quantitative results on Pix3D using the train/test split in [34] are shown in Table 3, where LIEN and MGN achieve better than ours. The major reason is that LIEN and MGN tend to be over-fitting on Pix3D which has only about 400 shapes. Because the split in [34] is based on different images, and all shapes in testing dataset also occur in training dataset. Also, the usage of pixel-aligned local feature makes our model achieve better generalization ability, but weaken the fitting performance. Nevertheless, our method still achieves comparable qualitative results (see the third row in Fig. 4).

Comparison of Generalization. To compare the generalization ability of the above three object reconstruction networks, we re-split Pix3D based on different

Table 2. Quantitative comparisons of object reconstruction on 3D-FUTURE (CD/F-Score). The values of CD are in units of 10^{-3}.

Method	Bed	Chair	Sofa	Table	Desk	Nightstand	Cabinet	Bookshelf	Mean ↓/↑
MGN [34]	**15.48**/46.81	**11.67**/57.49	8.72/64.61	**20.90**/49.80	**17.59**/46.82	17.11/47.91	13.13/54.18	10.21/54.55	**14.07**/55.64
LIEN [56]	16.81/44.28	41.40/31.61	9.51/61.40	35.65/43.22	26.63/37.04	16.78/50.76	7.44/69.21	11.70/55.33	28.52/45.63
Ours	18.17/**47.85**	14.06/**59.08**	**7.66**/**67.60**	23.25/**56.43**	33.33/**48.49**	**11.73**/**57.14**	**6.04**/**73.32**	**8.03**/**66.13**	14.46/**61.32**

Table 3. Quantitative comparisons of object reconstruction on Pix3D with split in [34] and non-overlapped split.

Split in [34]	Bed	Bookcase	Chair	Desk	Sofa	Table	Tool	Wardrobe	Misc	Mean ↓/↑
MGN [34]	5.99/**78.08**	6.56/62.98	**5.32**/**72.73**	**5.93**/75.04	**3.36**/**79.64**	14.19/65.27	3.12/**81.17**	3.83/85.51	26.93/46.76	6.84/**73.18**
LIEN [56]	**4.11**/65.26	**3.96**/46.05	5.45/59.84	7.85/76.03	5.61/64.02	**11.73**/**72.28**	**2.39**/36.09	4.31/58.59	**24.65**/**57.50**	6.72/63.96
Ours	9.52/59.47	4.38/**73.25**	14.40/48.26	13.70/64.24	8.21/57.17	22.6/57.52	7.76/69.36	**3.67**/**87.36**	30.32/35.05	13.60/56.07
Non-overlapped Split	Bed	Bookcase	Chair	Desk	Sofa	Table	Tool	Wardrobe	Misc	Mean ↓/↑
MGN [34]	22.91/34.69	36.61/28.42	56.47/**35.67**	33.95/34.90	9.27/51.15	81.19/17.05	94.70/57.16	10.43/52.04	137.5/10.41	44.32/36.20
LIEN [56]	11.88/37.13	29.61/15.51	40.01/25.70	65.36/26.01	10.54/49.71	146.13/21.16	29.63/5.85	4.88/59.46	144.06/11.04	51.31/31.45
Ours	**10.90**/**54.99**	**7.55**/**62.26**	**32.44**/35.30	**22.09**/**47.30**	**8.13**/**56.54**	**45.82**/**37.51**	**10.29**/**64.24**	**1.29**/**94.62**	**47.31**/**27.03**	**24.65**/**45.62**

shapes (70% for training and 30% for testing), which ensures that all shapes in testing dataset have not been seen when training (non-overlapped split). Quantitative results are shown in Table 3, where our method achieves the best result due to the use of local image features. In contrast, MGN and LIEN suffer from over-fitting caused by global image features. Qualitative results in Fig. 4 (the last row) give the same conclusion, where objects reconstructed by MGN and LIEN are coarse shapes. More results are shown in the supplementary material.

5.4 Qualitative Result of Holistic Scene Reconstruction

We compare our method with Total3D [34] and Im3D [56] in holistic indoor scene reconstruction on both 3D-FRONT [10] and SUN RGB-D [43] datasets. Qualitative comparisons shown in Fig. 5 demonstrate the superiority of our instance-aligned implicit representation. For fair comparison on SUN RGB-D, we first train the InstPIFu on 3D-FRONT and 3D-FUTURE and then finetune it on Pix3D. And we also use the predicted 3D object boxes by Im3D. Although our reconstructed scenes on SUN RGB-D may have some noisy patches due to the domain gap between the synthetic and the realistic datasets, the results are full of details in both the background and indoor objects, which reveals the good generalization ability of our method to some extend.

5.5 Ablation Study

To better study the effect of instance-aligned implicit representation for indoor object reconstruction, our method is ablated with five configurations:

- *Baseline*: only pixel-aligned feature is used in object reconstruction.
- C_0: pixel-aligned feature + global instance feature.
- C_1: C_0 + attentional channel filtering.
- C_2: C_0 + spatial-guided supervision.
- *Full*: C_0 + attentional channel filtering + spatial-guided supervision.

(a) Scene reconstruction results on 3D-FRONT

(b) Scene reconstruction results on SUN RGB-D

Fig. 5. Qualitative comparisons of holistic scene reconstruction. From the first row to the last: the input image, scene reconstruction results of Total3D, Im3D and ours. Note that the first four rows are compared on 3D-FRONT and the rest are on SUN RGB-D.

Fig. 6. Visual comparisons for ablation study. From left to right: the input image, results of *Baseline*, C_0, C_1, C_2 and *Full*.

Table 4. Ablation study for the network architecture.

Method	*Baseline*	C_0	C_1	C_2	*Full*
CD ↓	17.95	16.42(-1.53)	15.54(-2.41)	15.28(-2.67)	**14.46**(-3.49)
F-Score ↑	56.98	58.62($+1.64$)	60.23($+3.25$)	60.56($+3.58$)	**61.32**($+4.34$)

As the quantitative comparisons shown in Table 4, our *Full* model achieves the best results on metrics CD and F-Score, where we add the channel-wise attention together with the mask supervision to C_0. If we remove the anyone of these two modules from *Full*, that are C_1 and C_2, CD and F-Score both become worse. But C_1 and C_2 still perform better than C_0. This gives us the insight that both of channel-level filtering and spatial guidance help to decouple the feature ambiguity towards occluded objects. And the comparisons between the *Baseline* and C_0 show that concatenating global instance feature with pixel-aligned local feature is helpful for indoor object reconstruction. But from the comparisons of the whole table, we can see that only using global feature to tackle the ambiguity in occlusion region is not enough. Same conclusions can be drawn by the visual comparisons in Fig. 6.

6   Conclusion

We have introduced a new method based on implicit representation, called Inst-PIFu, for holistic and detailed 3D indoor scene reconstruction from a single image. To resolve the problem of ambiguous local features caused by object occlusions in an indoor scene, we proposed an instance-aligned attention module to effectively disentangle the mixed features for accurate instance shape reconstruction. Moreover, our method is the first to estimate the detailed room background via implicit representation, resulting in a more complete scene reconstruction. Extensive experiments on both synthetic and real datasets show that our method achieves state-of-the-art results for this problem.

Although our instance-aligned implicit function enables a more detailed and accurate indoor object reconstruction, the use of local feature makes the joint training of the 3D detection network and object reconstruction network not easy. Besides, real-world indoor scene datasets with high-quality 3D ground truth are scarce, and methods trained or finetuned with limited real data perform less well on real-world scenes compared with results on the synthetic scene (see Fig. 5). It would be interesting to explore how to take advantage of the existing large-scale and photo-realistic synthetic datasets for improving the generalization ability of the method.

Acknowledgement. The work was supported in part by the National Key R&D Program of China with grant No. 2018YFB1800800, the Basic Research Project No. HZQB-KCZYZ-2021067 of Hetao Shenzhen-HK S&T Cooperation Zone, by Shenzhen Outstanding Talents Training Fund 202002, by Guangdong Research Projects

No. 2017ZT07X152 and No. 2019CX01X104, and by the Guangdong Provincial Key Laboratory of Future Networks of Intelligence (Grant No. 2022B12 12010001). It was also supported by NSFC-62172348, NSFC-61902334 and Shenzhen General Project (No. JCYJ20190814112007258). Thanks to the ITSO in CUHKSZ for their High-Performance Computing Services.

References

1. Bhat, S.F., Alhashim, I., Wonka, P.: AdaBins: depth estimation using adaptive bins. In: Proceedings of the IEEE/CVF Conference on Computer Vision and Pattern Recognition, pp. 4009–4018 (2021)
2. Chen, Y., Huang, S., Yuan, T., Qi, S., Zhu, Y., Zhu, S.C.: Holistic++ scene understanding: single-view 3D holistic scene parsing and human pose estimation with human-object interaction and physical commonsense. arXiv preprint arXiv:1909.01507 (2019)
3. Chen, Z., Zhang, H.: Learning implicit fields for generative shape modeling. In: Proceedings of the IEEE Conference on Computer Vision and Pattern Recognition, pp. 5939–5948 (2019)
4. Choy, C.B., Xu, D., Gwak, J., Chen, K., Savarese, S.: 3D–R2N2: a unified approach for single and multi-view 3D object reconstruction. In: European Conference on Computer Vision, pp. 628–644. Springer (2016). https://doi.org/10.1007/978-3-319-46484-8_38
5. Dasgupta, S., Fang, K., Chen, K., Savarese, S.: DeLay: robust spatial layout estimation for cluttered indoor scenes. In: Proceedings of the IEEE Conference on Computer Vision and Pattern Recognition, pp. 616–624 (2016)
6. Deprelle, T., Groueix, T., Fisher, M., Kim, V.G., Russell, B.C., Aubry, M.: Learning elementary structures for 3D shape generation and matching. arXiv preprint arXiv:1908.04725 (2019)
7. Du, Y., et al.: Learning to exploit stability for 3D scene parsing. In: Advances in Neural Information Processing Systems, pp. 1726–1736 (2018)
8. Dupont, E., Martin, M.B., Colburn, A., Sankar, A., Susskind, J., Shan, Q.: Equivariant neural rendering. In: International Conference on Machine Learning, pp. 2761–2770. PMLR (2020)
9. Fan, H., Su, H., Guibas, L.J.: A point set generation network for 3D object reconstruction from a single image. In: Proceedings of the IEEE Conference on Computer Vision and Pattern Recognition, pp. 605–613 (2017)
10. Fu, H., et al.: 3D-FRONT: 3D furnished rooms with layouts and semantics. In: Proceedings of the IEEE/CVF International Conference on Computer Vision, pp. 10933–10942 (2021)
11. Fu, H., et al.: 3D-future: 3D furniture shape with texture. Int. J. Comput. Vis. 1–25 (2021). https://doi.org/10.1007/s11263-021-01534-z
12. Gkioxari, G., Malik, J., Johnson, J.: Mesh R-CNN. arXiv preprint arXiv:1906.02739 (2019)
13. Groueix, T., Fisher, M., Kim, V.G., Russell, B., Aubry, M.: AtlasNet: a Papier-Mâché approach to learning 3D surface generation. In: Proceedings of IEEE Conference on Computer Vision and Pattern Recognition (CVPR) (2018)
14. He, K., Gkioxari, G., Dollár, P., Girshick, R.: Mask R-CNN. In: Proceedings of the IEEE International Conference on Computer Vision, pp. 2961–2969 (2017)

15. Hedau, V., Hoiem, D., Forsyth, D.: Recovering the spatial layout of cluttered rooms. In: 2009 IEEE 12th International Conference on Computer Vision, pp. 1849–1856. IEEE (2009)
16. Hu, J., Shen, L., Sun, G.: Squeeze-and-excitation networks. In: Proceedings of the IEEE Conference on Computer Vision and Pattern Recognition, pp. 7132–7141 (2018)
17. Huang, S., Qi, S., Xiao, Y., Zhu, Y., Wu, Y.N., Zhu, S.C.: Cooperative holistic scene understanding: unifying 3D object, layout, and camera pose estimation. In: Advances in Neural Information Processing Systems, pp. 207–218 (2018)
18. Huang, S., Qi, S., Zhu, Y., Xiao, Y., Xu, Y., Zhu, S.C.: Holistic 3D scene parsing and reconstruction from a single RGB image. In: Proceedings of the European Conference on Computer Vision (ECCV), pp. 187–203 (2018)
19. Hueting, M., Reddy, P., Kim, V., Yumer, E., Carr, N., Mitra, N.: SeeThrough: finding chairs in heavily occluded indoor scene images. arXiv preprint arXiv:1710.10473 (2017)
20. Izadinia, H., Shan, Q., Seitz, S.M.: IM2CAD. In: Proceedings of the IEEE Conference on Computer Vision and Pattern Recognition, pp. 5134–5143 (2017)
21. Kato, H., Ushiku, Y., Harada, T.: Neural 3D mesh renderer. In: Proceedings of the IEEE Conference on Computer Vision and Pattern Recognition, pp. 3907–3916 (2018)
22. Kulkarni, N., Misra, I., Tulsiani, S., Gupta, A.: 3D-RelNet: joint object and relational network for 3D prediction. In: International Conference on Computer Vision (ICCV) (2019)
23. Kurenkov, A., et al.: DeformNet: free-form deformation network for 3D shape reconstruction from a single image. In: 2018 IEEE Winter Conference on Applications of Computer Vision (WACV), pp. 858–866. IEEE (2018)
24. Lee, D.C., Hebert, M., Kanade, T.: Geometric reasoning for single image structure recovery. In: 2009 IEEE Conference on Computer Vision and Pattern Recognition, pp. 2136–2143. IEEE (2009)
25. Li, L., Khan, S., Barnes, N.: Silhouette-assisted 3D object instance reconstruction from a cluttered scene. In: Proceedings of the IEEE International Conference on Computer Vision Workshops (2019)
26. Liao, Y., Donne, S., Geiger, A.: Deep marching cubes: learning explicit surface representations. In: Proceedings of the IEEE Conference on Computer Vision and Pattern Recognition, pp. 2916–2925 (2018)
27. Liu, C., Kim, K., Gu, J., Furukawa, Y., Kautz, J.: PlaneRCNN: 3D plane detection and reconstruction from a single image. In: Proceedings of the IEEE/CVF Conference on Computer Vision and Pattern Recognition, pp. 4450–4459 (2019)
28. Mallya, A., Lazebnik, S.: Learning informative edge maps for indoor scene layout prediction. In: Proceedings of the IEEE International Conference on Computer Vision, pp. 936–944 (2015)
29. Mandikal, P., KL, N., Babu, R.V.: 3D-PSRNet: part segmented 3D point cloud reconstruction from a single image. In: Proceedings of the European Conference on Computer Vision (ECCV) (2018)
30. Mescheder, L., Oechsle, M., Niemeyer, M., Nowozin, S., Geiger, A.: Occupancy networks: learning 3D reconstruction in function space. In: Proceedings of the IEEE Conference on Computer Vision and Pattern Recognition, pp. 4460–4470 (2019)
31. Michalkiewicz, M., Pontes, J.K., Jack, D., Baktashmotlagh, M., Eriksson, A.: Deep level sets: implicit surface representations for 3D shape inference. arXiv preprint arXiv:1901.06802 (2019)

32. Navaneet, K., Mandikal, P., Agarwal, M., Babu, R.V.: CAPNet: continuous approximation projection for 3D point cloud reconstruction using 2D supervision. In: Proceedings of the AAAI Conference on Artificial Intelligence, vol. 33, pp. 8819–8826 (2019)

33. Newell, A., Yang, K., Deng, J.: Stacked hourglass networks for human pose estimation. In: Leibe, B., Matas, J., Sebe, N., Welling, M. (eds.) ECCV 2016. LNCS, vol. 9912, pp. 483–499. Springer, Cham (2016). https://doi.org/10.1007/978-3-319-46484-8_29

34. Nie, Y., Han, X., Guo, S., Zheng, Y., Chang, J., Zhang, J.J.: Total3DUnderstanding: joint layout, object pose and mesh reconstruction for indoor scenes from a single image. In: Proceedings of the IEEE/CVF Conference on Computer Vision and Pattern Recognition, pp. 55–64 (2020)

35. Niemeyer, M., Mescheder, L., Oechsle, M., Geiger, A.: Differentiable volumetric rendering: learning implicit 3D representations without 3D supervision. In: Proceedings of the IEEE/CVF Conference on Computer Vision and Pattern Recognition, pp. 3504–3515 (2020)

36. Pan, J., Han, X., Chen, W., Tang, J., Jia, K.: Deep mesh reconstruction from single RGB images via topology modification networks. In: Proceedings of the IEEE International Conference on Computer Vision, pp. 9964–9973 (2019)

37. Park, J.J., Florence, P., Straub, J., Newcombe, R., Lovegrove, S.: DeepSDF: learning continuous signed distance functions for shape representation. arXiv preprint arXiv:1901.05103 (2019)

38. Paschalidou, D., Ulusoy, A.O., Geiger, A.: Superquadrics revisited: learning 3D shape parsing beyond cuboids. In: Proceedings of the IEEE Conference on Computer Vision and Pattern Recognition, pp. 10344–10353 (2019)

39. Ren, S., He, K., Girshick, R., Sun, J.: Faster R-CNN: towards real-time object detection with region proposal networks. In: Advances in Neural Information Processing Systems, pp. 91–99 (2015)

40. Ren, Y., Li, S., Chen, C., Kuo, C.-C.J.: A coarse-to-fine indoor layout estimation (CFILE) method. In: Lai, S.-H., Lepetit, V., Nishino, K., Sato, Y. (eds.) ACCV 2016. LNCS, vol. 10115, pp. 36–51. Springer, Cham (2017). https://doi.org/10.1007/978-3-319-54193-8_3

41. Riegler, G., Ulusoy, A.O., Geiger, A.: OctNet: learning deep 3D representations at high resolutions. In: Proceedings of the IEEE Conference on Computer Vision and Pattern Recognition, pp. 3577–3586 (2017)

42. Saito, S., Huang, Z., Natsume, R., Morishima, S., Kanazawa, A., Li, H.: PIFu: pixel-aligned implicit function for high-resolution clothed human digitization. In: Proceedings of the IEEE/CVF International Conference on Computer Vision, pp. 2304–2314 (2019)

43. Song, S., Lichtenberg, S.P., Xiao, J.: SUN RGB-D: A RGB-D scene understanding benchmark suite. In: Proceedings of the IEEE Conference on Computer Vision and Pattern Recognition, pp. 567–576 (2015)

44. Stekovic, S., Hampali, S., Rad, M., Sarkar, S.D., Fraundorfer, F., Lepetit, V.: General 3D room layout from a single view by render-and-compare. In: Vedaldi, A., Bischof, H., Brox, T., Frahm, J.-M. (eds.) ECCV 2020. LNCS, vol. 12361, pp. 187–203. Springer, Cham (2020). https://doi.org/10.1007/978-3-030-58517-4_12

45. Sun, X., et al.: Pix3d: dataset and methods for single-image 3D shape modeling. In: Proceedings of the IEEE Conference on Computer Vision and Pattern Recognition, pp. 2974–2983 (2018)

46. Tang, J., Han, X., Pan, J., Jia, K., Tong, X.: A skeleton-bridged deep learning approach for generating meshes of complex topologies from single RGB images. In: Proceedings of the IEEE Conference on Computer Vision and Pattern Recognition, pp. 4541–4550 (2019)

47. Tatarchenko, M., Dosovitskiy, A., Brox, T.: Octree generating networks: efficient convolutional architectures for high-resolution 3D outputs. In: Proceedings of the IEEE International Conference on Computer Vision, pp. 2088–2096 (2017)

48. Tian, Y., et al.: Learning to infer and execute 3D shape programs. arXiv preprint arXiv:1901.02875 (2019)

49. Tulsiani, S., Gupta, S., Fouhey, D.F., Efros, A.A., Malik, J.: Factoring shape, pose, and layout from the 2D image of a 3D scene. In: Proceedings of the IEEE Conference on Computer Vision and Pattern Recognition, pp. 302–310 (2018)

50. Tulsiani, S., Su, H., Guibas, L.J., Efros, A.A., Malik, J.: Learning shape abstractions by assembling volumetric primitives. In: Proceedings of the IEEE Conference on Computer Vision and Pattern Recognition, pp. 2635–2643 (2017)

51. Wallace, B., Hariharan, B.: Few-shot generalization for single-image 3D reconstruction via priors. In: Proceedings of the IEEE International Conference on Computer Vision, pp. 3818–3827 (2019)

52. Wang, N., Zhang, Y., Li, Z., Fu, Y., Liu, W., Jiang, Y.G.: Pixel2Mesh: generating 3D mesh models from single RGB images. In: Proceedings of the European Conference on Computer Vision (ECCV), pp. 52–67 (2018)

53. Wang, P.S., Sun, C.Y., Liu, Y., Tong, X.: Adaptive O-CNN: a patch-based deep representation of 3D shapes. In: SIGGRAPH Asia 2018 Technical Papers, p. 217. ACM (2018)

54. Xu, Q., Wang, W., Ceylan, D., Mech, R., Neumann, U.: DISN: deep implicit surface network for high-quality single-view 3D reconstruction. arXiv preprint arXiv:1905.10711 (2019)

55. Yang, C., Zheng, J., Dai, X., Tang, R., Ma, Y., Yuan, X.: Learning to reconstruct 3D non-cuboid room layout from a single RGB image. In: Proceedings of the IEEE/CVF Winter Conference on Applications of Computer Vision, pp. 2534–2543 (2022)

56. Zhang, C., Cui, Z., Zhang, Y., Zeng, B., Pollefeys, M., Liu, S.: Holistic 3D scene understanding from a single image with implicit representation. In: Proceedings of the IEEE/CVF Conference on Computer Vision and Pattern Recognition, pp. 8833–8842 (2021)

57. Zou, C., Colburn, A., Shan, Q., Hoiem, D.: LayoutNet: reconstructing the 3D room layout from a single RGB image. In: Proceedings of the IEEE Conference on Computer Vision and Pattern Recognition, pp. 2051–2059 (2018)

CompNVS: Novel View Synthesis with Scene Completion

Zuoyue Li[1]([✉]), Tianxing Fan[2], Zhenqiang Li[3], Zhaopeng Cui[2], Yoichi Sato[3], Marc Pollefeys[1,4], and Martin R. Oswald[1,5]

[1] ETH Zürich, Zürich, Switzerland
li.zuoyue@inf.ethz.ch, zuli@student.ethz.ch
[2] Zhejiang University, Hangzhou, China
[3] The University of Tokyo, Tokyo, Japan
[4] Microsoft, Redmond, USA
[5] University of Amsterdam, Amsterdam, Netherlands

Abstract. We introduce a scalable framework for novel view synthesis from RGB-D images with largely incomplete scene coverage. While generative neural approaches have demonstrated spectacular results on 2D images, they have not yet achieved similar photorealistic results in combination with scene completion where a spatial 3D scene understanding is essential. To this end, we propose a generative pipeline performing on a sparse grid-based neural scene representation to complete unobserved scene parts via a learned distribution of scenes in a 2.5D-3D-2.5D manner. We process encoded image features in 3D space with a geometry completion network and a subsequent texture inpainting network to extrapolate the missing area. Photorealistic image sequences can be finally obtained via consistency-relevant differentiable rendering. Comprehensive experiments show that the graphical outputs of our method outperform the state of the art, especially within unobserved scene parts.

Keywords: Novel view synthesis · Scene completion · Scene representation learning · 3D inpainting/extrapolation · 3D-aware generative modeling

1 Introduction

Recent advancements in 3D reconstruction and differentiable rendering have driven impressive progress in novel view synthesis (NVS) from single or multiple images. With such technology captured scenes can be easily virtually explored. However, with increasing scene complexity, the capturing effort for a complete scene increases substantially, and in practice captured scenes almost always contain holes due to occlusions and incomplete captures. Therefore, to synthesize such scenes immersively, the ability to extrapolate visual content into unseen areas in a spatially consistent manner is utterly important.

Supplementary Information The online version contains supplementary material available at https://doi.org/10.1007/978-3-031-19769-7_26.

Posed partial observations   No completion   pixelNeRF [25]   PixelSynth [17]   **Ours**

Fig. 1. Examplary input and output of our method. Our approach combines generative modeling with a sparse 3D feature representation to complete unobserved scene parts plausibly. (See supplementary material for an animated version of the figure.)

While GAN-based 2D inpainting approaches [26,27] can tackle the problem, they typically have trouble learning 3D correspondences and as a result, synthesized videos lack spatial and temporal consistency. More recent single-image NVS methods [17,24] combine 3D scene representations with generative modeling and become more 3D-aware. However, the multi-view inconsistency still occurs in the rendered sequences with large perspective changes due to the completion in 2D space. Grid-based 3D approaches can achieve better consistency, but may still fail when the input scene is incomplete due to sparse views [4].

In recent years, we have also seen the introduction of neural implicit representations [11,13] which further improved the (differentiable) rendering standards. These representations no longer strive for explicit 3D data structures but to "store" a single scene implicitly in the network, without being limited to a specific spatial resolution anymore. As a result, the quality of rendering images for the novel views is significantly improved and they can meanwhile hold very good multi-view consistency. However, these methods focus more on 3D reconstruction or NVS from multiple images with good coverage. Before inference for new query views, they have to optimize the corresponding scene individually, using its hundreds of observations. This kind of test-time optimization generally may not learn the priors across different scenes, and it is also difficult to directly use them for (conditional) scene generation or completion. There are also several applications of implicit representation combined with differentiable rendering in the field of scene or object generation. Most existing approaches [2,5,14–16] focus more on sampling from latent noise, and rarely involve conditions on some input. And they still rely heavily on the 2D network to ensure the generation quality, which may break the multi-view consistency from the 3D part.

In this project, we tackle the problem of 3D scene generation from limited partial observations. We present an approach that takes as input a partially scanned point cloud from a single or multiple posed RGB-D images and outputs novel synthesized views of a completed 3D model in a spatially consistent manner. Specifically, we perform geometry and texture synthesis directly in 3D space utilizing neural implicit representations with an explicit voxel grid. Exemplary results of our work are shown in Fig. 1. We hope that such an architecture provides stronger generation abilities, easier learns 3D priors from multiple scenes, and can thus generate high-quality multi-view consistent images via differentiable rendering, and benefit scene editing, or NVS from sparse observations.

In sum, we make the following contributions: (1) We propose a scalable NVS framework capable of generative modeling to complete larger unobserved scene parts with plausible and spatially consistent content. (2) We adapt NSVF [11] to a feed-forward version, in which the scene embeddings can be directly obtained by an encoder without test-time optimization. (3) We demonstrate superior experiment results on both synthetic and real-world datasets.

2   Related Work

Our work is at the cross-section of NVS, scene completion, and 3D generative modeling, which all have been studied with various scene representations implying crucial method properties. In the following, we discuss related works according to their underlying representations.

Explicit Scene Representations. Early 3D generative methods such as Holo-GAN [14] or BlockGAN [15] use the 3D representation of the voxel grid. The generation is performed in a coarse feature volume, which is further projected to each frame using the query camera poses, followed by upsampling and super-resolution through the 2D generation network to get the final result. The key idea is that the projected features naturally hold temporal consistency, but it is difficult to get high-quality results because of the limitation of the 3D volume resolution. Such a constraint also applies to scene completion methods like SPSG [4], which performs generation directly on a TSDF volume. On the other hand, in the field of single-image NVS, SynSin [24] and PixelSynth [17] use point cloud representation, which is also adopted by the recent works of cross-view generation Sat2Vid [10] and video translation WC-Vid2Vid [12] to encourage temporal consistency of the output video. The projection of point cloud is often finer than volumes, but the rendered frame usually has "holes", which also require a subsequent 2D network to make the completion, where the temporal consistency is hard to control especially when the "holes" are large. We see that either voxel grid or point cloud can obtain an initial multi-view consistency from a 3D explicit representation, however, such consistency could be easily destroyed by the 2D CNNs, which may still result in unsatisfactory video quality.

Implicit Scene Representations. The emergence of NeRF [13] allows scene information to be no longer stored explicitly like voxel grids or point clouds, but to map 3D coordinates (and view directions) to RGB/alpha values through a multi-layer perceptron (MLP), making it not limited by resolution. By introducing implicit representation to the differentiable rendering techniques, it can render high-quality photo-realistic images of an object or a scene from unseen points of view and can generate the video with very well temporal consistency.

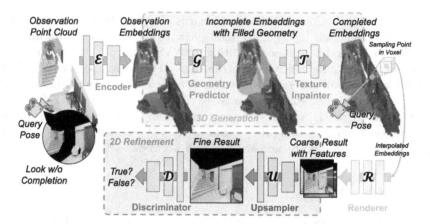

Fig. 2. Network architecture overview. Our pipeline generates a spatially consistent 3D representation for scene completion and novel view synthesis. The pipeline can be divided into four major steps: input feature encoding, generative scene completion (geometry & texture), differentiable rendering, and 2D upsampling and refinement.

However, NeRF optimization requires hundreds of images of the same scene or object. For different scenarios, different radiance fields are required to be trained individually. Such test-time optimization settings and the implicit representation make it inherently difficult for the generation. Thus, how to combine GAN techniques and implicit representations would be an attractive direction.

GRAF [20] is a pioneer work in the field of **generative NeRF**. It simply concatenates two global latent vectors representing the object shape and appearance respectively with the original input positions of MLP, which may not be powerful enough to handle complicated scenes. As NeRF's rendering speed is slow, it samples patches for the discriminator to speed-up training which may not work for a high-resolution generation. piGAN [2] further improves the visual quality by introducing StyleGAN-like [9] mapping and synthesis network with SIREN [21] activation, but is still far away from 2D GANs and the training and inference are slow. GIRAFFE [16] proposes a compositional NeRF followed by a 2D upsampler. The basic idea is similar to GRAF, but it further decomposes the 3D scene representation into several controllable local feature fields. It sacrifices the resolution of rendering in exchange for speed and then uses 2D CNN for super-resolution, which further impedes the multi-view consistency. In sum, direct generative NeRF training like GRAF and piGAN may suffer from low computation speed and is only acceptable at low resolution. Works like GIRAFFE using 2D CNNs can improve the visual quality but again have the problem of the multi-view consistency. More importantly, these works sample scenes from learned distributions, without conditioning on the given observations.

Hybrid Scene Representations. Based on the implicit radiance field, NSVF [11] further introduces an explicit representation of a sparse voxel grid to store the information of local neighborhoods, thus solving the problem of insufficient expression ability of MLP in NeRF. With the explicitly saved local scene embeddings, this hybrid approach can make the scene more scalable and no longer limited to the object level, and the generated results are more refined than NeRF. However, NSVF does not have a generative setting and still requires hundreds of images for test-time optimization. GSN [5] has similar ideas to NSVF and incorporates the GAN setting. It generates a 2D feature layout of a scene (vertical top view) from a latent code first, on which the rendering rays can thus extract features, followed by a general NeRF pipeline. Another work, pixelNeRF [25] can learn to predict a NeRF from a few posed images, allowing it to generate plausible novel views without test-time optimization. It assigns features from the input views to the sampling points via projection when rendering the target view, making the generation more applicable and image conditioned. However, such feature acquisition is naive and the points' spatial relationships are not well established, making the geometry difficult to learn, and further leading to blurry results.

3   Method

Pipeline Overview. A central goal of our work within the NVS setting is the completion of both the geometry and the texture of larger missing scene parts when observed from a novel query viewpoint. In contrast to 2D inpainting approaches, we aim to generate a completed 3D scene representation with texture information, which can be used to render videos with ensured spatial and temporal consistency. To this end, we propose the pipeline illustrated in Fig. 2 which takes as input an incomplete point cloud obtained by combining all known RGB-D input views. The input point cloud is encoded into a sparse voxel grid of 3D features, followed by two predictor networks that complete missing geometry by means of voxel occupancies and texture inpainting subsequently. Finally, for a given camera pose or trajectory, a single 2D image or video can be obtained via implicit differentiable rendering followed by a 2D upsampling and refinement module. We detail each component of the pipeline as follows.

Input and Output. The input to our method is a point cloud of a partially scanned scene and is reconstructed from given posed RGB-D images. The output is the synthesized image or video for a given query view or camera trajectory.

Point Cloud Encoder. Given a colored point cloud (P, C) with $P \in \mathbb{R}^{N \times 3}$ denoting the 3D positions of N points and $C \in \mathbb{R}^{N \times 3}$ representing their colors, we first employ a ResNet [7] module $\mathcal{E}$ based on a sparse voxel representation to encode them into M vertex embeddings $(V, F) = \mathcal{E}(P, C)$. Here the M embeddings $F \in \mathbb{R}^{M \times d}$ with d dimensional features correspond to all vertices of the occupied voxels, and $V \in \mathbb{R}^{M \times 3}$ records their coordinates.

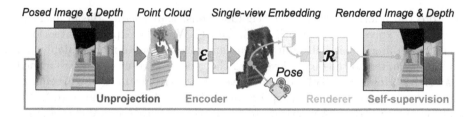

Fig. 3. The first step of training. We first train the encoder $\mathcal{E}$ and the renderer $\mathcal{R}$ jointly in a self-supervision way by forming them as a 2.5D-3D-2.5D autoencoder.

Geometry Predictor. In a second step we fill the geometry of the incomplete input scene. We adopt a geometry predictor $\mathcal{G}$, which yields the filled scene geometry $\widetilde{V} \in \mathbb{R}^{(M+m)\times 3}$ with m new occupied voxel vertices conditioned on the given V, i.e., $\widetilde{V} = \mathcal{G}(V)$. Please note that geometry filling is not directly based on the vertex coordinates but based on its associated occupied voxel grid. Here we omit the step of transformation between a voxel and its indices for simplicity. Also, F is not used in the geometry completion step. We implement the geometry predictor $\mathcal{G}$ as a sparse voxel U-Net [18] model with generative sparse convolutions [6] that can generate new coordinates.

Texture Inpainter. The newly added vertices from the previous step do not hold color embeddings yet and are required to be filled with the necessary texture information. We predict embeddings for the m vertices by composing a texture inpainter $\mathcal{T}$ which predicts the filled embeddings $\widehat{F} \in \mathbb{R}^{(M+m)\times d}$ by taking as input the intial voxel embeddings $F_0 = F \oplus 0^{m\times d} \in \mathbb{R}^{(M+m)\times d}$ which are formed by padding zeros on F for the m new vertices, here $\oplus$ denotes the concatenation operation. From the visibility information of the input we also indicate whether a vertex needs inpainting or not, so we have $v = 0^M \oplus 1^m \in \mathbb{R}^{(M+m)}$ and $\widehat{F} = \mathcal{T}(\widetilde{V}, F_0, v)$. We again employ a U-Net architecture with general sparse convolutions which only operate across the occupied locations.

Differentiable Renderer. In this step, we leverage the rendering pipeline of NSVF to generate a 2D image for the query view based on the explicit sparse voxel representation together with the implicit MLPs. We follow NSVF to get the feature of a sampling point inside one voxel (illustrated as a yellow cube in Fig. 2) by trilinearly interpolating the embeddings at its 8 vertices. In the original NSVF work, the features representing RGB and geometry are entangled in the same embedding, which is decoded to RGB and alpha value through the same MLP with two different tail branches. In contrast, we directly split the embeddings into two groups (1:3) in the feature dimension, and use two independent MLPs to decode the alpha value and RGB respectively, so as to make the embeddings disentangled. Another difference is that for each pixel, along its corresponding ray, we not only aggregate RGB values, but also compute a feature by assembling the embeddings based on the calculated alpha value. This feature will be used for subsequent refinement. In short, the renderer $\mathcal{R}$ takes as input the inpainted voxel embeddings $(\tilde{\boldsymbol{V}}, \tilde{\boldsymbol{F}})$, the query pose $\boldsymbol{Q}$ and outputs an image with features in half resolution, $i.e.$, $\tilde{\boldsymbol{I}} = \mathcal{R}(\tilde{\boldsymbol{V}}, \tilde{\boldsymbol{F}}, \boldsymbol{Q}) \in \mathbb{R}^{\frac{H}{2} \times \frac{W}{2} \times (3+d)}$.

Refinement Module. Inspired by Sat2Vid [10] we introduce a quality-enhancing refinement module after the rendering, which consists of a lightweight upsampler and a discriminator used during training. Generally, the paired discriminator improves the ability of generating visually plausible images through adversarial learning. The training with the discriminator $\mathcal{D}$ requires the images to be fully rendered. However, directly rendering the full-resolution image with NSVF is highly time-consuming. Additionally, the generated images tend to be blurry for larger voxel sizes. To overcome these challenges, our pipeline first renders a coarse half-resolution image and then uses an upsampler to double the resolution, $i.e.$, $\boldsymbol{I} = \mathcal{U}(\tilde{\boldsymbol{I}}) \in \mathbb{R}^{H \times W \times 3}$, which greatly shortens the rendering time. Please note that the upsampler is designed lightweight in order to not excessively destroy the temporal consistency obtained in rendering phase.

Network Training. The modules are trained step-by-step in four phases.
Step 1. We first train the encoder $\mathcal{E}$ jointly with the implicit renderer $\mathcal{R}$ (MLPs), which forms a structure similar to an autoencoder but in a 2.5D-3D-2.5D manner. This procedure is illustrated in Fig. 3. Different from NSVF which uses images to optimize the sparse voxel embeddings and the implicit renderer, we use an encoder to directly generate embeddings, and the parameters of the implicit renderer are also shared across scenes. Thus, a test-time optimization is not required here. The self-supervision loss is calculated on sampled rays during training instead of the entire image, using both RGB and depth values for training. To achieve good rendering quality, $\mathcal{E}$ here is required to have the ability to encode both geometry and texture information at the same time.
Step 2. Next we train the geometry predictor $\mathcal{G}$ separately. The decoder of $\mathcal{G}$ consists of several coarse-to-fine generative transposed convolution layers [6], where the binary cross-entropy loss is used for the occupancy prediction.
Step 3. Then we pre-train the texture inpainter $\mathcal{T}$. We use $\mathcal{E}$ trained on the first step to generate necessary paired incomplete and complete embeddings for training $\mathcal{T}$. During pre-training, the ground-truth geometry for the missing part is used, and the feature loss is only calculated in this area, $i.e.$, the orange part in the "incomplete embeddings with filled geometry" in Fig. 2.
Step 4. The final step is to train the 2D refinement module $\mathcal{U}$ and $\mathcal{D}$ and possibly refine $\mathcal{F}$ to adapt it to the predicted geometry in the full pipeline. We fix the parameters of the other modules during training. Here the losses are only calculated on the 2D part, $i.e.$, the RGB and depth losses for the half-resolution coarse results, as well as the general GAN losses for the upsampler and discriminator. There is no embedding supervision for $\mathcal{F}$ in this step.

4   Experiments

4.1   Configuration

Datasets. Experiments are conducted on both synthetic and real-world datasets.

Replica. Following GSN [5], we use Replica [22] for evaluation. The dataset contains 18 realistic synthetic scenes, and we generate the ground truth by the Habitat [19,23] renderer. Specifically, we first divide the 18 scenes into 48 disjoint sub-areas with similar space sizes, and randomly sample 300 camera poses that fit the human perspective for each sub-area. Then we collect the RGB and depth observations in 256×256 resolution for each camera pose. We finally select 15 scenes (42 sub-areas) for training and the rest 3 scenes (6 sub-areas) for validation. The generation is performed on a triplet of two input source views and one output query view. For each query view, the two source views are selected by satisfying: (1) no overlap; (2) less than 50% overlap with the query view; and (3) totally 65%–70% overlap with the query view, which means that the 30%–35% of the pixels in the query view need inpainting. For each sampled camera pose, we randomly select 3 triplets meeting the requirements, which results in ∼37k triplets for training and ∼5k for validation.

ARKitScenes. The real-world dataset, ARKitScenes [1], is one of the largest datasets for indoor scene understanding. The dataset consists of 5,047 captures of 1,661 unique scenes, with high-quality ground truth of registered RGB-D frames. Due to its huge amount of data, for each scene, we only select its longest capture and uniformly sample it with an interval of 10 frames (∼1 s). We adopt the same strategy described above in the Replica dataset to generate triplets but adjust the overlap rates of the query view with each individual source view and two total source views to 60% and 60%–80% respectively. Considering the temporal relationship between frames, the suitable source views are searched within a window of 9 frames centered by the query view for efficiency. We follow ARKitScenes' official train-val split and finally obtain ∼15k triplets for training and ∼2k triplets for validation. The image resolution is 256×192.

Implementation Details. Our framework is implemented in PyTorch and run on a single NVIDIA Tesla V100 GPU with 32 GB memory. The voxel size of all the scenes in our experiment is chosen to be 10 cm, and the embedding size is set as $d = 32$, (8 for alpha value and 24 for RGB). The point cloud encoder, geometry predictor, and texture inpainter are all implemented via the Minkowski Engine [3] and adopt their provided default network architectures with sparse voxel convolutions such as ResNet34, generative U-Net and the vanilla U-Net. The training takes around 6 days in total to go through all the training steps from scratch. The experiments are conducted on the validation sets of Replica and ARKitScenes. We not only compare the results of the target view, but also the short video generated based on the poses between the target view and the two source views. For Replica, the intermediate poses are linearly interpolated, and the ground-truth images are generated based on these poses using Habitat.

For ARKitScenes, we uniformly sample intermediate poses from the original scan sequence between the source views and the target view.

Metrics. For quantitative evaluation, we mainly follow pixelNeRF [25] and PixelSynth [17] to use SSIM as low-level metrics to measure the window differences between the ground truth and the prediction. The high-level measures such as FID [8] and LPIPS [28] (perceptual similarity) are also taken into account. For the commonly used PSNR, as explained also in PixelSynth [17] paper, it may not be a good metric in such an extrapolation task. Therefore, we report these numbers here for reference only. We also compare the generated geometry separately. Since the scene representations of each method are different, for convenience, we directly compare the rendered depth of the center frame by RMSE and $\delta^{1.25}$, which are commonly used in the field of depth estimation. The purpose of the video result comparison is to better evaluate the temporal consistency of the textures across frames, so we only report the image quality metrics for videos. The FID score of videos is also excluded because it uses globally pooled features for calculation, where both the temporal and spatial relationships are lost. Additionally, we incorporated a user study of 30 people to evaluate the quality of the generated videos, which consists of 20 multiple-choice questions (half from Replica and half from ARKitScenes). For each question, the participant was asked to select a single generated video from four options (baselines and ours) that they considered more visually plausible.

4.2   Baseline Comparison

Baselines. We select PixelSynth [17], pixelNeRF [25] and SPSG [4] as baselines. Since they are not original proposed for our task and the depths of the input views are required in the task, we make necessary adaptions on these baselines for a fair comparison, which are briefly described in the following.

PixelSynth [17] outpaints the missing parts in a chosen support view, which is then back-projected to the 3D space followed by the point cloud rendering technology with refinement. Here our target view is used as its support view. The depth of the input view in the original paper is predicted by a network, but in our task we directly provide the depth observations as input. The depth estimation module still exists in the adapted version for predicting the depth of the generated part. The rendering and refinement procedures follows the original. *pixelNeRF* [25] does not require depth inputs. The feature acquisition is done by projecting the sampling points on the rays of the target view to the source views and extracting encoded features of the corresponding pixel locations. Here we utilize the depths of the source views and back-project the features to the 3D space, where the k-NN search is afterward performed to more precisely obtain the features for the sampling points. The following procedures such as ray feature aggregation and rendering are the same as the original. Due to the additional searching phase, the runtime of this adapted version is slightly longer. *SPSG* [4] generates 3D models with both geometry and color from incomplete RGB-D scans, and raycasting is applied to render the target frames in specific

456   Z. Li et al.

Table 1. Quantitative baseline comparison on the generated query view for Replica [22] dataset. Numbers of each entry: full frame/unobserved regions.

Method	PSNR↑	SSIM↑	FID↓	LPIPS↓	Dep. RMSE↓	Dep. $\delta^{1.25}$ ↑
SPSG [4]	29.12/29.36	0.603/0.452	175.2/207.6	0.518/0.804	0.914/1.243	0.790/0.534
pixelNeRF [25]	28.78/28.90	0.650/0.583	257.3/215.8	0.610/0.592	0.713/0.793	0.560/0.491
PixelSynth [17]	31.28/30.44	0.633/0.587	80.58/**177.4**	**0.315**/0.576	**0.382**/0.534	0.869/0.664
Ours	29.65/29.42	**0.706/0.600**	**74.84**/182.6	0.332/**0.575**	0.397/**0.456**	**0.879/0.817**

Table 2. Quantitative baseline comparison on the generated query view for ARKitScenes [1] dataset. Numbers of each entry: full frame/unobserved regions.

Method	PSNR↑	SSIM↑	FID↓	LPIPS↓	Dep. RMSE↓	Dep. $\delta^{1.25}$ ↑
SPSG [4]	28.41/28.34	0.406/0.293	175.43/188.00	0.522/0.875	0.680/0.918	0.763/0.661
pixelNeRF [25]	28.12/28.10	0.504/**0.442**	262.65/189.01	0.724/1.065	0.546/0.541	0.589/0.553
PixelSynth [17]	28.42/28.40	0.385/0.260	**101.96/149.72**	0.496/0.920	0.445/0.696	0.779/0.397
Ours	28.38/28.10	**0.551**/0.414	141.40/160.57	**0.427/0.733**	**0.394/0.489**	**0.927/0.883**

poses. We turn the source views into the TSDF through volumetric fusion as the required input for SPSG, followed by the same original procedure. SPSG avoids memory issues by tessellating the scene into smaller chunks ($64 \times 64 \times 128$) for training, however, with a small voxel size (*e.g.*, 2 cm), the unobserved area is typically larger than such a chuck, thus unable to generate the desired data to train the baseline. To avoid the problem, the voxel scale is still similarly set to 10 cm to be consistent with our approach, and the scene is processed and completed as a whole for a fair comparison.

Quantitative Results. We show quantitative baseline comparisons for the *center frame (query view)* on Replica and ARKitScenes in Table 1 and Table 2, while the quantitative results in terms of *videos* are reported in Table 3.

Texture. On the low-level metric SSIM, our method outperforms all other baselines on both datasets and all the area levels (the unobserved part, the whole query view, and the video) except the unobserved part of the ARKitScenes dataset. This may indicate that although pixelNeRF generates blurry results in Fig. 5, the pixel color distribution of the local windows in the unobserved part could still be close to GT. In terms of perceptual similarity, our method outperforms all other baselines on both datasets and all the area levels except the full query view of the Replica dataset. Since PixelSynth's LPIPS score is similar to ours on the generated part, it could be inferred that PixelSynth may better perform on the observed part on the synthetic dataset. This may be because, in the case of a synthetic dataset, the geometry is rather simple and does not have many occlusions. Details could be recovered better through the point cloud projection they use in this case, rather than a voxel representation. For the FID score, ours performs on par with PixelSynth on the Replica dataset, but slightly worse on ARKitScenes, which means that PixelSynth may have a higher capacity

Table 3. Quantitative baseline comparison on the generated videos.

Method	Replica [22]				ARKitScenes [1]			
	PSNR↑	SSIM↑	LPIPS↓	Human↑	PSNR↑	SSIM↑	LPIPS↓	Human↑
SPSG [4]	27.91	0.521	0.694	1.3%	28.27	0.427	0.518	2.3%
pixelNeRF [25]	28.95	0.670	0.624	1.7%	28.96	0.524	0.727	0.0%
PixelSynth [17]	31.42	0.605	0.323	16.7%	28.95	0.393	0.463	43.7%
Ours	29.83	**0.728**	**0.301**	**80.3%**	28.66	**0.589**	**0.402**	**54.0%**

Input   GT   SPSG [4]   pixelNeRF [25] PixelSynth [17]   **Ours**

Fig. 4. Qualitative baseline comparison on the Replica [22] dataset. We show comparisons to the state of the art on a variety of examples. Our method generates more realistic videos with better temporal consistency and fewer artifacts. (See supplementary material for an animated version of the figure.)

to produce richer textures in the real-world dataset. However, since FID only represents the distribution between the overall features of the two sets of images, it generally cannot measure the spatial similarity of two images.

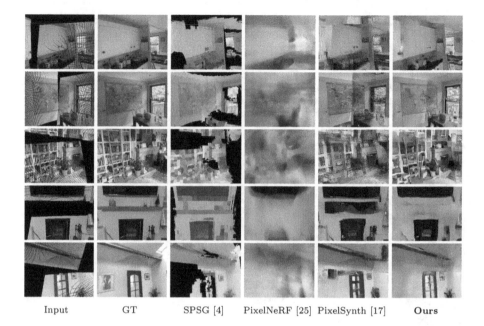

| Input | GT | SPSG [4] | PixelNeRF [25] | PixelSynth [17] | **Ours** |

Fig. 5. Qualitative baseline comparison on the ARKitScenes [1] dataset. We show comparisons to the state of the art on a variety of examples. Our method generates more realistic videos with better temporal consistency and fewer artifacts. (See supplementary material for an animated version of the figure.)

Geometry. In terms of geometric similarity, measured by depth RMSE and $\delta^{1.25}$, our method can predict more accurate depth in most cases. This could be the main reason that leads to a better 3D consistency of our generated videos and better performance in high-level metrics of videos in Table 4. On the other hand, the depths in the generated part of all methods are almost worse than that of the whole image, however, the drop of PixelSynth is relatively large. It can be inferred that the geometry predicted by PixelSynth of the unobserved part does not match the input very well. This is also reflected in the video results in Fig. 4 and Fig. 5, where its generated parts are disjointed from time to time.

User Study. We list the user study results in Table 3 ("Human" column). Compared to the baselines, we found that ours achieved higher average approval rates for the generated video quality. On the ARKitScenes dataset, PixelSynth scores second place with a much smaller gap. We believe this is mainly due to that the geometry generated by PixelSynth on ARKitScenes is generally better than that of Replica, and the gap between the generated part and the observation is much smaller, leading to a better 3D-consistency in the synthesized video.

Qualitative Results. Figure 4 and Fig. 5 visualize the generated videos on selected samples of the Replica and ARKitScenes datasets. It can be seen that the completion results by SPSG still have many empty spaces. We believe this is because SPSG is originally proposed for filling small holes on denser input and has difficulties adapting to our sparser input with large gaps. pixelNeRF's results are much more blurry than other columns. We attribute this to the low resolutions of features used by pixelNeRF for image rendering. PixelSynth is the most competitive baseline method for predicting relatively complete and sharp results. However, it tends to generate results that are less plausible in the visualization. For instance, on the second sample of Fig. 4, its prediction on the corner resembles the entrance of a new room, which has low geometrical plausibility with the neighboring observed walls. We believe that the absence of geometry information in the texture completion is the main reason for the implausible results of PixelSynth. Compared with the baseline methods, our method can better fill the gaps in scenes and produces more plausible outputs for both texture and geometry. For example, on the third sample of Fig. 5 with complicated texture, our method fills the gap with a ladder and bookshelf which are consistent with the contextual texture. On the final sample of Fig. 5 with high variance in the spatial structure, our method predicts better geometry of the wall and roof. However, compared with PixelSynth, our method comes short of the sharpness in the generated frames. This is mainly limited by the voxel embeddings used in our pipeline which is bound by the voxel resolution ($10\,cm^3$ size). Also, on many validation triplets, there are objects in the query view that are totally unobservable in the source views, *e.g.*, the door in the second sample of Fig. 4 and the shelf in the fourth sample of Fig. 5. In these empty regions, both methods tend to fill contents that are different from the ground truth. Hence, neither method can achieve an excellent score on the high-level metrics.

4.3   Ablation Study

To better evaluate the effectiveness of the individual components of our method, we also conduct an ablation study on Replica [22] by incrementally adding components to our basic framework (**B**). More specifically, we focus on the following three components: (**E**) the encoder used for processing the point cloud input; (**D**) the disentangled MLPs for RGB and alpha value; and (**R**) the 2D refinement module. In the basic framework, we directly voxelize the point cloud with RGB according to the defined voxel size as the input of the 3D generation pipeline, which outputs completed embeddings. All the ablations use the same geometry completion network which does not take color information as input.

Table 4 shows quantitative evaluation results of the ablation study and Fig. 6 illustrates the center frame generated by different ablative pipelines. It can be observed that frames generated by the full pipeline show the best performance in high-level measurements (FID and LPIPS). Comparing between results of B and B+E, we find that encoding the point cloud into the voxel embeddings via an encoder facilitates the texture generation and preserves more details, which is also consistent with the SSIM increase and LPIPS decrease. We still found a deviation between the generated colors and the GT. We believe this is due to the entangled embeddings having to take into account both geometry and texture

Table 4. Quantitative ablation study on the generated query view. The ablations are: **B**: basic framework; **+E**: point cloud encoder; **+D**: disentangled MLPs; **+R**: refinement module. Numbers of each entry: full frame/*unobserved* regions.

Method	PSNR↑	SSIM↑	FID↓	LPIPS↓
B	30.06/29.38	0.730/0.636	128.5/210.2	0.418/0.613
B+E	29.67/29.39	**0.746/0.644**	135.0/210.7	0.395/0.628
B+E+D	29.58/29.37	0.729/0.603	124.2/216.3	0.404/0.672
B+E+D+R (Ours)	29.65/29.42	0.706/0.600	**74.84/182.6**	**0.332/0.575**

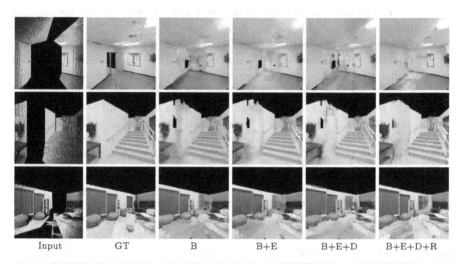

Input   GT   B   B+E   B+E+D   B+E+D+R

Fig. 6. Qualitative ablation study on the generated query view. We present exemplary qualitative results for various ablations of our method. The rendered videos visibly contain more details and achieve higher levels realism with our full method.

information. After integrating D, it can be seen that the recovered colors are further improved and closer to the ground truth. For example, the pink wall in the third sample of Fig. 6 is recovered to be pink after adding D. This indicates the disentangled prediction of alpha values and RGB values can improve the modeling of color patterns in scenes. Finally, the 2D refinement module (R) increases the details in the results as well as fills the small holes caused by the geometric incompleteness. One notable point shown in the table is that the SSIM decreases after adding disentangled MLPs (D) and the refinement module (R) even though they can produce sharper image results than B and B+E. Based on our experiments, we find that SSIM scores drop less when the image is blurred while the general structure is maintained. Likewise, sharpening an image does not improve SSIM scores much. We believe the pipeline scarifies structural similarity scores in favor of improving the visual quality.

4.4 Limitations

Although our method can efficiently complete a scene, generate a high-quality video from novel views, and outperform existing methods, there are three major limitations: (1) The completion of geometry can sometimes produce holes and incomplete structures, which result in visible artifacts. (2) The introduction of the 2D upsampling step may slightly break the strict 3D consistency obtained from rendering and frame-to-frame flickering is possible for individual pixels though this is rather unlikely due to local context consistency. (3) The voxel resolution and thus the voxel size is limited in our pipeline. A too-large voxel size may blur the result while a too-small value could lead to a huge number of voxels resulting in a slow rendering speed and excessive memory usage.

5 Conclusion

We proposed a scalable novel view synthesis pipeline that completes larger amounts of incomplete scene data with plausible information. Our method builds upon a sparse grid-based feature representation that jointly encodes local geometry and texture information. In contrast to NeRF or NSVF which requires test-time optimization, our pipeline directly encodes the few-shot input views in a feed-forward manner. The scene embeddings with geometry and texture for the missing area are generated during completion in the 3D domain. Compared with other approaches, our method effectively combines full 3D modeling which is crucial for spatial and temporal consistency with generative modeling.

Acknowledgments. This work was supported by JSPS Postdoctoral Fellowships for Research in Japan (Strategic Program) and JSPS KAKENHI Grant Number JP20H04205. Z. Li was supported by the Swiss Data Science Center Fellowship program. Z. Cui was affiliated with the State Key Lab of CAD & CG, Zhejiang University. M. R. Oswald was supported by a FIFA research grant.

References

1. Baruch, G., et al.: ARKitscenes - a diverse real-world dataset for 3D indoor scene understanding using mobile RGB-d data. In: Thirty-fifth Conference on Neural Information Processing Systems Datasets and Benchmarks Track (Round 1) (2021)
2. Chan, E., Monteiro, M., Kellnhofer, P., Wu, J., Wetzstein, G.: pi-GAN: periodic implicit generative adversarial networks for 3D-aware image synthesis. In: arXiv (2020)
3. Choy, C., Gwak, J., Savarese, S.: 4D spatio-temporal convnets: Minkowski convolutional neural networks. In: Proceedings of the IEEE Conference on Computer Vision and Pattern Recognition, pp. 3075–3084 (2019)
4. Dai, A., Siddiqui, Y., Thies, J., Valentin, J., Nießner, M.: SPSG: self-supervised photometric scene generation from RGB-D scans. In: Proceedings of the Computer Vision and Pattern Recognition (CVPR). IEEE (2021)
5. DeVries, T., Bautista, M.A., Srivastava, N., Taylor, G.W., Susskind, J.M.: Unconstrained scene generation with locally conditioned radiance fields. ICCV (2021)

6. Gwak, J.Y., Choy, C., Savarese, S.: Generative sparse detection networks for 3D single-shot object detection. In: Vedaldi, A., Bischof, H., Brox, T., Frahm, J.-M. (eds.) ECCV 2020. LNCS, vol. 12349, pp. 297–313. Springer, Cham (2020). https://doi.org/10.1007/978-3-030-58548-8_18

7. He, K., Zhang, X., Ren, S., Sun, J.: Deep residual learning for image recognition. In: Proceedings of the IEEE Conference on Computer Vision and Pattern Recognition (CVPR), June 2016

8. Heusel, M., Ramsauer, H., Unterthiner, T., Nessler, B., Hochreiter, S.: GANs trained by a two time-scale update rule converge to a local Nash equilibrium. In: Guyon, I., et al. (eds.) Advances in Neural Information Processing Systems, vol. 30. Curran Associates, Inc. (2017)

9. Karras, T., Laine, S., Aila, T.: A style-based generator architecture for generative adversarial networks. In: Proceedings of the IEEE/CVF Conference on Computer Vision and Pattern Recognition (CVPR), June 2019

10. Li, Z., Li, Z., Cui, Z., Qin, R., Pollefeys, M., Oswald, M.R.: Sat2Vid: street-view panoramic video synthesis from a single satellite image. In: Proceedings of the IEEE/CVF International Conference on Computer Vision (ICCV), pp. 12436–12445, October 2021

11. Liu, L., Gu, J., Lin, K.Z., Chua, T.S., Theobalt, C.: Neural sparse voxel fields. NeurIPS (2020)

12. Mallya, A., Wang, T.-C., Sapra, K., Liu, M.-Y.: World-consistent video-to-video synthesis. In: Vedaldi, A., Bischof, H., Brox, T., Frahm, J.-M. (eds.) ECCV 2020. LNCS, vol. 12353, pp. 359–378. Springer, Cham (2020). https://doi.org/10.1007/978-3-030-58598-3_22

13. Mildenhall, B., Srinivasan, P.P., Tancik, M., Barron, J.T., Ramamoorthi, R., Ng, R.: NeRF: representing scenes as neural radiance fields for view synthesis. In: The European Conference on Computer Vision (ECCV) (2020)

14. Nguyen-Phuoc, T., Li, C., Theis, L., Richardt, C., Yang, Y.L.: HoloGAN: unsupervised learning of 3D representations from natural images. In: Proceedings of the IEEE/CVF International Conference on Computer Vision (ICCV), October 2019

15. Nguyen-Phuoc, T., Richardt, C., Mai, L., Yang, Y.L., Mitra, N.: BlockGAN: learning 3D object-aware scene representations from unlabelled images. In: Advances in Neural Information Processing Systems, vol. 33, November 2020

16. Niemeyer, M., Geiger, A.: Giraffe: representing scenes as compositional generative neural feature fields. In: Proceedings of the IEEE Conference on Computer Vision and Pattern Recognition (CVPR) (2021)

17. Rockwell, C., Fouhey, D.F., Johnson, J.: PixelSynth: generating a 3D-consistent experience from a single image. In: ICCV (2021)

18. Ronneberger, O., Fischer, P., Brox, T.: U-Net: convolutional networks for biomedical image segmentation. In: Navab, N., Hornegger, J., Wells, W.M., Frangi, A.F. (eds.) MICCAI 2015. LNCS, vol. 9351, pp. 234–241. Springer, Cham (2015). https://doi.org/10.1007/978-3-319-24574-4_28

19. Savva, M., et al.: Habitat: a platform for embodied AI research. In: Proceedings of the IEEE/CVF International Conference on Computer Vision (ICCV) (2019)

20. Schwarz, K., Liao, Y., Niemeyer, M., Geiger, A.: GRAF: generative radiance fields for 3D-aware image synthesis. In: Advances in Neural Information Processing Systems (NeurIPS) (2020)

21. Sitzmann, V., Martel, J.N., Bergman, A.W., Lindell, D.B., Wetzstein, G.: Implicit neural representations with periodic activation functions. In: Advances in Neural Information Processing Systems (NeurIPS) (2020)

22. Straub, J., et al.: The Replica dataset: A digital replica of indoor spaces. arXiv preprint arXiv:1906.05797 (2019)
23. Szot, A., et al.: Habitat 2.0: training home assistants to rearrange their habitat. In: Advances in Neural Information Processing Systems (NeurIPS) (2021)
24. Wiles, O., Gkioxari, G., Szeliski, R., Johnson, J.: SynSin: end-to-end view synthesis from a single image. In: Proceedings of the IEEE/CVF Conference on Computer Vision and Pattern Recognition (CVPR), June 2020
25. Yu, A., Ye, V., Tancik, M., Kanazawa, A.: pixelNeRF: neural radiance fields from one or few images. In: Proceedings of the IEEE/CVF Conference on Computer Vision and Pattern Recognition (CVPR), pp. 4578–4587, June 2021
26. Yu, J., Lin, Z., Yang, J., Shen, X., Lu, X., Huang, T.S.: Free-form image inpainting with gated convolution. arXiv preprint arXiv:1806.03589 (2018)
27. Yu, J., Lin, Z., Yang, J., Shen, X., Lu, X., Huang, T.S.: Generative image inpainting with contextual attention. arXiv preprint arXiv:1801.07892 (2018)
28. Zhang, R., Isola, P., Efros, A.A., Shechtman, E., Wang, O.: The unreasonable effectiveness of deep features as a perceptual metric. In: CVPR (2018)

SketchSampler: Sketch-Based 3D Reconstruction via View-Dependent Depth Sampling

Chenjian Gao[1], Qian Yu[1(✉)], Lu Sheng[1], Yi-Zhe Song[2], and Dong Xu[3]

[1] School of Software, Beihang University, Beijing, China
{gaochenjian,qianyu,lsheng}@buaa.edu.cn
[2] SketchX, CVSSP, University of Surrey, Guildford, England
y.song@surrey.ac.uk
[3] Department of Computer Science, The University of Hong Kong, Hong Kong, China

Abstract. Reconstructing a 3D shape based on a single sketch image is challenging due to the large domain gap between a sparse, irregular sketch and a regular, dense 3D shape. Existing works try to employ the global feature extracted from sketch to directly predict the 3D coordinates, but they usually suffer from losing fine details that are not faithful to the input sketch. Through analyzing the 3D-to-2D projection process, we notice that the density map that characterizes the distribution of 2D point clouds (i.e., the probability of points projected at each location of the projection plane) can be used as a proxy to facilitate the reconstruction process. To this end, we first translate a sketch via an image translation network to a more informative 2D representation that can be used to generate a density map. Next, a 3D point cloud is reconstructed via a two-stage probabilistic sampling process: first recovering the 2D points (i.e., the x and y coordinates) by sampling the density map; and then predicting the depth (i.e., the z coordinate) by sampling the depth values at the ray determined by each 2D point. Extensive experiments are conducted, and both quantitative and qualitative results show that our proposed approach significantly outperforms other baseline methods.

1 Introduction

Sketching is an intuitive approach for humans to express their ideas, and it has been adopted for 3D modeling for decades. With the rapid development of deep learning and virtual reality (VR) techniques, sketch-based 3D modeling has attracted increasing attention from both academia and industry [4,12,13,15,29], showing great potential in designing, animation, and entertainment.

In recent years we have witnessed great progress in sketch-based 3D modeling. Motivated by the success of image-based single-view 3D reconstruction (SVR),

---

Supplementary Information The online version contains supplementary material available at https://doi.org/10.1007/978-3-031-19769-7_27.

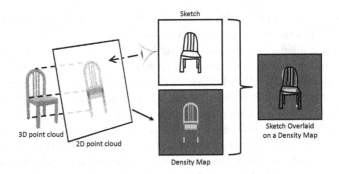

Fig. 1. The motivation of our work. We can see: 1) a 2D point cloud can be generated by projecting a 3D shape onto an image plane. On the projection plane, some locations have more than one 2D point with different depth values. 2) The distribution of 2D points can be characterized by a density map, where the value at each location indicates the probability of points projected at that location. 'Red' color indicates higher density. 3) The density map is spatially rough-aligned with the sketch. (Color figure online)

most sketch-based 3D modeling approaches follow a well-known pipeline of SVR [6,22], which firstly encodes a sketch into a feature vector with a convolution neural network (CNN), and then utilizes multilayer perception (MLP) based decoders to generate a fixed number of 3D coordinates that define the point cloud of a 3D shape. Considering that sketch is usually sparse and ambiguous, global feature is a reasonable sketch representation [31], as it summarizes the sketch in a coarse level (e.g., semantic category and its conceptual shape). However, it is hard for a model to reconstruct a 3D shape with fine details from a global feature due to the huge domain gap between a sketch and a 3D shape.

Figure 1 illustrates a 2D point cloud projected from a 3D shape. Note that on the projection plane, there could exist more than one 2D points with different depth values at the same location because of occlusions. The distribution of a 2D point cloud can be characterized by a density map, indicating the probability of points projected at each location. In other words, if we have a density map, we can infer the corresponding 2D point cloud. It is interesting to see that a sketch is spatially rough-aligned with the density map. Considering that both sketch and density map are 2D images, their domain gap is supposed to be smaller than that between sketch and 3D shape. This motivates us to introduce the density map as the proxy to facilitate sketch-to-3D reconstruction. Namely, given an input sketch, the reconstruction model first generates a density map to recover a 2D point cloud (i.e., x and y coordinates) and then predicts the depth value for each 2D point (i.e., z coordinate). *From a probabilistic view*, this reconstruction process can be interpreted as predicting the joint distribution of x, y, z coordinates from a 3D shape, which defines a two-stage sampling process. Specifically it *firstly* samples x, y coordinates from the distribution of $P(X, Y|I)$ generated from the sketch I, and *secondly* samples z coordinate for each (x, y) location from the distribution of $P(Z|x, y, I)$. Note that the distribution of $P(X, Y|I)$ is a 2D point cloud to be generated from a sketch.

However, considering the sparsity and ambiguity characteristics of sketch, there is also a domain gap between a sketch and a density map. As displayed in Fig. 1, the visible and occluded object surfaces, and the vacancy between surfaces can all be shown as blank in a sketch. We thus adopt an image translation model [10] to complete the missing information while preserving the spatial information in a sketch before predicting the density map.

In this work, we present a new method for sketch-based SVR, which consists of two components: a sketch translator and a point cloud generator. The sketch translator adopts a CNN-based encoder-decoder network, where the encoder network extracts features from the input sketch and the decoder network infers target 3D information and outputs a more informative 2D representation. Based on the output of the sketch translator, our point cloud generator aims to reconstruct a point cloud of the corresponding 3D shape. It first predicts the density map which can be used as guidance to recover 2D point clouds, and then samples along a ray determined by each 2D projected point to predict depth values, where the point with farther depth values means it is occluded by the point with nearer depth values.

It is worth noting that a sketch may exhibit different levels of deformation and abstraction. Here we focus on sketches with reliable shape and fine-grained details, i.e., sketches with significant deformation or only expressing conceptual ideas are not considered in this work. To demonstrate the effectiveness of our proposed model, we train and test it on a newly rendered dataset, *Synthetic-LineDrawing*. The contributions of this work are three-fold:

- First, we present a novel method for sketch-based single-view 3D reconstruction, in which a 3D shape is recovered in two easier but indispensable steps, sketch translation and point cloud generation;
- Second, we formulate the point generation process as a two-stage probabilistic sampling process, where the density map is introduced as a guidance. Besides, an image translation model is used for sketch translation to preserve the spatial information in a sketch and to further reduce domain gap;
- Third, the proposed model can reconstruct a 3D shape faithful to the sketched object. Its effectiveness has been demonstrated through extensive experiments on both synthetic and hand-drawn sketch datasets.

2   Related Works

2.1   Deep Sketch-Based 3D Modeling

Sketch-based modeling is a problem that has been studied for a long time. The earlier methods predicted local geometric properties from hand-crafted rules and then inferred the 3D shape from the geometric properties [32,33]. In recent years, some deep learning based methods have been proposed for sketch-based 3D modeling. Wang *et al.* introduced a method to reconstruct 3D shapes based on retrieval [23]. The work [22] proposed to generate point clouds from a single hand-drawn image. They enhanced the PSGN [6] method with a viewpoint estimation module. To alleviate the deformation of sketches, they proposed a sketch standardization module to alleviate the deformation problem of sketches. The work [32] discussed the additional challenges of line drawings in comparison with images in 3D reconstruction. In [33], sketches from two viewpoints were used as the input to perform 3D reconstruction, and they collected two datasets, ProSketch and AmateurSketch. Sketch2model [31] alleviated the ambiguity in sketch modeling by decoupling view code and shape code. Sketch2mesh [9] used an encoder/decoder architecture to learn a latent representation of an input sketch and refined it by matching the external contours of the reconstructed 3D mesh to the sketch during the inference process. While achieving good performance, this approach is time-consuming. Most deep sketch modeling methods encode a sketch as a latent code and then apply a decoder to convert the latent code to a 3D shape. However, these approaches fail to preserve spatial details in a sketch.

2.2   3D Reconstruction from Single RGB Image

3D reconstruction is a problem that has been widely studied in computer vision. Reconstructing a 3D shape from a single image is an ill-posed problem that requires strong prior knowledge. In recent years, with the development of deep learning, neural networks can be used to extract useful features for 3D reconstruction [6,19,27,28]. The early works focus on reconstructing 3D shapes represented by regular voxels [3,24,26]. MarrNet [26] and 3DensiNet [24] resorted to intermediate representations (i.e., 2.5D sketches and density heat-map) to facilitate reconstruction. In this work, we introduce the density map as a proxy, which reflects the probability of points projected at each location of the image plane. Unlike MarrNet and 3DensiNet that directly reconstruct 3D shapes based on the intermediate representations, we use the density map as the guidance for 2D points sampling, from which the depth value will be further predicted.

However, voxel reconstruction is inefficient because the information of 3D shape is distributed only on the surface of an object. Therefore, some methods [6,25] attempted to recover the surface information of 3D shapes, such as point clouds or meshes. Nevertheless, the surface of a 3D shape is sparse and irregular, posing great challenges for shape recovery. Some works [7,25] use the coarse-to-fine and feature pooling strategies to alleviate this problem. In addition to reconstructing a 3D shape based on explicit representations, recent approaches [1,16,30] explore 3D reconstruction based on implicit surface learning. But these methods require post-processing to obtain explicit 3D shapes.

3   Methodology

As shown in Fig. 2, our sketch-based modeling framework mainly consists of two components: a sketch translator and a point cloud generator. Given an input sketch I, the **sketch translator** first translates it to a feature map F. Next, the

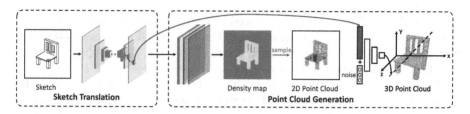

Fig. 2. Overview of our proposed method. It consists of sketch translation and point cloud generation. During sketch translation, missing information such as surface and occlusions are implicitly compensated. The resultant feature maps are used to predict the density map. When generating 3D points, a 2D point cloud is first sampled based on the density map, followed by sampling the depth values at each given 2D point.

point cloud generator produces a point cloud S based on the given feature map F. When recovering the point cloud, a 2D density map is first predicted, from which 2D points are sampled; then the depth of each 2D point is predicted by using the proposed conditional depth generator. Note that in line with [9,18], we adopt the commonly used "viewer-centered" setting [21], in which we assume the image space and the 3D space are aligned.

3.1 Sketch Translation

The goal of the sketch translator is to fully exploit the spatial information in a sketch and generate suitable features for 3D shape prediction. It is non-trivial because there is a large information discrepancy between a sketch and a 3D shape: 1) a sketch is sparse and mainly preserves structural framework of a corresponding 3D shape. The object surface, occluded object surface, and vacancy between surfaces can all be shown as blank in a sketch. 2) most depth information in a 3D shape is also lost in a sketch image. Therefore, the sketch translator aims to complement the missing information. For example, inferring whether a blank area belongs to an object, or which pixels are on the same surface.

Specifically, we adopt an encoder-decoder based CNN network for sketch translation. Firstly, an encoder network is used to extract features from the input sketch with multiple down-sampling blocks. This is to increase the receptive field of the neurons to acquire an overview of the input sketch. A decoder network consisting of multiple upsampling blocks is then used to gradually infer the information of the 3D shape with increased spatial resolution. Instead of using the last feature map F^n for point cloud generation, we use a similar idea [14] that leverages the feature maps at all scales by upsampling the feature map at each individual scale F^i to the size of F^n and concatenating them together to produce the final feature F.

After sketch translation, the response of the feature map F is much denser than the input sketch while the spatial alignment and resolution are roughly maintained. It will facilitate the prediction of point clouds of with fine details, which will be explained below.

3.2   Point Cloud Generation

The point cloud generator aims to recover the point cloud of the corresponding 3D shape S from the translated feature maps F. To utilize the spatial information of the input sketch, we decompose the point could generation process into two steps: 1) predicting the 2D point cloud which is the projection of the 3D point cloud into the image plane; 2) inferring the depth of each 2D point.

For generating a 2D point cloud, the point cloud generator predicts the joint distribution of the coordinates from the projected points $p(X, Y|I)$, where X, Y are random variables corresponding to the x, y axis respectively. Sampling from $P(X, Y|I)$ will generate the 2D point cloud. Figure 1 shows an example of projecting a 3D shape into the image plane and its corresponding density map. The probabilistic density at each location varies because it depends on how many surfaces are being passed by the ray centered at this location.

After a 2D point cloud is generated, the point cloud generator predicts the depth distribution of each point $p(Z_i|x_i, y_i, I)$, where x_i, y_i is the location of the i-th point in the image plane. Sampling from $P(Z_i|x_i, y_i, I)$ gives the depth of each point. Combining x, y coordinates from density map sampling and z coordinate from depth sampling, the overall 3D point cloud can be generated.

From a probabilistic view, this process actually models the shape of a 3D object as a joint distribution of x, y, z coordinates. Our generation process assumes a factorization process over projection and conditional independency between different locations for depth prediction, i.e., $P(X, Y, Z|I) = P(X, Y|I)P(Z|X, Y, I)$. The first term and the second respectively correspond to the process of generating a 2D point cloud and the process of predicting depth given a 2D point cloud and a sketch.

2D Point Cloud Generation. As all valid locations must lay inside the image, we firstly model the distribution of projected points in pixel coordinates. The image coordinates can be seen as quantizing the x, y location into $\mathcal{W} \times \mathcal{H}$ bins, where $P_{u,v} = P(X = u, Y = v)$ is the probability of a projected point inside the (u, v)-th bin.

We use a mask prediction head to directly predict the density map $M \in \mathcal{R}^{\mathcal{W} \times \mathcal{H}}$, where $M_{v,u} = P_{u,v}$. It takes the translated sketch feature F as the input, and resizes the feature map to the size of $\mathcal{W} \times \mathcal{H}$ by using bilinear interpolation. The interpolated feature map is then passed to three convolutional layers for density prediction. The hyper parameters $\mathcal{W}$ and $\mathcal{H}$ control the resolution of the point clouds.

To generate a 2D point cloud, we can see $P(X, Y)$ as a multinomial distribution over $\mathcal{W} \times \mathcal{H}$ locations. We firstly sample a specific number of locations with the probabilities defined by the density map M, and then use the column and row indices u, v as x, y coordinates. We normalize the coordinate to the range of $[-1, 1]^1$ to produce the coordinate in the image plane (x^I, y^I). It then converts the points to world coordinates by using the camera parameters. As we use the

---

[1] $x = \frac{2u}{\mathcal{W}-1} - 1$, $y = \frac{2v}{\mathcal{W}-1} - 1$, where $u = 0, 1, ..., \mathcal{W} - 1$, $v = 0, 1, ..., \mathcal{H} - 1$.

orthogonal projection model to produce the rendered sketches, the x, y coordinates are linearly mapped from that of the 3D point clouds. That is the x, y coordinates of a point in the raw 3D point cloud, and (x, y) can be computed as $(x^I/s, y^I/s)$, where s is a preset parameter of the projection model. Similar mapping functions can be drawn for other projection models.

Conditional Depth Estimation. After producing the 2D coordinates (x, y) of the 3D points, the next step is to predict their z coordinates. Given its x, y location, we assume estimating the depth for each individual point to be independent, so we predict the conditional depth distribution $P(Z_i|x_i, y_i)$ separately for each 2D location. Given a x, y location, the depth distribution $P(Z_i|x_i, y_i)$ can be multimodal and the number of modes tends to be varied, as there may be one or multiple points from different surfaces sharing the same 2D location. It is hard to explicitly define the probabilistic function of $P(Z_i|x_i, y_i)$.

Inspired by the Generative Adversarial Networks [8,10,17], we use an implicit approach and adopt the generator network design to model $P(Z_i|x_i, y_i)$. It takes a noise variable $N \in R^d$ and the local feature $f_{x,y}$ as input, and predict a scalar of depth z, where N is sampled from the uniform distribution $\mathcal{U}(0, 1)$ and $f_{x,y}$ is obtained by extracting from the feature map F at the corresponding location. Note that the depth generator can output different depth values given the same feature and different noise variables. It takes a multi-layer perceptron (MLP) as the backbone and its parameters are shared at all (x_i, y_i) locations.

For inference, we randomly sample a noise vector $\mathbf{n}_i$ by following the uniform distribution for each point (x_i, y_i) in the predicted 2D point cloud, and then predict the corresponding z_i. Putting the 2D location (x_i, y_i) and depth prediction z_i together will generate the final point cloud $\mathcal{S}$. Note that the sampled random noise $\mathbf{n}_i$ controls which mode the predicted depth z_i falls in if the corresponding $P(Z_i|x_i, y_i, I)$ is multimodal. Together with 2D point cloud sampling, the two-stage process can be seen as sampling from the joint distribution that defines the coordinates of a 3D shape. The detailed process of point cloud generation is listed in Algorithm 1. Note that our proposed method is compatible with both orthogonal projection and perspective projection. It can be controlled by the 'invproj' function in Algorithm 1.

3.3   Loss Function

A key role in our proposed approach is the density map. Fortunately, we can freely produce the ground-truth density map from a 3D shape by a customized renderer, i.e., counting the number of points that occurred when projecting a ray from a 3D point onto an image plane followed by normalization. To provide supervision information for the learning process of the density map, we use the L1 loss as a constraint, as shown in Eq. (1).

$$L_D = \sum_{x_i, y_i} \|\hat{p}(x_i, y_i) - p(x_i, y_i)\|_1. \tag{1}$$

Algorithm 1. Point Cloud Generation Process

Input: total number of points N, the predicted density map M, feature map F, and depth generator T.

Output: the reconstructed point clouds of the sketch S.

1: Let $S = \emptyset$
2: **while** $|S| \leq N$ **do**
3:    sample a location from the multinomial distribution defined by M, i.e. $(u, v) \sim Mult(x, y; M)$.
4:    convert u, v to the image plane coordinate x^I, y^I
5:    sample the noise vector $\mathbf{n} \sim \mathcal{U}(0, 1)$.
6:    inference the depth at u, v: $z_c = T(\mathbf{n}, F_{uv})$
7:    convert (x^I, y^I, z_c) to the world coordinate: $(x, y, z) = \text{invproj}(x^I, y^I, z_c)$
8:    $S = S \cup \{(x, y, z)\}$
9: **end while**
10: **return** S

To provide supervision information for the learning process of the conditional generator, we constrain the distance between the output point cloud and the ground-truth point cloud. We use the Chamfer distance as the loss function during the training process. Given two point clouds $S, \hat{S} \subseteq \mathcal{R}^3$, the Chamfer distance is defined as Eq. (2). The final loss function is shown in Eq. (3), and λ_1 and λ_2 are the weights of L_{CD} and L_D, respectively.

$$L_{CD} = \frac{1}{|S|} \sum_{p \in S} \min_{q \in \hat{S}} \|p - q\|_2^2 + \frac{1}{|\hat{S}|} \sum_{q \in \hat{S}} \min_{p \in S} \|q - p\|_2^2 \tag{2}$$

$$L = \lambda_1 L_{CD} + \lambda_2 L_D, \tag{3}$$

As shown in Fig. 2, during the training process, the feature maps from the encoder-decoder network are fed into two paths: 1) the convolutional layers to predict the density map; 2) the fully-connected layers to predict the depth value. Correspondingly, the L1 loss in Eq. (1) and the Chamfer loss in Eq. (2) are computed, and the gradients from these two losses will be separately backpropagated along two different paths back to the encoder-decoder network.

4   Experiment

In this section, we first introduce the datasets and evaluation metrics used in our experiments and provide implementation details (Sect. 4.1). We then compare our proposed method with the baseline methods on the Synthetic-LineDrawing dataset (Sect. 4.2) and three hand-drawn sketch datasets (Sect. 4.3). Ablation studies are conducted to show the effectiveness of individual modules (Sect. 4.4). We also evaluate our model on unseen classes (Sect. 4.4).

4.1   Experimental Setup and Evaluation Metrics

Synthetic-LineDrawing Dataset. Publicly available large-scale paired sketch-3D datasets are rare. So we contribute a new dataset, the Synthetic-LineDrawing dataset, by rendering sketch images from 3D models of the ShapeNet dataset [2]. Specifically, we use a subset of ShapeNet-core, which consists of around 50k 3D models spanning 13 classes. We select 5 random views for each object to render, resulting in 218,915 sketch images and corresponding 43,783 3D objects. We follow the conventional train/test splits as in [3], i.e., 4/5 and 1/5 for training and test, respectively.

Except the synthetic sketch dataset, we also conduct experiments on three hand-drawn sketch datasets:

- **ShapeNet-Sketch** [31] is a dataset consisting of $1,300$ free-hand sketches and their corresponding ground-truth 3D models, belonging to the same 13 categories of the ShapeNet dataset. All sketch images are drawn by volunteers with different drawing skills.
- **AmateurSketch** [33] contains $3,015$ sketch images of 500 chair models and $1,665$ sketch images of 555 lamp models. Each 3D model is drawn from 3 different viewpoints.
- **ProSketch-3DChair** [33] contains $1,500$ professional sketches of 500 chair models, and each 3D model is drawn from 3 different viewpoints: front, side and 45 degree.

Implementation Details. We use a CNN-based encoder-decoder network [10] as the sketch translator. For density map prediction, we use ReLU followed by normalization to ensure the sum of the values from all spatial positions of the density map is 1. When estimating the depth information, an MLP with residual connections is used. λ_1 and λ_2 are set to be 1 and 10^4, respectively. The model is trained for 30 epochs with an initial learning rate of 10^{-3}. Adam optimizer [11] is used for optimization. The ground-truth density map is obtained by counting the number of points that occurred when a ray projects from a 3D point onto the image plane.

Evaluation Metrics. We employ four evaluation metrics to measure the reconstruction performance on the above four datasets: Chamfer Distance (CD), Earth Mover's Distance (EMD), voxel Intersection over Union (Vox-IoU), and Fréchet Point cloud Distance (FPD). **CD** is a widely used as the evaluation metric for 3D generation and reconstruction tasks. Similar to CD, **EMD** is also used to evaluate the similarity between two point clouds. But it is more sensitive to the local details and density distribution. **FPD** is similar to FID, which calculates the 2-Wasserstein distance between the real and fake samples in the feature space extracted by a pre-trained PointNet. **Voxel-IoU** measures the coverage percentage of two volumetric models. Further details of these evaluation metrics are explained in Supplementary.

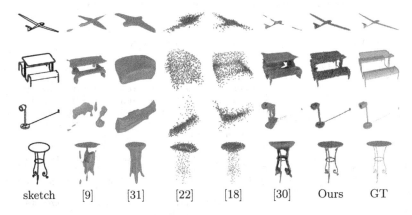

sketch   [9]   [31]   [22]   [18]   [30]   Ours   GT

Fig. 3. Reconstruction results on the Synthetic-LineDrawing Dataset.

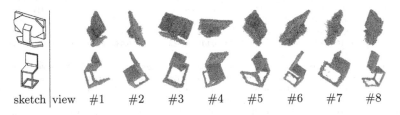

sketch | view   #1   #2   #3   #4   #5   #6   #7   #8

Fig. 4. Our reconstructed 3D shapes that are rendered under different viewpoints. The results show that the generated 3D point clouds are consistently accurate and faithful under all viewpoints.

4.2   Results on Synthetic-LineDrawing Dataset

Baseline Methods. We first compare our approach with three state-of-the-art methods for *sketch-based* single-view 3D reconstruction (SVR):

Sketch2Mesh [9]: Given an input sketch, this method also utilizes an encoder-decoder network to produce a 3D mesh estimate. It learns a compact feature representation and recovers the 3D shape by minimizing the 2D Chamfer distance between the 3D shape's projected contour and the input sketch.

Sketch2Model [31]: This method is proposed to reconstruct a 3D shape represented by a mesh. It employs an encoder-decoder network. To address the ambiguity problem of sketch, it introduces an additional encoder-decoder to decompose a sketch image to the view and shape space. During the inference process, each 3D shape is reconstructed based on an input sketch and the estimated viewpoint.

Sketch2Point [22]: This method is proposed to reconstruct a 3D point cloud from a sketch. It is built on PSGN [6], in which the key component is a standardization module used to handle sketches with various drawing styles.

Table 1. Results on the Synthetic-LineDrawing Dataset.

	Airplane	Bench	Cabinet	Car	Chair	Display	Lamp	Speaker	Rifle	Sofa	Table	Phone	Boat	Mean
Categories														
Chamfer Distance(↓) ×10⁻³														
Sketch2Mesh [9]	0.910	4.533	2.735	1.417	3.002	3.119	9.054	4.685	0.846	2.633	2.732	2.005	2.524	3.092
Sketch2Model [31]	1.814	6.404	3.010	2.720	3.997	6.976	6.617	5.579	1.495	5.721	4.632	2.723	2.755	4.188
Sketch2Point [22]	2.229	14.747	3.239	1.610	3.883	7.047	6.663	6.611	4.056	7.132	5.772	4.392	1.255	5.280
PCDNet [18]	0.571	1.151	1.480	1.002	1.664	1.389	3.104	2.120	0.621	1.416	1.647	1.207	1.051	1.417
DISN [30]	0.845	3.573	1.839	1.340	3.181	2.640	9.203	3.340	2.000	1.797	3.371	2.080	2.215	2.879
Ours	**0.389**	**0.729**	**1.153**	**0.866**	**1.033**	**0.959**	**1.907**	**1.561**	**0.428**	**1.050**	**0.949**	**0.957**	**0.780**	**0.982**
Earth Mover's Distance(↓) ×10⁻²														
Sketch2Mesh [9]	3.914	5.732	5.441	4.645	6.032	5.301	11.188	7.005	5.113	5.297	5.328	4.133	4.804	5.687
Sketch2Model [31]	5.587	7.460	5.852	5.662	6.867	7.436	10.615	7.109	6.297	7.035	7.206	4.679	6.529	6.795
Sketch2Point [22]	6.893	13.907	7.102	5.875	9.913	10.799	15.212	9.736	10.556	10.143	9.263	8.652	5.880	9.533
PCDNet [18]	7.114	8.723	9.745	7.420	10.948	9.493	16.054	10.465	7.464	10.121	10.450	7.880	7.255	9.472
DISN [30]	3.823	6.234	**4.911**	4.569	7.136	5.893	10.469	6.063	5.513	4.706	6.990	4.053	5.037	5.800
Ours	**3.178**	**3.978**	5.032	**4.240**	**5.293**	**4.553**	**6.722**	**5.690**	**3.436**	**4.662**	**4.421**	**3.762**	**3.969**	**4.534**
Fréchet Point Cloud Distance(↓) ×10														
Sketch2Mesh [9]	2.030	8.253	3.346	1.147	3.214	3.287	10.771	2.608	1.577	3.436	1.624	3.647	7.565	4.039
Sketch2Model [31]	1.524	13.900	5.546	1.121	3.220	9.622	2.887	6.364	3.494	20.001	4.393	3.188	5.490	6.212
Sketch2Point [22]	11.415	22.056	6.466	6.973	25.730	12.369	6.903	11.431	27.448	13.391	21.120	14.425	3.322	14.081
PCDNet [18]	0.991	1.117	0.760	1.107	0.846	0.892	1.657	1.177	0.916	0.957	1.050	0.760	1.313	1.042
DISN [30]	1.838	5.097	1.037	**0.285**	2.752	1.662	12.211	1.294	3.867	1.729	3.143	2.734	2.595	3.096
Ours	**0.516**	**0.542**	**0.358**	0.427	**0.454**	**0.519**	**1.082**	**0.734**	**0.635**	**0.561**	**0.633**	**0.347**	**0.729**	**0.580**
Voxel-IOU(↑)														
Sketch2Mesh [9]	0.693	0.506	0.383	0.515	0.442	0.469	0.355	0.280	0.691	0.418	0.493	0.596	0.553	0.492
Sketch2Model [31]	0.499	0.220	0.338	0.341	0.308	0.250	0.320	0.229	0.511	0.245	0.269	0.535	0.422	0.345
Sketch2Point [22]	0.427	0.174	0.172	0.335	0.204	0.231	0.209	0.125	0.263	0.184	0.137	0.293	0.514	0.251
PCDNet [18]	0.634	0.506	0.367	0.502	0.386	0.478	0.359	0.307	0.603	0.395	0.439	0.572	0.557	0.470
DISN [30]	0.698	0.464	0.407	0.521	0.397	0.462	0.332	0.325	0.683	0.437	0.426	0.627	0.547	0.487
Ours	**0.736**	**0.619**	**0.467**	**0.563**	**0.535**	**0.577**	**0.510**	**0.412**	**0.713**	**0.500**	**0.597**	**0.654**	**0.638**	**0.578**

We also compare our method with two *image-based* SVR methods:

PCDnet [18]: This method can generate a 3D point cloud of arbitrary size based on a single image. It extracts the global shape feature(i.e., a feature vector) from a sketch and predicts the 3D point cloud via a deformation network.

DISN [30]: This work proposes a signed distance fields (SDF) predictor, where both global and local features are used for prediction. It can produce a 3D shape with fine details since it exploits local features sampled from image feature maps. However, this approach works slowly during the inference process.

We follow their original works of Sketch2Model and Sketch2Mesh to train an individual model for each category. See more details in Supplementary.

Qualitative Results. The results of different methods are illustrated in Fig. 3. For point cloud based methods, **Sketch2Point** and **PCDNet** perform badly where the generated 3D shape can even fall into an incorrect category, e.g., the produced 3D point cloud of a plane is more like a rifle. For those whose class labels are correct, the shapes are still considerably different from the input sketches. These observations indicate that special designs are required for accurate and generalizable sketch-based SVR models. Note that the point cloud produced by PCDNet still exhibits more details than Sketch2Point, which demonstrates the benefit of using local features. **Sketch2Mesh** and **Sketch2Model**

Table 2. Results on the hand-drawn sketch datasets. *"ShapeNet-S." is short for ShapeNet-Sketch and "ProSketch" is short for ProSketch-3DChair.

	ShapeNet-S.	ProSketch	AmateurSketch	ShapeNet-S	ProSketch	AmateurSketch
	Chamfer Distance($\downarrow$) $\times 10^{-3}$			Fréchet Point Cloud Distance($\downarrow$) $\times 10$		
Sketch2Mesh [9]	12.324	6.317	14.739	14.997	6.089	14.785
Sketch2Model [31]	10.355	5.628	11.288	14.449	5.464	14.600
Sketch2Point [22]	11.176	8.019	10.547	35.864	38.991	24.794
Ours	**9.515**	**3.868**	**9.657**	**11.665**	**4.799**	**12.727**
	Earth Mover's Distance($\downarrow$) $\times 10^{-2}$			Voxel-IOU($\uparrow$)		
Sketch2Mesh [9]	9.947	7.921	13.164	0.195	0.283	0.217
Sketch2Model [31]	**9.256**	7.432	10.506	0.205	0.244	0.199
Sketch2Point [22]	13.443	13.179	16.506	0.163	0.185	0.166
Ours	9.626	**6.963**	**9.994**	**0.244**	**0.294**	**0.219**

are mesh based methods. As they are trained for each class separately, there is no confusion between different categories. The surface of Sketch2Mesh is often disconnected on sketch with thin strokes. Regularized by view constraint, the continuity of Sketch2Model becomes better, but it tends to generate over-smoothed meshes. **DISN** is a SDF based method and it could generate almost accurate 3D reconstruction results. However, limited by the resolution of 3D grids and the use of SDF, small object parts and thin lines in a sketch are missing in the reconstruction results.

Our method outperforms all competitors. The reconstructed point clouds are correct in terms of category labels and also exhibit notable level of details, even though our model is trained only once for all categories (i.e., class-agnostic). The overall layout of the point clouds are well aligned with the input sketches. Even small parts, e.g., the electric wire and the leg of the table, are depicted faithfully in the 3D point clouds. It suggests that our two-stage strategy is better in generalizing across different categories and capturing fine-details of sketch. The Reconstruction results from different views are also provided in Fig. 4.

Quantitative Results. The quantitative results are shown in Table 1 and we observe a similar trend with the qualitative results. Although the performance ranking of these models varies under different evaluation metrics, our method consistently performs the best in terms of all metrics. Moreover, our model even performs the best over almost all categories (except the *cabinet* class based on the EMD metric and the *car* class based on the FPD metric). It suggests that our reconstructed 3D shape captures both global structure and the local details. Notably, all methods perform considerably worse on the *lamp* class than other classes, where the sketches contain many thin strokes with fine structures(see Fig. 3). Nevertheless, our methods still achieves reasonable results. A possible explanation is that the proposed 2D point cloud generation strategy ensures the points can be sampled even from very thin strokes, with which the 3D point cloud could be successfully generated.

sketch   [9]   [31]   [22]   Ours   GT   sketch   [9]   [31]   [22]   Ours   GT

Fig. 5. Reconstruction results of different methods on hand-drawn sketches.

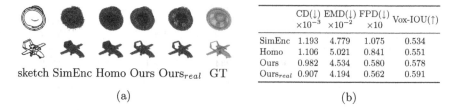

sketch SimEnc Homo Ours Ours$_{real}$ GT

	CD($\downarrow$) $\times 10^{-3}$	EMD($\downarrow$) $\times 10^{-2}$	FPD($\downarrow$) $\times 10$	Vox-IOU($\uparrow$)
SimEnc	1.193	4.779	1.075	0.534
Homo	1.106	5.021	0.841	0.551
Ours	0.982	4.534	0.580	0.578
Ours$_{real}$	0.907	4.194	0.562	0.591

(a)   (b)

Fig. 6. Reconstruction results of our method and different variants. Ours$_{real}$ is a variant method which uses the ground-truth density maps.

4.3 Results on Hand-Drawn Sketches

In this section, we test the generalization ability of our model on three hand-drawn sketch datasets ShapeNet-Sketch [31], AmateurSketch [33], and ProSketch-3DChair [33] without finetuning or domain adaptation. The three baseline methods which are specifically proposed for this task are used for comparison, including Sketch2Mesh, Sketch2Model, and Sketch2Point. The quantitative and qualitative results are shown in Table 2 and Fig. 5. Unlike the other baseline methods where the produced 3D shape and the input sketch are aligned only at the semantic level, the 3D shapes generated by our method are much more faithful to the input sketch. However, as shown in Fig. 5, a sketch can be inaccurate. For example, the second chair's leg seems to have shifted. Our method tends to be faithful to the sketch rather than generating an object by simply following categorical shape prior. We argue that there is often a trade-off between faithfulness and rationality. In this work, we focus on faithfulness.

4.4 Ablation Studies

Effectiveness of Sketch Translator. We use an encoder-decoder structure which is widely used for image translation to complement the information of an input sketch. Thanks to the translation module, we can produce a more informative feature map, so that a better 3D prediction can be achieved. To validate this design, we propose a comparative baseline model Simple-Encoder ('SimEnc'). When compared to our model, SimEnc removes the sketch decoder module. For a fair comparison, we increase the number of parameters of the depth estimator in SimEnc accordingly so that the amount of parameters in SimEnc is roughly the same as ours. As shown in Fig. 6, we can see the performance of

Fig. 7. Our reconstruction results on unseen classes.

SimEnc drops significantly. We suppose that the feature map produced by our sketch translation module is more informative. Using the encoder network, we cannot preserve much information to help predict reasonable 3D shape.

Effectiveness of Density-Guided Sampler. During point generation, our method uses a density map as guidance for sampling x, y coordinates. Taking into account that the inhomogeneous distribution of 3D information in 2D space can make modal transformation more effective, we compare our sampler with an alternative one denoted as homo-sampler, which treats 3D information as homogeneous distribution in two dimensions. Specifically, the homo-sampler only distinguishes between foreground and background, in which the same number of points is sampled at each position in the foreground. We use the points with $p(x, y)$ predicted by our method greater than 0 as the foreground for the homo-sampler method. The experimental results are shown in Fig. 6 (see the second row 'Homo'). A homogeneous sampling strategy does not allow the sampler to perceive the difference in the distribution of $p(Z_{xy} \mid X, Y)$ at different locations. We can see that the reconstruction performance when using a homo-sampler is worse than ours, as shown in Fig. 6.

Generalization on Unseen Classes. We evaluate the proposed method on unseen classes to verify its generalization ability. We randomly choose some sketches from the Sketchy database [20] (Fig. 7(Left)) and TU-Berlin sketch dataset [5] (Fig. 7(Right)) for testing. We can see that although our model is trained on rigid object classes, it also performs well on non-rigid objects.

5   Conclusion

In this work, we have proposed a new method for sketch-based single-view 3D reconstruction. During sketch translation, an informative feature map is derived from an input sketch via an image translation model, which is then used to predict a density map for point cloud generation. The point cloud generation process is implemented by two-stage sampling strategy: with the guidance of the density map, the x and y coordinates are first recovered; and then based on the conditions of x and y coordinates and the input sketch the z coordinate is further predicted by sampling. Experimental results have demonstrated that our proposed method can significantly outperform the baseline methods.

Acknowledgement. This work is supported by the National Natural Science Foundation of China (No. 62002012 and No. 62132001) and Key Research and Development Program of Guangdong Province, China (No. 2019B010154003).

References

1. Bian, W., Wang, Z., Li, K., Prisacariu, V.A.: Ray-ONet: efficient 3D reconstruction from a single RGB image. In: British Machine Vision Conference (BMVC) (2021)
2. Chang, A.X., et al.: ShapeNet: an information-rich 3D model repository. arXiv preprint arXiv:1512.03012 (2015)
3. Choy, C.B., Xu, D., Gwak, J.Y., Chen, K., Savarese, S.: 3D-R2N2: a unified approach for single and multi-view 3D object reconstruction. In: Leibe, B., Matas, J., Sebe, N., Welling, M. (eds.) ECCV 2016. LNCS, vol. 9912, pp. 628–644. Springer, Cham (2016). https://doi.org/10.1007/978-3-319-46484-8_38
4. Delanoy, J., Aubry, M., Isola, P., Efros, A.A., Bousseau, A.: 3D sketching using multi-view deep volumetric prediction. In: Proceedings of the ACM on Computer Graphics and Interactive Techniques(PACMCGIT) **1**(1), 1–22 (2018)
5. Eitz, M., Hays, J., Alexa, M.: How do humans sketch objects? ACM TOG **31**(4), 44:1–44:10 (2012)
6. Fan, H., Su, H., Guibas, L.J.: A point set generation network for 3D object reconstruction from a single image. In: CVPR (2017)
7. Gkioxari, G., Malik, J., Johnson, J.: Mesh R-CNN. In: ICCV (2019)
8. Goodfellow, I., et al.: Generative adversarial nets. In: NeurIPS (2014)
9. Guillard, B., Remelli, E., Yvernay, P., Fua, P.: Sketch2Mesh: reconstructing and editing 3D shapes from sketches. In: ICCV (2021)
10. Isola, P., Zhu, J.Y., Zhou, T., Efros, A.A.: Image-to-image translation with conditional adversarial networks. In: CVPR (2017)
11. Kingma, D.P., Ba, J.: Adam: a method for stochastic optimization (2015)
12. Li, C., Pan, H., Liu, Y., Tong, X., Sheffer, A., Wang, W.: BendSketch: modeling freeform surfaces through 2D sketching. ACM TOG **36**(4), 1–14 (2017)
13. Li, C., Pan, H., Liu, Y., Tong, X., Sheffer, A., Wang, W.: Robust flow-guided neural prediction for sketch-based freeform surface modeling. ACM TOG **37**(6), 1–12 (2018)
14. Lin, T.Y., Dollár, P., Girshick, R., He, K., Hariharan, B., Belongie, S.: Feature pyramid networks for object detection. In: CVPR (2017)
15. Lun, Z., Gadelha, M., Kalogerakis, E., Maji, S., Wang, R.: 3D shape reconstruction from sketches via multi-view convolutional networks. In: 3DV (2017)
16. Mescheder, L., Oechsle, M., Niemeyer, M., Nowozin, S., Geiger, A.: Occupancy networks: learning 3D reconstruction in function space. In: CVPR (2019)
17. Mirza, M., Osindero, S.: Conditional generative adversarial nets. arXiv preprint arXiv:1411.1784 (2014)
18. Nguyen, A.D., Choi, S., Kim, W., Lee, S.: GraphX-convolution for point cloud deformation in 2D-to-3D conversion. In: ICCV (2019)
19. Popov, S., Bauszat, P., Ferrari, V.: CoReNet: coherent 3D scene reconstruction from a single RGB image. In: Vedaldi, A., Bischof, H., Brox, T., Frahm, J.-M. (eds.) ECCV 2020. LNCS, vol. 12347, pp. 366–383. Springer, Cham (2020). https://doi.org/10.1007/978-3-030-58536-5_22
20. Sangkloy, P., Burnell, N., Ham, C., Hays, J.: The sketchy database: learning to retrieve badly drawn bunnies. ACM TOG **35**(4), 1–12 (2016)
21. Shin, D., Fowlkes, C.C., Hoiem, D.: Pixels, voxels, and views: a study of shape representations for single view 3D object shape prediction. In: CVPR (2018)
22. Wang, J., Lin, J., Yu, Q., Liu, R., Chen, Y., Yu, S.X.: 3D shape reconstruction from free-hand sketches. arXiv preprint arXiv:2006.09694 (2020)

23. Wang, L., Qian, C., Wang, J., Fang, Y.: Unsupervised learning of 3D model reconstruction from hand-drawn sketches. In: ACM MM (2018)
24. Wang, M., Wang, L., Fang, Y.: 3DensiNet: a robust neural network architecture towards 3D volumetric object prediction from 2D image. In: ACM MM (2017)
25. Wang, N., Zhang, Y., Li, Z., Fu, Y., Liu, W., Jiang, Y.G.: Pixel2Mesh: generating 3D mesh models from single RGB images. In: ECCV (2018)
26. Wu, J., Wang, Y., Xue, T., Sun, X., Freeman, B., Tenenbaum, J.: MarrNet: 3D shape reconstruction via 2.5D sketches. In: NeurIPS (2017)
27. Xie, H., Yao, H., Sun, X., Zhou, S., Zhang, S.: Pix2Vox: Context-aware 3D reconstruction from single and multi-view images. In: ICCV (2019)
28. Xie, H., Yao, H., Zhang, S., Zhou, S., Sun, W.: Pix2vox++: multi-scale context-aware 3D object reconstruction from single and multiple images. IJCV **128**(12), 2919–2935 (2020). https://doi.org/10.1007/s11263-020-01347-6
29. Xu, B., Chang, W., Sheffer, A., Bousseau, A., McCrae, J., Singh, K.: True2Form: 3D curve networks from 2D sketches via selective regularization. ACM TOG **33**(4), 1–13 (2014)
30. Xu, Q., Wang, W., Ceylan, D., Mech, R., Neumann, U.: DISN: deep implicit surface network for high-quality single-view 3D reconstruction. In: NeurIPS (2019)
31. Zhang, S.H., Guo, Y.C., Gu, Q.W.: Sketch2Model: view-aware 3D modeling from single free-hand sketches. In: CVPR (2021)
32. Zhong, Y., Gryaditskaya, Y., Zhang, H., Song, Y.Z.: Deep sketch-based modeling: tips and tricks. In: 3DV (2020)
33. Zhong, Y., Qi, Y., Gryaditskaya, Y., Zhang, H., Song, Y.Z.: Towards practical sketch-based 3d shape generation: the role of professional sketches. IEEE Trans. Circ. Syst. Video Technol. (T-CSVT) **31**(9), 3518–3528 (2020)

LocalBins: Improving Depth Estimation by Learning Local Distributions

Shariq Farooq Bhat[1]([✉])(iD), Ibraheem Alhashim[2](iD), and Peter Wonka[1](iD)

[1] KAUST, Thuwal, Saudi Arabia
shariq.bhat@kaust.edu.sa

[2] National Center for Artificial Intelligence (NCAI), Saudi Data and Artificial Intelligence Authority (SDAIA), Riyadh, Kingdom of Saudi Arabia

Abstract. We propose a novel architecture for depth estimation from a single image. The architecture itself is based on the popular encoder-decoder architecture that is frequently used as a starting point for all dense regression tasks. We build on AdaBins which estimates a global distribution of depth values for the input image and evolve the architecture in two ways. First, instead of predicting global depth distributions, we predict depth distributions of local neighborhoods at every pixel. Second, instead of predicting depth distributions only towards the end of the decoder, we involve all layers of the decoder. We call this new architecture LocalBins. Our results demonstrate a clear improvement over the state-of-the-art in all metrics on the NYU-Depth V2 dataset. Code and pretrained models will be made publicly available (https://github.com/shariqfarooq123/LocalBins).

Keywords: Single image depth estimation · Encoder-decoder architecture · Deep learning · Dense regression · Histogram prediction

1 Introduction

In this paper, we propose a new architecture for learning to estimate depth values given a single input image. In this line of work there are two main approaches that have been followed recently. Combining to train on multiple datasets at once while factoring out scale, e.g. [12,28] and training on a single dataset with consistent scale, e.g. [2,4,8,9,13,15,16,19,20,34,35]. Our approach falls in the second category. There are multiple published competing architectures, with AdaBins [4] currently being the most successful architecture on datasets such as NYU-Depth V2 [29]. Our newly proposed architecture, called *LocalBins* aims to improve upon this work.

The main idea of AdaBins is to predict adaptive bins that estimate one "global" depth distribution per image. This prediction works both as auxiliary supervision of depth estimation, but also directly influences the depth prediction.

Supplementary Information The online version contains supplementary material available at https://doi.org/10.1007/978-3-031-19769-7_28.

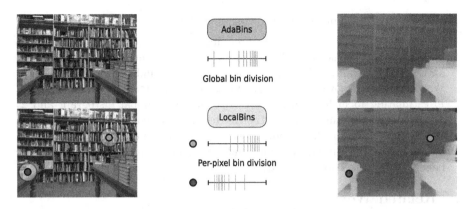

Fig. 1. Illustration of global adaptive bins vs local adaptive bins. While AdaBins predicts a depth distribution for a complete image, LocalBins predicts a depth distribution for the neighborhood of each pixel.

We initially formulated two objectives in evolving AdaBins. First, we wanted to see if predicting local distributions around each pixel can improve upon predicting one global distribution for the complete input image. Second, we wanted to design the architecture such that depth distribution supervision can be injected earlier in the network, preferably in a multi-scale fashion. AdaBins needs a special architecture design to work. Estimating global adaptive bins needs a transformer at "high resolution". Estimation of bins is done close to the output layer and most of the work is delegated to a specialized module based on a transformer. Even though this improves performance significantly, this offload of work may prevent earlier layers to fully exploit the "distribution supervision" to learn better representations. We call this the 'late injection problem' in our arguments. Any attempts to estimate global adaptive bins earlier in the network (e.g. near the bottleneck) or without a transformer leads to unstable training - divergence or convergence to a suboptimal point.

We realize these ideas in the following manner. To perform local predictions of depth distributions, we propose to use bin estimation at every pixel, and impose regularization on bin predictions via a query and response training scheme. Our proposed module is regularized to predict the depth distributions within the randomly selected bounding boxes within the image.

To perform multi-scale predictions of depth distributions, we let the network predict local depth distributions in a gradual step-wise manner throughout the decoder. Starting with a small N_{seed} number of bins, at the bottleneck each bin is subsequently split into two at every decoder layer, i.e., the ith layer of the decoder estimates $2^i N_{seed}$ bins at every pixel position. Together with locality, this coarse-to-fine construction lets us avoid unstable training and simultaneously solves the late injection problem.

Our proposed LocalBins module is lightweight (adding only ~1M params) and can be used in conjunction with any encoder-decoder network.

To summarize, we make the following contributions:

- We propose a new architecture for single image depth estimation that improves upon the state-of-the-art in all metrics on the NYU-Depth V2 [29] dataset. Models and code will be made publicly available.
- We propose two novel architecture ideas to single image depth estimation: 1) estimating local histograms instead of a single global histogram and 2) estimating histograms in a multi-scale fashion to benefit from distribution supervision earlier in the pipeline. Even beyond depth estimation, we are not aware of existing similar concepts and we believe that these ideas could be beneficial beyond depth estimation.

2   Related Work

There are multiple categories of depth estimation methods. The first category are unsupervised methods [5,10–12,22,32,33,40,41]. These methods do not use ground truth depth data, but use self-supervision generally by some form of 3D reconstruction to learn depth values. These methods typically use videos or stereo videos as input. The second category of methods learn depth estimation from multiple datasets [23,27,28] jointly. Combining multiple datasets requires considering the different depth scales of the scenes. Therefore, methods that train on multiple datasets are generally not comparable to methods that train on a single dataset, because the test protocol is different. The third category of algorithms are domain transfer methods [1,3,7,31,38]. These techniques assume the availability of ground truth data in one domain during training, but the images in the target domain do not have ground truth depth (or only a few of them have ground truth depth [39]). The fourth category are depth estimation methods that learn a depth estimation network for each dataset separately [2,4,8,9,13,15,16,19,20,34,35]. Our method belongs to this category of supervised monocular depth estimation. These methods formulate the task as the regression of a depth map from a single RGB input image. The current dominant architecture follows an encoder-decoder network. Most high performing methods apply such architecture with some variations on the process of extracting of relevant feature maps during encoding and the fusion of these features with the intermediate maps produced during the decoding stage. Finally, we mention two very recent arXiv submissions that are concurrent to our work for the sake of completeness. The first is GLP-depth [17] which proposes a hierarchical transformer encoder that captures global features and a simple decoder that considers the local context. The second method [37] employs a neural window fully-connected Conditional Random Fields (CRFs) module for the decoder and a vision transformer for the encoder.

3   Methodology

3.1   Background

AdaBins [4] divides the depth interval (d_{min}, d_{max}) into bins. This bin-division is global (one proposed bin-division per image) and adaptive (varies from image

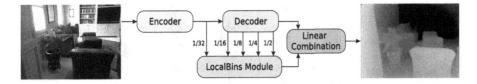

Fig. 2. Architecture overview. LocalBins module contains pixel-wise operations and estimate local neighborhood bin density for each pixel location.

to image), and reflects the global depth distribution for the input image. Each bin can have a different size and the predicted bin centers are closer to each other near more frequently occurring depth values. AdaBins employs an encoder-decoder architecture followed by a transformer based module to predict the adaptive bins. The final depth estimation is obtained by predicting the pixel-wise probability distribution over the bins and computing an expectation over the predicted global bin centers. This can also be seen as expressing the final depth value as a linear combination of bin centers. The reader is referred to [4] for more details.

We build upon the basic idea of AdaBins to estimate depth distributions, but we change the architecture design to incorporate two novel ideas. First, instead of predicting a single global depth bin-division, our architecture estimates a bin-division at every pixel, reflecting the depth distribution in the local neighborhood. Thus, the bin-divisions not only can vary from image to image (adaptiveness) but also from pixel to pixel (locality). Second, we do not use a transformer as a subsequent separate architecture block but integrate the depth prediction more tightly in the decoder. We utilize all the layers of the convolutional decoder to gradually refine the bin-division proposed for each pixel. The details of our architecture are described in the next section.

3.2 Architecture

Our architecture has two major components (see Fig. 2): 1) a standard encoder-decoder block and 2) our proposed LocalBins module.

Encoder-Decoder. We use the same encoder-decoder architecture as AdaBins to facilitate a fair comparison (EfficientNet-B5 with skip connections).

LocalBins Module. The LocalBins module uses the bottleneck features and the decoder features from the encoder-decoder block to estimate the local distribution of depth values at every pixel. As in AdaBins, the estimated distributions are encoded as the adaptive bin-divisions of the depth range interval, with the density of resulting bin-centers directly reflecting the density of the depth values in the local neighborhood. In practice, the bin-divisions are formulated as a vector of normalized bin-widths at every pixel from which the bin-centers can be easily obtained via Eq. 5. To estimate the bin-divisions at every pixel, we employ a *coarse-to-fine binning* strategy. Starting with N_{seed} number of bins for every pixel at the bottleneck, the number of bins doubles at every decoder layer.

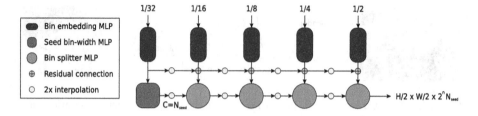

Fig. 3. Design of the LocalBins module. See Sect. 3.2 for more details.

The LocalBins module consists of three main types of layers: a) Bin embedding layers b) Seed bin width estimators c) Bin splitters. All these layers are composed of pointwise MLPs (a.k.a 1×1 convolutional blocks) with two hidden layers with hidden dimension h.

a) Bin embedding layers. All the feature blocks input to the LocalBins module (the bottleneck features from the encoder and the decoder features of all scales) are first fed to bin embedding layers. These layers project the features of varying channel dimensionality into the same space (dim = 128) that we refer to as the 'bin embedding space'. The further layers in the LocalBins module 'decide' the bin-division for each pixel based on their bin-embeddings.

b) Seed bin width estimator. This layer takes bin embeddings from the bottleneck as input and predicts N_{seed} number of bins at each pixel of the bottleneck. This bin-division estimate is taken as the seed and subsequently each bin is divided into two bins for every subsequent decoder layer by the bin splitters (along with the spatial 2x interpolation).

c) Bin splitters. These pointwise MLPs are used to realize our coarse-to-fine binning strategy. Loosely speaking, bin splitters 'decide' where to put more bin-centers for each pixel based on their bin-embeddings. As illustrated in the Fig. 3, bin-embeddings and bin-widths from the previous layer are first bilinearly upsampled to match the spatial resolution of the current layer. A bin splitter MLP at layer k, denoted as $\mathcal{S}^k$, takes as input the 'current' layer bin-embeddings after a residual connection with the upsampled previous layer bin-embeddings. The output is then used to split each bin-width from the previous layer into two. Specifically, let $\boldsymbol{b}_{ij} \in \mathbb{R}^m$ be the normalized m-bin-widths at pixel location (i, j) (after 2x interpolation). Then, the new $2m$-bin-widths $\boldsymbol{b}'_{ij} \in \mathbb{R}^{2m}$ are given by:

$$\boldsymbol{\alpha}_{ij} = \sigma(\mathcal{S}^k(\mathbf{e}_{ij}^{k-1} + \mathbf{e}_{ij}^k)) \tag{1}$$

$$\boldsymbol{b}'_{ij} = \{\alpha_{ij}^0 b_{ij}^0, (1 - \alpha_{ij}^0)b_{ij}^0, \alpha_{ij}^1 b_{ij}^1, (1 - \alpha_{ij}^1)b_{ij}^1, \ldots, \alpha_{ij}^m b_{ij}^m, (1 - \alpha_{ij}^m)b_{ij}^m\} \tag{2}$$

where, $\sigma(\cdot)$ represents the splitter activation function that outputs values $\alpha \in (0, 1)$, $\mathbf{e}_{ij}^k$ is the bin-embedding of the pixel at (i, j) at the kth layer and v^a represent the components of a vector $\mathbf{v}$.

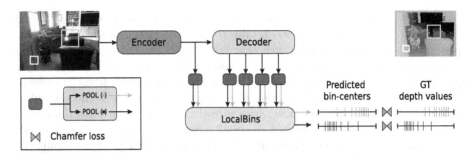

Fig. 4. Illustration of Query-Response training. See Sect. 3.3 for more details.

We explore three designs of the splitter activation function $\sigma(\cdot)$, namely:

1. Constant splitter: $\sigma(x) = 0.5 \; \forall x$, that divides a bin in half irrespective of the splitter MLP output (and indirectly the bin-embeddings).
2. Sigmoid splitter: where $\sigma(\cdot)$ represents the sigmoid activation function.
3. Linear norm splitter: In this case, we let the splitter MLP $\mathcal{S}^k$ output two positive values (x_1, x_2) (via ReLU) for each bin. Then the linear norm split is given by:

$$\sigma(x_1, x_2) = \frac{x_1}{x_1 + x_2 + \epsilon} \tag{3}$$

where $\epsilon = 1e^{-4}$ is used for numerical stability.

Refer to Sect. 5.3 for their comparison.

Since the number of bins doubles every layer, we have $2^n N_{seed}$ bins at the end of an n-layer decoder. Therefore we use $output\_channels = 2^n N_{seed}$ for the last convolutional layer in the decoder. We then use the hybrid regression as in AdaBins, to obtain the final depth map. The difference is that now the bins change from pixel-to-pixel:

$$c(b_{ij}^k) = d_{min} + (d_{max} - d_{min})(b_{ij}^k/2 + \sum_{s=1}^{k-1} b_{ij}^s) \tag{4}$$

$$\tilde{d}_{ij} = \sum_{k=1}^{2^n N_{seed}} c(b_{ij}^k) p_{ij}^k \tag{5}$$

where $\tilde{d}_{ij}$ is the final estimated depth value, b_{ij} and p_{ij} are the bin-widths at the output layer and softmax scores respectively at location (i, j).

3.3   Training

Our network needs supervision in two forms. First, we need a pixel-wise loss ($\mathcal{L}_{pixel}$) to provide supervision for the final estimated depth values. For this we use the Scale-Invariant Loss as used in recent works [4,20]. Second, we need to supervise our network such that the bin predictions at a pixel actually reflect the density of depth values in its local neighborhood.

A naive way would be to just choose a fixed-sized local window around every pixel (say a 5 × 5 window) and directly impose the regularization on bin-predictions at a pixel location such that they reflect the corresponding ground truth depth distributions within the window. However, there are two major problems with this approach: 1) It is not clear how one would choose the size of the window. Empirical determination is not a scalable solution as the amount of detail within a fixed sized window varies with the spatial resolution of the image. Alternatively, choosing different sizes simultaneously would lead to inconsistent regularization (e.g. bin predictions at the same pixel location would be compared to GT distributions within, say, 5 × 5 and 7 × 7 windows at the same time) 2) Chamfer loss, which is used in AdaBins to implement the distribution loss, is not computationally scalable. In AdaBins, it is computationally feasible because there is only one bin-division proposal per image. While in our case, we have a bin-division proposal at every pixel. This would mean we would have to compute Chamfer loss between the point sets at ~300K locations per image (= total number of pixels) at the highest resolution, which is not feasible in terms of memory or computation. One can, in practice, subsample the number of locations. For example, for a batch size of 16, we could fit around ~2% of randomly selected pixel locations on four NVIDIA A100 GPUs. However, as expected, this leads to inferior performance (Sect. 5.3).

One obvious reason is the significant loss of the spatial coverage of locations at which the loss is computed. We initially identified two possible workarounds for this problem. Either investigate efficient ways for subsampling or let the gradients from the loss computation at a given pixel location directly flow to neighboring regions to increase the coverage. This work focuses on designing the latter solution. To increase the coverage, we propose to involve all the pixel locations within the window to compute the loss (instead of just the center one), while keeping the loss computation feasible. This means that instead of regularizing the bin-predictions at individual pixel locations, we propose to regularize the bin-predictions of the entire local window together, potentially solving the coverage problem. We achieve this via the following formulation that we call 'Query-Response' training.

Query-Response Training. We train the network via the following locality constraint regularization:

Consider a bounding box B at any given location in the input image. The LocalBins module, when applied on the spatial average of the features within B, must predict the density of depth values within B.

This is illustrated in Fig. 4. By using bounding boxes of different sizes, we can enforce the features to contain the local distributional information at multiple scales. Furthermore, since the spatial averaging operation covers the entire window, we can potentially achieve the complete coverage with a relatively smaller number of bounding boxes per-image.

In order to implement such a regularization we make use of ROIAlign aggregation [14], a popular operation used in object detection pipelines. ROIAlign aggregation allows us to extract and pool the features at different layers (bottleneck and decoder features) corresponding to a given bounding box. Details on how ROIAlign works is found in [14].

Given a bounding box $\mathcal{B}$ (a.k.a query), we apply ROIAlign aggregation and use the pointwise MLPs (Bin embbeding, seed bin width estimator and bin splitter layers from the LocalBins module) on the pooled features to get the bins $b(\mathcal{B})$ (a.k.a the response). The resulting bins $b(\mathcal{B})$ are then 'forced' to match the ground truth depth distribution within that bounding box. We use the 1D bi-directional Chamfer loss as in [4] as the matching loss: $chamfer(b(\mathcal{B}), Depth(\mathcal{B}))$.

"Foveated" Loss. We now have a few choices on how to aggregate losses from different bounding boxes to calculate the final loss. We choose to have smaller bounding boxes to have more influence than the larger ones. We therefore use the loss weights that exponentially decay with the bounding box size. Suppose we generate N different sets of bounding boxes $\mathcal{Q}_1, \ldots, \mathcal{Q}_N$ with different box sizes such that $size(\mathcal{B}_a \in \mathcal{Q}_i) < size(\mathcal{B}_b \in \mathcal{Q}_j)$ $\forall a, b$ and $i < j$, the total loss is given by:

$$\mathcal{L}_{bins} = \sum_{L=1}^{n} \gamma_l^{n-L} \sum_{k=1}^{N} \gamma_b^{k-1} \sum_{\mathcal{B} \in \mathcal{Q}_k} chamfer(b_L(\mathcal{B}), Depth(\mathcal{B})) \qquad (6)$$

where n is the number of layers and $b_L(\mathcal{B})$ is the response at layer L (running from bottleneck to output layer). We use $\gamma_l = \gamma_b = 0.3$ in our experiments and use 5 different sizes of bounding boxes as described below. In summary, we follow the steps:

1. Generate five sets of random bounding boxes of sizes, 3×3, 7×7, 15×15, 31×31, 63×63, each containing $M = 200$ boxes.
2. Extract the depth values from the ground truth depth map corresponding to these bboxes to use in the Chamfer loss calculation.
3. Use ROIAlign with average pooling to get the corresponding pooled features at the bottleneck and decoder layers.
4. Use bin embedding MLPs to get the corresponding bin-embeddings. At this stage, we have one bin-embedding vector at each layer corresponding to each bbox.
5. Use seed bin estimator and bin splitter MLPs as usual to get the resulting bins.
6. Calculate the unweighted Chamfer loss for the predicted bins against their ground truth (step 2).
7. Calculate the weights and compute the weighted sum to get the final Chamfer loss, with the weighting scheme in Eq. 6.

Finally, we define the total loss as:

$$\mathcal{L}_{total} = \mathcal{L}_{pixel} + \beta \mathcal{L}_{bins}, \qquad (7)$$

where we set $\beta = 0.02$ in all our experiments.

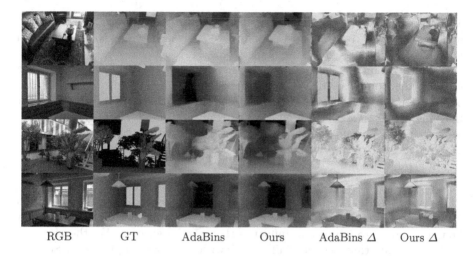

RGB   GT   AdaBins   Ours   AdaBins Δ Ours Δ

Fig. 5. Qualitative results on iBims-1 benchmark without fine-tuning.

4   Implementation Details

We implement the model in PyTorch [25]. We use hidden dimension $h = 256$ for the Seed bin width MLP and $h = 128$ for Bin embedding MLPs and Bin splitter MLPs. We train the network with batch size of 16 and use the AdamW [24] optimizer with weight decay of 10^{-1}. We use a learning rate of 3.57×10^{-4} which is decayed by a factor of 10^4 in the last 30% iterations using a cosine decay schedule. We train the models for 10 epochs in all our experiments.

5   Experiments and Results

5.1   Comparison to State-of-the-Art

NYU-Depth V2. We use NYU-Depth V2 [29] as the most important dataset for evaluation. Table 1 presents the performance comparison on the official test set of NYU-Depth V2. Our proposed model shows the state-of-the-art performance across all metrics, with ~4% reduction in the absolute relative error metric. Qualitatively, as shown in Fig. 7, our model is better at predicting depths of thin objects as well as planar surfaces. Note that our final model is able to beat the current published state-of-the-art model AdaBins while having the same encoder-decoder backbone. In total, our model has even fewer parameters since we do not use global attention or transformers. We attribute this performance improvement to the proposed LocalBins design and better utilization of the depth statistics via our novel training scheme.

5.2   Zero-Shot Performance

To evaluate the generalization performance of our model, we use the model pretrained on the NYU-Depth V2 dataset and evaluate it on other datasets without fine-tuning.

Table 1. Comparison of performance on the NYU-Depth V2 dataset. The reported numbers are from the corresponding original papers. Best results are in bold, second best are underlined.

Method	Encoder	#p	$\delta_1\uparrow$	$\delta_2\uparrow$	$\delta_3\uparrow$	REL ↓	RMS ↓	log_{10} ↓
Eigen et al. [8]	–	141	0.769	0.950	0.988	0.158	0.641	–
Laina et al. [19]	ResNet-50	64	0.811	0.953	0.988	0.127	0.573	0.055
Hao et al. [13]	ResNet-101	60	0.841	0.966	0.991	0.127	0.555	0.053
Lee et al. [21]	–	119	0.837	0.971	0.994	0.131	0.538	–
Fu et al. [9]	ResNet-101	110	0.828	0.965	0.992	0.115	0.509	0.051
SharpNet [26]	–	–	0.836	0.966	0.993	0.139	0.502	0.047
Hu et al. [15]	SENet-154	157	0.866	0.975	0.993	0.115	0.530	0.050
Chen et al. [6]	SENet	210	0.878	0.977	0.994	0.111	0.514	0.048
Yin et al. [36]	ResNeXt-101	114	0.875	0.976	0.994	0.108	0.416	0.048
BTS [20]	DenseNet-161	47	0.885	0.978	0.994	0.110	0.392	0.047
AdaBins [4]	EfficientNet-B5	78	<u>0.903</u>	<u>0.984</u>	<u>0.997</u>	<u>0.103</u>	<u>0.364</u>	<u>0.044</u>
LocalBins (Ours)	EfficientNet-B5	74	**0.907**	**0.987**	**0.998**	**0.099**	**0.357**	**0.042**

Table 2. Quantitative results on the iBims benchmark without fine-tuning.

Method	$\delta_1\uparrow$	$\delta_2\uparrow$	$\delta_3\uparrow$	REL↓	RMS↓	log10↓
BTS	0.54	0.86	0.95	0.23	0.93	0.11
AdaBins	0.55	0.87	0.96	**0.21**	0.90	0.11
LocalBins (Ours)	**0.56**	**0.88**	**0.97**	0.21	**0.88**	**0.10**

iBims-1 Benchmark. [18] (independent Benchmark images and matched scans version 1) is a high quality RGB-D dataset acquired using a digital single-lens reflex (DSLR) camera and a high-precision laser scanner. Table 2 lists the performance on this benchmark, with our proposed model outperforming prior state-of-the-art methods. In addition, we show qualitative results in Fig. 5. AdaBins noticeably underestimates the depth range of the scenes with relative error growing with distance, whereas LocalBins more consistently predicts scale-accurate depths across varying depth ranges. This further emphasizes the generalization capability of our model.

SUN-RGBD. [30] is an indoor dataset characterized by high scene diversity. We evaluate our model without fine-tuning on the official test set of 5050 images and report the results in Table 5.

5.3   Analysis and Ablation Studies

Here, we present the results of the extensive experiments we performed to analyse the properties and the importance of various components in our proposed model.

Table 3. Ablation experiments showing the importance of various components in our proposed model. **Enc-Dec**: Base encoder-decoder model, **LBM**: LocalBins module, **Naive**: Naive training strategy discussed in Sect. 3.3, **QR**: Query-Response training, $\mathcal{L}_{bins}$: Chamfer loss supervision, **Foveated**: Foveated weighting in Chamfer loss. Data in the 'Naive' column indicates the bbox size used on GT to compute density.

	Enc-Dec	LBM	Naive	QR	Lbins	Foveated	**REL**	**RMS**
1	✓	✗	✗	✗	✗	✗	0.111	0.419
2	✓	✓	✗	✗	✗	✗	0.106	0.375
3	✓	✓	3×3	✗	✓	✗	0.108	0.381
4	✓	✓	5×5	✗	✓	✗	0.107	0.375
5	✓	✓	15×15	✗	✓	✗	0.108	0.377
8	✓	✓	✗	✓	✓	✗	0.099	0.364
9	✓	✓	✗	✓	✓	✓	0.099	0.357

Table 4. Quantitative demonstration of the efficiency of Query-Response Training. **PSCI**: Point Set Comparisons per Image. **#px**: Total number of pixels covered for loss computation. **Coverage**: Percentage of #px with respect to image resolution.

	PSCI	#px	Coverage (%) ↑	REL ↓
Naive	4096	4096	1.33	0.1075
	8192	8192	2.67	0.1071
Query-Response	250	52,130	16.97	0.1043
	500	104,260	33.94	0.1002
	1000	208,520	67.88	0.0992

LocalBins Module. We first evaluate the importance of our LocalBins module. We remove the LocalBins module from the network and evaluate our base encoder-decoder architecture. We use the pixel-wise loss ($\mathcal{L}_{pixel}$) to train the network. We also evaluate our proposed model (with LocalBins module) without the Chamfer loss to study the capacity of our design in absence of extra supervision. Results are reported in Table 3.

Query-Response Training and Foveated Loss. We evaluate our proposed 'Query-Response' training scheme against the naive implementation discussed in Sect. 3.3. As listed in Table 3, Query-Response training gives a significant boost to the performance, improving the absolute relative error by ~7%. We believe the Query-Response training scheme is a general, powerful regularization technique that can be directly used in tasks beyond depth estimation. The foveated weighting scheme further improves the squared error based metrics.

In order to demonstrate the power of Query-Response training, we compare it with the Naive scheme in terms of computation of Chamfer loss against the coverage (total number of pixel locations involved in loss computation - higher the better). We define 'PSCI' as the total number of *Point Set Comparisons*

Table 5. Results of models trained on the NYU-Depth V2 dataset and tested on the SUN RGB-D dataset [30] without fine-tuning.

Method	$\delta_1 \uparrow$	$\delta_2 \uparrow$	$\delta_3 \uparrow$	REL↓	RMS↓	$log_{10} \downarrow$
Chen [6]	0.757	0.943	0.984	0.166	0.494	0.071
Yin [36]	0.696	0.912	0.973	0.183	0.541	0.082
BTS [20]	0.740	0.933	0.980	0.172	0.515	0.075
AdaBins [4]	0.771	0.944	0.983	0.159	0.476	0.068
Ours	**0.777**	**0.949**	**0.985**	**0.156**	**0.470**	**0.067**

Table 6. Types of splitters.

Splitter activation	REL	RMS
Constant	0.117	0.454
Sigmoid	0.100	0.361
Linear norm	0.099	0.364

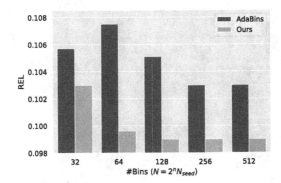

Fig. 6. REL vs #Bins.

performed per Image and use it as an indicator of computational efficiency. Note that in practice, the computations are batched, and PSCI indicates the total number of 'samples' in the batch contributed per image. For the Naive scheme, PSCI is equal to the total number of subsampled locations. For the Query-Response training scheme, PSCI is equal to the total number of bounding box queries per image. We take random bounding boxes of sizes $\{3 \times 3, 7 \times 7, 15 \times 15, 31 \times 31, 63 \times 63\}$ and compute their average total area. The results are given in Table 4. Our proposed training scheme performs ~7.6% better with 8× fewer point set comparisons compared to the naive scheme.

Splitter Activation Function. We evaluate the three types of splitter activation functions as discussed in Sect. 3.2. The results are given in Table 6.

RGB   GT   AdaBins [4]   Ours   AdaBins Δ Ours Δ

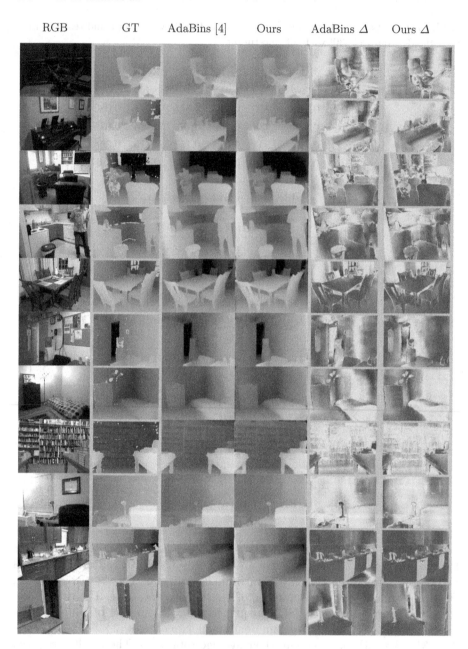

Fig. 7. Qualitative results on NYU-Depth V2.

Effect of N_{seed}. We analyse the effect of varying the N_{seed} and hence the total number of bins, and compare with AdaBins. The results are plotted in Fig. 6. We find that our model is more robust to the total number of bins used and generally has better performance.

6   Conclusions

We introduced a new network architecture for depth estimation from a single image. We build on an encoder-decoder architecture and evolve the current state-of-the-art model AdaBins in two aspects. First, we add three building blocks to estimate local depth distributions in the neighborhood of a pixel. These three building blocks, bin embedding layers, seed bin width estimator, and bin splitters are tightly integrated with the decoder in a multi-scale fashion. Second, we propose a query - response acceleration strategy for training, since a naive implementation of the idea would be highly time and memory consuming. In future work, we would like to adapt the LocalBins concept to other dense regression algorithms, such as image segmentation or inpainting.

Acknowledgements. This work was supported by the KAUST Office of Sponsored Research (OSR) under Award No. OSR-CRG2018-3730.

References

1. Akada, H., Bhat, S.F., Alhashim, I., Wonka, P.: Self-supervised learning of domain invariant features for depth estimation. In: IEEE/CVF Winter Conference on Applications of Computer Vision, WACV 2022, Waikoloa, HI, USA, 3–8 January 2022, pp. 997–1007. IEEE (2022). https://doi.org/10.1109/WACV51458.2022.00107
2. Alhashim, I., Wonka, P.: High quality monocular depth estimation via transfer learning. CoRR abs/1812.11941 (2018). http://arxiv.org/abs/1812.11941
3. Atapour-Abarghouei, A., Breckon, T.P.: Real-time monocular depth estimation using synthetic data with domain adaptation via image style transfer. In: Proceedings of the IEEE Conference on Computer Vision and Pattern Recognition, pp. 2800–2810 (2018)
4. Bhat, S.F., Alhashim, I., Wonka, P.: AdaBins: depth estimation using adaptive bins. In: 2021 IEEE/CVF Conference on Computer Vision and Pattern Recognition (CVPR), pp. 4008–4017. IEEE Computer Society, Los Alamitos, CA, USA, June 2021. https://doi.org/10.1109/CVPR46437.2021.00400
5. Casser, V., Pirk, S., Mahjourian, R., Angelova, A.: Unsupervised monocular depth and ego-motion learning with structure and semantics. In: CVPR Workshop on Visual Odometry and Computer Vision Applications Based on Location Cues (VOCVALC) (2019)
6. Chen, X., Chen, X., Zha, Z.J.: Structure-aware residual pyramid network for monocular depth estimation. In: Proceedings of the Twenty-Eighth International Joint Conference on Artificial Intelligence, IJCAI-19, pp. 694–700. International Joint Conferences on Artificial Intelligence Organization, July 2019. https://doi.org/10.24963/ijcai.2019/98

7. Chen, Y.C., Lin, Y.Y., Yang, M.H., Huang, J.B.: CrDoCo: pixel-level domain transfer with cross-domain consistency. In: IEEE Conference on Computer Vision and Pattern Recognition (CVPR) (2019)

8. Eigen, D., Puhrsch, C., Fergus, R.: Depth map prediction from a single image using a multi-scale deep network. In: NIPS (2014)

9. Fu, H., Gong, M., Wang, C., Batmanghelich, N., Tao, D.: Deep ordinal regression network for monocular depth estimation. In: 2018 IEEE/CVF Conference on Computer Vision and Pattern Recognition, pp. 2002–2011 (2018)

10. Godard, C., Aodha, O.M., Brostow, G.J.: Unsupervised monocular depth estimation with left-right consistency. In: 2017 IEEE Conference on Computer Vision and Pattern Recognition (CVPR), pp. 6602–6611 (2017)

11. Godard, C., Aodha, O.M., Brostow, G.J.: Digging into self-supervised monocular depth estimation. CoRR abs/1806.01260 (2018)

12. Gordon, A., Li, H., Jonschkowski, R., Angelova, A.: Depth from videos in the wild: unsupervised monocular depth learning from unknown cameras. In: Proceedings of the IEEE/CVF International Conference on Computer Vision (ICCV), October 2019

13. Hao, Z., Li, Y., You, S., Lu, F.: Detail preserving depth estimation from a single image using attention guided networks. In: 2018 International Conference on 3D Vision (3DV), pp. 304–313 (2018)

14. He, K., Gkioxari, G., Dollár, P., Girshick, R.: Mask R-CNN. In: 2017 IEEE International Conference on Computer Vision (ICCV), pp. 2980–2988 (2017). https://doi.org/10.1109/ICCV.2017.322

15. Hu, J., Ozay, M., Zhang, Y., Okatani, T.: Revisiting single image depth estimation: toward higher resolution maps with accurate object boundaries. In: 2019 IEEE Winter Conference on Applications of Computer Vision (WACV), pp. 1043–1051 (2018)

16. Huynh, L., Nguyen-Ha, P., Matas, J., Rahtu, E., Heikkila, J.: Guiding monocular depth estimation using depth-attention volume. arXiv preprint arXiv:2004.02760 (2020). https://doi.org/10.1007/978-3-030-58574-7_35

17. Kim, D., Ga, W., Ahn, P., Joo, D., Chun, S., Kim, J.: Global-local path networks for monocular depth estimation with vertical cutdepth. arXiv preprint arXiv:2201.07436 (2022)

18. Koch, T., Liebel, L., Fraundorfer, F., Körner, M.: Evaluation of CNN-based single-image depth estimation methods. In: Leal-Taixé, L., Roth, S. (eds.) ECCV 2018. LNCS, vol. 11131, pp. 331–348. Springer, Cham (2019). https://doi.org/10.1007/978-3-030-11015-4_25

19. Laina, I., Rupprecht, C., Belagiannis, V., Tombari, F., Navab, N.: Deeper depth prediction with fully convolutional residual networks. In: 2016 Fourth International Conference on 3D Vision (3DV), pp. 239–248 (2016)

20. Lee, J.H., Han, M.K., Ko, D.W., Suh, I.H.: From big to small: multi-scale local planar guidance for monocular depth estimation. arXiv preprint arXiv:1907.10326 (2019)

21. Lee, W., Park, N., Woo, W.: Depth-assisted real-time 3D object detection for augmented reality. In: ICAT 2011, vol. 2, pp. 126–132 (2011)

22. Li, H., Gordon, A., Zhao, H., Casser, V., Angelova, A.: Unsupervised monocular depth learning in dynamic scenes. arXiv preprint arXiv:2010.16404 (2020)

23. Li, Z., Snavely, N.: MegaDepth: learning single-view depth prediction from internet photos. In: Computer Vision and Pattern Recognition (CVPR) (2018)

24. Loshchilov, I., Hutter, F.: Decoupled weight decay regularization. In: 7th International Conference on Learning Representations, ICLR 2019, New Orleans, LA, USA, 6–9 May 2019. OpenReview.net (2019). https://openreview.net/forum?id=Bkg6RiCqY7

25. Paszke, A., et al.: PyTorch: an imperative style, high-performance deep learning library. In: Wallach, H., Larochelle, H., Beygelzimer, A., d'Alché-Buc, F., Fox, E., Garnett, R. (eds.) Advances in Neural Information Processing Systems, vol. 32, pp. 8026–8037. Curran Associates, Inc. (2019). https://proceedings.neurips.cc/paper/2019/file/bdbca288fee7f92f2bfa9f7012727740-Paper.pdf

26. Ramamonjisoa, M., Lepetit, V.: SharpNet: fast and accurate recovery of occluding contours in monocular depth estimation. In: Proceedings of the IEEE/CVF International Conference on Computer Vision (ICCV) Workshops, October 2019

27. Ranftl, R., Bochkovskiy, A., Koltun, V.: Vision transformers for dense prediction. In: Proceedings of the IEEE/CVF International Conference on Computer Vision (ICCV), pp. 12179–12188, October 2021

28. Ranftl, R., Lasinger, K., Hafner, D., Schindler, K., Koltun, V.: Towards robust monocular depth estimation: mixing datasets for zero-shot cross-dataset transfer. IEEE Trans. Patt. Anal. Mach. Intell. (TPAMI) (2020)

29. Silberman, N., Hoiem, D., Kohli, P., Fergus, R.: Indoor segmentation and support inference from RGBD images. In: Fitzgibbon, A., Lazebnik, S., Perona, P., Sato, Y., Schmid, C. (eds.) ECCV 2012. LNCS, vol. 7576, pp. 746–760. Springer, Heidelberg (2012). https://doi.org/10.1007/978-3-642-33715-4_54

30. Song, S., Lichtenberg, S.P., Xiao, J.: Sun RGB-D: A RGB-D scene understanding benchmark suite. In: 2015 IEEE Conference on Computer Vision and Pattern Recognition (CVPR), pp. 567–576 (2015). https://doi.org/10.1109/CVPR.2015.7298655

31. Tonioni, A., Poggi, M., Mattoccia, S., di Stefano, L.: Unsupervised domain adaptation for depth prediction from images. CoRR abs/1909.03943 (2019). http://arxiv.org/abs/1909.03943

32. Watson, J., Mac Aodha, O., Prisacariu, V., Brostow, G., Firman, M.: The temporal opportunist: self-supervised multi-frame monocular depth. In: Proceedings of the IEEE/CVF Conference on Computer Vision and Pattern Recognition (CVPR), pp. 1164–1174, June 2021

33. Xie, J., Girshick, R., Farhadi, A.: Deep3D: fully automatic 2D-to-3D video conversion with deep convolutional neural networks. In: Leibe, B., Matas, J., Sebe, N., Welling, M. (eds.) ECCV 2016. LNCS, vol. 9908, pp. 842–857. Springer, Cham (2016). https://doi.org/10.1007/978-3-319-46493-0_51

34. Xu, D., Ricci, E., Ouyang, W., Wang, X., Sebe, N.: Multi-scale continuous CRFs as sequential deep networks for monocular depth estimation. In: Proceedings of the IEEE Conference on Computer Vision and Pattern Recognition, pp. 5354–5362 (2017)

35. Xu, D., Wang, W., Tang, H., Liu, H.W., Sebe, N., Ricci, E.: Structured attention guided convolutional neural fields for monocular depth estimation. In: 2018 IEEE/CVF Conference on Computer Vision and Pattern Recognition, pp. 3917–3925 (2018)

36. Yin, W., Liu, Y., Shen, C., Yan, Y.: Enforcing geometric constraints of virtual normal for depth prediction. In: Proceedings of the IEEE/CVF International Conference on Computer Vision (ICCV), October 2019

37. Yuan, W., Gu, X., Dai, Z., Zhu, S., Tan, P.: NeW CRFs: neural window fully-connected CRFs for monocular depth estimation. arXiv e-prints arXiv:2203.01502, March 2022

38. Zhao, S., Fu, H., Gong, M., Tao, D.: Geometry-aware symmetric domain adaptation for monocular depth estimation. In: Proceedings of the IEEE/CVF Conference on Computer Vision and Pattern Recognition, pp. 9788–9798 (2019)
39. Zhao, Y., Kong, S., Shin, D., Fowlkes, C.: Domain decluttering: simplifying images to mitigate synthetic-real domain shift and improve depth estimation. In: Proceedings of the IEEE/CVF Conference on Computer Vision and Pattern Recognition (CVPR), June 2020
40. Zhou, H., Greenwood, D., Taylor, S.: Self-supervised monocular depth estimation with internal feature fusion. In: British Machine Vision Conference (BMVC) (2021)
41. Zhou, T., Brown, M.R., Snavely, N., Lowe, D.G.: Unsupervised learning of depth and ego-motion from video. In: 2017 IEEE Conference on Computer Vision and Pattern Recognition (CVPR), pp. 6612–6619 (2017)

2D GANs Meet Unsupervised Single-View 3D Reconstruction

Feng Liu$^{(\boxtimes)}$ and Xiaoming Liu

Michigan State University, Computer Science and Engineering, East Lansing, USA
{liufeng6,liuxm}@msu.edu

Abstract. Recent research has shown that controllable image generation based on pre-trained GANs can benefit a wide range of computer vision tasks. However, less attention has been devoted to 3D vision tasks. In light of this, we propose a novel image-conditioned neural implicit field, which can leverage 2D supervisions from GAN-generated multi-view images and perform the single-view reconstruction of generic objects. Firstly, a novel offline StyleGAN-based generator is presented to generate plausible pseudo images with full control over the viewpoint. Then, we propose to utilize a neural implicit function, along with a differentiable renderer to learn 3D geometry from pseudo images with object masks and rough pose initializations. To further detect the unreliable supervisions, we introduce a novel uncertainty module to predict uncertainty maps, which remedy the negative effect of uncertain regions in pseudo images, leading to a better reconstruction performance. The effectiveness of our approach is demonstrated through superior single-view 3D reconstruction results of generic objects. Code is available at http://cvlab.cse.msu.edu/project-gansvr.html.

Keywords: 2D GANs · Multi-view Pseudo Images · Unsupervised · Single-view 3D Reconstruction · Generic objects · Uncertainty

1 Introduction

Realistic image synthesis is an important research area of computer vision. There has been remarkable progress in this field with the advent of 2D Generative Adversarial Networks (GANs) [9], such as StyleGAN [23] and its variations [21,22,24], which can generate high-fidelity images of diverse object categories with a wide variety of attributes (*e.g.,* pose, identity). Such superior capabilities of modeling the semantic image manifold enable the generated photorealistic images to be leveraged for many vision tasks, such as image editing [20,55], domain translation [73], face recognition [30], and video generation [60]. However, it remains much less explored in 3D vision tasks, *e.g.,* 3D reconstruction [31,32].

The 2D GAN manifolds appear to learn 3D geometrical properties implicitly, where recent GAN interpretation methods [14,51] have shown that manipulating

---

Supplementary Information The online version contains supplementary material available at https://doi.org/10.1007/978-3-031-19769-7_29.

S. Avidan et al. (Eds.): ECCV 2022, LNCS 13661, pp. 497–514, 2022.
https://doi.org/10.1007/978-3-031-19769-7_29

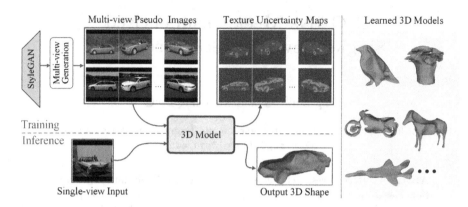

Fig. 1. Our approach leverages StyleGAN-generated multi-view pseudo images to learn a 3D model without 3D supervision, which can perform single-view 3D reconstruction for a variety of generic objects, *e.g.*, airplanes, birds, cars, horses, motorbikes, potted plants, *etc.*. In addition, our framework produces uncertainty maps, indicating the unreliable local areas in the pseudo images.

the latent code of the pre-trained GAN models can produce images of the same object under different viewpoints. Our work aims to answer the following question. *Using the GAN-generated multi-view images, can we learn a category-specific multi-view stereo system without 3D supervision that can reconstruct 3D shapes from a single image?*

Early attempts [44,52,63] are made to mine 3D geometric cues from the pre-trained 2D GAN models in an unsupervised manner. However, without modeling objects in the 3D space, these methods only recover 2.5D representations (depth or normals). Recently, StyleGANRender [71] integrates StyleGAN to generate multi-view images, which may be used to train an inverse graphics network for 3D reconstruction. However, the method focuses more on performing independent manipulation of 3D properties in GAN's latent space by fine-tuning the GAN models. In addition, the unreliable texture existing in GAN-generated multi-view images is a common issue, and has not been investigated.

It remains a challenge to leverage GAN-generated multi-view supervision for single-view 3D reconstruction. First of all, the pre-trained 2D GAN models lack explicit and precise camera pose to control over generated images, which is a necessity for classic multi-view stereo. Second, the GAN-generated multi-view images often suffer from *local* distortion and low perceptual quality, which severely breaks the *consistency* of either object shape or texture across views, and thereby ruins the *cornerstone* of multi-view stereo.

To address these challenges, we propose a novel framework to leverage GAN-generated multi-view images (termed '*pseudo images*') in learning generic object shape models, for the purpose of 3D reconstruction from a single image (Fig. 1). To first generate multi-view imagery by a pre-trained GAN, *e.g.*, StyleGAN, we carefully study the latent space of StyleGAN and devise a simple but effective technique, which generates plausible images with an azimuth range of $0-360°$. Consequently, during training, given a realistic image generated by StyleGAN, we can produce a set of pseudo images of the same object under different viewpoints. We then introduce a neural implicit network to simultaneously learn the unknown geometry, texture, and camera parameters for the objective of reconstructing the pseudo images, by incorporating a differentiable renderer.

A key component is that we introduce a learning framework that enables the neural implicit network to be conditioned on a single image. Specifically, we adopt an image encoder as a hypernetwork to predict the network parameters of the implicit function. This image conditioning allows the framework to be trained on multi-view images, where it learns object geometry priors within the category to perform single-view reconstruction. Moreover, to address the unreliable texture supervision issue in pseudo images, we devise an uncertainty prediction module, together with an uncertainty-aware photometric loss to estimate uncertainty maps, which can effectively filter out the unreliable supervision signals/inconsistencies within/across multi-view pseudo images, leading to a more precise reconstruction. Comprehensive experiments show the superiority of our method over existing methods in unsupervised single-view 3D reconstruction.

In summary, the contributions of this work include:

⋄ We propose a novel image-conditioned neural implicit network, which can exploit 2D supervision from GAN-generated multi-view pseudo images and performs single-view 3D reconstruction of generic objects.
⋄ We introduce a multi-view image generation mechanism based on the pre-trained StyleGAN models, which can produce plausible images with full control over viewpoints.
⋄ We propose an uncertainty prediction module to ignore unreliable texture supervision in pseudo images, enabling a reliable self-supervised learning.
⋄ Our method shows superior single-view 3D reconstruction for rigid and non-rigid generic objects in the wild.

2 Prior Work

Application of Pre-trained 2D GANs. While research on GANs is rapidly growing, our review mainly focuses on the pre-trained unconditional 2D GAN models. The capability to produce high-quality images makes 2D GANs applicable to many vision tasks, *e.g.*, image restoration [61,67], image editing (inpainting, super-resolution, semantic manipulation) [12,45], segmentation [72], and DeepFake attack and defense [2,6,7,48]. Further, the pre-trained 2D GAN models have been applied to data augmentation to reduce overfitting and bias in deep models [47,54]. To expand to 3D vision applications, prior works [36,56,62,63,74] adopt GANs to learn 3D shapes from images but rely on either 3D supervision or a 3D generator, which suffers from heavy memory consumption or extra training difficulties. Recently, LiftedGAN [52] lifts a pre-trained StyleGAN and distill it into a 3D aware generator, producing depth maps as a by-product. Similarly, GAN2Shape [44] produces an unsupervised decomposition by using a GAN model as supervision. However, those methods require inefficient online

500   F. Liu and X. Liu

Table 1. Comparison of *unsupervised* shape learning methods. [Keys: Cam. = camera poses per training sample, Requ. or Cons. = requirement or constraint, Real data = whether can train on real-world images, ⊛ = camera poses for a set of reference images]

Method	Output representation	Required template	Required cam	Additional Requ. or Cons	Real data
LiftedGAN [52]	2.5D, depth	✗	✗	GAN models, pre-trained	✓
GAN2Shape [44]	2.5D, depth	✗	✗	GAN models, pre-trained	✓
StyleGANRender [71]	3D, mesh	✗	✗	GAN models, fine-tuning	✓
DVR [42]	3D, implicit	✗	✓	Multi-view	✗
DIST [34]	3D, implicit	✗	✓	Multi-view	✗
SDFDiff [17]	3D, implicit	✗	✓	Multi-view	✗
CSDM [58]	3D, mesh	✓	✓	2D semantic	✓
CMR [18]	3D, mesh	✓	✗	2D semantic	✓
U-CMR [8]	3D, mesh	✓	✗	Viewpoint distribution	✓
UMR [28]	3D, mesh	✗	✗	3D semantic	✓
CSM [27]	3D, mesh	✓	✗	–	✓
A-CSM [26]	3D, mesh	✓	✗	–	✓
DRC [59]	3D, voxel	✗	✓	Multi-view	✓
SRN [53]	3D, implicit	✗	✓	Multi-view	✗
NeRF [38]	3D, implicit	✗	✓	Multi-view	✓
SDF-SRN [29]	3D, implicit	✗	✓	–	✓
ShSMesh [69]	3D, volumetric	✗	✗	–	✓
Proposed	3D, implicit	✗	⊛	GAN models, pre-trained	✓

image generation during training and infer 2.5D representations only. StyleGAN-Render [71] exploits StyleGAN as a multi-view generator to learn an *mesh-based* inverse graphics network to turn the StyleGAN into a controllable render. However, they require a *fine-tuning* step for the entire StyleGAN model, which is not desirable in this work. Moreover, they do not tackle the unreliability in pseudo multi-view images. In contrast, our method focuses on leveraging *pre-trained* 2D GANs for single-view 3D reconstruction. Despite both methods utilizing GAN-generated pseudo images for 3D modeling, we step forward in more plausible multi-view generation, robust shape and texture representation, and uncertainty-aware photometric supervision mechanism.

3D-Aware Generative Models. Understanding the latent representation of GANs has resulted in a body of works disentangling various factors of generated objects in a 3D-controllable manner, *e.g.*, viewpoint. These approaches can be classified into two groups. One adds additional modules or losses in training to explicitly disentangle 3D factors [37]. For example, HoloGAN [39] controls the object pose by rigid-body transformations via a 3D feature module. StyleFlow [1] learns non-linear paths in the latent space by normalizing flows conditioned on the attribute. Recently there has been a trend combining of the neural radiance fields (NeRF) [5,11,40,41,49] with GANs to devise 3D-aware generators. Another line of works, such as InterFaceGAN [50], SeFa [51], GANSpace [14], discover the latent semantic directions of a pre-trained GAN model that can manipulate object rotation unaware of its underlying 3D model. It is preferable to exploit the knowledge contained in a pre-trained GAN image manifold for the goal of recovering 3D object shapes without retraining the GAN models.

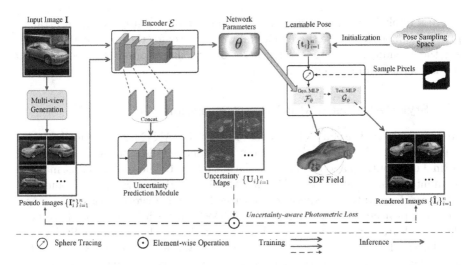

Fig. 2. Overview. The proposed framework is composed of two key modules: an offline StyleGAN-based multi-view generation and an image-conditioned neural implicit network. During training, the neural implicit network learns unknown geometry, texture, and camera poses for the objective of approximating the multi-view pseudo images. At inference time, the learned neural implicit function performs 3D reconstruction for the object from a single image.

Shape Learning without 3D Supervision. While 3D reconstruction, especially for faces [3,33,57], is a long-standing topic, we focus our review on shape learning from real-world images of generic objects without 3D supervision. Recent neural networks tackle this ill-posed problem via a differentiable renderer along with a choice of 3D shape representation [17,34,35,42,53,59]. However, in these works, multiple views of the same object with known cameras are required, which limits their learning from real-world images. Another branch of works show promising reconstruction from real-world images [8,26–28,64]. SDF-SRN [29] mines more supervision from 2D silhouette for superior reconstruction, yet still requires camera pose. ShSMesh [69] further discards the need for former constraints, but their reconstructions are of lower quality. NeRF [38] and its variations are scene- or object-specific models, which limit their applications for 3D reconstruction from unseen objects or scenes. In contrast, our models are category-specific, and can perform single-view 3D reconstruction for novel instances. Table 1 summarizes the differences between our method and prior work.

3   Proposed Method

We start with an overview of the proposed framework (Fig. 2) and then present the individual modules in detail. We first introduce an offline and effective multi-view generator based on the pre-trained StyleGAN models, which produce plausible multi-view images with full control over viewpoints. Then, we detail the proposed image-conditioned neural implicit field learning framework, including a neural implicit network, differentiable rendering procedure, and an uncertainty prediction module. These three modules work jointly for the objective of exploiting the pseudo images to learn generic object shape priors and perform 3D reconstruction from a single input image.

3.1 StyleGAN Based Multi-view Generation

We briefly review the embedding space of the StyleGAN [23, 24]. Typically, a generator $G(\cdot)$ samples a latent code $\mathbf{z}$ from a pre-defined distribution $\mathcal{Z}$ such as the normal distribution, and produces an output image $\mathbf{I}$. The code $\mathbf{z}$ is first mapped to an intermediate latent space $\mathcal{W}$ via a Multi-layer Perceptron (MLP), and then $\mathcal{W}$ is transformed to $\mathcal{W}^+$ space by 16 learned affine transformations. The generator $G(\cdot)$ projects $\mathbf{W}$ to the final image: $G(\mathbf{W}) = \mathbf{I}$. Such latent codes have been shown to learn various disentangled semantics [14, 51]. For instance, StyleGAN-Render [71] finds that the latent codes $\mathbf{W}_v := (\mathbf{w}_1, \mathbf{w}_2, \mathbf{w}_3, \mathbf{w}_4) \in \mathbb{R}^{4 \times 512}$ in the first 4 layers control camera viewpoints. That is, given a source and reference generated image pair $(\mathbf{I}^S, \mathbf{I}^R)$ with their latent codes $(\mathbf{W}^S, \mathbf{W}^R)$, we can generate an image of the source object with the reference viewpoint by swapping $(\mathbf{W}_v^S, \mathbf{W}_v^R)$ and keeping the rest dimensions of $\mathbf{W}^S$. We denote this multi-view generation strategy as **Baseline**. However, while $\mathbf{W}_v$ indeed alters the object viewpoint, it still perceives the shape of the reference object, as shown in Fig. 3a.

It is difficult to develop a multi-view stereo system from noisy multi-view images with inconsistent shapes. To tackle this issue, inspired by SeFa [51], we propose a novel offline multi-view generation mechanism that generates plausible images with enhanced cross-view consistency. SeFa suggests that the weight parameters in early affine transformations contain essential knowledge of image variations. One can obtain interpretable directions in the latent space by computing eigenvectors of their weight matrices and selecting eigenvectors with the largest eigenvalues. With this observation, we propose to filter the viewpoint-irrelevant features in $\mathbf{W}_v^R$ guided by the eigenvectors of the k largest eigenvalues, which are computed from the weight parameters in the first 4 transformations. It can be formulated as:

$$\arg\min_{\alpha} ||\hat{\mathbf{W}}_v - \mathbf{W}_v^R||^2, \quad \hat{\mathbf{W}}_v = \alpha \mathbf{V} + \mathbf{W}_v^S, \tag{1}$$

where $\hat{\mathbf{W}}_v$ is the enhanced viewpoint latent code. α is a k-dim viewpoint coefficient. $\mathbf{V} \in \mathbb{R}^{k \times 512}$ denotes the eigenvectors of $\mathbf{A}^T \mathbf{A}$ associated with the k largest eigenvalues. $\mathbf{A} \in \mathbb{R}^{m \times 512}$ are the weights of transformations. $\alpha \mathbf{V}$ is duplicated into four rows in Eq. 1 and α can be solved by gradient descent. Finally, we generate a new image by combining $\hat{\mathbf{W}}_v$ and remaining the dimensions of $\mathbf{W}^S$.

Training Pseudo Images. Given a pre-trained StyleGAN model, we first synthesize training images and filter out images that have more than one instance or an unrealistic instance, resulting in N training samples $\{\mathbf{I}^j\}_{j=1}^N$. Then, we manually select n reference view samples, which roughly cover the common object viewpoints ranging from 0–360° in azimuth. Finally, for each training sample $\mathbf{I}^j$, we produce n multi-view images of the same object with fixed camera poses

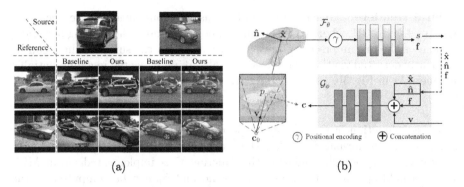

(a)   (b)

Fig. 3. (a) Multi-view generator comparisons of **Baseline** [71] and our approach (**Ours**). As can be observed, besides the viewpoint, the baseline perceives shape cues from the references (yellow circle, best view in zoom in). While our generated images show more consistency in object shape across views, which benefits shape learning. Please refer to **Supp** for more categories. (b) The architecture of our neural implicit fields and differentiable renderer. (Color figure online)

as the pseudo images $\{\mathbf{I}_i^{*j}\}_{i=1}^n \in \mathbb{R}^{W \times H \times 3}$. We further apply the work [46] to obtain corresponding instance segmentation $\{\mathbf{M}_i^j\}_{i=1}^n$ of pseudo images.

Pose Sampling Space. We assume a pinhole camera model $\mathcal{C} = (\mathbf{t}, \mathbf{K})$, where $\mathbf{K} \in \mathbb{R}^{3 \times 3}$ is the intrinsic parameter. We represent the camera pose/extrinsic parameters $\mathbf{t} = (\mathbf{c}_0, \mathbf{q})$ based on its 3D position $\mathbf{c}_0 \in \mathbb{R}^3$ and its rotation from a canonical view. $\mathbf{q} \in \mathbb{R}^4$ is the quaternion vector representing the camera rotation. We assume the observed object is approximately inside the unit sphere. We further assume a known intrinsic and the principal point at the image center, as commonly assumed in stereo systems [19]. For the pose initialization, we manually annotate the camera poses $\{\tilde{\mathbf{t}}_i\}_{i=1}^n$ for the n reference images, which initializes the pseudo images' poses, $\mathbf{t}_i^j = \tilde{\mathbf{t}}_i, j \in [1, N]$. *Since we only need to annotate n reference samples per object category, it is far more practical than prior works that require camera pose label per training sample* (see Table 1).

3.2 Image-Conditioned Neural Implicit Field

3D Geometry Representation. As illustrated in Fig. 3b, the object geometry to be reconstructed is represented by the function [68]: $\mathcal{F} : \gamma(\mathbf{x}) \rightarrow (s, \mathbf{f})$ that maps a point $\mathbf{x} \in \mathbb{R}^3$ to its signed distance value s to the object surface and a local geometry feature $\mathbf{f} \in \mathbb{R}^{d_f}$. $\gamma(\cdot)$ denotes a positional encoding operator on $\mathbf{x}$ with 6 exponentially increasing frequencies introduced in NeRF [38]. The surface $\mathcal{S}$ is represented as the zero level set of MLP $\mathcal{F}$ with learnable parameters θ:

$$\mathcal{S} = \{\mathbf{x} \in \mathbb{R}^3 | \mathcal{F}_\theta^{(s)}(\gamma(\mathbf{x})) = 0\}. \tag{2}$$

Neural Renderer. Given a pixel p of a masked input image, we march a ray $\mathbf{r} = \{\mathbf{c}_0 + t\mathbf{v} | t \geqslant 0\}$, where $\mathbf{c}_0$ is the camera position and $\mathbf{v}$ the viewing direction. $\hat{\mathbf{x}}$ denotes the first intersection between the ray $\mathbf{r}$ and the surface, which can be efficiently detected via the sphere tracing [15] and implemented in a differentiable manner. As in Fig. 3b, the rendered color of the pixel p is encoded as [68]

$$\mathcal{G} : (\hat{\mathbf{x}}, \hat{\mathbf{n}}, \mathbf{f}, \mathbf{v}) \to \mathbf{c}, \tag{3}$$

which is a function of the surface properties at $\hat{\mathbf{x}}_p$, including the surface point $\hat{\mathbf{x}}_p$, the surface normal $\hat{\mathbf{n}}_p \in \mathbb{R}^3$, the local geometry feature $\mathbf{f}_p$, and a viewing direction $\mathbf{v}_p \in \mathbb{R}^3$. Similarly, the function $\mathcal{G}$ is implemented as an MLP with learnable parameters ϕ. The surface normal $\hat{\mathbf{n}}_p$ can be computed by the spatial derivative $\frac{\delta \mathcal{F}^{(s)}}{\delta \mathbf{x}_p}$ via back-propagation through the network $\mathcal{F}$. Incorporating the surface normal and view direction enables $\mathcal{G}$ to represent the light reflected from a surface point $\mathbf{x}$ across different viewpoints [38,68]. Also, according to [68], introducing the local geometry feature vector $\mathbf{f}$ allows the renderer to handle more complex appearances, which might appear in GAN-generated images. The neural implicit field can be learned through back-propagation without any 3D supervision by comparing the rendered images with the multi-view pseudo images. Minimizing this error encourages multi-view photo consistency because only when the point is on the actual surface, the MLPs $(\mathcal{F}, \mathcal{G})$ will predict accurate color with fewer multi-view variations.

Image Encoder. We propose to utilize architecture to condition on a single image, such that the learned neural implicit field can generalize to a new object instance without re-training. Specifically, as shown in Fig. 2, we use an image encoder $\mathcal{E}$ as a hyper network [13,29] to predict θ (the parameters of $\mathcal{F}$), written as $\theta = \mathcal{E}_\Phi(\mathbf{I})$, where Φ is the neural network weights.

3.3   Model Learning

Given N sets of masked pseudo images $\{\mathbf{I}^j, \{\mathbf{I}_i^{*j}\}_{i=1}^n\}_{j=1}^N$, along with their initial camera poses $\{\{\mathbf{t}_i^j\}_{i=1}^n\}_{j=1}^N$, we optimize the encoder parameters Φ, texture MLP parameters ϕ and camera poses $\{\{\mathbf{t}_i^j\}_{i=1}^n\}_{j=1}^N$ by minimizing the loss:

$$\mathcal{L} = \sum_{j=1}^N (\mathcal{L}_{RGB} + \lambda_{mask}\mathcal{L}_{Mask} + \lambda_{eik}\mathcal{L}_{Eik}), \tag{4}$$

where $\mathcal{L}_{RGB}$ is photometric loss, $\mathcal{L}_{Mask}$ is silhouette loss, $\mathcal{L}_{Eik}$ is Eikonal regularization, and λ_* are loss weights.

Photometric Loss. Let $\mathbf{I}_p^*$, $\mathbf{M}_p \in \{0,1\}$ be the RGB and silhouette values of pixel p in an image sample $\mathbf{I}_i^*$ taken at the view direction $\mathbf{v}_p$ associated with camera $\mathcal{C}_i$. $p \in P$ indexes all pixels in the input image set $\{\mathbf{I}_i^*\}_{i=1}^n$. The photometric loss is defined on mini-batches of pixels in P:

$$\mathcal{L}_{RGB} = \frac{1}{|P|} \sum_{p \in P^{in}} |\mathbf{I}_p^* - \mathcal{G}(\hat{\mathbf{x}}_p, \hat{\mathbf{n}}_p, \mathbf{f}_p, \mathbf{v}_p)|, \tag{5}$$

where $\mathbf{f}_p$, $\hat{\mathbf{x}}_p$, $\hat{\mathbf{n}}_p$ are defined in Eq. 3. $P^{in} \subset P$ represents the subset of pixels P where intersection has been found and $\mathbf{M}_p = 1$. $|\cdot|$ denotes the L_1 loss.

Silhouette Loss. We define the silhouette loss as

$$\mathcal{L}_{Mask} = \frac{1}{|P|} \sum_{p \in P^{out}} CE(\mathbf{M}_p, \hat{\mathbf{M}}_p), \tag{6}$$

where $\hat{\mathbf{M}}$ is the masked rendering. $P^{out} = P - P^{in}$ represents the indices in the mini-batch for which there is no ray-geometry intersection or $\mathbf{M}_p = 0$. $CE(\cdot, \cdot)$ denotes the cross-entropy loss. Conventionally, given the ray $\mathbf{r}_p = \{\mathbf{c}_0 + t\mathbf{v}_p | t \geq 0\}$ of pixel p, $\hat{\mathbf{M}}_p$ is defined as:

$$\hat{\mathbf{M}}_p = \begin{cases} 1 & \mathbf{r}_p \cap \mathcal{S} \\ 0 & \text{otherwise.} \end{cases} \tag{7}$$

To make this differentiable, we follow [68] and compute

$$\hat{\mathbf{M}}_p = \text{sigmoid}(-\beta \min_{t \geq 0} \mathcal{F}^{(s)}(\mathbf{c}_0 + t\mathbf{v}_p)). \tag{8}$$

When $\beta \to \infty$, $\mathcal{F}^{(s)} < 0$ means inside the surface and $\mathcal{F}^{(s)} > 0$ outside.

Eikonal Regularization. A special property of signed distance functions is their differentiability with a gradient of unit norm, satisfying the Eikonal equation $||\nabla \mathcal{F}||_2 = 1$ [10,43]. We thus encourage our implicit geometry representation to satisfy the Eikonal property:

$$\mathcal{L}_{Eik} = \sum_{\tilde{\mathbf{x}}} \left| \, ||\nabla_{\tilde{\mathbf{x}}} \mathcal{F}^{(s)}(\tilde{\mathbf{x}})||_2 - 1 \right| \Big|_2^2, \tag{9}$$

where $\tilde{\mathbf{x}}$ is uniformly sampled at the 3D region of interest.

3.4 Uncertainty Prediction Module

The pseudo images are inherently aleatoric, *i.e.*, there might be areas with either notable artifacts in one image or with inconsistent shape/texture across images. To adaptively treat these problematic areas in learning, we propose to use Bayesian learning [25,66] to model this aleatoric uncertainty. Specifically, we exploit the feature space of the image encoder $\mathcal{E}$ and train a shallow decoder to estimate an uncertainty map $\mathbf{U}$, which has the same size as pseudo images. Formally, we model the observed color $\mathbf{I}_p^*$ at pixel p with a likelihood function $\mathcal{P}(\mathbf{I}_p^*)$ that follows the Laplacian distribution with ray-dependent variance σ_p:

$$\mathcal{P}(\mathbf{I}_p^*) = \frac{1}{2\sigma_p} \exp\left(-\frac{|\mathbf{I}_p^* - \mathcal{G}(\hat{\mathbf{x}}_p, \hat{\mathbf{n}}_p, \mathbf{f}_p, \mathbf{v}_p)|}{\sigma_p}\right), \tag{10}$$

where σ_p denotes the uncertainty. Since L_1 distance is less sensitive to outliers, which is more suitable for optimizing the rendered appearance against the pseudo

Fig. 4. The uncertainty maps produced by our proposed method and humans.

RGB values. Thus, we adopt Laplacian likelihood to model the inconsistent uncertainty distribution. To find the parameters best explaining the model, we maximize the likelihood function, *i.e.*, minimizing the negative log-likelihood:

$$-\log(\mathcal{P}(\mathbf{I}_p^*)) = \frac{|\mathbf{I}_p^* - \mathcal{G}(\hat{\mathbf{x}}_p, \hat{\mathbf{n}}_p, \mathbf{f}_p, \mathbf{v}_p)|}{\sigma_p} + \log\sigma_p + \log 2. \qquad (11)$$

Therefore, we update the photometric loss as:

$$\mathcal{L}_{RGB} = \frac{1}{|P|} \sum_{p \in P_{in}} \left(e^{-\mathbf{U}_p} |\mathbf{I}_p^* - \mathcal{G}(\hat{\mathbf{x}}_p, \hat{\mathbf{n}}_p, \mathbf{f}_p, \mathbf{v}_p)| + \mathbf{U}_p \right). \qquad (12)$$

We train this loss on mini-batches of pixels in P. Here P indexes all pixels in the input multi-view image set $\{\mathbf{I}_{i=1}^*\}_{i=1}^n$, which contribute to the same object's depth, texture and uncertainty learning. During training, for the surface point $\hat{\mathbf{x}}_p$, the first term $e^{-\mathbf{U}_p}$ can be seen as a weighted distance which assigns larger weights to less uncertain pixels. The second term $\mathbf{U}_p$ is a penalty term.

$\mathcal{L}_{RGB}$ encourages the neural implicit representation to bring together the texture information of all cross-view corresponding pixels in all images. Consequently, the pixel-wise uncertainty value is able to mine the multi-view inconsistencies among those pixels which contribute to the same 3D surface point. In practice, we train a 2-layer convolutional network to predict the log variance $\mathbf{U}_p := \log\sigma_p$ (please refer to **Supp** for more details).

3.5 Implementation Details

The encoder $\mathcal{E}$ is implemented as a ResNet-18 [16] followed by fully-connected layers. Both $\mathcal{F}$ and $\mathcal{G}$ have 4 fully-connected layers. For the main experiment, we set $N = 2{,}000$, $n = 40$, $k = 5$, $W = H = 256$, $d_f = 256$, $\beta = 50$, $\lambda_{mask} = 0.01$, $\lambda_{eik} = 0.1$. We implement our model in Pytorch and use the Adam optimizer with a learning rate of $1e-4$ for both network and camera pose parameters.

4 Experimental Results

We evaluate our approach on six category-specific StyleGAN models, including rigid objects such as airplanes, cars, motorbikes, as well as non-rigid objects such as birds, horses, and potted plants. We use the official car and horse models from

StyleGAN2 repo[1], trained on the LSUN dataset [70]. For other 4 categories, we train the StyleGAN models with StyleGAN2-ADA-Pytorch library[2] on LSUN airplanes, birds, motorbikes, and potted plants with $200K$ images respectively.

4.1 Human Study vs. Our Method on the Uncertainty Prediction

Although GAN interpretation methods have shown that manipulating the latent code of StyleGAN produces multi-view images of the same object [14,51], no studies have quantitatively evaluated the unreliable/inconsistent object shape or texture in/across the multi-view pseudo images. Thanks to our multi-view-stereo-like neural implicit network, our uncertainty map can serve as a means to detect the unreliable/inconsistent areas in the GAN-generated multi-view images. On the other hand, volunteers were asked to label the potentially problematic areas in the pseudo images. As shown in Fig. 4, given an image, the multi-view generator is able to generate pseudo images with varying viewpoints. We believe humans are able to reason unreliable/inconsistent regions in/across the pseudo images. Specifically, this is accomplished using a random set of 100 images from the PASCAL3D+ car category. For each image, we generate a pseudo image with a different viewpoint (Sect. 3.1). Then, the volunteers manually labels polygon-based regions of interest (uncertainty region) on the pseudo images, using the Matlab Image Labeler app. As can be observed in Fig. 4, human labels mainly focus on the global object shape inconsistency across views, which might not have the granularity to evaluate the pixel-level inconsistency. Nevertheless, we quantify the detection ability of our uncertainty maps by using the human labels as the ground-truth. We achieve 34.6% Intersection over Union (IoU), which shows our capability in detecting the inconsistent object shape in GAN-generated multi-view pseudo images. To our knowledge, this is the first method that tries to investigate the unreliable supervision across the GAN-generated pseudo multi-view images in 3D object modeling.

4.2 Quantitative 3D Reconstruction Evaluation

We quantitatively evaluate on the PASCAL3D+ dataset [65], a 3D reconstruction benchmark of real-world images with (approximate) CAD model annotations. Similar to prior work [29,59], we use annotations of airplane and car categories on the test set for evaluation.

Evaluation Metrics. We adopt standard 3D reconstruction metrics: IoU and Chamfer-L_1 Distance (CD). Following [59], we compute 3D IoU between ground truth and prediction with the resolution of 32^3. Following [29], we uniformly sample 3D points from the ground truth and prediction to compute CD.

Baselines. We compare against SoTA unsupervised single-view 3D reconstruction baselines: CSDM [58], DRC [59], CMR [18], U-CMR [8] and SDF-SRN [29].

---

[1] https://github.com/NVlabs/stylegan2.
[2] https://github.com/NVlabs/stylegan2-ada-pytorch.

Table 2. (a) Quantitative 3D reconstruction results on PASCAL3D+. During training, CSDM, DRC and SDF-SRN require ground-truth camera pose per training sample, CMR uses 2D keypoints and object-specific templates as additional constraints, and U-CMR only relies on object-specific templates. [Keys: Requ. or Cons. = requirement or constraint in training, T = category-specific templates, C = poses per training sample, C^* = poses for reference images, K = 2D keypoints, S = semantic information, GANs = pre-trained GAN models]. All CD values are scaled by 10 following [29]. (b) Ablation of uncertainty prediction and the number of pseudo images, using the PASCAL3D+ car category.

(a)

Category	Requ. or Cons.	Airplane CD (↓)	Airplane IoU (↑)	Car CD (↓)	Car IoU (↑)
CSDM [58]	T, C	–	0.400	–	0.600
DRC [59]	C	–	0.420	–	0.670
CMR [18]	T, K	0.625	–	0.474	0.640
U-CMR [8]	T	–	–	–	0.646
UMR [28]	S	–	–	–	0.620
SDF-SRN [29]	C	0.303	0.405	0.233	0.653
SDF-SRN* [29]	C	0.297	0.412	0.230	0.661
Proposed	GANs, C^*	**0.286**	**0.473**	**0.195**	**0.702**

(b)

	Proposed w/o Uncertainty ($n=40$)	Proposed ($n=$) 10	20	40	50
CD (↓)	0.208	0.336	0.243	0.195	**0.191**
IoU (↑)	0.681	0.612	0.643	0.702	**0.714**

As detailed in Tables 1 and 2a, some baselines require additional unsupervised constraints, *e.g.*, DRC (implicit) and SDF-SRN (implicit) both require ground-truth camera pose for each training sample. The mesh-based methods such as CSDM, CMR, and U-CMR need expert object-specific templates as additional constraints. Here, we do not compare with ShSMesh [69] as it neither quantitatively evaluates on PASCAL3D+, nor trains on real-world car/airplane images. Also, we do not quantitatively compare with StyleGANRender [71] as the code or trained model is not publicly available, and StyleGANRender does not report full 3D shape reconstruction errors as our baselines did.

Results. We present the comparisons of our approach in Table 2a and visualize sample predictions in Fig. 5. It can be observed that the **Proposed** model is significantly better than baselines in both CD (10.1% relative over SDF-SRN) and IoU (7.8% relative over DRC). It is worth noting that our models trained with Proposed setting only require ground-truth pose annotations for the reference images. It is more practical for real-world scenarios than DRC and SDF-SRN, which require ground-truth camera pose per training sample. Further, we retrain SDF-SRN with our GAN-generated training data and report the results in Table 2a (SDF-SRN*). It can be observed, despite the minor improvement over the original SDF-SRN due to our pseudo images, the new model still performs worse than ours. Figure 5 shows visual comparisons to SDF-SRN [29] and StyleGANRender [71] results. As can be observed, our approach suffers slightly from shape ambiguity, *e.g.*, windows of cars tend to be concave. Nonetheless, our predictions more closely resemble the ground truth.

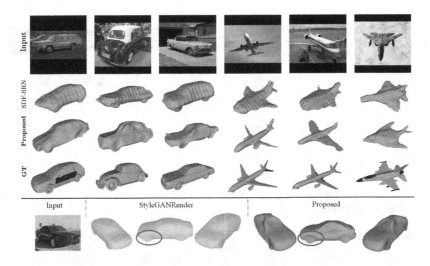

Fig. 5. Qualitative comparisons with SDF-SRN [29] (SOTA baseline) and StyleGAN-Render [71] on PASCAL3D+ car or airplane categories. Our approach recovers significantly more accurate 3D shapes and topologies from the images.

4.3   Ablation Study

All ablations use the models trained on the car category.

Effect on Uncertainty Prediction. We evaluate single-view 3D reconstruction on a model trained without uncertainty prediction, *i.e.* using Eq. 5 instead of Eq. 12. As shown in Table 2b, our uncertainty prediction module can remedy the negative impact of uncertain texture in GAN-generated multi-view images, leading to improved 3D reconstruction (CD: 0.208 → 0.195).

Effect on n. The key assumption, as well as motivation of our work, is that GAN-generated multi-view images can be leveraged to learn a multi-view stereo system. To validate the impact of the amount of pseudo images, we train models with different numbers of image viewpoints, $n = 10, 20, 40, 50$. Table 2b shows that the model trained with $n = 40$ images significantly outperforms the ones with $n = 10, 20$, and saturates when $n = 40 \rightarrow 50$. Considering the tradeoff between reconstruction accuracy and training cost, we use $n = 40$ for all experiments.

Effect on Multi-view Generation. Following the same setting, we re-train a model with pseudo imaged produced by *Baseline* method (Sect. 3.1). Quantitatively, such a model only obtain the IoU of 0.679, much worse than ours (0.702). The comparisons show the superior quality of our multi-view generation method.

4.4   Qualitative Evaluation

Comparison with U-CMR [8] and DRC [59]. We show qualitative comparisons with U-CMR and DRC on the PASCAL3D+ cars, motorbikes or airplanes,

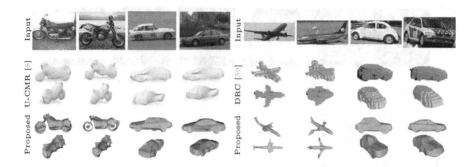

Fig. 6. Additional qualitative comparisons with U-CMR [8] and DRC [59] on cars and motorcycles or airplanes.

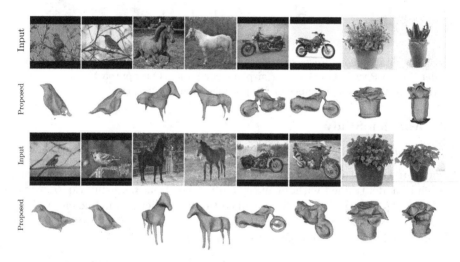

Fig. 7. Qualitative results of birds, horses, motorbikes, and potted plants.

in Fig. 6. Our approach achieves more faithful reconstructions than U-CMR. Note that U-CMR requires expert templates while our approach does not.

Results on More Categories. While our quantitative evaluation is on the PASCAL3D+ airplane and car, we provide more qualitative results for the birds, horses, motorbikes and potted plants in Fig. 7. As can be seen, our approach can effectively capture the thin structure present in 3D shapes from single-view images, *e.g.* horses' legs and motorbikes' hand clutch. All testing images are from the LSUN dataset, which never appear in the training set of our GAN models.

5   Conclusions

To leverage pre-trained GAN models for 3D vision tasks, we propose an image-conditioned neural implicit network that can learn the shape priors from GAN-generated multi-view images and perform single-view 3D reconstruction. Moreover, we naturally introduce a novel uncertainty prediction module, which can avoid invalid supervisions for better single-view 3D reconstruction. Experimentally, our approach significantly outperforms the SoTA unsupervised single-view 3D reconstruction methods, while requiring less supervision during training. We believe this work opens up a path for improving the ability to semantically control GAN generation and facilitates 2D GAN priors for 3D vision tasks.

Limitations. For some categories (*e.g.*, chair), StyleGAN is unable to converge to satisfying results, partially due to chairs' large topology variations. We believe the rapid development of GANs will extend to these challenging categories, and thus our method can leverage them for 3D reconstruction of more categories. Also, similar to most prior works, our model is category-specific. One future direction is to develop a single model for multiple categories based on universal GAN models, *e.g.*, BigGAN [4], improving generalization to unseen categories.

References

1. Abdal, R., Zhu, P., Mitra, N.J., Wonka, P.: StyleFlow: attribute-conditioned exploration of styleGAN-generated images using conditional continuous normalizing flows. TOG (2021)
2. Asnani, V., Yin, X., Hassner, T., Liu, X.: Reverse engineering of generative models: Inferring model hyperparameters from generated images. arXiv:2106.07873 (2021)
3. Bai, Z., Cui, Z., Rahim, J.A., Liu, X., Tan, P.: Deep facial non-rigid multi-view stereo. In: CVPR (2020)
4. Brock, A., Donahue, J., Simonyan, K.: Large scale GAN training for high fidelity natural image synthesis. In: ICLR (2019)
5. Chan, E.R., Monteiro, M., Kellnhofer, P., Wu, J., Wetzstein, G.: pi-GAN: periodic implicit generative adversarial networks for 3D-aware image synthesis. In: CVPR (2021)
6. Dang, H., Liu, F., Stehouwer, J., Liu, X., Jain, A.K.: On the detection of digital face manipulation. In: CVPR (2020)
7. Dolhansky, B., et al.: The deepfake detection challenge (DFDC) dataset. arXiv preprint arXiv:2006.07397 (2020)
8. Goel, S., Kanazawa, A., Malik, J.: Shape and viewpoint without keypoints. In: Vedaldi, A., Bischof, H., Brox, T., Frahm, J.-M. (eds.) ECCV 2020. LNCS, vol. 12360, pp. 88–104. Springer, Cham (2020). https://doi.org/10.1007/978-3-030-58555-6_6
9. Goodfellow, I., et al.: Generative adversarial nets. In: NeurIPS (2014)
10. Gropp, A., Yariv, L., Haim, N., Atzmon, M., Lipman, Y.: Implicit geometric regularization for learning shapes. arXiv preprint arXiv:2002.10099 (2020)
11. Gu, J., Liu, L., Wang, P., Theobalt, C.: StyleNeRF: a style-based 3D-aware generator for high-resolution image synthesis. arXiv preprint arXiv:2110.08985 (2021)
12. Gu, J., Shen, Y., Zhou, B.: Image processing using multi-code GAN prior. In: CVPR (2020)
13. Ha, D., Dai, A., Le, Q.V.: Hypernetworks. arXiv preprint arXiv:1609.09106 (2016)
14. Härkönen, E., Hertzmann, A., Lehtinen, J., Paris, S.: GANSpace: discovering interpretable GAN controls. In: NeurIPS (2020)

15. Hart, J.C.: Sphere tracing: a geometric method for the antialiased ray tracing of implicit surfaces. Vis. Comput. **12**, 527–545 (1996)
16. He, K., Zhang, X., Ren, S., Sun, J.: Deep residual learning for image recognition. In: CVPR (2016)
17. Jiang, Y., Ji, D., Han, Z., Zwicker, M.: SDFDiff: differentiable rendering of signed distance fields for 3D shape optimization. In: CVPR (2020)
18. Kanazawa, A., Tulsiani, S., Efros, A.A., Malik, J.: Learning category-specific mesh reconstruction from image collections. In: ECCV (2018)
19. Kar, A., Häne, C., Malik, J.: Learning a multi-view stereo machine. In: NeurIPS (2017)
20. Karras, T., Aila, T., Laine, S., Lehtinen, J.: Progressive growing of GANs for improved quality, stability, and variation. In: ICLR (2018)
21. Karras, T., Aittala, M., Hellsten, J., Laine, S., Lehtinen, J., Aila, T.: Training generative adversarial networks with limited data. In: NeurIPS (2020)
22. Karras, T., et al.: Alias-free generative adversarial networks. In: NeurIPS (2021)
23. Karras, T., Laine, S., Aila, T.: A style-based generator architecture for generative adversarial networks. In: CVPR (2019)
24. Karras, T., Laine, S., Aittala, M., Hellsten, J., Lehtinen, J., Aila, T.: Analyzing and improving the image quality of StyleGan. In: CVPR (2020)
25. Kendall, A., Gal, Y.: What uncertainties do we need in Bayesian deep learning for computer vision? In: NeurIPS (2017)
26. Kulkarni, N., Gupta, A., Fouhey, D.F., Tulsiani, S.: Articulation-aware canonical surface mapping. In: CVPR (2020)
27. Kulkarni, N., Gupta, A., Tulsiani, S.: Canonical surface mapping via geometric cycle consistency. In: ICCV (2019)
28. Li, X., et al.: Self-supervised Single-View 3D Reconstruction via Semantic Consistency. In: Vedaldi, A., Bischof, H., Brox, T., Frahm, J.-M. (eds.) ECCV 2020. LNCS, vol. 12359, pp. 677–693. Springer, Cham (2020). https://doi.org/10.1007/978-3-030-58568-6_40
29. Lin, C.H., Wang, C., Lucey, S.: SDF-SRN: learning signed distance 3D object reconstruction from static images. In: NeurIPS (2020)
30. Liu, F., Kim, M., Jain, A., Liu, X.: Controllable and guided face synthesis for unconstrained face recognition. In: ECCV (2022)
31. Liu, F., Liu, X.: Voxel-based 3D detection and reconstruction of multiple objects from a single image. NeurIPS (2021)
32. Liu, F., Tran, L., Liu, X.: Fully understanding generic objects: modeling, segmentation, and reconstruction. In: CVPR (2021)
33. Liu, F., Zhao, Q., Liu, X., Zeng, D.: Joint face alignment and 3D face reconstruction with application to face recognition. TPAMI (2018)
34. Liu, S., Zhang, Y., Peng, S., Shi, B., Pollefeys, M., Cui, Z.: DIST: rendering deep implicit signed distance function with differentiable sphere tracing. In: CVPR (2020)
35. Liu, S., Chen, W., Li, T., Li, H.: Soft rasterizer: differentiable rendering for unsupervised single-view mesh reconstruction. In: ICCV (2019)
36. Lunz, S., Li, Y., Fitzgibbon, A., Kushman, N.: Inverse graphics GAN: learning to generate 3D shapes from unstructured 2D data. In: NeurIPS (2020)
37. Medin, S.C., et al.: MOST-GAN: 3D morphable StyleGan for disentangled face image manipulation. In: AAAI (2022)
38. Mildenhall, B., Srinivasan, P.P., Tancik, M., Barron, J.T., Ramamoorthi, R., Ng, R.: NeRF: representing scenes as neural radiance fields for view synthesis. In: ECCV (2020)

39. Nguyen-Phuoc, T., Li, C., Theis, L., Richardt, C., Yang, Y.L.: HoloGAN: unsupervised learning of 3D representations from natural images. In: ICCV (2019)
40. Niemeyer, M., Geiger, A.: CAMPARI: camera-aware decomposed generative neural radiance fields. arXiv preprint arXiv:2103.17269 (2021)
41. Niemeyer, M., Geiger, A.: GIRAFFE: representing scenes as compositional generative neural feature fields. In: CVPR (2021)
42. Niemeyer, M., Mescheder, L., Oechsle, M., Geiger, A.: Differentiable volumetric rendering: learning implicit 3D representations without 3D supervision. In: CVPR (2020)
43. Osher, S., Fedkiw, R., Piechor, K.: Level set methods and dynamic implicit surfaces. Appl. Mech. Rev. **57**, B15 (2004)
44. Pan, X., Dai, B., Liu, Z., Loy, C.C., Luo, P.: Do 2D GANs know 3D shape? Unsupervised 3D shape reconstruction from 2D image GANs. In: ICLR (2021)
45. Pan, X., Zhan, X., Dai, B., Lin, D., Loy, C.C., Luo, P.: Exploiting deep generative prior for versatile image restoration and manipulation. TPAMI (2021)
46. Qin, X., Zhang, Z., Huang, C., Dehghan, M., Zaiane, O.R., Jagersand, M.: U2-Net: going deeper with nested u-structure for salient object detection. Patt. Recogn. **106**, 107404 (2020)
47. Rojtberg, P., Pöllabauer, T., Kuijper, A.: Style-transfer GANs for bridging the domain gap in synthetic pose estimator training. In: AIVR (2020)
48. Rossler, A., Cozzolino, D., Verdoliva, L., Riess, C., Thies, J., Nießner, M.: FaceForensics++: learning to detect manipulated facial images. In: ICCV (2019)
49. Schwarz, K., Liao, Y., Niemeyer, M., Geiger, A.: GRAF: generative radiance fields for 3D-aware image synthesis. In: NeurIPS (2020)
50. Shen, Y., Yang, C., Tang, X., Zhou, B.: InterfaceGAN: interpreting the disentangled face representation learned by GANs. TPAMI (2020)
51. Shen, Y., Zhou, B.: Closed-form factorization of latent semantics in GANs. In: CVPR (2021)
52. Shi, Y., Aggarwal, D., Jain, A.K.: Lifting 2D StyleGAN for 3D-aware face generation. In: CVPR (2021)
53. Sitzmann, V., Zollhöfer, M., Wetzstein, G.: Scene representation networks: Continuous 3D-structure-aware neural scene representations. In: NeurIPS (2019)
54. Su, K., Zhou, E., Sun, X., Wang, C., Yu, D., Luo, X.: Pre-trained StyleGAN based data augmentation for small sample brain CT motion artifacts detection. In: Yang, X., Wang, C.-D., Islam, M.S., Zhang, Z. (eds.) ADMA 2020. LNCS (LNAI), vol. 12447, pp. 339–346. Springer, Cham (2020). https://doi.org/10.1007/978-3-030-65390-3_26
55. Suzuki, R., Koyama, M., Miyato, T., Yonetsuji, T., Zhu, H.: Spatially controllable image synthesis with internal representation collaging. arXiv preprint arXiv:1811.10153 (2018)
56. Szabó, A., Meishvili, G., Favaro, P.: Unsupervised generative 3D shape learning from natural images. arXiv preprint arXiv:1910.00287 (2019)
57. Tran, L., Liu, X.: On learning 3D face morphable model from in-the-wild images. TPAMI (2019)
58. Tulsiani, S., Kar, A., Carreira, J., Malik, J.: Learning category-specific deformable 3D models for object reconstruction. TPAMI (2016)
59. Tulsiani, S., Zhou, T., Efros, A.A., Malik, J.: Multi-view supervision for single-view reconstruction via differentiable ray consistency. In: CVPR (2017)
60. Tulyakov, S., Liu, M.Y., Yang, X., Kautz, J.: MoCoGAN: decomposing motion and content for video generation. In: CVPR (2018)

61. Wang, X., Li, Y., Zhang, H., Shan, Y.: Towards real-world blind face restoration with generative facial prior. In: CVPR (2021)
62. Wu, J., Zhang, C., Xue, T., Freeman, W.T., Tenenbaum, J.B.: Learning a probabilistic latent space of object shapes via 3D generative-adversarial modeling. In: NeurIPS (2016)
63. Wu, S., Rupprecht, C., Vedaldi, A.: Unsupervised learning of probably symmetric deformable 3D objects from images in the wild. In: CVPR (2020)
64. Wu, Y., Sun, Z., Song, Y., Sun, Y., Zhong, Y., Shi, J.: Shape-pose ambiguity in learning 3D reconstruction from images. In: AAAI (2021)
65. Xiang, Y., Mottaghi, R., Savarese, S.: Beyond pascal: a benchmark for 3D object detection in the wild. In: WACV (2014)
66. Xu, H., et al.: Digging into uncertainty in self-supervised multi-view stereo. In: ICCV (2021)
67. Yang, T., Ren, P., Xie, X., Zhang, L.: GAN prior embedded network for blind face restoration in the wild. In: CVPR (2021)
68. Yariv, L., et al.: Multiview neural surface reconstruction by disentangling geometry and appearance. In: NeurIPS (2020)
69. Ye, Y., Tulsiani, S., Gupta, A.: Shelf-supervised mesh prediction in the wild. In: CVPR (2021)
70. Yu, F., Seff, A., Zhang, Y., Song, S., Funkhouser, T., Xiao, J.: LSUN: construction of a large-scale image dataset using deep learning with humans in the loop. arXiv preprint arXiv:1506.03365 (2015)
71. Zhang, Y., et al.: Image GANs meet differentiable rendering for inverse graphics and interpretable 3D neural rendering. In: ICLR (2021)
72. Zhang, Y., et al.: DatasetGAN: efficient labeled data factory with minimal human effort. In: CVPR (2021)
73. Zhu, J.Y., Park, T., Isola, P., Efros, A.A.: Unpaired image-to-image translation using cycle-consistent adversarial networks. In: ICCV (2017)
74. Zhu, J.Y., et al.: Visual object networks: image generation with disentangled 3D representation. In: NeurIPS (2018)

InfiniteNature-Zero: Learning Perpetual View Generation of Natural Scenes from Single Images

Zhengqi Li[1(✉)], Qianqian Wang[1,2], Noah Snavely[1], and Angjoo Kanazawa[3]

[1] Google Research, Mountain View, USA
zhengqili@google.com
[2] Cornell Tech, Cornell University, Ithaca, USA
[3] UC Berkeley, Berkeley, USA

Abstract. We present a method for learning to generate unbounded fly-through videos of natural scenes starting from a single view. This capability is learned from a collection of *single photographs*, without requiring camera poses or even multiple views of each scene. To achieve this, we propose a novel self-supervised view generation training paradigm where we sample and render virtual camera trajectories, including cyclic camera paths, allowing our model to learn stable view generation from a collection of single views. At test time, despite never having seen a video, our approach can take a single image and generate long camera trajectories comprised of hundreds of new views with realistic and diverse content. We compare our approach with recent state-of-the-art supervised view generation methods that require posed multi-view videos and demonstrate superior performance and synthesis quality. Our project webpage, including video results, is at https://infinite-nature-zero.github.io.

1 Introduction

There are millions of photos of natural landscapes on the Internet, capturing breathtaking scenery across the world. Recent advances in vision and graphics have led to the ability to turn such images into compelling 3D photos [30,38,70]. However, most prior work can only extrapolate scene content within a limited range of views corresponding to a small head movement. What if, instead, we could step into the picture and fly through the scene like a bird and explore the world in 3D, and see diverse elements like mountain, lakes, and forests appear naturally as we move through the landscape? This challenging new task was recently proposed by Liu *et al.* [43], who called it *perpetual view generation*: given a single RGB image, the goal is to synthesize a video depicting a scene captured from a moving camera with an arbitrary long camera trajectory. Methods that tackle this problem have applications in content creation and virtual reality.

Supplementary Information The online version contains supplementary material available at https://doi.org/10.1007/978-3-031-19769-7_30.

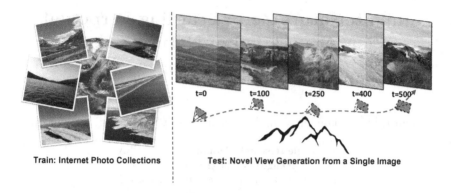

Fig. 1. Learning perpetual view generation from single images. Given a single RGB image input, our approach generates novel views corresponding to a continuous long camera trajectory, without ever seeing a video during training.

However, perceptual view generation is a very challenging problem: as the camera travels through the world, we must fill in unseen missing regions in a harmonious manner, and must resolve new details as scene content approaches the camera, all the while maintaining photo-realism and diversity. Liu *et al.* [43] proposed a supervised solution that generates view sequences in an auto-regressive manner. To train their model, Liu *et al.* (which we will refer to as *Infinite Nature*), require a large dataset of video clips of nature scenes along with per-frame camera poses. In essence, perpetual view generation is a video synthesis task, but the requirement of *posed* video makes data collection very challenging. Obtaining large amounts of diverse, high-quality, and long videos of nature scenes is difficult enough, let alone estimating accurate camera poses on these videos at scale. In contrast, Internet photos of nature landscapes are much easier to collect, and have spurred research into panorama synthesis [42,74], image extrapolation [10,63], image editing [57], and multi-model image synthesis [17,29].

Can we use existing *single*-image datasets for perpetual 3D view generation? In other words, can we learn view generation by simply observing many photos, without requiring video or camera poses? Training with less powerful supervision would seemingly make this already challenging synthesis task even harder. And doing so is not a straightforward application of prior methods. For instance, prior single-image view synthesis methods either require posed multi-view data [36,61, 87], or can only extrapolate within a limited range of viewpoints [28,30,38,70]. Other methods for video synthesis [1,40,79,92] require videos spanning multiple views as training data, and can only generate a limited number of novel frames with no ability to control camera motion at runtime.

In this work, we present a new method for learning perpetual view generation from only a collection of single photos, without requiring multiple views of each scene or camera information. Despite using much less information, our approach improves upon the visual quality of prior methods that require multi-view data. We do so by utilizing *virtual* camera trajectories and computing losses that enable high-quality perpetual view generation results. Specifically, we first introduce a self-supervised view synthesis strategy that utilizes *cyclic* virtual camera trajectories, where we know that the synthesized end frame should be identical to the starting frame. This idea provides the network a training signal for generating a single view synthesis step without multi-view data. Second, to learn to generate a long sequence of novel views we employ an adversarial perpetual view generation training technique, encouraging views along a long virtual camera trajectory to be realistic and generation to be stable. The only requirement for our approach is an off-the-shelf monocular depth network to obtain disparity for the initial frame, but this depth network does not need to be trained on our data. In this sense, our method is self-supervised, leveraging underlying pixel statistics from single-image collections. Because we train with no video data whatsoever, we call our approach *InfiniteNature-Zero*.

We show that training our model using prior video/view generation methods leads to training divergence or mode collapse. We therefore introduce balanced GAN sampling and progressive trajectory growing strategies that stabilize model training. In addition, to prevent artifacts and drift during inference, we propose a global sky correction technique that yields more consistent and realistic synthesis results along long camera trajectories.

We evaluate our method on two public nature scene datasets, and compare to recent supervised video synthesis and view generation methods. We demonstrate superior performance compared to state-of-the-art baselines trained on multi-view collections, even though our model only requires single-view photos during training. To our knowledge, our work is the first to tackle unbounded 3D view generation for natural scenes trained on 2D image collections, and believe this capability will enable new methods for generative 3D synthesis that leverage more limited supervision.

2   Related Work

Image Extrapolation. An inspiring early approach to infinite view extrapolation, called *Infinite Images* was proposed by Kaneva *et al.* [32], which continually retrieves, transforms, and blends imagery from a database to create an infinite 2D landscape. We revisit this idea in a 3D context, which requires inpainting, i.e., filling missing content within an image [25,44,90,91,95], as well as *outpainting*, extending the image and inferring unseen content outside the image boundaries [4,61,63,75,85,88] in order to generate images from novel camera viewpoints. Super-resolution [21,39] is also an important aspect of perpetual view generation, as approaching a distant object requires synthesizing additional high-resolution detail. Image-specific GAN methods demonstrate super-resolution of textures and natural images as a form of image extrapolation [67,72,73,97]. In contrast to the above methods that address these problems individually, our methods handles inpainting, outpainting, and superresolution jointly.

Generative View Synthesis. View synthesis is the problem of generating novel views of a scene from existing views. Many view synthesis methods require multiple views of a scene as input [7,11,19,41,45,47,49,50,60,84,96], though recent works can also generate novel views from a single image [9,31,37,56,62, 69,71,77,78,87]. These methods often require multi-view posed datasets such as RealEstate10k [96]. However, empowered by advances in neural rendering, recent works show that one can unconditionally generate 3D scene representations for 3D-consistent image synthesis [5,16,23,52–54,64]. Many of these methods only require unstructured 2D images for training. When GAN inversion is possible, these methods can also be used for single-image view synthesis, although they have only been demonstrated on specific object categories like faces [5,6]. All of the works above only allow for a limited range of output viewpoints. In contrast, our method can generate new views perpetually, eventually reaching an entirely new distant view, from a single input image. Most related to our work is Liu *et al.* [43], which also performs perpetual view generation. However, Liu *et al.* require posed videos during training. Our method can be trained with unstructured 2D images, and also experimentally achieves better view generation diversity and quality.

Video Synthesis. Our work is also related to video synthesis [13,76], which can be roughly divided into three categories: 1) unconditional video generation [20,46,51,79], which produces a video sequence from an input noise; 2) video prediction [27,40,81–83,86], which generates a video sequence from one or more initial observations; and 3) video-to-video synthesis, which maps a video from a source domain to a target domain. Most video prediction methods focus on generating videos of dynamic objects under a static camera [15,18,40,82,83,89,93], e.g., human motion [3] or the movement of robot arms [18]. In contrast, we focus on generating new views of static nature scenes with a moving camera. Several video prediction methods can also simulate moving cameras [1,14,40,80], but unlike our approach, they require long video sequences for training, do not reason about underlying 3D scene geometry, and do not allow for explicit control over camera viewpoint. More recently, Koh *et al.* propose to navigate and synthesize indoor scenes with controllable camera motion [36]. However, they require ground truth RGBD panoramas as supervision and can only generate novel frames up to 6 steps. Many prior methods in this vein also require 3D inputs, such as voxel grids [24] or dense point clouds [48], whereas we require only a single RGB image.

3   Learning View Generation from Single-Image Collections

We formulate the task of perpetual view generation as follows: given an starting RGB image I_0, generate an image sequence $(\hat{I}_1, \hat{I}_2, ..., \hat{I}_t, ...)$ corresponding to an arbitrary camera trajectory $(c_1, c_2, ..., c_t, ...)$ starting from I_0, where the camera viewpoints c_t can be specified either algorithmically or via user input.

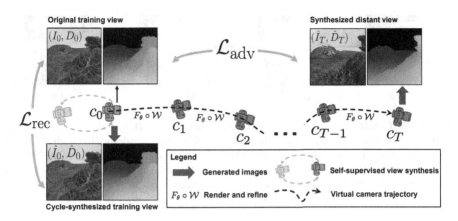

Fig. 2. Self-supervised view generation via virtual cameras. Given a starting RGBD image (I_0, D_0) at viewpoint c_0, our training procedure samples two virtual camera trajectories: 1) a cycle to and back from a single virtual view (dashed orange arrows), creating a *self-supervised view synthesis* signal enforced by the reconstruction loss $\mathcal{L}_{\mathrm{rec}}$. 2) a longer virtual camera path for which we generate corresponding images via the render-refine-repeat process (black dashed arrows and gray cameras). An adversarial loss $\mathcal{L}_{\mathrm{adv}}$ between the final view $(\hat{I}_T, \hat{D}_T)$ and the real image (I_0, D_0) enables the network to learn long-range view generation. (Color figure online)

The prior Infinite Nature method tackles this problem by decomposing it into three phases: **render**, **refine** and **repeat** [43]. Given an RGBD image $(\hat{I}_{t-1}, \hat{D}_{t-1})$ at camera c_{t-1}, the **render** phase renders a new view $(\tilde{I}_t, \tilde{D}_t)$ at c_t by transforming and warping $(\hat{I}_{t-1}, \hat{D}_{t-1})$ using a differentiable 3D renderer $\mathcal{W}$. This yields a warped view $(\tilde{I}_t, \tilde{D}_t) = \mathcal{W}\big((I_{t-1}, D_{t-1}), T_{t-1}^t\big)$, where T_{t-1}^t is an $SE(3)$ transformation from c_{t-1} to c_t. In the **refine** phase, the warped RGBD image $(\tilde{I}_t, \tilde{D}_t)$ is fed into a refinement network F_θ to fill in missing content and add details: $(\hat{I}_t, \hat{D}_t) = F_\theta(\tilde{I}_t, \tilde{D}_t)$. The refined outputs $(\hat{I}_t, \hat{D}_t)$ are then treated as a starting view for the next iteration of the **repeat** step, from which the process iterates. We refer readers to the original work for more details [43].

To supervise a view generation model, Infinite Nature trains on video clips of natural scenes, where each video frame has camera pose derived from structure from motion (SfM) [96]. During training, it randomly chooses one frame in a video clip as the starting view I_0, and performs the render-refine-repeat process along the provided SfM camera trajectory. At a camera viewpoint c_t along the trajectory, a reconstruction loss and an adversarial loss are computed between the image predicted by the network $(\hat{I}_t, \hat{D}_t)$ and the corresponding real RGBD frame (I_t, D_t). However, obtaining long nature videos with accurate camera poses is difficult due to often distant or non-Lambertian contents of landscape scenes (e.g., sea, mountain, and sky). In contrast, our method does not require videos at all, whether with camera poses or not.

We show that 2D photo collections alone provide sufficient supervision signals to learn perceptual view generation, given an off-the-shelf monocular depth

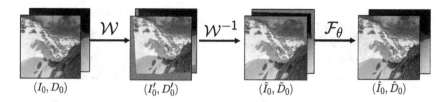

$$(I_0, D_0) \qquad (I_0', D_0') \qquad (\tilde{I}_0, \tilde{D}_0) \qquad (\hat{I}_0, \hat{D}_0)$$

Fig. 3. Self-supervised view synthesis. From a real RGBD image (I_0, D_0), we synthesize an input $(\tilde{I}_0, \tilde{D}_0)$ to a refinement model by cycle-rendering through a virtual viewpoint. From left to right: input image; input rendered to a virtual "previous" view; virtual view rendered *back* to the starting viewpoint; final image $(\hat{I}_0, \hat{D}_0)$ refined with refinement network $\mathcal{F}_\theta$, trained to match the starting image.

prediction network. Our key idea is to sample and render *virtual* camera trajectories starting from the training image, using the refined depth at each frame to warp it to the next view. We generate two kinds of camera trajectories, illustrated in Fig. 2: First, we produce *cyclic* camera trajectories that start and end at the training image. Since the start and end frame should be identical, we can use a reconstruction loss on the initial frame as a self-supervised loss (Sect. 3.1). This self-supervision trains our network to do geometry-aware view refinement during view generation. Second, we synthesize longer virtual camera paths and compute an adversarial loss $\mathcal{L}_{\mathrm{adv}}$ on the final rendered image (Sect. 3.2). This signal trains our network to learn stable view generation over long camera trajectories. The rest of this section describes the two training signals in detail, as well as a sky correction component (Sect. 3.3) that prevents drift in sky regions at test time, yielding more realistic and stable long-range trajectories for nature scenes.

3.1   Self-supervised View Synthesis

In Infinite Nature's supervised learning framework, a reconstruction loss is applied between predicted and corresponding real RGBD images to train the network to refine the inputs rendered from a previous viewpoint. Note that unlike the task of free-form image inpainting [95], this next-view supervision provides a crucial signal for the network to learn to add suitable details and to fill in missing regions around disocclusions using background context, while preserving 3D perspective. Accordingly, we cannot fully simulate the necessary 3D training signals from standard 2D inpainting supervision alone. Instead, our idea is to treat the known real image as the held-out "next" view, and simulate a rendered image input from a virtual "previous" viewpoint. We implement this idea by rendering a *cyclic* virtual camera trajectory starting and ending at the known input training view, then comparing the final rendered image at the end of the cycle to the known ground truth input view. In practice, we find that a cycle including just one other virtual view (i.e., warping to a sampled viewpoint, then rendering back to the input viewpoint) is sufficient. Figure 3 shows an example sequence of views produced in such a cyclic rendering step.

To implement this idea, we first predict the depth D_0 from a real image I_0 using a standard monocular depth network [58]. We randomly sample a nearby viewpoint with relative camera pose T within a set of maximum values for each camera parameter. We then synthesize the view at virtual pose T by rendering (I_0, D_0) to a new image $(I_0', D_0') = \mathcal{W}((I_0, D_0), T)$. Next, to encourage the network to learn to fill in missing content at disocclusions, we create a per-pixel binary mask M_0' derived from the rendered depth D_0' at the virtual viewpoint [30,43]. Finally, we render this virtual view with mask (I_0', D_0', M_0') back to the starting viewpoint via transform T^{-1}: $(\tilde{I}_0, \tilde{D}_0, \tilde{M}_0) = \mathcal{W}((I_0', D_0', M_0'), T^{-1})$ where the rendered mask is element-wise multiplied with the rendered RGBD image. Intuitively, this strategy constructs inputs whose pixel statistics, including blur and missing content, are similar to those produced by warping a view forward to a next viewpoint, yielding naturalistic input to view refinement.

The cycle-rendered images $(\tilde{I}_0, \tilde{D}_0)$ are then fed into the refinement network F_θ, whose outputs $(\hat{I}_0, \hat{D}_0) = F_\theta(\tilde{I}_0, \tilde{D}_0)$ are compared to the original RGBD image (I_0, D_0) to yield a reconstruction loss $\mathcal{L}_{\text{rec}}$. Because this method does not require actual multiple views or SfM camera poses, we can generate an effectively infinite set of virtual camera motions during training. Because the target view is always an input training view we seek to reconstruct, this approach can be thought of as a self-supervised way of training view synthesis.

3.2   Adversarial Perpetual View Generation

Although the insight above enables the network to learn to refine a rendered image, directly applying such a network iteratively during inference over multiple steps quickly degenerates (see third row of Fig. 4). As observed by prior work [43], we must train a synthesis model through multiple recurrently-generated camera viewpoints in order for the view generation to be stable. Therefore, in addition to the self-supervised training in Sect. 3.1, we also train on longer virtual camera trajectories. In particular, during training, for a given input RGBD image (I_0, D_0), we randomly sample a virtual camera trajectory $(c_1, c_2, ..., c_T)$ starting from (I_0, D_0) by iteratively performing render-refine-repeat T times, yielding a sequence of generated views $(\hat{I}_1, \hat{I}_2, ..., \hat{I}_T)$. To prevent the camera from traversing out-of-distribution viewpoints (e.g., crashing into mountains or water) we adopt the auto-pilot algorithm from [43] to sample the camera path. The auto-pilot algorithm determines the pose of the next view based on the proportion of sky and foreground elements as determined by the estimated disparity map at the current viewpoint (see supplemental material for more details). Next, we discuss how we train our model using such sampled virtual camera trajectories.

Balanced GAN Sampling. We now have a generated sequence of views along a virtual camera trajectory from the input image, but we do not have the ground truth sequence corresponding to these views. How can we train the model without such ground truth? We find that it is sufficient to compute an adversarial loss that trains a discriminator to distinguish between real images and the synthesized "fake" images along the virtual camera path. One straightforward implementation of this idea is to treat all T predictions $\{\hat{I}_t, \hat{D}_t\}_{t=1}^T$, along the virtual path as fake samples, and sample T real images randomly from the dataset. However, this strategy leads to unstable training, because there is a significant discrepancy in pixel statistics between the generated view sequence and the set of sampled real photos: a generated sequence along a camera trajectory has frames with similar content with smoothly changing viewpoints, whereas randomly sampled real images from the dataset exhibit completely different content and viewpoints. This vast difference in the distribution of images that the discriminator observes leads to unstable training in conditional GAN settings [24]. To address this issue, we propose a simple but effective technique to stabilize the training. Specifically, for a generated sequence, we only feed the discriminator the generated image $(\hat{I}_T, \hat{D}_T)$ at the last camera c_T as the fake sample, and use its corresponding input image (I_0, D_0) at the starting view as the real sample, as shown in Fig. 2. In this case, the real and fake sample in each batch will exhibit similar content and viewpoint variations. Further, during each training iteration, we randomly sample the length of virtual camera trajectory T between 1 and a predefined maximum length $T_{\max}$, so that the prediction at any viewpoint and step will be sufficiently trained.

Progressive Trajectory Growing. We observe that without the guidance of ground truth sequences, the discriminator quickly gains an overwhelming advantage over the generator at the beginning of training. Similarly to issues explored in prior work on 2D GANs [33,34,68], we find that it takes longer for the network to predict plausible views at more distant viewpoints. As a result, the discriminator will easily distinguish real images from fake ones generated at distant views, and hence offer meaningless gradients to the generator. To address this issue, we propose to progressively grow the length of the virtual camera trajectory. We begin with self-supervised view synthesis as described in Sect. 3.1 and pretrain the model for 200K steps. We then increase the maximum length of the virtual camera trajectory T by 1 every 25K iterations until reaching a predefined maximum length $T_{\max}$. This progressive growing strategy ensures that images rendered at a previous viewpoint c_{t-1} have been sufficiently initialized before being fed to the refinement network to generate the view at the next viewpoint c_t.

3.3   Global Sky Correction

The sky is an indispensable visual element of nature scenes with unique characteristics—it should change much more slowly than the foreground content, since the sky is at infinity. However, we found that the sky synthesized by Infinite Nature can contain unrealistic artifacts after multiple steps. We also found that monocular depth predictions can be inaccurate in sky regions, leading to sky contents to quickly approach the camera in an unrealistic manner.

Therefore, we devise a method to correct the sky regions of refined RGBD images at each test time iteration by leveraging the sky content from the starting view. In particular, we use an off-the-shelf semantic segmentation method [8] and

I_0 w/o BGS   w/o repeat   w/o PTG   w/o SVS   w/o sky   Full

Fig. 4. Generated views after 50 steps with different settings. Each row shows results for a different input image. From left to right: input view; results without balanced GAN sampling; without the adversarial perpetual view generation strategy; without progressive trajectory growing; without self-supervised view synthesis; without global sky correction; full approach.

the predicted disparity map to determine soft sky masks for the starting and for each generated view, which we found to be effective in identifying sky pixels. We then correct the sky texture and disparity at every step by alpha blending the homography-warped sky content from the starting view (warped according to the camera rotation's effect on the plane at infinity) with the foreground content in the current generated view. To avoid redundantly outpainting the same sky regions, we expand the input image and disparity through GAN inversion [10,12] to seamlessly create a canvas of higher resolution and field of view. We refer readers to the supplemental material for more details. As shown in the penultimate column of Fig. 4, when applying global sky correction at test time, sky regions exhibit significantly fewer artifacts, resulting in more realistic generated views.

3.4   Network and Supervision Losses

We adopt a variant of the CoMod-GAN conditional StyleGAN model [95] as our backbone refinement module F_θ. Specifically, F_θ consists of a global encoder and a StyleGAN generator, where the encoder produces a global latent code z_0 from the input view. At each refine step, we co-modulate intermediate feature layers of the StyleGAN generator via concatenation of z_0 and a latent code z mapped from Gaussian noise. The training loss for the generator and discriminator is:

$$\mathcal{L}^F = \mathcal{L}^F_{\text{adv}} + \lambda_1 \mathcal{L}_{\text{rec}}, \quad \mathcal{L}^D = \mathcal{L}^D_{\text{adv}} + \lambda_2 \mathcal{L}_{R_1} \tag{1}$$

where $\mathcal{L}^F_{\text{adv}}$ and $\mathcal{L}^D_{\text{adv}}$ are non-saturated GAN losses [22], applied on the last view from the camera trajectory and the corresponding training image. $\mathcal{L}_{\text{rec}}$ is a reconstruction loss between real images (and depth maps) and their corresponding cycle-synthesized views described in Sec 3.1: $\mathcal{L}_{\text{rec}} = \sum_l ||\phi^l(\hat{I}_0) - \phi^l(I_0)||_1 + ||\hat{D}_0 - D_0||_1$, where ϕ^l is a feature map at scale l from different layers of a pretrained VGG network [65]. $\mathcal{L}_{R_1}$ is a gradient regularization term that is applied to the discriminator during training [35].

Table 1. Quantitative comparisons on the ACID test set. "MV?" indicates whether a method requires (posed) multi-view data for training. We report view synthesis results with two different types of ground truth (shown as X/Y): sequences rendered with 3D Photos [71] (left), and real sequences (right). KID and Style are scaled by 10 and 10^5 respectively. See Sect. 4.4 for descriptions of baselines.

Method	MV?	View Synthesis			View Generation			
		PSNR↑	SSIM↑	LPIPS↓	FID↓	FID$_{sw}$ ↓	KID↓	Style↓
GFVS [62]	Yes	11.3/11.9	0.68/0.69	0.33/0.34	109	117	0.87	14.6
PixelSynth [61]	Yes	20.0/19.7	0.73/0.70	0.19/0.20	111	119	1.12	10.54
SLAMP [1]	Yes	-	-	-	114	138	1.91	15.2
DIGAN [92]	Yes	-	-	-	53.4	57.6	0.43	5.85
Liu *et al.* [43]	Yes	23.0/**21.1**	**0.83/0.74**	0.14/0.18	32.4	37.2	0.22	9.37
Ours	No	**23.5/21.1**	0.81/0.71	**0.10/0.15**	**19.3**	**25.1**	**0.11**	**5.63**

4 Experiments

4.1 Datasets and Baselines

We evaluate our approach on two public datasets of nature scenes: the Landscape High Quality (LHQ) dataset [74], a collection of 90K landscapes photos collected from the Internet, and the Aerial Coastline Imagery Dataset (ACID) [43], a video dataset of nature scenes with SfM camera poses.

On the ACID dataset, where posed video data is available, we compare with several state-of-the-art supervised learning methods. Our main baseline is Infinite Nature, the recent state-of-the-art view generation method designed for natural scenes [43]. We also compare with other recent view and video synthesis methods, including geometry-free view synthesis (GFVS) [62] and PixelSynth [61], both of which are based on VQ-VAE [17,59] for long-range view synthesis. Additionally, we compare with two recent video synthesis methods, SLAMP [1] and DIGAN [92]. Following their original protocols, we train both methods with video clips of 16 frames from the ACID dataset until convergence.

For the LHQ dataset, since there is no multi-view training data and we are unaware of prior methods that can train on single images, we show results from our approach with different configurations, described in more detail in Sect. 4.5.

4.2 Metrics

We evaluate synthesis quality on two tasks that we refer to as *short-range view synthesis* and *long-range view generation*. By (short-range) view synthesis, we mean the ability to render high fidelity views near a source view, and we report standard error metrics between predicted and ground truth views, including PSNR, SSIM and LPIPS [94]. Since there is no multi-view data for LHQ, we create pseudo ground truth images over a trajectory of length 5 from a global LDI mesh [66] computed using 3D Photos [71]; please see the supplemental

Table 2. Ablation study on the LHQ test set. KID and Style are scaled by 10 and 10^5 respectively. See Sect. 4.5 for a description of each baseline.

Method	Configurations					View Synthesis			View Generation			
	$\mathcal{L}_{rec}$	$\mathcal{L}_{adv}$	PTG	BGS	Sky	PSNR↑	SSIM↑	LPIPS↓	FID ↓	FID$_{sw}$ ↓	KID ↓	Style↓
Naive	✓	✓				28.0	0.87	0.07	38.1	52.1	0.25	6.36
w/o BGS	✓	✓	✓		✓	28.0	0.89	0.08	34.9	41.1	0.20	6.45
w/o PTG	✓	✓		✓	✓	28.1	0.90	0.07	35.3	42.6	0.21	6.04
w/o repeat	✓				✓	26.8	0.86	0.15	61.3	85.5	0.40	8.15
w/o SVS		✓	✓	✓	✓	26.6	0.85	0.08	23.4	30.2	0.12	6.37
w/o sky	✓	✓	✓	✓		28.3	0.90	0.07	24.8	31.3	0.11	6.43
Ours (full)	✓	✓	✓	✓	✓	**28.4**	**0.91**	**0.06**	**19.4**	**25.8**	**0.09**	**5.91**

material for more details. On the ACID dataset, we report errors on real video sequences where we use SfM-aligned depth maps to render images from each method. We also report results from ground truth sequences created with 3D Photos, since we observe that in real video sequences, pixel misalignments can also be caused by factors like scene motion and errors in monocular depth or camera poses.

For the task of (long-range) view generation, following prior work [43] we adopt the Fréchet Inception Distance (FID), sliding window FID (FID$_{sw}$) (with window size $\omega = 20$), and Kernel Inception Distance (KID) [2] to measure the synthesis quality of different methods. We also introduce a style consistency metric that computes an average style loss between the starting image and each generated view along a camera trajectory. This metric reflects how much the style of a generated sequence deviates from the original image; we evaluate it over a trajectory of length 50. For FID and KID calculations, we compute real statistics from 50K images randomly sampled from each dataset, and calculate fake statistics from 70K and 100K generated images on ACID and LHQ respectively, where 700 and 1000 test images are used as starting images evaluated over 100 steps. Note that since SLAMP and DIGAN do not support camera viewpoint control, we only evaluate them on the view generation task.

4.3 Implementation Details

We set the maximum camera trajectory length $T_{max} = 10$. The weight of R_1 regularization λ_2 is set to 0.15 and 0.004 for the LHQ and ACID datasets, respectively. During training, we found that treating a predicted view along a long virtual trajectory as ground truth and adding a small self-supervised view synthesis loss over these predictions yields more stable view generation results. Therefore we set the reconstruction weight $\lambda_1 = 1$ for the input training image at the starting viewpoint, and $\lambda_1 = 0.05$ for frames predicted on a camera trajectory. Following [35], we apply lazy regularization to the discriminator gradient regularization every 16 training steps and adopt gradient clipping and exponential moving averaging to update the parameters of the refinement network. For all experiments, we train on centrally cropped images of size 128×128 for 1.8 M

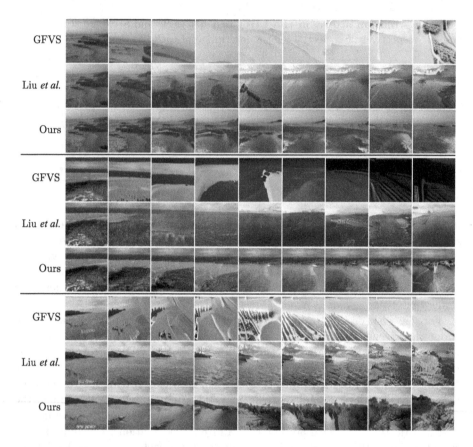

GFVS

Liu *et al.*

Ours

GFVS

Liu *et al.*

Ours

GFVS

Liu *et al.*

Ours

Fig. 5. Qualitative comparisons on the ACID test set. From left to right, we show generated views over trajectories of length 100 for three methods: GFVS [62], Liu *et al.* [43] and Ours.

steps with batch size 32 using 8 NVIDIA A100 GPUs, which takes ~6 days to converge. During rendering, we use softmax splatting [55] to 3D render images via their depth maps. Our method can also generate higher resolution 512×512 views. Rather than directly training the model at high resolution, which would take an estimate of 3 weeks, we train an extra super-resolution module that takes one day to converge using the same self-supervised learning idea. We refer readers to the supplementary material for more details and high-resolution results.

4.4 Quantitative Comparisons

Table 1 shows quantitative comparisons between our approach and other baselines on the ACID test set. Although the model only observes single images, our approach outperforms the other baselines in view generation on all error metrics, while achieving competitive performance on the view synthesis task. Specifically,

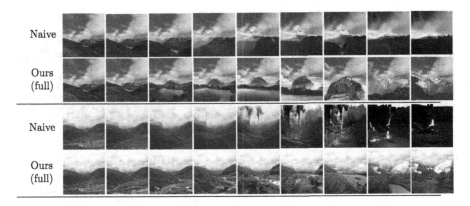

Fig. 6. Qualitative comparisons on the LHQ test set. On two starting views, from left to right, we show generated views over trajectories of length 100 from a naive baseline and our full approach. See Sect. 4.5 for more details.

our approach demonstrates the best FID and KID scores, indicating better realism and diversity for our generated views. Our method also achieves the best style consistency score. For the view synthesis task, we achieve the best LPIPS score over the baselines, suggesting higher perceptual quality for our rendered images. We also obtain PSNR and SSIM errors on the ACID test set that are competitive with the supervised learning method from Infinite Nature, which uses a supervised reconstruction loss computed on real sequences.

4.5   Ablation Study

We perform an ablation study on the LHQ test set to analyze the effectiveness of each component in our proposed system. We test the following configurations: (1) a naive baseline where we apply an adversarial loss between all the predictions along a camera trajectory and a set of randomly sampled real photos, and apply geometry re-grounding as introduced in Infinite Nature [43] at test time (Naive); and configurations without: (2) using balanced GAN sampling (w/o BGS), (3) progressive trajectory growing (w/o PTG), (4) GAN training via long camera trajectories (w/o repeat), (5) applying self-supervised view synthesis (w/o SVS), and (6) employing global sky correction (w/o sky). Quantitative and qualitative comparisons are shown in Table 2 and Fig. 4 respectively. Our full system achieves the best view synthesis and view generation performance of these configurations. In particular, adding self-supervised view synthesis significantly improves view synthesis performance. Training via virtual camera trajectories, adopting introduced GAN sampling/training strategies, and applying global sky correction all improve view generation performance by a large margin.

| t=0 | t=50 | t=100 | t=200 | t=250 | t=300 | t=400 | t=500 |

Fig. 7. Perpetual view generation. Given a single RGB image, we show the results of our method generating sequences of 500 realistic new views of natural scenes without suffering significant drift. Please see video for animated results. (Color figure online)

4.6 Qualitative Comparisons

Figure 5 shows visual comparisons between our approach, Infinite Nature [43], and GFVS [62] on the ACID test set. GFVS quickly degenerates due to the large distance between the input and generated viewpoints. Infinite Nature can generate plausible views over multiple steps, but the content and style of generated views quickly diverge into an unrelated unimodal scene. Our approach, in contrast, not only generates more consistent views with respect to starting images, but also demonstrates significantly improved synthesis quality and realism.

Figure 6 shows visual comparisons between the naive baseline described in Sect. 4.5 and our full approach. The generated views from the baseline quickly deviate from realism due to ineffective training/inference strategies. In contrast, our full approach can generate much more realistic, consistent, and diverse results over long camera trajectories. For example, the views generated by our approach cover diverse and realistic natural elements such as lakes, trees, and mountains.

4.7 Single-Image Perpetual View Generation

Finally, we visualize our model's ability to generate long view trajectories from a single RGB image in Fig. 7. Although our approach only sees single images during training, it learns to generate long sequences of 500 new views depicting realistic natural landscapes, without suffering significant drift or degeneration. We refer readers to the supplemental video for the full effect and to see results generated from different types of camera trajectories.

5 Discussion

Limitations and Future Directions. Our method inherits some limitations from prior video and view generation methods. For example, although our method produces globally consistent background sky, it does not ensure global consistency of foreground content. Addressing this issue potentially requires generating an entire 3D world model, which is an exciting future direction to explore.

In addition, as with Infinite Nature, our method can generate unrealistic views if the desired camera trajectory is not seen during training (e.g., in-place rotation). Alternative generative methods such as VQ-VAE [59] and diffusion models [26] may provide promising paths towards addressing this limitation.

Conclusion. We presented a method for learning perpetual view generation of natural scenes solely from single-view photos, without requiring camera poses and multi-view data. At test time, given a single RGB image, our approach allows for generating hundreds of new views covering realistic natural scenes along a long camera trajectory. We conduct extensive experiments and demonstrate the improved performance and synthesis quality of our approach over prior supervised approaches. We hope this work demonstrates a new step towards unbounded generative view synthesis from Internet photo collections.

References

1. Akan, A.K., Erdem, E., Erdem, A., Guney, F.: SLAMP: stochastic latent appearance and motion prediction. In: Proceedings of the International Conference on Computer Vision (ICCV), pp. 14728–14737 (2021)
2. Bińkowski, M., Sutherland, D.J., Arbel, M., Gretton, A.: Demystifying MMD GANs. In: Proceedings of the International Conference on Learning Representations (ICLR) (2018)
3. Blank, M., Gorelick, L., Shechtman, E., Irani, M., Basri, R.: Actions as space-time shapes. In: Proceedings of the International Conference on Computer Vision (ICCV), vol. 2, pp. 1395–1402. IEEE (2005)
4. Bowen, R.S., Chang, H., Herrmann, C., Teterwak, P., Liu, C., Zabih, R.: OCONet: image extrapolation by object completion. In: Proceedings of the Computer Vision and Pattern Recognition (CVPR), pp. 2307–2317 (2021)
5. Chan, E.R., et al.: Efficient geometry-aware 3D generative adversarial networks. In: Proceedings of the Computer Vision and Pattern Recognition (CVPR) (2022)
6. Chan, E.R., Monteiro, M., Kellnhofer, P., Wu, J., Wetzstein, G.: pi-GAN: periodic implicit generative adversarial networks for 3d-aware image synthesis. In: Proceedings of the Computer Vision and Pattern Recognition (CVPR), pp. 5799–5809 (2021)
7. Chaurasia, G., Duchene, S., Sorkine-Hornung, O., Drettakis, G.: Depth synthesis and local warps for plausible image-based navigation. ACM Trans. Graphics **32**(3), 1–12 (2013)
8. Chen, L.C., Papandreou, G., Schroff, F., Adam, H.: Rethinking atrous convolution for semantic image segmentation. arXiv preprint arXiv:1706.05587 (2017)
9. Chen, X., Song, J., Hilliges, O.: Monocular neural image based rendering with continuous view control. In: Proceedings of the International Conference on Computer Vision (ICCV), pp. 4090–4100 (2019)
10. Cheng, Y.C., Lin, C.H., Lee, H.Y., Ren, J., Tulyakov, S., Yang, M.H.: In&out: diverse image outpainting via GAN inversion. In: Proceedings of the Computer Vision and Pattern Recognition (CVPR) (2022)
11. Choi, I., Gallo, O., Troccoli, A., Kim, M.H., Kautz, J.: Extreme view synthesis. In: Proceedings of the International Conference on Computer Vision (ICCV), pp. 7781–7790 (2019)

12. Chong, M.J., Lee, H.Y., Forsyth, D.: StyleGAN of all trades: image manipulation with only pretrained StyleGAN. arXiv preprint arXiv:2111.01619 (2021)
13. Clark, A., Donahue, J., Simonyan, K.: Adversarial video generation on complex datasets. arXiv preprint arXiv:1907.06571 (2019)
14. Clark, A., Donahue, J., Simonyan, K.: Efficient video generation on complex datasets. arXiv preprint arXiv:1907.06571 (2019)
15. Denton, E., Fergus, R.: Stochastic video generation with a learned prior. In: Proceedings of the International Conference on Machine Learning (ICML), pp. 1174–1183. PMLR (2018)
16. DeVries, T., Bautista, M.A., Srivastava, N., Taylor, G.W., Susskind, J.M.: Unconstrained scene generation with locally conditioned radiance fields. In: Proceedings of the International Conference on Computer Vision (ICCV), pp. 14304–14313 (2021)
17. Esser, P., Rombach, R., Ommer, B.: Taming transformers for high-resolution image synthesis. In: Proceedings of the Computer Vision and Pattern Recognition (CVPR), pp. 12873–12883 (2021)
18. Finn, C., Goodfellow, I., Levine, S.: Unsupervised learning for physical interaction through video prediction. In: Neural Information Processing Systems (2016)
19. Flynn, J., Broxton, M., et al.: Deepview: view synthesis with learned gradient descent. In: Proceedings of the Computer Vision and Pattern Recognition (CVPR), pp. 2367–2376 (2019)
20. Fox, G., Tewari, A., Elgharib, M., Theobalt, C.: StyleVideoGAN: a temporal generative model using a pretrained styleGAN. In: Proceedings of the British Machine Vision Conference (BMVC) (2021)
21. Glasner, D., Bagon, S., Irani, M.: Super-resolution from a single image. In: Proceedings of the International Conference on Computer Vision (ICCV), pp. 349–356. IEEE (2009)
22. Goodfellow, I., et al.: Generative adversarial nets. In: Neural Information Processing Systems (2014)
23. Gu, J., Liu, L., Wang, P., Theobalt, C.: StyleNeRF: a style-based 3D-aware generator for high-resolution image synthesis. arXiv preprint arXiv:2110.08985 (2021)
24. Hao, Z., Mallya, A., Belongie, S., Liu, M.Y.: Gancraft: unsupervised 3D neural rendering of minecraft worlds. In: Proceedings of the International Conference on Computer Vision (ICCV), pp. 14072–14082 (2021)
25. Hays, J., Efros, A.A.: Scene completion using millions of photographs. In: ACM Transactions on Graphics (SIGGRAPH North America) (2007)
26. Ho, J., Jain, A., Abbeel, P.: Denoising diffusion probabilistic models. In: Neural Information Processing Systems. vol. 33, pp. 6840–6851 (2020)
27. Hsieh, J.T., Liu, B., Huang, D.A., Fei-Fei, L.F., Niebles, J.C.: Learning to decompose and disentangle representations for video prediction. In: Neural Information Processing Systems, vol. 31 (2018)
28. Hu, R., Ravi, N., Berg, A.C., Pathak, D.: Worldsheet: wrapping the world in a 3D sheet for view synthesis from a single image. In: Proceedings of the International Conference on Computer Vision (ICCV) (2021)
29. Huang, X., Mallya, A., Wang, T.C., Liu, M.Y.: Multimodal conditional image synthesis with product-of-experts GANs. arXiv preprint arXiv:2112.05130 (2021)
30. Jampani, V., et al.: SLIDE: single image 3D photography with soft layering and depth-aware inpainting. In: Proceedings of the International Conference on Computer Vision (ICCV), pp. 12518–12527 (2021)

31. Jang, W., Agapito, L.: CodeNeRF: disentangled neural radiance fields for object categories. In: Proceedings of the International Conference on Computer Vision (ICCV), pp. 12949–12958 (2021)
32. Kaneva, B., Sivic, J., Torralba, A., Avidan, S., Freeman, W.T.: Infinite images: creating and exploring a large photorealistic virtual space. In: Proceedings of the IEEE (2010)
33. Karnewar, A., Wang, O.: Msg-GAN: multi-scale gradients for generative adversarial networks. In: Proceedings of the Computer Vision and Pattern Recognition (CVPR), pp. 7799–7808 (2020)
34. Karras, T., Aila, T., Laine, S., Lehtinen, J.: Progressive growing of GANs for improved quality, stability, and variation. In: Proceedings of the International Conference on Learning Representations (ICLR) (2018)
35. Karras, T., Laine, S., Aittala, M., Hellsten, J., Lehtinen, J., Aila, T.: Analyzing and improving the image quality of styleGAN. In: Proceedings of the Computer Vision and Pattern Recognition (CVPR), pp. 8110–8119 (2020)
36. Koh, J.Y., Lee, H., Yang, Y., Baldridge, J., Anderson, P.: Pathdreamer: a world model for indoor navigation. In: Proceedings of the International Conference on Computer Vision (ICCV), pp. 14738–14748 (2021)
37. Kopf, J., et al.: One shot 3D photography. ACM Trans. Graph. (Proc. ACM SIGGRAPH) **39**(4), 1–13 (2020)
38. Kopf, J., et al.: One shot 3D photography. In: ACM Transactions on Graphics (SIGGRAPH North America) (2020)
39. Ledig, C., et al.: Photo-realistic single image super-resolution using a generative adversarial network. In: Proceedings of the Computer Vision and Pattern Recognition (CVPR), pp. 4681–4690 (2017)
40. Lee, W., et al.: Revisiting hierarchical approach for persistent long-term video prediction. arXiv preprint arXiv:2104.06697 (2021)
41. Levoy, M., Hanrahan, P.: Light field rendering. In: ACM Transactions on Graphics (SIGGRAPH North America) (1996)
42. Lin, C.H., Cheng, Y.C., Lee, H.Y., Tulyakov, S., Yang, M.H.: InfinityGAN: towards infinite-pixel image synthesis. In: Proceedings of the International Conference on Learning Representations (ICLR) (2022)
43. Liu, A., Tucker, R., Jampani, V., Makadia, A., Snavely, N., Kanazawa, A.: Infinite nature: perpetual view generation of natural scenes from a single image. In: Proceedings of the International Conference on Computer Vision (ICCV), pp. 14458–14467 (2021)
44. Liu, H., Wan, Z., Huang, W., Song, Y., Han, X., Liao, J.: PD-GAN: probabilistic diverse GAN for image inpainting. In: Proceedings of the Computer Vision and Pattern Recognition (CVPR), pp. 9371–9381 (2021)
45. Liu, L., Gu, J., Lin, K.Z., Chua, T.S., Theobalt, C.: Neural sparse voxel fields. NeurIPS (2020)
46. Liu, Y., Shu, Z., Li, Y., Lin, Z., Perazzi, F., Kung, S.Y.: Content-aware GAN compression. In: Proceedings of the Computer Vision and Pattern Recognition (CVPR), pp. 12156–12166 (2021)
47. Lombardi, S., Simon, T., Saragih, J., Schwartz, G., Lehrmann, A., Sheikh, Y.: Neural volumes: learning dynamic renderable volumes from images. ACM Trans. Graph. **38**(4), 1–14 (2019)
48. Mallya, A., Wang, T.-C., Sapra, K., Liu, M.-Y.: World-consistent video-to-video synthesis. In: Vedaldi, A., Bischof, H., Brox, T., Frahm, J.-M. (eds.) ECCV 2020. LNCS, vol. 12353, pp. 359–378. Springer, Cham (2020). https://doi.org/10.1007/978-3-030-58598-3_22

49. Mildenhall, B., et al.: Local light field fusion: practical view synthesis with prescriptive sampling guidelines. In: ACM Transactions on Graphics (SIGGRAPH North America) (2019)
50. Mildenhall, B., Srinivasan, P.P., Tancik, M., Barron, J.T., Ramamoorthi, R., Ng, R.: NeRF: representing scenes as neural radiance fields for view synthesis. In: Vedaldi, A., Bischof, H., Brox, T., Frahm, J.-M. (eds.) ECCV 2020. LNCS, vol. 12346, pp. 405–421. Springer, Cham (2020). https://doi.org/10.1007/978-3-030-58452-8_24
51. Munoz, A., Zolfaghari, M., Argus, M., Brox, T.: Temporal shift GAN for large scale video generation. In: Proceedings Winter Conference on Computer Vision (WACV), pp. 3179–3188 (2021)
52. Nguyen-Phuoc, T., Li, C., Theis, L., Richardt, C., Yang, Y.L.: Hologan: unsupervised learning of 3D representations from natural images. In: The IEEE International Conference on Computer Vision (ICCV) (2019)
53. Niemeyer, M., Geiger, A.: CAMPARI: camera-aware decomposed generative neural radiance fields. In: 2021 International Conference on 3D Vision (3DV), pp. 951–961. IEEE (2021)
54. Niemeyer, M., Geiger, A.: GIRAFFE: representing scenes as compositional generative neural feature fields. In: Proceedings of the Computer Vision and Pattern Recognition (CVPR), pp. 11453–11464 (2021)
55. Niklaus, S., Liu, F.: Softmax splatting for video frame interpolation. In: Proceedings of the Computer Vision and Pattern Recognition (CVPR), pp. 5437–5446 (2020)
56. Niklaus, S., Mai, L., Yang, J., Liu, F.: 3D Ken Burns effect from a single image. ACM Trans. Graphics 38(6), 1–15 (2019)
57. Park, T., et al.: Swapping autoencoder for deep image manipulation. In: Neural Information Processing Systems, pp. 7198–7211 (2020)
58. Ranftl, R., Lasinger, K., Hafner, D., Schindler, K., Koltun, V.: Towards robust monocular depth estimation: mixing datasets for zero-shot cross-dataset transfer. In: Transactions Pattern Analysis and Machine Intelligence (2020)
59. Razavi, A., Van den Oord, A., Vinyals, O.: Generating diverse high-fidelity images with VQ-VAE-2. Neural Information Processing Systems 32 (2019)
60. Riegler, G., Koltun, V.: Free view synthesis. In: Vedaldi, A., Bischof, H., Brox, T., Frahm, J.-M. (eds.) ECCV 2020. LNCS, vol. 12364, pp. 623–640. Springer, Cham (2020). https://doi.org/10.1007/978-3-030-58529-7_37
61. Rockwell, C., Fouhey, D.F., Johnson, J.: Pixelsynth: Generating a 3D-consistent experience from a single image. In: Proceedings of the International Conference on Computer Vision (ICCV), pp. 14104–14113 (2021)
62. Rombach, R., Esser, P., Ommer, B.: Geometry-free view synthesis: transformers and no 3D priors. In: Proceedings of the International Conference on Computer Vision (ICCV), pp. 14356–14366 (2021)
63. Saharia, C., et al.: Palette: image-to-image diffusion models. arXiv preprint arXiv:2111.05826 (2021)
64. Schwarz, K., Liao, Y., Niemeyer, M., Geiger, A.: Graf: Generative radiance fields for 3D-aware image synthesis. Neural Inf. Process. Syst. 33, 20154–20166 (2020)
65. Sengupta, A., Ye, Y., Wang, R., Liu, C., Roy, K.: Going deeper in spiking neural networks: VGG and residual architectures. Front. Neurosci. 13, 95 (2019)
66. Shade, J., Gortler, S., He, L.W., Szeliski, R.: Layered depth images. In: ACM Transactions Graphics (SIGGRAPH North America), pp. 231–242 (1998)

67. Shaham, T.R., Dekel, T., Michaeli, T.: Singan: learning a generative model from a single natural image. In: Proceedings of the International Conference on Computer Vision (ICCV), pp. 4570–4580 (2019)
68. Shaham, T.R., Dekel, T., Michaeli, T.: Singan: learning a generative model from a single natural image. In: Proceedings of the International Conference on Computer Vision (ICCV), pp. 4569–4579 (2019)
69. Shi, L., Hassanieh, H., Davis, A., Katabi, D., Durand, F.: Light field reconstruction using sparsity in the continuous fourier domain. In: ACM Trans. Graphics (SIGGRAPH North America) (2014)
70. Shih, M.L., Su, S.Y., Kopf, J., Huang, J.B.: 3D photography using context-aware layered depth inpainting. In: Proceedings of the Computer Vision and Pattern Recognition (CVPR), pp. 8028–8038 (2020)
71. Shih, M.L., Su, S.Y., Kopf, J., Huang, J.B.: 3D photography using context-aware layered depth inpainting. In: Proceedings of the Computer Vision and Pattern Recognition (CVPR) (2020)
72. Shocher, A., Bagon, S., Isola, P., Irani, M.: InGAN: capturing and remapping the "DNA" of a natural image. In: Proceedings International Conference on Computer Vision (ICCV) (2019)
73. Shocher, A., Cohen, N., Irani, M.: "zero-shot" super-resolution using deep internal learning. In: Proceedings of the Computer Vision and Pattern Recognition (CVPR), pp. 3118–3126 (2018)
74. Skorokhodov, I., Sotnikov, G., Elhoseiny, M.: Aligning latent and image spaces to connect the unconnectable. In: Proceedings of the International Conference on Computer Vision (ICCV), pp. 14144–14153 (2021)
75. Teterwak, P., et al.: Boundless: generative adversarial networks for image extension. In: Proceedings of the International Conference on Computer Vision (ICCV), pp. 10521–10530 (2019)
76. Tian, Y., et al.: A good image generator is what you need for high-resolution video synthesis. arXiv preprint arXiv:2104.15069 (2021)
77. Tucker, R., Snavely, N.: Single-view view synthesis with multiplane images. In: Proceedings of the Computer Vision and Pattern Recognition (CVPR) (2020)
78. Tulsiani, S., Tucker, R., Snavely, N.: Layer-structured 3D scene inference via view synthesis. In: Ferrari, V., Hebert, M., Sminchisescu, C., Weiss, Y. (eds.) ECCV 2018. LNCS, vol. 11211, pp. 311–327. Springer, Cham (2018). https://doi.org/10.1007/978-3-030-01234-2_19
79. Tulyakov, S., Liu, M.Y., Yang, X., Kautz, J.: MoCoGAN: decomposing motion and content for video generation. In: Proceedings of the Computer Vision and Pattern Recognition (CVPR), pp. 1526–1535 (2018)
80. Villegas, R., Pathak, A., Kannan, H., Erhan, D., Le, Q.V., Lee, H.: High fidelity video prediction with large stochastic recurrent neural networks. In: Neural Information Processing Systems (2019)
81. Villegas, R., Yang, J., Hong, S., Lin, X., Lee, H.: Decomposing motion and content for natural video sequence prediction. arXiv preprint arXiv:1706.08033 (2017)
82. Vondrick, C., Pirsiavash, H., Torralba, A.: Generating videos with scene dynamics. In: Neural Information Processing Systems (2016)
83. Vondrick, C., Torralba, A.: Generating the future with adversarial transformers. In: Proceedings of the Computer Vision and Pattern Recognition (CVPR), pp. 1020–1028 (2017)
84. Wang, Q., et al.: IBRNet: learning multi-view image-based rendering. In: Proceedings of the Computer Vision and Pattern Recognition (CVPR), pp. 4690–4699 (2021)

85. Wang, Y., Tao, X., Shen, X., Jia, J.: Wide-context semantic image extrapolation. In: Proceedings of the Computer Vision and Pattern Recognition (CVPR), pp. 1399–1408 (2019)
86. Wang, Y., Long, M., Wang, J., Gao, Z., Yu, P.S.: PredRNN: recurrent neural networks for predictive learning using spatiotemporal LSTMs. In: Neural Information Processing Systems, pp. 879–888 (2017)
87. Wiles, O., Gkioxari, G., Szeliski, R., Johnson, J.: SynSin: end-to-end view synthesis from a single image. In: Proceedings of the Computer Vision and Pattern Recognition (CVPR), pp. 7467–7477 (2020)
88. Yang, Z., Dong, J., Liu, P., Yang, Y., Yan, S.: Very long natural scenery image prediction by outpainting. In: Proceedings of the International Conference on Computer Vision (ICCV), pp. 10561–10570 (2019)
89. Ye, Y., Singh, M., Gupta, A., Tulsiani, S.: Compositional video prediction. In: Proceedings of the International Conference on Computer Vision (ICCV) (2019)
90. Yu, J., Lin, Z., Yang, J., Shen, X., Lu, X., Huang, T.S.: Generative image inpainting with contextual attention. In: Proceedings of the Computer Vision and Pattern Recognition (CVPR), pp. 5505–5514 (2018)
91. Yu, J., Lin, Z., Yang, J., Shen, X., Lu, X., Huang, T.S.: Free-form image inpainting with gated convolution. In: Proceedings of the International Conference on Computer Vision (ICCV), pp. 4471–4480 (2019)
92. Yu, S., et al.: Generating videos with dynamics-aware implicit generative adversarial networks. In: Proceedings of the International Conference on Learning Representations (ICLR) (2022)
93. Yu, S., et al.: Generating videos with dynamics-aware implicit generative adversarial networks. In: The Tenth International Conference on Learning Representations (2022)
94. Zhang, R., Isola, P., Efros, A.A., Shechtman, E., Wang, O.: The unreasonable effectiveness of deep features as a perceptual metric. In: Proceedings of the Computer Vision and Pattern Recognition (CVPR) (2018)
95. Zhao, S., et al.: Large scale image completion via co-modulated generative adversarial networks. In: Proceedings of the International Conference on Learning Representations (ICLR) (2021)
96. Zhou, T., Tucker, R., Flynn, J., Fyffe, G., Snavely, N.: Stereo magnification: learning view synthesis using multiplane images. In: ACM Transactions on Graphics (SIGGRAPH North America) (2018)
97. Zhou, Y., Zhu, Z., Bai, X., Lischinski, D., Cohen-Or, D., Huang, H.: Non-stationary texture synthesis by adversarial expansion. In: ACM Transactions on Graphics (SIGGRAPH North America) (2018)

Semi-supervised Single-View 3D Reconstruction via Prototype Shape Priors

Zhen Xing[1,2], Hengduo Li[3], Zuxuan Wu[1,2]($\boxtimes$), and Yu-Gang Jiang[1,2]

[1] Shanghai Key Laboratory of Intelligent Information Processing, School of CS,
Fudan University, Shanghai, China
zxwu@fudan.edu.cn
[2] Shanghai Collaborative Innovation Center on Intelligent Visual Computing,
Shanghai, China
[3] University of Maryland, College Park, USA

Abstract. The performance of existing single-view 3D reconstruction methods heavily relies on large-scale 3D annotations. However, such annotations are tedious and expensive to collect. Semi-supervised learning serves as an alternative way to mitigate the need for manual labels, but remains unexplored in 3D reconstruction. Inspired by the recent success of semi-supervised image classification tasks, we propose SSP3D, a semi-supervised framework for 3D reconstruction. In particular, we introduce an attention-guided prototype shape prior module for guiding realistic object reconstruction. We further introduce a discriminator-guided module to incentivize better shape generation, as well as a regularizer to tolerate noisy training samples. On the ShapeNet benchmark, the proposed approach outperforms previous supervised methods by clear margins under various labeling ratios, (*i.e.*, 1%, 5% , 10% and 20%). Moreover, our approach also performs well when transferring to real-world Pix3D datasets under labeling ratios of 10%. We also demonstrate our method could transfer to novel categories with few novel supervised data. Experiments on the popular ShapeNet dataset show that our method outperforms the zero-shot baseline by over 12% and we also perform rigorous ablations and analysis to validate our approach. Code is available at https://github.com/ChenHsing/SSP3D.

Keywords: Semi-supervised learning · 3D Reconstruction · Shape priors

1 Introduction

Reconstructing 3D shape from RGB images plays an important role in many applications, such as 3D printing, virtual reality and 3D scene understanding.

Supplementary Information The online version contains supplementary material available at https://doi.org/10.1007/978-3-031-19769-7_31.

Human can easily infer 3D shape and scene object from single-view images mainly because of the powerful shape priors of human visual systems, yet it remains challenging to model such strong priors for accurate single-view 3D reconstruction. While Structure From Motion(SFM) [24] and Simultaneous Localization and Mapping (SLAM) [3] are feasible solutions, they require abundant data annotations and inferring camera parameters.

Recently, with the growing interest in deep learning, great success has been achieved in predicting 3D shape from a single image with deep Convolutional Neural Networks (CNNs) [6,34,39]. But there are still limitations of these methods: (i) The astounding performance comes at the cost of massive amount of labeled images with fine-grained 3D shape, which is time-consuming and labour-intensive to obtain. (ii) Inferring 3D shape from a single image is an ill-posed problem because there are multiple plausible shapes given a 2D image.

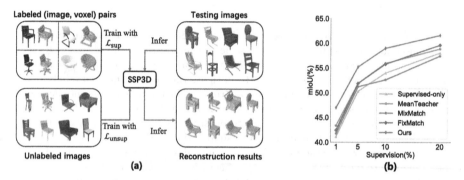

Fig. 1. (a) Illustration of semi-supervised single-view 3D reconstruction. Our SSP3D can predict 3D shape for an unlabeled image after training with a mixture of labeled data and unlabeled data. (b) Our proposed model can efficient leverage the unlabeled data and outperform supervised-only method and state-of-the-art semi-supervised image recognition extended methods.

Semi-Supervised learning (SSL) is a popular strategy to learn in the low-data regime by leveraging the readily available unlabeled data, which has demonstrated great success for image classification [1,29,35] and object detection [19]. Generalizing best practices [27,29] that work well in the 2D domain to 3D reconstruction, while appealing, is challenging. On one hand, it remains unclear how to evaluate the quality of 3D shape pseudo labels, which are the core for SSL. On the other hand, inferring the actual 3D shape of an object from a single image requires strong shape priors, yet existing single-view 3D reconstruction methods [6,39] require a large amount of annotated data to learn the shape priors implicitly with the model parameters. As a result, the 3D reconstruction network trained with limited annotations will likely produce low-quality reconstruction results, especially for the images with heavy occlusion.

To tackle these challenges, we propose a semi-supervised learning framework with several components specially designed for single-view 3D reconstruction as shown in Fig. 1. Inspired by the recent advances in SSL for image classification [27,29], we use the teacher-student pseudo labeling method as the training paradigm of our framework. In order to generate more reliable pseudo labels for the unlabeled images, we use a Prototype Attentive Module for providing shape priors explicitly. In particular, we first obtain 3D prototype shape as candidate shape priors through clustering algorithms (e.g., KMeans). For a given image, we extract the image feature through a 2D encoder. The relationship of image feature and 3D prototype is captured with the help of the attention mechanism to obtain the shape priors, which serve as a bridge to encourage perceptually realistic reconstruction and prevent mode collapses [6,39].

In addition, we introduce a module named Shape Naturalness Module that serves as a discriminator distinguishing predicted 3D shapes from ground-truth 3D shapes. During training, an additional loss is used to penalize unnatural reconstruction results from the model in a generative adversarial learning manner such that the model is incentivized to generate more realistic 3D shapes. Meanwhile, the output of the discriminator can be directly used as an approximation of the quality of pseudo labels such that the inaccurate pseudo labels can be ignored or down-weighted accordingly when training the student model. In conclusion, the main contributions of this paper is summarized as follows:

- We propose a semi-supervised prototype 3D reconstruction network (SSP3D) to reconstruct 3D shapes from a single RGB image. Our work is the first attempt to reconstruct 3D volume in semi-supervised learning with only 1% labeled data of train set.
- Without additional information, an effective yet lightweight shape prior fusion module is proposed, which can be easily incorporated into 3D reconstruction networks with similar architecture. In addition, the discriminator module we proposed guides the generation of natural shapes and serves as a scorer to filter out noisy training samples for the student model.
- We are the first to establish a semi-supervised benchmark to measure the single-view 3D reconstruction network. Experiments show that our model achieves the state-of-the-art on two datasets and settings under various labeling ratios. We hope that our results serve a strong baseline to encourage future research in more robust semi-supervised 3D reconstruction methods.

2 Related Work

Deep Learning for 3D Reconstruction. Recently, deep learning techniques have been widely used for 3D reconstruction. 3D-R2N2 [6] is among the earliest work exploring the 3D reconstruction based on Recurrent Neural Network. It establishes a benchmark for 3D reconstruction with a synthetic ShapeNet dataset. 3D-VAE-GAN [37] builds upon Variational Autoencoders (VAE) and Generative Adversarial Networks (GANs) to reconstruct 3D shapes. OGN [30] and Matryoshka Networks [22] use octree and nested shape layers to represent 3D volumes of objects, respectively. Marrnet [36], ShapeHD [38] and GenRe [46] adopt 2.5D information such as depth, silhouette and surface normal of RGBs as intermediate shape priors to reconstruct 3D shapes. Pix2Vox [39] and Pix2Vox++ [40] build robust backbones for 3D volume reconstruction and achieve state-of-the-art results with encoder-decoder architectures. Mem3D [43] requires a great extra storage space to provide shape priors, which limits its applicability. EVolT [33] and 3D-RETR [25] leverage transformers as backbone networks to reconstruct 3D shapes. Unlike most existing work that are trained in a supervised manner, we explore semi-supervised learning for 3D reconstruction.

Deep Semi-supervised Learning. The overall purpose of semi-supervised learning (SSL) is to effectively use unlabeled data without relying on any manual supervision to expand supervised learning when the labeled training data is scarce. Recent semi-supervised methods mainly contain two principles: data augmentations and consistency regularization. The model is expected to be consistent and robust to data augmentations—producing consistent outputs for the original and augmented inputs. Many methods use different data augmentations [1,16,23] or dropout [29] of models to generate images of different transformations. Researchers also use multiple networks to generate different views of the same input data [21], or mix input data to generate training data and labels [10,13,44,45]. In single-view 3D reconstruction, Semi-supervised Soft Rasterizer (SSR) [17] and [42] try to reconstruct 3D objects with few amount of annotation data, but they all rely on the annotations of additional camera pose or silhouette. To the best of knowledge, the settings of SSL with only single-view image have not been studied in 3D reconstruction, a complex and challenging task that depends on fine-grained human annotations.

3   Method

Problem Definition. For a single-view image x of any object, the goal is to reconstruct the 3D shape y of the object. As discussed earlier, current methods for single-view 3D reconstruction typically require large amount of annotations that are time-consuming and labour-intensive to obtain. We thus explore developing a semi-supervised learning framework for the task to alleviate the need of annotated data during training.

Suppose we have N training samples, including N_L labeled image-3D pairs $(x_l, y_l) \in D_L$ and N_U unlabeled image data $(x_u) \in D_U$. As in prior work, D_L and D_U are sampled from the same data distribution (*e.g.*, either synthetic or real-world). Our purpose is to leverage D_L and D_U together to train the model for an improved performance on reconstructing the 3D shapes of objects.

Overview. As shown in Fig. 2, our framework SSP3D contains two training stages: Warm-up stage and Teacher-student mutual learning stage. In the Warm-up stage, the available labeled set D_L is used to train a "teacher" model; in the Teacher-student mutual learning stage, the teacher model first generates pseudo labels (*i.e.*, predicted 3D shapes) for the unlabeled set D_U, and then a

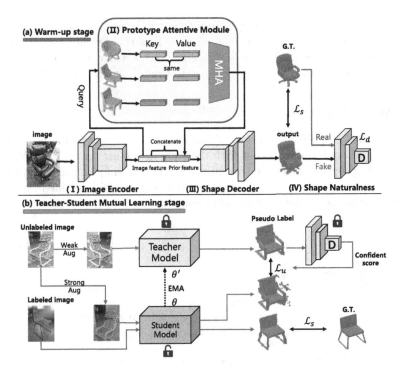

Fig. 2. Overview of Our SSP3D. SSP3D consists of two stages. **Warm-up**: we use available supervised data to train 3D reconstruction network. **Teacher-Student mutual learning stage**: for unsupervised data, *Teacher* with fixed parameters generate pseudo-labels to train *Student*. At the same time, *Teacher* and *Student* are given weakly and strongly augmented inputs respectively. In order to avoid the interference of pseudo-labels noise generated by the *Teacher*, we give a confidence score weight to the unsupervised loss by discriminator. The knowledge learned by *Student* online is slowly transferred to the weight replication mode of *Teacher* through exponential moving average (EMA). When the reconstruction network is trained and converged in the Warm-up stage, we switch to the Teacher-Student mutual learning stage.

"student" model – initialized from the pre-trained teacher model – is trained on D_L and D_U for an improved performance. For effective distillation, strong data augmentation is applied on the input to student model. The teacher model also temporally aggregates the weights from the student model to produce more refined pseudo labels.

While appealing, directly extending existing SSL methods like Mean-Teacher [29] and FixMatch [27] for single-view 3D reconstruction is challenging since the pseudo labels from the teacher model can be quite noisy for two main reasons: 1) inferring accurate 3D shape from single-view image requires strong prior that is difficult to learn without massive annotated data; 2) it is unclear how to evaluate the quality of the predicted pseudo 3D shapes to filter out inaccurate predictions. To this end, we propose two modules namely Prototype Attention Module and Shape Naturalness Module to address these challenges.

In the following text, we first introduce our proposed model components in Warm-up stage (Sect. 3.1), and then we show how the Teacher-student mutual learning stage works with pseudo labelling and teacher refinement methods (Sect. 3.2). Finally, we elaborate the optimization of our framework in Sect. 3.3.

3.1 Warm-Up Stage

As shown in Fig. 2, SSP3D consists of four modules, among which image encoder and shape decoder are consistent with the state-of-the-art method Pix2Vox [39], whereas the proposed prototype attentive module and shape naturalness module will be presented below. At this stage, the teacher model is trained on labeled set D_L in standard supervised learning manner.

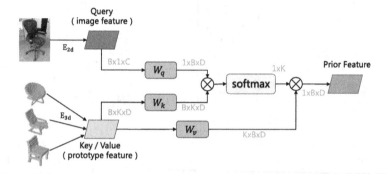

Fig. 3. Overview of our prototype attentive module.

Prototype Attentive Module. In most existing work on single-view 3D reconstruction, the shape priors are learned implicitly with the parameters of the model [6,33,39], which may lead to poor performance for some noisy or occluded images [6], especially when annotated training data is not abundant. Therefore, standard 3D reconstruction models are likely to produce noisy and inaccurate pseudo labels when used as the teacher model directly, resulting in poor performance when training the student model under such semi-supervised learning setting.

To tackle this problem, we propose to augment the image features with learned category-specific shape priors explicitly, so that the strong priors learned from labeled data could help infer more realistic and natural object shapes.

For supervised data, we obtain the shape prototype $P_i{}^k$ of the specified categories by doing K-Means clustering on the features learned by a 3D Autoencoder[1], and we take the clustering center of these categories as the prototype shape priors. The designed attention-based shape priors acquisition mechanism is shown in Fig. 3. Firstly, image encoder extracts the 2D feature of the image

---

[1] Please refer to Appendix for more details.

as query in Eq. 1. Secondly, we extract feature of the prototype through 3D encoder to obtain the prototype feature in Eq. 2, which is used as the key and value in the attention mechanism [31]. We then use three separate linear layers parameterized by W_q, W_k and W_v to extract query, key, value embedding Q, K and V in Eq. 3. Formally, the shape prior feature can be obtained by multi-head attention (MHA) [31] in Eq. 4.

$$\text{Image features: } Query = \texttt{Encoder2d}(I_q), \tag{1}$$

$$\text{Prototype features: } Key, Value = \texttt{Encoder3d}(P_i), \tag{2}$$

$$Q = Query \cdot W_q, \quad K = Key \cdot W_k, \quad V = Value \cdot W_v, \tag{3}$$

$$\text{Prior features} = \texttt{MHA}(Q, K, V). \tag{4}$$

Here, I_q is the query image, P_i indicates the prototype 3D voxel, $W_q \in \mathbb{R}^{C \times D}$, W_k, $W_v \in \mathbb{R}^{D \times D}$ are learnable matrices. In the previous work using shape priors [32,43], 3D voxel can be directly used as shape priors in the form of additional inputs, however they can not capture the correlations between the images and multiple prototype shape priors. In contrast, we use the attention-based module to extract the shape priors by exploring the association between image features and 3D prototypes.

Shape Naturalness Module. The shape reconstruction network typically uses only one supervised loss during training, yet the inherent uncertainty of the loss will lead to unrealistic and inaccurate prediction of object shapes especially on object surface.

Inspired by [38], we develop a shape naturalness module that servers as a discriminator distinguishing predicted shape and the corresponding ground-truth shape, and penalizes the network in an adversarial learning manner when unnatural shapes are generated.

Unlike [37] and [38] which use a pre-trained 3D-GAN as a discriminator to judge whether a shape is real, our framework is learned in an end-to-end generative adversarial training manner. In particular, we take parts (I)–(III) in Fig. 2 as the generator, and the shape naturalness module is used to distinguish the generated shape from the real shape. The optimization is achieved by minimizing the following loss $\mathcal{L}_d$:

$$\mathcal{L}_d = \mathbb{E}_{y_p \sim D_p} \log D(\mathbf{y_p}) + \mathbb{E}_{y_g \sim D_g} \log(1 - D(\mathbf{y_g})), \tag{5}$$

where D_p and D_g are predicted and groundtruth distributions, y_p and y_g are samples in D_p and D_g respectively, and D is the discriminator here.

3.2 Teacher-Student Mutual Learning Stage

Overview. After the teacher model converges in the Warm-up stage, it is used to produce pseudo labels on unlabeled images to supervise the student model. For effective and efficient distillation, we initialize the student model with the weights of the teacher model and apply strong data augmentations on input images to the student following the common SSL paradigm [27]. On the other hand, the teacher takes weakly augmented images as inputs and aggregates the weights of the student temporally throughout the Teacher-Student Mutual Learning stage to generate more reliable pseudo labels.

Student Learning. To utilize the readily available unlabeled images D_u, we use the pseudo-labeling method to generate labels for D_u to train the student model, which has been shown effective for semi-supervised image classification [27,29] and object detection [18,19].

Formally, for unsupervised data, the teacher model first generates the soft label $\hat{I}$ in voxel, in which each voxel entry belongs to $[0,1]$. We first binarize it into hard labels, where each entry in the 3D voxel is binarized as follows:

$$
I(i,j,k) = \begin{cases} 1, & \hat{I}(i,j,k) > \delta \qquad\qquad (6) \\ 0, & \text{otherwise} \qquad\qquad (7) \end{cases}
$$

We then train the student model by taking the binarized pseudo label I as the ground truth. In addition, we jointly train students with the same amount of unsupervised and supervised data in each mini-batch to ensure that the model is not biased by pseudo labels.

Confidence Scores for Pseudo Label. The predictions from the teacher model are more or less inaccurate compared with the ground-truth shapes. Therefore, a filtering mechanism is desired to keep only the mostly accurate predictions as pseudo labels to train the student model. Existing semi-supervised classification methods often use the confidence scores predicted by the network as a proxy and only keep the confident predictions as pseudo labels through applying pre-defined thresholds [41] or using Top-k selection [27]. However, such confidence scores are missing in 3D reconstruction, and a new solution is needed to measure the quality of the generated 3D pseudo labels.

To this end, the shape naturalness module is also designed to serve as a naturalness "scorer" directly. In particular, the sigmoid-normalized output of the discriminator naturally indicates the possibility that an output sample is real or fake since the discriminator is optimized by a binary cross entropy loss using label 1 for real ground truth, and 0 for generated shape. We therefore use this output as the confidence score to measure the quality of generated pseudo label. The confidence score can be used to reweight the unsupervised loss, which will be described in detail in Sect. 3.3.

Teacher Refinement. In order to obtain more refined pseudo labels, we use exponential moving average (EMA) to gradually update the teacher model with the weights of the student model. The slow updating process of teacher model can be considered as an ensemble of student models at different training time stamps. The update rule is defined below:

$$
\theta_t \leftarrow \alpha\theta_t + (1-\alpha)\theta_s, \qquad\qquad (8)
$$

where α is momentum coefficient. In order to make the training process more stable, we slowly increase α to 1 through cosine design as in [7]. This method has been proved to be effective in many existing works, such as self-supervised learning [9,11], SSL image classification [29] and SSL object detection method [18,19]. Here, we are the first to introduce it and validate its effectiveness in semi-supervised 3D reconstruction to the best of our knowledge.

3.3 Training Paradigm

The training process is completed in two stages. In the Warm-up stage, we adopt reconstruction loss and GAN loss jointly and train the teacher model on D_L. In the Teacher-Student mutual learning stage, the generator part is duplicated as two models (Teacher and Student). The parameters of teacher and discriminator are fixed in this stage. We only optimize students through supervised and unsupervised losses.

Reconstruction Loss. For the 3D reconstructions network, both the reconstruction prediction and the ground truth are in the form of voxels. We follow previous works [20,32,39,40] that adopt binary cross entropy loss as the reconstruction loss function:

$$\mathcal{L}_{rec} = \frac{1}{r_v^3} \sum_{i=1}^{r_v^3} [gt_i \log(pr_i) + (1 - gt_i) \log(1 - pr_i)], \tag{9}$$

where r_v represents the resolution of the voxel space, pr and gt represent the predict and the ground truth volume.

Warm-Up Loss. In the Warm-up stage, all parts of the models are end-to-end trained on labeled set D_L. The objective function is:

$$\min_{\theta_f} \max_{\theta_d} \mathcal{L}_{rec}(\theta_f) + \lambda_d \mathcal{L}_d(\theta_d). \tag{10}$$

where θ_f and θ_d are the parameter of generator and discriminator, respectively. λ_d is the balance parameter of loss terms. We set λ_d to 1e-3 here.

Teacher-Student Mutual Loss. At the second stage, for supervised data, we use the BCE loss function as in Eq. 9. For unlabeled data, we use the loss function below:

$$\mathcal{L}_{unsup} = \sum_{i=1}^{n} \text{score}_i (\hat{y}_i - y_i)^2, \tag{11}$$

where y_i and $\hat{y}_i$ are the target and predicted shapes, respectively. $score_i$ denotes the confidence score of $\hat{y}_i$ output by the discriminator. Note that we used squared L2 loss or the Birer score [2] instead of binary cross entry loss in the optimization of unsupervised data. The Brier score is widely used in semi-supervised literature because it is bounded and does not severely penalize the probability of being far away from the ground truth. Our initial experiments show that square L2 loss results in slightly better performance than binary cross entropy.

The loss function for training the student model is shown below:

$$\mathcal{L} = \mathcal{L}_{rec} + \lambda_u \mathcal{L}_{unsup}. \tag{12}$$

where λ_u is the balance parameter of loss terms, which is set as 5 here. Through the joint training of supervised loss and unsupervised loss, we can make full use of labeled and unlabeled data to achieve better performance.

4 Experiments

4.1 Experimental Setup

Datasets. We use ShapeNet [4] and Pix3D [28] in our experiments. The ShapeNet [4] is described in 3D-R2N2 [6], which has 13 categories and 43,783 3D models. Following the split defined in Pix2Vox [39], we randomly divide the training set into supervised data and unlabeled data based on the ratio of labeled samples, *i.e.*, 1%, 5%, 10% and 20%. The voxel resolution of ShapeNet is 32^3. Pix3D [28] is a large-scale benchmark with image-shape pairs and pixel level 2D-3D alignment containing 9 categories. We follow the standard S1-split, which contains 7,539 train images and 2,530 test images as in Mesh R-CNN [8]. Because Pix3D is loosely annotated (*i.e.*, an image may contain more than one object but only one object is labeled), we use ground-truth bounding boxes to cut all the images as [39,40]. Similarly, we randomly sample 10% of the training set as labeled data and use the remaining samples as unlabeled data. The voxel resolution of Pix3D is 128^3 and we have also changed the network parameters accordingly following common practice [40].

Evaluation Metric. We used Intersection over Union (IoU) for the evaluation metric as in [6,39]. It is defined as follows:

$$\text{IoU} = \frac{\sum_{i,j,k} \mathcal{F}(\hat{p}_{(i,j,k)} > t)\mathcal{F}(p_{(i,j,k)})}{\sum_{i,j,k} \mathcal{F}[\mathcal{F}(\hat{p}_{(i,j,k)} > t) + \mathcal{F}(p_{(i,j,k)})]}, \tag{13}$$

where $\hat{p}_{(i,j,k)}$ and $p_{(i,j,k)}$ represent the predicted possibility and the value of ground truth at voxel entry (i,j,k), respectively. $\mathcal{F}$ is a shifted unit step function and t represents the threshold, which is set to 0.3 in our experiments.

Implementation Details. In both stages, the batch size is set to 32, and the learning rate decays from $1e-3$ to $1e-4$. We use Adam [15] as the optimizer. We set α to 0.9996, the number of clusters for prototypes to 3, and the number of multi-head of attention to 2. The δ is set to 0.3. We train the network for 250 epochs in the Warm-up stage and 100 epochs in the Teacher-Student mutual learning stage.

4.2 Main Results

Baseline. We compare our approach with various baselines and direct extensions of popular semi-supervised approaches for 2D image classification. Firstly,

Table 1. Comparisons of single-view 3D object reconstruction on ShapeNet at 32^3 resolution with different labeling ratios. We report the mean IoU (%) of all categories. The best number for each category is highlighted in bold.

Approach\ split	1% 301 labels 30596 images	5% 1527 labels 30596 images	10% 3060 labels 30596 images	20% 6125 labels 30596 images
Supervised (ICCV'19)	41.13	50.32	53.99	58.06
Mean-Teacher (NeurIPS'17)	43.36 (↑2.23)	51.92 (↑1.60)	55.93 (↑1.94)	58.88 (↑0.82)
MixMatch (NeurIPS'19)	41.77 (↑0.64)	51.23 (↑0.91)	52.62 (↓1.37)	57.43 (↓0.63)
FixMatch (NeurIPS'20)	42.44 (↑1.31)	51.89 (↑1.57)	55.79 (↑1.80)	59.63 (↑1.57)
SSP3D (ours)	**46.99 (↑5.86)**	**55.23 (↑4.91)**	**58.98 (↑4.99)**	**61.64 (↑3.58)**

we consider the encoder-decoder architecture of Pix2Vox [39] as our supervised baseline. Note that we change the backbone from VGG19 [26] to ResNet-50 [12] for decreasing parameters following Pix2Vox++ [40]. Secondly, we extend state-of-the-art SSL methods for image classification such as MeanTeacher [29], Mix-Match [1] and FixMatch [27], to the task of 3D reconstruction, which serve as strong semi-supervised baselines. We use the same backbone and experimental settings for all the baselines and our approach for fair comparisons. More details of implementation could be found in Appendix.

Results on ShapeNet and Pix3D. As shown in Table 1, we compare our method with the supervised-only models under the settings of 1%, 5%, 10% and 20% labeled data. The experimental results show that our model outperforms supervised baselines by clear margins, especially under the setting with only 1% labels where our model outperforms the supervised model by 5.86%. Notably, our model outperforms the latest SOTA method FixMatch [27] by 4.55% with only 1% labeled data, demonstrating that directly extending existing SSL methods is sub-optimal for the task of single-view 3D reconstruction and that the proposed prototype attentive module and shape naturalness module are effective.

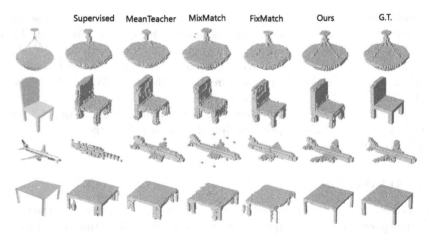

Fig. 4. Examples of single-view 3D Reconstruction on ShapeNet with 5% labels.

546   Z. Xing et al.

Table 2. Comparisons of single-view 3D object reconstruction on Pix3D at 128^3 resolution with 10% supervised data. We report the mean IoU (%) of all categories. The best number for each category is highlighted in bold.

Approach\ split	Chair 267/2672	Bed 78/781	Bookcase 28/282	Desk 54/546	Misc 4/48	Sofa 153/1532	Table 145/1451	Tool 3/36	Wardrobe 18/189	Mean
Supervised [39]	19.27	32.10	23.99	25.32	18.37	62.29	22.77	11.32	81.88	29.80
MeanTeacher [29]	21.66 (↑2.39)	35.04 (↑2.94)	18.88 (↓5.11)	26.17 (↑0.85)	22.37 (↑4.00)	64.19 (↑1.90)	24.03 (↑1.26)	9.18 (↓2.14)	84.34 (↑2.46)	31.40 (↑1.60)
FixMatch [27]	21.95 (↑2.68)	26.69 (↓5.41)	16.06 (↓7.93)	22.12 (↓3.20)	17.87 (↓0.50)	63.74 (↑1.45)	20.64 (↓2.13)	6.89 (↓4.43)	84.45 (↑2.57)	30.35 (↑0.55)
SSP3D (ours)	**23.97** (↑3.04)	**46.33** (↑13.36)	**32.77** (↑7.01)	**32.89** (↑4.76)	**24.35** (↑2.96)	**68.32** (↑5.76)	**23.84** (↑2.13)	**39.06** (↑27.58)	**89.59** (↑6.88)	**35.39** (↑5.59)

We further conduct experiments on Pix3D [28]. The experimental results are shown in Table 2. Considering that the 3D voxel resolution of Pix3D is 128^3, which increases the model complexity, we compare it with the supervised method and the two state-of-the-art methods of MeanTeacher [29] and FixMatch [27] only under the setting of 10% labeled data due to limited GPU resources. We report the reconstruction performance of each category. For some categories with few labeled data (e.g., tool and bed), our model outperforms supervised models by 27.58% and 13.36%. Due to the scarcity of annotated training data and that the other two methods (MeanTeacher and FixMatch) do not have the guidance of strong shape priors, they do not perform as well as supervised methods. Overall, our method outperforms supervised methods by 5.59% measured by on the mean IoU of all categories. Compared with MeanTeacher [29], it is also better by 4.99%.

We further provide qualitative results on both datasets in Fig. 4 and Fig. 5. As can be seen from Fig. 4, for images with clean background, our method produces a smoother object surface than baseline methods. For data with complex background and heavy occlusions in Fig. 5, shapes generated by our method are much better than alternative methods.

Transferring to Novel Category Results. Wallace et al. [32] propose a few-shot setting for single-view 3D reconstruction via shape priors. We also train the model with seven base categories and finetune the model with only 10 labeled data in the novel categories. During inference, we report the performance on novel categories. We also compare our method with CGCE [20] and PADMix [5], as shown in Table 3. Under the 10-shot setting, our method outperforms the zero-shot baseline by 12%. We hypothesize that the improvement is mainly due to our more reasonable shape prior module design as well as the usage of a large number of unlabeled data.

4.3 Ablation Study

In this section, we evaluate the effectiveness of proposed modules and the impact of hyper-parameters. The experiments are under the 1% ShapeNet setting and 10% Pix3D setting if not mentioned elsewhere.

Fig. 5. Examples of single-view 3D Reconstruction on Pix3D with 10% labels.

Prototype Attentive Module. Here we analyze the effectiveness of shape prior module in 3D reconstruction. To do this, we compare our method with various prior aggregation methods including totally removing the prototype attention module (w/o PAM), fusing class-specific prototypes through averaging (w. average) and using LSTM [14] fusion for prototype shape priors. As shown in Table 4, removing the prototype attentive module results in a large drop of 3.55% and 4.17% in performance on both datasets, demonstrating the effectiveness of using class-specific shape priors for single-view 3D reconstruction. Our prior module also outperforms all other prior aggregation methods, indicating that self-attention mechanism is better at capturing the relationships between input images and class-specific prototype shape priors.

Shape Naturalness Module. In order to demonstrate the effectiveness of the shape naturalness module, we remove the module (w/o SNM), that is, remove the GAN loss $\mathcal{L}_d$ and only use $\mathcal{L}_{rec}$ to optimize the network in the warm-up stage. In addition, we also verify the effectiveness of the confidence scores generated

Table 3. Comparison of single-view 3D object reconstruction on novel categories of ShapeNet at 32^3 resolution under 10-shot setting. We report the mean IoU(%) per category. The best number for each category is highlighted in bold.

	Cabinet	Sofa	Bench	Watercraft	Lamp	Firearm	**Mean**
Zero-shot	69	52	37	28	19	13	36
Wallace (ICCV'19) [32]	69 (↑0)	54 (↑2)	36 (↓1)	36 (↑8)	19 (↑0)	24 (↑11)	39 (↑3)
CGCE (ECCV'20) [20]	71 (↑2)	54 (↑2)	37 (↑0)	41 (↑13)	20 (↑1)	23 (↑10)	41 (↑5)
PADMiX (AAAI'22) [5]	66 (↓3)	57 (↑5)	41 (↑4)	46 (↑18)	31 (↑12)	39 (↑26)	47 (↑11)
SSP3D(ours)	**72** (↑3)	**61** (↑9)	**43** (↑6)	**49** (↑21)	**31** (↑12)	34 (↑21)	**48** (↑12)

Table 4. Ablation study of different modules and losses. We report the mean IoU(%) of both datasets.

	PAM	Average	LSTM	SNM	$\mathcal{L}_{unsup}$	$\mathcal{L}_{BCE}$	$\mathcal{L}_{rec}$	Score	ShapeNet	Pix3D
Baseline								✓	41.13 (↓5.86)	29.80 (↓5.59)
w/o PAM				✓	✓		✓	✓	43.44 (↓3.55)	31.32 (↓4.17)
w average		✓		✓	✓		✓	✓	43.80 (↓3.19)	32.64 (↓2.75)
w LSTM			✓	✓	✓		✓	✓	44.61 (↓2.38)	33.42 (↓1.97)
w/o SNM	✓				✓		✓	-	45.87 (↓1.12)	33.90 (↓1.49)
w/o score	✓			✓	✓		✓		46.32 (↓0.67)	34.62 (↓0.77)
w $\mathcal{L}_{BCE}$	✓			✓		✓	✓	✓	45.85 (↓1.14)	34.02 (↓1.37)
SSP3D(ours)	✓			✓	✓		✓	✓	**46.99**	**35.39**

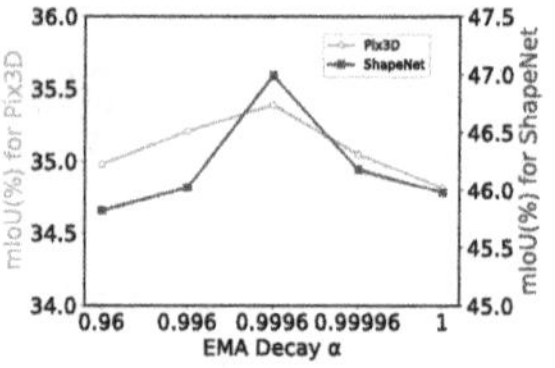

Fig. 6. Ablation study of different EMA decay α.

by the discriminator through replacing all the confidence scores as 1 (w/o score) and check the performance. Experiments show that the performance drops 1.12% and 0.67% on ShapeNet without SNM and scorer respectively, indicating that SNM plays an important role in our framework, which may avoid unnatural 3D shape generation, and the confidence score could avoid the negativeness of noisy or biased labels.

EMA and Loss. We also verify the effect of EMA. In our experiments, we find that EMA decay coefficient $\alpha = 0.9996$ gives the best validation performance. As shown in Fig. 6, the performance slightly drops at different decay rates. For the unsupervised loss function in the Teacher-student mutual learning stage, if the unsupervised squared L2 loss is replaced by binary cross entropy (w $\mathcal{L}_{BCE}$), the performance of the model will also drop 1.14% and 1.37% in performance on two datasets shown in Table 4.

5   Conclusion

We introduced SSP3D, which is the first semi-supervised approach for single-view 3D reconstruction. We presented an effective prototype attentive module for semi-supervised setting to cope with limited annotation data. We also used a discriminator to evaluate the quality of pseudo-labels so as to generate better shapes. We conducted extensive experiments on multiple benchmarks and the results demonstrate the effectiveness of the proposed approach. In future work, we would like to explore the semi-supervised setting on other 3D representation, such as mesh or implicit function.

Acknowledgement. Y.-G. Jiang was sponsored in part by "Shuguang Program" supported by Shanghai Education Development Foundation and Shanghai Municipal Education Commission (No. 20SG01). Z. Wu was supported by NSFC under Grant No. 62102092.

References

1. Berthelot, D., Carlini, N., Goodfellow, I., Papernot, N., Oliver, A., Raffel, C.A.: Mixmatch: a holistic approach to semi-supervised learning. In: NeurIPS (2019)
2. Brier, G.W., et al.: Verification of forecasts expressed in terms of probability. Monthly Weather Rev. **78**, 1–3 (1950)
3. Cadena, C., et al.: Past, present, and future of simultaneous localization and mapping: toward the robust-perception age. IEEE Trans. Rob. **32**, 1309–1332 (2016)
4. Chang, A.X., et al.: Shapenet: an information-rich 3D model repository. arXiv preprint arXiv:1512.03012 (2015)
5. Cheng, T.Y., Yang, H.R., Trigoni, N., Chen, H.T., Liu, T.L.: Pose adaptive dual mixup for few-shot single-view 3D reconstruction. In: AAAI (2022)
6. Choy, C.B., Xu, D., Gwak, J.Y., Chen, K., Savarese, S.: 3D-R2N2: a unified approach for single and multi-view 3D object reconstruction. In: Leibe, B., Matas, J., Sebe, N., Welling, M. (eds.) ECCV 2016. LNCS, vol. 9912, pp. 628–644. Springer, Cham (2016). https://doi.org/10.1007/978-3-319-46484-8_38
7. Ge, C., Liang, Y., Song, Y., Jiao, J., Wang, J., Luo, P.: Revitalizing CNN attention via transformers in self-supervised visual representation learning. In: NeurIPS (2021)
8. Gkioxari, G., Malik, J., Johnson, J.: Mesh r-cnn. In: ICCV (2019)
9. Grill, J.B., et al.: Bootstrap your own latent-a new approach to self-supervised learning. In: NeurIPS (2020)
10. Guo, H., Mao, Y., Zhang, R.: Mixup as locally linear out-of-manifold regularization. In: AAAI (2019)
11. He, K., Fan, H., Wu, Y., Xie, S., Girshick, R.: Momentum contrast for unsupervised visual representation learning. In: CVPR (2020)
12. He, K., Zhang, X., Ren, S., Sun, J.: Deep residual learning for image recognition. In: CVPR (2016)
13. Hendrycks, D., Mu, N., Cubuk, E.D., Zoph, B., Gilmer, J., Lakshminarayanan, B.: Augmix: a simple data processing method to improve robustness and uncertainty. In: ICLR (2020)
14. Hochreiter, S., Schmidhuber, J.: Long short-term memory. Neural Comput. **9**, 1735–1780 (1997)
15. Kingma, D.P., Ba, J.: Adam: a method for stochastic optimization. In: ICLR (2015)
16. Laine, S., Aila, T.: Temporal ensembling for semi-supervised learning. In: ICLR (2017)
17. Laradji, I., Rodríguez, P., Vazquez, D., Nowrouzezahrai, D.: SSR: semi-supervised soft rasterizer for single-view 2D to 3D reconstruction. In: ICCVW (2021)
18. Li, H., Wu, Z., Shrivastava, A., Davis, L.S.: Rethinking pseudo labels for semi-supervised object detection. In: AAAI (2022)
19. Liu, Y.C., et al.: Unbiased teacher for semi-supervised object detection. In: ICLR (2021)

20. Michalkiewicz, M., Parisot, S., Tsogkas, S., Baktashmotlagh, M., Eriksson, A., Belilovsky, E.: Few-shot single-view 3-D object reconstruction with compositional priors. In: Vedaldi, A., Bischof, H., Brox, T., Frahm, J.-M. (eds.) ECCV 2020. LNCS, vol. 12370, pp. 614–630. Springer, Cham (2020). https://doi.org/10.1007/978-3-030-58595-2_37

21. Qiao, S., Shen, W., Zhang, Z., Wang, B., Yuille, A.: Deep co-training for semi-supervised image recognition. In: ECCV (2018)

22. Richter, S.R., Roth, S.: Matryoshka networks: predicting 3D geometry via nested shape layers. In: CVPR (2018)

23. Sajjadi, M., Javanmardi, M., Tasdizen, T.: Regularization with stochastic transformations and perturbations for deep semi-supervised learning. In: NeurIPS (2016)

24. Schonberger, J.L., Frahm, J.M.: Structure-from-motion revisited. In: CVPR (2016)

25. Shi, Z., Meng, Z., Xing, Y., Ma, Y., Wattenhofer, R.: 3D-RETR: end-to-end single and multi-view 3D reconstruction with transformers. In: BMVC (2021)

26. Simonyan, K., Zisserman, A.: Very deep convolutional networks for large-scale image recognition. In: ICLR (2015)

27. Sohn, K., et al.: Fixmatch: simplifying semi-supervised learning with consistency and confidence. In: NeurIPS (2020)

28. Sun, X., et al.: Pix3D: dataset and methods for single-image 3D shape modeling. In: CVPR (2018)

29. Tarvainen, A., Valpola, H.: Mean teachers are better role models: weight-averaged consistency targets improve semi-supervised deep learning results. In: NeurIPS (2017)

30. Tatarchenko, M., Dosovitskiy, A., Brox, T.: Octree generating networks: efficient convolutional architectures for high-resolution 3D outputs. In: ICCV (2017)

31. Vaswani, A., et al.: Attention is all you need. In: NeurIPS (2017)

32. Wallace, B., Hariharan, B.: Few-shot generalization for single-image 3D reconstruction via priors. In: ICCV (2019)

33. Wang, D., et al.: Multi-view 3D reconstruction with transformers. In: ICCV (2021)

34. Wang, N., Zhang, Y., Li, Z., Fu, Y., Liu, W., Jiang, Y.G.: Pixel2mesh: generating 3D mesh models from single RGB images. In: ECCV (2018)

35. Weng, Z., Yang, X., Li, A., Wu, Z., Jiang, Y.G.: Semi-supervised vision transformers. In: ECCV (2022)

36. Wu, J., Wang, Y., Xue, T., Sun, X., Freeman, B., Tenenbaum, J.: Marrnet: 3D shape reconstruction via 2.5D sketches. In: NeurIPS (2017)

37. Wu, J., Zhang, C., Xue, T., Freeman, W.T., Tenenbaum, J.B.: Learning a probabilistic latent space of object shapes via 3D generative-adversarial modeling. In: NeurIPS (2016)

38. Wu, J., Zhang, C., Zhang, X., Zhang, Z., Freeman, W.T., Tenenbaum, J.B.: Learning shape priors for single-view 3D completion and reconstruction. In: ECCV (2018)

39. Xie, H., Yao, H., Sun, X., Zhou, S., Zhang, S.: Pix2vox: context-aware 3D reconstruction from single and multi-view images. In: ICCV (2019)

40. Xie, H., Yao, H., Zhang, S., Zhou, S., Sun, W.: Pix2vox++: multi-scale context-aware 3D object reconstruction from single and multiple images. IJCV 128, 2919–2935 (2020)

41. Yalniz, I.Z., Jégou, H., Chen, K., Paluri, M., Mahajan, D.: Billion-scale semi-supervised learning for image classification. arXiv preprint arXiv:1905.00546 (2019)

42. Yang, G., Cui, Y., Belongie, S., Hariharan, B.: Learning single-view 3D reconstruction with limited pose supervision. In: ECCV (2018)

43. Yang, S., Xu, M., Xie, H., Perry, S., Xia, J.: Single-view 3D object reconstruction from shape priors in memory. In: CVPR (2021)
44. Yun, S., Han, D., Oh, S.J., Chun, S., Choe, J., Yoo, Y.: Cutmix: regularization strategy to train strong classifiers with localizable features. In: ICCV (2019)
45. Zhang, H., Cisse, M., Dauphin, Y.N., Lopez-Paz, D.: mixup: beyond empirical risk minimization. In: ICLR (2018)
46. Zhang, X., Zhang, Z., Zhang, C., Tenenbaum, J., Freeman, B., Wu, J.: Learning to reconstruct shapes from unseen classes. In: NeurIPS (2018)

Bilateral Normal Integration

Xu Cao[1]([✉])[iD], Hiroaki Santo[1][iD], Boxin Shi[2,3][iD], Fumio Okura[1][iD], and Yasuyuki Matsushita[1][iD]

[1] Osaka University, Suita, Japan
{cao.xu,santo.hiroaki,okura,yasumat}@ist.osaka-u.ac.jp
[2] NERCVT, School of Computer Science, Peking University, Beijing, China
shiboxin@pku.edu.cn
[3] Peng Cheng Laboratory, Shenzhen, China

Abstract. This paper studies the discontinuity preservation problem in recovering a surface from its surface normal map. To model discontinuities, we introduce the assumption that the surface to be recovered is *semi-smooth*, *i.e.*, the surface is one-sided differentiable (hence one-sided continuous) everywhere in the horizontal and vertical directions. Under the semi-smooth surface assumption, we propose a bilaterally weighted functional for discontinuity preserving normal integration. The key idea is to relatively weight the one-sided differentiability at each point's two sides based on the definition of one-sided depth discontinuity. As a result, our method effectively preserves discontinuities and alleviates the under- or over-segmentation artifacts in the recovered surfaces compared to existing methods. Further, we unify the normal integration problem in the orthographic and perspective cases in a new way and show effective discontinuity preservation results in both cases (Source code is available at https://github.com/hoshino042/bilateral_normal_integration.).

Keywords: Normal integration · Discontinuity preservation · Semi-smooth surface · One-sided differentiability · Photometric shape recovery

1 Introduction

Photometric shape recovery aims at high-fidelity three-dimensional (3D) surface reconstruction by exploiting the shading information. Representative methods include photometric stereo [29] and shape from polarization [17]. These methods typically estimates shape in the form of a surface normal map (Fig. 1(a)). To recover the 3D surface, it is needed to integrate the surface normals, which is called the normal integration problem [24]. Therefore, the normal integration plays a key role in photometric surface recovery.

Despite the importance, most normal integration methods are limited to recovering smooth surfaces. Namely, the target surface is assumed differentiable

---

Supplementary Information The online version contains supplementary material available at https://doi.org/10.1007/978-3-031-19769-7_32.

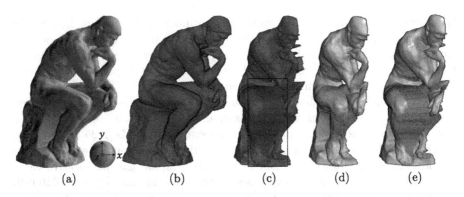

Fig. 1. The discontinuity preservation problem in normal integration. **(a)** RGB color-coded normal map rendered by a perspective camera ("The Thinker by Auguste Rodin" CC BY 4.0. https://sketchfab.com/3d-models/the-thinker-by-auguste-rodin-08a1e693c9674a3292dec2298b09e0ae.). **(b,c)** Corresponding ground truth surface (front and side views). The red box highlights the depth gap at occlusion boundary. **(d)** Under the smooth surface assumption [3], the integrated surface is wrongly connected at occlusion boundary. **(e)** Under the semi-smooth surface assumption, our method preserves large depth gaps at occlusion boundary even in the perspective case.

(hence continuous) everywhere [3, 24, 33]. However, this assumption is violated when the depths abruptly change at the occlusion boundaries, as shown in Fig. 1(b) and (c). In such a case, applying a method with the smooth surface assumption [3] yields distorted surfaces, as shown in Fig. 1(d).

Unfortunately, preserving the discontinuities in normal integration remains an open problem. So far, several assumptions on the discontinuities have been introduced to ease the problem. By assuming that discontinuities exist sparsely, robust estimators-based methods have been studied [2,7]. By assuming that discontinuity locations are short curves in the integration domain, the Mumford-Shah weighted approach has been proposed [25]. However, these methods can be fragile depending on scenes as they only statistically model the discontinuities, while the distribution of discontinuity locations of real surfaces can be arbitrary.

This paper instead introduces the *semi-smooth* surface assumption, which models the depth discontinuity directly from its definition. We assume that even if the surface is discontinuous at a point, it is discontinuous at only one side but not both sides of the point. To determine the discontinuous side for each point, we design a weight function based on the definition of one-sided depth discontinuity. Intuitively, if the depth gap at one side of a point is much larger than the other side, then the side with a larger depth gap is more likely to be discontinuous. We will show that this weight function design idea can be naturally derived from the semi-smooth surface assumption.

Under the semi-smooth surface assumption, we propose a bilaterally weighted functional for discontinuity preserving normal integration. In our functional, we approximate the normal vector observed at a point from that point's two sides in each of the horizontal and vertical directions. Using our weight functions, the unreliable approximation from the discontinuous side is ignored, while the reliable approximation from the continuous side is kept. In this way, the surface can be accurately recovered without being affected by discontinuities. As our functional considers and compares the two sides of each point, we term our method as "bilateral" normal integration (BiNI).

Experimental results show that our method can faithfully locate and preserve discontinuities over the iterative optimization. Our method also reduces the number of hyperparameters that are needed in existing methods [2,25] to only one, which makes it easy to use in practice. In addition, we present a new unification of the normal integration problem in the orthographic and perspective cases, which allows us to treat the two cases in the same manner by our bilaterally weighted functional. While existing discontinuity preservation methods focus on the orthographic case [2,25,31], we confirm effective discontinuity preservation results for the first time in the perspective case, as shown in Fig. 1(e).

In summary, this paper's contributions are

- a bilaterally weighted functional under the semi-smooth surface assumption,
- its numerical solution method, which can effectively preserve the discontinuities, and
- new unified normal integration equations that cover both orthographic and perspective cases.

2   Proposed Method

The goal of normal integration is to estimate the depth or height map of a surface given its normal map and corresponding camera projection parameters. In this section, we first derive the unified partial differential equations (PDEs) relating the depth map to its normal map in the orthographic and perspective cases in Sect. 2.1. We then describe our semi-smooth surface assumption and present the bilaterally weighted functional in Sect. 2.2. Finally, we present in Sect. 2.3 the solution method for the proposed functional. We will discuss our method's differences to the related work in Sect. 3.

2.1   Unified Normal Integration Equations

Let $\mathbf{p} = [x, y, z]^\top \in \mathbb{R}^3$ be a surface point in a 3D space, and $\mathbf{n}(\mathbf{p}) = [n_x, n_y, n_z]^\top \in \mathcal{S}^2 \subset \mathbb{R}^3$ be the unit surface normal vector at the surface point $\mathbf{p}$. When the surface is observed by a camera, the surface point and its normal vector are projected in the image plane with coordinates $\mathbf{u} = [u, v]^\top \in \mathbb{R}^2$. Therefore, we can parameterize the surface and its normal map as vector-valued functions $\mathbf{p}(\mathbf{u}) = [x(\mathbf{u}), y(\mathbf{u}), z(\mathbf{u})]^\top$ and $\mathbf{n}(\mathbf{u}) = [n_x(\mathbf{u}), n_y(\mathbf{u}), n_z(\mathbf{u})]^\top$ respectively. By definition, the normal vector $\mathbf{n}(\mathbf{u})$ is orthogonal to the tangent plane to the surface at the point $\mathbf{p}(\mathbf{u})$. Hence, $\mathbf{n}(\mathbf{u})$ is orthogonal to the two tangent vectors in the tangent plane at $\mathbf{p}(\mathbf{u})$:

$$\mathbf{n}^\top \partial_u \mathbf{p} = 0 \quad \text{and} \quad \mathbf{n}^\top \partial_v \mathbf{p} = 0. \tag{1}$$

Here, ∂_u and ∂_v are partial derivatives with respect to u and v, and we omit the dependencies of $\mathbf{p}$ and $\mathbf{n}$ on $\mathbf{u}$ for brevity.

We consider the normal maps observed on a closed and connected subset Ω in the image plane (i.e., $\mathbf{u} \in \Omega \subset \mathbb{R}^2$) under orthographic or perspective projection. We now discuss the problem formulation in these two cases.

Orthographic Case: Under orthographic projection,

$$\mathbf{p}(\mathbf{u}) = \begin{bmatrix} u \\ v \\ z(\mathbf{u}) \end{bmatrix}, \quad \partial_u \mathbf{p} = \begin{bmatrix} 1 \\ 0 \\ \partial_u z \end{bmatrix}, \quad \text{and} \quad \partial_v \mathbf{p} = \begin{bmatrix} 0 \\ 1 \\ \partial_v z \end{bmatrix}. \tag{2}$$

Inserting Eq. (2) into Eq. (1) results in a pair of PDEs

$$n_z \partial_u z + n_x = 0 \quad \text{and} \quad n_z \partial_v z + n_y = 0. \tag{3}$$

Perspective Case: Let f be the camera's focal length and $[c_u, c_v]^\top$ be the coordinates of the principal point in the image plane, the surface is then $\mathbf{p}(\mathbf{u}) = z(\mathbf{u}) [(u - c_u)/f, (v - c_v)/f, 1]^\top$. The two tangent vectors are

$$\partial_u \mathbf{p} = \begin{bmatrix} \frac{1}{f}((u-c_u)\partial_u z + z) \\ \frac{1}{f}(v-c_v)\partial_u z \\ \partial_u z \end{bmatrix} \quad \text{and} \quad \partial_v \mathbf{p} = \begin{bmatrix} \frac{1}{f}(u-c_u)\partial_v z \\ \frac{1}{f}((v-c_v)\partial_v z + z) \\ \partial_v z \end{bmatrix}. \tag{4}$$

Similar to [6,7,24], we introduce a log depth map $\tilde{z}(\mathbf{u})$ satisfying $z(\mathbf{u}) = \exp(\tilde{z}(\mathbf{u}))$ to unify the formulations. By chain rule, we have

$$\partial_u z = z \partial_u \tilde{z} \quad \text{and} \quad \partial_v z = z \partial_v \tilde{z}. \tag{5}$$

Plugging Eq. (5) into Eq. (4) leads to

$$\partial_u \mathbf{p} = z \begin{bmatrix} \frac{1}{f}((u-c_u)\partial_u \tilde{z} + 1) \\ \frac{1}{f}(v-c_v)\partial_u \tilde{z} \\ \partial_u \tilde{z} \end{bmatrix} \quad \text{and} \quad \partial_v \mathbf{p} = z \begin{bmatrix} \frac{1}{f}(u-c_u)\partial_v \tilde{z} \\ \frac{1}{f}((v-c_v)\partial_v \tilde{z} + 1) \\ \partial_v \tilde{z} \end{bmatrix}. \tag{6}$$

Further plugging Eq. (6) into Eq. (1) cancels out z. Rearranging the remaining terms yields

$$\begin{cases} (n_x(u - c_u) + n_y(v - c_v) + n_z f)\partial_u \tilde{z} + n_x = 0 \\ (n_x(u - c_u) + n_y(v - c_v) + n_z f)\partial_v \tilde{z} + n_y = 0 \end{cases}. \tag{7}$$

Denoting $\tilde{n}_z = n_x(u - c_u) + n_y(v - c_v) + n_z f$ simplifies Eq. (7) as

$$\tilde{n}_z \partial_u \tilde{z} + n_x = 0 \quad \text{and} \quad \tilde{n}_z \partial_v \tilde{z} + n_y = 0, \tag{8}$$

which are in the same form as the orthographic counterpart Eq. (3). We can pre-compute $\tilde{n}_z$ from the normal map and camera parameters. Once the log dpeth map $\tilde{z}$ is estimated, we can exponentiate it to obtain the depth map z.

We have unified the PDEs in the orthographic and perspective cases as Eq. (3) and (8). In Sect. 2.2 and 2.3, we will not distinguish between orthographic and perspective cases.

2.2 Bilaterally Weighted Functional

We now describe our bilaterally weighted functional for discontinuity preserving normal integration. The proposed functional can be applied to either Eq. (3) or (8) depending on the camera projection model. Without loss of generality, we will use the notations in Eq. (3) hereafter. To pave the way for the bilaterally weighted functional under the semi-smooth surface assumption, we first discuss the traditional quadratic functional under the smooth surface assumption.

When assuming a smooth surface, the target surface is differentiable everywhere, $i.e.$, the partial derivatives $\partial_u z$ and $\partial_v z$ exist everywhere. We can therefore minimize the quadratic functional to find the depth map

$$\min_z \iint_\Omega (n_z \partial_u z + n_x)^2 + (n_z \partial_v z + n_y)^2 \, du \, dv. \tag{9}$$

When a function is differentiable at a point, it is also one-sided differentiable at the point's two sides. The one-sided partial derivatives hence exist at both sides of the point horizontally ($\partial_u^+ z$ and $\partial_u^- z$) and vertically ($\partial_v^+ z$ and $\partial_v^- z$), and are equal to the partial derivative, $i.e.$, $\partial_u^+ z = \partial_u^- z = \partial_u z$ and $\partial_v^+ z = \partial_v^- z = \partial_v z$. Therefore, the quadratic functional (9) under the smooth surface assumption is equivalent to

$$\min_z \iint_\Omega 0.5(n_z \partial_u^+ z + n_x)^2 + 0.5(n_z \partial_u^- z + n_x)^2$$
$$+ 0.5(n_z \partial_v^+ z + n_y)^2 + 0.5(n_z \partial_v^- z + n_y)^2 \, du \, dv, \tag{10}$$

where the one-sided partial derivatives are defined as

$$\partial_u^+ z = \lim_{h \to 0^+} \frac{z(u+h,v) - z(u,v)}{h}, \quad \partial_u^- z = \lim_{h \to 0^-} \frac{z(u+h,v) - z(u,v)}{h},$$
$$\partial_v^+ z = \lim_{h \to 0^+} \frac{z(u,v+h) - z(u,v)}{h}, \quad \partial_v^- z = \lim_{h \to 0^-} \frac{z(u,v+h) - z(u,v)}{h}. \tag{11}$$

Considering the one-sided differentiability at a point's two sides leads to our semi-smooth surface assumption. We assume a semi-smooth surface can be indifferentiable (hence discontinuous) at and only at one side of a point in each of the horizontal and vertical directions. As illustrated in Fig. 2, this assumption contains three cases. At differentiable points, a semi-smooth surface is also guaranteed to be both left- and right-differentiable (Fig. 2(a), both $\partial_u^+ z$ and $\partial_u^- z$ exist). Unlike a smooth surface, a semi-smooth surface can be one-sided indifferentiable at one-sided discontinuous points (Fig. 2(b) and (c), either $\partial_u^+ z$ or $\partial_u^- z$ exists)[1]. On the other hand, a semi-smooth surface does not contain any point that is indifferentiable from its both sides in the horizontal or vertical direction (Fig. 2(d) and (e), the case neither $\partial_u^+ z$ nor $\partial_u^- z$ exists is not allowed).

---

[1] This requirement is stricter than jump discontinuity, which requires the one-sided limits exist but are unequal at a point's two sides. Figure 2(b), (c), and (e) are jump discontinuity examples, but a semi-smooth surface allows only Fig. 2(b) and (c).

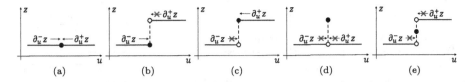

Fig. 2. A semi-smooth surface allows and only allows a point to be indifferentiable at one side. At a point, a semi-smooth surface can be **(a)** left and right differentiable, **(b,c)** only left or right differentiable, but *cannot* be **(d,e)** neither left nor right differentiable. Black dots indicate the function values; black circles indicate the one-sided limits; red cross indicate the non-existence of the one-sided partial derivatives.

Under the semi-smooth surface assumption, we propose the bilaterally weighted functional

$$
\min_z \iint_\Omega w_u(n_z\partial_u^+ z + n_x)^2 + (1 - w_u)(n_z\partial_u^- z + n_x)^2 \\
+ w_v(n_z\partial_v^+ z + n_y)^2 + (1 - w_v)(n_z\partial_v^- z + n_y)^2 \, du \, dv,
\tag{12}
$$

where w_u and w_v indicate the one-sided differentiability at each point's two sides:

$$
w_u = \begin{cases} 1 & \text{(only right diff.)} \\ 0.5 & \text{(left \& right diff.)} \\ 0 & \text{(only left diff.)} \end{cases} \text{ and } w_v = \begin{cases} 1 & \text{(only upper diff.)} \\ 0.5 & \text{(upper \& lower diff.)} \\ 0 & \text{(only lower diff.)} \end{cases}. \tag{13}
$$

The bilaterally weighted functional (12) states that, for example, when the depth map is left but not right differentiable at a point, the data term is kept at the left side but ignored at the right side. When the depth map is differentiable at a point, the data terms at the two sides are equally weighted. This relative weighting thus covers all possible cases at every point in a semi-smooth surface.

Now, the problem is how to determine the surface's one-sided differentiability at each point. To this end, we use the fact that one-sided differentiability requires one-sided continuity. For example, a function being right (in)differentiable at a point must be right (dis)continuous at that point. A function being one-sided continuous at a point requires the function value at that point to be equal to the limit approached from the corresponding side. Formally, denote the differences between the function value and one-sided limits at a point as

$$
\Delta_u^+ z = z(u, v) - \lim_{h\to 0^+} z(u + h, v), \quad \Delta_u^- z = z(u, v) - \lim_{h\to 0^-} z(u + h, v), \\
\Delta_v^+ z = z(u, v) - \lim_{h\to 0^+} z(u, v + h), \quad \Delta_v^- z = z(u, v) - \lim_{h\to 0^-} z(u, v + h).
\tag{14}
$$

The function is right continuous if $\Delta_u^+ z = 0$ or right discontinuous if $\Delta_u^+ z \neq 0$; so for the left, upper, and lower continuity. To further judge which side of a point is discontinuous by one function, we can compare the one-sided continuity at the point's two sides. As detailed in Fig. 3, the surface is continuous at a point if $(\Delta_u^- z)^2 - (\Delta_u^+ z)^2 = 0$, *only* left continuous if $(\Delta_u^- z)^2 - (\Delta_u^+ z)^2 < 0$, and *only*

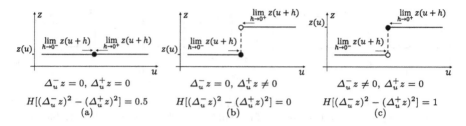

$$\Delta_u^- z = 0, \Delta_u^+ z = 0 \qquad \Delta_u^- z = 0, \Delta_u^+ z \neq 0 \qquad \Delta_u^- z \neq 0, \Delta_u^+ z = 0$$

$$H[(\Delta_u^- z)^2 - (\Delta_u^+ z)^2] = 0.5 \qquad H[(\Delta_u^- z)^2 - (\Delta_u^+ z)^2] = 0 \qquad H[(\Delta_u^- z)^2 - (\Delta_u^+ z)^2] = 1$$

$$\text{(a)} \qquad\qquad\qquad \text{(b)} \qquad\qquad\qquad \text{(c)}$$

Fig. 3. Our weight function $w_u = H[(\Delta_u^- z)^2 - (\Delta_u^+ z)^2]$ compares the one-sided continuity at a point's two sides and can indicate all three cases in a semi-smooth surface. **(a)** $w_u = 0.5$ when left and right continuous. **(b)** $w_u = 0$ when left continuous and right discontinuous. **(c)** $w_u = 1$ when left discontinuous and right continuous.

right continuous if $(\Delta_u^- z)^2 - (\Delta_u^+ z)^2 > 0$. By wrapping this difference with a Heaviside step function $H(x) = \{1 \text{ if } x > 0; 0.5 \text{ if } x = 0; 0 \text{ if } x < 0\}$, we have our weight functions in the horizontal and vertical directions:

$$w_u = H((\Delta_u^- z)^2 - (\Delta_u^+ z)^2) \quad \text{and} \quad w_v = H((\Delta_v^- z)^2 - (\Delta_v^+ z)^2). \qquad (15)$$

It can be verified that the weight functions Eq. (15) take values as Eq. (13).

2.3 Solution Method

This section presents the discretization of the bilaterally weighted functional (12) and the weight functions Eq. (15) and derives a solution method given the normal map observed in the discrete pixel domain, *i.e.*, $\Omega \subset \mathbb{Z}^2$. In the following, we will not distinguish the symbols between the continuous and discrete cases.

Discretization: We first discretize the one-sided partial derivatives Eq. (11) by forward or backward differences, *i.e.*, plugging $h = 1$ or -1 into Eq. (11):

$$\partial_u^+ z \approx z(u+1, v) - z(u, v), \quad \partial_u^- z \approx z(u, v) - z(u-1, v),$$
$$\partial_v^+ z \approx z(u, v+1) - z(u, v), \quad \partial_v^- z \approx z(u, v) - z(u, v-1). \qquad (16)$$

We then approximate the one-sided limits by the depth values at adjacent pixels, and Eq. (14) becomes

$$\Delta_u^+ z \approx n_z(z(u, v) - z(u+1, v)), \quad \Delta_u^- z \approx n_z(z(u, v) - z(u-1, v)),$$
$$\Delta_v^+ z \approx n_z(z(u, v) - z(u, v+1)), \quad \Delta_v^- z \approx n_z(z(u, v) - z(u, v-1)). \qquad (17)$$

Here, the depth differences are scaled by n_z to measure the difference along the normal direction at the point. To avoid the step function always taking binary values in the discrete domain (*i.e.*, treating every pixel as one-sided discontinuous), we approximate the step function by a sigmoid function:

$$H(x) \approx \sigma_k(x) = \frac{1}{1 + e^{-kx}}, \qquad (18)$$

where the parameter k controls the sharpness of the sigmoid function. Combining Eq. (16) to (18) together, we have the discretized bilaterally weighted functional

$$
\begin{aligned}
\min_{z(u,v)} \sum_{\Omega} &\sigma_k \left((\Delta_u^- z)^2 - (\Delta_u^+ z)^2\right) (n_z \partial_u^+ z + n_x)^2 \\
&+ \sigma_k \left((\Delta_u^+ z)^2 - (\Delta_u^- z)^2\right) (n_z \partial_u^- z + n_x)^2 \\
&+ \sigma_k \left((\Delta_v^- z)^2 - (\Delta_v^+ z)^2\right) (n_z \partial_v^+ z + n_y)^2 \\
&+ \sigma_k \left((\Delta_v^+ z)^2 - (\Delta_v^- z)^2\right) (n_z \partial_v^- z + n_y)^2.
\end{aligned}
\tag{19}
$$

In (19), we use the sigmoid function's property $1 - \sigma_k(x) = \sigma_k(-x)$ to make it more compact. Intuitively, the optimization problem (19) states that if the depth difference at one side of a pixel is much larger than the other side, then the larger side is more likely to be discontinuous, and correspondingly the quadratic data term is less weighted at the discontinuous side.

Optimization: The optimization problem (19) is non-convex due to the weights being non-linear sigmoid functions of unknown depths. To solve (19), we use iteratively re-weighted least squares (IRLS) [12]. To describe the iteration process, we first prepare the matrix form of (19).

Let $\mathbf{z}$, $\mathbf{n}_x$, $\mathbf{n}_y$, and $\mathbf{n}_z \in \mathbb{R}^m$ be the vectors of $z(\mathbf{u})$, $n_x(\mathbf{u})$, $n_y(\mathbf{u})$, and $n_z(\mathbf{u})$ from all $|\Omega| = m$ pixels serialized in the same order. Let $\mathrm{diag}(\mathbf{x})$ be the diagonal matrix whose i-th diagonal entry is the i-th entry of $\mathbf{x}$, and denote $\mathbf{N}_z = \mathrm{diag}(\mathbf{n}_z)$. We can write the optimization problem (19) in the matrix form as

$$
\min_{\mathbf{z}} (\mathbf{Az} - \mathbf{b})^\top \mathbf{W}(\mathbf{z})(\mathbf{Az} - \mathbf{b})
\tag{20}
$$

with

$$
\mathbf{A} = \begin{bmatrix} \mathbf{N}_z \mathbf{D}_u^+ \\ \mathbf{N}_z \mathbf{D}_u^- \\ \mathbf{N}_z \mathbf{D}_v^+ \\ \mathbf{N}_z \mathbf{D}_v^- \end{bmatrix}, \quad
\mathbf{b} = \begin{bmatrix} -\mathbf{n}_x \\ -\mathbf{n}_x \\ -\mathbf{n}_y \\ -\mathbf{n}_y \end{bmatrix}, \quad \text{and} \quad
\mathbf{W}(\mathbf{z}) = \mathrm{diag}\left(\begin{bmatrix} \mathbf{w}_u(\mathbf{z}) \\ 1 - \mathbf{w}_u(\mathbf{z}) \\ \mathbf{w}_v(\mathbf{z}) \\ 1 - \mathbf{w}_v(\mathbf{z}) \end{bmatrix} \right), \tag{21}
$$

where

$$
\begin{aligned}
\mathbf{w}_u(\mathbf{z}) &= \sigma_k \left[(\mathbf{N}_z \mathbf{D}_u^- \mathbf{z})^{\circ 2} - (\mathbf{N}_z \mathbf{D}_u^+ \mathbf{z})^{\circ 2} \right], \\
\mathbf{w}_v(\mathbf{z}) &= \sigma_k \left[(\mathbf{N}_z \mathbf{D}_v^- \mathbf{z})^{\circ 2} - (\mathbf{N}_z \mathbf{D}_v^+ \mathbf{z})^{\circ 2} \right].
\end{aligned}
\tag{22}
$$

Here, the four matrices $\mathbf{D}_u^+$, $\mathbf{D}_u^-$, $\mathbf{D}_v^+$, and $\mathbf{D}_v^- \in \mathbb{R}^{m \times m}$ are discrete partial derivative matrices. The i-th row either contains only two non-zero entries -1 and 1 or is a zero vector if the adjacent pixel of i-th pixel is outside the domain Ω; more details can be found in [25]. Besides, $\mathbf{1} \in \mathbb{R}^m$ is an all-one vector, $\sigma_k(\cdot)$ is now the element-wise version of the sigmoid function Eq. (18), and $(\cdot)^{\circ 2}$ is an element-wise square function on a vector.

At each step t during the optimization, we first fix the weight matrix $\mathbf{W}(\mathbf{z}^{(t)})$ and then solve for the depths $\mathbf{z}$:

$$
\mathbf{z}^{(t+1)} = \underset{\mathbf{z}}{\arg\min} (\mathbf{Az} - \mathbf{b})^\top \mathbf{W}(\mathbf{z}^{(t)})(\mathbf{Az} - \mathbf{b}).
\tag{23}
$$

When $\mathbf{W}(\mathbf{z}^{(t)})$ is fixed, Eq. (23) boils down to a convex weighted least-squares problem. We can find $\mathbf{z}^{(t+1)}$ by solving the normal equation of Eq. (23)

$$\mathbf{A}^\top \mathbf{W}(\mathbf{z}^{(t)})\mathbf{A}\mathbf{z} = \mathbf{A}^\top \mathbf{W}(\mathbf{z}^{(t)})\mathbf{b}. \tag{24}$$

The matrix $\mathbf{A}$ is rank 1 deficient; the 1D nullspace basis is an all-one vector, corresponding to the offset ambiguity in the result. In the perspective case, the offset ambiguity becomes scale ambiguity after exponentiating the result. In our implementation, we use a conjugate gradient method [11] to solve Eq. (24).

We initialize $\mathbf{z}^{(0)}$ as a plane, or equivalently, initialize all weights as 0.5. The estimated depth map at the first step $\mathbf{z}^{(1)}$ is then the same as the one estimated under the smooth surface assumption, *i.e.*, using the functional (10). Denoting the energy of the objective function at step t as $E_t = (\mathbf{A}\mathbf{z}^{(t)} - \mathbf{b})^\top \mathbf{W}(\mathbf{z}^{(t)})(\mathbf{A}\mathbf{z}^{(t)} - \mathbf{b})$, we terminate the iteration once the relative energy $|E_t - E_{t-1}|/E_{t-1}$ is smaller than the user-provided tolerance or the maximum number of iterations is exceeded.

3   Related Work

This section briefly reviews related works and discusses the differences between our method and existing methods.

Unified Normal Integration Equations: Since the emergence of the normal integration problem [13,16], the majority of the methods estimates the depth map based on the PDEs

$$\partial_u z - p = 0 \quad \text{and} \quad \partial_v z - q = 0, \tag{25}$$

where $[p, q]^\top = [-\frac{n_x}{n_z}, -\frac{n_y}{n_z}]^\top$ is the gradient field computed from the normal map. By introducing the log depth map $\tilde{z}$, we can unify the PDEs in the perspective case as the same form as Eq. (25), with a different gradient field $[\tilde{p}, \tilde{q}]^\top = [-\frac{n_x}{\tilde{n}_z}, -\frac{n_y}{\tilde{n}_z}]^\top$ [6,7,24]. The normal integration problem is therefore also called shape/height/depth from gradient [2,8–10,14,19,21,26].

Equation (25) is derived from the constraint that a normal vector should be parallel to the cross product of the two tangent vectors $\mathbf{n} \parallel \partial_u z \times \partial_v z$. Alternatively, Zhu and Smith [33] derived the PDEs from the orthogonal constraint Eq. (1) and found it benefits numerical stability. However, Zhu and Smith [33] derived inconsistent PDEs in the orthographic and perspective cases. In the perspective case, Zhu and Smith [33] solves a homogeneous system using singular value decomposition, which can be more time consuming than solving an inhomogeneous system in the orthographic case.

We combine the strength of both derivations. Like Zhu and Smith [33], we derive the PDEs from the orthogonal constraint for numerical stability. Like [6, 7,24], we introduce the log depth map to unify the formulations. In this way, we can solve the normal integration problem in the two cases in the same manner while being numerically more stable[2].

---

[2] See experiments in the supplementary material.

Discontinuity Preserving Surface Recovery: We now discuss two strategies for discontinuity preservation: Robust estimator-based and weighted approaches.

As the residuals of Eq. (25) become large at discontinuous points, robust estimator-based methods apply robust functions ρ to the data terms as

$$\min_z \iint_\Omega \rho\left(\partial_u z - p\right) + \rho\left(\partial_v z - q\right) \, du \, dv. \tag{26}$$

Properly designed ρ-functions that can suppress the influence of large residuals are expected to preserve the discontinuities. Lorentzian function [6], total variation [23], and triple sparsity [2] have been studied.

Instead of applying robust functions, weighted approaches assign the weights to quadratic residuals of PDEs to eliminates the effects of discontinuities:

$$\min_z \iint_\Omega w_u \left(\partial_u z - p\right)^2 + w_v \left(\partial_v z - q\right)^2 \, du \, dv. \tag{27}$$

If the data terms at discontinuous points are appropriately assigned smaller weights, then the discontinuities are expected to be preserved.

A class of weighted approaches detects discontinuous points as a preprocessing step before optimizing (27). The major differences among these works are the clues used for discontinuity detection. Karacali and Snyder [18] detect the discontinuity based on the residuals of Eq. (25). Wu and Tang [30] use the expectation-maximization algorithm to estimate a discontinuity map from the normal map. Wang *et al.* [28] detect a binary discontinuity map using both photometric stereo images and the normal map. Xie *et al.* [31] handcraft features from the normal map. The one-time detection can be fragile as there is no scheme to correct possibly wrong detection in the optimization afterward.

A more effective type of weighted approach iteratively updates the weights. Alpha-surface method [1] first creates a minimal spanning tree from the integration domain, then iteratively adds to the spanning-tree the edges that are treated continuous. Anisotropic diffusion [1,25] applies diffusion tensors to the gradient field. Quéau *et al.* [25] design the diffusion tensors as functions of depths, and the diffusion tensors are iteratively updated during optimization. Mumford-Shah integrator [25] bypasses the detection by jointly optimizing for the weights and depths by assuming that discontinuities are short curves in the domain.

Our method can be categorized as the weighted approach. Unlike previous methods, we assume a semi-smooth target surface and *relatively weight* the one-sided differentiability at each point's two sides. Our weights are iteratively updated during the optimization, which is different to the methods determining the weights once before the optimization [28,30,31]. Further, unlike most methods determining the weights without depth information, our weights are functions of unknown depths and thus are adaptively determined during the optimization.

4   Comparison

To verify our method's effectiveness, this section compares our method to existing ones using synthetic and real-world normal maps in orthographic and perspective cases. Readers can find more experimental analysis and the discussion on the limitations of our method in the supplementary material.

4.1   Experimental Settings

Baselines: We compare our method to six methods. The first one assumes smooth surfaces and uses inverse plane fitting (IPF) [3]. The remaining five methods all aim at discontinuity preservation using triple sparsity (TS) [2], total variation (TV) [25], robust estimator (RE)[3] [25], Mumford-Shah (MS) [25], and anisotropic diffusion (AD) [25].

Implementation: We use the publicly available official implementations[4] of IPF [3] and the four discontinuity preservation methods presented in [25]. We use the five-point version of IPF, and there is no hyperparameter. For the hyperparameters of TV, RE, MS, AD methods, we follow the suggestions in [25]: $\alpha = 0.1$ in TV, $\gamma = 0.5$ and $\beta = 0.8$ in RE, $\mu = 45$ and $\epsilon = 0.01$ in MS, and $\mu = 0.2$ and $\nu = 10$ in AD. We implement TS [2] by ourselves and following the hyperparameter setup in the paper [2]. For our method, there is one hyperparameter k in the objective function, and we set $k = 2$. The maximum iteration number and stopping tolerance of IRLS are set as 100 and 1×10^{-5}, respectively.

Metric: When the ground-truth (GT) surfaces are available, we show the absolute depth error maps and report the mean absolute depth error (MADE) between the integrated and GT depth maps. To remove the offset ambiguity in the orthographic case, we shift the integrated surfaces such that the $L1$ norm between the shifted and the GT depth maps is minimal. We similarly remove the scale ambiguity by scaling the integrated surfaces in the perspective case.

4.2   Results in the Orthographic Case

Figure 4 shows quantitative comparisons on synthetic orthographic normal maps. The first normal map (top rows) is analytically computed, while the second one (bottom rows) is rendered from the object "Reading" in DiLiGenT-MV dataset [20] by the Mitsuba renderer [22]. Restricted by the smooth surface assumption, the IPF method [3] only recovers smooth surfaces as expected. When the target surface contains large depth gaps, the recovered surfaces by the IPF method [3] are heavily distorted. Compared to the IPF method [3], the TS [2] and TV method [25] marginally improve the result. The TS [2] and TV method [25] under-segments the surface and cannot faithfully locate all discontinuities. On the other hand, both the RE and MS methods [25] can identify unnecessary or incorrect discontinuity locations and thus introduce the over-segmentation artifacts into the recovered surfaces. The AD method [25] performs

---

[3] The method we call robust estimator is called non-convex estimator in [25].

[4] https://github.com/hoshino042/NormalIntegrationhttps://github.com/yqueau/normal_integration.

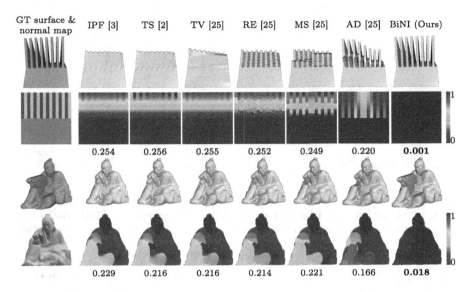

GT surface &
normal map | IPF [3] | TS [2] | TV [25] | RE [25] | MS [25] | AD [25] | BiNI (Ours)

0.254   0.256   0.255   0.252   0.249   0.220   **0.001**

0.229   0.216   0.216   0.214   0.221   0.166   **0.018**

Fig. 4. Quantitative comparison using an analytically computed and a Mitsuba-rendered normal map as input. The odd and even rows display the integrated surfaces and absolute depth error maps, respectively. Numbers underneath are MADEs.

better at identifying discontinuity locations but suffers from the distortion problem. The distortion is clear in Fig. 4 (top). The ideally straight stripes recovered by the AD method [25] are still bent, although the discontinuities are well located. In contrast, our method locates discontinuity and reduces the under- or over-segmentation artifacts in the surfaces. As a result, our method achieve the smallest MADEs among all compared methods on both surfaces.

Figure 5 shows a qualitative comparison on real-world orthographic normal maps obtained in three applications. The first normal map is estimated by photometric stereo [15] on the real-world images from Light Stage Data Gallery [4]; the second one is estimated by shape from polarization [5]; and the third one is inferred from a single RGB human image by learning-based method [32]. Consistent with the trends for synthetic normal maps, the results by the baseline methods suffer from under- or over-segmentation artifacts. Our method can still preserves discontinuities reasonably well for noisy real-world normal maps. This experiment demonstrates the wide application of our method.

In addition, our method can be easier to use in practice. Compared to existing discontinuity preservation methods, our method reduce the number of hyperparameter from six [2] or two [25] to only one.

4.3 Results in the Perspective Case

Figure 6 shows the quantitative comparison on a Mitsuba-rendered normal map with a perspective camera model. The TV, RE, MS, and AD methods [25] all initialize the surface by solving the quadratic Poisson equation [25], as shown in

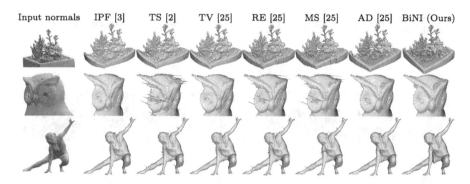

Fig. 5. Qualitative comparison of surfaces integrated from real-world noisy normal maps, which are estimated by (1st row) photometric stereo [15] (2nd row) shape from polarization [5]. (3rd row) a deep network inferred from a single RGB human image [32]. Best viewed on screen.

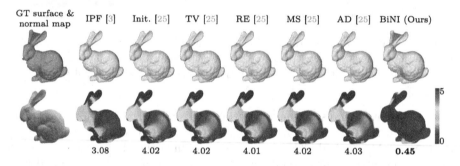

Fig. 6. Quantitative comparison using synthetic perspective normal maps rendered from Stanford Bunny. We additionally show the initialization (3rd column) for the TV, RE, MS, and AD methods [25]. The four methods do not improve the initialization and do not preserve discontinuities in the perspective case.

the third column in Fig. 6. However, we can barely see the difference between the initialization and final results of the four methods. As the four methods [25] are all based on the traditional unified formulations Eq. (25), it is likely that the traditional formulation Eq. (25) is unsuitable for the discontinuity preservation methods in the perspective case. In contrast, based on our unified formulation Eq. (8), our bilaterally weighted functional still preserves discontinuities in the perspective case.

Figure 7 shows integration results from the GT perspective normal maps in DiLiGenT benchmark [27]. We again observe that the TV, RE, MS, and AD methods [25] do not improve the initialization; therefore, we only display the results from the IPF method [3] and our method. Compared to the IPF method [33], our method preserves the discontinuities and largely reduces the distortion. The MADEs are within 1 mm except for two objects ("Harvest" and "Goblet"). Especially, we achieve 0.07 mm MADE for the object "Cow." To our

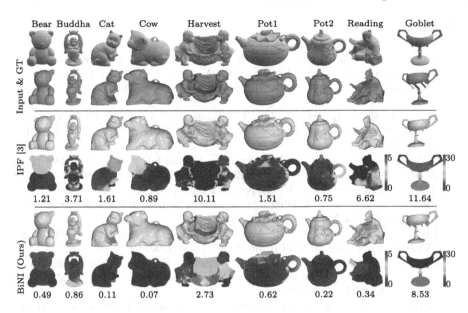

Fig. 7. Quantitative comparison on DiLiGenT benchmark [27]. (**1st & 2nd rows**) The input perspective normal maps and GT surfaces. Viewpoints of surfaces are adjusted to emphasize discontinuities. (**3rd & 4th rows**) Surfaces integrated by IPF method [3] and absolute depth error maps. The colormap scale is the same for the first eight objects. (**5th & 6th rows**) Surfaces integrated by our method and absolute depth error maps. Numbers underneath are MADEs [mm].

knowledge, our discontinuity preservation method is the first to be effective in the perspective case.

5 Conclusions

We have presented and evaluated bilateral normal integration for discontinuity preserving normal integration. Compared to existing methods, our method preserves discontinuities more effectively and alleviates the under- or over-segmentation artifacts. The effectiveness of our method relies on the bilaterally weighted functional under the semi-smooth surface assumption. Further, we have unified the normal integration formulations in the orthographic and perspective cases appropriately. As a result, we have first shown effective discontinuity preservation results in the perspective case.

Acknowledgments. This work was supported by JSPS KAKENHI Grant Number JP19H01123, and National Natural Science Foundation of China under Grant No. 62136001, 61872012.

References

1. Agrawal, A., Raskar, R., Chellappa, R.: What is the range of surface reconstructions from a gradient field? In: Leonardis, A., Bischof, H., Pinz, A. (eds.) ECCV 2006. LNCS, vol. 3951, pp. 578–591. Springer, Heidelberg (2006). https://doi.org/10.1007/11744023_45
2. Badri, H., Yahia, H., Aboutajdine, D.: Robust surface reconstruction via triple sparsity. In: Proceedings of Computer Vision and Pattern Recognition (CVPR), pp. 2283–2290 (2014)
3. Cao, X., Shi, B., Okura, F., Matsushita, Y.: Normal integration via inverse plane fitting with minimum point-to-plane distance. In: Proceedings of Computer Vision and Pattern Recognition (CVPR), pp. 2382–2391 (2021)
4. Chabert, C.F., et al.: Relighting human locomotion with flowed reflectance fields. In: Proceedings of ACM SIGGRAPH (2006)
5. Deschaintre, V., Lin, Y., Ghosh, A.: Deep polarization imaging for 3D shape and SVBRDF acquisition. In: CVPR, pp. 15567–15576 (2021)
6. Durou, J.D., Aujol, J.F., Courteille, F.: Integrating the normal field of a surface in the presence of discontinuities. In: Workshops of the Computer Vision and Pattern Recognition (CVPRW) (2009)
7. Durou, J.D., Courteille, F.: Integration of a normal field without boundary condition. In: Workshops of the International Conference on Computer Vision (ICCVW) (2007)
8. Harker, M., O'Leary, P.: Least squares surface reconstruction from measured gradient fields. In: Proceedings of Computer Vision and Pattern Recognition (CVPR) (2008)
9. Harker, M., O'Leary, P.: Least squares surface reconstruction from gradients: direct algebraic methods with spectral, Tikhonov, and constrained regularization. In: Proceedings of Computer Vision and Pattern Recognition (CVPR) (2011)
10. Harker, M., O'leary, P.: Regularized reconstruction of a surface from its measured gradient field. J. Math. Imaging Vision **51**, 46–70 (2015)
11. Hestenes, M.R., Stiefel, E.: Methods of conjugate gradients for solving linear systems. J. Res. Natl. Bureau Stand. **49**(6), 409 (1952)
12. Holland, P.W., Welsch, R.E.: Robust regression using iteratively reweighted least-squares. Commun. Stat.-theory Methods **6**(9), 813–827 (1977)
13. Horn, B.K., Brooks, M.J.: The variational approach to shape from shading. In: Computer Vision, Graphics, and Image Processing(1986)
14. Horovitz, I., Kiryati, N.: Depth from gradient fields and control points: bias correction in photometric stereo. Image Vision Comput. **22**, 681–694 (2004)
15. Ikehata, S.: CNN-PS: CNN-based photometric stereo for general non-convex surfaces. In: Proceedings of European Conference on Computer Vision (ECCV), pp. 3–18 (2018)
16. Ikeuchi, K.: Constructing a depth map from images. Technical report, Massachusetts Inst of Tech Cambridge Artificial Intelligence Lab (1983)
17. Kadambi, A., Taamazyan, V., Shi, B., Raskar, R.: Polarized 3D: high-quality depth sensing with polarization cues. In: Proceedings of International Conference on Computer Vision (ICCV) (2015)
18. Karacali, B., Snyder, W.: Reconstructing discontinuous surfaces from a given gradient field using partial integrability. Comput. Vision Image Underst. **92**, 78–111 (2003)

19. Klette, R., Schluens, K.: Height data from gradient maps. In: Machine Vision Applications, Architectures, and Systems Integration V, vol. 2908, pp. 204–215. SPIE (1996)
20. Li, M., Zhou, Z., Wu, Z., Shi, B., Diao, C., Tan, P.: Multi-view photometric stereo: a robust solution and benchmark dataset for spatially varying isotropic materials. IEEE Trans. Image Process. **29**, 4159–4173 (2020)
21. Ng, H.S., Wu, T.P., Tang, C.K.: Surface-from-gradients without discrete integrability enforcement: a Gaussian kernel approach. IEEE Trans. Pattern Anal. Mach. Intell. (PAMI) **32**(11), 2085–2099 (2009)
22. Nimier-David, M., Vicini, D., Zeltner, T., Jakob, W.: Mitsuba 2: a retargetable forward and inverse renderer. ACM Trans. Graph. (TOG) **38**(6), 1–17 (2019)
23. Quéau, Y., Durou, J.-D.: Edge-preserving integration of a normal field: weighted least-squares, TV and L^1 approaches. In: Aujol, J.-F., Nikolova, M., Papadakis, N. (eds.) SSVM 2015. LNCS, vol. 9087, pp. 576–588. Springer, Cham (2015). https://doi.org/10.1007/978-3-319-18461-6_46
24. Quéau, Y., Durou, J.D., Aujol, J.F.: Normal integration: a survey. J. Math. Imaging Vision **60**, 576–593 (2018)
25. Quéau, Y., Durou, J.D., Aujol, J.F.: Variational methods for normal integration. J. Math. Imaging Vision **60**, 609–632 (2018)
26. Saracchini, R.F., Stolfi, J., Leitão, H.C., Atkinson, G.A., Smith, M.L.: A robust multi-scale integration method to obtain the depth from gradient maps. Comput. Vision Image Underst. **116**(8), 882–895 (2012)
27. Shi, B., Mo, Z., Wu, Z., Duan, D., Yeung, S.K., Tan, P.: A benchmark dataset and evaluation for non-Lambertian and uncalibrated photometric stereo. IEEE Trans. Pattern Anal. Mach. Intell. (PAMI) (2019)
28. Wang, Y., Bu, J., Li, N., Song, M., Tan, P.: Detecting discontinuities for surface reconstruction. In: Proceedings of International Conference on Pattern Recognition (ICPR), pp. 2108–2111. IEEE (2012)
29. Woodham, R.J.: Photometric stereo: a reflectance map technique for determining surface orientation from image intensity. In: Proceedings of Image Understanding Systems and Industrial Applications I (1979)
30. Wu, T.P., Tang, C.K.: Visible surface reconstruction from normals with discontinuity consideration. In: Proceedings of Computer Vision and Pattern Recognition (CVPR), vol. 2, pp. 1793–1800. IEEE (2006)
31. Xie, W., Wang, M., Wei, M., Jiang, J., Qin, J.: Surface reconstruction from normals: a robust DGP-based discontinuity preservation approach. In: Proceedings of Computer Vision and Pattern Recognition (CVPR) (2019)
32. Xiu, Y., Yang, J., Tzionas, D., Black, M.J.: ICON: implicit clothed humans obtained from normals. In: Proceedings of Computer Vision and Pattern Recognition (CVPR), pp. 13296–13306 (2022)
33. Zhu, D., Smith, W.A.P.: Least squares surface reconstruction on arbitrary domains. In: Vedaldi, A., Bischof, H., Brox, T., Frahm, J.-M. (eds.) ECCV 2020. LNCS, vol. 12367, pp. 530–545. Springer, Cham (2020). https://doi.org/10.1007/978-3-030-58542-6_32

S²Contact: Graph-Based Network for 3D Hand-Object Contact Estimation with Semi-supervised Learning

Tze Ho Elden Tse[1], Zhongqun Zhang[1(✉)], Kwang In Kim[2], Aleš Leonardis[1], Feng Zheng[3], and Hyung Jin Chang[1]

[1] University of Birmingham, Birmingham, UK
ZXZ064@student.bham.ac.uk
[2] UNIST, Ulsan, Korea
[3] SUSTech, Shenzhen, China

Abstract. Despite the recent efforts in accurate 3D annotations in hand and object datasets, there still exist gaps in 3D hand and object reconstructions. Existing works leverage contact maps to refine inaccurate hand-object pose estimations and generate grasps given object models. However, they require explicit 3D supervision which is seldom available and therefore, are limited to constrained settings, e.g., where thermal cameras observe residual heat left on manipulated objects. In this paper, we propose a novel semi-supervised framework that allows us to learn contact from monocular images. Specifically, we leverage visual and geometric consistency constraints in large-scale datasets for generating pseudo-labels in semi-supervised learning and propose an efficient graph-based network to infer contact. Our semi-supervised learning framework achieves a favourable improvement over the existing supervised learning methods trained on data with 'limited' annotations. Notably, our proposed model is able to achieve superior results with less than half the network parameters and memory access cost when compared with the commonly-used PointNet-based approach. We show benefits from using a contact map that rules hand-object interactions to produce more accurate reconstructions. We further demonstrate that training with pseudo-labels can extend contact map estimations to out-of-domain objects and generalise better across multiple datasets. Project page is available (https://eldentse.github.io/s2contact/).

1 Introduction

Understanding hand-object interactions have been an active area of study in recent years [3,16–18,20,26,31,49,60]. Besides common practical applications in augmented and virtual reality [15,36,52], it is a key ingredient to advanced

---

T. H. E. Tse and Z. Zhang—Equal contribution.

Supplementary Information The online version contains supplementary material available at https://doi.org/10.1007/978-3-031-19769-7_33.

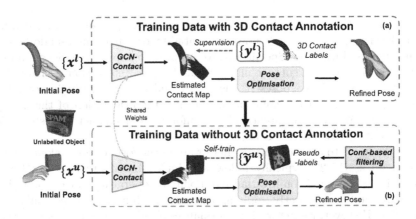

Fig. 1. Overview of our semi-supervised learning framework, S²Contact. (a) The model is pre-trained on a small annotated dataset. (b) Then, it is deployed on unlabelled datasets to collect pseudo-labels. The pseudo-labels are filtered with confidence-based on visual and geometric consistencies. Upon predicting the contact map, the hand and object poses are jointly optimised to achieve target contact via a contact model [12].

human-computer interaction [50] and imitation learning in robotics [62]. In this paper, as illustrated in Fig. 1, we tackle the problem of 3D reconstruction of the hand and manipulated object with the focus on contact map estimation.

Previous works in hand-object interactions typically formulate this as a joint hand and object pose estimation problem. Along with the development of data collection and annotation methods, more accurate 3D annotations for real datasets [4,11,14] are available for learning-based methods [10,49]. Despite the efforts, there still exist gaps between hand-object pose estimation and contact as ground-truth in datasets are not perfect. Recent works attempt to address this problem with interaction constraints (attraction and repulsion) under an optimisation framework [3,17,60]. However, inferred poses continue to exhibit sufficient error to cause unrealistic hand-object contact, making downstream tasks challenging [12]. In addition, annotations under constrained laboratory environments rely on strong priors such as limited hand motion which prevents the trained model from generalising to novel scenes and out-of-domain objects.

To address the problem of hand-object contact modelling, Brahmbhatt *et al.* [1] used thermal cameras observing the heat transfer from hand to object after the grasp to capture detailed ground-truth contact. Their follow-up work contributed a large grasp dataset (*ContactPose*) with contact maps and hand-object pose annotations. Recent works are able to leverage contact maps to refine inaccurate hand-object pose estimations [12] and generate grasps given object model [20]. Therefore, the ability to generate an accurate contact map is one of the key elements to reasoning physical contact. However, the number of annotated objects is incomparable to manipulated objects in real life and insufficient to cover a wide range of human intents. Furthermore, obtaining annotations for contact maps is non-trivial as it requires thermal sensors during data collection.

To enable the wider adoption of contact maps, we propose a unified framework that leverages existing hand-object datasets for generating pseudo-labels in semi-supervised learning. Specifically, we propose to exploit the visual and geometric consistencies of contact maps in hand-object interactions. This is built upon the idea that the poses of the hands and objects are highly-correlated where the 3D pose of the hand often indicates the orientation of the manipulated object. We further extend this by enforcing our contact consistency loss for the contact maps across a video.

As the input to contact map estimator are in the form of point clouds, recent related works [12,20] typically follow a PointNet-based architectures [38,39]. This achieves permutation invariance of points by operating on each point independently and subsequently applying a symmetric function to accumulate features [53]. However, the network performances are limited as points are treated independently at a local scale to maintain permutation invariance. To overcome this fundamental limitation, many recent approaches adopt graph convolutional networks (GCN) [9,25] and achieve state-of-the-art performances in 3D representation learning on point clouds for classification, part segmentation and semantic segmentation [28,30,53]. The ability to capture local geometric structures while maintaining permutation invariance is particularly important for estimating contact maps. However, it comes at the cost of high computation and memory usage for constructing a local neighbourhood with K-nearest neighbour (K-NN) search on point clouds at each training epoch. For this reason, we design a graph-based neural network that demonstrates superior results with less than half the learning parameters and faster convergence.

Our contributions are three-fold:

- We propose a novel semi-supervised learning framework that combines pseudo-label with consistency training. Experimental results demonstrate the effectiveness of this training strategy.
- We propose a novel graph-based network for processing hand-object point clouds, which is at least two times more efficient than PointNet-based architecture for estimating contact between hand and object.
- We conduct comprehensive experiments on three commonly-used hand-object datasets. Experiments show that our proposed framework S^2Contact outperforms recent semi-supervised methods.

2   Related Work

Our work tackles the problem of hand and object reconstruction from monocular RGB videos, exploiting geometric and visual consistencies on contact maps for semi-supervised learning. To the best of our knowledge, we are the first to apply such consistencies on hand-object scenarios. We first review the literature on *hand-object reconstruction*. Then, we review *point cloud analysis* with the focus of graph-based methods. Finally, we provide a brief review on *semi-supervised learning in 3D hand-object pose estimation*.

2.1   Hand-Object Reconstruction

Previous works mainly tackle 3D pose estimations on hands [35,42,44,47,58, 58,64] and objects [5,6,27,29,56] separately. Joint reconstruction of hands and objects has been receiving increasing attention [3,16–18]. Hasson *et al.* [18] introduces an end-to-end model to regress MANO hand parameters jointly with object mesh vertices deformed from a sphere and incorporates contact losses which encourages contact surfaces and penalises penetrations between hand and object. A line of works [3,10,12,16,17,19,49,60] assume known object models and regress a 6DoF object pose instead. Other works focus on grasp synthesis [8,20,21,46]. In contrast, our method is in line with recent optimisation-based approaches for modelling 3D hand-object contact. ContactOpt [12] proposes a contact map estimation network and a contact model to produce realistic hand-object interaction. ContactPose dataset [2] is unique in capturing ground-truth thermal contact maps. However, 3D contact labels are seldom available and limited to constrained labratory settings. In this work, we treat contact maps as our primary learning target and leverage unannotated datasets.

2.2   Point Cloud Analysis

Since point cloud data is irregular and unordered, early works tend to project the original point clouds to intermediate voxels [33] or images [61], *i.e.* translating into a well-explored 2D image problem. As information loss caused by projection degrades the representational quality, PointNet [38] is proposed to directly process unordered point sets and PointNet++ [39] is extends on local point representation in multi-scale. As PointNet++ [39] can be view as the generic point cloud analysis network framework, the research focus has been shifted to generating better regional points representation. Methods can be divided into convolution [55,57], graph [28,30,53] and attention [13,63] -based.

Graph-Based Methods. GCNs have been gaining much attention in the last few years. This is due to two reasons: 1) the rapid increase of non-Euclidean data in real-world applications and 2) the limited performance of convolutional neural networks when dealing with such data. As the unstructured nature of point clouds poses a representational challenge in the community, graph-based methods treat points as nodes of a graph and formulate edges according to their spatial/feature relationships. MoNet [34] defines the convolution as Gaussian mixture models in a local pseudo-coordinate system. 3D-GCN [30] proposes a deformable kernels which has shift and scale-invariant properties for point cloud processing. DGCNN [53] proposes to gather nearest neighbouring points in feature space and follow by the EdgeConv operators for feature extraction. The EdgeConv operator dynamically computes node adjacency at each graph layer using the distance between point features. In this paper, we propose a computationally efficient network for contact map estimation which requires less than half the parameters of PointNet [38] and GPU memory of DGCNN [53].

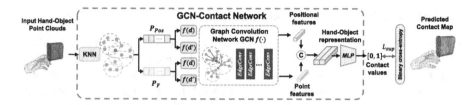

Fig. 2. Framework of GCN-Contact. The network takes hand-object point clouds $\mathbf{P} = (\mathbf{P}_{hand}, \mathbf{P}_{obj})$ as input and perform K-NN search separately on 3D position $\mathbf{P}_{pos}$ and point features $\mathbf{P}_F$, *i.e.* $\mathbf{P} = \{\mathbf{P}_{pos}, \mathbf{P}_F\}$. Different dilation factors d, d' are used to enlarge the receptive field for graph convolution $f(\cdot)$. Finally, features are concatenated and pass to MLP to predict contact map.

2.3   Semi-supervised Learning in 3D Hand-Object Pose Estimation

Learning from both labelled and unlabelled data simultaneously has recently attracted growing interest in 3D hand pose estimation [7,23,43,48,59]. They typically focus on training models with a small amount of labelled data as well as a relatively larger amount of unlabelled data. After training on human-annotated datasets, pseudo-labelling and consistency training can be used to train further and a teacher-student network with exponential moving average (EMA) strategy [51] is common to accelerate the training. For instance, So-HandNet [7] leverages the consistency between the recovered hand point cloud and the original hand point cloud for semi-supervised training. SemiHand [59] is the first to combine pseudo-labelling and consistency learning for hand pose estimation. Liu *et al.* [31] is the only prior work on 3D hand-object pose estimation with semi-supervised learning. They proposed spatial and temporal constraints for selecting the pseudo-labels from videos. However, they are limited to pseudo hand labels and did not account for physical contact with manipulated objects. In contrast, our work is the first to explore pseudo-labelling for 3D hand-object contact map with geometric and visual consistency constraints.

3   Methodology

Given a noisy estimate of hand and object meshes from an image-based algorithm, we seek to learn a hand-object contact region estimator by exploiting real-world hand and object video datasets without contact ground-truths. Figure 1 shows an overview of our approach. In the following section, we describe our learned contact map estimation network (GCN-Contact) in Sect. 3.1 and our newly proposing semi-supervised training pipeline (S²Contact) in Sect. 3.2 that utilise a teacher-student mutual learning framework.

3.1   GCN-Contact: 3D Hand-Object Contact Estimation

As pose estimates from an image-based algorithm can be potentially inaccurate, GCN-Contact learns to infer contact maps $\mathbf{C} = (\mathbf{C}_{hand}, \mathbf{C}_{obj})$ from hand and object point clouds $\mathbf{P} = (\mathbf{P}_{hand}, \mathbf{P}_{obj})$. We adopted the differential MANO [41] model from [18]. It maps pose ($\boldsymbol{\theta} \in \mathbb{R}^{51}$) and shape ($\boldsymbol{\beta} \in \mathbb{R}^{10}$) parameters to a mesh with $N = 778$ vertices. Pose parameters ($\boldsymbol{\theta}$) consists of 45 DoF (*i.e.* 3 DoF for each of the 15 finger joints) plus 6 DoF for rotation and translation of the wrist joint. Shape parameters ($\boldsymbol{\beta}$) are fixed for a given person. We sample 2048 points randomly from object model to form object point cloud. Following [12], we include F-dimensional point features for each point: binary per-point feature indicating whether the point belongs to the hand or object, distances from hand to object and surface normal information. With network input $\mathbf{P} = (\mathbf{P}_{hand}, \mathbf{P}_{obj})$ where $\mathbf{P}_{hand} \in \mathbb{R}^{778 \times F}$ and $\mathbf{P}_{obj} \in \mathbb{R}^{2048 \times F}$, GCN-Contact can be trained to infer discrete contact representation ($\mathbf{C} = (\mathbf{C}_{hand}, \mathbf{C}_{obj}) \in [0, 1]$) [2] using binary cross-entropy loss. Similarly to [12], the contact value range $[0, 1]$ is evenly split into 10 bins and the training loss is weighted to account for class imbalance.

Revisiting PointNet-Based Methods. Recent contact map estimators are based on PointNet [20] and PoinetNet++ [12]. PointNet [38] directly processes unordered point sets using shared multi-layer perceptron (MLP) networks. Point-Net++ [39] learns hierarchical features by stacking multiple learning stages and recursively capturing local geometric structures. At each learning stage, farthest point sampling (FPS) algorithm is used to re-sample a fixed number of points and K neighbours are obtained from ball query's local neighbourhood for each sampled point to capture local structures. The kernel operation of PointNet++ for point $p_i \in \mathbb{R}^F$ with F-dimensional features can be described as:

$$\dot{p}_i = \sigma\big(\Phi(p_j | j \in \mathcal{N}(i))\big), \tag{1}$$

where the updated point $\dot{p}_i$ is formed by max-pooling function $\sigma(\cdot)$ and Point-Net as the basic building block for local feature extractor $\Phi(\cdot)$ around point neighbourhood $\mathcal{N}(i)$ of point p_i. The kernel of the point convolution can be implemented with MLPs. However, MLPs are unnecessarily performed on the neighbourhood features which causes a considerable amount of latency in Point-Net++ [40]. This motivates us to employ advanced local feature extractors such as convolution [55,57], graph [28,30,53] or self-attention mechanisms [13,63].

Local Geometric Information. While contact map estimation can take advantage of detailed local geometric information, they usually suffer from two major limitations. First, the computational complexity is largely increased with delicate extractors which leads to low inference latency. For instance, in graph-based methods, neighbourhood information gathering modules are placed for better modelling of the locality on point clouds. This is commonly established by K-nearest neighbour (K-NN) search which increases the computational cost quadratically with the number of points and even further for dynamic feature representation [53]. For reference on ModelNet40 point cloud classification task [40], the inference speed of PointNet [38] is 41 times faster than DGCNN [53]. Second, Liu *et al.*'s investigation on local aggregation operators reveals that advanced local feature extractors make surprisingly similar contributions to the network performance under the same network input [32]. For these reasons, we are encouraged to develop a computationally efficient design while maintaining comparable accuracy for learning contact map estimation.

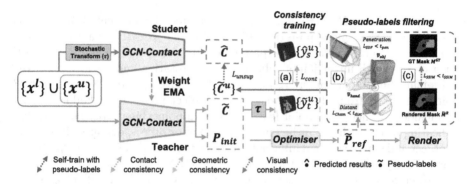

Fig. 3. S^2Contact pipeline. We adopt our proposed graph-based network GCN-Contact as backbone. We utilise a teacher-student mutual learning framework which is composed of a learnable student and an EMA teacher. The student network is trained with labelled data $\{\mathbf{x}^l, \mathbf{y}^l\}$. For unlabelled data $\mathbf{x}^u$, the student network takes pseudo contact labels $\widetilde{\mathbf{C}}^u$ from its EMA teacher and compares with its predictions $\widehat{\mathbf{C}}$. Please note that pseudo contact labels $\widetilde{\mathbf{C}}^u$ is a subset of $\widetilde{\mathbf{C}}$, *i.e.* $\widetilde{\mathbf{C}}^u \in \widetilde{\mathbf{C}}$. (a) refers to contact consistency constraint for consistency training. To improve the quality of pseudo-label, we adopt a confidence-based filtering mechanism to geometrically (b) and visually (c) filter out predictions that violate contact constraints.

Proposed Method. To overcome the aforementioned limitations, we present a simple yet effective graph-based network for contact map estimation. We use EdgeConv [53] to generate edge features that describe the relationships between a point and its neighbours:

$$\Phi(p_i, p_j) = \text{ReLU}\big(\text{MLP}(p_j - p_i, p_i)\big), \quad j \in \mathcal{N}(i), \qquad (2)$$

where neighbourhood $\mathcal{N}(i)$ is obtained by K-NN search around the point p_i. As shown in Fig. 2, we only compute K-NN search once at each network pass to improve computational complexity and reduce memory usage. In addition, we apply dilation on the K-NN results to increase the receptive field without loss of resolution. To better construct local regions when hand and object are perturbed, we propose to perform K-NN search on 3D position and point features separately. Note that [12] perform ball query on $0.1 - 0.2\,m$ radius and [53] combine both position and features. Finally, we take inspirations from the Inception model [45] in which they extract multi-scale information by using different kernel sizes in different paths of the architecture. Similarly, we process spatial information at various dilation factors and then aggregates. The experiment demonstrates the effectiveness of our proposed method and is able to achieve constant memory access cost regardless of the size of dilation factor d (see Table 1).

3.2 S^2Contact: Semi-supervised Training Pipeline

Collecting ground-truth contact annotation for hand-object dataset can be both challenging and time-consuming. To alleviate this, we introduce a semi-supervised learning framework to learn 3D hand-object contact estimation by leveraging large-scale unlabelled videos. As shown in Fig. 1, our proposed framework relies on two training stages: 1) pre-training stage where the model is pre-trained on the existing labelled data [2]; 2) semi-supervised stage where the model is trained by the pseudo-labels from unlabelled hand-object datasets [4,11,14]. As pseudo-labels are often noisy, we propose confidence-based filtering with geometric and visual consistency constraints to improve the quality of pseudo-labels.

Pre-training. As good initial contact estimate enables semi-supervision, we pre-train our graph-based contact estimator using a small labelled dataset $\{\mathbf{x}^l, \mathbf{y}^l\}$. We followed [12] and optimise hand-object poses to achieve target contact. Upon convergence, we clone the network to create a pair of student-teacher networks.

Pseudo-Label Generation. To maintain a reliable performance margin over the student network throughout the training, we adopt an EMA teacher which is commonly used in semi-supervised learning. The output of the student network is the predicted contact map $\widehat{\mathbf{C}}$. The teacher network generates pseudo-labels which includes pre-filter contact map $\widetilde{\mathbf{C}}$ and refined hand-object pose $\widetilde{P}_{ref}$. As it is crucial for the teacher network to generate high-quality pseudo-labels under a semi-supervised framework, we propose a confidence-based filtering mechanism that leverages geometric and visual consistency constraints.

Contact Consistency Constraint for Consistency Training. We propose a contact consistency loss to encourage robust and stable predictions for unlabelled data $\mathbf{x}^u$. As shown in Fig. 3a, we first apply stochastic transformations $\mathcal{T}$ which includes flipping, rotation and scaling on the input hand-object point clouds $\mathbf{x}^u$ for the student network. The predictions of the student network $\widehat{\mathbf{y}}_s^u \in \widehat{\mathbf{C}}$ are compared with the teacher predictions $\widetilde{\mathbf{y}}_t^u \in \widetilde{\mathbf{C}}$ processed by the same transformation $\mathcal{T}$ using contact consistency loss:

$$\mathcal{L}_{cont}(\mathbf{x}^u) = \|\Omega(\mathcal{T}(\mathbf{x}^u)) - (\mathcal{T}(\Omega(\mathbf{x}^u)))\|_1$$
$$= \|\widehat{\mathbf{y}}_s^u - \widetilde{\mathbf{y}}_t^u\|_1, \qquad (3)$$

where $\Omega(\cdot)$ represents the predicted contact map.

Geometric Consistency Constraint for Pseudo-Labels Filtering. As shown in Fig. 3b, we propose a geometric consistency constraint to the hand and object pseudo pose label $\widetilde{P}_{ref}$. Concretely, we allow the Chamfer distance $\mathcal{L}_{Cham}$ between hand and object meshes to be less than threshold t_{dist}:

$$\mathcal{L}_{Cham}(\widetilde{\mathbf{v}}_{hand}, \widetilde{\mathbf{v}}_{obj}) = \frac{1}{|\widetilde{\mathbf{v}}_{obj}|} \sum_{x \in \widetilde{\mathbf{v}}_{obj}} d_{\widetilde{\mathbf{v}}_{hand}}(x) + \frac{1}{|\widetilde{\mathbf{v}}_{hand}|} \sum_{y \in \widetilde{\mathbf{v}}_{hand}} d_{\widetilde{\mathbf{v}}_{obj}}(y), \quad (4)$$

where $\widetilde{\mathbf{v}}_{hand}$ and $\widetilde{\mathbf{v}}_{obj}$ refers to hand and object point sets, $d_{\widetilde{\mathbf{v}}_{hand}}(x) = \min_{y \in \widetilde{\mathbf{v}}_{hand}} \|x - y\|_2^2$, and $d_{\widetilde{\mathbf{v}}_{obj}}(y) = \min_{x \in \widetilde{\mathbf{v}}_{obj}} \|x - y\|_2^2$. Similarly for interpenetration, we use $\mathcal{L}_{SDF}(\widetilde{\mathbf{v}}_{obj}) = \sum_{hand,obj} \sum_i \Psi_h\left(\widetilde{\mathbf{v}}_{obj}^i\right) \le t_{pen}$ to ensure object is being manipulated by hand. Ψ_h is the Signed Distance Field (SDF) from the hand mesh (*i.e.*, $\Psi_h(\widetilde{\mathbf{v}}_{obj}) = -\min(\text{SDF}(\widetilde{\mathbf{v}}_{obj}), 0)$) to detect object penetrations.

Visual Consistency Constraint for Pseudo-Labels Filtering. We observed that geometric consistency is insufficient to correct hand grasp (see Table 5). To address this, we propose a visual consistency constraint to filter out the pseudo-labels whose rendered hand-object image $\widetilde{I}_{ho}$ does not match the input image. We first use a renderer [22] to render the hand-object image from the refined pose $\widetilde{P}_{ref}$ and obtain the hand-object segment of the input image I by applying the segmentation mask M_{gt}. Then, the structural similarity (SSIM) [54] between two images can be computed. We keep pseudo-labels when $\mathcal{L}_{SSIM} \leq t_{SSIM}$:

$$\mathcal{L}_{SSIM}(I, M_{gt}, \widetilde{I}_{ho}) = 1 - SSIM\left(I \odot M_{gt}, \widetilde{I}_{ho}\right), \qquad (5)$$

where $\odot$ denotes element-wise multiplication.

Self-training with Pseudo-Labels. After filtering pseudo-labels, our model is trained with the union set of the human-annotated dataset and the remaining pseudo-labels. The total loss $\mathcal{L}_{semi}$ can be described as:

$$\mathcal{L}_{semi}(\widehat{\mathbf{C}}, \mathbf{y}^l, \widetilde{\mathbf{C}}^u, \mathbf{x}^u) = \mathcal{L}_{sup}\left(\widehat{\mathbf{C}}, \mathbf{y}^l\right) + \mathcal{L}_{unsup}\left(\widehat{\mathbf{C}}, \widetilde{\mathbf{C}}^u\right) + \lambda_c \mathcal{L}_{cont}\left(\mathbf{x}^u\right), \quad (6)$$

where $\mathcal{L}_{sup}$ is a supervised contact loss, $\mathcal{L}_{unsup}$ is a unsupervised contact loss with pseudo-labels and λ_c is a hyperparameter. Note that $\mathcal{L}_{sup}$ (see Fig. 2) and $\mathcal{L}_{unsup}$ (see Fig. 3) are both binary cross-entropy loss.

4   Experiments

Implementation Details. We implement our method in PyTorch [37]. All experiments are run on an Intel i9-CPU @ 3.50 GHZ, 16 GB RAM, and one NVIDIA RTX 3090 GPU. For pseudo-labels filtering, $t_{dist} = 0.7$, $t_{pen} = 6$ and $t_{SSIM} = 0.25$ are the constant thresholds and stochastic transformations includes flipping ($\pm 20\%$), rotation ($\pm 180°$) and scaling ($\pm 20\%$). We train all parts of the network simultaneously with Adam optimiser [24] at a learning rate 10^{-3} for 100 epochs. We empirically fixed $K = 10, d = 4$ to produce the best results.

Datasets and Evaluation Metrics. *ContactPose* is the first dataset [2] of hand-object contact paired with hand pose, object pose and RGB-D images. It contains 2,306 unique grasps of 25 household objects grasped with 2 functional intents by 50 participants, and more than 2.9M RGB-D grasp images. For fair comparisons with ContactOpt [12], we follow their *Perturbed Contact-Pose* dataset where hand meshes are modified by additional noise to MANO parameters. This results in 22,624 training and 1,416 testing grasps. *DexYCB* is a recent real dataset for capturing hand grasping of objects [4]. It consists of a total of 582,000 image frames on 20 objects from the YCB-Video dataset [56]. We present results on their default official dataset split settings. *HO-3D* [14] is similar to *DexYCB* where it consists of 78,000 images frames on 10 objects. We present results on the official dataset split (version 2). The hand mesh error is reported after procrustes alignment and in *mm*.

Table 1. Hand error rates (*mm*) on *Perturbed ContactPose* and *ContactPose* datasets.

	Baseline		DGCNN [53]		Ours	
	joint ↓	mesh ↓	joint ↓	mesh ↓	joint ↓	mesh ↓
Perturbed ContactPose	32.988	33.147	32.592	32.762	**29.442**	**30.635**
ContactPose	8.880	8.769	8.767	32.988	**5.878**	**5.765**

- *Hand error:* We report the mean end-point error (*mm*) over 21 joints and mesh error in *mm*.
- *Object error:* We report the percentage of average object 3D vertices error within 10% of object diameter (ADD-0.1D).
- *Hand-object interaction:* We report the intersection volume (cm^3) and contact coverage (%). Intersection volume is obtained by voxelising the hand and object using a voxel size of 0.5 *cm*. Contact coverage refers to the percentage of hand points between ±2 *mm* of the object surface [12].

Baseline. For refining image-based pose estimates, we use the baseline pose estimation network from Hasson *et al.* [16] and retrain it on the training split of the respecting dataset. We filter out frames where the minimum distance between the ground truth hand and object surfaces is greater than 2 *mm*. We also use the contact estimation network DeepContact from Grady *et al.* [12] which takes ground-truth object class and pose. For semi-supervised learning, we use the baseline method from Liu *et al.* [31], a semi-supervised learning pipeline for 3D hand-object pose estimation from large-scale hand-object interaction videos.

4.1 Comparative Results

Refining Small and Large Inaccuracies. We use *ContactPose* to evaluate GCN-Contact for refining poses with small (*ContactPose*) and large (*Perturbed ContactPose*) inaccuracies. Table 1 shows the results for both cases. For *Perturbed ContactPose*, the mean end-point error over 21 joints is 82.947 *mm* before refinement. This is aimed at testing the ability to improve hand poses with large errors. In contrast, *ContactPose* is used to evaluate *mm*-scale refinement. As shown, our method consistently outperforms baseline and DGCNN [53]. We attribute the performance gain to multi-scale feature aggregation with dilation. Qualitative comparison with ContactOpt [12] is provided in Fig. 4.

Refining Image-Based Pose Estimates. We evaluate S²Contact in refining poses from an image-based pose estimator. We use the baseline image-based pose estimation network from Hasson *et al.* [16] and retrain it on the training split of the respecting dataset. Unlike [12], we do not rely on ground-truth object class and pose. In particular, we compare with the current state-of-the-art [31] which is also a semi-supervised framework for 3D hand-object pose estimation. Liu *et al.* [31] proposes spatial-temporal consistency in large-scale hand-object videos to

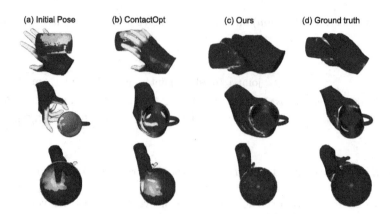

Fig. 4. Qualitative comparison with ContactOpt [12] on *ContactPose*. We observed that penetrations across hand and object (can be seen in (b)) is likely to be caused by contact predictions appearing on both object surfaces. Our model, trained only on *ContactPose*, shows robustness to various hand poses and objects.

Table 2. Error rates on *HO-3D*. Note that Liu *et al.* [31] is the current state-of-the-art semi-supervised method. **ave**, **inter** and **cover** refers to average, intersection volume and contact coverage, respectively.

Methods	Hand Error		Object ADD-0.1D(↑)				Contact	
	joint ↓	mesh ↓	bottle	can	bleach	avg	cover ↑	inter ↓
Initial Pose [16]	11.1	11.0	–	–	–	74.5	4.4	15.3 ± 21.1
ContactOpt [12]	9.7	9.7	–	–	–	75.5	14.7	6.0 ± 6.7
Liu *et al.* [31]	9.9	9.5	69.6	53.2	86.9	69.9	–	–
Ours	**8.7**	**8.9**	**79.1**	**71.8**	**93.3**	**81.4**	**19.2**	$\mathbf{3.5 \pm 1.8}$

generate pseudo-labels for hand. In contrast, we leverage physical contact and visual consistency constraints to generate pseudo contact labels which can be optimised jointly with hand and object poses. As shown in Table 2, our method outperforms [31] by 11.5% in average object ADD-0.1D score. Besides, we also compare with our baseline contact model ContactOpt [12]. As shown in Table 2, we are able to further improve hand error by 1 *mm* and 0.8 *mm* over joints and mesh. In addition to hand-object pose performance, our method is able to better reconstruct hand and object with less intersection volume and higher contact coverage. The above demonstrates that our method provides a more practical alternative to alleviate the reliance on heavy dataset annotation in hand-object. In addition, we provide qualitative comparison on *HO-3D* in Fig. 5. We also report the cross-dataset generalisation performance of our model on *DexYCB* in Table 3. We select three objects (*i.e.* , mustard bottle, potted meat can and bleach cleanser), to be consistent with *HO-3D*. As shown, our method consistently shown improvements across all metrics.

Table 3. Error rates of the cross-dataset generalisation performance on *DexYCB*.

Models	Hand Error		Object ADD-0.1D($\uparrow$)				Contact	
	joint $\downarrow$	mesh $\downarrow$	bottle	can	bleach	avg	cover $\uparrow$	inter $\downarrow$
w/o semi-supervised	13.0	13.7	76.9	46.2	68.4	63.8	4.1	16.0 ± 12.3
semi-supervised	**11.8**	**12.1**	**83.6**	**53.7**	**74.2**	**70.5**	**9.3**	$\mathbf{10.5 \pm 7.9}$

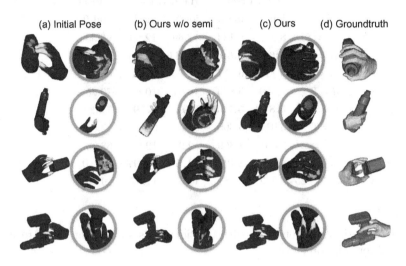

Fig. 5. Qualitative comparison with Initial Pose [31] on *HO-3D*. (a) We observed that the reconstructions of hand and object are more likely to be distant in 3D space due to depth and scale ambiguity with only RGB image as an input. (b) In contrast, when without semi-supervised learning, there exists more interpenetration or incorrect grasps between hand and object. (c) As shown, together with visual cues and contact map estimations, our method is able correct the above failure cases and generate more physically-plausible reconstructions.

4.2   Ablation Study

Number of K Neighbours. Table 4a shows the results of varying number of K neighbours without dilation. As shown, increasing K improves immediately at $K = 10$ but does not gain performance further. ***Size of Dilation Factor d.*** As shown above that performance saturates at $K = 10$, we now fix K and vary the size of dilation factor d in Table 4b. We find that the combination $K = 10, d = 4$ produce the best performance and do not improve by further increasing d. This demonstrates the effectiveness of increasing the receptive field for contact map estimation.

Combining K-NN Computation. To further study the effect of separately computing K-NN, we experiment with combined K-NN computation with $d = 4$ in Table 4c. It can be seen that the performance exceed the lower bound of (a) and similar to DGCNN's performance in Table 1. This is expected as this is

Table 4. Performances of different GCN-Contact design choices on *ContactPose* and *HO-3D*. **semi** refers to semi-supervised learning. We experiment on (a) number of K neighbours without dilation, (b) size of dilation factor d with $K = 10$ and (c) combining K-NN computation (denoted with *) with $d = 4$. Full results are in supplementary.

	Models	*ContactPose*		*HO-3D* w/o semi		
		Hand Error		Hand Error		Object Error
		joint ↓	mesh ↓	joint ↓	mesh ↓	add-0.1d (↑)
(a)	$K = 5$	8.252	8.134	12.69	12.86	66.33
	$K = 10$	**6.691**	**6.562**	**11.81**	**11.91**	**68.71**
	$K = 15$	6.715	6.617	11.85	11.92	68.70
(b)	$d = 2$	5.959	5.865	10.91	10.86	70.25
	$d = 4$	**5.878**	**5.765**	**9.92**	**9.79**	**72.81**
	$d = 6$	5.911	5.805	9.95	9.79	72.81
(c)	$K^* = 5$	8.451	8.369	12.91	12.86	68.86
	$K^* = 10$	**8.359**	**8.251**	**11.55**	**11.97**	**69.10**
	$K^* = 25$	8.369	8.286	11.57	11.97	69.06

similar to static EdgeConv [53] with dilation. This shows that separate K-NN is crucial for this framework.

Impact of Our Components. We study the impact of semi-supervised learning on *HO-3D*. Since the hand model (MANO) is consistent across datasets, the contact estimator can easily transfer hand contact to new datasets without re-training. However, it is insufficient to adapt to unlabelled dataset due to diverse object geometries. Therefore, we propose a semi-supervised learning method to generate high-quality pseudo-labels. As shown in Table 5, our method enables performance boost on both hand and object. The hand joint error is $8.74mm$ while it is $9.92mm$ without semi-supervised training. Also, the average object ADD-0.1D has a significant 8.56% improvement under S^2Contact.

Table 5 shows a quantitative comparison of S^2Contact with various filtering constraints disabled demonstrating that constraints from both visual and geometry domains are essential for faithful training. We also observed that disabling $\mathcal{L}_{cont}$ can easily lead to unstable training and 5.45% performance degradation in object error. In contrast, geometric consistencies ($\mathcal{L}_{Cham}$ and $\mathcal{L}_{SDF}$) have a comparably smaller impact on hand and object pose. Despite that they account for less than 2% performance drop to object error, geometric consistencies are important for contact (*i.e.*, more than 5% for contact coverage). The remaining factor, measuring visual similarity, has a more significant impact. Disabling visual consistency constraint $\mathcal{L}_{SSIM}$ results in hand joint error and object error increase by $1mm$ and 8.04%, respectively. We validate that the combination of our pseudo-label filtering constraints are critical for generating high-quality pseudo-labels and improving hand-object pose estimation performance. Finally, we provide qualitative examples on out-of-domain objects in supplementary.

Table 5. Performances of different filtering constraints under semi-supervised learning on *HO-3D*. **semi** refers to semi-supervised learning.

Models	Hand Error		Object Error	Contact	
	joint ↓	mesh ↓	add-0.1d (↑)	cover ↑	inter ↓
W/o semi	9.92	9.79	72.81	12.1	8.3 ± 10.5
w/ semi	**8.74**	**8.86**	**81.37**	**19.2**	**3.5 ± 1.8**
w/o $\mathcal{L}_{Cham}$	8.53	8.48	80.11	13.1	6.9 ± 6.2
w/o $\mathcal{L}_{SDF}$	8.61	8.59	80.90	14.7	5.5 ± 6.0
w/o $\mathcal{L}_{cont}$	9.28	9.19	75.92	13.9	4.8 ± 3.1
w/o $\mathcal{L}_{SSIM}$	9.71	9.57	73.33	16.3	3.7 ± 2.6

Computational Analysis. We report the model parameters and GPU memory cost in Table 4 of supplementary material. For fair comparisons, all models are tested using a batch size of 64. As shown, our model has 2.4X less the number of learnable model parameters and 2X less the GPU memory cost when compared to baseline and DGCNN [53], respectively. We alleviate the need to keep a high density of points across the network (DGCNN) while gaining performance.

5 Conclusion

In this paper, we have proposed a novel semi-supervised learning framework which enables learning contact with monocular videos. The main idea behind this study was to demonstrate that this can successfully be achieved with visual and geometric consistency constraints for pseudo-label generation. We designed an efficient graph-based network for inferring contact maps and shown benefits of combining visual cues and contact consistency constraints to produce more physically-plausible reconstructions. In the future, we would like to explore more consistencies over time and or multiple views to further improve the accuracy.

Acknowledgements. This research was supported by the MSIT (Ministry of Science and ICT), Korea, under the ITRC (Information Technology Research Center) support program (IITP–2022–2020–0–01789) supervised by the IITP (Institute of Information & Communications Technology Planning & Evaluation) and the Baskerville Tier 2 HPC service (https://www.baskerville.ac.uk/) funded by the Engineering and Physical Sciences Research Council (EPSRC) and UKRI through the World Class Labs scheme (EP/T022221/1) and the Digital Research Infrastructure programme (EP/W032244/1) operated by Advanced Research Computing at the University of Birmingham. KIK was supported by the National Research Foundation of Korea (NRF) grant (No. 2021R1A2C2012195) and IITP grants (IITP–2021–0–02068 and IITP–2020–0–01336). ZQZ was supported by China Scholarship Council (CSC) Grant No. 202208060266. AL was supported in part by the EPSRC (grant number EP/S032487/1). FZ was supported by the National Natural Science Foundation of China under Grant No. 61972188 and 62122035.

References

1. Brahmbhatt, S., Ham, C., Kemp, C.C., Hays, J.: ContactDB: analyzing and predicting grasp contact via thermal imaging. In: CVPR (2019)
2. Brahmbhatt, S., Tang, C., Twigg, C.D., Kemp, C.C., Hays, J.: ContactPose: a dataset of grasps with object contact and hand pose. In: ECCV (2020)
3. Cao, Z., Radosavovic, I., Kanazawa, A., Malik, J.: Reconstructing hand-object interactions in the wild. In: ICCV (2021)
4. Chao, Y.W., et al.: DexYCB: a benchmark for capturing hand grasping of objects. In: CVPR (2021)
5. Chen, W., Jia, X., Chang, H.J., Duan, J., Leonardis, A.: G2L-Net: global to local network for real-time 6D pose estimation with embedding vector features. In: CVPR (2020)
6. Chen, W., Jia, X., Chang, H.J., Duan, J., Shen, L., Leonardis, A.: FS-Net: fast shape-based network for category-level 6D object pose estimation with decoupled rotation mechanism. In: CVPR (2021)
7. Chen, Y., Tu, Z., Ge, L., Zhang, D., Chen, R., Yuan, J.: SO-HandNet: self-organizing network for 3D hand pose estimation with semi-supervised learning. In: CVPR (2019)
8. Corona, E., Pumarola, A., Alenya, G., Moreno-Noguer, F., Rogez, G.: GanHand: predicting human grasp affordances in multi-object scenes. In: CVPR (2020)
9. Defferrard, M., Bresson, X., Vandergheynst, P.: Convolutional neural networks on graphs with fast localized spectral filtering. In: NeurIPS (2016)
10. Doosti, B., Naha, S., Mirbagheri, M., Crandall, D.J.: HOPE-Net: a graph-based model for hand-object pose estimation. In: CVPR (2020)
11. Garcia-Hernando, G., Yuan, S., Baek, S., Kim, T.K.: First-person hand action benchmark with RGB-D videos and 3D hand pose annotations. In: CVPR (2018)
12. Grady, P., Tang, C., Twigg, C.D., Vo, M., Brahmbhatt, S., Kemp, C.C.: ContactOpt: optimizing contact to improve grasps. In: CVPR (2021)
13. Guo, M.-H., Cai, J.-X., Liu, Z.-N., Mu, T.-J., Martin, R.R., Hu, S.-M.: PCT: point cloud transformer. Computational Visual Media **7**(2), 187–199 (2021). https://doi.org/10.1007/s41095-021-0229-5
14. Hampali, S., Rad, M., Oberweger, M., Lepetit, V.: Honnotate: A method for 3D annotation of hand and object poses. In: CVPR (2020)
15. Han, S., et al.: MEgATrack: monochrome egocentric articulated hand-tracking for virtual reality. In: SIGGRAPH (2020)
16. Hasson, Y., Tekin, B., Bogo, F., Laptev, I., Pollefeys, M., Schmid, C.: Leveraging photometric consistency over time for sparsely supervised hand-object reconstruction. In: CVPR (2020)
17. Hasson, Y., Varol, G., Laptev, I., Schmid, C.: Towards unconstrained joint hand-object reconstruction from RGB videos. In: 3DV (2021)
18. Hasson, Y., et al.: Learning joint reconstruction of hands and manipulated objects. In: CVPR (2019)
19. Huang, L., Tan, J., Meng, J., Liu, J., Yuan, J.: HOT-Net: non-autoregressive transformer for 3D hand-object pose estimation. In: ACM MM (2020)
20. Jiang, H., Liu, S., Wang, J., Wang, X.: Hand-object contact consistency reasoning for human grasps generation. In: ICCV (2021)
21. Karunratanakul, K., Yang, J., Zhang, Y., Black, M.J., Muandet, K., Tang, S.: Grasping field: learning implicit representations for human grasps. In: 3DV (2020)
22. Kato, H., Ushiku, Y., Harada, T.: Neural 3D mesh renderer. In: CVPR (2018)

23. Kaviani, S., Rahimi, A., Hartley, R.: Semi-Supervised 3D hand shape and pose estimation with label propagation. arXiv preprint arXiv:2111.15199 (2021)
24. Kingma, D.P., Ba, J.: Adam: a method for stochastic optimization. In: ICLR (2015)
25. Kipf, T.N., Welling, M.: Semi-supervised classification with graph convolutional networks. In: ICLR (2017)
26. Kwon, T., Tekin, B., Stühmer, J., Bogo, F., Pollefeys, M.: H2O: two hands manipulating objects for first person interaction recognition. In: ICCV (2021)
27. Labbé, Y., Carpentier, J., Aubry, M., Sivic, J.: CosyPose: consistent multi-view multi-object 6D pose estimation. In: ECCV (2020)
28. Li, G., Muller, M., Thabet, A., Ghanem, B.: DeepGNSs: can GCNs go as deep as CNNs? In: ICCV (2019)
29. Li, Y., Wang, G., Ji, X., Xiang, Y., Fox, D.: DeepIM: deep iterative matching for 6D pose estimation. In: ECCV (2018)
30. Lin, Z.H., Huang, S.Y., Wang, Y.C.F.: Convolution in the cloud: learning deformable kernels in 3D graph convolution networks for point cloud analysis. In: CVPR (2020)
31. Liu, S., Jiang, H., Xu, J., Liu, S., Wang, X.: Semi-supervised 3D hand-object poses estimation with interactions in time. In: CVPR (2021)
32. Liu, Z., Hu, H., Cao, Y., Zhang, Z., Tong, X.: A closer look at local aggregation operators in point cloud analysis. In: ECCV (2020)
33. Maturana, D., Scherer, S.: VoxNet: a 3D convolutional neural network for real-time object recognition. In: IROS (2015)
34. Monti, F., Boscaini, D., Masci, J., Rodola, E., Svoboda, J., Bronstein, M.M.: Geometric deep learning on graphs and manifolds using mixture model CNNs. In: CVPR (2017)
35. Mueller, F., et al.: GANerated hands for real-time 3D hand tracking from monocular RGB. In: CVPR (2018)
36. Mueller, F., et al.: Real-time pose and shape reconstruction of two interacting hands with a single depth camera. In: SIGGRAPH (2019)
37. Paszke, A., et al.: Automatic Differentiation in Pytorch. In: NeurIPS (2017)
38. Qi, C.R., Su, H., Mo, K., Guibas, L.J.: PointNet: deep learning on point sets for 3D classification and segmentation. In: CVPR (2017)
39. Qi, C.R., Yi, L., Su, H., Guibas, L.J.: PointNet++: deep hierarchical feature learning on point sets in a metric space. In: NeurIPS (2017)
40. Qian, G., Hammoud, H., Li, G., Thabet, A., Ghanem, B.: ASSANet: an anisotropic separable set abstraction for efficient point cloud representation learning. NeurIPS (2021)
41. Romero, J., Tzionas, D., Black, M.J.: Embodied hands: modeling and capturing hands and bodies together. ACM Trans. Graph. (ToG) 36(6), 1–17 (2017)
42. Simon, T., Joo, H., Matthews, I., Sheikh, Y.: Hand keypoint detection in single images using multiview bootstrapping. In: CVPR (2017)
43. Spurr, A., Molchanov, P., Iqbal, U., Kautz, J., Hilliges, O.: Adversarial motion modelling helps semi-supervised hand pose estimation. arXiv preprint arXiv:2106.05954 (2021)
44. Spurr, A., Song, J., Park, S., Hilliges, O.: Cross-modal deep variational hand pose estimation. In: CVPR (2018)
45. Szegedy, C., et al.: Going deeper with convolutions. In: CVPR (2015)
46. Taheri, O., Ghorbani, N., Black, M.J., Tzionas, D.: GRAB: a dataset of whole-body human grasping of objects. In: ECCV (2020)
47. Tang, D., Chang, H.J., Tejani, A., Kim, T.K.: Latent regression forest: structured estimation of 3D articulated hand posture. In: CVPR (2014)

48. Tang, D., Yu, T.H., Kim, T.K.: Real-time articulated hand pose estimation using semi-supervised transductive regression forests. In: ICCV (2013)
49. Tekin, B., Bogo, F., Pollefeys, M.: H+O: unified egocentric recognition of 3D hand-object poses and interactions. In: CVPR (2019)
50. Ueda, E., Matsumoto, Y., Imai, M., Ogasawara, T.: A hand-pose estimation for vision-based human interfaces. IEEE Trans. Ind. Electron. 50(4), 676–684 (2003)
51. Wang, H., Cong, Y., Litany, O., Gao, Y., Guibas, L.J.: 3DIoUMatch: leveraging IoU prediction for semi-supervised 3D object detection. In: CVPR (2021)
52. Wang, J., et al.: RGB2Hands: real-time tracking of 3D hand interactions from monocular RGB video. In: SIGGRAPH (2020)
53. Wang, Y., Sun, Y., Liu, Z., Sarma, S.E., Bronstein, M.M., Solomon, J.M.: Dynamic graph CNN for learning on point clouds. In: SIGGRAPH (2019)
54. Wang, Z., Bovik, A.C., Sheikh, H.R., Simoncelli, E.P.: Image quality assessment: from error visibility to structural similarity. IEEE Trans. Image Process. **13**(4), 600–612 (2004)
55. Wu, W., Qi, Z., Fuxin, L.: PointConv: deep convolutional networks on 3D point clouds. In: CVPR (2019)
56. Xiang, Y., Schmidt, T., Narayanan, V., Fox, D.: PoseCNN: a convolutional neural network for 6D object pose estimation in cluttered scenes. In: RSS (2018)
57. Xu, M., Ding, R., Zhao, H., Qi, X.: PAConv: position adaptive convolution with dynamic kernel assembling on point clouds. In: CVPR (2021)
58. Yang, J., Chang, H.J., Lee, S., Kwak, N.: SeqHAND: RGB-sequence-based 3D hand pose and shape estimation. In: ECCV (2020)
59. Yang, L., Chen, S., Yao, A.: SemiHand: semi-supervised hand pose estimation with consistency. In: ICCV (2021)
60. Yang, L., Zhan, X., Li, K., Xu, W., Li, J., Lu, C.: CPF: learning a contact potential field to model the hand-object interaction. In: ICCV (2021)
61. You, H., Feng, Y., Ji, R., Gao, Y.: PVNet: a joint convolutional network of point cloud and multi-view for 3D shape recognition. In: ACM Multimedia (2018)
62. Zhang, T., et al.: Deep imitation learning for complex manipulation tasks from virtual reality teleoperation. In: ICRA (2018)
63. Zhao, H., Jiang, L., Jia, J., Torr, P.H., Koltun, V.: Point transformer. In: ICCV (2021)
64. Zimmermann, C., Brox, T.: Learning to estimate 3D hand pose from single RGB images. In: ICCV (2017)

SC-wLS: Towards Interpretable Feed-forward Camera Re-localization

Xin Wu[1,2(✉)], Hao Zhao[1,3], Shunkai Li[4], Yingdian Cao[1,2], and Hongbin Zha[1,2]

[1] Key Laboratory of Machine Perception (MOE), School of AI, Peking University, China, China
{wuxin1998,zhao-hao,yingdianc}@pku.edu.cn, zha@cis.pku.edu.cn
[2] PKU-SenseTime Machine Vision Joint Lab, China, China
[3] Intel Labs China, Beijing, China
[4] Kuaishou Technology, Beijing, China
lishunkai@pku.edu.cn
https://github.com/XinWu98/SC-wLS

Abstract. Visual re-localization aims to recover camera poses in a known environment, which is vital for applications like robotics or augmented reality. Feed-forward absolute camera pose regression methods directly output poses by a network, but suffer from low accuracy. Meanwhile, scene coordinate based methods are accurate, but need iterative RANSAC post-processing, which brings challenges to efficient end-to-end training and inference. In order to have the best of both worlds, we propose a feed-forward method termed SC-wLS that exploits all scene coordinate estimates for weighted least squares pose regression. This differentiable formulation exploits a weight network imposed on 2D-3D correspondences, and requires pose supervision only. Qualitative results demonstrate the interpretability of learned weights. Evaluations on 7Scenes and Cambridge datasets show significantly promoted performance when compared with former feed-forward counterparts. Moreover, our SC-wLS method enables a new capability: self-supervised test-time adaptation on the weight network. Codes and models are publicly available.

Keywords: Camera re-localization · Differentiable optimization

1 Introduction

Visual re-localization [10,16,32,38] determines the global 6-DoF poses (*i.e.*, position and orientation) of query RGB images in a known environment. It is a fundamental computer vision problem and has many applications in robotics and augmented reality. Recently there is a trend to incorporate deep neural networks into various 3D vision tasks, and use differentiable formulations that optimize

---

X. Wu and H. Zhao—Equal contribution.

Supplementary Information The online version contains supplementary material available at https://doi.org/10.1007/978-3-031-19769-7_34.

S. Avidan et al. (Eds.): ECCV 2022, LNCS 13661, pp. 585–601, 2022.
https://doi.org/10.1007/978-3-031-19769-7_34

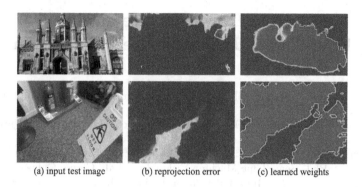

(a) input test image   (b) reprojection error   (c) learned weights

Fig. 1. For input images (a), our network firstly regresses their scene coordinates, then predicts correspondence-wise weights (c). With these weights, we can use all 2D-3D correspondences for end-to-end differentiable least squares pose estimation. We use re-projection errors (b) to illustrate scene coordinate quality. Our weights select high-quality scene coordinates. A higher color temperature represents a higher value.

losses of interest to learn result-oriented intermediate representation. Following this trend, many learning-based absolute pose regression (APR) methods [8,22] have been proposed for camera re-localization, which only need a single feed-forward pass to recover poses. However, they treat the neural network as a black box and suffer from low accuracy [33]. On the other hand, scene coordinate based methods learn pixel-wise 3D scene coordinates from RGB images and solve camera poses using 2D-3D correspondences by Perspective-n-Point (PnP) [24]. In order to handle outliers in estimated scene coordinates, the random sample consensus (RANSAC) [11] algorithm is usually used for robust fitting. Compared to the feed-forward APR paradigm, scene coordinate based methods achieve state-of-the-art performance on public camera re-localization datasets. However, RANSAC-based post-processing is an iterative procedure conducted on CPUs, which brings engineering challenges for efficient end-to-end training and inference [4,7].

In order to get the best of both worlds, we develop a new feed-forward method based upon the state-of-the-art (SOTA) pipeline DSAC* [7], thus enjoying the strong representation power of scene coordinates. We propose an alternative option to RANSAC that exploits all 3D scene coordinate estimates for weighted least squares pose regression (SC-wLS). The key to SC-wLS is a weight network that treats 2D-3D correspondences as 5D point clouds and learns weights that capture geometric patterns in this 5D space, with only pose supervision. Our learned weights can be used to interpret how much each scene coordinate contributes to the least squares solver.

Our SC-wLS estimates poses using only tensor operators on GPUs, which is similar to APR methods due to the feed-forward nature but out-performs APR methods due to the usage of scene coordinates. Furthermore, we show that a self-supervised test-time adaptation step that updates the weight network can lead to further performance improvements. This is potentially useful in scenarios like a robot vacuum adapts to specific rooms during standby time. Although we focus on comparisons with APR methods, we also equip SC-wLS with the LM-Refine post-processing module provided in DSAC* [7] to explore the limits of SC-wLS, and show that it out-performs SOTA on the outdoor dataset Cambridge.

Our major contributions can be summarized as follows: (1) We propose a new feed-forward camera re-localization method, termed SC-wLS, that learns interpretable scene coordinate quality weights (as in Fig. 1) for weighted least squares pose estimation, with only pose supervision. (2) Our method combines the advantages of two paradigms. It exploits learnt 2D-3D correspondences while still allows efficient end-to-end training and feed-forward inference in a principled manner. As a result, we achieve significantly better results than APR methods. (3) Our SC-wLS formulation allows test-time adaptation via self-supervised fine-tuning of the weight network with the photometric loss.

2   Related Works

Camera Re-localization. In the following, we discuss camera re-localization methods from the perspective of map representation.

Representing image databases with global descriptors like thumbnails [16], BoW [39], or learned features [1] is a natural choice for camera re-localization. By retrieving poses of similar images, localization can be done in the extremely large scale [16,34]. Meanwhile, CNN-based absolute pose regression methods [8,22,42,49] belong to this category, since their final-layer embeddings are also learned global descriptors. They regress camera poses from single images in an end-to-end manner, and recent work primarily focuses on sequential inputs [48] and network structure enhancement [13,37,49]. Although the accuracies of this line of methods are generally low due to intrinsic limitations [33], they are usually compact and fast, enabling pose estimation in a single feed-forward pass.

Maps can also be represented by 3D point cloud [46] with associated 2D descriptors [26] via SfM tools [35]. Given a query image, feature matching establishes sparse 2D-3D correspondences and yields very accurate camera poses with RANSAC-PnP pose optimization [32,53]. The success of these methods heavily depends on the discriminativeness of features and the robustness of matching strategies. Inspired by feature based pipelines, scene coordinate regression learns a 2D-3D correspondence for each pixel, instead of using feature extraction and matching separately. The map is implicitly encoded into network parameters. [27] demonstrates impressive localization performance using stereo initialization and sequence input. Recently, [2] shows that the algorithm used to create pseudo ground truth has a significant impact on the relative ranking of above methods.

Apart from random forest based methods using RGB-D inputs [28,38,40], scene coordinate regression on RGB images is seeing steady progress [4-7,56]. This line of work lays the foundation for our research. In this scheme, predicted scene coordinates are noisy due to single-view ambiguity and domain gap during inference. As such, [4,5] use RANSAC and non-linear optimization to deal with outliers, and NG-RANSAC [6] learns correspondence-wise weights to guide RANSAC sampling. [6] conditions weights on RGB images, whose statistics is often influenced by factors like lighting, weather or even exposure time. Object

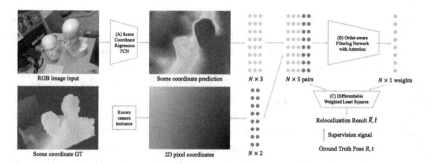

Fig. 2. The overview of SC-wLS. Firstly, a fully convolutional network (A) regresses pixel-wise scene coordinates from an input RGB image. Scene coordinate predictions are flattened to the shape of $N \times 3$, with N being pixel count. We concatenate it with normalized $N \times 2$ 2D pixel coordinates, forming $N \times 5$ correspondence inputs. The correspondences are fed into the weight learning network (B), producing $N \times 1$ weights indicating scene coordinate quality. The architecture of B is an order-aware filtering network [52] with graph attention modules [31]. Thirdly, correspondences and weights are sent into a differentiable weighted least squares layer (C), directly outputting camera poses. The scene coordinate ground truth is not used during training.

pose and room layout estimation [17,18,23,50,54,55] can also be addressed with similar representations [3,44].

Differentiable Optimization. To enhance their compatibility with deep neural networks and learn result-oriented feature representation, some recent works focus on re-formulating geometric optimization techniques into an end-to-end trainable fashion, for various 3D vision tasks. [20] proposes several standard differentiable optimization layers. [30,51] propose to estimate fundamental/essential matrix via solving weighted least squares problems with spectral layers. [12] further shows that the eigen-value switching problem when solving least squares can be avoided by minimizing linear system residuals. [14] develops generic black-box differentiable optimization techniques with implicit declarative nodes.

3   Method

Given an RGB image I, we aim to find an estimate of the absolute camera pose consisting of a 3D translation and a 3D rotation, in the world coordinate system. Towards this goal, we exploit the scene coordinate representation. Specifically, for each pixel i with position $\mathbf{p}_i$ in an image, we predict the corresponding 3D scene coordinate $\mathbf{s}_i$. As illustrated in Fig. 2, we propose an end-to-end trainable deep network that directly calculates global camera poses via weighted least squares. The method is named as **SC-wLS**. Figure 2-A is a standard fully convolutional network for scene coordinate regression, as used in former works [7]. Our innovation lies in Fig. 2-B/C, as elaborated below.

3.1   Formulation

Given learned 3D scene coordinates and corresponding 2D pixel positions, our goal is to determine the absolute poses of calibrated images taking all correspondences into account. This would inevitably include outliers, and we need to give them proper weights. Ideally, if all outliers are rejected by zero weights, calculating the absolute pose can be formulated as a linear least squares problem. Inspired by [51] (which solves an essential matrix problem instead), we use all of the N 2D-3D correspondences as input and predict N respective weights $\mathbf{w}_i$ using a neural network (Fig. 2-B). $\mathbf{w}_i$ indicates the uncertainty of each scene coordinate prediction. As such the ideal least squares problem is turned into a weighted version for pose recovery.

Specifically, the input correspondence $\mathbf{c}_i$ to Fig. 2-B is

$$\mathbf{c}_i = [x_i, y_i, z_i, u_i, v_i] \tag{1}$$

where x_i, y_i, z_i are the three components of the scene coordinate $\mathbf{s}_i$, and u_i, v_i denote the corresponding pixel position. u_i, v_i are generated by normalizing $\mathbf{p}_i$ with the known camera intrinsic matrix. The absolute pose is written as a transformation matrix $\mathbf{T} \in \mathbb{R}^{3 \times 4}$. It projects scene coordinates to the camera plane as below:

$$\begin{bmatrix} u_i \\ v_i \\ 1 \end{bmatrix} = \mathbf{T} \begin{bmatrix} x_i \\ y_i \\ z_i \\ 1 \end{bmatrix} = \begin{bmatrix} p_1 & p_2 & p_3 & p_4 \\ p_5 & p_6 & p_7 & p_8 \\ p_9 & p_{10} & p_{11} & p_{12} \end{bmatrix} \begin{bmatrix} x_i \\ y_i \\ z_i \\ 1 \end{bmatrix} \tag{2}$$

When $N > 6$, the transformation matrix $\mathbf{T}$ can be recovered by Direct Linear Transform (DLT) [15], which converts Eq. 2 into a linear system:

$$\mathbf{X}\text{Vec}(\mathbf{T}) = 0 \tag{3}$$

Vec$(\mathbf{T})$ is the vectorized $\mathbf{T}$. $\mathbf{X}$ is a $\mathbb{R}^{2N \times 12}$ matrix whose $2i - 1$ and $2i$ rows $\mathbf{X}^{(2i-1)}$, $\mathbf{X}^{(2i)}$ are as follows:

$$\begin{bmatrix} x_i, y_i, z_i, 1, & 0, & 0, & 0, & 0, & -u_i x_i, -u_i y_i, -u_i z_i, -u_i \\ 0, & 0, & 0, & 0, & x_i, y_i, z_i, 1, & -v_i x_i, -v_i y_i, -v_i z_i, -v_i \end{bmatrix} \tag{4}$$

As such, pose estimation is formulated as a least squares problem. Vec$(\mathbf{T})$ can be recovered by finding the eigenvector associated to the smallest eigenvalue of $\mathbf{X}^\top \mathbf{X}$.

Note that in SC-wLS, each correspondence contributes differently according to $\mathbf{w}_i$, so $\mathbf{X}^\top \mathbf{X}$ can be rewritten as $\mathbf{X}^\top \text{diag}(\mathbf{w})\mathbf{X}$ and Vec$(\mathbf{T})$ still corresponds to its smallest eigenvector. As the rotation matrix $\mathbf{R}$ needs to be orthogonal and has determinant 1, we further refine the DLT results by the generalized Procrustes algorithm [36], which is also differentiable. More details about this post-processing step can be found in the supplementary material.

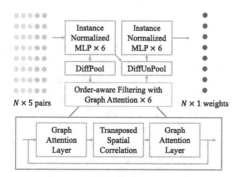

Fig. 3. The architecture of Fig. 2-B, which is inherited from OANet [52]. Here we use two self-attention graph layers from [31] while in the original OANet they are instance normazlied MLPs.

3.2   Network Design

Then we describe the architecture of Fig. 2-B. We treat the set of correspondences $\{\mathbf{c}_i\}$ as unordered 5-dimensional point clouds and resort to a PointNet-like architecture [52] dubbed OANet. This hierarchical network design consists of multiple components including DiffPool layers, Order-Aware filtering blocks and DiffUnpool layers. It can guarantee the permutation-invariant property of input correspondences. Our improved version enhances the Order-Aware filtering block with self-attention.

The network is illustrated in Fig. 3. Specifically speaking, the DiffPool layer firstly clusters inputs into a particularly learned canonical order. Then the clusters are spatially correlated to retrieve the permutation-invariant context and finally recovered back to original size through the DiffUnpool operator. Inspired by the success of transformer [41] and its extension in 3D vision [31], we propose to introduce the self-attention mechanism into OANet to better reason about the underlying relationship between correspondences. As shown in Fig. 3, there are two instance normalized MLP modules before and after the transposed spatial correlation module, in original OANet. We replace them with attention-based message passing modules, which can exploit the spatial and contextual information simultaneously.

The clusters can be regarded as nodes $\mathcal{V} = \{v_i\}$ in a graph $\mathcal{G}$ and the edges $\mathcal{E} = \{e_i\}$ are constructed between every two nodes. Since the number of clusters is significantly less than that of original inputs, calculating self-attention on these fully connected weighted edges would be tractable, in terms of computation speed and memory usage. Let $^{(in)}\mathbf{f}_i$ be the intermediate representation for node v_i at input, and the self-attention operator can be described as:

$$^{(out)}\mathbf{f}_i = {}^{(in)}\mathbf{f}_i + \mathrm{MLP}\left(\left[{}^{(in)}\mathbf{f}_i \,\|\, \mathbf{m}_{\mathcal{E}\to i}\right]\right), \qquad (5)$$

where $\mathbf{m}_{\mathcal{E}\to i}$ is the message aggregated from all other nodes $\{j : (i,j) \in \mathcal{E}\}$ using self-attention described in [31], and $[\cdot \,\|\, \cdot]$ denotes concatenation.

Finally, we apply two sequential activation functions (a ReLU followed by a Tanh) upon the outputs of Fig. 3 to get the weights $\mathbf{w}_i$ in the range $[0, 1)$, for indoor scenes. We use the log-sigmoid activation for outdoor environments following [6], for more stable gradient back-propagation.

3.3   Loss Functions

Generating ground-truth scene coordinates for supervision is time-consuming for real-world applications. To this end, our whole framework is solely supervised by ground-truth poses without accessing any 3D models. Successfully training the network with only pose supervision requires three stages using different loss functions. Before describing the training protocol, we first define losses here.

Re-projection loss is defined as follows:

$$r_i = ||\mathbf{K}\mathbf{T}^{-1}\mathbf{s}_i - \mathbf{p}_i||_2, \tag{6}$$

$$\mathcal{L}_p^{(i)} = \begin{cases} r_i & \text{if } \mathbf{s}_i \in V \\ ||\bar{\mathbf{s}}_i - \mathbf{s}_i||_1 & \text{otherwise.} \end{cases} \tag{7}$$

$$\mathcal{L}_p = \frac{1}{N}\sum_{i=1}^{N}\mathcal{L}_p^{(i)} \tag{8}$$

$\mathbf{K}$ is the known camera intrinsic matrix, and V is a set of predicted $\mathbf{s}_i$ that meet some specific validity constraints following DSAC* [7]. If $\mathbf{s}_i \in V$, we use the re-projection error r_i in Eq. 6. Otherwise, we generate a heuristic $\bar{\mathbf{s}}_i$ assuming a depth of 10 m. This hyper-parameter is inherited from [7].

Classification loss $\mathcal{L}_c$ Without knowing the ground truth for scene coordinate quality weights, we utilize the re-projection error r_i for weak supervision:

$$l_i = \begin{cases} 0, & \text{if } r_i > \tau \\ 1, & \text{otherwise} \end{cases} \tag{9}$$

$$\mathcal{L}_c = \frac{1}{N}\sum_{i=1}^{N} H\left(l_i, \mathbf{w}_i\right), \tag{10}$$

where H is the binary cross-entropy function. τ is empirically set to 1 pixel to reject outliers. Ablation studies about $\mathcal{L}_c$ can be found in the supplementary.

Regression loss $\mathcal{L}_r$ Since our SC-wLS method is fully differentiable using the eigen-decomposition (ED) technique [20], imposing L_2 or other losses on the transform matrix $\mathbf{T}$ is straightfoward. However, as illustrated in [12], the eigen-decomposition operations would result in gradient instabilities due to the eigen vector switching phenomenon. Thus we utilize the eigen-decomposition free loss [12] to constrain the pose output, which avoids explicitly performing ED so as to guarantee convergence:

$$\mathcal{L}_r = \mathbf{t}^\top \mathbf{X}^\top \text{diag}(\mathbf{w})\mathbf{X}\mathbf{t} + \alpha e^{-\beta tr(\bar{\mathbf{X}}^\top \text{diag}(\mathbf{w})\bar{\mathbf{X}})} \tag{11}$$

where $\mathbf{t}$ is the flattened ground-truth pose, $\bar{\mathbf{X}} = \mathbf{X}(\mathbf{I} - \mathbf{t}\mathbf{t}^\top)$, α and β are positive scalars. Two terms in Eq. 11 serve as different roles. The former tends to minimize the error of pose estimation while the latter tries to alleviate the impact of null space. The trace value in the second term can change up to thousands for different batches, thus the hyper-parameter β should be set to balance this effect.

3.4   Training Protocol

We propose a three-stage protocol with different objective functions to train our network Fig. 2. Note that ground truth poses are known in all three stages.

Scene Coordinate Initialization. Firstly, we train our scene coordinate regression network (Fig. 2-A) by optimizing the re-projection error $\mathcal{L}_p$ in Eq. 8, using ground truth poses. This step gives us reasonable initial scene coordinates, which are critical for the convergence of $\mathcal{L}_c$ and $\mathcal{L}_r$.

Weight Initialization. Secondly, we exploit $\mathcal{L} = \mathcal{L}_c + \gamma\mathcal{L}_r$ to train Fig. 2-B, where γ is a balancing weight. The classification loss $\mathcal{L}_c$ is used to reject outliers. The regression loss $\mathcal{L}_r$ is used to constrain poses, making a compromise between the accuracy of estimated weights and how much information is reserved in the null-space. $\mathcal{L}_c$ is important to the stable convergence of this stage.

End-to-End Optimization. As mentioned above, our whole pipeline is fully differentiable thus capable of learning both scene coordinates and quality weights, directly from ground truth poses. In the third stage, we train both the scene coordinate network (Fig. 2-A) and quality weight network (Fig. 2-B) using $\mathcal{L}$, forcing both of them to learn task-oriented feature representations.

3.5   Optional RANSAC-like Post-processing

The SC-wLS method allows us to directly calculate camera poses with DLT. Although we focus on feed-forward settings, it is still possible to use RANSAC-like post-processing for better results. Specifically, we adopt the Levenberg-Marquardt solver developed by [7] to post-process DLT poses as an optional step (shortened as **LM-Refine**). It brings a boost of localization accuracy.

Specifically, this LM-Refine module is an iterative procedure. It first determines a set of inlier correspondences according to current pose estimate then optimizes poses using the Levenberg-Marquardt algorithm over the inlier set w.r.t. re-projection errors in Eq. 6. This process stops when the number of inlier set converges or until the maximum iteration, which we set to 100 following [7].

3.6   Self-supervised Adaptation

At test time, we do not have ground truth pose for supervision. However, if the test data is given in the form of image sequences, we could use the photometric loss in co-visible RGB images to supervise our quality weight network Fig. 2-B. We sample two consecutive images I^s and I^t from the test set and synthesize $\tilde{I}^t$ by warping I^s to I^t, similar to self-supervised visual odometry methods [25,57]:

Table 1. Median errors on the 7Scenes dataset [38], with translational and rotational errors measured in m and $°$. Our results are evaluated directly using weighted DLT, without the (optional) LM-Refine step (Sect. 3.5) and self-supervised adaptation step (Sect. 3.6). We compare with methods that directly predict poses with neural networks. Ours *(dlt)* and Ours *(dlt+e2e)* show results evaluated w/o and w/ the third stage mentioned in Sect. 3.4. Our results are significantly better.

Methods	Chess	Fire	Heads	Office	Pumpkin	Kitchen	Stairs
PoseNet17 [21]	0.13/4.5	0.27/11.3	0.17/13.0	0.19/5.6	0.26/4.8	0.23/5.4	0.35/12.4
LSTM-Pose [42]	0.24/5.8	0.34/11.9	0.21/13.7	0.30/8.1	0.33/7.0	0.37/8.8	0.40/13.7
BranchNet [47]	0.18/5.2	0.34/9.0	0.20/14.2	0.30/7.1	0.27/5.1	0.33/7.4	0.38/10.3
GPoseNet [9]	0.20/7.1	0.38/12.3	0.21/13.8	0.28/8.8	0.37/6.9	0.35/8.2	0.37/12.5
MLFBPPose [45]	0.12/5.8	0.26/12.0	0.14/13.5	0.18/8.2	0.21/7.1	0.22/8.1	0.38/10.3
AttLoc [43]	0.10/4.1	0.25/11.4	0.16/11.8	0.17/5.3	0.21/4.4	0.23/5.4	0.26/10.5
MapNet [8]	0.08/3.3	0.27/11.7	0.18/13.3	0.17/5.2	0.22/4.0	0.23/4.9	0.30/12.1
LsG [48]	0.09/3.3	0.26/10.9	0.17/12.7	0.18/5.5	0.20/3.7	0.23/4.9	0.23/11.3
GL-Net [49]	0.08/2.8	0.26/8.9	0.17/11.4	0.18/5.1	0.15/2.8	0.25/4.5	0.23/8.8
MS-Transformer [37]	0.11/4.7	0.24/9.6	0.14/12.2	0.17/5.7	0.18/4.4	0.17/5.9	0.26/8.5
Ours *(dlt)*	0.029/0.78	0.051/1.04	**0.026/2.00**	0.063/0.93	0.084/1.28	0.099/1.60	0.179/3.61
Ours *(dlt+e2e)*	**0.029/0.76**	**0.048/1.09**	0.027/1.92	**0.055/0.86**	**0.077/1.27**	**0.091/1.43**	**0.123/2.80**

$$p^t \sim \mathbf{K}\mathbf{T}_t^{-1}\mathbf{s}^s \qquad (12)$$

$$\mathcal{L}_{\mathrm{ph}} = \sum_p ||I^t(p) - \tilde{I}^t(p)||_1 + \mathcal{L}_{\mathrm{SSIM}} \qquad (13)$$

where p indexes over pixel coordinates, and $\tilde{I}^t$ is the source view I^s warped to the target frame based on the predicted global scene coordinates $\mathbf{s}^s$. $\mathcal{L}_{\mathrm{SSIM}}$ is the structured similarity loss [58] imposed on $I^t(p)$ and $\tilde{I}^t(p)$.

Scene coordinates can be trained by $\mathcal{L}_{\mathrm{ph}}$ via two gradient paths: (1) the sampling location generation process in Eq. 12. (2) the absolute pose $\mathbf{T}_t$ which is calculated by DLT on scene coordinates. However, we observe that fine-tuning scene coordinates with $\mathcal{L}_{\mathrm{ph}}$ results in divergence. To this end, we detach scene coordinates from computation graphs and only fine-tune the quality weight net (Fig. 2-B) with self supervision. As will be shown later in experiments, this practice greatly improves re-localization performance. A detailed illustration of this adaptation scheme is provided in the supplementary material.

To clarity, our test-time adaptation experiments exploit all frames in the test set for self-supervision. An ideal online formulation would only use frames before the one of interest. The current offline version can be useful in scenarios like robot vacuums adapts to specific rooms during standby time.

4 Experiments

4.1 Experiment Setting

Datasets. We evaluate our SC-wLS framework for camera re-localization from single RGB images, on both indoor and outdoor scenes. Following [7], we choose the publicly available indoor 7Scenes dataset [38] and outdoor Cambridge dataset

Table 2. Median errors on the Cambridge dataset [22], with translational and rotational errors measured in m and $°$. Settings are the same as Table 1. Our results are significantly better than other methods. Note that GL-Net uses a sequence as input.

Methods	Greatcourt	King's College	Shop Facade	Old Hospital	Church
ADPoseNet [19]	N/A	1.30/1.7	1.22/6.7	N/A	2.28/4.8
PoseNet17 [21]	7.00/3.7	0.99/1.1	1.05/4.0	2.17/2.9	1.49/3.4
GPoseNet [9]	N/A	1.61/2.3	1.14/5.7	2.62/3.9	2.93/6.5
MLFBPPose [45]	N/A	0.76/1.7	0.75/5.1	1.99/2.9	1.29/5.0
LSTM-Pose [42]	N/A	0.99/3.7	1.18/7.4	1.51/4.3	1.52/6.7
SVS-Pose [29]	N/A	1.06/2.8	0.63/5.7	1.50/4.0	2.11/8.1
MapNet [8]	N/A	1.07/1.9	1.49/4.2	1.94/3.9	2.00/4.5
GL-Net [49]	6.67/2.8	0.59/0.7	0.50/2.9	1.88/2.8	1.90/3.3
MS-Transformer [37]	N/A	0.83/1.5	0.86/3.1	1.81/2.4	1.62/4.0
Ours (*dlt*)	1.81/1.2	0.22/0.9	0.15/1.1	0.46/1.9	0.50/1.5
Ours (*dlt+e2e*)	**1.64/0.9**	**0.14/0.6**	**0.11/0.7**	**0.42/1.7**	**0.39/1.3**

[22], which have different scales, appearance and motion patterns. These two datasets have fairly different distributions of scene coordinates. We would show later that, in both cases, SC-wLS successfully predicts interpretable scene coordinate weights for feed-forward camera re-localization.

Implementation. The input images are proportionally resized so that the heights are 480 pixels. We randomly zoom and rotate images for data augmentation following [7]. As mentioned in Sect. 3.3, α and β in Eq. 11 are sensitive to specific scenes. We set them to 5 and 1e−4 for indoor 7Scenes while 5 and 1e−6 for outdoor Cambridge, respectively. The balancing hyper-parameter γ is set to 5 empirically. We train our model on one nVIDIA GeForce RTX 3090 GPU. The ADAM optimizer with initial learning rate 1e−4 is utilized in the first two training stages and the learning rate is set to 1e−5 in the end-to-end training stage. The batch size is set to 1 as [7]. The scene coordinate regression network architecture for 7Scenes is adopted from [7]. As for Cambridge, we add residual connections to the early layers of this network. Architecture details can be found in the supplementary material.

4.2   Feed-Forward Re-localization

An intriguing property of the proposed SC-wLS method is that we can re-localize a camera directly using weighted least squares. This inference scheme seamlessly blends into the forward pass of a neural network. So we firstly compare with other APR re-localization methods that predict camera poses directly with a neural network during inference. Quantitative results on 7Scenes and Cambridge are summarized in Table 1 and Table 2, respectively.

As for former arts, we distinguish between two cases: single frame based and sequence based. They are separated by a line in Table 1 and Table 2. As for our results, we report under two settings: Ours (*dlt*) and Ours (*dlt + e2e*). Ours (*dlt + e2e*) means all three training stages are finished (see Sect. 3.4), while Ours (*dlt*) means only the first two stages are used.

Table 3. Results on the 7Scenes dataset [38] and the Cambridge dataset [22], with translational and rotational errors measured in m and $°$. Here ref means LM-Refine. Note DSAC* [7] w/o model and NG-RANSAC [6] also use the LM pose refinement process. Note that with ref used, our method loses the feed-forward nature. NG-RANSAC w/o model† is retrained using DSAC* RGB initialization.

7Scenes	Chess	Fire	Heads	Office	Pumpkin	Kitchen	Stairs
DSAC* [7] w/o model	0.019/1.11	**0.019/1.24**	**0.011/1.82**	**0.026**/1.18	**0.042**/1.42	**0.030/1.70**	**0.041/1.42**
Ours ($dlt+ref$)	**0.018/0.63**	0.026/**0.84**	0.015/**0.87**	0.038/0.85	0.045/**1.05**	0.051/1.17	0.067/1.93
Ours ($dlt+e2e+ref$)	0.019/**0.62**	0.025/0.88	0.014/0.90	0.035/**0.78**	0.051/1.07	0.054/1.15	0.058/1.57
Cambridge		Greatcourt	King's College	Shop Facade	Old Hospital	Church	
DSAC* [7] w/o model		0.34/0.2	0.18/0.3	0.05/1.3	0.21/0.4	0.15/0.6	
NG-RANSAC [6]		0.35/0.2	0.13/0.2	0.06/0.3	0.22/0.4	0.10/0.3	
NG-RANSAC w/o model†		0.31/0.2	0.15/0.3	0.05/0.3	0.20/0.4	0.12/0.4	
Ours ($dlt+ref$)		0.32/0.2	0.09/0.3	0.04/0.3	0.12/0.4	0.12/0.4	
Ours ($dlt+e2e+ref$)		**0.29/0.2**	**0.08/0.2**	**0.04/0.3**	**0.11/0.4**	**0.09/0.3**	

Table 4. Recalls (%) on the 7Scenes dataset [38] and the Cambridge dataset [22].

7Scenes	Chess	Fire	Heads	Office	Pumpkin	Kitchen	Stairs	Avg
DSAC* [7] w/o model	96.7	92.9	**98.2**	87.1	60.7	65.3	64.1	**80.7**
Active Search [32]	86.4	86.3	95.7	65.6	34.1	45.1	**67.8**	68.7
Ours ($dlt + e2e + ref$)	93.7	82.2	65.7	73.5	49.6	45.3	43.0	64.7
Ours ($dlt + e2e + dsac*$)	**96.8**	**97.4**	94.3	**88.6**	**62.4**	**65.5**	55.3	80.0
Cambridge		Great Court	King's College	Shop Facade	Old Hospital	Church	Avg	
DSAC* [7] w/o model		62.9	80.8	86.4	55.5	88.9	74.9	
NG-RANSAC w/o model†		67.6	81.9	88.3	56.0	93.8	77.5	
Ours ($dlt + e2e + ref$)		73.3	**97.3**	**90.3**	**75.3**	82.2	**83.7**	
Ours ($dlt + e2e + dsac*$)		**74.3**	87.2	87.4	53.8	**99.1**	80.4	

On 7Scenes, it is clear that our SC-wLS method out-performs APR camera re-localization methods by significant margins. Note we only take a single image as input, and still achieve much lower errors than sequence-based methods like GLNet [49]. On Cambridge, SC-wLS also reports significantly better results including the Greatcourt scene. Note that this scene has extreme illumination conditions and most former solutions do not even work.

Why SC-wLS performs much better than former APR methods that directly predict poses without RANSAC? Because they are black boxes, failing to model explicit geometric relationships. These black-box methods suffer from hard memorization and dataset bias. By contrast, our formulation is based upon 2D-3D correspondences, while still allowing feed-forward inference.

Lastly, Ours ($dlt + e2e$) shows lower errors than Ours (dlt) in most scenes, validating the effectiveness of jointly fine-tuning Fig. 2-A and Fig. 2-B.

4.3   Optional LM-Refine

The most interesting part about SC-wLS is the strong re-localization ability without using iterative pose estimation methods like RANSAC or LM-Refine.

Table 5. Results on the 7Scenes dataset [38] and the Cambridge dataset [22], with translational and rotational errors measured in m and $°$. Here *self* means self-supervised adaptation. Note these results are evaluated without LM-Refine.

7Scenes	Chess	Fire	Heads	Office	Pumpkin	Kitchen	Stairs
Ours ($dlt + e2e$)	0.029/0.76	0.048/1.09	0.027/1.92	0.055/0.86	0.077/1.27	0.091/1.43	0.123/2.80
Ours ($dlt + e2e + self$)	**0.021/0.64**	**0.023/0.80**	**0.013/0.79**	**0.037/0.76**	**0.048/1.06**	**0.051/1.08**	**0.055/1.48**
Cambridge		Greatcourt	King's College	Shop Facade	Old Hospital	Church	
Ours ($dlt + e2e$)		1.64/0.9	0.14/0.6	0.11/0.7	0.42/1.7	0.39/1.3	
Ours ($dlt + e2e + self$)		**0.94/0.5**	**0.11/0.3**	**0.05/0.4**	**0.18/0.7**	**0.17/0.8**	

However, although optional, incorporating such methods does lead to lower errors. We report quantitative comparisons in Table 3. We compare with the best published method DSAC* in the 'w/o model' setting for two reasons: 1) Our training protocol does not involve any usage of 3D models; 2) DSAC* also uses LM-Refine as a post-processing step. Note LM-Refine needs pose initialization, so we use DLT results as initial poses. Similar to above experiments, Ours ($dlt + e2e + ref$) means all three training stages are used. For Cambridge, we also compare to NG-RANSAC [6], which predicts weights from RGB images.

On the Cambridge dataset, SC-wLS outperforms state-of-the-art methods. On the Old Hospital scene, we reduce the translational error from 0.21 m to 0.11 m, which is a 47.6% improvement. Meanwhile, Ours ($dlt + e2e + ref$) consistently achieves lower errors than Ours ($dlt + ref$), showing the benefits of end-to-end joint learning. On the 7Scenes dataset, our results under-perform DSAC* and the third training stage does not bring clear margins.

We also report average recalls with pose error below 5 cm, 5deg (7Scenes) and translation error below 0.5% of the scene size (Cambridge) in Table 4. It is demonstrated that the recall value of Ours ($dlt+e2e+ref$) under-performs SOTA on 7Scenes, which is consistent with the median error results. We also evaluate DSAC*'s exact post-processing, denoted as Our ($dlt+e2e+dsac*$). The difference is that for Ours ($dlt + e2e + ref$) we use DLT results for LM-Refine initialization and for Ours ($dlt + e2e + dsac*$) we use RANSAC for LM-Refine initialization. It is shown that using DSAC*'s exact post-processing can compensate for the performance gap on 7Scenes.

4.4 Self-supervised Adaptation

We show the effectiveness of self-supervised weight network adaptation during test time, which is a potentially useful new feature of SC-wLS (Sect. 3.6). Results are summarized in Table 5, which is evaluated under the weighted DLT setting (same as Sect. 4.2). Obviously, for all sequences in 7Scenes and Cambridge, Ours ($dlt + e2e + self$) outperforms Ours ($dlt + e2e$) by clear margins. On *Stairs* and *Old Hospital*, translational and rotational error reductions are both over 50%. In these experiments, self-supervised adaptation runs for about 600k iterations. Usually, this adaptation process converges to a reasonably good state, within only 150k iterations. More detailed experiments are in the supplementary.

Fig. 4. Visualizations on 7Scenes and Cambridge test sets, demonstrating that our network learns interpretable scene coordinate weights consistent with re-projection errors, by solely considering the intrinsic structure of input 2D-3D correspondences. A higher color temperature represents a higher value.

4.5 Visualization

Firstly, we demonstrate more learnt weights on the test sets of 7Scenes and Cambridge, in Fig. 4. The heatmaps for reprojection error and learnt weight are highly correlated. Pixels with low quality weights usually have high reprojection errors and occur in sky or uniformly textured regions.

Secondly, We show learnt 3D maps with and without quality filtering, in Fig. 5. Since scene coordinates are predicted in the world frame, we directly show the point clouds generated by aggregating scene coordinate predictions on test frames. It is shown that only predicting scene coordinates results in noisy point clouds, especially in outdoor scenes where scene coordinate predictions on sky regions are only meaningful in term of their 2D projections. We show good scene coordinates by filtering out samples with a quality weight lower than 0.9.

4.6 Training and Inference Efficiency

As shown in Table 6, our method, as a feed-forward (fw) one, is faster than the well-engineered iterative ($iter$) method DSAC* and the transformer-based method MS-Transformer [37]. We could make a tradeoff using OANet w/o attention for even faster speed. Thus it's reasonable to compare our feed-forward method with APR methods. We also evaluate end-to-end training efficiency.

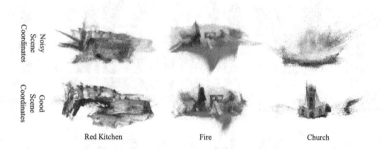

Noisy Scene Coordinates

Good Scene Coordinates

Red Kitchen   Fire   Church

Fig. 5. Map visualization on 7Scenes and Cambridge. More comparisons are provided in the supplementary material.

Table 6. Average end-to-end training and inference time comparisons.

Inference time (ms/frame)							End-to-end training time (s/frame)			
Method	AtLoc [43]	MapNet [8]	Ours (OANet)	Ours [37]	MS-T [7]	DSAC*	Setting	Ours	DSAC* [7]	NG-RANSAC [6]
Type	fw	fw	fw	fw	fw	$iter$	1-GPU	**0.09**	0.20	0.31
Time	**6**	9	13	19	30	32	5-GPU	**0.09**	0.85	1.05

When training 5 network instances for 5 different scenes on 5 GPUs, the average time of ours stays unchanged, while that of DSAC* increases due to limited CPU computation/scheduling capacity. We believe large-scale training of many re-localization network instances on the cloud is an industry demand.

5   Conclusions

In this study, we propose a new camera re-localization solution named SC-wLS, which combines the advantages of feed-forward formulations and scene coordinate based methods. It exploits the correspondences between pixel coordinates and learnt 3D scene coordinates, while still allows direct camera re-localization through a single forward pass. This is achieved by a correspondence weight network that finds high-quality scene coordinates, supervised by poses only. Meanwhile, SC-wLS also allows self-supervised test-time adaptation. Extensive evaluations on public benchmarks 7Scenes and Cambridge demonstrate the effectiveness and interpretability of our method. In the feed-forward setting, SC-wLS results are significantly better than APR methods. When coupled with LM-Refine post-processing, our method out-performs SOTA on outdoor scenes and under-performs SOTA on indoor scenes.

Acknowledgement. This work was supported by the National Natural Science Foundation of China under Grant 62176010.

References

1. Arandjelovic, R., Gronat, P., Torii, A., Pajdla, T., Sivic, J.: NetVLAD: CNN architecture for weakly supervised place recognition. In: Proceedings of the IEEE Conference on Computer Vision and Pattern Recognition, pp. 5297–5307 (2016)
2. Brachmann, E., Humenberger, M., Rother, C., Sattler, T.: On the limits of pseudo ground truth in visual camera re-localisation. In: Proceedings of the IEEE/CVF International Conference on Computer Vision, pp. 6218–6228 (2021)
3. Brachmann, E., Krull, A., Michel, F., Gumhold, S., Shotton, J., Rother, C.: Learning 6D object pose estimation using 3D object coordinates. In: Fleet, D., Pajdla, T., Schiele, B., Tuytelaars, T. (eds.) ECCV 2014. LNCS, vol. 8690, pp. 536–551. Springer, Cham (2014). https://doi.org/10.1007/978-3-319-10605-2_35
4. Brachmann, E., et al.: DSAC-differentiable RANSAC for camera localization. In: CVPR (2017)
5. Brachmann, E., Rother, C.: Learning less is more - 6D camera localization via 3D surface regression. In: CVPR (2018)
6. Brachmann, E., Rother, C.: Neural-guided RANSAC: learning where to sample model hypotheses. In: ICCV (2019)
7. Brachmann, E., Rother, C.: Visual camera re-localization from RGB and RGB-D images using DSAC. IEEE Trans. Pattern Anal. Mach. Intell. (2021)
8. Brahmbhatt, S., Gu, J., Kim, K., Hays, J., Kautz, J.: Geometry-aware learning of maps for camera localization. In: CVPR (2018)
9. Cai, M., Shen, C., Reid, I.: A hybrid probabilistic model for camera relocalization (2019)
10. Cao, S., Snavely, N.: Graph-based discriminative learning for location recognition. In: IJCV (2015)
11. Choi, S., Kim, T., Yu, W.: Performance evaluation of RANSAC family. J. Comput. Vision **24**(3), 271–300 (1997)
12. Dang, Z., Yi, K.M., Hu, Y., Wang, F., Fua, P., Salzmann, M.: Eigendecomposition-free training of deep networks for linear least-square problems. TPAMI (2020)
13. Ding, M., Wang, Z., Sun, J., Shi, J., Luo, P.: CamNet: coarse-to-fine retrieval for camera re-localization. In: Proceedings of the IEEE/CVF International Conference on Computer Vision, pp. 2871–2880 (2019)
14. Gould, S., Hartley, R., Campbell, D.J.: Deep declarative networks. TPAMI (2021)
15. Hartley, R., Zisserman, A.: Multiple view geometry in computer vision: N-view geometry (2004)
16. Hays, J., Efros, A.A.: IM2GPS: estimating geographic information from a single image. In: CVPR (2008)
17. Hedau, V., Hoiem, D., Forsyth, D.: Recovering the spatial layout of cluttered rooms. In: 2009 IEEE 12th International Conference on Computer Vision, pp. 1849–1856. IEEE (2009)
18. Hirzer, M., Lepetit, V., Roth, P.: Smart hypothesis generation for efficient and robust room layout estimation. In: Proceedings of the IEEE/CVF Winter Conference on Applications of Computer Vision, pp. 2912–2920 (2020)
19. Huang, Z., Xu, Y., Shi, J., Zhou, X., Bao, H., Zhang, G.: Prior guided dropout for robust visual localization in dynamic environments. In: ICCV (2019)
20. Ionescu, C., Vantzos, O., Sminchisescu, C.: Matrix backpropagation for deep networks with structured layers. In: ICCV (2015)
21. Kendall, A., Cipolla, R.: Geometric loss functions for camera pose regression with deep learning. In: CVPR (2017)

22. Kendall, A., Grimes, M., Cipolla, R.: PoseNet: a convolutional network for real-time 6-DOF camera relocalization. In: ICCV (2015)
23. Lepetit, V., Fua, P., et al.: Monocular model-based 3D tracking of rigid objects: a survey. Found. Trends® Comput. Graph. Vision **1**(1), 1–89 (2005)
24. Li, S., Xu, C., Xie, M.: A robust o (n) solution to the perspective-n-point problem. IEEE Trans. Pattern Anal. Mach. Intell. **34**(7), 1444–1450 (2012)
25. Li, S., Wu, X., Cao, Y., Zha, H.: Generalizing to the open world: deep visual odometry with online adaptation. In: Proceedings of the IEEE/CVF Conference on Computer Vision and Pattern Recognition, pp. 13184–13193 (2021)
26. Lowe, D.G.: Distinctive image features from scale-invariant keypoints. IJCV (2004)
27. Mair, E., Strobl, K.H., Suppa, M., Burschka, D.: Efficient camera-based pose estimation for real-time applications. In: 2009 IEEE/RSJ International Conference on Intelligent Robots and Systems, pp. 2696–2703. IEEE (2009)
28. Meng, L., Tung, F., Little, J.J., Valentin, J., de Silva, C.W.: Exploiting points and lines in regression forests for RGB-D camera relocalization. In: IROS (2018)
29. Naseer, T., Burgard, W.: Deep regression for monocular camera-based 6-DOF global localization in outdoor environments. In: IROS (2017)
30. Ranftl, R., Koltun, V.: Deep fundamental matrix estimation. In: Ferrari, V., Hebert, M., Sminchisescu, C., Weiss, Y. (eds.) ECCV 2018. LNCS, vol. 11205, pp. 292–309. Springer, Cham (2018). https://doi.org/10.1007/978-3-030-01246-5_18
31. Sarlin, P.E., DeTone, D., Malisiewicz, T., Rabinovich, A.: Superglue: learning feature matching with graph neural networks. In: CVPR (2020)
32. Sattler, T., Leibe, B., Kobbelt, L.: Efficient & effective prioritized matching for large-scale image-based localization. TPAMI (2016)
33. Sattler, T., Zhou, Q., Pollefeys, M., Leal-Taixe, L.: Understanding the limitations of CNN-based absolute camera pose regression. In: CVPR (2019)
34. Schindler, G., Brown, M., Szeliski, R.: City-scale location recognition. In: CVPR (2007)
35. Schonberger, J.L., Frahm, J.M.: Structure-from-motion revisited. In: Proceedings of the IEEE Conference on Computer Vision and Pattern Recognition, pp. 4104–4113 (2016)
36. Schönemann, P.H.: A generalized solution of the orthogonal procrustes problem. Psychometrika **31**(1), 1–10 (1966)
37. Shavit, Y., Ferens, R., Keller, Y.: Learning multi-scene absolute pose regression with transformers. In: Proceedings of the IEEE/CVF International Conference on Computer Vision, pp. 2733–2742 (2021)
38. Shotton, J., Glocker, B., Zach, C., Izadi, S., Criminisi, A., Fitzgibbon, A.: Scene coordinate regression forests for camera relocalization in RGB-D images. In: CVPR (2013)
39. Sivic, J., Zisserman, A.: Video Google: a text retrieval approach to object matching in videos. In: ICCV (2003)
40. Valentin, J., Nießner, M., Shotton, J., Fitzgibbon, A., Izadi, S., Torr, P.H.: Exploiting uncertainty in regression forests for accurate camera relocalization. In: CVPR (2015)
41. Vaswani, A., et al.: Attention is all you need (2017)
42. Walch, F., Hazirbas, C., Leal-Taixe, L., Sattler, T., Hilsenbeck, S., Cremers, D.: Image-based localization using LSTMs for structured feature correlation. In: ICCV (2017)
43. Wang, B., Chen, C., Lu, C.X., Zhao, P., Trigoni, N., Markham, A.: Atloc: attention guided camera localization. In: Proceedings of the AAAI Conference on Artificial Intelligence, vol. 34, pp. 10393–10401 (2020)

44. Wang, H., Sridhar, S., Huang, J., Valentin, J., Song, S., Guibas, L.J.: Normalized object coordinate space for category-level 6d object pose and size estimation. In: Proceedings of the IEEE/CVF Conference on Computer Vision and Pattern Recognition, pp. 2642–2651 (2019)
45. Wang, X., Wang, X., Wang, C., Bai, X., Wu, J., Hancock, E.R.: Discriminative features matter: multi-layer bilinear pooling for camera localization. In: BMVC (2019)
46. Wu, C.: Towards linear-time incremental structure from motion. In: 3DV (2013)
47. Wu, J., Ma, L., Hu, X.: Delving deeper into convolutional neural networks for camera relocalization. In: ICRA (2017)
48. Xue, F., Wang, X., Yan, Z., Wang, Q., Wang, J., Zha, H.: Local supports global: deep camera relocalization with sequence enhancement. In: ICCV (2019)
49. Xue, F., Wu, X., Cai, S., Wang, J.: Learning multi-view camera relocalization with graph neural networks. In: 2020 IEEE/CVF Conference on Computer Vision and Pattern Recognition (CVPR), pp. 11372–11381. IEEE (2020)
50. Yan, C., Shao, B., Zhao, H., Ning, R., Zhang, Y., Xu, F.: 3D room layout estimation from a single RGB image. IEEE Trans. Multimedia **22**(11), 3014–3024 (2020)
51. Yi, K., Trulls, E., Ono, Y., Lepetit, V., Salzmann, M., Fua, P.: Learning to find good correspondences. In: CVPR (2018)
52. Zhang, J., et al.: Learning two-view correspondences and geometry using order-aware network. In: ICCV (2019)
53. Zhang, W., Kosecka, J.: Image based localization in urban environments. In: 3DPTV (2006)
54. Zhao, H., Lu, M., Yao, A., Guo, Y., Chen, Y., Zhang, L.: Physics inspired optimization on semantic transfer features: an alternative method for room layout estimation. In: Proceedings of the IEEE Conference on Computer Vision and Pattern Recognition, pp. 10–18 (2017)
55. Zhong, L., et al.: Seeing through the occluders: robust monocular 6-DOF object pose tracking via model-guided video object segmentation. IEEE Robot. Autom. Lett. **5**(4), 5159–5166 (2020)
56. Zhou, L., et al.: KfNet: learning temporal camera relocalization using Kalman filtering. In: Proceedings of the IEEE/CVF Conference on Computer Vision and Pattern Recognition, pp. 4919–4928 (2020)
57. Zhou, T., Brown, M., Snavely, N., Lowe, D.G.: Unsupervised learning of depth and ego-motion from video. In: CVPR (2017)
58. Wang, Z., Bovik, A.C., Sheikh, H.R., Simoncelli, E.P.: Image quality assessment: from error visibility to structural similarity. IEEE Trans. Image Process. **13**(4), 600–612 (2004). https://doi.org/10.1109/TIP.2003.819861

FloatingFusion: Depth from ToF and Image-Stabilized Stereo Cameras

Andreas Meuleman[1], Hakyeong Kim[1], James Tompkin[2], and Min H. Kim[1(✉)]

[1] KAIST, Daejeon, South Korea
{ameuleman,hkkim,minhkim}@vclab.kaist.ac.kr
[2] Brown University, Providence, USA

Abstract. High-accuracy per-pixel depth is vital for computational photography, so smartphones now have multimodal camera systems with time-of-flight (ToF) depth sensors and multiple color cameras. However, producing accurate high-resolution depth is still challenging due to the low resolution and limited active illumination power of ToF sensors. Fusing RGB stereo and ToF information is a promising direction to overcome these issues, but a key problem remains: to provide high-quality 2D RGB images, the main color sensor's lens is optically stabilized, resulting in an unknown pose for the floating lens that breaks the geometric relationships between the multimodal image sensors. Leveraging ToF depth estimates and a wide-angle RGB camera, we design an automatic calibration technique based on dense 2D/3D matching that can estimate camera extrinsic, intrinsic, and distortion parameters of a stabilized main RGB sensor from a single snapshot. This lets us fuse stereo and ToF cues via a correlation volume. For fusion, we apply deep learning via a real-world training dataset with depth supervision estimated by a neural reconstruction method. For evaluation, we acquire a test dataset using a commercial high-power depth camera and show that our approach achieves higher accuracy than existing baselines.

Keywords: Online camera calibration · 3D imaging · Depth estimation · Multi-modal sensor fusion · Stereo imaging · Time of flight

1 Introduction

Advances in computational photography allow many applications such as 3D reconstruction [18], view synthesis [23,43], depth-aware image editing [49,53], and augmented reality [20,48]. Vital to these algorithms is *high-accuracy per-pixel depth*, e.g., to integrate virtual objects by backprojecting high-resolution camera color into 3D. To this end, smartphones now have camera systems with

Supplementary Information The online version contains supplementary material available at https://doi.org/10.1007/978-3-031-19769-7_35.

Fig. 1. (a) Multi-modal smartphone imaging. (b) Reference RGB image. (c) ToF depth reprojected to the reference. (d) Our depth from floating fusion.

multiple sensors, lenses of different focal lengths, and active-illumination time-of-flight (ToF). For instance, correlation-based ToF provides depth by measuring the travel time of infrared active illumination with a gated infrared sensor.

We consider two challenges in providing high-accuracy per-pixel depth: (1) ToF sensor spatial resolution is orders of magnitude less than that of its compatriot color cameras. RGB spatial resolution has increased dramatically on smartphones—12–64 million pixels is common—whereas ToF is often 0.05–0.3 million pixels. One might correctly think that fusing depth information from ToF with depth information from color camera stereo disparity is a good strategy to increase our depth resolution. Fusion might also help us overcome the low signal-to-noise ratio in ToF signals that arises from the low-intensity illumination of a battery-powered device. For fusion, we need to accurately know the geometric poses of all sensors and lenses in the camera system.

This leads to our second challenge: (2) As RGB spatial resolution has increased, smartphones now use optical image stabilization [40,50]: a *floating* lens compensates for camera body motion to avoid motion blur during exposure. Two low-power actuators suspend the lens body vertically and horizontally to provide a few degrees of in-plane rotation or translation, similar to how a third actuator translates the lens along the optical axis for focus. The magnetic actuation varies with focus and even with the smartphone's orientation due to gravity, and the pose of the stabilizer is not currently possible to measure or read out electronically. As such, we can only use a fusion strategy if we can automatically optically calibrate the geometry of the floating lens for each exposure taken.

This work proposes a *floating fusion* algorithm to provide high accuracy per pixel depth estimates from an optically-image-stabilized camera, a second RGB camera, and a ToF camera (Fig. 1). We design an online calibration approach for the floating lens that uses ToF measurements and dense optical flow matching between the RGB camera pair. This lets us form 2D/3D correspondences to recover intrinsic, extrinsic, and lens distortion parameters in an absolute manner (not 'up to scale'), and for every snapshot. This makes it suitable for dynamic environments. Then, to fuse multi-modal sensor information, we build a correlation volume that integrates both ToF and stereo RGB cues, then predict disparity via a learned function. There are few large multi-modal datasets to train this function, and synthetic data creation is expensive and retains a domain gap to the real world. Instead, we capture real-world scenes with multiple views and optimize a neural radiance field [6] with ToF supervision. The resulting depth maps are lower noise and higher detail than those of a depth camera, and provide us with high-quality training data. For validation, we build a test dataset using a Kinect Azure and show that our method outperforms other traditional and data-driven approaches for snapshot RGB-D imaging.

2   Related Work

ToF and RGB Fusion. Existing data-driven approaches [1,2,36] heavily rely on synthetic data, creating a domain gap. This is exacerbated when using imperfect low-power sensors such on mobile phones. In addition, current stereo-ToF fusion [10,14,15] typically estimates disparity from stereo and ToF separately before fusion. One approach is to estimate stereo and ToF confidence to merge the disparity maps [1,2,33,37]. In contrast, our ToF estimates are directly incorporated into our disparity pipeline before depth selection. Fusion without stereo [22] tackles more challenging scenarios than direct ToF depth estimation. However, Jung et al.'s downsampling process can blur over occlusion edges, producing incorrect depth at a low resolution that is difficult to fix after reprojection at finer resolutions.

Phone and Multi-sensor Calibration. DiVerdi and Barron [12] tackle per shot stereo calibration up to scale in the challenging mobile camera environment; however, absolute calibration is critical for stereo/ToF fusion. We leverage coarse ToF depth estimates for absolute stereo calibration. Gil et al. [16] estimate two-view stereo calibration by first estimating a monocular depth map in one image before optimizing the differentiable projective transformation (DPT) that maximizes the consistency between the stereo depth and the monocular depth. The method refines the DPT parameters, handling camera pose shift after factory calibration and improving stereo depth quality, but it still requires the initial transformation to be sufficiently accurate for reasonable stereo depth estimation. In addition, to allow for stable optimization, a lower degree of freedom model is selected, which can neglect camera distortion and lens shift. Works tackling calibration with phone and ToF sensors are not common. Gao et al. [15] use Kinect RGB-D inputs, match RGB to the other camera, use depth to lift points to 3D, then solves a PnP problem to find the transformation. Since it matches sparse keypoints, it is not guaranteed that depth is available where a keypoint is, leading to too few available keypoints. In addition, the method does not account for intrinsic or distortion refinement.

Data-Driven ToF Depth Estimation. Numerous works [3,17,32,39,44,46] attempt to tackle ToF depth estimation via learned approaches. While these approaches have demonstrated strong capabilities in handling challenging artifacts (noise, multi-path interference, or motion), our approach does not strictly require a dedicated method for ToF depth estimation as we directly merge ToF samples in our stereo fusion pipeline.

Conventional Datasets. Accurate real-world datasets with ground-truth depth maps are common for stereo depth estimation [34,41,45]. However, the variety of fusion systems makes it challenging to acquire large-high-quality, real-world datasets. A majority of ToF-related works leverage rendered data [17,32], particularly for fusion datasets [1,2]. These datasets enable improvement over conventional approaches, but synthesizing RGB and ToF images accurately is challenging. A domain gap is introduced as the noise profile and imaging artifacts are different from the training data. Notable exceptions are Son et al. [44], and Gao and Fan et al. [39], where an accurate depth camera provides training data for a lower-quality ToF module. The acquisition is partially automated thanks to a robotic arm. However, this bulky setup limits the variety of the scenes: all scenes are captured on the same table, with similar backgrounds across the dataset. In addition, the use of a single depth camera at a different location from the ToF module introduces occlusion, with some areas in the ToF image having no supervision. In addition, this method only tackles ToF depth estimation, and the dataset does not feature RGB images.

Multiview Geometry Estimation. Several approaches are capable of accurate depth estimation from multiview images [42], even in dynamic environments [24,26,31]. Despite their accuracy, including ToF data to these approaches is not obvious. Scene representations optimized from a set of images [4,6,21] have recently shown good novel view synthesis and scene geometry reconstruction, including to refine depth estimates in the context of multiview stereo [51]. Since the optimization can accept supervision from varied sources, including ToF measurements is straightforward. For this reason, we select a state-of-the-art neural representation that has the advantage to handle heterogeneous resolutions [6] for our training data generation. TöRF [5] renders phasor images from a volume representation to optimize raw ToF image reconstruction. While efficiently improving NeRF's results and tackling ToF phase wrapping, this approach is not necessary for our context as our device is not prone to phase wrapping due to its low illumination range (low power) and thanks to the use of several modulation frequencies. We also observe that, in the absence of explicit ToF confidence, erroneous ToF measurements tend to be more present in depth maps rendered from a TöRF. Finally, approaches based on ICP registration [18] cannot be applied directly to our data since depth maps from the low-power ToF module are too noisy to be registered through ICP.

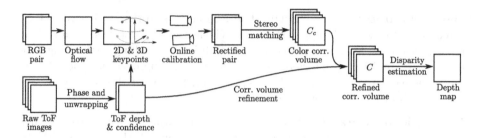

Fig. 2. Overview of our method. The first step is to estimate ToF depth via phase unwrapping (Sect. 3.1). Then, and after dense matching between the two RGB cameras, we use ToF depth to estimate the floating camera's intrinsic and extrinsic parameters (Sect. 3.2). Once the stereo pair is calibrated and rectified, we extract features in each image to build a correlation volume C_c. This volume is refined using the ToF samples (Sect. 3.3) before the final disparity estimation.

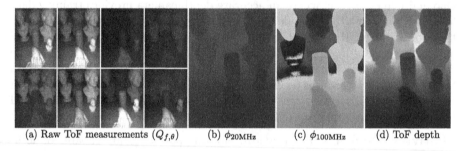

(a) Raw ToF measurements $(Q_{f,\theta})$ (b) $\phi_{20\text{MHz}}$ (c) $\phi_{100\text{MHz}}$ (d) ToF depth

Fig. 3. ToF depth estimation from raw measurements. From raw ToF images for two different frequencies (a), we estimate a coarse but unwrapped phase map (b) and a finer but wrapped phase map (c). By unwrapping $\phi_{100\text{MHz}}$ using the lower frequency phase $\phi_{20\text{MHz}}$, we estimate a more accurate depth map.

3 Method

We use an off-the-shelf Samsung Galaxy S20+ smartphone. This has the main camera with a 12MP color sensor and a magnetic mount 79° lens for stabilization and focusing, a secondary 12MP color camera with a fixed ultrawide 120° lens, and a 0.3MP ToF system with an infrared fixed 78° lens camera and infrared emitter (Fig. 1a). As the ultrawide camera and the ToF module are rigidly fixed, we calibrate their intrinsics K_{UW}, K_{ToF}, extrinsics $[R|t]_{\text{UW}}$, $[R|t]_{\text{ToF}}$, and lens distortion parameters using an offline method based on checkerboard corner estimation. We use a checkerboard with similar absorption in the visible spectrum as in infrared. However, calibrating the floating main camera (subscript $_{\text{FM}}$) is not possible offline, as its pose changes from snapshot to snapshot. OIS introduces lens shift in x, y for stabilization and in z for focus, with the z direction inducing additional lens distortion changes. The lens also tilts (pitch/yaw rotations) depending on the phone's orientation because of gravity. As such, we must estimate per snapshot a new intrinsic matrix K_{FM}, new extrinsic matrix $[R|t]_{FM}$, and three radial and two tangential distortion coefficients $\{k_1, k_2, k_3, p_1, p_2\}_{FM}$ from the Brown-Conrady model [7,9] for the main floating camera. To tackle this challenge, we present a method to estimate these parameters at an absolute scale (not relative or 'up to scale').

3.1 ToF Depth Estimation

The ToF system modulates its infra-red light source by 20 MHz and 100 MHz square waves, and alternates between both frequencies sequentially. The sensor captures four shots per frequency, with the same modulation as the light source shifted by 90°. With two frequencies, we obtain eight raw ToF measurements per snapshot:

$$Q_{f,\theta}, \quad f \in \{20\,\text{MHz}, 100\,\text{MHz}\}, \quad \theta \in \{0, \pi/2, \pi, 3\pi/2\}, \tag{1}$$

where f is the modulation frequency and θ is the sensor phase shift. Figure 3(a) shows an example of captured ToF measurements. To estimate depth, the first step is to estimate phase (Figs. 3(b) and 3(c)) for each modulation frequency:

$$\phi_f = \text{arctan2}\left(Q_{f,\pi} - Q_{f,0}, Q_{f,3\frac{\pi}{2}} - Q_{f,\frac{\pi}{2}}\right). \tag{2}$$

From these phases, we can estimate the distance:

$$d_f = \frac{c}{4\pi f}\phi_f + k_f\frac{c}{2f}, \quad k_f \in \mathbb{N}, \tag{3}$$

with c the speed of light. This equation shows that, given a phase, the depth is known up to a $k_f\frac{c}{2f}$ shift. That is, we observe phase wrapping. The ToF depth variance is inversely proportional to the modulation frequency [25]. Therefore, the higher frequency tends to be more accurate; however, the wrapping range is shorter ($\approx$1.5 m for 100 MHz against $\approx$7.5 m for 20 MHz). Since the ToF illumination is low power, the signal becomes weak for further objects, making the phase estimation highly unreliable beyond the wrapping point of the lower frequency. Based on this, we assume that $\phi_{20\text{MHz}}$ does not show phase wrapping: $k_{20\text{MHz}} := 0$. Therefore, we unwrap $\phi_{100\text{MHz}}$ using $\phi_{20\text{MHz}}$ to benefit from the lower depth variance associated with a higher frequency without the phase wrapping ambiguity. In detail, we find $\hat{k}_{100\text{MHz}}$ that minimizes the depth difference between the two frequencies:

$$\hat{k}_{100\text{MHz}} = \underset{k}{\text{argmin}} \left| d_{20\text{MHz}} - \frac{c}{4\pi \cdot 10^8}\phi_{100\text{MHz}} + k\frac{c}{2 \cdot 10^8} \right|. \tag{4}$$

From this, we can compute the distance $d_{100\text{MHz}}$, which will be used to obtain the depth d_{ToF} (Fig. 3d). For more details on ToF, refer to [19,25].

With the estimated depth, we assign a confidence map ω based on the signal's amplitude, the concordance between $d_{20\text{MHz}}$ and $d_{100\text{MHz}}$ and local depth changes. First, we assign lower scores when the signal is weak:

$$\omega_A = \exp\left(-1/\sum_f A_f/(2\sigma_A^2)\right), \tag{5}$$

With $A_f = \sqrt{(Q_{f,0} - Q_{f,\frac{\pi}{2}})^2 + (Q_{f,\pi} - Q_{f,3\frac{\pi}{2}})^2}/2$ and $f \in \{20\,\text{MHz}, 100\,\text{MHz}\}$. We also take into account the difference between the estimated distance from the two frequencies:

$$\omega_d = \exp\left(-|d_{20\text{MHz}} - d_{100\text{MHz}}|^2/(2\sigma_d^2)\right). \tag{6}$$

Since ToF is less reliable at depth discontinuities, we deem areas with large depth gradient to be less reliable:

$$\omega_\nabla = \exp\left(-\|\nabla d_{\text{ToF}}\|^2/(2\sigma_\nabla^2)\right) \cdot \exp\left(-\|\nabla(1/d_{\text{ToF}})\|^2/(2\sigma_\nabla^2)\right). \tag{7}$$

The complete confidence is: $\omega = \omega_A \omega_d \omega_\nabla$. For our experiments, we set $\sigma_A = 20$, $\sigma_d = 0.05$, and $\sigma_\nabla = 0.005$.

3.2 Online Calibration

To calibrate our floating main camera, we need to find sufficient correspondences between known 3D world points and projections of those points in 2D. We must find a way to correspond our 3D points from ToF with the main camera even though it does not share a spectral response. Our overall strategy is to use the known fixed relationship between the ToF and ultrawide cameras and additionally exploit 2D color correspondences from the optical flow between the ultrawide and floating cameras. In this way, we can map from main camera 2D coordinates to ultrawide camera 2D coordinates to ToF camera 3D coordinates. While ToF can be noisy, it still provides sufficient points to robustly calibrate all intrinsic, extrinsic, and lens distortion parameters of the main camera.

From ToF to Ultrawide. The first step is to reproject the ToF depth estimates d_{ToF} to the ultrawide camera. We transform the depth map to a point cloud P:

$$P_{\text{ToF}} = K_{\text{ToF}}^{-1}[u, v, d_{\text{ToF}}]^\top, \tag{8}$$

where K_{ToF} is the known ToF camera matrix, and (u, v) are pixel coordinates with corresponding depth d_{ToF}. Then, the point cloud can be transformed to the ultrawide camera's space:

$$P_{\text{UW}} = [R|t]_{\text{ToF}\to\text{UW}}[P_{\text{ToF}}^T|1]^\top, \tag{9}$$

where $[R|t]_{ToF\to\text{UW}}$ is the relative transformation from the ToF camera space to the ultrawide camera space. From point cloud P_{UW}, we obtain the pixel coordinate $[u_{\text{UW}}, v_{\text{UW}}]$ of the ToF point cloud reprojected to the ultrawide camera:

$$d_{\text{UW}}[u_{\text{UW}}, v_{\text{UW}}, 1] = K_{\text{UW}}P_{\text{UW}}. \tag{10}$$

The reprojected ToF points P_{UW} and their subpixel coordinates $[u_{\mathrm{UW}}, v_{\mathrm{UW}}]$ will be used to estimate calibration for our main camera in a later stage.

From Ultrawide to Floating Main. Next, we match the ultrawide camera to the floating main camera to be accurately calibrated. Since both cameras are located near to each other, they share a similar point of view, thus making sparse scale and rotation invariant feature matching unnecessary. As such, we use dense optical flow [47] to find correspondences. To use flow, we first undistort the ultrawide camera image given its calibration, then rectify it *approximately* to the floating main camera given an initial *approximate* offline calibration. This calibration will be wrong, but as flow is designed for small unconstrained image-to-image correspondence, this rectification approach will still find useful correspondences. As output, we receive a 2D vector field $\mathcal{F}_{\mathrm{UW} \rightarrow \mathrm{FM}}$.

From ToF to Floating Main. We can now use the optical flow to form 2D/3D matches. For each ToF point reprojected to the ultrawide camera P_{UW}, we find the corresponding pixel in the floating camera:

$$[u'_{\mathrm{FM}}, v'_{\mathrm{FM}}] = [u_{\mathrm{UW}}, v_{\mathrm{UW}}] + \mathcal{F}_{\mathrm{UW} \rightarrow \mathrm{FM}}([u_{\mathrm{UW}}, v_{\mathrm{UW}}]). \tag{11}$$

We sample the flow using bilinear interpolation since $[u_{\mathrm{UW}}, v_{\mathrm{UW}}]$ are estimated with subpixel precision.

From this 2D/3D matching between the 2D points in the floating camera $[u'_{\mathrm{FM}}, v'_{\mathrm{FM}}]$ and the 3D points in the ultrawide camera space P_{UW}, we can estimate the floating camera's calibration. We first solve the optimization through RANSAC for outlier removal, followed by Levenberg-Marquardt optimization, obtaining the transformation between the two RGB cameras $[R, t]_{\mathrm{FM} \rightarrow \mathrm{UW}}$, as well as the camera matrix K_{FM} and its distortion coefficients. Once the calibration is achieved, we rectify the two RGB images, enabling stereo/ToF fusion.

Discussion. Both Eqs. (10) and (11) are not occlusion-aware but, due to the small baseline, occluded world points are only a small portion of the total number of matched points. RANSAC helps us to avoid outlier correspondences from occlusion, incorrect flow estimates, and noisy ToF estimates for calibration.

Further, while we rely on a fixed RGB camera and ToF module, extending the approach to scenarios without fixed cameras is possible. Reliable feature matching between spectral domains has been demonstrated [8,13,55] as well as RGB/IR optical flow [38]. Using matching between the ToF module's IR camera and the RGB cameras, calibration can likely be achieved even if no second RGB camera exists as fixed with respect to the ToF module.

3.3   Fusing ToF and Stereo

Given the now-calibrated color stereo pair, and the ToF depth samples, we will fuse these into an accurate high-resolution depth map for a color camera. The first step is to build a correlation volume $\mathcal{C}_c$ from our RGB pair. A point in the volume $\mathcal{C}_c$ at coordinate $[u, v, u']$ represents the correlation between a pixel $[u, v]$ in the reference image and a pixel $[u', v]$ in the target image at some disparity.

Thus, the correlation volume's shape is (width×height×width) since disparity is horizontal along the width direction. We compute correlation volume values by extracting 256-dim. image features from each view using RAFT-stereo [28]'s learned feature encoder, then by taking the dot product of the feature vectors from each RGB camera for each disparity amount. Note that the feature extraction process downsamples the images four times and a disparity map at original resolution is recovered through RAFT-stereo's convex upsampling. For ease of evaluation, we use the fixed ultrawide camera as a reference, although the floating camera can be chosen without algorithmic change.

A 3D world point P_{ToF} corresponds to a 2D sensor coordinate $[u_{\text{UW}}, v_{\text{UW}}]$ in the ultrawide camera and in the floating main camera $[u_{\text{FM}}, v_{\text{FM}}]$ (see reprojection Eqs. (8),(9), and (10))). A ToF point also has a confidence ω estimated along with the ToF depth maps. We leverage those points by increasing the correlation at the location of the sample points in the volume. For a point, the corresponding location in the volume is $[u_{\text{UW}}, v_{\text{UW}}, u_{\text{FM}}]$. Coordinates given by a ToF point $[u_{\text{UW}}, v_{\text{UW}}, u_{\text{FM}}]$ have eight integer neighbors with defined values in the correlation volume. We inject the ToF point into RAFT-stereo's correlation volume using linear-like weights:

$$
\begin{aligned}
\mathcal{C}(\lfloor u_{\text{UW}} \rfloor, \lfloor v_{\text{UW}} \rfloor, \lfloor u_{\text{FM}} \rfloor) &= \mathcal{C}_c(\lfloor u_{\text{UW}} \rfloor, \lfloor v_{\text{UW}} \rfloor, \lfloor u_{\text{FM}} \rfloor) \\
&+ \tau \cdot \omega \left((\lceil u_{\text{UW}} \rceil - u_{\text{UW}})(\lceil v_{\text{UW}} \rceil - v_{\text{UW}})(\lceil u_{\text{FM}} \rceil - u_{\text{FM}}) \right).
\end{aligned}
\tag{12}
$$

$\lfloor u \rfloor$ is the nearest integer $\leq u$, and $\lceil \rceil$ is the nearest integer $\geq u$. We perform the same operation for all eight integer points with coordinates around $[u, v, u']$. If several ToF points affect the same point in the correlation, their contributions accumulate. τ is a scalar parameter that is optimized during the training process. We then use the updated correlation volume $\mathcal{C}$ with the next steps of RAFT-stereo's pipeline to estimate disparity. Thanks to this approach, we efficiently and robustly combine stereo correlation cues and ToF measurements before estimating the depth map.

3.4   Dataset Generation

Background. Learning-based stereo/ToF fusion methods require training data. For this, we optimize neural radiance fields [29, 30, 35] from RGB images. These allow querying density and appearance for every point in the scene, allowing us to render estimated depth maps. A typical volume rendering is as follows:

$$
\hat{C}(r) = \sum_{i=1}^{N} T_i (1 - \exp(-\sigma_i \delta_i)) c_i,
\tag{13}
$$

where $\hat{C}(r)$ is the rendered color for the ray r, σ_i and c_i are the density and color of the representation at i point. N points are sampled along the ray r, where δ_i is the distance between neighboring samples, and $T_i = \exp\left(-\sum_{i' < i} \sigma_{i'} \delta_{i'}\right)$ is the approximated transparency between the ray origin and the sample.

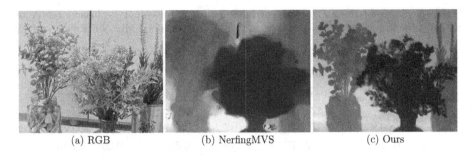

(a) RGB   (b) NerfingMVS   (c) Ours

Fig. 4. NerfingMVS [51]'s results before filtering show more artifacts than ours.

This rendering is differentiable on the density and color of the sample points, allowing gradient-based optimization. To render depth maps from these representations, we swap the color term c_i in Eq. (13) for the depth of the point w.r.t. the ray origin.

Design Choices and Our Approach. Despite improving on numerous aspects such as optimization time and novel view quality over neural representations, Plenoxels [4] produces fuzzy depth maps (see the supplemental material).We also observe that NerfingMVS's [51] guided NeRF optimization based on reducing the sampling range around the expected depth tends to create artifacts (Fig. 4) that require an additional filtering step [48] for accurate depth maps. Since filtering at novel views is impossible, this approach is not suitable for creating training data for unseen views. Thus, we use MipNeRF [6] as a basis for our multiview depth estimation pipeline. This can naturally handle the resolution difference between the ToF and RGB cameras. For depth supervision [11], we use a straightforward approach similar to VideoNeRF [52]'s supervision on inverse depth maps computed from multi-view stereo: we add a loss using depth samples to the optimization:

$$\mathcal{L}_{depth} = \omega \left\| d_{\text{rendered}} - d_{\text{ToF}} \right\|_2^2, \tag{14}$$

with the ToF confidence ω (Sect. 3.1). In addition to this supervision, we implement extrinsic and intrinsic refinement [21] following a 6D continuous rotation representation [54]. Note that, since Mip-NeRF prevents too high frequencies in positional encoding and poses and that the intrinsics are well initialized, a coarse-to-fine positional encoding [27] is not required. Our training and validation datasets have eight scenes with around 100 snapshots per scene. The scenes feature varied depth range, background, objects, and materials.

4   Results

4.1   Evaluation Dataset

To evaluate our method, we build a real-world dataset with ground-truth depth obtained using a Kinect Azure. Since the depth camera is higher power, noise is

Table 1. We compare our multiview fusion against a ToF-supervised scene representation [5] and multiview-stereo approaches [24,31,51]. Original Mip-NeRF [6] on which our implementation is based is given for reference.

	Bad ratio (%)		Depth error			
	>0.2	>0.05	MAE Rel.	RMSE Rel.	MAE	RMSE
TöRF [5]	1.82	26.48	0.055	0.075	0.041	0.062
NerfingMVS [51]	3.37	20.11	0.047	0.069	0.039	0.071
RCVD [24]	35.12	75.68	0.249	0.326	0.208	0.315
CVD [31]	1.68	11.67	0.033	0.056	0.028	0.056
Mip-NeRF [6]	17.47	61.71	0.156	0.199	0.115	0.170
Ours	**0.81**	**7.07**	**0.028**	**0.047**	**0.022**	**0.044**

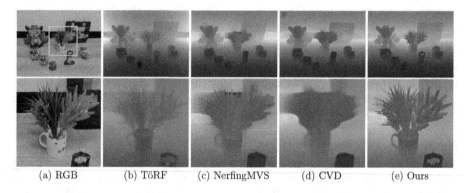

(a) RGB   (b) TöRF   (c) NerfingMVS   (d) CVD   (e) Ours

Fig. 5. Multiview depth estimation with ToF-supervised scene representation TöRF [5] and the multiview-stereo approaches NerfingMVS [51] and CVD [31].

reduced, and depth quality is much better than our phone's ToF module. After securing the phone and the Azure Kinect on a joint mount, we calibrate the phone's ToF module and the ultrawide camera w.r.t. the depth camera. Once the calibration is estimated, depth maps can be reprojected to the ultrawide and ToF cameras for comparison. We capture four scenes for a total of 200 snapshots.

In addition to this RGB-D dataset, we calibrate the floating camera using the conventional multi-shot offline pipeline with chessboards. This provides ground truth for our online calibration evaluation on four scenes.

4.2   Depth Estimation for Training

Figure 5 shows that our method can preserve thin structures, and Table 1 confirms that our approach can efficiently merge ToF and stereo data from multiple views. While TöRF [5] is designed to handle ToF inputs, it performs worse than our method when ToF and RGB resolutions differ. We run CVD and RCVD at

Table 2. Our online calibration versus Gao et al. [15] on our real dataset.

	Pose error (mm)		Rotation error (deg.)	
	MAE	RMSE	MAE	RMSE
Gao et al. [15]	5.664	5.997	1.292	1.431
[15] + DFM [13]	5.174	5.803	1.344	1.499
Ours	**3.346**	**3.989**	**1.264**	**1.411**

Table 3. Fusion evaluation. In the first rows, we evaluate other approaches for RGB/ToF fusions. Since their stereo matching method are not robust against noise and imaging artifacts, their fusion results are highly inaccurate. In the next rows, we replace their less robust stereo matching with a state-of-the-art method [28]. We use our calibration for all methods, except for "Ours (ignoring OIS)" to highlight the importance of our per-shot calibration. The results show that our fusion approach outperforms existing methods.

	Bad ratio (%)		Depth error			
	>0.2	>0.05	MAE Rel.	RMSE Rel.	MAE	RMSE
Marin et al. [33]	61.25	96.80	0.545	0.708	0.410	0.610
Agresti et al. [1,2]	91.29	98.29	0.962	1.172	0.722	1.012
Gao et al. [15]	14.95	40.82	2.262	6.493	1.394	5.066
[33] (RAFT-stereo [28])	2.01	13.87	0.043	0.092	0.034	0.085
[1,2] (RAFT-stereo [28])	1.48	13.96	0.037	0.065	0.031	0.065
[15] (RAFT-stereo [28])	1.67	7.71	0.031	1.021	0.026	0.854
Stereo only [28]	2.36	14.18	0.041	1.300	0.035	1.078
Ours (ignoring OIS)	9.06	29.54	0.082	0.163	0.073	0.164
Ours	**1.40**	**7.17**	**0.028**	**0.050**	**0.024**	**0.051**

their default resolution since we observed degradation in accuracy when increasing image size. Note that progressive geometry integration methods such as [18] fail on our data: the raw ToF depth maps from our smartphone are not accurate enough for ICP registration.

4.3   Snapshot RGB-D Imaging

Calibration. Table 2 details the accuracy of our calibration method. Gao et al. [15] calibrate two GoPro cameras w.r.t. a Kinect RGB-D camera. In our comparison, we substitute the RGB-D camera by the ultrawide RGB plus the ToF module, and substitute the GoPro camera by the main camera. Gao et al. [15]'s calibration shows much lower accuracy than ours, even when the method is paired with a state-of-the-art feature matcher [13]. In addition, only our method is able to refine the camera matrix and distortion parameters.

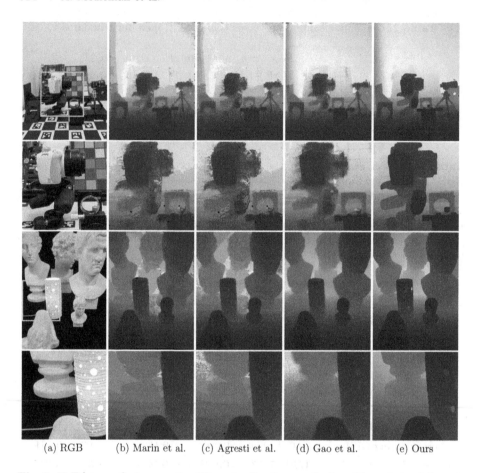

(a) RGB   (b) Marin et al.   (c) Agresti et al.   (d) Gao et al.   (e) Ours

Fig. 6. ToF/stereo fusion results. We pair all other methods with a state-of-the-art stereo approach for fair comparison [28]. Marin et al. [33] and Agresti et al. [1] suffer from quantization as their sub-pixel resolution approach is not suitable for small phone camera baselines. Gao et al. [15] relies heavily on ToF measurements, degrading its performance when ToF is inaccurate (e.g., black parts of the camera). Refer to the supplemental material for additional results.

Stereo/ToF Fusion. We evaluate our fusion approach against our real-world RGB-D dataset. For comparison, we implement [1,15,33] and we train Agresti et al.'s method [1] using their rendered SYNTH3 dataset. Since Gao et al. [15]'s calibration is too inaccurate, rectification fails severely on some snapshots. Total calibration failure occurred for 34 of the 200 snapshots (17%) in our test dataset. We show examples of poorly rectified stereo pairs in the supplemental material.

Thus, evaluate all methods using our calibration. We also evaluate if we can ignore OIS: we calibrate the main camera using a checkerboard while the phone is fixed, then we move the phone to capture our test scenes. We report the results in Table 3 under "Ours (ignoring OIS)", showing a large decrease in depth accuracy. Thus, online calibration is both necessary and effective. Figure 6 shows that our method allows for robust depth estimation with better edge and hole preservation. The low RMSE in Table 3 suggests that our method is robust against strong outliers. While other methods suffer from a less robust stereo matching—swapping theirs for RAFT-stereo [28] significantly improving their results—our approach maintains higher accuracy.

4.4 Dependency on the Device

Under the assumption of a narrow baseline on smartphones, the method should generalize as it allows accurate optical flow estimation between the two RGB cameras. In addition, we show the results of our fusion of datasets based on different hardware: a ZED stereo camera and a Microsoft Kinect v2 ToF depth camera for REAL3 [2], and two calibrated BASLER scA1000 RGB cameras and a MESA SR4000 ToF camera for LTTM5 [10] in the supplemental material.

4.5 Limitations

While our approach applies to indoor environments, the reliance on ToF and stereo prevents application in some scenarios. First, the ToF module cannot estimate depth accurately at large distances due to its low power. Second, ToF depth estimation is not reliable within strong IR ambient illumination (e.g., direct daylight). Since our calibration relies directly on ToF measurements, it becomes inaccurate if no ToF depth can be estimated. In addition, some materials—particularly translucent or specular materials—are challenging for both ToF and stereo depth estimation and cannot be tackled by our fusion approach.

5 Conclusion

Optical-image-stabilized lenses are now common but present problems for pose estimation when wanting to fuse information across multiple sensors in a camera system. This limits our ability to estimate high-quality depth maps from a single snapshot. Our method is designed for consumer devices, tackling calibration and robust sensor fusion for indoor environments. As our approach uses only a single snapshot and does not exploit camera motion for pose estimation, the acquisition is quick and could be used on dynamic scenes. Evaluated on real-world inputs, our method estimates more accurate depth maps than state-of-the-art ToF and stereo fusion methods.

Acknowledgement. Min H. Kim acknowledges funding from Samsung Electronics, in addition to the partial support of the MSIT/IITP of Korea (RS-2022-00155620, 2022-0-00058, and 2017-0-00072), the NIRCH of Korea (2021A02P02-001), Microsoft Research Asia, and the Samsung Research Funding Center (SRFC-IT2001-04) for developing 3D imaging algorithms. James Tompkin thanks US NSF CAREER-2144956.

References

1. Agresti, G., Minto, L., Marin, G., Zanuttigh, P.: Deep learning for confidence information in stereo and ToF data fusion. In: ICCV Workshops (2017)
2. Agresti, G., Minto, L., Marin, G., Zanuttigh, P.: Stereo and ToF data fusion by learning from synthetic data. Inf. Fus. **49**, 161–173 (2019)
3. Agresti, G., Zanuttigh, P.: Deep learning for multi-path error removal in ToF sensors. In: ECCV Workshops (2018)
4. Yu, A., Fridovich-Keil, S., Tancik, M., Chen, Q., Recht, B., Kanazawa, A.: Plenoxels: radiance fields without neural networks (2021)
5. Attal, B., et al.: Törf: time-of-flight radiance fields for dynamic scene view synthesis. In: NeurIPS (2021)
6. Barron, J.T., Mildenhall, B., Tancik, M., Hedman, P., Martin-Brualla, R., Srinivasan, P.P.: Mip-NeRF: a multiscale representation for anti-aliasing neural radiance fields. In: ICCV (2021)
7. Brown, D.C.: Decentering distortion of lenses. Photogramm. Eng. (1966)
8. Brown, M.A., Süsstrunk, S.: Multi-spectral sift for scene category recognition. In: CVPR (2011)
9. Conrady, A.E.: Decentred lens-systems. Monthly Notices of the Royal Astronomical Society (1919)
10. Dal Mutto, C., Zanuttigh, P., Cortelazzo, G.M.: Probabilistic ToF and stereo data fusion based on mixed pixels measurement models. IEEE Trans. Patt. Anal. Mach. Intell. (TPAMI) **37**, 2260–2272 (2015)
11. Deng, K., Liu, A., Zhu, J.Y., Ramanan, D.: Depth-supervised nerf: fewer views and faster training for free. arXiv preprint arXiv:2107.02791 (2021)
12. DiVerdi, S., Barron, J.T.: Geometric calibration for mobile, stereo, autofocus cameras. In: WACV (2016)
13. Efe, U., Ince, K.G., Alatan, A.: Dfm: A performance baseline for deep feature matching. In: CVPR Workshops (2021)
14. Evangelidis, G.D., Hansard, M.E., Horaud, R.: Fusion of range and stereo data for high-resolution scene-modeling. IEEE Trans. Patt. Anal. Mach. Intell. (TPAMI) **37**, 2178–2192 (2015)
15. Gao, Y., Esquivel, S., Koch, R., Keinert, J.: A novel self-calibration method for a stereo-ToF system using a Kinect V2 and two 4K GoPro cameras. In: 3DV (2017)
16. Gil, Y., Elmalem, S., Haim, H., Marom, E., Giryes, R.: Online training of stereo self-calibration using monocular depth estimation. IEEE Trans. Comput. Imaging **7**, 812–823 (2021)
17. Guo, Q., Frosio, I., Gallo, O., Zickler, T., Kautz, J.: Tackling 3D ToF artifacts through learning and the flat dataset. In: ECCV (2018)
18. Ha, H., Lee, J.H., Meuleman, A., Kim, M.H.: NormalFusion: real-time acquisition of surface normals for high-resolution RGB-D scanning. In: CVPR (2021)
19. Hansard, M., Lee, S., Choi, O., Horaud, R.: Time of Flight Cameras: Principles, Methods, and Applications. Springer Briefs in Computer Science, Springer (2012). https://doi.org/10.1007/978-1-4471-4658-2
20. Holynski, A., Kopf, J.: Fast depth densification for occlusion-aware augmented reality. ACM Trans. Graph. (Proc. SIGGRAPH Asia) **37**, 1–11 (2018)
21. Jeong, Y., Ahn, S., Choy, C., Anandkumar, A., Cho, M., Park, J.: Self-calibrating neural radiance fields. In: ICCV (2021)
22. Jung, H., Brasch, N., Leonardis, A., Navab, N., Busam, B.: Wild ToFu: improving range and quality of indirect time-of-flight depth with RGB fusion in challenging environments. In: 3DV (2021)

23. Kopf, J., et al.: One shot 3D photography. In: ACM Transactions on Graphics (Proceedings of ACM SIGGRAPH) (2020)
24. Kopf, J., Rong, X., Huang, J.B.: Robust consistent video depth estimation. In: CVPR (2021)
25. Li, L.: Time-of-flight camera – an introduction (2014). https://www.ti.com/lit/wp/sloa190b/sloa190b.pdf
26. Li, Z., et al.: Learning the depths of moving people by watching frozen people. In: CVPR (2019)
27. Lin, C.H., Ma, W.C., Torralba, A., Lucey, S.: BARF: bundle-adjusting neural radiance fields. In: ICCV (2021)
28. Lipson, L., Teed, Z., Deng, J.: RAFT-Stereo: multilevel recurrent field transforms for stereo matching. In: 3DV (2021)
29. Lombardi, S., Simon, T., Saragih, J., Schwartz, G., Lehrmann, A., Sheikh, Y.: Neural volumes: learning dynamic renderable volumes from images. ACM Trans. Graph. (2019)
30. Lombardi, S., Simon, T., Schwartz, G., Zollhoefer, M., Sheikh, Y., Saragih, J.: Mixture of volumetric primitives for efficient neural rendering. ACM Trans. Graph. (2021)
31. Luo, X., Huang, J., Szeliski, R., Matzen, K., Kopf, J.: Consistent video depth estimation. ACM Trans. Graph. (Proceedings of ACM SIGGRAPH) (2020)
32. Marco, J., et al.: DeepToF: off-the-shelf real-time correction of multipath interference in time-of-flight imaging. ACM Trans. Graph. 36, 1–12 (2017)
33. Marin, G., Zanuttigh, P., Mattoccia, S.: Reliable fusion of ToF and stereo depth driven by confidence measures. In: Leibe, B., Matas, J., Sebe, N., Welling, M. (eds.) ECCV 2016. LNCS, vol. 9911, pp. 386–401. Springer, Cham (2016). https://doi.org/10.1007/978-3-319-46478-7_24
34. Menze, M., Geiger, A.: Object scene flow for autonomous vehicles. In: CVPR (2015)
35. Mildenhall, B., Srinivasan, P.P., Tancik, M., Barron, J.T., Ramamoorthi, R., Ng, R.: NeRF: representing scenes as neural radiance fields for view synthesis. In: ECCV (2020)
36. Pham, F.: Fusione di dati stereo e time-of-flight mediante tecniche di deep learning (2019). https://github.com/frankplus/tof-stereo-fusion
37. Poggi, M., Agresti, G., Tosi, F., Zanuttigh, P., Mattoccia, S.: Confidence estimation for ToF and stereo sensors and its application to depth data fusion. IEEE Sens. J. 20, 1411–1421(2020)
38. Qiu, D., Pang, J., Sun, W., Yang, C.: Deep end-to-end alignment and refinement for time-of-flight RGB-D modules. In: ICCV (2019)
39. Gao, R., Fan, N., Li, C., Liu, W., Chen, Q.: Joint depth and normal estimation from real-world time-of-flight raw data. In: IROS (2021)
40. Sachs, D., Nasiri, S., Goehl, D.: Image stabilization technology overview. InvenSense Whitepaper (2006)
41. Scharstein, D., et al.: High-resolution stereo datasets with subpixel-accurate ground truth. In: Jiang, X., Hornegger, J., Koch, R. (eds.) GCPR 2014. LNCS, vol. 8753, pp. 31–42. Springer, Cham (2014). https://doi.org/10.1007/978-3-319-11752-2_3
42. Schönberger, J.L., Frahm, J.M.: Structure-from-motion revisited. In: CVPR (2016)
43. Shih, M.L., Su, S.Y., Kopf, J., Huang, J.B.: 3D photography using context-aware layered depth inpainting. In: CVPR (2020)
44. Son, K., Liu, M.Y., Taguchi, Y.: Learning to remove multipath distortions in time-of-flight range images for a robotic arm setup. In: ICRA (2016)
45. Song, S., Lichtenberg, S.P., Xiao, J.: Sun RGB-d: A RGB-d scene understanding benchmark suite. In: CVPR (2015)

46. Su, S., Heide, F., Wetzstein, G., Heidrich, W.: Deep end-to-end time-of-flight imaging. In: CVPR (2018)
47. Teed, Z., Deng, J.: RAFT: recurrent all-pairs field transforms for optical flow. In: Vedaldi, A., Bischof, H., Brox, T., Frahm, J.-M. (eds.) ECCV 2020. LNCS, vol. 12347, pp. 402–419. Springer, Cham (2020). https://doi.org/10.1007/978-3-030-58536-5_24
48. Valentin, J., et al.: Depth from motion for smartphone ar. SIGGRAPH Asia (2018)
49. Wadhwa, N., et al.: Synthetic depth-of-field with a single-camera mobile phone. SIGGRAPH (2018)
50. Wang, J., Qiu, K.F., Chao, P.: Control design and digital implementation of a fast 2-degree-of-freedom translational optical image stabilizer for image sensors in mobile camera phones. Sensors **17**, 2333 (2017)
51. Wei, Y., Liu, S., Rao, Y., Zhao, W., Lu, J., Zhou, J.: NerfingMVS: guided optimization of neural radiance fields for indoor multi-view stereo. In: ICCV (2021)
52. Xian, W., Huang, J.B., Kopf, J., Kim, C.: Space-time neural irradiance fields for free-viewpoint video. In: CVPR (2021)
53. Zhang, X., Matzen, K., Nguyen, V., Yao, D., Zhang, Y., Ng, R.: Synthetic defocus and look-ahead autofocus for casual videography. SIGGRAPH (2019)
54. Zhou, Y., Barnes, C., Jingwan, L., Jimei, Y., Hao, L.: On the continuity of rotation representations in neural networks. In: CVPR (2019)
55. Zhu, R., Yu, D., Ji, S., Lu, M.: Matching RGB and infrared remote sensing images with densely-connected convolutional neural networks. Remote Sens. **11**, 2836 (2019)

DELTAR: Depth Estimation from a Light-Weight ToF Sensor and RGB Image

Yijin Li[1], Xinyang Liu[1], Wenqi Dong[1], Han Zhou[1], Hujun Bao[1],
Guofeng Zhang[1], Yinda Zhang[2](✉), and Zhaopeng Cui[1](✉)

[1] State Key Laboratory of CAD & CG, Zhejiang University, Hangzhou, China
zhpcui@zju.edu.cn
[2] Google, Mountain View, USA
yindaz@gmail.com

Abstract. Light-weight time-of-flight (ToF) depth sensors are small, cheap, low-energy and have been massively deployed on mobile devices for the purposes like autofocus, obstacle detection, etc. However, due to their specific measurements (depth distribution in a region instead of the depth value at a certain pixel) and extremely low resolution, they are insufficient for applications requiring high-fidelity depth such as 3D reconstruction. In this paper, we propose DELTAR, a novel method to empower light-weight ToF sensors with the capability of measuring high resolution and accurate depth by cooperating with a color image. As the core of DELTAR, a feature extractor customized for depth distribution and an attention-based neural architecture is proposed to fuse the information from the color and ToF domain efficiently. To evaluate our system in real-world scenarios, we design a data collection device and propose a new approach to calibrate the RGB camera and ToF sensor. Experiments show that our method produces more accurate depth than existing frameworks designed for depth completion and depth super-resolution and achieves on par performance with a commodity-level RGB-D sensor. Code and data are available on the project webpage.

Keywords: Light-weight ToF Sensor · Depth estimation

1 Introduction

The depth sensor is a game changer in computer vision, especially with commodity-level products being widely available [6,22,29,33,53]. As the main player, time-of-flight (ToF) sensors have competitive features , e.g., compact and less sensitive to mechanical alignment and environmental lighting conditions. and thus have become one of the most popular classes in the depth sensor market. However, the price and power consumption, though already significantly lower than other technologies such as structured light (Microsoft Kinect V1) ,

---

Supplementary Information The online version contains supplementary material available at https://doi.org/10.1007/978-3-031-19769-7_36.

S. Avidan et al. (Eds.): ECCV 2022, LNCS 13661, pp. 619–636, 2022.
https://doi.org/10.1007/978-3-031-19769-7_36

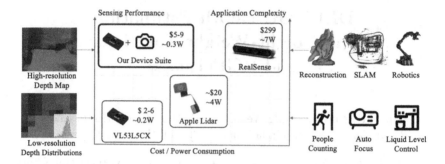

Fig. 1. Comparison between different depth sensors. Low-cost and low-power-consumed sensors like VL53L5CX are designed for simple applications such as people counting and autofocus. In this paper, we show how to improve the depth quality to be on par with a commodity-level RGB-D sensor by our DELTAR algorithm. (Icon credit: Iconfinder [20].)

are still one to two orders of magnitudes higher than a typical RGB camera when reaching a similar resolution due to a large number of photons needs to be emitted, collected, and processed. On the other hand, light-weight ToF sensors are designed to be low-cost, small, and low-energy, which have been massively deployed on mobile devices for the purposes like autofocus, obstacle detection, etc. [41]. However, due to the light-weight electronic design, the depth measured by these sensors has more uncertainty (i.e., in a distribution instead of single depth value) and low spatial resolution (e.g., $< 10 \times 10$), and thus cannot support applications like 3D reconstruction or SLAM [6,22], that require high-fidelity depth (see Fig. 1).

In contrast, RGB cameras are also widely deployed in modern devices with the advantage of capturing rich scene context at high resolution, but they are not able to estimate accurate depth with a single capture due to the inherent scale ambiguity of monocular vision. We observe that these two sensors sufficiently complement each other and thus propose a new setting, i.e., estimating accurate dense depth maps from paired sparse depth distributions (by the light-weight ToF sensor) and RGB image. The setting is essentially different from previous depth super-resolution and completion in terms of the input depth signal. Specifically, the task of super-resolution targets relatively low-resolution consumer depth sensors, (e.g., 256×192 for the Apple LIDAR and 240×180 for the ToF sensor on the Huawei P30). In contrast, our task targets light-weight ToF sensors with several orders of magnitude lower resolution (e.g., 8×8), but provides a depth distribution per zone (see Fig. 2). Depth completion, on the other hand, aims to densify incomplete dense high-resolution maps (e.g., given hundreds of depth samples), which is not available for light-weight ToF sensors. Therefore, our task is unique and challenging due to the extremely low resolution of the input depth but accessibility to the rich depth distribution.

To demonstrate, we use ST VL53L5CX [42] (denoted as L5) ToF sensor, which outputs 8×8 zones, each with a depth distribution, covering a total of $63°$ diagonal field-of-view (FoV) and runs at a power consumption of about

200 mW (vs. 4W of an Apple Lidar). To fully exploit the L5 depth signals, we design DELTAR (**D**epth **E**stimation from **L**ight-weight **T**oF **A**nd **R**GB image), a neural network architecture tailored with respect to the underlying physics of the L5 sensors. Specifically, we first build the depth hypothesis map sampled from the distribution reading of L5, and then use cross-domain attention to exchange the information between the RGB image and the depth hypothesis. A self-attention is also run on image domain to exchange the information between regions covered by L5 and beyond, hence the output aligns with the RGB image and covers the whole FoV. Experiments show that DELTAR outperforms existing architectures designed for depth completion and super-resolution, and improve the raw depth readings of L5 to maintain the quality on par with commodity-level depth sensors, such as Intel RealSense D435i.

Moreover, as no public datasets are available for this new task, we build a capturing system by mounting an L5 sensor and a RealSense RGB-D sensor on a frame-wire with reasonable field-of-view overlap. To align the RGB image and L5's zones, we need to calibrate the sensors, which is challenging as the correspondence cannot be trivially built between two domains. To this end, we propose a new calibration method. An EM-like algorithm is first designed to estimate the plane from L5 signals and then the extrinsic parameters between the L5 sensor and the color camera are optimized by solving point-to-plane alignment in a natural scene with multiple planes. With this capturing system, we create a dataset called ZJU-L5, which includes about 1000 L5-image pairs from 15 real-world scenes with pixel-aligned RGB and ToF signals for training and evaluation purposes. Besides the real-world data, we also simulate synthetic L5 signals using depth from NYU-Depth V2 dataset and use them to augment the training data. The dataset is publicly available to facilitate and inspire further research in the community.

Our contributions can be summarized as follows. First, we demonstrate that light-weight ToF sensors designed for autofocus can be empowered for high-resolution and accurate depth estimation by cooperating with a color image. Second, we prove the concept with a hardware setup and design a cross-domain calibration method to align RGB and low-resolution depth distributions, which enables us to collect large-scale data. The dataset is released to motivate further research. Third, we propose DELTAR, a novel end-to-end transformer-based architecture, based on the sensors' underlying physics, can well utilize the captured depth distribution from the sensor and the color image for dense depth estimation. Experiments show that DELTAR performs better than previous architectures designed for depth completion or super-resolution and achieves more accurate depth prediction results.

2 Related Work

Monocular Depth Estimation. These methods predict a dense depth map for each pixel with a single RGB image. Early approaches [37–40] use hand-crafted features or graphical models to estimate a depth map. More recent methods employ deep CNN [9,13,15,23,48,49] due to its strong feature representation. Among them, some methods [19,24] exploit assumptions about indoor

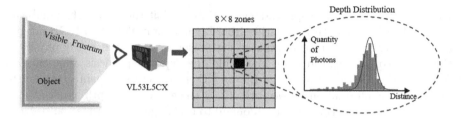

Fig. 2. L5 Sensing Principle. L5 has an extremely low resolution (8 × 8 zones) and provides depth distribution per zone.

environments, e.g., plane constraints, to regularize the network. Other methods [35,36] try to benefit from more large-scale and diverse data samples by designing loss functions and mixing strategies. Besides, [1,11] propose to model depth estimation as a classification task or hybrid regression to improve accuracy and generalization. Nonetheless, these methods cannot generalize well on different scenes due to their lack of metric scale innately.

Depth Completion. Depth completion aims to recover a high-resolution depth map given some sparse depth samples and an RGB image. Spatial propagation network(SPN) series methods [4,5,27,31] are one of the most popular methods which learned local affinities to refine depth predictions. Recently, some works [3,34,50] attempt to introduce 3D geometric cues in the depth completion task, e.g., by introducing surface normals as the intermediate representation, or learn a guided network [44] to utilize the RGB image better. More recently, PENet [16] propose an elaborate two-branch framework, which reaches the state-of-the-art. This type of method, however, is not suitable for our task because it assumes the pixel-wise depth-to-RGB alignment, while light-weight ToF sensors only provide a coarse depth distribution in each zone area without exact pixel-wise correspondence.

Depth Super-Resolution. This task aims to boost the consumer depth sensor to a higher spatial resolution to match the resolution of RGB images. Most early works are based on filtering [26,51] or formulate depth super-resolution as an optimization problem [7,30]. Later researches focus more on learning-based method [18,45–47]. Among them, Xia et al. [47] propose a task-agnostic network which can be used to process depth information from different sources. Wang et al. [45] iteratively updates the intermediate feature map to be consistent with the given low-resolution depth. In contrast to these methods which usually take a depth map with more than 10 thousand pixels as input, our task targets light-weight ToF sensors with several orders of magnitude lower resolution (e.g., 8×8), but provides a depth distribution per region.

3 Hybrid Sensor Setup

This paper aims to predict a high-resolution depth image from a light-weight ToF sensor (e.g., L5) guided by a color image. While no public datasets are available,

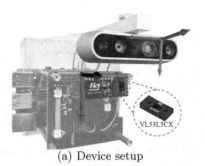

(a) Device setup   (b) Aligned L5's zones and color image

Fig. 3. Hybrid sensor setup. (a) We mount a L5 with an Intel RealSense 435i on a metal frame. (b) Blending color images with L5' depth. White color represents close range, black color represents long range. According to the valid status returned by L5, we hide all invalid zones which may receive too less photons or fail in measurement consistency.

we build a device with hybrid sensors and propose the calibration method for this novel setup.

3.1 L5 Sensing Principle

L5 is a light-weight ToF-based depth sensor. In conventional ToF sensors, the output is typically in a resolution higher than 10 thousand pixels and measures the per-pixel distance along the ray from the optical center to the observed surfaces. In contrast, L5 provides multiple depth distributions with an extremely low resolution of 8×8 zones, covering $63°$ diagonal FoV in total. The distribution is originally measured by counting the number of photons returned in each discretized range of time, and then fitted with a Gaussian distribution (see Fig. 2) in order to reduce the broadband load and energy consumption since only mean and variance needs to be transmitted. Due to the low resolution and high uncertainty of L5, it cannot be directly used for indoor dense depth estimation. Please refer to supplementary materials for more details about L5.

3.2 Device Setup

Figure 3-(a) shows our proposed device suite. An L5 and an Intel RealSense D435i are mounted on a metal frame facing in the same direction. It is worth noting that we only used RealSense's color camera along with L5 when estimating depth, and the depth output by RealSense is used as the ground truth to measure the quality of our estimation. The horizontal and vertical FoV of the L5 are both $45°$, while the RealSense's color camera has $55°$ horizontal FoV and $43°$ vertical FoV. As a result, the L5 sensor and the color camera share most of the FoV but not all in our setup.

3.3   Calibration

In order to align the L5 outputs with the color image, we need to calibrate the multi-sensor setup, i.e., computing the relative rotation and translation between the color camera and the L5 sensor. Similar to the calibration between LIDAR and camera [12], we also calibrate our device suite by solving a point-to-plane fitting problem. However, it is not trivial to fit a plane with raw L5 signals since it does not provide the pixel position of the depth value. We observe that, when facing a plane, in each zone $k \in Z$, there must be a location (x_k, y_k), though unknown, whose depth is equal to the mean of the corresponding distribution m_k returned by L5. Therefore we can optimize both the plane parameter $\{n, d\}$ (the frame subscript is omitted for brevity) and the pixel position (x_k, y_k) through:

$$\{n, d, x_k, y_k \mid k \in Z\} = \arg \min \sum_{k \in Z} \|n \cdot K^{-1}(x_k, y_k, m_k)^T + d\|^2 \tag{1}$$

$$\text{s.t. } x^k_{\min} \leq x_k \leq x^k_{\max}, y^k_{\min} \leq y_k \leq y^k_{\max},$$

where $(x, y)^k_{(\min,\max)}$ is the boundary of the zone k in L5 coordinates, and K is the intrinsic matrix. Clearly, Eq. 1 is non-convex thus we solve it by an EM-like algorithm. Specifically, we first initialize all 2D positions at the center of the zone. In the E-step, we back-project these 2D points with measured mean depth, and then fit a 3D plane. In the M-step, we adjust the 2D positions within each zone by minimizing the distance of the 3D points to the plane. The steps run iteratively until convergence. During the iteration, the points that are too far from the plane are discarded from the optimization.

We then obtain the extrinsic transformation matrix by solving a point-to-plane fitting problem. We use our device suite to scan three planes that are not parallel to each other, and ensure that we only observe one plane most of the time. We employ an RGB-D SLAM [28], which recovers from color images a set of camera poses and point cloud P in real-world metric scale, and each point belongs to a certain plane.

For each time stamp $i \in F$, we use P_i to represent the subset of P that are visible in frame i and transformed from the world coordinate system to current RGB camera's. We also have the planar parameters $\{n_i \in \mathbb{R}^3, d_i \in \mathbb{R}\}$ (normal and offset to the origin) in the current L5's coordinate system by solving Eq. 1, then the extrinsic parameters can be solved by minimizing the point to plane distance:

$$\{R, t\} = \arg \min \sum_{i \in F} \sum_{p \in P_i} \|n_i \cdot (R \cdot p + t) + d_i\|^2, \tag{2}$$

where $[R, t]$ are the transformation that map 3D points from the RGB camera's coordinate system to L5's.

For the device setup shown in Fig. 3-(a), we are able to recover the rotation transformation of the two sensors close to $90°$. The mean distances between the L5 measurement and the 3D point cloud before and after calibration are 7.5 cm and 1.5 cm, respectively. An example of aligned L5's zones and color image is shown in Fig. 3-(b).

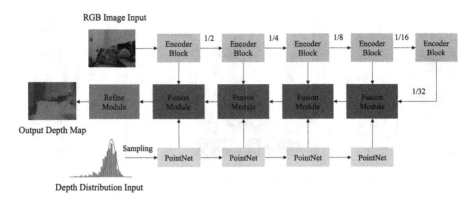

Fig. 4. Overview architecture of the fusion network. Our model takes depth distributions and color image as input, and fuse them at multiple-scale with attention based module before predicting the final depth map.

4   The DELTAR Model

With the calibration, we are able to align the L5'zones with the color image. Based on the characteristic of each modality, we propose a novel attention-based network to predict a high-resolution depth image given the aligned L5 signals and color image. We first design a module to extract features from the distribution (Sect. 4.1), and then propose a cross-domain Transformer-based module to fuse with the color image features at different resolutions (Sect. 4.2). Finally, we predict the final depth values through a refinement module (Sect. 4.3). An overview of the proposed method is shown in Fig. 4.

4.1   Hybrid Feature Extraction

Many works have been designed for fine-grained observations, such as RGB images, depth and point clouds. In contrast, how to extract features from distributions is barely studied. A straightforward idea is to encode the mean and variance directly. However, the depth variance is often smaller than the mean by several magnitudes, which may make it difficult to train the network because of internal covariate shift [21]. In Sect. 5.3 we show that directly encoding the mean and the variance does not work well in our experiments. Therefore, we propose to discretize the distribution by sampling depth hypotheses. Instead of uniform sampling [2,14], we uniformly sample on the inverse cumulative distribution function of the distribution, so the density of the sampling follows the distribution. We utilize PointNet [32] without T-Net to extract features from the sampled depth hypotheses. Multiple pointnets are stacked to extract multi-level distribution features. We use a standard convolutional architecture for the color image, i.e., Efficient B5 [43], to extract multi-level features. Unlike image feature extraction, we do not conduct a down-sample operation when distilling multi-level distribution features.

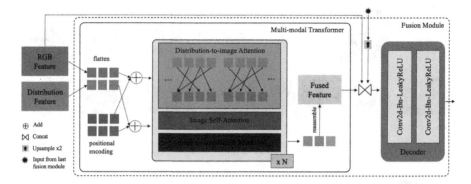

Fig. 5. Details of fusion module. The image and distribution features are flattened into 1-D vectors and added with the positional encoding. The added features are then fused by three different attention mechanisms and concatenated with last layer's feature and the skip-connected image features. Finally, a decoder decodes the concatenated feature and outputs it to the next fusion layer.

4.2   Transformer-Based Fusion Module

The fusion module takes two multi-modal data, including image features and distribution features as input, and outputs fused features. In the task of depth completion and super-resolution, depth features and RGB features are usually concatenated or summed in the fusion step [16,34]. This may be sufficient for fine-grained observations which provide the pixel-wise depth-to-RGB alignment, but it is not suitable for our task since pixel-wise alignment is not available between the depth hypothesis map and the RGB features. Inspired by the recent success of Transformer [8,17,25], we adopt attention mechanisms, which process the entire input all at once and learn to focus on sub-components of the cross-modal information and retrieve useful information from each other.

Cross-Attention Considering Patch-Distribution Correspondence. The Transformer adopts an attention mechanism with the Query-Key-Value model. Similar to information retrieval, the query vector Q retrieves information from the value vector V, according to the attention scores computed from the dot product of queries Q and keys K corresponding to each value. The vanilla version of Transformer contains only self-attention, in which the key, query, and value vectors are from the same set. In multi-modal learning, researchers use cross-attention instead, in which the key and value vectors are from one modal data, and the query vectors are from the other. We first conduct Distribution-to-image attention, that is, taking the key and value vectors from the distribution's feature, and the query vector from the image features, so that the network learns to retrieve information from the candidate depth space. Considering that each distribution from L5 signals corresponds to a specific region in the image, we only conduct cross-attention between the corresponding patch image and the distribution (see Fig. 3-(b)). In Sect. 5.3, we show that conducting cross-attention

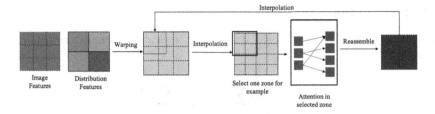

Fig. 6. Interpolation for solving misalignment. When the boundary of the L5's zones and the image feature cannot be aligned precisely, simply quantizing the floating-number boundary could introduce a large negative effect, so we propose to fuse feature after interpolation.

without considering the patch-distribution correspondence leads to severe performance degradation. Empirically, we find that adding image-to-distribution attention leads to better performance.

Propagation by Self-attention. It is not enough to conduct cross-attention as many regions on the image are not covered by the L5's FoV, and these regions cannot benefit from the distribution features. To propagate the depth information further, we also include image self-attention. This step helps the learned depth information propagate to a global context. Besides, the fused feature can be blended to make the feature map smoother. We conduct cross-attention between the two modal data and self-attention over the image feature alternatively for N times, as shown in Fig. 5. In our experiment, we set $N = 2$.

Solving Misalignment by Interpolation. Misalignment occurs when warping L5 zones to an image. See Fig. 6 for a toy example. Simply quantizing the floating-number boundary could introduce a largely negative effect, especially when the fusion is operated on low-resolution feature maps. Moreover, the image resolution corresponding to each zone should be the same to facilitate putting them into a batch. To this end, we fuse on the interpolated feature and then interpolate the fused image features back.

4.3 Refinement Module

We employ the mViT proposed in Adabins [1] as our refinement module to generate the final depth map. Unlike directly regressing depth, the refinement module predicts the depth as a linear combination of multiple depth bins. Specifically, the refinement module predicts a bin-width vector b per image and linear coefficient l at each pixel. The depth-bin's centers $c(b)$ can be calculated from b. Suppose the depth range is divided into N bins, the depth at pixel k can be formulated as:

$$d_k = \sum_{n=1}^{N} c(b_n) l_n. \tag{3}$$

4.4   Supervision

Following [1, 24], we use a scaled version of the Scale-Invariant loss (SI) introduced by [10]:

$$\mathcal{L} = \alpha \sqrt{\frac{1}{T} \sum_i g_i^2 - \frac{\lambda}{T^2} (\sum_i g_i)^2}, \qquad (4)$$

where $g_i = \log \tilde{d}_i - \log d_i$ defined by the estimated depth $\tilde{d}_i$ and ground truth depth $\log d_i$, and T denotes the number of pixels with valid ground truth values. We use $\lambda = 0.85$ and $\alpha = 10$ for all our experiments.

5   Experiment

5.1   Datasets and Evaluation Metrics

NYU-Depth V2 for Training. We use the NYU-Depth V2 dataset to simulate and generate the training data containing L5 signals and color images, from which we select a 24K subset following [1, 24]. For each image, we select a set of zones and according to the L5 sensing principle, we count the histograms of the ground true depth map in each zone and fit them with Gaussian distributions. The fitted mean and variance are used together with the color images as the input for network training. We exclude the depths beyond the L5 measurement range during the histogram statistics.

ZJU-L5 Dataset for Testing. Since the current datasets do not contain the L5 signals, we create an indoor depth dataset using the device suit in Fig. 3-(a) to evaluate our method. This dataset contains 1027 L5-image pairs from 15 scenes, of which the test set contains 527 pairs and the other 500 pairs are used for fine-tuning network. We show the results after fine-tuning in the supplementary material.

Evaluation Metrics. We report the results in terms of standard metrics including thresholded accuracy (δ_i), mean absolute relative error (REL), root mean square error (RMSE) and average ($\log_{10}$) error. The detailed definitions of the metrics are provided in the supplementary material.

5.2   Comparison with State-of-the-Art

Since we are the first to utilize L5 signals and color images to predict depth, there is no existing method for a direct comparison. Therefore, we pick three types of existing methods and let them make use of information from L5 as fully as possible. The first method is monocular depth estimation, where we use the depth information of L5 to align the predicted depth globally. The second method is depth completion, where we assume that each zone's mean depth lies at the zone's centroid to construct a sparse depth map as the input. The third method is depth super-resolution, where we consider the L5 signals as an 8×8

Table 1. Quantitative evaluation on the ZJU-L5 dataset. Our method outperforms all baselines for monocular depth estimation, depth completion, and depth super-resolution.

Comparison with monocular depth estimation						
Methods	$\delta_1 \uparrow$	$\delta_2 \uparrow$	$\delta_3 \uparrow$	REL↓	RMSE↓	$\log_{10} \downarrow$
VNL [52]	0.661	0.861	0.928	0.225	0.653	0.104
BTS [24]	0.739	0.914	0.964	0.174	0.523	0.079
AdaBins [1]	0.770	0.926	0.970	0.160	0.494	0.073
Ours	**0.853**	**0.941**	**0.972**	**0.123**	**0.436**	**0.051**
Comparison with depth completion						
Methods	$\delta_1 \uparrow$	$\delta_2 \uparrow$	$\delta_3 \uparrow$	REL↓	RMSE↓	$\log_{10} \downarrow$
PrDepth [47]	0.161	0.395	0.660	0.409	0.937	0.249
NLSPN [31]	0.583	0.784	0.892	0.345	0.653	0.120
PENet [16]	0.807	0.914	0.954	0.161	0.498	0.065
Ours	**0.853**	**0.941**	**0.972**	**0.123**	**0.436**	**0.051**
Comparison with depth super resolution						
Methods	$\delta_1 \uparrow$	$\delta_2 \uparrow$	$\delta_3 \uparrow$	REL↓	RMSE↓	$\log_{10} \downarrow$
PnP-Depth [45]	0.805	0.904	0.948	0.144	0.560	0.068
PrDepth [47]	0.800	0.926	0.969	0.151	0.460	0.063
Ours	**0.853**	**0.941**	**0.972**	**0.123**	**0.436**	**0.051**

low-resolution depth map, with each pixel (zone) corresponding to a region of the image. Since the state-of-the-art RGB-D method is sensitive to the sparsity of the input points, we re-trained these methods for a fair comparison.

Table 1 summarizes the comparison between ours and these three types of methods. For all metrics, our method achieves the best performance among all methods, which indicates that our network customized with respect to the underlying physics of the L5 is effective in learning from the depth distribution.

Figure 7 shows the qualitative comparison of our method with other solutions for our task [1,16,47]. Overall, our method produces the most accurate depth as reflected by the error map. The monocular estimation method [1] can produce sharp object boundaries, however tends to make mistakes for regions with ambiguous textures. Guided depth super-resolution [47] and completion [16] by design are easier to maintain plausible depth measurement, but the output depths are often overly blurry and lack geometry details. In contrast, our method learns to leverage the high-resolution color image and low-quality L5 readings, producing the most accurate depths that are rich of details.

5.3 Ablation Studies

To understand the impact of each model component on the final performance, we conduct a comprehensive ablation study by disabling each component respec-

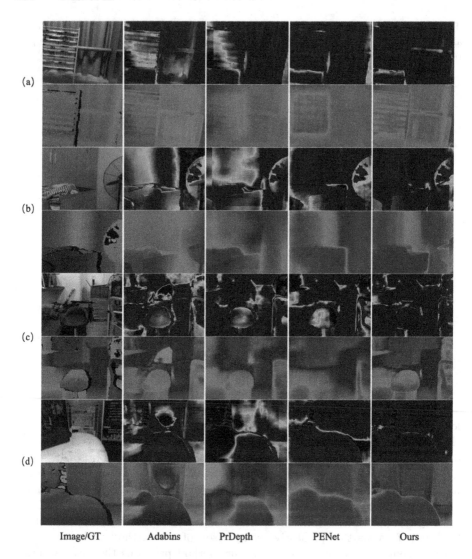

Fig. 7. Qualitative comparison on ZJU-L5 dataset with error map. Monocular estimation method [1] tends to make mistakes on some misleading textures. Guided depth super-resolution [47] and completion [16] produce overly blurry depths that are lack of geometry details. In contrast, our method learns to leverage the high resolution color image and low quality L5 reading, and produces the most accurate depths with sharp object boundaries.

tively. The quantitative results are shown in Table 2. There is a reasonable drop in performance with each component disabled, while the full model works the best.

Table 2. Ablation studies. We evaluate our method with each design or network component turned off. Overall, our full model achieves the best performance, which indicates the positive contribution from all design choices.

Models	$\delta_1 \uparrow$	REL$\downarrow$	RMSE$\downarrow$
Mean-Var PointNet	0.434	0.298	0.669
Five-channel Input	0.619	0.251	0.583
Feature Concat	0.825	0.140	0.454
w/o Patch-Dist-Corr	0.749	0.182	0.512
w/o Img-Self-Attn	0.835	0.133	0.456
w/o Img-Dist-Attn	0.840	0.135	0.446
Uniform Sampling	0.849	0.127	0.439
w/o Refine	0.850	0.126	0.462
Full	**0.853**	**0.123**	**0.436**

Learning Directly from the Mean/Variance. We implement two baseline methods which learn directly from the mean and variance of the depth distribution. For the first one, we change the input to a five-dimensional tensor, which consists of RGB, mean depth and depth variance, named "Five-channel Input". For the second one, we extract features directly from the mean and variance instead of sampled depth, named "Mean-Var PointNet". The performance of these two baselines drops significantly compared to our full model, which indicates the effectiveness of our distribution feature extractor and the fusion module.

Compared with Direct Feature Concatenation. We also replace our Transformer-based fusion module with direct feature concatenation (but retain our proposed feature extractor). It shows that our fusion module performs better than the direct concatenation, which benefits from the fact that our strategy can better gather features from totally different modalities and boost the overall accuracy by propagating features in a global receptive field.

Cross-Attention Without Considering Patch-Distribution Correspondence. We re-train a model by relaxing the constrain on cross-attention, name "w/o Patch-Dist-Corr". Specifically, we conduct cross-attention between all distribution features and image features without considering patch-distribution correspondence. The performance degradation shows the importance of considering this correspondence.

Impact of Multiple Attention Mechanisms. We train models without image self-attention ("w/o Img-Self-Attn") or image-to-distribution attention ("w/o Img-Dist-Attn") respectively that are proposed in Sect. 4.2. The performance drop indicates that the attention modules positively contribute to our fusion model.

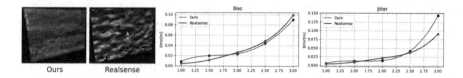

Fig. 8. Quantitative comparison with RealSense. Our method improves the raw L5 reading to a quality on par with commodity-level RGB-D sensor as reflected by the bias and jitter.

Impact of Probability-Driven Sampling. We compare our methods trained with uniform sampling and probability-related sampling. The experiment indicates it brings an improvement of 0.3 cm in terms of RMSE by considering the distribution probability. We report the impact of sampling points' number in the supplementary.

Impact of Refinement Module. We also study the impact of the refinement module by replacing it with a simpler decoder consisting of two convolutional layers that output bin-widths vector and linear coefficient respectively. The overall performance drops but not much, which indicates, though the refiner helps, the majority improvements are brought by our distribution feature extractor and the fusion network.

5.4   Quantitative Comparison with RealSense

In this section, we compare our methods with RealSense quantitatively using traditional metrics in the area of stereo matching [54], such as jitter and bias. Specifically, we recorded multiple frames with the device in front of a flat wall at distances ranging from 1000 mm to 3000 mm. In this case, we evaluate by comparing to "ground truth" obtained with robust plane fitting. We compute bias as the average L1 error between the predicted depth and the ground truth plane to characterize the precision and compute the jitter as the standard deviation of the depth error to characterize the noise. Figure 8 shows the comparison between our method and RealSense, together with visualizations of point clouds colored by surface normals. It can be seen that at a close range (less than three meters), our method achieves a similar and even better performance than RealSense. But as it approaches the upper range limit of L5, the jitter of our method increases dramatically. Overall, it indicates that our method improves the raw depth readings of the L5 to a quality (both resolution and accuracy) on par with a commodity-level depth sensor (i.e., the Intel RealSense D435i) in the working range of L5.

6   Conclusion and Future Work

In this work, we show that it is feasible to estimate high-quality depth, on par with commodity-level RGB-D sensors, using a color image and low-quality depth from a light-weight ToF depth sensor. The task is non-trivial due to the extremely low resolution and specific measurements of depth distribution, thus requiring a customized model to effectively extract features from depth distribution and fuse them with RGB image. One limitation of our method is that it is not fast enough for real-time performance. It is promising to further optimize the network complexity such that the system can run without much extra cost of energy consumption, or to further extend the system for more applications such as 3D reconstruction or SLAM.

Acknowledgment. This work was partially supported by Zhejiang Lab (2021PE 0AC01) and NSF of China (No. 62102356).

References

1. Bhat, S.F., Alhashim, I., Wonka, P.: AdaBins: depth estimation using adaptive bins. In: Proceedings of the IEEE/CVF Conference on Computer Vision and Pattern Recognition, pp. 4009–4018 (2021)
2. Bleyer, M., Rhemann, C., Rother, C.: PatchMatch stereo-stereo matching with slanted support windows. In: British Machine Vision Conference, vol. 11, pp. 1–11 (2011)
3. Chen, Y., Yang, B., Liang, M., Urtasun, R.: Learning joint 2D–3D representations for depth completion. In: Proceedings of the IEEE/CVF International Conference on Computer Vision, pp. 10023–10032 (2019)
4. Cheng, X., Wang, P., Guan, C., Yang, R.: CSPN++: learning context and resource aware convolutional spatial propagation networks for depth completion. In: Proceedings of the AAAI Conference on Artificial Intelligence, vol. 34, pp. 10615–10622 (2020)
5. Cheng, X., Wang, P., Yang, R.: Depth estimation via affinity learned with convolutional spatial propagation network. In: Ferrari, V., Hebert, M., Sminchisescu, C., Weiss, Y. (eds.) ECCV 2018. LNCS, vol. 11220, pp. 108–125. Springer, Cham (2018). https://doi.org/10.1007/978-3-030-01270-0_7
6. Dai, A., Nießner, M., Zollhöfer, M., Izadi, S., Theobalt, C.: BundleFusion: real-time globally consistent 3d reconstruction using on-the-fly surface reintegration. ACM Trans. Graph. (ToG) **36**(4), 1 (2017)
7. Diebel, J., Thrun, S.: An application of Markov random fields to range sensing. Adv. Neural Inf. Process. Syst. **18**, 291–298 (2005)
8. Dosovitskiy, A., et al.: An image is worth 16 × 16 words: transformers for image recognition at scale. arXiv preprint arXiv:2010.11929 (2020)
9. Eigen, D., Puhrsch, C., Fergus, R.: depth map prediction from a single image using a multi-scale deep network. In: Advances in Neural Information Processing Systems, vol. 27. Curran Associates, Inc. (2014)
10. Eigen, D., Puhrsch, C., Fergus, R.: Depth map prediction from a single image using a multi-scale deep network. Adv. Neural Inf. Process. Syst. **27** (2014)
11. Fu, H., Gong, M., Wang, C., Batmanghelich, K., Tao, D.: Deep ordinal regression network for monocular depth estimation. In: Proceedings of the IEEE Conference on Computer Vision and Pattern Recognition, pp. 2002–2011 (2018)
12. Geiger, A., Moosmann, F., Car, Ö., Schuster, B.: Automatic camera and range sensor calibration using a single shot. In: 2012 IEEE International Conference on Robotics and Automation, pp. 3936–3943. IEEE (2012)

13. Hao, Z., Li, Y., You, S., Lu, F.: Detail preserving depth estimation from a single image using attention guided networks. In: 2018 International Conference on 3D Vision (3DV), pp. 304–313. IEEE (2018)
14. Hirschmuller, H.: Stereo processing by semiglobal matching and mutual information. IEEE Trans. Pattern Anal. Mach. Intell. **30**(2), 328–341 (2007)
15. Hu, J., Ozay, M., Zhang, Y., Okatani, T.: Revisiting single image depth estimation: toward higher resolution maps with accurate object boundaries. In: 2019 IEEE Winter Conference on Applications of Computer Vision (WACV), pp. 1043–1051. IEEE (2019)
16. Hu, M., Wang, S., Li, B., Ning, S., Fan, L., Gong, X.: PENet: towards precise and efficient image guided depth completion. In: 2021 IEEE International Conference on Robotics and Automation (ICRA), pp. 13656–13662. IEEE (2021)
17. Huang, Z., et al.: FlowFormer: a transformer architecture for optical flow. arXiv preprint arXiv:2203.16194 (2022)
18. Hui, T.-W., Loy, C.C., Tang, X.: Depth map super-resolution by deep multi-scale guidance. In: Leibe, B., Matas, J., Sebe, N., Welling, M. (eds.) ECCV 2016. LNCS, vol. 9907, pp. 353–369. Springer, Cham (2016). https://doi.org/10.1007/978-3-319-46487-9_22
19. Huynh, L., Nguyen-Ha, P., Matas, J., Rahtu, E., Heikkilä, J.: Guiding monocular depth estimation using depth-attention volume. In: Vedaldi, A., Bischof, H., Brox, T., Frahm, J.-M. (eds.) ECCV 2020. LNCS, vol. 12371, pp. 581–597. Springer, Cham (2020). https://doi.org/10.1007/978-3-030-58574-7_35
20. Iconfinder: Iconfinder. www.iconfinder.com/ (2022). Accessed 19 Jul 2022
21. Ioffe, S., Szegedy, C.: Batch normalization: accelerating deep network training by reducing internal covariate shift. In: International Conference on Machine Learning, pp. 448–456. PMLR (2015)
22. Izadi, S., et al.: KinectFusion: real-time 3D reconstruction and interaction using a moving depth camera. In: Proceedings of the 24th Annual ACM Symposium on User Interface Software and Technology, pp. 559–568 (2011)
23. Laina, I., Rupprecht, C., Belagiannis, V., Tombari, F., Navab, N.: Deeper depth prediction with fully convolutional residual networks. In: 2016 Fourth International Conference on 3D Vision (3DV), pp. 239–248 (2016). https://doi.org/10.1109/3DV.2016.32
24. Lee, J.H., Han, M.K., Ko, D.W., Suh, I.H.: From big to small: multi-scale local planar guidance for monocular depth estimation. arXiv preprint arXiv:1907.10326 (2019)
25. Lin, T., Wang, Y., Liu, X., Qiu, X.: A survey of transformers. arXiv preprint arXiv:2106.04554 (2021)
26. Liu, M.Y., Tuzel, O., Taguchi, Y.: Joint geodesic upsampling of depth images. In: Proceedings of the IEEE Conference on Computer Vision and Pattern Recognition, pp. 169–176 (2013)
27. Liu, S., De Mello, S., Gu, J., Zhong, G., Yang, M.H., Kautz, J.: SPN: learning affinity via spatial propagation networks. In: Advances in Neural Information Processing Systems, vol. 30. Curran Associates, Inc. (2017)
28. Mur-Artal, R., Tardós, J.D.: ORB-SLAM2: an open-source slam system for monocular, stereo, and RGB-D cameras. IEEE Trans. Rob. **33**(5), 1255–1262 (2017)
29. Newcombe, R.A., Fox, D., Seitz, S.M.: DynamicFusion: reconstruction and tracking of non-rigid scenes in real-time. In: Proceedings of the IEEE Conference on Computer Vision and Pattern Recognition, pp. 343–352 (2015)

30. Park, J., Kim, H., Tai, Y.W., Brown, M.S., Kweon, I.: High quality depth map upsampling for 3D-ToF cameras. In: 2011 International Conference on Computer Vision, pp. 1623–1630. IEEE (2011)
31. Park, Jinsun, Joo, Kyungdon, Hu, Zhe, Liu, Chi-Kuei., So Kweon, In.: Non-local spatial propagation network for depth completion. In: Vedaldi, Andrea, Bischof, Horst, Brox, Thomas, Frahm, Jan-Michael. (eds.) ECCV 2020. LNCS, vol. 12358, pp. 120–136. Springer, Cham (2020). https://doi.org/10.1007/978-3-030-58601-0_8
32. Qi, C.R., Su, H., Mo, K., Guibas, L.J.: PointNet: deep learning on point sets for 3D classification and segmentation. In: Proceedings of the IEEE Conference on Computer Vision and Pattern Recognition, pp. 652–660 (2017)
33. Qian, C., Sun, X., Wei, Y., Tang, X., Sun, J.: Realtime and robust hand tracking from depth. In: Proceedings of the IEEE Conference on Computer Vision and Pattern Recognition, pp. 1106–1113 (2014)
34. Qiu, J., et al.: DeepLiDAR: deep surface normal guided depth prediction for outdoor scene from sparse lidar data and single color image. In: Proceedings of the IEEE/CVF Conference on Computer Vision and Pattern Recognition, pp. 3313–3322 (2019)
35. Ranftl, R., Bochkovskiy, A., Koltun, V.: Vision transformers for dense prediction. In: Proceedings of the IEEE/CVF International Conference on Computer Vision, pp. 12179–12188 (2021)
36. Ranftl, R., Lasinger, K., Hafner, D., Schindler, K., Koltun, V.: Towards robust monocular depth estimation: mixing datasets for zero-shot cross-dataset transfer. IEEE Trans. Patt. Anal. Mach. Intell. **44**(3), 1623–1637 (2020)
37. Ranftl, R., Vineet, V., Chen, Q., Koltun, V.: Dense monocular depth estimation in complex dynamic scenes. In: Proceedings of the IEEE Conference on Computer Vision and Pattern Recognition, pp. 4058–4066 (2016)
38. Saxena, A., Chung, S., Ng, A.: Learning depth from single monocular images. In: Advances in Neural Information Processing Systems, vol. 18. MIT Press (2005)
39. Saxena, A., Sun, M., Ng, A.Y.: Make3D: learning 3D scene structure from a single still image. IEEE Trans. Pattern Anal. Mach. Intell. **31**(5), 824–840 (2008)
40. Shi, J., Tao, X., Xu, L., Jia, J.: Break Ames room illusion: depth from general single images. ACM Trans. Graph. (TOG) **34**(6), 1–11 (2015)
41. STMicroelectronics: STMicroelectronics Ships 1 Billionth Time-of-Flight Module. www.st.com/content/st_com/en/about/media-center/press-item.html/t4210.html. Accessed 19 Jul 2022
42. STMicroelectronics: Time-of-Flight 8 × 8 multizone ranging sensor with wide field of view. https://www.st.com/en/imaging-and-photonics-solutions/vl53l5cx.html. Accessed 19 Jul 2022
43. Tan, M., Le, Q.: EfficientNet: rethinking model scaling for convolutional neural networks. In: International Conference on Machine Learning, pp. 6105–6114. PMLR (2019)
44. Tang, J., Tian, F.P., Feng, W., Li, J., Tan, P.: Learning guided convolutional network for depth completion. IEEE Trans. Image Process. **30**, 1116–1129 (2020)
45. Wang, T.H., Wang, F.E., Lin, J.T., Tsai, Y.H., Chiu, W.C., Sun, M.: Plug-and-play: improve depth prediction via sparse data propagation. In: 2019 International Conference on Robotics and Automation (ICRA), pp. 5880–5886. IEEE (2019)
46. Wang, Z., Ye, X., Sun, B., Yang, J., Xu, R., Li, H.: 40 Depth upsampling based on deep edge-aware learning. Patt. Recogn. **103**, 107274 (2020)
47. Xia, Z., Sullivan, P., Chakrabarti, A.: Generating and exploiting probabilistic monocular depth estimates. In: Proceedings of the IEEE/CVF Conference on Computer Vision and Pattern Recognition, pp. 65–74 (2020)

48. Xu, D., Ricci, E., Ouyang, W., Wang, X., Sebe, N.: Multi-scale continuous CRFs as sequential deep networks for monocular depth estimation. In: Proceedings of the IEEE Conference on Computer Vision and Pattern Recognition, pp. 5354–5362 (2017)
49. Xu, D., Wang, W., Tang, H., Liu, H., Sebe, N., Ricci, E.: Structured attention guided convolutional neural fields for monocular depth estimation. In: Proceedings of the IEEE Conference on Computer Vision and Pattern Recognition, pp. 3917–3925 (2018)
50. Xu, Y., Zhu, X., Shi, J., Zhang, G., Bao, H., Li, H.: Depth completion from sparse lidar data with depth-normal constraints. In: Proceedings of the IEEE/CVF International Conference on Computer Vision, pp. 2811–2820 (2019)
51. Yang, Q., Yang, R., Davis, J., Nistér, D.: Spatial-depth super resolution for range images. In: 2007 IEEE Conference on Computer Vision and Pattern Recognition, pp. 1–8. IEEE (2007)
52. Yin, W., Liu, Y., Shen, C.: Virtual normal: enforcing geometric constraints for accurate and robust depth prediction. IEEE Trans. Patt. Anal. Mach. Intell. (2021)
53. Zhang, Y., Funkhouser, T.: Deep depth completion of a single RGB-D image. In: Proceedings of the IEEE Conference on Computer Vision and Pattern Recognition, pp. 175–185 (2018)
54. Zhang, Y., et al.: ActiveStereoNet: end-to-end self-supervised learning for active stereo systems. In: Proceedings of the European Conference on Computer Vision (ECCV), pp. 784–801 (2018)

3D Room Layout Estimation
from a Cubemap of Panorama Image
via Deep Manhattan Hough Transform

Yining Zhao[1], Chao Wen[2], Zhou Xue[2], and Yue Gao[1]($\boxtimes$)

[1] BNRist, THUIBCS, BLBCI, KLISS, School of Software, Tsinghua University,
Beijing, China
gaoyue@tsinghua.edu.cn
[2] Pico IDL, ByteDance, Beijing, China

Abstract. Significant geometric structures can be compactly described by global wireframes in the estimation of 3D room layout from a single panoramic image. Based on this observation, we present an alternative approach to estimate the walls in 3D space by modeling long-range geometric patterns in a learnable Hough Transform block. We transform the image feature from a cubemap tile to the Hough space of a Manhattan world and directly map the feature to the geometric output. The convolutional layers not only learn the local gradient-like line features, but also utilize the global information to successfully predict occluded walls with a simple network structure. Unlike most previous work, the predictions are performed individually on each cubemap tile, and then assembled to get the layout estimation. Experimental results show that we achieve comparable results with recent state-of-the-art in prediction accuracy and performance. Code is available at https://github.com/Starrah/DMH-Net.

Keywords: Panorama images · Room layout estimation · Holistic scene structure

1 Introduction

Recovering 3D geometry from a single image is one of the most studied topics in computer vision. This ill-posed task is generally solved under specific scenarios, with certain assumptions or with prior knowledge. The goal of our work is to reconstruct the 3D room layout from a single panorama image under the Manhattan world assumption.

As described in [4], humans perceive 3D world with great efficiency and robustness by using geometrically salient global structures. For our task, an intuitive way of representing the room layout is to use the wireframe for a typical

---

Supplementary Information The online version contains supplementary material available at https://doi.org/10.1007/978-3-031-19769-7_37.

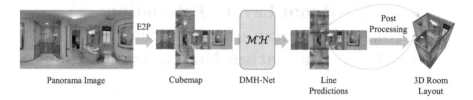

Panorama Image   Cubemap   DMH-Net   Line   3D Room
                                                        Predictions   Layout

Fig. 1. The processing pipeline of our method. Taking panorama image as input, we first apply Equirectangular-to-Perspective transform to generate cubemap, then utilize Deep Manhattan Hough Transform to predict the positions of the wall-wall, wall-floor and wall-ceiling intersection line in each cubemap tile, and recover 3D room layouts by with post-processing procedures.

3D room, consisting of lines on the horizontal plane and vertical lines denoting the junctions between vertical walls. Although using wireframes to estimate 3D room structure is compact and sparse, it could be challenging for vision algorithms to detect long and thin lines with few appearance clues, especially when lines are heavily occluded in a cluttered room.

Existing methods either model the structure estimation as a segmentation problem in panoramic or perspective [39] or decompose the estimation of the geometry elements into a series of regression problems [33]. The motivation of our work is to take advantage of the compact and efficient representation in the VR area and provide an alternative perspective on this problem. We introduce the cubemap [12] of the panorama image and obtain appropriate visual cues in each cubemap tile. To increase the robustness of the line description, we resort to the Hough transform, which is widely used in line segment detection. It uses two geometric terms, an offset, and an angle, to parameterize lines. As the parameterization of lines is global, the estimation is less prone to be affected by noisy observation and partial occlusions.

In our task, we introduce the Manhattan world assumption to the Hough transform, making the representation even simpler. The wireframe of a room has three types of lines. The first two types are wall-ceiling intersection lines and wall-floor intersection lines, which live on horizontal planes and are perpendicular to each other under the Manhattan world assumption. The third type is vertical lines in the 3D space, which represent the junctions between walls and are perpendicular to the first two types of lines. By taking some pre-processing steps to align the room with the camera coordinate system, the first two types of lines can always be aligned with either the x-axis or the y-axis in the floor/ceiling view.

With the aligned input images, after we adopt the equirectangular to perspective (E2P) transform [12,39] to get the cubemap from the panorama images, it can be proven that a straight line which is along one of the coordinate axes in the camera 3D space is either a horizontal line, a vertical line or a line passing the center of the image in the cubemap tiles. This simplifies the estimation of the wireframe lines greatly as only lines with specific characteristics are needed for detection so that the network can be more concentrated to learn a better line detection model suitable for Manhattan room layout estimation.

In this work, we add the Manhattan world line priors into deep neural networks to overcome the challenge of lacking appearance features and occlusion for 3D room wireframes by relying on Hough transform. We embed the Hough transform into trainable neural networks in order to combine Manhattan world wireframe priors with local learned appearances.

The main contributions of this paper can be summarized as follows:

– We introduce the Manhattan world assumption through Deep Hough Transform to capture the long-range geometric pattern of room layouts.
– We propose a novel framework estimating layouts on each cubemap tile individually, which is distortion-free for standard CNN.
– We directly predict Manhattan lines with explicit geometric meaning, which achieves comparable performance towards recent state-of-the-art works.

2 Related Work

Room Layout Estimation. 3D room layout estimation from a single image attracted a lot of research over the past decade. Most previous studies exploit the Manhattan world assumption [7] which means that all boundaries are aligned with the global coordinate system. Moreover, vanishing points detection can be used to inferring the layout based on the assumption.

Traditional methods extract geometric cues and formalize this task as an optimization problem. Since the images may differ in the FoV (field of view), ranging from perspective to 360° panoramas, the methods vary with types of input images. In terms of perspective images, Delage et al. [10] propose a dynamic Bayesian network model to recognize "floor-wall" geometry of the indoor scene. Lee et al. [23] using Orientation Map (OM) while Hedau et al. [16] using Geometric Context (GC) for geometry reasoning to tackle the problem. The strategies have been employed by other approaches leveraging enhanced scoring function [30,31], or modeling the objects-layout interaction [9,13,45].

On the other hand, since the 360° panoramas provides more information, there are multiple papers exploiting in this direction. Zhang et al. [43] proposes to estimate layout and 3D objects by combining the OM and GC on a panoramic image. Yang et al. [38] takes line segments and superpixel facets as features and iteratively optimize the 3D layout. Xu et al. [36] estimate layout use detected objects, their pose, and context in the scene. In order to recover the spatial layout, Yang et al. [40] use more geometric cues and semantic cues as input, while Pintore et al. [27] utilize the gradient map.

With the astonishing capability, neural network based methods leverage data prior to improve layout estimation. Most studies in this field focused on dense prediction, which train deep classification network to pixel-wise estimate boundary probability map [25,29,44], layout surface class [8,19] or corner keypoints heatmaps [11,22]. The panoramas-based approach has recently attracted wide interest to 3D room layout estimation problems. Zou *et al.* [46] predict the corner and boundary probability map directly from a single panorama. Yang *et al.* [39] propose DuLa-Net which leverages both the equirectangular view and the perspective floor/ceiling view images to produce 2D floor plan. Pintore *et al.* [26] follows the similar approach, which directly adopts equirectangular to perspective (E2P) conversion to infer 2D footprint. Fernandez-Labrador *et al.* [11] present EquiConvs, a deformable convolution kernel which is specialized for equirectangular image. Zeng *et al.* [42] jointly learns layout prediction and depth estimation from a single indoor panorama image. Wang *et al.* [35] propose a differentiable depth rendering procedure which can learn depth estimation without depth ground truth. Although extensive research has been carried out on 3D room layout estimation, few studies exist which try to produce a more compact representation to infer the layout. Sun *et al.* propose HorizonNet [33] and HoHoNet [34], which encodes the room layout as boundary and corner probability vectors and propose to recover the 3D room layouts from 1D predictions. LGT-Net [20] is a recent work which represents the room layout by horizon-depth and room height. NonCuboidRoom [37] takes lines into account and recovers partial structures from a perspective image, but cannot estimate the whole room. Although these existing methods are sound in predicting layout, they still have limitations in representing layout. Our approach also assembles compact results but is different from the existing methods. Rather than regressing boundaries values, our approach combines the cubemap and parametric representation of lines.

Hough Transform Based Detectors. Hough *et al.* [18] devise Hough transform to detect line or generalized shapes [2] *e.g.* circle from images. Through the extensive use of Hough transform, traditional line detectors apply edge detection filter at the beginning (*e.g.* Canny [5] and Sobel [32]), then identify significant peaks in the parametric spaces. Qi *et al.* [28] using the Hough voting schemes for 3D object detection. Beltrametti *et al.* [3] adopt voting conception for curve detection. Recently, learning-based approaches demonstrate the representation capability of the Hough transform. Han *et al.* [14] propose Hough Transform to aggregate line-wise features and detect lines in the parametric space. Lin *et al.* [24] consider adding geometric line priors through Hough Transform in networks. Our work use Deep Hough features in a new task to tackle the room layout estimation problem.

3   Method

3.1   Overview

Our goal is to estimate the Manhattan room layout from a single 360° panoramic image. However, panoramic images have distortion, *i.e.* a straight line in the 3D space may not be straight in the equirectangular view of panoramic images. Instead of make predictions directly on the panoramic image like previous works

[33–35,46], we get a cubemap [12] which contains six tiles by adopting E2P transform [12,39].

Given a single RGB panoramic image as input, first we take some pre-processing steps to align the image, get the cubemap and transform ground truth labels. Then we use our proposed Deep Manhattan Hough Network (DMH-Net) to detect three types of straight lines on the tiles of the cubemap. Finally, with an optimization-based post-processing procedure, the line detection results can be assembled and fully optimized to generate the 3D room layout.

In Sect. 3.2, we introduce our pre-processing procedure. The Deep Manhattan Hough Transform for room layout estimation is presented in Sect. 3.3. We summarize the network architecture of the proposed DMH-Net in Sect. 3.4. Finally, in Sect. 3.5, we introduce our optimization-based post-processing method.

3.2 Pre-processing

Aligning the Image. Receiving a single panorama which covers a 360° H-FoV, we first align the image based on the LSD algorithm and vanishing points mentioned in [33,46]. Our approach exploits both the Manhattan world assumption and the properties of aligned panoramas. After the alignment, the cubemap tiles are aligned with the three principal axes [7], *i.e.*, the optical axis of the front camera is perpendicular to a wall. Aligning the panorama makes the vanishing point precisely the center of each cubemap tile.

Cubemap Generation. E2P transform is conducted six times on the equirectangular image with different azimuth angle and elevation angle to generate cubemap tiles $I_{front}, I_{back}, I_{left}, I_{right}, I_{ceil}, I_{floor}$, as Fig. 1 shown. For all of the cubemap tiles, FoV is set to 90° on both horizontal and vertical direction, and image size is set to 512×512.

Ground Truth Transformation. After adopting the aligning method described above for all the datasets we use, the ground truth of the room's layout is provided in the format of corner coordinates in the panoramic image. Using the same E2P transform as described previously, we can also transform the coordinates of the ground truth corner coordinate from the panoramic image to the cubemap and then get lines in the cubemap by connecting the points by their connectivity relation in the original panoramic image. The lines can be categorized into three types: horizontal lines, vertical lines and lines passing the center of the image. Since Deep Manhattan Hough Transform can only detect straight lines rather than line segments, we do not care about the specific position of the line's start point and end point. So for a horizontal (or vertical) line, we only use the y (or x) coordinate to represent the line, and for a line passing the center, we only use the orientation angle θ to represent it.

3.3 Deep Manhattan Hough Transform

In the case of line detection, the traditional Hough Transform [18] parameterizes lines in images to polar coordinates with two parameters, an orientation θ and a distance ρ. Every image pixel votes to a discretized parameter bin, which represents the corresponding polar coordinates. The parameter space is denoted as the Hough space, and the maximum local peaks represent the lines in the image. Specifically, given an single channel input $\mathbf{X} \in \mathbb{R}^{h_o \times w_o}$, the Hough Transform $\mathcal{H}$ can be represented as:

$$\mathcal{H}(\rho, \theta) = \sum_{i \in l} \mathbf{X}(x_i, y_i) \tag{1}$$

in which l is a line whose orientation angle is θ and distance to the coordinate origin is ρ, and i is each of the point in line l.

A key concept of our network is detecting all possible positions of the layout boundaries of the room in each cubemap tile. We propose to incorporate the deep network with Hough Transform for layout boundaries detection. Specifically, we present the Deep Manhattan Hough Transform (DMHT) combining the deep CNN feature and Manhattan world assumption. It is based on the following two assumption:

1. The Manhattan world assumption, *i.e.*, all of the walls, the ceiling and the floor must be perpendicular to each other, and all of the intersection lines of them must be parallel with one of the coordinate axes of some orthogonal coordinate space (named Manhattan space).
2. The input image must be aligned, *i.e.*, the camera of each cubemap tile faces precisely to one of the walls, and its optical axis is parallel with one of the coordinate axes of the Manhattan space.

In practice, the two assumptions are quite straightforward since most of the rooms in human buildings obey the Manhattan world assumption, and the second assumption can be implemented with the pre-processing steps described in Sect. 3.2. Under these two assumptions, it can be proven that any line in the wireframe of the room, including wall-wall line, wall-ceiling line and wall-floor line, must be either a horizontal line ($\theta = 0$), a vertical line ($\theta = \pi/2$) or a line passing the center ($\rho = 0$) in the cubemap tiles, which is a special case of single-point perspective [17] (more details in supplementary materials).

As shown in Fig. 2, given the feature maps of a cubemap tile extracted by the encoder network as input, for each channel $\mathbf{X} \in \mathbb{R}^{h \times w}$ of the feature maps, the Deep Manhattan Hough Transform $\mathcal{MH}$ output three vectors $\mathbf{H}, \mathbf{V}$ and $\mathbf{C}$, corresponding to bins in the Hough space representing horizontal lines, vertical lines and lines passing the center, which is defined as:

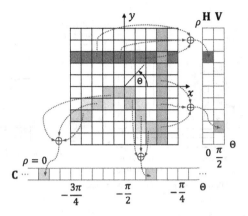

Fig. 2. Deep Manhattan Hough Transform Overview. Each feature map of the CNN can be regarded as many discretized bins. According to the Hough Transform, the bins with the same polar coordinate parameters vote for the same single bin in the Hough space. A bin in the vector **H** (green) is calculated by aggregating horizontal features in the 2D feature map, while bins in **V**(orange) and **C**(blue) are calculated by aggregate vertical features and features passing the center respectively. (Color figure online)

$$\mathbf{H}(\rho) = \mathcal{MH}_H(\rho) = \mathcal{H}(\rho, 0) = \sum_{x_i = -\frac{w}{2}}^{\frac{w}{2}} \mathbf{X}(x_i, \rho),$$

$$\mathbf{V}(\rho) = \mathcal{MH}_V(\rho) = \mathcal{H}(\rho, \frac{\pi}{2}) = \sum_{y_i = -\frac{h}{2}}^{\frac{h}{2}} \mathbf{X}(\rho, y_i), \tag{2}$$

$$\mathbf{C}(\theta) = \mathcal{MH}_C(\theta) = \mathcal{H}(0, \theta)$$

$$= \begin{cases} \sum_{x_i=0}^{\frac{w}{2}} \mathbf{X}(x_i, x_i \cdot \tan(\theta)), & |\tan(\theta)| \leq 1 \\ \sum_{y_i=0}^{\frac{h}{2}} \mathbf{X}(y_i \cdot \cot(\theta), y_i), & |\tan(\theta)| > 1 \end{cases}$$

in which $\rho \in [-\frac{h}{2}, \frac{h}{2}]$ for **H** and $\rho \in [-\frac{w}{2}, \frac{w}{2}]$ for **V**, and $0 \leq \theta \leq 2\pi$. $\mathbf{H}(\rho)$ is the bin of **H** with Hough space parameter ρ, and similar for **V** and **C**.

To calculate the proposed DMHT efficiently, effective discretization is necessary. It is natural to discretize ρ to be integers so that each bin in **H** and **V** represents to a 1-pixel-wide line in the image, thus $\mathbf{H} \in \mathbb{R}^h$ and $\mathbf{V} \in \mathbb{R}^w$. In our experiment, we discretize θ so that $\mathbf{C} \in \mathbb{R}^{2(h+w)}$ for each bin in **C**, the corresponding line intersects the image's border at positions whose coordinates are integer. With the discretization technique above, the process of DMHT can be implemented with matrix addition and multiplications, which is highly parallelizable and suitable for GPU calculation. See the supplementary materials for more detail.

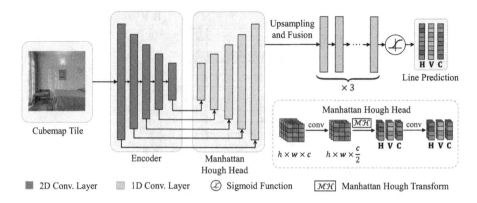

Fig. 3. Overview of the architecture of Deep Manhattan Hough Network (DMH-Net). Given a cubemap tile image as input, a CNN encoder is adopted to extract multi-level image features. Manhattan Hough Head are utilized to handle multi-scale 2D features and perform the Manhattan Hough Transform to get three feature vectors in the Hough space, $\mathbf{H} \in \mathbb{R}^{h \times \frac{c}{2}}$, $\mathbf{V} \in \mathbb{R}^{w \times \frac{c}{2}}$ and $\mathbf{C} \in \mathbb{R}^{2(h+w) \times \frac{c}{2}}$. Finally, the feature vectors are fused to generate lines prediction result, in the format of line position probability vector.

3.4 Network Architecture

Feature Extractor. We employ Dilated Residual Networks [15,41] as our encoder, which utilizes dilated convolution to improve spatial acuity for the images to learn better feature for thin line prediction. The input shape of panorama in equirectangular view I_{equi} is $512 \times 1024 \times 3$. For each of the six tiles of the cubemap, the input shape is $512 \times 512 \times 3$. To capture both low-level and high-level features, we extract the intermediate features of the encoder network. In particular, we gather the perceptual feature from each block before downsample layers. Then, all the features are fed into five independent **Manhattan Hough Head** with different input and output size.

Manhattan Hough Head. As shown in Fig. 3, Manhattan Hough Head receives feature maps of cubemap tile and produces feature vectors in Hough space for line detection. Specifically, given input feature maps with channel number c, the module first applies 2D convolution to reduce the feature channels to $\frac{c}{2}$, then transform the feature through DMHT described in Sect. 3.3 to get results in Hough space. The results are then filtered by a 1D convolution layer with kernel size = 3 to capture context information surrounding certain Hough bins. The output of this module is three multi-channel feature vectors in the Hough space: $\mathbf{H} \in \mathbb{R}^{h \times \frac{c}{2}}$, $\mathbf{V} \in \mathbb{R}^{w \times \frac{c}{2}}$ and $\mathbf{C} \in \mathbb{R}^{2(h+w) \times \frac{c}{2}}$.

Upsampling, Fusion and Generating Line Prediction. Since the size of the feature maps extracted by the feature extractor varies with the depth of the layer, the size of the output feature vectors of the five Manhattan Hough Heads differs. We upsample all the feature vectors to the same size which equals to the width and height of the original image, $h = 512$ and $w = 512$, with bilinear interpolation. Then, feature vectors of the same type are fused by channel-wise concatenation, and filtered by three 1D convolution layers. Finally, the Sigmoid function is applied to generate three single-channel 1D vectors $\mathbf{H} \in \mathbb{R}^h$, $\mathbf{V} \in \mathbb{R}^w$ and $\mathbf{C} \in \mathbb{R}^{2(h+w)}$, which represents the predicted probability of horizontal lines, vertical lines and lines passing the center of the image.

Loss Functions. If we define the ground truth of the probability vectors $\mathbf{H}, \mathbf{V}$ and $\mathbf{C}$ as binary-valued vectors with 0/1 labels simply, it would be too sparse to train, e.g. there would be only less than two non-zero values out of 512 in a cubemap tile. Similar as HorizonNet [33], we smooth the ground truth based on the exponential function of the distance to nearest ground truth line position. Then, for $\mathbf{H}$, $\mathbf{V}$ and $\mathbf{C}$, we can apply binary cross entropy loss:

$$\mathcal{L}_{bce}(\mathbf{X}, \mathbf{X}^*) = -\sum_i x_i^* \log(x_i) + (1 - x_i^*) \log(1 - x_i) \tag{3}$$

in which x_i denotes the i-th element of $\mathbf{X}$.

The overall loss for layout estimation is defined by the sum of the loss of the three type of lines:

$$\mathcal{L} = \mathcal{L}_{bce}(\mathbf{H}, \mathbf{H}^*) + \mathcal{L}_{bce}(\mathbf{V}, \mathbf{V}^*) + \mathcal{L}_{bce}(\mathbf{C}, \mathbf{C}^*) \tag{4}$$

where the notation $*$ indicates ground truth data.

3.5   Post-processing

Since the output of our network is line existence confidence on the cubemap, post-processing is necessary to generate the final 3D layout result. Our post-processing procedure consists of two stages: initialization and optimization (more details in supplementary materials).

Parametric Room Layout Representation. In 3D space, take the camera as the coordinate origin and z-axis points to the ceiling, and regard the distance from the camera to the floor can be set as a known value similar to Zou et al. [46]. The layout of a Manhattan room with n vertical walls can be represents with $n + 1$ parameters: n for the distance on coordinate axis from the camera to each wall, the other for the height of the room.

Layout Parameters Initialization. Given the lines predictions in the cube-map, we can calculate the elevation angle of each wall-ceiling and wall-floor intersection line, and since the distance from the camera to the floor is known, the distance to each of the wall-floor lines can be directly estimated, which is used as initial values for the first n parameters of the layout representation. Though the distance from the camera to the ceiling is unknown, by assuming and optimizing the ratio of the distance from the camera to the ceiling and the floor so that the ceiling 2D frame has the highest 2DIoU with the floor 2D frame, we can get the initial estimated value of the room height.

Gradient Descent Based Optimization. Given the initialized layout parameters, we convert the them into corner point coordinates in the panoramic image and then line positions in the cubemap. For each line position in the cubemap tiles, we can get line existence confidence from the output H, V and C of the network. We add the confidence of all the line position together to get a score representing the overall confidence of the layout parameters, define the loss as the negative of the score and optimize the loss using SGD, with learning rate 0.01, no momentum and 100 optimization steps. With the optimization process, the line predictions on different tiles of the cubemap can be effectively integrated together to get a better layout estimation from the global aspect.

4 Experiments

4.1 Experimental Setup

Dataset. We use three different datasets. The PanoContext dataset [43] and the Stanford 2D-3D dataset [1,46] are used for cuboid layout estimation, while the Matterport 3D dataset [6] is used for non-cuboid Manhattan layout estimation. For fair comparison, we follow the train/validation/test split adopted by other works [33,46,47].

Evaluation Metric. We employ standard evaluation metrics for 3D room layout estimation. For both cuboid and non-cuboid estimation, we use **(i) 3D IoU**, which is defined as the volumetric intersection over the union between the prediction and the ground truth. In addition, for cuboid layout estimation, we also use **(ii) Corner Error** (CE), which measures the average Euclidean distance between predicted corners and ground truth, and **(iii) Pixel Error** (PE), which is the pixel-wise accuracy. For non-cuboid estimation, we also use **(iv) 2D IoU**, which project 3D IoU to the 2D plane, and **(v)** δ_i, which is defined as the percentage of pixels where the ratio between the prediction and the ground truth is no more than 1.25. For 3D IoU, 2DIoU and δ_i, the larger is better. For Corner Error and Pixel Error, the smaller is better.

Baselines. We compare to previous works for both cuboid and non-cuboid room layout estimation and evaluate quantitative performance. The methods we compare to include PanoContext [43], LayoutNet [46], CFL [11], DuLa-Net [39], HorizonNet [33], AtlantaNet [26], HoHoNet [34] and LED^2-Net [35]. In addition, we also compared the improved version of the algorithm proposed in the recent work of Zou *et al.* [47], *i.e.*, Layoutnet-v2, DuLa-Net-v2 and HorizonNet$^+$. We compare with all these methods based on published comparable results. More specifically, DuLa-Net [39] and CFL [11] take 256×512 resolution images as input while others using the input resolution of 512 × 1024. Additionally, PanoContext [43] and LayoutNet [46] need additional input (*i.e.* Line Segment or Orientation Map) besides a single RGB panorama image.

Implementation Details. The network is implemented in PyTorch and optimized using Adam [21]. The network is trained for 75 epochs with batch size 8

Table 1. Quantitative results of cuboid room layout estimation evaluated on the testset of the PanoContext dataset [43] and the Stanford 2D-3D dataset [1,46]. CE means Corner Error, and PE means Pixel Error.

Method	PanoContext			Stanford 2D-3D		
	3DIoU	CE	PE	3DIoU	CE	PE
PanoContext [43]	67.23	1.60	4.55	-	-	-
LayoutNet [46]	74.48	1.06	3.34	76.33	1.04	2.70
DuLa-Net [39]	77.42	-	-	79.36	-	-
CFL [11]	78.79	0.79	2.49	-	-	-
HorizonNet [33]	82.17	0.76	2.20	79.79	0.71	2.39
AtlantaNet [11]	-	-	-	82.43	0.70	2.25
LED^2-Net [35]	82.75	-	-	83.77	-	-
DMH-Net (ours)	**85.48**	**0.73**	**1.96**	**84.93**	**0.67**	**1.93**

and learning rate 1e−4. The model is trained on a single NVIDIA A100 GPU. For both cuboid and non-cuboid estimation, we use the Pano Stretch [33], left-right flipping for data augmentation. In addition, for non-cuboid estimation, we also apply vertical flip and random 90° rotation augmentation on the cubemap tiles generated by the E2P transform.

4.2   Cuboid Room Results

Table 1 show the quantitative comparison on the PanoContext dataset [43] and Stanford 2D-3D dataset [1,46], respectively. Blank fields in the tables indicate metrics which is not reported in the corresponding publications. As can be seen, our approach achieves the state-of-the-art results under these two datasets in cuboid 3D room layout estimation. The qualitative results of the PanoContext and the Stanford 2D-3D dataset are shown in Fig. 4.

4.3   Non-cuboid Room Results

Table 2 shows the quantitative comparison on the Matterport 3D dataset [6]. Blank fields in the table indicate metrics which is not reported in the corresponding publications. As can be seen, our approach achieves competitive overall performance with the state-of-the-art methods in non-cuboid 3D layout estimation. The qualitative results of the Matterport 3D dataset are shown in Fig. 4. We present results from different methods aligned with ground truth. Please see supplemental materials for more qualitative results.

4.4   Ablation Study

In this section, we perform an ablation study to demonstrate the effect of the major designs in our method.

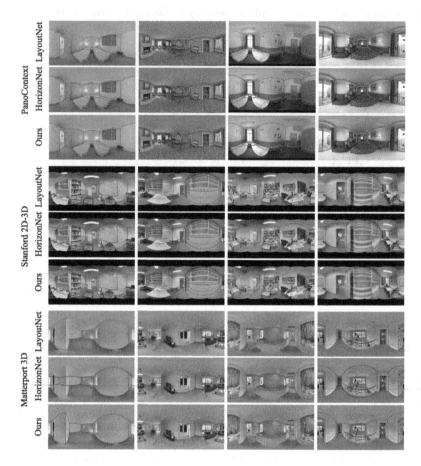

Fig. 4. Qualitative results of both cuboid and non-cuboid layout estimation. The results are selected from four folds that comprise results with the best 0–25%, 25–50%, 50–75% and 75–100% 3DIoU (displayed from left to right). The green lines are ground truth layout while the pink, blue and orange lines are estimated by LayoutNet [46], HorizonNet [33] and our DMH-Net. (Color figure online)

Table 2. Quantitative results of non-cuboid room layout estimation evaluated on the Matterport 3D dataset [6].

Metrics	3DIoU					2DIoU					δ_i				
# of corners	4	6	8	10+	Overall	4	6	8	10+	Overall	4	6	8	10+	Overall
LayoutNet-v2 [47]	81.35	72.33	67.45	63	75.82	84.61	75.02	69.79	65.14	78.73	0.897	0.827	0.877	0.8	0.871
DuLa-Net-v2 [39,47]	77.02	78.79	71.03	63.27	75.05	81.12	82.69	74	66.12	78.82	0.818	0.859	0.823	0.741	0.818
HorizonNet+ [33,47]	81.88	82.26	71.78	68.32	79.11	84.67	84.82	73.91	70.58	81.71	0.945	0.938	0.903	0.861	0.929
AtlantaNet [26]	82.64	80.1	71.79	**73.89**	**81.59**	85.12	82.00	74.15	**76.93**	**84.00**	**0.950**	0.815	**0.911**	**0.915**	**0.945**
HoHoNet [34]	82.64	82.16	73.65	69.26	79.88	85.26	84.81	75.59	70.98	82.32	-	-	-	-	-
LED²-Net [35]	84.22	**83.22**	**76.89**	70.09	81.52	86.91	**85.53**	**78.72**	71.79	83.91	-	-	-	-	-
DMH-Net (ours)	**84.39**	80.22	66.15	64.46	78.97	**86.94**	82.31	67.99	66.2	81.25	0.949	**0.951**	0.838	0.864	0.925

Choice of Manhattan Line Types. We first study the necessity of the three types of lines in the DMHT. We conduct experiments with each of the types of line detection omitted, respectively. The results are summarized in Table 3. All of the types of line detection are important as disabling any of each results in an obvious performance loss. Among them, horizontal lines detection is more important, as horizontal lines make up the ceiling and floor wireframe which is essential to the initialization in the post-processing procedure.

Cubemap and DMHT. We then verify whether the cubemap representation and DMHT helps to recover 3D room layout. We analyze the effectiveness of our full pipeline (*i.e.* III) with another two ablation variants in Table 4. The variation (I) encodes feature from cubemap as ours but replaces decoder from DMHT to HorizonNet-like LSTM [33]. The significant performance drop without DMHT in (I) indicates the necessity of DMHT in our full pipeline. We also compare the deep Hough feature against classical Hough Transforms line detectors, please refer to supplementary materials for more details. The variation (II) directly extracts features from the panorama image then apply E2P transform to feed features into DMHT. It shows the effectiveness of cubemap representation since the distortion in equirectangular features can degrade performance. These ablations between (I, II) and (III) provide evidence that the performance gain owe to the use of cubemap and DMHT.

Table 3. Ablation study on the impact of different types of line detection on the PanoContext dataset, in which **H** means horizontal line detection, **V** means vertical line detection and **C** means detection of lines passing the center.

Line detection			3DIoU	CE	PE
H	V	C			
×	✓	✓	43.58	3.69	11.91
✓	×	✓	83.28	0.89	2.44
✓	✓	×	84.15	0.79	**1.96**
✓	✓	✓	**85.48**	**0.73**	**1.96**

4.5 Model Analysis

Encoder Network. We first investigate how our cubemap representation and DMHT are model agnostic. We choose four different encoder backbones: ResNet-18,34,50 [15] and DRN-38 [41]. We change backbone for both our network and HorizonNet [33] baseline, and then train and evaluate the networks on the PanoContext dataset. As shown in Table 5, our method is not only effective for the DRN-38 architecture we use, but also consistently outperforms the competitors quantitatively. Ours with Res-50 is even slightly higher than DRN, may owing to deeper layers.

Table 4. Ablation study on encoder feature representation and decoder choice. We analyze the effectiveness of our pipeline (*i.e.* III) with another two ablation variants in Table 4. Our pipeline (*i.e.* III) achieves best quantitative results.

Method	Encoder		Decoder		3D IoU
	Cuebmap	equirectangular	DMHT	RNN	
I	✓			✓	81.44
II		✓	✓		83.99
III (Ours)	✓		✓		**85.48**

Table 5. Model analysis of the model agnostic generalization capability of DMH-Net on the PanoContext dataset. (Bold and underline indicates better than other competitors and the best among all, respectively.) Our DMH-Net consistently outperforms the competitors quantitatively when changing different backbones.

Method	Ours			HorizonNet [33]		
Backbone	3DIoU	CE	PE	3DIoU	CE	PE
ResNet-18	**82.96**	**0.83**	**2.18**	79.02	1.52	2.77
ResNet-34	**83.84**	**0.87**	**2.13**	79.97	0.91	2.49
ResNet-50	**85.90**	<u>**0.64**</u>	<u>**1.75**</u>	82.17	0.76	2.20
DRN-38	**85.48**	**0.73**	**1.96**	81.52	0.79	2.19

Robustness to Occlusion. We then test how our method outperforms competitors w.r.t. occlusion scenarios. We provide qualitative comparison in Fig. 5(a). As illustrated, our method has more accurate prediction results in regions where walls are occluded by other clutter. We further set up a more challenging experiment. We manually add challenging noise patches, and then test the network without re-training. As shown in Fig. 5(b), the raw output of ours is more reasonable than competitors in noisy regions. Moreover, quantitative results on the PanoContext dataset also show that our method achieves a 3D IoU of 84.39 while HorizonNet is 78.74.

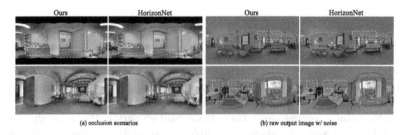

Fig. 5. Qualitative comparison in occlusion scenarios. (a) shows the result after post-processing, (b) shows the raw output of the network with challenging noise.

Limitations. We show two examples for analyzing our performance limitations on non-cuboid data. Figure 6(a) shows that our method may suffer from discriminating two very close walls. Figure 6(b) shows low line confidence due to "Vote Splitting". The probability peaks of short lines (in leftmost tile) are less significant than long ones. The short ones may not have enough feature bins to vote for. These cases may be improved by exploring adding interactions between tiles and a improved version of DMHT, which will be our future works.

(a) close parallel walls   (b) short walls

Fig. 6. Non-cuboid results explanations.

5 Conclusion

In this paper, we introduced a new method for estimating 3D room layouts from a cubemap of panorama image by leveraging DMHT. The proposed method detects horizontal lines, vertical lines and lines passing the image center on the cubemap and applies post-processing steps to combine the line prediction results to get the room layout. The learnable Deep Hough transform enables the network to capture the long-range geometric pattern and precisely detect the lines in the wireframe of the room layout. Both quantitative and qualitative results demonstrate that our method achieves better results for cuboid room estimation and comparable results for non-cuboid room estimation with recent state-of-the-art methods in prediction accuracy and performance. For future work, combining with fully differentiable post-processing or Atlanta world assumption are some of the practical directions to explore.

Acknowledgements. This work was supported by National Natural Science Funds of China (No. 62088102, 62021002) and ByteDance Research Collaboration Project.

References

1. Armeni, I., Sax, S., Zamir, A.R., Savarese, S.: Joint 2D–3D-semantic data for indoor scene understanding. arXiv preprint arXiv:1702.01105 (2017)
2. Ballard, D.H.: Generalizing the Hough transform to detect arbitrary shapes. Pattern Recogn. **13**(2), 111–122 (1981)
3. Beltrametti, M.C., Campi, C., Massone, A.M., Torrente, M.: Geometry of the Hough transforms with applications to synthetic data. Math. Comput. Sci. 1–23 (2020)

4. Bertamini, M., Helmy, M., Bates, D.: The visual system prioritizes locations near corners of surfaces (not just locations near a corner). Attention Percept. Psychophys. **75**(8), 1748–1760 (2013). https://doi.org/10.3758/s13414-013-0514-1

5. Canny, J.: A computational approach to edge detection. IEEE Trans. Pattern Anal. Mach. Intell. **8**(6), 679–698 (1986)

6. Chang, A., et al.: Matterport3D: learning from RGB-D data in indoor environments. In: Proceedings of the International Conference on 3D Vision (3DV), pp. 667–676 (2017)

7. Coughlan, J.M., Yuille, A.L.: Manhattan world: compass direction from a single image by Bayesian inference. In: Proceedings of the IEEE International Conference on Computer Vision (ICCV), pp. 941–947 (1999)

8. Dasgupta, S., Fang, K., Chen, K., Savarese, S.: Delay: robust spatial layout estimation for cluttered indoor scenes. In: Proceedings of IEEE Conference on Computer Vision and Pattern Recognition (CVPR), pp. 616–624 (2016)

9. Del Pero, L., Bowdish, J., Kermgard, B., Hartley, E., Barnard, K.: Understanding Bayesian rooms using composite 3D object models. In: Proceedings IEEE Conference on Computer Vision and Pattern Recognition (CVPR), pp. 153–160 (2013)

10. Delage, E., Lee, H., Ng, A.Y.: A dynamic Bayesian network model for autonomous 3D reconstruction from a single indoor image. In: Proceedings of IEEE Conference on Computer Vision and Pattern Recognition (CVPR), pp. 2418–2428 (2006)

11. Fernandez-Labrador, C., Facil, J.M., Perez-Yus, A., Demonceaux, C., Civera, J., Guerrero, J.J.: Corners for layout: End-to-end layout recovery from 360 images. IEEE Robot. Autom. Lett. (RA-L) **5**(2), 1255–1262 (2020)

12. Greene, N.: Environment mapping and other applications of world projections. IEEE Comput. Graph. Appl. **6**(11), 21–29 (1986)

13. Gupta, A., Hebert, M., Kanade, T., Blei, D.: Estimating spatial layout of rooms using volumetric reasoning about objects and surfaces. In: Advances in Neural Information Processing Systems (NeurIPS), pp. 1288–1296 (2010)

14. Han, Q., Zhao, K., Xu, J., Cheng, M.M.: Deep Hough transform for semantic line detection. IEEE Trans. Pattern Anal. Mach. Intell. 249–265 (2021)

15. He, K., Zhang, X., Ren, S., Sun, J.: Deep residual learning for image recognition. In: Proceedings of IEEE Conference on Computer Vision and Pattern Recognition (CVPR), pp. 770–778 (2016)

16. Hedau, V., Hoiem, D., Forsyth, D.: Recovering the spatial layout of cluttered rooms. In: Proceedings of the IEEE International Conference on Computer Vision (ICCV), pp. 1849–1856 (2009)

17. Horry, Y., Anjyo, K.I., Arai, K.: Tour into the picture: using a spidery mesh interface to make animation from a single image. In: Proceedings of the 24th Annual Conference on Computer Graphics and Interactive Techniques, pp. 225–232 (1997)

18. Hough, P.V.: Method and means for recognizing complex patterns, US Patent 3,069,654, 18 December 1962

19. Izadinia, H., Shan, Q., Seitz, S.M.: IM2CAD. In: Proceedings of IEEE Conference on Computer Vision and Pattern Recognition (CVPR), pp. 2422–2431 (2017)

20. Jiang, Z., Xiang, Z., Xu, J., Zhao, M.: LGT-Net: indoor panoramic room layout estimation with geometry-aware transformer network. In: Proceedings of IEEE Conference on Computer Vision and Pattern Recognition (CVPR), pp. 1654–1663 (2022)

21. Kingma, D.P., Ba, J.: Adam: a method for stochastic optimization. In: Proceedings of the International Conference on Learning Representations (ICLR) (2015)

22. Lee, C.Y., Badrinarayanan, V., Malisiewicz, T., Rabinovich, A.: RoomNet: end-to-end room layout estimation. In: Proceedings of the IEEE International Conference on Computer Vision (ICCV), pp. 4875–4884 (2017)
23. Lee, D.C., Hebert, M., Kanade, T.: Geometric reasoning for single image structure recovery. In: Proceedings of IEEE Conference on Computer Vision and Pattern Recognition (CVPR), pp. 2136–2143 (2009)
24. Lin, Y., Pintea, S.L., van Gemert, J.C.: Deep Hough-transform line priors. In: Vedaldi, A., Bischof, H., Brox, T., Frahm, J.-M. (eds.) ECCV 2020. LNCS, vol. 12367, pp. 323–340. Springer, Cham (2020). https://doi.org/10.1007/978-3-030-58542-6_20
25. Mallya, A., Lazebnik, S.: Learning informative edge maps for indoor scene layout prediction. In: Proceedings of the IEEE International Conf. on Computer Vision (ICCV), pp. 936–944 (2015)
26. Pintore, G., Agus, M., Gobbetti, E.: AtlantaNet: inferring the 3D indoor layout from a single 360° image beyond the Manhattan world assumption. In: Vedaldi, A., Bischof, H., Brox, T., Frahm, J.-M. (eds.) ECCV 2020. LNCS, vol. 12353, pp. 432–448. Springer, Cham (2020). https://doi.org/10.1007/978-3-030-58598-3_26
27. Pintore, G., Garro, V., Ganovelli, F., Gobbetti, E., Agus, M.: Omnidirectional image capture on mobile devices for fast automatic generation of 2.5 d indoor maps. In: Proceedings of IEEE Winter Conference on Applications of Computer Vision (WACV), pp. 1–9 (2016)
28. Qi, C.R., Litany, O., He, K., Guibas, L.J.: Deep Hough voting for 3D object detection in point clouds. In: Proceedings of the IEEE International Confernece on Computer Vision (ICCV), pp. 9276–9285 (2019)
29. Ren, Y., Li, S., Chen, C., Kuo, C.C.J.: A coarse-to-fine indoor layout estimation (cfile) method. In: Asia Conference on Computer Vision (ACCV), pp. 36–51 (2016)
30. Schwing, A.G., Hazan, T., Pollefeys, M., Urtasun, R.: Efficient structured prediction for 3d indoor scene understanding. In: Proceedings of IEEE Conference on Computer Vision and Pattern Recognition (CVPR), pp. 2815–2822 (2012)
31. Schwing, A.G., Urtasun, R.: Efficient exact inference for 3D indoor scene understanding. In: Fitzgibbon, A., Lazebnik, S., Perona, P., Sato, Y., Schmid, C. (eds.) ECCV 2012. LNCS, vol. 7577, pp. 299–313. Springer, Heidelberg (2012). https://doi.org/10.1007/978-3-642-33783-3_22
32. Sobel, I., Feldman, G.: A 3x3 isotropic gradient operator for image processing. A talk at the Stanford Artificial Project, pp. 271–272 (1968)
33. Sun, C., Hsiao, C.W., Sun, M., Chen, H.T.: Horizonnet: learning room layout with 1D representation and pano stretch data augmentation. In: Proceedings of IEEE Conference on Computer Vision and Pattern Recognition (CVPR), pp. 1047–1056 (2019)
34. Sun, C., Sun, M., Chen, H.T.: Hohonet: 360 indoor holistic understanding with latent horizontal features. In: Proceedings of IEEE Conference on Computer Vision and Pattern Recognition (CVPR), pp. 2573–2582 (2021)
35. Wang, F.E., Yeh, Y.H., Sun, M., Chiu, W.C., Tsai, Y.H.: LED2-Net: monocular 360deg layout estimation via differentiable depth rendering. In: Proceedings of IEEE Conference on Computer Vision and Pattern Recognition (CVPR), pp. 12956–12965 (2021)
36. Xu, J., Stenger, B., Kerola, T., Tung, T.: Pano2CAD: room layout from a single panorama image. In: Proceedings of IEEE Winter Conference on Applications of Computer Vision (WACV), pp. 354–362 (2017)

37. Yang, C., Zheng, J., Dai, X., Tang, R., Ma, Y., Yuan, X.: Learning to reconstruct 3D non-cuboid room layout from a single RGB image. In: Proceedings of
the IEEE/CVF Winter Conference on Applications of Computer Vision, pp. 2534–
2543 (2022)
38. Yang, H., Zhang, H.: Efficient 3D room shape recovery from a single panorama.
In: Proceedings of IEEE Conference on Computer Vision and Pattern Recognition
(CVPR), pp. 5422–5430 (2016)
39. Yang, S.T., Wang, F.E., Peng, C.H., Wonka, P., Sun, M., Chu, H.K.: Dula-net: a
dual-projection network for estimating room layouts from a single RGB panorama.
In: Proceedings of IEEE Conference on Computer Vision and Pattern Recognition
(CVPR), pp. 3363–3372 (2019)
40. Yang, Y., Jin, S., Liu, R., Bing Kang, S., Yu, J.: Automatic 3d indoor scene modeling from single panorama. In: Proceedings of IEEE Conference on Computer
Vision and Pattern Recognition (CVPR), pp. 3926–3934 (2018)
41. Yu, F., Koltun, V., Funkhouser, T.: Dilated residual networks. In: Proceedings
of IEEE Conference on Computer Vision and Pattern Recognition (CVPR), pp.
636–644 (2017)
42. Zeng, W., Karaoglu, S., Gevers, T.: Joint 3D layout and depth prediction from
a single indoor panorama image. In: Vedaldi, A., Bischof, H., Brox, T., Frahm,
J.-M. (eds.) ECCV 2020. LNCS, vol. 12361, pp. 666–682. Springer, Cham (2020).
https://doi.org/10.1007/978-3-030-58517-4_39
43. Zhang, Y., Song, S., Tan, P., Xiao, J.: PanoContext: a whole-room 3D context
model for panoramic scene understanding. In: Fleet, D., Pajdla, T., Schiele, B.,
Tuytelaars, T. (eds.) ECCV 2014. LNCS, vol. 8694, pp. 668–686. Springer, Cham
(2014). https://doi.org/10.1007/978-3-319-10599-4_43
44. Zhao, H., Lu, M., Yao, A., Guo, Y., Chen, Y., Zhang, L.: Physics inspired optimization on semantic transfer features: an alternative method for room layout
estimation. In: Proceedings of IEEE Conference on Computer Vision and Pattern
Recognition (CVPR), pp. 870–878 (2017)
45. Zhao, Y., Zhu, S.C.: Scene parsing by integrating function, geometry and appearance models. In: Proceedings of IEEE Conference on Computer Vision and Pattern
Recognition (CVPR), pp. 3119–3126 (2013)
46. Zou, C., Colburn, A., Shan, Q., Hoiem, D.: Layoutnet: reconstructing the 3D room
layout from a single RGB image. In: Proceedings of IEEE Conference on Computer
Vision and Pattern Recognition (CVPR), pp. 2051–2059 (2018)
47. Zou, C., et al.: Manhattan room layout reconstruction from a single 360° image:
a comparative study of state-of-the-art methods. Int. J. Comput. Vision (IJCV)
129(5), 1410–1431 (2021)

RBP-Pose: Residual Bounding Box Projection for Category-Level Pose Estimation

Ruida Zhang[1]([✉]), Yan Di[2], Zhiqiang Lou[1], Fabian Manhardt[3], Federico Tombari[2,3], and Xiangyang Ji[1]

[1] Tsinghua University, Beijing, China
{zhangrd21,lzq20}@mails.tsinghua.edu.cn, xyji@tsinghua.edu.cn
[2] Technical University of Munich, Munich, Germany
tombari@in.tum.de
[3] Google, Zurich, Switzerland
fabianmanhardt@google.com

Abstract. Category-level object pose estimation aims to predict the 6D pose as well as the 3D metric size of arbitrary objects from a known set of categories. Recent methods harness shape prior adaptation to map the observed point cloud into the canonical space and apply Umeyama algorithm to recover the pose and size. However, their shape prior integration strategy boosts pose estimation indirectly, which leads to insufficient pose-sensitive feature extraction and slow inference speed. To tackle this problem, in this paper, we propose a novel geometry-guided **R**esidual Object **B**ounding Box **P**rojection network **RBP-Pose** that jointly predicts object pose and residual vectors describing the displacements from the shape-prior-indicated object surface projections on the bounding box towards the real surface projections. Such definition of residual vectors is inherently zero-mean and relatively small, and explicitly encapsulates spatial cues of the 3D object for robust and accurate pose regression. We enforce geometry-aware consistency terms to align the predicted pose and residual vectors to further boost performance. Finally, to avoid overfitting and enhance the generalization ability of RBP-Pose, we propose an online non-linear shape augmentation scheme to promote shape diversity during training. Extensive experiments on NOCS datasets demonstrate that RBP-Pose surpasses all existing methods by a large margin, whilst achieving a real-time inference speed.

Keywords: Category-level pose estimation · 3D Object Detection · Scene understanding

R. Zhang and Y. Di—Equal contributions.
Codes are released at https://github.com/lolrudy/RBP_Pose.

Supplementary Information The online version contains supplementary material available at https://doi.org/10.1007/978-3-031-19769-7_38.

1 Introduction

Category-level object pose estimation describes the task of estimating the full 9 degrees-of-freedom (DoF) object pose (consisting of the 3D rotation, 3D translation and 3D metric size) for objects from a given set of categories. The problem has gained wide interest in research due to its essential role in many applications, such as augmented reality [34], robotic manipulation [5] and scene understanding [29,50]. In comparison to conventional instance-level pose estimation [13,14], which assumes the availability of a 3D CAD model for each object of interest, the category-level task puts forward a higher requirement for adaptability to various shapes and textures within each category.

Noteworthy, category-level pose estimation has recently experienced a large leap forward in recent past, thanks to novel deep learning architecture that can directly operate on point clouds [1,3,43,44]. Thereby, most of these works try to establish 3D-3D correspondences between the input point cloud and either a predefined normalized object space [43] or a deformed shape prior to better address intra-class shape variability [2,38] (Fig. 1). Eventually, the 9DoF pose is commonly recovered using the Umeyama algorithm [39]. Nonetheless, despite achieving great performance, these methods typically still suffer from two shortcomings. First, their shape prior integration only boosts pose estimation indirectly, which leads to insufficient pose-sensitive feature extraction and slow inference speed. Second, due to the relatively small amount of available real-world data [43], these works tend to overfit as they are directly trained on these limited datasets.

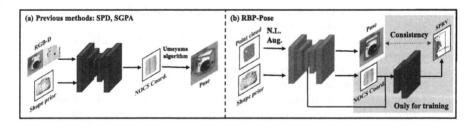

Fig. 1. Comparison of RBP-Pose and previous shape prior adaptation methods. Previous methods [2,38] predict NOCS coordinates (NOCS Coord.) and recover the pose using the Umeyama algorithm [39]. In comparison, we use NOCS coordinates within our point-wise bounding box projection and predict Shape Prior Guided Residual Vectors (SPRV), which encapsulates the pose explicitly. Moreover, we propose Non-Linear Shape Augmentation (**N.L. Aug.**) to increase shape diversity during training.

As for the lack of real labeled data, we further propose an online non-linear shape augmentation scheme for training to avoid overfitting and enhance the generalization ability of RBP-Pose. In FS-Net [3], the authors propose to stretch or compress the object bounding box to generate new instances. However, the proportion between different parts of the object basically remains unchanged, as shown in Fig. 4. Therefore, we propose a category-specific non-linear shape augmentation technique. In particular, we deform the object shape by adjusting its scale via a truncated parabolic function along the direction of a selected axis. To this end, we either choose the symmetry axis for symmetric objects or select the axis corresponding to the facing direction to avoid unrealistic distortions for non-symmetric objects. In this way we are able to increase the dataset size while preserving the representative shape characteristic of each category.

Interestingly, to tackle the former limitation, the authors of GPV-Pose [7] have proposed to leverage **D**isplacement **V**ectors from the observed points to the corresponding **P**rojections on the **B**ounding box (DVPB), in an effort to explicitly encapsulate the spatial cues of the 3D object and, thus, improve direct pose regression. While this performs overall well, the representation still exhibits weaknesses. In particular, DVPB is not necessarily a small vector with zero-mean. In fact, the respective values can become very large (as for large objects like laptops), which can make it very difficult for standard networks to predict them accurately. Based on these grounds, in this paper we propose to overcome this limitation by means of integrating shape priors into DVPB. We essentially describe the displacement field from the shape-prior-indicated projections towards the real projections onto the object bounding box. We dub the residual vectors in this displacement field as SPRV for **S**hape **P**rior Guided **R**esidual **V**ectors. SPRV is inherently zero-centered and relatively small, allowing robust estimation with a deep neural network. In practice, we adopt a fully convolutional decoder to directly regress SPRV and then establish geometry-aware consistency with the predicted pose to enhance feature extraction (Fig. 1). We experimentally show that our novel geometry-guided **R**esidual **B**ounding Box **P**rojection network RBP-Pose provides state-of-the-art results and clearly outperforms the DVPB representation. Overall, our main contributions are summarized as follows,

1. We propose a **R**esidual **B**ounding Box **P**rojection network (RBP-Pose) that jointly predicts 9DoF pose and shape prior guided residual vectors. We demonstrate that these nearly zero-mean residual vectors can be effectively predicted from our network and well encapsulate the spatial cues of the pose whilst enabling geometry-guided consistency terms.
2. To enhance the robustness of our method, we additionally propose a non-linear shape augmentation scheme to improve shape diversity during training whilst effectively preserving the commonality of geometric characteristics within categories.
3. RBP-Pose runs at inference speed 25 Hz and achieves state-of-the-art performance on both synthetic and real-world datasets.

2   Related Works

Instance-Level 6D Pose Estimation. Instance-level pose estimation tries to estimate the 6DoF object pose, composed of the 3D rotation and 3D translation, for a known set of objects with associated 3D CAD models. The majority of monocular methods falls into three groups. The first group of methods [16,18,20,26,27,47] regresses the pose directly, whereas the second group instead establishes 2D-3D correspondences via keypoint detection or dense pixel-wise prediction of 3D coordinates [12,21,30,31,33,49]. The pose can be then obtained by adopting the PnP algorithm. Noteworthy, a few methods [6,15,42] adopt a neural network to learn the optimization step instead of relying on PnP. The last group of methods [35,36] attempt to learn a pose-sensitive latent embedding for subsequent pose retrieval. As for RGB-D based methods, most works [10,11,19,41] again regress the pose directly, while a few methods [17,46] resort to latent embedding similar to [35,36]. In spite of great advance in recent years, the practical use of instance-level methods is limited as they can typically only deal with a handful of objects and additionally require CAD models.

Category-Level Pose Estimation. In the category-level setting, the goal is to predict the 9DoF pose for previously seen or unseen objects from a known set of categories [28,43]. The setting is fairly more challenging due to the large intra-class variations of shape and texture within categories. To tackle this issue, Wang *et al.* [43] derive the *Normalized Object Coordinate Space* (NOCS) as a unified representation. They map the observed point cloud into NOCS and then apply the Umeyama algorithm [39] for pose recovery. CASS [1] introduces a learned canonical shape space instead. FS-Net [3] proposes a decoupled representation for rotation and directly regresses the pose. DualPoseNet [23] adopts two networks for explicit and implicit pose prediction and enforces consistency between them for pose refinement. While 6-PACK [40] tracks the object's pose by means of semantic keypoints, CAPTRA [45] instead combines coordinate prediction with direct regression. GPV-Pose [7] harnesses geometric insights into bounding box projection to enhance the learning of category-level pose-sensitive features. To explicitly address intra-class shape variation, a certain line of works make use of shape priors [2,8,22,38]. Thereby, SPD [38] extracts the prior point cloud for each category as the mean of all shapes adopting a PointNet [32] autoencoder. SPD further deforms the shape prior to fit the observed instance and assigns the observed point cloud to the reconstructed shape model. SGPA [2] dynamically adapts the shape prior to the observed instance in accordance with its structural similarity. DO-Net [22] also utilizes shape prior, yet, additionally harnesses the geometric object properties to enhance performance. ACR-Pose [8] adopts a shape prior guided reconstruction network and a discriminator network to learn high-quality canonical representations. Noteworthy, as shape prior integration only improves pose estimation indirectly, all these methods commonly suffer from insufficient pose-sensitive feature extraction and slow inference speed.

3   Methodology

In this paper, we aim at tackling the problem of category-level 9DoF pose estimation. In particular, given an RGB-D image with objects from a set of known categories, our objective is to detect all present object instances in the scene and recover their 9DoF object poses, including the 3DoF rotation as rotation matrix $R \in \mathbb{R}^{3\times3}$, the 3DoF translation $t \in \mathbb{R}^3$ and 3DoF metric size $s \in \mathbb{R}^3$.

3.1 Overview

As illustrated in Fig. 2, RBP-Pose consists of 5 modules, responsible for i) input preprocessing, ii) feature extraction from the input and prior point cloud, iii) 9DoF pose regression, iv) adaptation of the shape prior given the extracted features, and, finally, v) Shape Prior Guided Residual Vectors (SPRV) prediction.

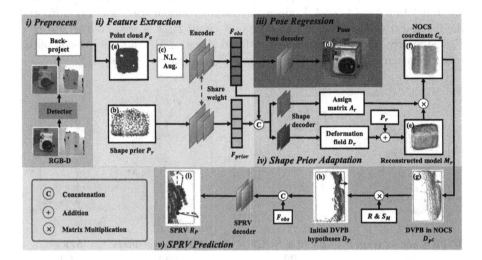

Fig. 2. An overview of RBP-Pose. RBP-Pose takes RGB-D image and shape prior as inputs. We perform a non-linear shape augmentation (c) after extracting point cloud of the object of interest (a). It deforms shape prior (e) and predicts NOCS coordinates (f) to retrieve DVPB in NOCS (g). Integrating the predicted rotation R, the pre-computed category mean size S_M, we compute the initial DVPB hypotheses (h) as the input of SPRV decoder. Finally, RBP-Pose predicts pose (d) and SPRV (i) and enforces consistency between them. During inference, only the Preprocessing, Feature Extraction and Pose Regression modules are needed.

Preprocessing. Given an RGB-D image, we first leverage an off-the-shelf object detector (*e.g.* Mask-RCNN [9]) to segment objects of interest and then back-project their corresponding depth values to generate the associated object point clouds. We then uniformly sample $N = 1024$ points from each detected object and feed it as the input P_o to the following modules, as shown in Fig. 2 (a).

Feature Extraction. Since 3DGC [24] is insensitive to shift and scale of the given point cloud, we adopt it as our feature extractor to respectively obtain pose-sensitive features F_{obs} and F_{prior} from P_o and a pre-computed mean shape prior P_r (with $M = 1024$ points) as in SPD [38]. We introduce a non-linear shape augmentation scheme to increase the diversity of shapes and promote robustness, which will be discussed in detail in Sect. 3.5. Finally, F_{obs} is fed to the Pose Regression module for direct pose estimation and to the Shape Prior Adaptation module after concatenation with F_{prior}.

9DoF Object Pose Regression. To represent the 3DoF rotation, we follow GPV-Pose [7] and decompose the rotation matrix R into 3 columns $r_x, r_y, r_z \in \mathbb{R}^3$, each representing a plane normal of the object bounding box. We predict the first two columns r_x, r_y along with their uncertainties u_x, u_y, and calculate the calibrated plane normals r'_x, r'_y via uncertainty-aware averaging [7]. Eventually, the predicted rotation matrix is recovered as $R' = [r'_x, r'_y, r'_x \times r'_y]$. For translation and size, we follow FS-Net, adopting their residual representation [3]. Specifically, for translation $t = \{t_x, t_y, t_z\}$, given the output residual translation $t_r \in \mathbb{R}^3$ and the mean P_M of the observed visible point cloud P_o, t is recovered as $t = t_r + P_M$. Similarly, given the estimated residual size $s_r \in \mathbb{R}^3$ and the pre-computed category mean size S_M, we have $s = s_r + S_M$, where $s = \{s_x, s_y, s_z\}$.

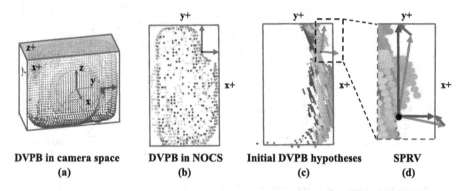

DVPB in camera space	DVPB in NOCS	Initial DVPB hypotheses	SPRV
(a)	(b)	(c)	(d)

Fig. 3. Illustration of DVPB and SPRV. We show DVPB and SPRV of a point to $x+, y+$ plane. The target is to predict DVPB in the camera space (purple vectors in (a)). For better demonstration, we project the point cloud to $z+$ plane (the blue plane in (a)) in (b)–(d). Using the predicted coordinates in (b), we recover DVPB in NOCS (blue vectors in (b)). We then transform it into the camera space and compute the initial DVPB hypotheses (brown vectors in (c)). RBP-Pose predicts the residual vector from hypotheses to ground truth DVPB (SPRV, red vectors in (d)). (Color figure online)

Shape Prior Adaptation. We first concatenate the feature maps F_{obs} from the observed point cloud P_o and F_{prior} from shape prior P_r in a channel-wise manner. Subsequently, we use two sub-decoders from SPD [38] to predict the row-normalized assignment matrix $A_r \in \mathbb{R}^{N \times M}$ and the deformation field $D_r \in \mathbb{R}^{M \times 3}$, respectively. Given the shape prior $\mathcal{P}_r$, D_r deforms $\mathcal{P}_r$ to reconstruct the normalized object shape model M_r with $M_r = P_r + D_r$ (Fig. 2 (e)). Further, A_r associates each point in P_o with M_r. Thereby, the NOCS coordinate $C_o \in \mathbb{R}^{N \times 3}$ of the input point cloud P_o is computed as $C_o = A_r M_r$ (Fig. 2 (f)).

Shape Prior Guided Residual Vectors (SPRV) Prediction. The main contribution of our work resides in the use of Shape Prior Guided Residual Vectors (SPRV) to integrate shape priors into the direct pose regression network, enhancing the performance whilst keeping a fast inference speed. In the following section we will now introduce this module in detail.

3.2   Residual Bounding Box Projection

Preliminaries. In GPV-Pose [7], the authors propose a novel confidence-aware point-wise voting method to recover the bounding box. For each observed object point P, GPV-Pose thereby predicts its **D**isplacement **V**ector towards its **P**rojections on each of the 6 **B**ounding Box faces (DVPB), as shown in Fig. 3 (a). Exemplary, when considering the $x+$ plane, the DVPB of the observed point P onto the $x+$ face of the bounding box is defined as,

$$D_{P,x+} = (s_x/2 - \langle r_x, P \rangle + \langle r_x, t_x \rangle)r_x, \tag{1}$$

where $\langle *, * \rangle$ denotes the inner product and, as before, r_x denotes the first column of the rotation matrix R. Thus, each point P provides 6 DVPBs with respect to all 6 bounding box faces $\mathcal{B} = \{x\pm, y\pm, z\pm\}$. Notice that, since symmetries lead to ambiguity in bounding box faces around the corresponding symmetry axis, GPV-Pose only compute the DVPB on the ambiguity-free faces.

Although GPV-Pose reports great results when leveraging DVPB, it still suffers from two important shortcomings. First, DVPB is not necessarily a small vector with zero-mean. In fact, the respective values can become very large (as for large objects like laptops), which can make it very difficult for standard networks to predict them accurately. Second, DVPB is not capable of conducting automatic outlier filtering, hence, noisy point cloud observations may significantly deteriorate the predictions of DVPB. On that account, we propose to incorporate shape prior into DVPB in the form of Shape Prior Guided Residual Vectors (SPRV) to properly address the aforementioned shortcomings.

Shape Prior Guided Residual Vectors (SPRV). As illustrated in Fig. 2 (e), we predict the deformation field D_r that deforms the shape prior P_r to the shape of the observed instance with $M_r = P_r + D_r$. Thereby, during experimentation we made two observations. First, as P_r is outlier-free and D_r is regularized to be small, similar to SPD [38], we can safely assume that M_r contains no outliers, allowing us to accurately recover its bounding box in NOCS by selecting the outermost points along the x, y and z axis, respectively. Second, since A_r is the row-normalized assignment matrix, we know that M_r shares the same bounding box with $C_o = A_r M_r$, which is assumed to be inherently outlier-free and accurate. Based on the above two observations, we can utilize M_r and C_o to provide initial hypotheses for DVPB with respect to each point in P_o.

Specifically, for a point P^C in C_o, as it is in NOCS, its DVPB $D_{P^C,x+}$ (Fig. 3 (b)) can be represented as,

$$D_{P^C,x+} = (s_x^M/2 - P^C)n_x, \tag{2}$$

where s_x^M denotes the size of M_r along the x axis and $n_x = [1, 0, 0]^T$ is the normalized bounding box face normal. We then transform $D_{P^C,x+}$ from NOCS to the camera coordinate to obtain the initial DVPB hypotheses for the corresponding point P in P_o (Fig. 3 (c)) as,

$$D_{P,x+} = LD_{P^C,x+}r_x \tag{3}$$

where $L = \sqrt{s_x^2 + s_y^2 + s_z^2}$ is the diagonal length of the bounding box. Note that L and r_x are calculated from the category mean size S_M and the rotation prediction of our Pose Regression Module respectively.

Given the ground truth DVPB $D_{P,x+}^{gt}$ and initial DVPB hypotheses $D_{P,x+}$, the SPRV of P to the $x+$ bounding box face (Fig. 3 (d)) is calculated as,

$$R_{P,x+} = D_{P,x+}^{gt} - D_{P,x+}. \tag{4}$$

The calculation of SPRV with respect to the other bounding box faces in $\mathcal{B}$ follows the same principal. By this means, SPRV can be approximately modelled with zero-mean Laplacian distribution, which enables effective prediction with a simple network. In the SPRV Prediction module, we feed the estimated initial DVPB hypotheses together with the feature map F_{obs} into a fully convolutional decoder to directly regress SPRV. As this boils down to a multi-task prediction problem, we employ the Laplacian aleatoric uncertainty loss from [4] to weight the different contributions within SPRV according to

$$
\begin{aligned}
\mathcal{L}_{SPRV}^{data} &= \sum_{P \in P_o} \sum_{j \in \mathcal{B}} \frac{\sqrt{2}}{\sigma_{s_j}} |R_{P_j} - R_{P_j}^{gt}| + log(\sigma_{s_j}) \\
\mathcal{L}_{SPRV}^{reg} &= \sum_{P \in P_o} \sum_{j \in \mathcal{B}} \frac{\sqrt{2}}{\sigma'_{s_j}} |R_{P_j}| + log(\sigma'_{s_j}) \\
\mathcal{L}_{SPRV} &= \mathcal{L}_{SPRV}^{data} + \lambda_0 \mathcal{L}_{SPRV}^{reg}
\end{aligned} \tag{5}
$$

Thereby, $R_{P_j}^{gt}$ refers to the ground truth SPRV as calculated by the provided ground truth NOCS coordinates and respective pose annotations. Further, σ_{s_j}, σ'_{s_j} denote the standard variation of Laplacian distribution that are utilized to model the uncertainties. Note that the first term $\mathcal{L}_{SPRV}^{data}$ is fully supervised using the respective ground truth, while $\mathcal{L}_{SPRV}^{reg}$ is a regularization term that enforces the SPRV network to predict small displacements. In addition, λ_0 is a weighting parameter to balance the two terms. Note that we do not apply Gaussian-distribution-based losses. We follow GPV-Pose [7] to supervise other branches with $\mathcal{L}_1$ loss for stability. Thus we adopt Eq. 5 for convenient adjustment of the weight of each term.

3.3   SPRV for Pose Consistency

Since SPRV explicitly encapsulates pose-related cues, we utilize it to enforce geometric consistency between the SPRV prediction and the pose regression. To this end. we first employ the predicted pose to estimate DVPB D^{Pose} according to Eq. 1. We then recover D^{SPRV} via adding the predicted SPRV to the initial hypotheses. Finally, our consistency loss term is defined as follows,

$$\mathcal{L}_{con} = \sum_{P \in P_o} \sum_{j \in \mathcal{B}} |D_{P,b}^{Pose} - D_{P,b}^{SPRV}|, \tag{6}$$

where $| * |$ denotes the $\mathcal{L}_1$ distance.

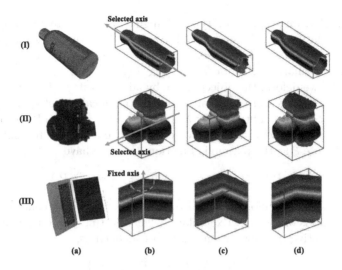

Fig. 4. Demonstration of non-linear data augmentation. (I), (II) and **(III)** show non-linear data augmentation for *bottle, camera* and *laptop*. For the instances **(a)**, we augment the object shape in our non-linear manner from **(b)** to **(c)**. FS-Net [3] adopts the linear bounding box deformation in data augmentation **(d)**, which can be regarded as a special case of our non-linear shape augmentation.

3.4 Overall Loss Function

The overall loss function is defined as follows,

$$\mathcal{L} = \lambda_1 \mathcal{L}_{pose} + \lambda_2 \mathcal{L}_{shape} + \lambda_3 \mathcal{L}_{SPRV} + \lambda_4 \mathcal{L}_{con} \tag{7}$$

For $\mathcal{L}_{pose}$, we utilize the loss terms from GPV-Pose [7] to supervise R, t, s with the ground truth. For $\mathcal{L}_{shape}$, we adopt the loss terms from SPD [38] to supervise the prediction of the deformation field D_r and the assignment matrix A_r. Further, $\mathcal{L}_{SPRV}$ and $\mathcal{L}_{con}$ are defined in Eq. 5 and Eq. 6. Finally, $\lambda_1, \lambda_2, \lambda_3, \lambda_4$ denote the utilized weights to balance the individual loss contributions, and are chosen empirically.

3.5 Non-linear Shape Augmentation

To tackle the intra-class shape variations and improve the robustness and generalizability of RBP-Pose, we propose a category-specific non-linear shape augmentation scheme (Fig. 4). In FS-Net [3], the authors augment the shape by stretching or compression of the object bounding box. Their augmentation is linear and unable to cover the large shape variations within a category, since the proportions between different parts of the object basically remain unchanged (Fig. 4 (d)). In contrast, we propose a novel non-linear shape augmentation method which is designed to generate diverse unseen instances, whilst preserving the representative shape features of each category (Fig. 4 (c)).

In particular, we propose two types of augmentation strategies for categories provided by the REAL275 dataset [43]: axis-based non-linear scaling transformation ($A1$) for *camera, bottle, can, bowl, mug* (Fig. 4 (I, II)) and plane-based rotation transformation ($A2$) for *laptop* (Fig. 4 (III)).

As for $A1$, we deform the object shape by adjusting its scale along the direction of a selected axis. For each point P in the canonical object space, its deformation scale $\mathcal{S}_{A1}(P)$ is obtained by $\mathcal{S}_{A1}(P) = \xi(P_*)$, where $\xi(P_*)$ is a random non-linear function and P_* is the projection of P on the selected axis. In this paper, we choose ξ as the parabolic function, thus, we have

$$\mathcal{S}_{A1}(P) = \xi(P_*) = \gamma_{min} + 4(\gamma_{max} - \gamma_{min})(P_*)^2, \tag{8}$$

where $\gamma_{max}, \gamma_{min}$ are uniformly sampled random variables that control the upper and lower bounds of $\mathcal{S}_{A1}(P)$. Exemplary, when selecting y as our augmentation axis, the respective transformation function is defined as,

$$\mathcal{T}_{A1}(P) = \{\gamma P_x, \mathcal{S}_{A1}(P)P_y, \gamma P_z\}, \tag{9}$$

where γ is the random variable that controls the scaling transformation along x and z axis. In practice, we select the symmetry axis (y-axis) for *bottle, can, bowl* and *mug* as the transformation axis. Moreover, for *camera*, we select the axis that passes through the camera lens (x-axis), to keep its roundish shape after augmentation. The corresponding transformation function is then defined as in Eq. 8 and Eq. 9, yet, $\mathcal{S}_{A1}(P)$ is only applied to P_x and γ is applied to P_y, P_z.

As for $A2$, since *laptop* is an articulated object consisting of two movable planes, we conduct shape augmentation by modifying the angle between the upper and lower plane (Fig. 4 (III)). Thereby, we rotate the upper plane by a certain angle along the fixed axis, while the lower plane remains static. Please refer to the Supplementary Material for details of $A2$ transformation.

4   Experiments

Datasets. We employ the common REAL275 and CAMERA25 [43] benchmark datasets for evaluation. Thereby, REAL275 is a real-world dataset consisting of 7 scenes with 4.3K images for training and 6 scenes with 2.75K images for testing. It covers 6 categories, including *bottle, bowl, camera, can, laptop* and *mug*. Each category contains 3 unique instances in both training and test set. On the other hand, CAMERA25 is a synthetic dataset generated by rendering virtual objects on real background. CAMERA25 contains 275k images for training and 25k for testing. Note that CAMERA25 shares the same six categories with REAL275.

Implementation Details. Following [23,43], we use Mask-RCNN [9] to generate 2D segmentation masks for a fair comparison. As for our category-specific non-linear shape augmentation, we uniformly sample $\gamma_{max} \sim \mathcal{U}(1, 1.3)$, $\gamma_{min} \sim \mathcal{U}(0.7, 1)$ and $\gamma \sim \mathcal{U}(0.8, 1.2)$. Besides our non-linear shape augmentation, we add random Gaussian noise to the input point cloud, and employ random rotational and translational perturbations as well as random scaling of the

Table 1. Comparison with state-of-the-art methods on REAL275 dataset.

Method	Prior	IoU_{75}	5° 2 cm	5° 5 cm	10° 2 cm	10° 5 cm	Speed (FPS)
NOCS [43]		30.1	7.2	10.0	13.8	25.2	5
CASS [1]		-	-	23.5	-	58.0	-
DualPoseNet [23]		62.2	29.3	35.9	50.0	66.8	2
FS-Net [3]		-	-	28.2	-	60.8	20
FS-Net(Ours)		52.0	19.9	33.9	46.5	69.1	20
GPV-Pose [7]		<u>64.4</u>	32.0	<u>42.9</u>	-	<u>73.3</u>	20
SPD [38]	✓	53.2	19.3	21.4	43.2	54.1	4
CR-Net [44]	✓	55.9	27.8	34.3	47.2	60.8	-
DO-Net [22]	✓	63.7	24.1	34.8	45.3	67.4	10
SGPA [2]	✓	61.9	<u>35.9</u>	39.6	<u>61.3</u>	70.7	-
Ours	✓	**67.8**	**38.2**	**48.1**	**63.1**	**79.2**	25

Overall best results are in bold and the second best results are underlined. **Prior** denotes whether the method makes use of shape priors. We reimplement FS-Net as **FS-Net (Ours)** for a fair comparison since FS-Net uses different detection results.

Table 2. Comparison with state-of-the-art methods on CAMERA25 dataset.

Method	Prior	IoU_{50}	IoU_{75}	5° 2 cm	5° 5 cm	10° 2 cm	10° 5 cm
NOCS [43]		83.9	69.5	32.3	40.9	48.2	64.6
DualPoseNet [23]		92.4	86.4	64.7	70.7	77.2	84.7
GPV-Pose [7]		<u>93.4</u>	<u>88.3</u>	<u>72.1</u>	<u>79.1</u>	-	<u>89.0</u>
SPD [38]	✓	93.2	83.1	54.3	59.0	73.3	81.5
CR-Net [44]	✓	**93.8**	88.0	72.0	76.4	81.0	87.7
SGPA [2]	✓	93.2	88.1	70.7	74.5	**82.7**	88.4
Ours	✓	93.1	**89.0**	**73.5**	**79.6**	<u>82.1</u>	**89.5**

Overall best results are in bold and the second best results are underlined. **Prior** denotes whether the method utilizes shape priors.

object. Unless specified, we set the employed balancing factors $\{\lambda_1, \lambda_2, \lambda_3, \lambda_4\}$ to $\{8.0, 10.0, 3.0, 1.0\}$. Finally, the parameter λ_0 in Eq. 5 is set to 0.01. We train RBP-Pose in a two-stage manner to stabilize the training process. In the first stage, we only train the pose decoder and the shape decoder employing only $\mathcal{L}_{pose}$ and $\mathcal{L}_{shape}$. In the second stage we train all the modules except the Pre-processing as explained in Eq. 7. This strategy ensures that our two assumptions in Sect. 3.2 are reasonable and enables smooth training. Notice that similar to other works [2,38,43], we train a single model for all categories. Unlike [2,23] that train with both synthetic and real data for evaluation on REAL275, we only use the real data for training. We train RBP-Pose for 150 epochs in each stage and employ a batch size of 32. We further employ the Ranger optimizer [25,48,52] with a base learning rate of 1e−4, annealed at 72% of the training phase using a cosine schedule. Our experiments are conducted on a single NVIDIA-A100 GPU.

Evaluation Metrics. Following the widely adopted evaluation scheme [2, 23, 43], we utilize the two standard metrics for quantitative evaluation of the performance. In particular, we report the mean precision of 3D IoU, which computes intersection over union for two bounding boxes under the predicted and the ground truth pose. Thereby, a prediction is considered correct if the IoU is larger than the employed threshold. On the other hand, to directly evaluate rotation and translation errors, we use the 5° 2 cm, 5° 5 cm, 10° 2 cm and 10° 5 cm metrics. A pose is hereby considered correct if the translational and rotational errors are less than the respective thresholds.

4.1   Comparison with State-of-the-Art

Performance on NOCS-REAL275. In Table 1, we compare RBP-Pose with 9 state-of-the-art methods, among which 4 methods utilize shape priors. It can be easily observed that our method outperforms all other competitors by a large margin. Specifically, under IoU_{75}, we achieve a mAP of 67.8%, which exceeds the second best method DO-Net [22] by 4.1%. Regarding the rotation and translation accuracy, RBP-Pose outperforms SGPA [2] by 2.3% in 5° 2 cm, 8.5% in 5° 5 cm, 1.8% in 10° 2 cm and 8.5% in 10° 5 cm. Moreover, when comparing with GPV-Pose [7], we can outperform them by 6.2% in 5° 2 cm, 5.2% in 5° 5 cm and 5.9% in 10° 5 cm. Noteworthy, despite achieving significant accuracy improvements, RBP-Pose still obtains a real-time frame rate 25 Hz when using YOLOv3 [37] and ATSA [51] for object detection. Moreover, we present a detailed per-category comparison for 3D IoU, rotation and translation accuracy of RBP-Pose and SGPA [2] in Fig. 6. It can be deduced that our method obtains superior results over SGPA in terms of mean precision for all metrics, especially in rotation. Moreover, our method is superior in dealing with complex categories with significant intra-class shape variations, *e.g. camera* (green line in Fig. 6).

Performance on NOCS-CAMERA25. The results for CAMERA25 are shown in Table 2. Our method outperforms all competitors for stricter metrics IoU_{75}, 5° 2 cm, 5° 5 cm and 10° 5 cm, and is on par with the best methods for IoU_{50} and 10° 2 cm. Specifically, our method exceeds the second best methods for IoU_{75}, 5° 2 cm, 5° 5 cm and 10° 5 cm by 0.7%, 1.4%, 0.5% and 0.5%, respectively.

4.2   Ablation Study

Effect of Shape Prior Guided Residual Vectors. In Table 3, we evaluate the performance of our method under different configurations. From E1 to E3, we compare three variants of RBP-Pose w.r.t the integration of DVPB: removing the DVPB related modules, predicting DVPB directly and predicting SPRV. By directly predicting DVPB like in GPV-Pose [7], the mAP improves by 0.8% under 5° 2 cm and 2.1% in 5° 5 cm, which indicates that DVPB explicitly encapsulates pose information, helping the network to extract pose-sensitive features.

Fig. 5. Qualitative results of our method (green line) and SGPA [2] (blue line). Images (a)–(f) demonstrate 2D segmentation results. (Color figure online)

By utilizing shape priors to generate initial hypothesis of DVPB and additionally predicting SPRV, the performance improves 2.6% under 5° 2 cm and 2.5% under 10° 2 cm, while the mAP of 5° 5 cm and 10° 5 cm decreases. In general, by solely adopting the auxiliary task of predicting SPRV, the translation accuracy rises while the rotation accuracy falls. This, however, can be solved using our consistency loss between SPRV and pose. E7 adopts the consistency term in Eq. 6 based on E3, and boosts the performance by a large margin under all metrics. This shows that the consistency term is able to guide the network to align predictions from different decoders by jointly optimizing them. E4 enforces the consistency term on DVPB without residual reasoning. Performance deteriorates since initial DVPB hypotheses in are typically inaccurate. SPRV decoder refines the hypotheses by predicting residuals, and thus enhances overall performance.

Effect of Non-linear Shape Augmentation. In Table 3 E5, we remove the non-linear shape augmentation and preserve all other components. Comparing E5 and E7, it can be deduced that the performance degrades dramatically without non-linear shape augmentation, where the mAP of 5° 2 cm and 5° 5 cm drops by 15.6% and 18.4%, respectively. The main reason is that we only train the network on real-world data containing only 3 objects for each category with 4k images, leading to severe overfitting. The non-linear data augmentation mitigates this problem and enhances the diversity of shapes in the training data.

Table 3. Ablation study on DVPB and data augmentation.

	DVPB	SPRV	Con.	L. Aug.	N.L. Aug.	IoU_{75}	5° 2 cm	5° 5 cm	10° 2 cm	10° 5 cm
E1					✓	65.8	32.6	43.8	57.9	75.6
E2	✓				✓	66.6	33.4	45.9	59.1	77.5
E3		✓			✓	67.7	36.0	44.5	61.6	77.1
E4	✓		✓		✓	66.2	34.4	44.8	61.3	77.5
E5		✓	✓			61.3	23.8	29.7	53.8	66.4
E6		✓	✓	✓		66.2	36.1	47.0	62.2	78.8
E7		✓	✓		✓	**67.8**	**38.2**	**48.1**	**63.1**	**79.2**

DVPB denotes predicting DVPB directly, *i.e.*, the decoder only takes F_{obs} as input and outputs DVPB instead of SPRV. **SPRV** denotes SPRV prediction introduced in Sect. 3.2. **Con.** denotes the loss term $\mathcal{L}_{con}$. **L. Aug.** denotes the linear bounding-box-based shape augmentation from FS-Net [3] and **N.L. Aug.** denotes our non-linear shape augmentation.

Non-linear *vs* Linear Shape Augmentation. We compare our non-linear shape augmentation with the linear bounding-box-based augmentation from FS-Net [3] in Table 3 E6 and E7. Our non-linear shape augmentation boosts the mAP w.r.t. all metrics. Specifically, the accuracy improves by 1.6% for IoU_{75}, 2.1% for 5° 2 cm, 1.1% for 5° 5 cm and 0.9% for 10° 2 cm. The main reason is that our non-linear shape augmentation covers more kinds of shape variations than the linear counterpart, which improves the diversity of training data and mitigates the problem of overfitting.

4.3   Qualitative Results

We provide a qualitative comparison between RBP-Pose and SGPA [2] in Fig. 5. Comparative advantage of our method over SGPA is significant, especially in the accuracy of the rotation estimation. Moreover, our method consistently outperforms SGPA when estimating the pose for the *camera* category, which supports our claim that we can better handle categories with large intra-class variations. We discuss **Failure Cases** and **Limitations** in the supplemental material.

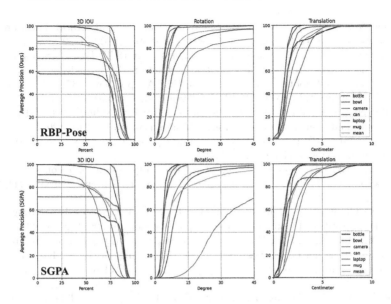

Fig. 6. Quantitative comparison with SGPA [2] on REAL275 in terms of average precision in 3D IoU, Rotation and Translation.

5 Conclusion

In this paper, we propose RBP-Pose, a novel method that leverages Residual Bounding Box Projection for category-level object pose estimation. RBP-Pose jointly predicts 9DoF pose and shape prior guided residual vectors. We illustrate that these nearly zero-mean residual vectors encapsulate the spatial cues of the pose and enable geometry-guided consistency terms. We also propose a non-linear data augmentation scheme to improve shape diversity of the training data. Extensive experiments on the common public benchmark demonstrate the effectiveness of our design and the potential of our method for future real-time applications such as robotic manipulation and augmented reality.

References

1. Chen, D., Li, J., Wang, Z., Xu, K.: Learning canonical shape space for category-level 6d object pose and size estimation. In: Proceedings of the IEEE/CVF Conference on Computer Vision and Pattern Recognition, pp. 11973–11982 (2020)
2. Chen, K., Dou, Q.: SGPA: structure-guided prior adaptation for category-level 6D object pose estimation. In: Proceedings of the IEEE/CVF International Conference on Computer Vision, pp. 2773–2782 (2021)
3. Chen, W., Jia, X., Chang, H.J., Duan, J., Linlin, S., Leonardis, A.: FS-net: fast shape-based network for category-level 6d object pose estimation with decoupled rotation mechanism. In: Proceedings of the IEEE/CVF Conference on Computer Vision and Pattern Recognition (CVPR), pp. 1581–1590 (2021)

4. Chen, Y., Tai, L., Sun, K., Li, M.: Monopair: Monocular 3D object detection using pairwise spatial relationships. In: CVPR, pp. 12093–12102 (2020)
5. Deng, X., Xiang, Y., Mousavian, A., Eppner, C., Bretl, T., Fox, D.: Self-supervised 6D object pose estimation for robot manipulation. In: 2020 IEEE International Conference on Robotics and Automation (ICRA), pp. 3665–3671. IEEE (2020)
6. Di, Y., Manhardt, F., Wang, G., Ji, X., Navab, N., Tombari, F.: So-pose: exploiting self-occlusion for direct 6D pose estimation. In: Proceedings of the IEEE/CVF International Conference on Computer Vision, pp. 12396–12405 (2021)
7. Di, Y., et al.: GPV-pose: category-level object pose estimation via geometry-guided point-wise voting. arXiv preprint (2022)
8. Fan, Z., et al.: ACR-pose: adversarial canonical representation reconstruction network for category level 6D object pose estimation. arXiv preprint arXiv:2111.10524 (2021)
9. He, K., Gkioxari, G., Dollár, P., Girshick, R.: Mask R-CNN. In: Proceedings of the IEEE International Conference on Computer Vision, pp. 2961–2969 (2017)
10. He, Y., Huang, H., Fan, H., Chen, Q., Sun, J.: FFB6D: a full flow bidirectional fusion network for 6d pose estimation. In: Proceedings of the IEEE/CVF Conference on Computer Vision and Pattern Recognition (CVPR), pp. 3003–3013, June 2021
11. He, Y., Sun, W., Huang, H., Liu, J., Fan, H., Sun, J.: PVN3D: a deep point-wise 3D keypoints voting network for 6dof pose estimation. In: Proceedings of the IEEE/CVF Conference on Computer Vision and Pattern Recognition, pp. 11632–11641 (2020)
12. Hodan, T., Barath, D., Matas, J.: Epos: estimating 6D pose of objects with symmetries. In: CVPR, pp. 11703–11712 (2020)
13. Hodaň, T., et al.: BOP: benchmark for 6D object pose estimation. In: Ferrari, V., Hebert, M., Sminchisescu, C., Weiss, Y. (eds.) ECCV 2018. LNCS, vol. 11214, pp. 19–35. Springer, Cham (2018). https://doi.org/10.1007/978-3-030-01249-6_2
14. Hodaň, T., et al.: BOP challenge 2020 on 6D object localization. In: Bartoli, A., Fusiello, A. (eds.) ECCV 2020. LNCS, vol. 12536, pp. 577–594. Springer, Cham (2020). https://doi.org/10.1007/978-3-030-66096-3_39
15. Hu, Y., Fua, P., Wang, W., Salzmann, M.: Single-stage 6D object pose estimation. In: Proceedings of the IEEE/CVF Conference on Computer Vision and Pattern Recognition, pp. 2930–2939 (2020)
16. Kehl, W., Manhardt, F., Tombari, F., Ilic, S., Navab, N.: SSD-6D: making RGB-based 3D detection and 6D pose estimation great again. In: The IEEE International Conference on Computer Vision (ICCV), October 2017
17. Kehl, W., Milletari, F., Tombari, F., Ilic, S., Navab, N.: Deep learning of local RGB-D patches for 3D object detection and 6D pose estimation. In: Leibe, B., Matas, J., Sebe, N., Welling, M. (eds.) ECCV 2016. LNCS, vol. 9907, pp. 205–220. Springer, Cham (2016). https://doi.org/10.1007/978-3-319-46487-9_13
18. Labbé, Y., Carpentier, J., Aubry, M., Sivic, J.: CosyPose: consistent multi-view multi-object 6D pose estimation. In: Vedaldi, A., Bischof, H., Brox, T., Frahm, J.-M. (eds.) ECCV 2020. LNCS, vol. 12362, pp. 574–591. Springer, Cham (2020). https://doi.org/10.1007/978-3-030-58520-4_34
19. Li, C., Bai, J., Hager, G.D.: A unified framework for multi-view multi-class object pose estimation. In: Ferrari, V., Hebert, M., Sminchisescu, C., Weiss, Y. (eds.) ECCV 2018. LNCS, vol. 11220, pp. 263–281. Springer, Cham (2018). https://doi.org/10.1007/978-3-030-01270-0_16
20. Li, Y., Wang, G., Ji, X., Xiang, Y., Fox, D.: DeepIM: deep iterative matching for 6d pose estimation. IJCV, 1–22 (2019)

21. Li, Z., Wang, G., Ji, X.: CDPN: coordinates-based disentangled pose network for real-time RGB-based 6-DoF object pose estimation. In: ICCV, pp. 7678–7687 (2019)
22. Lin, H., Liu, Z., Cheang, C., Zhang, L., Fu, Y., Xue, X.: Donet: learning category-level 6d object pose and size estimation from depth observation. arXiv preprint arXiv:2106.14193 (2021)
23. Lin, J., Wei, Z., Li, Z., Xu, S., Jia, K., Li, Y.: Dualposenet: category-level 6d object pose and size estimation using dual pose network with refined learning of pose consistency. arXiv preprint arXiv:2103.06526 (2021)
24. Lin, Z.H., Huang, S.Y., Wang, Y.C.F.: Convolution in the cloud: learning deformable kernels in 3d graph convolution networks for point cloud analysis. In: Proceedings of the IEEE/CVF Conference on Computer Vision and Pattern Recognition, pp. 1800–1809 (2020)
25. Liu, L., et al.: On the variance of the adaptive learning rate and beyond. In: International Conference on Learning Representations (2019)
26. Manhardt, F., et al.: Explaining the ambiguity of object detection and 6D pose from visual data. In: Proceedings of the IEEE International Conference on Computer Vision, pp. 6841–6850 (2019)
27. Manhardt, F., Kehl, W., Navab, N., Tombari, F.: Deep model-based 6D pose refinement in RGB. In: Ferrari, V., Hebert, M., Sminchisescu, C., Weiss, Y. (eds.) Computer Vision – ECCV 2018. LNCS, vol. 11218, pp. 833–849. Springer, Cham (2018). https://doi.org/10.1007/978-3-030-01264-9_49
28. Manhardt, F., et al.: Cps++: improving class-level 6D pose and shape estimation from monocular images with self-supervised learning. arXiv preprint arXiv:2003.05848v3 (2020)
29. Nie, Y., Han, X., Guo, S., Zheng, Y., Chang, J., Zhang, J.J.: Total3dunderstanding: joint layout, object pose and mesh reconstruction for indoor scenes from a single image. In: CVPR, pp. 55–64 (2020)
30. Park, K., Patten, T., Vincze, M.: Pix2pose: pixel-wise coordinate regression of objects for 6D pose estimation. In: ICCV (2019)
31. Peng, S., Liu, Y., Huang, Q., Zhou, X., Bao, H.: PvNet: pixel-wise voting network for 6dof pose estimation. In: CVPR (2019)
32. Qi, C.R., Su, H., Mo, K., Guibas, L.J.: Pointnet: deep learning on point sets for 3D classification and segmentation. In: Proceedings of the IEEE Conference on Computer Vision and Pattern Recognition (CVPR), July 2017
33. Song, C., Song, J., Huang, Q.: Hybridpose: 6D object pose estimation under hybrid representations. In: Proceedings of the IEEE/CVF Conference on Computer Vision and Pattern Recognition, pp. 431–440 (2020)
34. Su, Y., Rambach, J., Minaskan, N., Lesur, P., Pagani, A., Stricker, D.: Deep multi-state object pose estimation for augmented reality assembly. In: 2019 IEEE International Symposium on Mixed and Augmented Reality Adjunct (ISMAR-Adjunct), pp. 222–227. IEEE (2019)
35. Sundermeyer, M., et al.: Multi-path learning for object pose estimation across domains. In: CVPR, pp. 13916–13925 (2020)
36. Sundermeyer, M., Marton, Z.-C., Durner, M., Brucker, M., Triebel, R.: Implicit 3D orientation learning for 6D object detection from RGB images. In: Ferrari, V., Hebert, M., Sminchisescu, C., Weiss, Y. (eds.) ECCV 2018. LNCS, vol. 11210, pp. 712–729. Springer, Cham (2018). https://doi.org/10.1007/978-3-030-01231-1_43
37. Tekin, B., Sinha, S.N., Fua, P.: Real-time seamless single shot 6D object pose prediction. In: CVPR, pp. 292–301 (2018)

38. Tian, M., Ang, M.H., Lee, G.H.: Shape prior deformation for categorical 6D object pose and size estimation. In: Vedaldi, A., Bischof, H., Brox, T., Frahm, J.-M. (eds.) ECCV 2020. LNCS, vol. 12366, pp. 530–546. Springer, Cham (2020). https://doi.org/10.1007/978-3-030-58589-1_32

39. Umeyama, S.: Least-squares estimation of transformation parameters between two point patterns. IEEE Trans. Pattern Anal. Mach. Intell. **13**(04), 376–380 (1991). https://doi.org/10.1109/34.88573

40. Wang, C., et al.: 6-pack: Category-level 6D pose tracker with anchor-based keypoints. In: 2020 IEEE International Conference on Robotics and Automation (ICRA), pp. 10059–10066. IEEE (2020)

41. Wang, C., et al.: DenseFusion: 6D object pose estimation by iterative dense fusion. In: CVPR, pp. 3343–3352 (2019)

42. Wang, G., Manhardt, F., Tombari, F., Ji, X.: GDR-net: geometry-guided direct regression network for monocular 6D object pose estimation. In: CVPR, June 2021

43. Wang, H., Sridhar, S., Huang, J., Valentin, J., Song, S., Guibas, L.J.: Normalized object coordinate space for category-level 6d object pose and size estimation. In: Proceedings of the IEEE/CVF Conference on Computer Vision and Pattern Recognition, pp. 2642–2651 (2019)

44. Wang, J., Chen, K., Dou, Q.: Category-level 6D object pose estimation via cascaded relation and recurrent reconstruction networks. In: IEEE/RSJ International Conference on Intelligent Robots and Systems (IROS) (2021)

45. Weng, Y., et al.: Captra: category-level pose tracking for rigid and articulated objects from point clouds. arXiv preprint arXiv:2104.03437 (2021)

46. Wohlhart, P., Lepetit, V.: Learning descriptors for object recognition and 3D pose estimation. In: CVPR (2015)

47. Xiang, Y., Schmidt, T., Narayanan, V., Fox, D.: PoseCNN: a convolutional neural network for 6D object pose estimation in cluttered scenes. In: RSS (2018)

48. Yong, H., Huang, J., Hua, X., Zhang, L.: Gradient centralization: a new optimization technique for deep neural networks. In: Vedaldi, A., Bischof, H., Brox, T., Frahm, J.-M. (eds.) ECCV 2020. LNCS, vol. 12346, pp. 635–652. Springer, Cham (2020). https://doi.org/10.1007/978-3-030-58452-8_37

49. Zakharov, S., Shugurov, I., Ilic, S.: Dpod: dense 6d pose object detector in RGB images. In: ICCV (2019)

50. Zhang, C., Cui, Z., Zhang, Y., Zeng, B., Pollefeys, M., Liu, S.: Holistic 3D scene understanding from a single image with implicit representation. In: Proceedings of the IEEE/CVF Conference on Computer Vision and Pattern Recognition, pp. 8833–8842 (2021)

51. Zhang, M., Fei, S.X., Liu, J., Xu, S., Piao, Y., Lu, H.: Asymmetric two-stream architecture for accurate RGB-D saliency detection. In: Vedaldi, A., Bischof, H., Brox, T., Frahm, J.-M. (eds.) ECCV 2020. LNCS, vol. 12373, pp. 374–390. Springer, Cham (2020). https://doi.org/10.1007/978-3-030-58604-1_23

52. Zhang, M., Lucas, J., Ba, J., Hinton, G.E.: Lookahead optimizer: k steps forward, 1 step back. In: Wallach, H., Larochelle, H., Beygelzimer, A., d' Alché-Buc, F., Fox, E., Garnett, R. (eds.) Advances in Neural Information Processing Systems, vol. 32. Curran Associates, Inc. (2019)

Monocular 3D Object Reconstruction with GAN Inversion

Junzhe Zhang[1,3], Daxuan Ren[1,3], Zhongang Cai[1,3], Chai Kiat Yeo[2], Bo Dai[4], and Chen Change Loy[1(✉)]

[1] S-Lab, Nanyang Technological University, Singapore, Singapore
{junzhe001,daxuan001,caiz0023}@e.ntu.edu.sg
[2] Nanyang Technological University, Singapore, Singapore
{asckyeo,ccloy}@ntu.edu.sg
[3] SenseTime Research, Hong Kong, China
[4] Shanghai AI Laboratory, Shanghai, China
daibo@pjlab.org.cn

Abstract. Recovering a textured 3D mesh from a monocular image is highly challenging, particularly for in-the-wild objects that lack 3D ground truths. In this work, we present **MeshInversion**, a novel framework to improve the reconstruction by exploiting the *generative prior* of a 3D GAN pre-trained for 3D textured mesh synthesis. Reconstruction is achieved by searching for a latent space in the 3D GAN that best resembles the target mesh in accordance with the single view observation. Since the pre-trained GAN encapsulates rich 3D semantics in terms of mesh geometry and texture, searching within the GAN manifold thus naturally regularizes the realness and fidelity of the reconstruction. Importantly, such regularization is directly applied in the 3D space, providing crucial guidance of mesh parts that are unobserved in the 2D space. Experiments on standard benchmarks show that our framework obtains faithful 3D reconstructions with consistent geometry and texture across both observed and unobserved parts. Moreover, it generalizes well to meshes that are less commonly seen, such as the extended articulation of deformable objects. Code is released at https://github.com/junzhezhang/mesh-inversion.

1 Introduction

We consider the task of recovering the 3D shape and texture of an object from its monocular observation. A key challenge in this task is the lack of 3D or multi-view supervision due to the prohibitive cost of data collection and annotation for object instances in the wild.

Prior attempts resort to weak supervision based on 2D silhouette annotations of monocular images to solve this task. For instance, Kanazawa *et al.* [19] propose the use of more readily available 2D supervisions including keypoints as the

---

Bo Dai completed this work when he was with S-Lab, NTU.

Supplementary Information The online version contains supplementary material available at https://doi.org/10.1007/978-3-031-19769-7_39.

S. Avidan et al. (Eds.): ECCV 2022, LNCS 13661, pp. 673–689, 2022.
https://doi.org/10.1007/978-3-031-19769-7_39

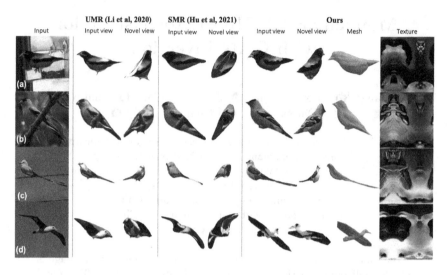

Fig. 1. We propose an alternative approach to monocular 3D reconstruction by exploiting generative prior encapsulated in a pre-trained GAN. Our method has three major advantages: **1)** It reconstructs highly faithful and realistic 3D objects, even when observed from novel views; **2)** The reconstruction is robust against occlusion in (b); **3)** The method generalizes reasonably well to less common shapes, such as birds with (c) extended tails or (d) open wings.

supervision. To further relax the supervision constraint, several follow-up studies propose to learn the 3D manifold in a self-supervised manner, only requiring single-view images and their corresponding masks for training [4,10,15,24]. Minimizing the reconstruction error in the 2D domain tends to ignore the overall 3D geometry and back-side appearance, leading to a shortcut solution that may look plausible only from the input viewpoint, *e.g.*, SMR [15] in Fig. 1 (a)(c). While these methods compensate the relaxed supervision by exploiting various forms of prior information, *e.g.*, categorical semantic invariance [24] and interpolated consistency of the predicted 3D attributes [15], this task remains challenging.

In this work, we propose a new approach, **MeshInversion**, that is built upon generative prior possessed by Generative Adversarial Networks (GANs) [11]. GANs are typically known for their exceptional ability to capture comprehensive knowledge [5,21,41], empowering the success of GAN inversion in image restoration [12,37] and point cloud completion [52]. We believe that by training a GAN to synthesize 3D shapes in the form of a topology-aligned texture and deformation map, one could enable the generator to capture rich prior knowledge of a certain object category, including high-level semantics, object geometries, and texture details.

We propose to exploit the appealing generative prior through GAN inversion. Specifically, our framework finds the latent code of the pre-trained 3D GAN that best recovers the 3D object in accordance with the single-view observation. Given the RGB image and its associated silhouette mask estimated by an off-the-shelf segmentation model, the latent code is optimized towards minimizing

2D reconstruction losses by rendering the 3D object onto the 2D image plane. Hence, the latent manifold of the 3D GAN *implicitly* constrains the reconstructed 3D shape within the realistic boundaries, whereas minimization of 2D losses *explicitly* drives the 3D shape towards a faithful reflection of the input image.

Searching for the optimal latent code in the GAN manifold for single-view 3D object reconstruction is non-trivial due to following challenges: 1) Accurate camera poses are not always available for real-world applications. Inaccurate camera poses easily lead to reprojection misalignment and thus erroneous reconstruction. 2) Existing geometric losses that are computed between 2D masks inevitably discretize mesh vertices into a grid of pixels during rasterization. Such discretization typically makes the losses less sensitive in reflecting the subtle geometric variations in the 3D space. To address the misalignment issue, we propose a **Chamfer Texture Loss**, which relaxes the one-to-one pixel correspondences in existing losses and allows the match to be found within a local region. By jointly considering the appearance and positions of image pixels or feature vectors, it provides a robust texture distance despite inaccurate camera poses and in the presence of high-frequency textures. To improve the geometric sensitivity, we propose a **Chamfer Mask Loss**, which intercepts the rasterization process and computes the Chamfer distance between the projected vertices before discretization to retain information, with the foreground pixels of the input image projected to the same continuous space. Hence, it is more sensitive to small variations in shape and offers a more accurate gradient for geometric learning.

MeshInversion demonstrates compelling performance for 3D reconstruction from real-world monocular images. Even with the assumption of inaccurate masks and camera poses, our method still gives highly plausible and faithful 3D reconstruction in terms of both appearance and 3D shape, as depicted in Fig. 1. It achieves state-of-the-art results on the perceptual metric, *i.e.*, FID, when evaluating the textured mesh from various viewpoints, and is on-par with the existing CMR-based frameworks in terms of geometric accuracy. In addition, while its holistic understanding of the objects benefits from the generative prior, it not only gives a realistic recovery of the back-side texture but also generalizes well in the presence of occlusion, *e.g.*, Fig. 1 (b). Furthermore, MeshInversion also demonstrates significantly better generalization for 3D shapes that are less commonly seen, such as birds with open wings and long tails, as shown in Fig. 1 (d) and (c) respectively.

2   Related Work

Single-View 3D Reconstruction. Many methods have been proposed to recover the 3D information of an object, such as its shape and texture, from a single-view observation. Some methods use image-3D object pairs [32, 35, 39, 46] or multi-view images [28, 33, 34, 47, 51] for training, which limit the scenarios to synthetic data. Another line of work fits the parameters of a 3D prior morphable model, *e.g.*, SMPL for humans and 3DMM for faces [8, 18, 40], which are typically expensive to build and difficult to extend to various natural object categories.

To relax the constraints on supervision, CMR [19] reconstructs category-specific textured mesh by training with a collection of monocular images and associated 2D supervisions, *i.e.*, 2D key-points, camera poses, and silhouette masks. Thereafter, several follow-up studies further relax the supervision, *e.g.*, masks only, and improves the reconstruction results by exploiting different forms of prior. Specifically, they incorporate the prior by enforcing various types of cycle consistencies, such as texture cycle consistency [4, 24], rotation adversarial cycle consistency [4], and interpolated consistency [15]. Some of these methods also leverage external information, *e.g.*, category-level mesh templates [4, 10], and semantic parts provided by an external SCOPS model [24]. In parallel, Shelf-Sup [7] first gives a coarse volumetric prediction, and then converts the coarse volume into a mesh followed by test-time optimization. Without categorical mesh templates in existing approaches, this design demonstrates its scalability to categories with high-genus meshes, *e.g.*, chairs and backpacks.

For texture modeling, direct regression of pixel values in the UV texture map often leads to blurry images [10]. Therefore, the mainstream approach is to regress pixel coordinates, *i.e.*, learning *texture flow* from the input image to the texture map. Although texture flow is easier to regress and usually provides a vivid front view result, it often fails to generalize well to novel views or occluded regions. Our approach directly predicts the texture pixel values by incorporating a pre-trained GAN. In contrast to the texture flow approach, it benefits from a holistic understanding of the objects given the generative prior and offers high plausibility and fidelity at the same time.

GAN Inversion. A well-trained GAN usually captures useful statistics and semantics underlying the training data. In the 2D domain, GAN prior has been explored extensively in various image restoration and editing tasks [3, 12, 37]. GAN inversion, the common method in this line of work, finds a latent code that best reconstructs the given image using the pre-trained generator. Typically, the target latent code can be obtained via gradient descent [27, 29], projected by an additive encoder that learns the inverse mapping of a GAN [2], or a combination of them [53]. There are recent attempts to apply GAN inversion in the 3D domain. Zhang *et al.* [52] use a pre-trained point cloud GAN to address shape completion in the canonical pose, giving remarkable generalization for out-of-domain data such as real-world partial scans. Pan *et al.* [36] recover the geometric cues from pre-trained 2D GANs and achieve exceptional reconstruction results, but the reconstructed shapes are limited to 2.5D due to limited poses that 2D GANs can synthesize. In this work, we directly exploit the prior from a 3D GAN to reconstruct the shape and texture of complete 3D objects.

Textured Mesh Generation. 3D object generation approaches that use voxels [9, 43, 48, 50, 54] or point clouds [1, 41] typically require some form of 3D supervision and are unfriendly for modeling texture. Chen *et al.* [6] propose DIB-R, a GAN framework for textured mesh generation, where 3D meshes are differentiably rendered into 2D images and discriminated with multi-view images of synthetic objects. Later on, Henderson *et al.* [13] relax the multi-view restriction and propose a VAE framework [22] that leverages a collection of single-view

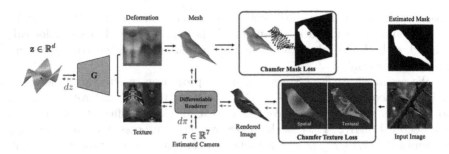

Fig. 2. Overview of the MeshInversion framework. We reconstruct a 3D object from its monocular observation by incorporating a pre-trained textured 3D GAN **G**. We search for the latent code **z** and fine-tune the imperfect camera π that minimizes 2D reconstruction losses via gradient descent. To address the intrinsic challenges associated with 3D-to-2D degradation, we propose two Chamfer-based losses: 1) **Chamfer Texture Loss** (Sect. 3.2) relaxes the pixel-wise correspondences between two RGB images or feature maps, and factorizes the pairwise distance into spatial and textural distance terms. We illustrate the distance maps between one anchor point from the rendered image to the input image, where brighter regions correspond to smaller distances. 2) **Chamfer Mask Loss** (Sect. 3.3) intercepts the discretization process and computes the Chamfer distance between the projected vertices and the foreground pixels projected to the same continuous space. No labeled data is assumed during inference: the mask and camera are estimated by off-the-shelf pre-trained models.

natural images. The appearance is parameterized by face colors instead of texture maps, limiting the visual detail of generated objects. Under the same setting, ConvMesh [38] achieves more realistic 3D generations by generating 3D objects in the form of topology-aligned texture maps and deformation maps in the UV space, where discrimination directly takes place in the UV space against pseudo ground truths. Pseudo deformation maps are obtained by overfitting a mesh reconstruction baseline on the training set. Subsequently, the associated pseudo texture maps can then be obtained by projecting natural images on the UV space. Our proposed method is built upon a pre-trained ConvMesh model to incorporate its generative prior in 3D reconstruction (Fig. 2).

3   Approach

Preliminaries. We represent a 3D object as a textured triangle mesh $\mathbf{O} \equiv (\mathbf{V}, \mathbf{F}, \mathbf{T})$, where $\mathbf{V} \in \mathbb{R}^{|v| \times 3}$ represents the location of the vertices, $\mathbf{F}$ represents the faces that define the fixed connectivity of vertices in the mesh, and $\mathbf{T}$ represents the texture map. An individual mesh is iso-morphic to a 2-pole sphere, and thus we model the deformation $\Delta\mathbf{V}$ from the initial sphere template, and then obtain the final vertex positions by $\mathbf{V} = \mathbf{V}_{sphere} + \Delta\mathbf{V}$. Previous methods [10,19,24] mostly regress deformation of individual vertices via a fully connected network (MLP). In contrast, recent studies have found that using a 2D convolutional neural network (CNN) to learn a deformation map in the UV space would benefit from consistent semantics across the entire category [4,38]. In addition, the deformation map $\mathbf{S}$ and the texture map $\mathbf{T}$ are topologically aligned, so both the values can be mapped to the mesh via the same predefined mapping function.

We assume a weak-perspective camera projection, where the camera pose π is parameterized by scale $\mathbf{s} \in \mathbb{R}$, translation $\mathbf{t} \in \mathbb{R}^2$, and rotation in the form of quaternion $\mathbf{r} \in \mathbb{R}^4$. We use DIB-R [6] as our differentiable renderer. We denote $\mathbf{I} = R(\mathbf{S}, \mathbf{T}, \pi)$ as the image rendering process. Similar to previous baselines [10,15,19,24], we enforce reflectional symmetry along the x axis, which both benefits geometric performance and reduces computation cost.

3.1 Reconstruction with Generative Prior

Our study presents the first attempt to explore the effectiveness of generative prior in monocular 3D construction. Our framework assumes a pre-trained textured 3D GAN. In this study, we adopt ConvMesh [38], which is purely trained with 2D supervisions from single-view natural images. With the help of GAN prior, our goal is to recover the geometry and appearance of a 3D object from a monocular image and its associated mask. Unlike SMR [15] that uses ground truth masks, our method takes silhouettes estimated by an off-the-shelf segmentation model [23].

Next, we will detail the proposed approach to harness the meaningful prior, such as high-level semantics, object geometries, and texture details, from this pre-trained GAN to achieve plausible and faithful recovery of 3D shape and appearance. Note that our method is not limited to ConvMesh, and other pretrained GANs that generate textured meshes are also applicable. More details of ConvMesh can be found in the supplementary materials.

Pre-training Stage. Prior to GAN inversion, we first pre-train the textured GAN on the training split to capture desirable prior knowledge for 3D reconstruction. As discussed in Sect. 2, the adversarial training of ConvMesh takes place in the UV space, where generated deformation maps and texture maps are discriminated against their corresponding pseudo ground truth. In addition to the UV space discrimination, we further enhance the photorealism of the generated 3D objects by introducing a discriminator in the image space, following the architecture of PatchGAN as in [16]. The loss functions for the pre-training stage are shown as follows, where D_{uv} and D_I refer to the discriminators in the UV space and image space respectively, and λ_{uv} and λ_I are the corresponding weights. We use least-squares losses following [30]. An ablation study on image space discrimination can be found in the supplementary materials.

$$\mathcal{L}_G = \lambda_{uv}\mathbb{E}_{\mathbf{z} \sim P_{\mathbf{z}}}[(D_{uv}(G(\mathbf{z})) - 1)^2] + \lambda_I\mathbb{E}_{\mathbf{z} \sim P_{\mathbf{z}}(\mathbf{z})}[(D_I(R(G(\mathbf{z})), \pi) - 1)^2]. \quad (1)$$

$$\mathcal{L}_{D_{uv}} = \mathbb{E}_{\mathbf{S},\mathbf{T} \sim P_{pseudo}}[(D_{uv}(\mathbf{S}, \mathbf{T}) - 1)^2] + \mathbb{E}_{\mathbf{z} \sim P_{\mathbf{z}}(\mathbf{z})}[(D_{uv}(G(\mathbf{z})))^2]. \quad (2)$$

$$\mathcal{L}_{D_I} = \mathbb{E}_{\mathbf{I} \sim P_{data}}[(D_I(\mathbf{I}) - 1)^2] + \mathbb{E}_{\mathbf{z} \sim P_{\mathbf{z}}(\mathbf{z})}[(D_I(R(G(\mathbf{z})), \pi)^2]. \quad (3)$$

Inversion Stage. We now formally introduce GAN inversion for single-view 3D reconstruction. Given a pre-trained ConvMesh that generates a textured mesh from a latent code, $\mathbf{S}, \mathbf{T} = G(\mathbf{z})$, we aim to find the $\mathbf{z}$ that best recovers the 3D object from the input image $\mathbf{I}_{in}$ and its silhouette mask $\mathbf{M}_{in}$. Specifically, we search for such $\mathbf{z}$ via gradient descent towards minimizing the overall reconstruction loss $\mathcal{L}_{inv}$, which can be denoted by

$$\mathbf{z}^* = \arg\min_{\mathbf{z}} \mathcal{L}_{inv}(R(G(\mathbf{z}), \pi), \mathbf{I}_{in}). \tag{4}$$

Given the single-view image and the associated mask, we would need to project the reconstructed 3D object to the observation space for computing $\mathcal{L}_{inv}$. However, such 3D-to-2D degradation is non-trivial. Unlike existing image-based GAN inversion tasks where we can always assume pixel-wise image correspondence in the observation space, rendering 3D objects in the canonical frame onto the image space is explicitly controlled via camera poses. For real-world applications, unfortunately, perfect camera poses are not always available to guarantee such pixel-wise image correspondence. While concurrently optimizing the latent code and the camera pose from scratch seems a plausible approach, this often suffers from camera-shape ambiguity [24] and leads to erroneous reconstruction. To this end, we initialize the camera with a camera pose estimator (CMR [19]), which can be potentially inaccurate, and jointly optimize it in the course of GAN inversion, for which we have:

$$\mathbf{z}^*, \pi^* = \arg\min_{\mathbf{z}, \pi} \mathcal{L}_{inv}(R(G(\mathbf{z}), \pi), \mathbf{I}_{in}). \tag{5}$$

As the 3D object and cameras are constantly optimized throughout the inversion stage, it is infeasible to assume perfect image alignment. In addition, the presence of high-frequency textures, *e.g.*, complex bird feathers, often leads to blurry appearance even with slight discrepancies in pose. Consequently, it calls for a robust form of texture loss in Sect. 3.2.

3.2   Chamfer Texture Loss

To facilitate searching in the GAN manifold without worrying about blurry reconstructions, we reconsider the appearance loss by relaxing the pixel-aligned assumption in existing low-level losses. Taking inspiration from the point cloud data structure, we treat a 2D image as a set of 2D colored points, which have both appearance attributes, *i.e.*, RGB values, and spatial attributes, the values of which relate to their coordinates in the image grid. Thereafter, we aim to measure the dissimilarity between the two colored point sets via Chamfer distance,

$$\mathcal{L}_{CD}(\mathbb{S}_1, \mathbb{S}_2) = \frac{1}{|\mathbb{S}_1|} \sum_{x \in \mathbb{S}_1} \min_{y \in \mathbb{S}_2} \mathbf{D}_{xy} + \frac{1}{|\mathbb{S}_2|} \sum_{y \in \mathbb{S}_2} \min_{x \in \mathbb{S}_1} \mathbf{D}_{yx}. \tag{6}$$

Intuitively, defining the pairwise distance between pixel x and pixel y in the two respective images should jointly consider their appearance and location. In this regard, we factorize the overall pairwise distance $\mathbf{D}_{xy}$ into an appearance term $\mathbf{D}_{xy}^a$ and a spatial term $\mathbf{D}_{xy}^s$, both of which are L2 distance. Like conventional Chamfer distance, single-sided pixel correspondences are determined by column-wise or row-wise minimum in the distance matrix $\mathbf{D}$.

Importantly, we desire the loss to be tolerant and only tolerant of local mis-alignment, as large misalignment will potentially introduce noisy pixel corre-spondences that may jeopardize appearance learning. Inspired by the focal loss for detection [25], we introduce an exponential operation in the spatial term to penalize those spatially distant pixel pairs. Therefore, we define the overall distance matrix $\mathbf{D} \in \mathbb{R}^{|S_1| \times |S_2|}$ as follows

$$\mathbf{D} = \max((\mathbf{D}^s + \epsilon_s)^\alpha, 1) \otimes (\mathbf{D}^a + \epsilon_a), \qquad (7)$$

where $\mathbf{D}^a$ and $\mathbf{D}^s$ are the appearance distance matrix and spatial distance matrix respectively; $\otimes$ denotes element-wise product; ϵ_s and ϵ_a are residual terms to avoid incorrect matches with identical location or identical pixel value respec-tively; α is the scaling factor for flexibility. Specifically, we let $\epsilon_s < 1$ so that the spatial term remains one when two pixels are slightly misaligned. Note that the spatial term is not differentiable and it only serves as a weight matrix for appearance learning. By substituting the resulting $\mathbf{D}$ into Eq. 6, we thus have the final formulation of our proposed **Chamfer Texture Loss**, denoted as $\mathcal{L}_{CT}$.

The proposed relaxed formulation provides a robust measure of texture dis-tance, which effectively eases searching of the target latent code while preventing blurry reconstructions; in return, although $\mathcal{L}_{CT}$ only concerns about local patch statistics but not photorealism, the use of GAN prior is sufficient to give realistic predictions. Besides, the GAN prior also allows computing $\mathcal{L}_{CT}$ with a down-sampled size of colored points. In practice, we randomly select 8096 pixels from each image as a point set.

The proposed formulation additionally gives flexible control between appear-ance and spatial attributes. The appearance term is readily extendable to accept misaligned feature maps to achieve more semantically faithful 3D reconstruction. Specifically, we apply the Chamfer texture loss between the (foreground) feature maps extracted with a pre-trained VGG-19 network [42] from the rendered image and the input image. It is worth noting that the feature-level Chamfer texture loss is somewhat related to the contextual loss [31], which addresses the misalign-ment issue for image transfer. The key difference is that the contextual loss only considers the feature distances but ignores their locations. We compare against the contextual loss in the experiment.

3.3   Chamfer Mask Loss

Conventionally, the geometric distance is usually computed between two binary masks in terms of L1 or IoU loss [10,15,19,24]. However, obtaining the mask of the reconstructed mesh usually involves rasterization that discretizes the mesh into a grid of pixels. This operation inevitably introduces information loss and thus inaccurate supervision signals. This is particularly harmful to a well-trained

ConvMesh, the shape manifold of which is typically smooth. Specifically, a small perturbation in $\mathbf{z}$ usually corresponds to a slight variation in the 3D shape, which may translate to an unchanged binary mask. This usually leads to an insensitive gradient for back-propagation, which undermines geometric learning. We analyze the sensitivity of existing losses in the experiment.

To this end, we propose a **Chamfer Mask Loss**, or $\mathcal{L}_{CM}$, to compute the geometric distance in an unquantized 2D space. Instead of rendering the mesh into a binary mask, we directly project the 3D vertices of the mesh onto the image plane, $\mathbb{S}_v = P(\mathbf{S}, \mathbf{T}, \pi)$. For the foreground mask, we obtain the positions of the foreground pixels by normalizing their pixel coordinates in the range of $[-1, 1]$, denoted as $\mathbb{S}_f$. Thereafter, we compute the Chamfer distance between $\mathbb{S}_v$ and $\mathbb{S}_f$ as the Chamfer mask loss. Note that one does not need to distinguish visible and occluded vertices, as eventually they all fall within the rendered silhouette. The bidirectional Chamfer distance between the sparse set $\mathbb{S}_v$ and dense set $\mathbb{S}_f$ would regularize the vertices from highly uneven deformation.

3.4   Overall Objective Function

We apply the pixel-level Chamfer texture loss $\mathcal{L}_{pCT}$ and the feature-level one $\mathcal{L}_{fCT}$ as our appearance losses, and the Chamfer mask loss $\mathcal{L}_{CM}$ as our geometric loss. Besides, we also introduce two regularizers: the smooth loss $\mathcal{L}_{smooth}$ that encourages neighboring faces to have similar normals, *i.e.*, low cosine; the latent space loss $\mathcal{L}_z$ that regularizes the L2 norm of $\mathbf{z}$ to ensure Gaussian distribution. In summary, the overall objective function is shown in Eq. 8.

$$\mathcal{L}_{inv} = \mathcal{L}_{pCT} + \mathcal{L}_{fCT} + \mathcal{L}_{CM} + \mathcal{L}_{smooth} + \mathcal{L}_z. \tag{8}$$

4   Experiments

Datasets and Experimental Setting. We primarily evaluate MeshInversion on CUB-200-2011 dataset [45]. It consists of 200 species of birds with a wide range of shapes and feathers, making it an ideal benchmark to evaluate 3D reconstruction in terms of both geometric and texture fidelity. Apart from the organic shapes like birds, we also validate our method on 11 man-made rigid car categories from PASCAL3D+ [49].

We use the same train-validation-test split as provided by CMR [19]. The images in both datasets are annotated with foreground masks and camera poses. Specifically, we pre-train ConvMesh on the pseudo ground truths derived from the training split following a class conditional setting [38]. During inference, we conduct GAN inversion on the test split without assuming additional labeled data compared to existing methods. We use the silhouette masks predicted by an off-the-shelf instance segmentation method PointRend [23] pre-trained on COCO [26], which gives an IoU of 0.886 against ground truth masks. We use camera poses predicted by CMR [19], which can be inaccurate. In particular, the poses estimated yields 6.03 degree of azimuth error and 4.33 degree of elevation

Table 1. Quantitative results on CUB show the effectiveness of applying generative prior in 3D reconstruction. As all the baseline methods are regression-based whereas our method involves optimization during inference, we report both baselines and test-time optimization (TTO) results for existing methods, if applicable, with access to masks estimated by PointRend [23]. SMR baseline uses ground truth mask, and it shows noticeable IoU drop with estimated mask. [†]: We report results from [4] since no implementation released; the results are based on ground truth cameras, whereas our method optimizes from imperfect cameras.

Methods	TTO	Input mask	IoU $\uparrow$	FID$_1$ $\downarrow$	FID$_{10}$ $\downarrow$	FID$_{12}$ $\downarrow$
CMR [19]		-	0.703	140.9	176.2	180.1
UMR [24]		-	0.734	40.0	72.8	86.9
U-CMR [10]		-	0.701	65.0	314.9	315.2
View-gen [4] [†]		-	0.629	-	-	70.3
SMR [15]		Estimated	0.751	55.9	65.7	85.6
SMR		Ground truth	**0.800**	52.9	63.2	79.3
CMR	✓	Estimated	0.717	121.6	150.5	158.4
UMR	✓	Estimated	0.739	38.8	78.2	91.3
Ours	✓	Estimated	0.752	**37.3**	**38.7**	**56.8**

error compared to the ground truth cameras via structure-from-motion (SfM). During evaluation, we report quantitative results based on ground truth masks.

Evaluation Strategy. Since there are no 3D ground truths available for CUB, we evaluate MeshInversion against various baselines from three aspects: 1) We evaluate the geometry accuracy in the 2D domain by IoU between the rendered masks and the ground truths. 2) We evaluate the appearance quality by the image synthesis metric FID [14], which compares the distribution of test set images and the render of reconstruction. Since a plausible 3D shape should look photo-realistic observed from multiple viewpoints, we report both single-view FID (**FID$_1$**) and multi-view FIDs. Following SMR [15] and View-gen [4], we render our reconstructed 3D shape from 12 different views (**FID$_{12}$**), which covers azimuth from $0°$ to $360°$ at an interval of $30°$. We additionally report **FID$_{10}$** since the exact front view ($90°$) and the exact back view ($270°$) are rarely seen in CUB. Note that this is in favour of existing methods that do not use any GAN prior as ours. 3) Apart from extensive qualitative results, we conduct a user study to evaluate human preferences in terms of both shape and appearance. For PASCAL3D+, it provides approximated 3D shapes using a set of 10 CAD models, which allows us to evaluate geometric performance in terms of 3D IoU.

4.1   Comparison with Baselines

We compare MeshInversion with various existing methods on the CUB dataset, and report quantitative results in Table 1. Overall, MeshInversion achieves state-of-the-art results on perceptual metrics, particularly multi-view FIDs, and is on par with existing methods in terms of IoU. The qualitative results in Fig. 1, Fig. 3 and Fig. 4 show that MeshInversion achieves highly faithful and realistic

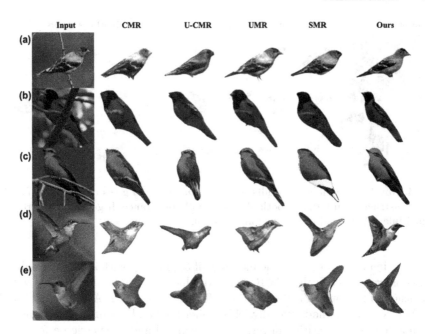

Fig. 3. Qualitative results on CUB. Our method achieves highly faithful and realistic 3D reconstruction. In particular, it exhibits superior generalization under various challenging scenarios, including with occlusion (b, c) and extended articulation (d, e).

3D reconstruction, particularly when observed from novel views. Moreover, our method generalizes reasonably well to highly articulated shapes, such as birds with long tails and open wings, where many of the existing methods fail to give satisfactory reconstructions, as Fig. 1 (c)(d) and Fig. 3 (d)(e) show. We note that although SMR gives the competitive IoU with estimated mask, it lacks fine geometric details and looks less realistic, *e.g.*, the beak in Fig. 1 (c) and Fig. 3.

Texture Flow vs. Texture Regression. Texture flow is extensively adopted in existing methods, except for U-CMR. Although texture flow-based methods are typically easier to learn and give superior texture reconstruction for visible regions, they tend to give incorrect predictions for invisible regions, *e.g.*, abdomen or back as shown in Fig. 1. In contrast, MeshInversion, which performs direct regression of textures, benefits from a holistic understanding of the objects and gives remarkable performance in the presence of occlusion, while texture flow-based methods only learn to copy from the foreground pixels including the obstacles, *e.g.*, twig, from the bird, as shown in Fig. 1 (a) and Fig. 3 (b)(c). Due to the same reason, these methods also tend to copy background pixels onto the reconstructed object when the shape prediction is inaccurate, as shown in Fig. 1 (d) and Fig. 3 (d). More qualitative results and multi-view comparisons can be found in the supplementary materials.

Test-Time Optimization. While existing methods mostly adopt an auto-encoder framework and perform inference with a single forward pass, MeshInversion is optimization-based. For a fair comparison, we also introduce test-time

Input	Texture	Novel-view images	Novel-view meshes

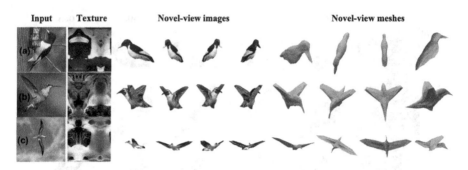

Fig. 4. Novel-view rendering results on CUB. Our method gives realistic and faithful 3D reconstruction in terms of both 3D shape and appearance. It generalizes fairly well to invisible regions and challenging articulations.

optimization (TTO) for baseline methods, if applicable, with access to predicted masks as well. Specifically, CMR and UMR have a compact latent code with a dimension of 200, which is desirable for efficient fine-tuning. As shown in Tab. 1, TTO of existing methods overall yields higher fidelity, but our proposed method remains highly competitive in terms of perceptual and geometric performance. Interestingly, UMR with TTO achieves marginal improvement in terms of IoU and single-view FID at the cost of worsening novel-view FID. This further shows the superiority of generative prior captured through adversarial training over that captured in an auto-encoder, including UMR that is coupled with adversarial training, and its effectiveness of such appealing prior in 3D reconstruction.

User Study. We further conduct a user preference study on multi-view renderings of 30 randomly selected birds, and ask 40 users to choose the most realistic and faithful reconstruction in terms of texture, shape, and overall 3D reconstruction. Table 2 shows that MeshInversion gives the most preferred results, whereas all texture flow-based methods give poor results mainly due to their incorrect prediction for unseen regions.

Table 2. User preference study on CUB in terms of the quality and faithfulness of texture, shape, and overall 3D reconstruction.

Criterion	CMR	U-CMR	UMR	SMR	Ours
Texture	2.7%	15.7%	13.2%	6.6%	**61.8%**
Shape	2.7%	19.6%	14.6%	4.2%	**58.9%**
Overall	2.5%	19.8%	12.8%	3.3%	**61.5%**

4.2 Texture Transfer

As the shape and texture are topologically and semantically aligned in the UV space, it allows us to modify the surface appearance across bird instances. In Fig. 5, we sample pairs of instances and swap their texture maps. Thanks to the categorical semantic consistency, the resulting new 3D objects remain highly realistic even for extended articulations like open wings and long tails.

Input   Reconstruction   Texture transfer   Input   Reconstruction   Texture transfer

Fig. 5. Our method enables faithful and realistic texture transfer between bird instances even with highly articulated shapes.

| Input | Novel-view images | Input | Novel-view images |

Fig. 6. Qualitative results on PASCAL3D+ Car. Our method gives reasonably good performance across different car models and appearances.

4.3   Evaluation on PASCAL3D+ Car

We also evaluate MeshInversion on the man-made rigid car category. As demonstrated in Fig. 6, it performs reasonably well across different car models and appearances. Unlike [4] and [10] that explicitly use one or multiple mesh templates provided by PASCAL3D+, the appeal-

Table 3. 3D IoU on PASCAL3D+ Car. Both the deformable model fitting-based method CSDM [20] and volume-based method DRC [44] do not predict object texture.

	CSDM	DRC	CMR	Ours
3D IoU	0.60	**0.67**	0.64	0.66

ing GAN prior implicitly provides a rich number of templates that makes it possible to reconstruct cars of various models. Given the approximated 3D ground truths in PASCAL3D+, we show in Table 3 that MeshInversion performs comparably to baselines in terms of 3D IoU.

4.4   Ablation Study

Effectiveness of Chamfer Mask Loss. As compared to Chamfer mask loss in Table 4, both conventional mask losses give significantly worse reconstruction results. As existing mask losses are obtained through rasterization, the induced information loss often makes them less accurate in capturing subtle shape variations, which undermines geometry recovery. A detailed sensitivity study of various mask losses can be found in the supplementary materials.

Effectiveness of Chamfer Texture Loss. In the presence of imperfect poses and high-frequency details in textures, we show in Table 4 that our proposed

Table 4. Ablation study. Compared to conventional texture and mask losses, our proposed pixel-level Chamfer texture losses $\mathcal{L}_{pCT}$ and feature-level one $\mathcal{L}_{fCT}$, and Chamfer mask loss $\mathcal{L}_{CM}$ are effective to address the challenges due to misalignment and quantization during rendering. Despite using not-so-accurate camera poses, our method gives compelling geometric and perceptual performance by jointly optimizing 3D shape and camera during GAN inversion.

Mask loss	Texture loss	Camera	IoU ↑	FID$_1$ ↓	FID$_{10}$ ↓	FID$_{12}$ ↓
IoU loss	L1 + perceptual loss	Fine-tuned	0.580	82.8	78.3	92.01
$\mathcal{L}_{CM}$	L1 loss	Fine-tuned	0.732	58.9	78.3	74.7
$\mathcal{L}_{CM}$	L1 + perceptual loss	Fine-tuned	0.741	56.3	44.0	64.9
$\mathcal{L}_{CM}$	Contextual loss	Fine-tuned	0.718	69.0	57.7	75.5
L1 loss	$\mathcal{L}_{pCT} + \mathcal{L}_{fCT}$	Fine-tuned	0.589	71.8	73.2	74.1
IoU loss	$\mathcal{L}_{pCT} + \mathcal{L}_{fCT}$	Fine-tuned	0.604	72.1	71.8	70.5
$\mathcal{L}_{CM}$	$\mathcal{L}_{pCT} + \mathcal{L}_{fCT}$	Fixed	0.695	40.9	41.38	60.0
$\mathcal{L}_{CM}$	$\mathcal{L}_{pCT} + \mathcal{L}_{fCT}$	Fine-tuned	0.752	**37.3**	**38.7**	**56.8**

pixel- and feature-level Chamfer texture losses are highly effective compared to existing losses. In particular, pixel-to-pixel L1 loss tends to give blurry reconstructions. Feature-based losses, perceptual loss [17] and contextual loss [31], are generally more robust to misalignment, but they are usually not discriminative enough to reflect the appearance details between two images. Although the contextual loss is designed to address the misalignment issue for image-to-image translation, it only considers feature distances while ignoring their positions in the image.

Effectiveness of Optimizing Camera Poses. In Table 4, we show that compared to only optimizing latent code, jointly fine-tuning imperfect camera poses effectively improves geometric performance.

5   Discussion

We have presented a new approach for monocular 3D object reconstruction. It exploits generative prior encapsulated in a pre-trained GAN and reconstructs textured shapes through GAN inversion. To address reprojection misalignment and discretization-induced information loss due to 3D-to-2D degradation, we propose two Chamfer-based losses in the 2D space, $i.e.$, Chamfer texture loss and Chamfer mask loss. By efficiently incorporating the GAN prior, MeshInversion achieves highly realistic and faithful 3D reconstruction, and exhibits superior generalization power for challenging cases, such as in the presence of occlusion or extended articulations. However, this challenging problem is far from being solved. In particular, although we can faithfully reconstruct flying birds with open wings, the wings are only represented by a few vertices due to semantic consistency across the entire category, which strictly limits the representation power in terms of geometry and texture details. Therefore, future work may explore more flexible solutions, for instance, an adaptive number of vertices can be assigned to articulated regions to accommodate richer details.

Acknowledgement. This study is supported under the RIE2020 Industry Alignment Fund – Industry Collaboration Projects (IAF-ICP) Funding Initiative, Singapore MOE AcRF Tier 2 (MOE-T2EP20221-0011), Shanghai AI Laboratory, as well as cash and in-kind contribution from the industry partners.

References

1. Achlioptas, P., Diamanti, O., Mitliagkas, I., Guibas, L.: Learning representations and generative models for 3D point clouds. In: ICML (2018)
2. Bau, D., et al.: Semantic photo manipulation with a generative image prior. In: SIGGRAPH (2019)
3. Bau, D., et al.: Seeing what a GAN cannot generate. In: ICCV (2019)
4. Bhattad, A., Dundar, A., Liu, G., Tao, A., Catanzaro, B.: View generalization for single image textured 3D models. In: CVPR (2021)
5. Brock, A., Donahue, J., Simonyan, K.: Large scale GAN training for high fidelity natural image synthesis. In: ICLR (2019)
6. Chen, W., et al.: Learning to predict 3D objects with an interpolation-based differentiable renderer. In: NeurIPS (2019)
7. Ye, Y., et al.: Shelf-supervised mesh prediction in the wild. In: CVPR (2021)
8. Gecer, B., Ploumpis, S., Kotsia, I., Zafeiriou, S.: GANFIT: generative adversarial network fitting for high fidelity 3D face reconstruction. In: CVPR (2019)
9. Girdhar, R., Fouhey, D.F., Rodriguez, M., Gupta, A.: Learning a predictable and generative vector representation for objects. In: Leibe, B., Matas, J., Sebe, N., Welling, M. (eds.) ECCV 2016. LNCS, vol. 9910, pp. 484–499. Springer, Cham (2016). https://doi.org/10.1007/978-3-319-46466-4_29
10. Goel, S., Kanazawa, A., Malik, J.: Shape and viewpoint without keypoints. In: Vedaldi, A., Bischof, H., Brox, T., Frahm, J.-M. (eds.) ECCV 2020. LNCS, vol. 12360, pp. 88–104. Springer, Cham (2020). https://doi.org/10.1007/978-3-030-58555-6_6
11. Goodfellow, I., et al.: Generative adversarial nets. In: NeurIPS (2014)
12. Gu, J., Shen, Y., Zhou, B.: Image processing using multi-code GAN prior. In: CVPR (2020)
13. Henderson, P., Tsiminaki, V., Lampert, C.H.: Leveraging 2D data to learn textured 3D mesh generation. In: CVPR (2020)
14. Heusel, M., Ramsauer, H., Unterthiner, T., Nessler, B., Hochreiter, S.: GANs trained by a two time-scale update rule converge to a local NASH equilibrium. In: NeurIPS (2017)
15. Hu, T., Wang, L., Xu, X., Liu, S., Jia, J.: Self-supervised 3D mesh reconstruction from single images. In: CVPR (2021)
16. Isola, P., Zhu, J.Y., Zhou, T., Efros, A.A.: Image-to-image translation with conditional adversarial networks. In: CVPR (2017)
17. Johnson, J., Alahi, A., Fei-Fei, L.: Perceptual losses for real-time style transfer and super-resolution. In: Leibe, B., Matas, J., Sebe, N., Welling, M. (eds.) ECCV 2016. LNCS, vol. 9906, pp. 694–711. Springer, Cham (2016). https://doi.org/10.1007/978-3-319-46475-6_43

18. Kanazawa, A., Black, M.J., Jacobs, D.W., Malik, J.: End-to-end recovery of human shape and pose. In: CVPR (2018)
19. Kanazawa, A., Tulsiani, S., Efros, A.A., Malik, J.: Learning category-specific mesh reconstruction from image collections. In: Ferrari, V., Hebert, M., Sminchisescu, C., Weiss, Y. (eds.) ECCV 2018. LNCS, vol. 11219, pp. 386–402. Springer, Cham (2018). https://doi.org/10.1007/978-3-030-01267-0_23
20. Kar, A., Tulsiani, S., Carreira, J., Malik, J.: Category-specific object reconstruction from a single image. In: CVPR (2015)
21. Karras, T., Laine, S., Aila, T.: A style-based generator architecture for generative adversarial networks. In: CVPR (2019)
22. Kingma, D.P., Welling, M.: Auto-encoding variational Bayes. In: ICLR (2014)
23. Kirillov, A., Wu, Y., He, K., Girshick, R.: PointRend: image segmentation as rendering. In: CVPR (2020)
24. Li, X., et al.: Self-supervised single-view 3D reconstruction via semantic consistency. In: Vedaldi, A., Bischof, H., Brox, T., Frahm, J.-M. (eds.) ECCV 2020. LNCS, vol. 12359, pp. 677–693. Springer, Cham (2020). https://doi.org/10.1007/978-3-030-58568-6_40
25. Lin, T.Y., Goyal, P., Girshick, R., He, K., Dollár, P.: Focal loss for dense object detection. In: ICCV (2017)
26. Lin, T.-Y., et al.: Microsoft COCO: common objects in context. In: Fleet, D., Pajdla, T., Schiele, B., Tuytelaars, T. (eds.) ECCV 2014. LNCS, vol. 8693, pp. 740–755. Springer, Cham (2014). https://doi.org/10.1007/978-3-319-10602-1_48
27. Lipton, Z.C., Tripathi, S.: Precise recovery of latent vectors from generative adversarial networks. CoRR arXiv:1702.04782 (2017)
28. Liu, S., Chen, W., Li, T., Li, H.: Soft rasterizer: differentiable rendering for unsupervised single-view mesh reconstruction. In: ICCV (2019)
29. Ma, F., Ayaz, U., Karaman, S.: Invertibility of convolutional generative networks from partial measurements. In: NeurIPS (2018)
30. Mao, X., Li, Q., Xie, H., Lau, R.Y., Wang, Z., Paul Smolley, S.: Least squares generative adversarial networks. In: ICCV (2017)
31. Mechrez, R., Talmi, I., Zelnik-Manor, L.: The contextual loss for image transformation with non-aligned data. In: Ferrari, V., Hebert, M., Sminchisescu, C., Weiss, Y. (eds.) Computer Vision – ECCV 2018. LNCS, vol. 11218, pp. 800–815. Springer, Cham (2018). https://doi.org/10.1007/978-3-030-01264-9_47
32. Mescheder, L., Oechsle, M., Niemeyer, M., Nowozin, S., Geiger, A.: Occupancy networks: learning 3D reconstruction in function space. In: CVPR (2019)
33. Niemeyer, M., Mescheder, L., Oechsle, M., Geiger, A.: Differentiable volumetric rendering: learning implicit 3D representations without 3D supervision. In: CVPR (2020)
34. Oechsle, M., Peng, S., Geiger, A.: UNISURF: unifying neural implicit surfaces and radiance fields for multi-view reconstruction. In: ICCV (2021)
35. Pan, J., Han, X., Chen, W., Tang, J., Jia, K.: Deep mesh reconstruction from single RGB images via topology modification networks. In: ICCV (2019)
36. Pan, X., Dai, B., Liu, Z., Loy, C.C., Luo, P.: Do 2D GANs know 3D shape? Unsupervised 3D shape reconstruction from 2D image GANs. In: ICLR (2021)
37. Pan, X., Zhan, X., Dai, B., Lin, D., Loy, C.C., Luo, P.: Exploiting deep generative prior for versatile image restoration and manipulation. PAMI (2021)
38. Pavllo, D., Spinks, G., Hofmann, T., Moens, M.F., Lucchi, A.: Convolutional generation of textured 3D meshes. In: NeurIPS (2020)
39. Rematas, K., Martin-Brualla, R., Ferrari, V.: ShaRF: shape-conditioned radiance fields from a single view. In: ICML (2021)

40. Sanyal, S., Bolkart, T., Feng, H., Black, M.J.: Learning to regress 3D face shape and expression from an image without 3D supervision. In: CVPR (2019)
41. Shu, D.W., Park, S.W., Kwon, J.: 3D point cloud generative adversarial network based on tree structured graph convolutions. In: ICCV (2019)
42. Simonyan, K., Zisserman, A.: Very deep convolutional networks for large-scale image recognition. In: ICLR (2015)
43. Smith, E.J., Meger, D.: Improved adversarial systems for 3D object generation and reconstruction. In: CoRL (2017)
44. Tulsiani, S., Zhou, T., Efros, A.A., Malik, J.: Multi-view supervision for single-view reconstruction via differentiable ray consistency. In: CVPR (2017)
45. Wah, C., Branson, S., Welinder, P., Perona, P., Belongie, S.: The Caltech-UCSD birds-200-2011 dataset (2011)
46. Wang, N., Zhang, Y., Li, Z., Fu, Y., Liu, W., Jiang, Y.-G.: Pixel2Mesh: generating 3D mesh models from single RGB images. In: Ferrari, V., Hebert, M., Sminchisescu, C., Weiss, Y. (eds.) ECCV 2018. LNCS, vol. 11215, pp. 55–71. Springer, Cham (2018). https://doi.org/10.1007/978-3-030-01252-6_4
47. Wang, P., Liu, L., Liu, Y., Theobalt, C., Komura, T., Wang, W.: NeuS: learning neural implicit surfaces by volume rendering for multi-view reconstruction. In: NeurIPS (2021)
48. Wu, J., Zhang, C., Xue, T., Freeman, B., Tenenbaum, J.: Learning a probabilistic latent space of object shapes via 3D generative-adversarial modeling. In: NeurIPS (2016)
49. Xiang, Y., Mottaghi, R., Savarese, S.: Beyond PASCAL: a benchmark for 3D object detection in the wild. In: WACV (2014)
50. Xie, J., Zheng, Z., Gao, R., Wang, W., Zhu, S.C., Wu, Y.N.: Learning descriptor networks for 3D shape synthesis and analysis. In: CVPR (2018)
51. Yariv, L., et al.: Multiview neural surface reconstruction by disentangling geometry and appearance. In: NeurIPS (2020)
52. Zhang, J., et al.: Unsupervised 3D shape completion through GAN inversion. In: CVPR (2021)
53. Zhu, J., Shen, Y., Zhao, D., Zhou, B.: In-domain GAN inversion for real image editing. In: Vedaldi, A., Bischof, H., Brox, T., Frahm, J.-M. (eds.) ECCV 2020. LNCS, vol. 12362, pp. 592–608. Springer, Cham (2020). https://doi.org/10.1007/978-3-030-58520-4_35
54. Zhu, J.Y., et al.: Visual object networks: image generation with disentangled 3D representations. In: NeurIPS (2018)

Map-Free Visual Relocalization: Metric Pose Relative to a Single Image

Eduardo Arnold[1,2(✉)], Jamie Wynn[1], Sara Vicente[1],
Guillermo Garcia-Hernando[1], Áron Monszpart[1], Victor Prisacariu[1,3],
Daniyar Turmukhambetov[1], and Eric Brachmann[1]

[1] Niantic, London, UK
eduardoarnold@nianticlabs.com
[2] University of Warwick, Coventry, UK
[3] University of Oxford, Oxford, UK
http://github.com/nianticlabs/map-free-reloc

Abstract. Can we relocalize in a scene represented by a single reference image? Standard visual relocalization requires hundreds of images and scale calibration to build a scene-specific 3D map. In contrast, we propose *Map-free Relocalization*, *i.e.*, using only one photo of a scene to enable instant, metric scaled relocalization. Existing datasets are not suitable to benchmark map-free relocalization, due to their focus on large scenes or their limited variability. Thus, we have constructed a new dataset of 655 small places of interest, such as sculptures, murals and fountains, collected worldwide. Each place comes with a reference image to serve as a relocalization anchor, and dozens of query images with known, metric camera poses. The dataset features changing conditions, stark viewpoint changes, high variability across places, and queries with low to no visual overlap with the reference image. We identify two viable families of existing methods to provide baseline results: relative pose regression, and feature matching combined with single-image depth prediction. While these methods show reasonable performance on some favorable scenes in our dataset, map-free relocalization proves to be a challenge that requires new, innovative solutions.

1 Introduction

Given not more than a single photograph we can imagine what a depicted place looks like, and where we, looking through the lens, would be standing relative to that place. Visual relocalization mimics the human capability to estimate a camera's position and orientation from a single query image. It is a well-researched task that enables exciting applications in augmented reality (AR) and robotic navigation. State-of-the-art relocalization methods surpass human rule-of-thumb estimates by a noticeable margin [10,33,53–55,58], allowing centimeter accurate predictions of a camera's pose. But this capability comes with a price:

E. Arnold—Work done during internship at Niantic.

S. Avidan et al. (Eds.): ECCV 2022, LNCS 13661, pp. 690–708, 2022.
https://doi.org/10.1007/978-3-031-19769-7_40

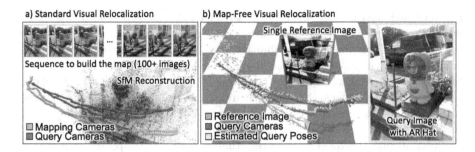

Fig. 1. Standard relocalization methods build a scene representation from hundreds of mapping images (a). For map-free relocalization (b), only a single photo (cyan) of the scene is available to relocalize queries. We show ground truth poses (purple) and estimated poses (yellow) in (b), and we use one estimate to render a virtual hat on the statue. We achieve these results with SuperGlue [54] feature matching and DPT [48] depth estimation. (Color figure online)

each scene has to be carefully pre-scanned and reconstructed. First, images need to be gathered from hundreds of distinct viewpoints, ideally spanning different times of day and even seasons. Then, the 3D orientation and position of these images needs to be estimated, *e.g.*, by running structure-from-motion (SfM) [60, 64,81,82] or simultaneous-localization-and-mapping (SLAM) [20,43] software. Oftentimes, accurate multi-camera calibration, alignment against LiDAR scans, high-definition maps or inertial sensor measurements are needed to recover poses in metric units, *e.g.*, [58,59]. Finally, images and their camera poses are fed to a relocalization pipeline. For traditional structure-based systems [33,53,55,57], the final scene representation consists of a point cloud triangulated from feature correspondences, and associated feature descriptors, see Fig. 1a).

The requirement for systematic pre-scanning and mapping restricts how visual relocalization can be used. For example, AR immersion might break if a user has to record an entire image sequence of an unseen environment first, gathering sufficient parallax by walking sideways, all in a potentially busy public space. Furthermore, depending on the relocalization system, the user then has to wait minutes or hours until the scene representation is built. We propose a new flavour of relocalization, termed *Map-free Relocalization*. We ask whether the mapping requirement can be relaxed to the point where a single reference image is enough to relocalize new queries in a metric coordinate system. Map-free relocalization enables instant AR capabilities at new locations: User A points their camera at a structure, takes a photo, and any user B can instantly relocalize w.r.t. user A. Map-free relocalization constitutes a systematic, task-oriented benchmark for two-frame relative pose estimation, namely between the reference image and a query image, see Fig. 1 b).

Relocalization by relative pose estimation is not new. For example, neural networks have been trained to regress metric relative poses directly from two images [4,42]. Thus far, such systems have been evaluated on standard relocalization benchmarks where structure-based methods rule supreme, to the extent where the accuracy of the ground truth is challenged [9]. We argue that we should strive towards enabling new capabilities that traditional structure-based methods cannot provide. Based on a single photo, a scene cannot be reconstructed by SfM or SLAM. And while feature matching still allows to estimate the relative pose between two images, the reference and the query, there is no notion of absolute scale [32]. To recover a *metric* estimate, some heuristic or world knowledge has to be applied to resolve the scale ambiguity which we see as the key problem.

Next to pose regression networks, that predict metric poses by means of supervised learning, we recognize a second family of methods as suitable for map-free relocalization. We show that a combination of deep feature matching [54, 66] and deep single-image depth prediction [40,48] currently achieves highest relative pose accuracy. To the best of our knowledge, this variant of relative pose estimation has not gained attention in relocalization literature thus far.

While we provide evidence that existing methods can solve map-free relocalization with acceptable precision, such results are restricted to a narrow window of situations. To stimulate further research in map-free relocalization, we present a new benchmark and dataset. We have gathered images of 655 places of interest worldwide where each place can be represented well by a single reference image. All frames in each place of interest have metric ground truth poses. There are 522,921 frames for training, 36,998 query frames across 65 places for validation, and 14,778 query frames (subsampled from 73,902 frames) across 130 places in the test set. Following best practice in machine learning, we provide a public validation set while keeping the test ground truth private, accessed through an online evaluation service. This dataset can serve as a test bed for advances in relative pose estimation and associated sub-problems such as wide-baseline feature matching, robust estimation and single-image depth prediction.

We summarize our **contributions** as follows:

- Map-free relocalization, a new flavor of visual relocalization that dispenses with the need for creating explicit maps from extensive scans of a new environment. A single reference image is enough to enable relocalization.
- A dataset that provides reference and query images of over 600 places of interest worldwide, annotated with ground truth poses. The dataset includes challenges such as changing conditions, stark viewpoint changes, high variability across places, and queries with low to no visual overlap with the reference image.
- Baseline results for map-free relocalization using relative pose regression methods, and feature matching on top of single image-depth prediction. We expose the primary problems of current approaches to guide further research.
- Additional experiments and ablation studies on ScanNet and 7Scenes datasets, allowing comparisons to related, previous research on relative pose estimation and visual relocalization.

2   Related Work

Scene Representations in Visual Relocalization: In the introduction, we have discussed traditional structure-based relocalizers that represent a scene by an explicit SfM or SLAM reconstruction. As an alternative, recent learning-based relocalizers encode the scene *implicitly* in the weights of their neural networks by training on posed mapping images. This is true for both scene coordinate regression [10–12,14,38,62] and absolute pose regression (APR) [15,36,37,61,77]. More related to our map-free scenario, some relative pose regression (RPR) methods avoid training scene specific networks [4,73,80]. Given a query, they use image retrieval [3,33,45,46,51,59,72] to look up the closest database image and its pose. A generic relative pose regression network estimates the pose between query and database images to obtain the absolute pose of the query. RPR methods claim to avoid creating costly scene-specific representations but ultimately these works do not discuss how posed database images would be obtained without running SfM or SLAM. ExReNet [80], a recent RPR method, shows that the database of posed images can be extremely sparse, keeping as little as four strategically placed reference images to cover an indoor room. Although only a few images make up the final representation, continuous pose tracking is required when recording them. In contrast, map-free relocalization means keeping only a single image to represent a scene without any need for pose tracking or pose reconstruction. The reference image has the identity pose.

Relative Pose by Matching Features: The pose between two images with known intrinsics can be recovered by decomposing the essential matrix [32]. This yields the relative rotation, and a *scaleless* translation vector. The essential matrix is classically estimated by matching local features, followed by robust estimation, such as using a 5-point solver [44] inside a RANSAC [26] loop. This basic formula has been improved by learning better features [8,21,41,52,74], better matching [54,66] and better robust estimators [5,6,13,47,49,67,85], and progress has been measured in wide-baseline feature matching challenges [35] and small overlap regimes.

In the relocalization literature, scaleless pairwise relative poses between the query and multiple reference images have been used to triangulate the scaled, metric pose of a query [80,86,87]. However, for map-free relocalization only two images (reference and query) are available at any time, making query camera pose triangulation impossible. Instead, we show that estimated depth can be used to resolve the scale ambiguity of poses recovered via feature matching.

Relative Pose Regression (RPR): Deep learning methods that predict the relative pose from two input images bypass explicit estimation of 2D correspondences [1,4,24,42,75,80]. Some methods recover pose up to a scale factor [42,80] and rely on pose triangulation, while others aim to estimate metric relative pose [1,4,24]. Both RelocNet [4] and ExReNet [80] show generalization of RPR across datasets by training on data different from the test dataset.

Recently, RPR was applied in scenarios that are challenging for correspondence-based approaches. Cai et al. [16] focus on estimating the relative rotation between two images in extreme cases, including when there is no overlap between the two images. Similarly, the method in [18] estimates scaleless relative pose for pairs of images with very low overlap. We take inspiration from the methods above to create baselines and discuss in more detail the different architectures and output parameterizations in Sect. 3.2.

Single-Image Depth Prediction: Advances in deep learning have allowed practical methods for single-image depth estimation, *e.g.*, [23,29]. There are two versions of the problem: relative and absolute depth prediction. Relative, also called scaleless, depth prediction aims at estimating depth maps up to an unknown linear or affine transformation, and can use scaless training data such as SfM reconstructions [39], uncalibrated stereo footage [50,83] or monocular videos [30,88]. Absolute depth prediction methods (*e.g.*, [29,40,48,79]) aim to predict depth in meters by training or fine-tuning on datasets that have absolute metric depth such as the KITTI [28], NYUv2 [63] and ScanNet [19] datasets. Generalizing between domains (*e.g.*, driving scenes vs. indoors) is challenging as collecting metric depth in various conditions can be expensive. Moreover, generalization of a single network that is robust to different input image resolutions, aspect ratios and camera focal lengths is also challenging [25].

Recently, single-image depth prediction was leveraged in some pose estimation problems. In [71], predicted depth maps are used to rectify planar surfaces before local feature computation for improved relative pose estimation under large viewpoint changes. However, that work did not use metric depth information to estimate the scale of relative poses. Depth prediction was incorporated into monocular SLAM [69,70] and Visual Odometry [17,84] pipelines to combat scale drift and improve camera pose estimation. Predicted depths were used as a soft constraint in multi-image problem, while we use depth estimates to scale relative pose between two images.

3   Map-Free Relocalization

Our aim is to obtain the camera pose of a query image given a single RGB reference image of a scene. We assume intrinsics of both images are known, as they are generally reported by modern devices. The absolute pose of a query image Q is parameterized by $R \in SO(3)$, $t \in \mathbb{R}^3$, which maps a world point $\mathbf{y}$ to point $\mathbf{x}$ in the camera's local coordinate system as $\mathbf{x} = R\mathbf{y} + t$. Assuming the global coordinate system is anchored to the reference image, the problem of estimating the absolute pose of the query becomes one of estimating a scaled relative pose between two images. Next, we discuss different approaches for obtaining a metric relative pose between a pair of RGB images. The methods are split into two categories: methods based on feature matching with estimated depth, and methods based on direct relative pose regression.

3.1   Feature Matching and Scale from Estimated Depth

The relative pose from 2D correspondences is estimated up to scale via the Essential matrix [32]. We consider SIFT [41] as a traditional baseline as well

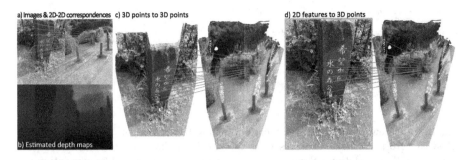

Fig. 2. Given the reference and query images, we obtain 2D-2D correspondences using the feature matching method in [54] (a). Inlier correspondences for the robust RANSAC-based essential matrix computation are visualized in green and outlier correspondences in red. Estimated monocular depth maps using [48] are shown in (b). The depth maps can be coupled with the 2D-2D correspondences to obtain 3D-3D correspondences (c) or 2D-3D correspondences (d), which are used in the geometric methods discussed in Sect. 3.1.

as more recent learning-based matchers such as SuperPoint + SuperGlue [54] and LoFTR [66]. To recover the missing scale, we utilize monocular depth estimation. For indoors, we experimented with DPT [48] fine-tuned on the NYUv2 dataset [63] and PlaneRCNN [40], which was trained on ScanNet [19]. For outdoors, we use DPT [48] fine-tuned on KITTI [28]. Given estimated depth and 2D correspondences we compute scaled relative poses in the following variants. See also Fig. 2 for an illustration.

(2D-2D) Essential matrix + depth scale (*Ess.Mat. + D.Scale*): We compute the Essential matrix using a 5-point solver [44] with MAGSAC++ [6] and decompose it into a rotation and a unitary translation vector. We back-project MAGSAC inlier correspondences to 3D using the estimated depth. Each 3D-3D correspondence provides one scale estimate for the translation vector, and we select the scale estimate with maximum consensus across correspondences, see the supplemental material for details.

(2D-3D) Perspective-n-Point (PnP): Using estimated depth, we back-project one of the two images to 3D, giving 2D-3D correspondences. This allows us to use a PnP solver [27] to recover a metric pose. We use PnP within RANSAC [26] and refine the final estimate using all inliers. We use 2D features from the query image and 3D points from the reference image.

(3D-3D) Procrustes: Using estimated depth, we back-project both images to 3D, giving 3D-3D correspondences. We compute the relative pose using Orthogonal Procrustes [22] inside a RANSAC loop [26]. Optionally, we can refine the relative pose using ICP [7] on the full 3D point clouds. This variant performs significantly worse compared to the previous two, so we report its results in the supplemental material.

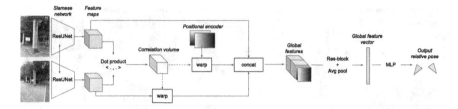

Fig. 3. Overview of the network architecture for RPR. We use a Siamese network (ResUNet [16]) to extract features from the two input images. Following [16,80], we compute a 4D correlation volume to mimic soft feature matching. The correlation volume is used to warp the features of the second image and a regular grid of coordinates (positional encoding). These are concatenated channel-wise with the first image's feature map to create the global feature map. The global volume is fed to four ResNet blocks followed by global average pooling, resulting in a single 512-dimensional global feature vector. Finally, an MLP generates the output poses. See supplement for details.

3.2   Relative Pose Regression

Relative pose regression (RPR) networks learn to predict metric relative poses in a forward pass. We implement a baseline architecture following best practices reported in the literature [80,87,89] – see Fig. 3, and the supplement for more details. In the following, we focus on the different output parameterizations and leave a discussion about losses and other design choices to the supplement.

RPR networks often parameterize rotations as **quaternions** [24,42,80] (denoted as $R(q)$). [89] argues that a **6D parameterization** of rotation avoids discontinuities of other representations: the network predicts two 3D vectors and creates an orthogonal basis through a partial Gram-Schmidt process (denoted as $R(6D)$). Finally, for rotation, we experiment with **Discrete Euler angles** [16], denoted as $R(\alpha, \beta, \gamma)$. Following [16], we use 360 discrete values for yaw and roll, and 180 discrete values for the pitch angle. For the translation vector we investigate three parameterization options: predicting the **scaled translation** (denoted as t), predicting a **scale and unitary translation** separately (denoted as $s \cdot \hat{t}$), and **scale and discretized unitary translation**. For the latter we predict translation in spherical coordinates ϕ, θ with quantized bins of 1deg as well as a 1D scale (denoted as $s \cdot \hat{t}(\phi, \theta)$). As an alternative which model rotation and translation jointly, we adapt the method of [68] which predicts 3D-3D correspondences for predefined keypoints of specific object classes. Here, we let the network predict **three 3D-3D correspondences** (denoted as $[3D - 3D]$). We compute the transformation that aligns these two sets of point triplets using Procrustes, which gives the relative rotation and translation between the two images. The models are trained end-to-end until convergence by supervising the output pose with the ground truth relative pose. We experimented with different loss functions and weighting between rotation and translation losses. For details, see supplemental material.

4 Map-Free Relocalization Datasets

In this section, we first discuss popular relocalization datasets and their limitations for map-free relocalization. Then, we introduce the Niantic map-free relocalization dataset which was collected specifically for the task. Finally, we define evaluation metrics used to benchmark baseline methods.

4.1 Existing Relocalization Datasets

One of the most commonly used datasets for visual relocalization is 7Scenes [62], consisting of seven small rooms scanned with KinectFusion [34]. 12Scenes [76] provides a few more, and slightly larger environments, while RIO10 [78] provides 10 scenes focusing on condition changes between mapping and query images. For outdoor relocalization, Cambridge Landmarks [37] and Aachen Day-Night [58], both consisting of large SfM reconstructions, are popular choices.

We find existing datasets poorly suited to benchmark map-free relocalization. Firstly, their scenes are not well captured by a single image which holds true for both indoor rooms and large-scale outdoor reconstructions. Secondly, the variability across scenes is extremely limited, with 1–12 distinct scenes in each single dataset. For comparison, our proposed dataset captures 655 distinct outdoor places of interest with 130 reserved for testing alone. Despite these issues, we have adapted the 7Scenes dataset to our map-free relocalization task.

Regarding relative pose estimation, ScanNet [19] and MegaDepth [39] have become popular test beds, *e.g.*, for learning-based 2D correspondence methods such as SuperGlue [54] and LoFTR [66]. However, both datasets do not feature distinctive mapping and query sequences as basis for a relocalization benchmark. Furthermore, MegaDepth camera poses do not have metric scale. In our experiments, we use ScanNet [19] as a training set for scene-agnostic relocalization methods to be tested on 7Scenes. In the supplemental material, we also provide ablation studies on metric relative pose accuracy on ScanNet.

4.2 Niantic Map-Free Relocalization Dataset

We introduce a new dataset for development and evaluation of map-free relocalization. The dataset consists of 655 outdoor scenes, each containing a small 'place of interest' such as a sculpture, sign, mural, etc., such that the place can be well-captured by a single image. Scenes of the dataset are shown in Fig. 4.

The scenes are split into 460 training scenes, 65 validation scenes, and 130 test scenes. Each training scene has two sequences of images, corresponding to two different scans of the scene. We provide the absolute pose of each training image, which allows determining the relative pose between any pair of training images. We also provide overlap scores between any pair of images (intra- and inter-sequence), which can be used to sample training pairs. For validation and test scenes, we provide a single reference image obtained from one scan and a sequence of query images and absolute poses from a different scan. Camera intrinsics are provided for all images in the dataset.

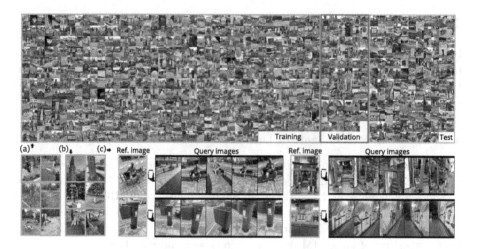

Fig. 4. Niantic map-free relocalisation dataset. (a) Dataset overview. Training (460 scenes), validation (65) and test (130) thumbnails. Better seen in color and magnified in electronic format. (b) Examples of training pairs sampled from training scenes. (c) Reference frame (enclosed in blue) and an example of query images. Query sequences have been sampled at relative temporal frames: 0%, 25%, 50%, 75% and 100% of the sequence duration. (Color figure online)

The Niantic map-free dataset was crowdsourced from members of the public who scanned places of interest using their mobile phones. Each scan contains video frames, intrinsics and (metric) poses estimated by ARKit (iOS) [2] or ARCore (Android) [31] frameworks and respective underlying implementations of Visual-Inertial Odometry and Visual-Inertial SLAM. We use automatic anonymization software to detect and blur faces and car license plates in frames. Scans were registered to each other using COLMAP [60]. First, we bundle adjust the scans individually by initializing from raw ARKit/ARCore poses. Then, the two 3D reconstructions are merged into a single reconstruction by matching features between scans, robustly aligning the two scans and bundle adjusting all frames jointly. We then compute a scale factor for each scan, so that the frames of the 3D reconstructions of each scan would (robustly) align to the raw ARKit/ARCore poses. Finally, the 3D reconstruction is rescaled using the average scale factor of the two scans. Further details are provided in supplemental material. Poses obtained via SfM constitute only a *pseudo* ground truth, and estimating their uncertainty bounds has recently been identified as an open problem in relocalization research [9]. However, as we will discuss below, given the challenging nature of map-free relocalization, we evaluate at much coarser error threshold than standard relocalization works. Thus, we expect our results to be less susceptible to inaccuracies in SfM pose optimization.

The places of interest in the Niantic map-free dataset are drawn from a wide variety of locations around the world and captured by a large number of people. This leads to a number of interesting challenges, such as variations in the

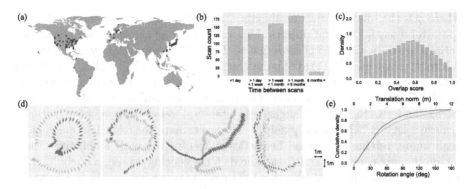

Fig. 5. Niantic map-free dataset statistics. (a) Geographic location of scans. (b) Time elapsed between two different scans from the same scene. (c) Visual overlap between training frames estimated using co-visible SfM points, inspired by [54]. (d) Sample of different dataset trajectories seen from above. Each plot represents one scene and shows two different trajectories corresponding to two different scans, one in each color. The direction of the arrows represent the camera viewing direction. Each trajectory has been subsampled for visualization. (e) Relative pose distribution between reference image and query images in the test set.

capture time, illumination, weather, season, and cameras, and even the geometry of the scene; and variations in the amount of overlap between the scans. Figure 5 summarizes these variations.

4.3 Evaluation Protocol

Our evaluation protocol consists of rotation, translation and reprojection errors computed using ground truth and estimated relative poses that are predicted for each query and reference image pair. Given estimated (R, t) and ground truth (R_{gt}, t_{gt}) poses, we compute the rotation error as the angle (in degrees) between predicted and ground truth rotations, $\angle(R, R_{gt})$. We measure the translation error as the Euclidean distance between predicted c and ground truth c_{gt} camera centers in world coordinate space, where $c = -R^T t$.

Our proposed reprojection error provides an intuitive measure of AR content misalignment. We were inspired by the Dense Correspondence Reprojection Error (DCRE) [78] which measures the average Euclidean distance between corresponding original pixel positions and reprojected pixel positions obtained via back-projecting depth maps. As our dataset does not contain depth maps we cannot compute the DCRE. Hence, we propose a Virtual Correspondence Reprojection Error (VCRE): ground truth and estimated transformations are used to project virtual 3D points, located in the query camera's local coordinate system. VCRE is the average Euclidean distance of the reprojection errors:

$$\text{VCRE} = \frac{1}{|\mathcal{V}|} \sum_{\mathbf{v} \in \mathcal{V}} \left\| \pi(\mathbf{v}) - \pi(T T_{gt}^{-1} \mathbf{v}) \right\|_2 \quad \text{with} \quad T = [R|t], \tag{1}$$

where π is the image projection function, and $\mathcal{V}$ is a set of 3D points in camera space representing virtual objects. For convenience of notation, we assume all entities are in homogeneous coordinates. To simulate an arbitrary placement of AR content, we use a 3D grid of points for $\mathcal{V}$ (4 in height, 7 in width, 7 in depth) with equal spacing of 30 cm and with an offset of 1.8 m along the camera axis. See supplemental material for a video visualisation, and an ablation showing that DCRE and VCRE are well-aligned. In standard relocalization, best methods achieve a DCRE below a few pixels [9]. However, map-free relocalization is more challenging, relying on learned heuristics to resolve the scale ambiguity. Thus, we apply more generous VCRE thresholds for accepting a pose, namely 5% and 10% of the image diagonal. While a 10% offset means a noticeable displacement of AR content, we argue that it can still yield an acceptable AR experience.

Our evaluation protocol also considers the confidence of pose estimates. Confidence enables the relocalization system to flag and potentially reject unreliable predictions. This is a crucial capability for a map-free relocalization system to be practical since a user might record query images without any visual overlap with the reference frame. A confidence can be estimated as the number of inlier correspondences in feature matching baselines. Given a confidence threshold, we can compute the ratio of query images with confidence greater-or-equal to the threshold, *i.e.*, the ratio of confident estimates or the ratio of non-rejected samples. Similarly, we compute the precision as the ratio of non-rejected query images for which the pose error (translation, rotation) or the reprojection error is acceptable (below a given threshold). Each confidence threshold provides a different trade-off between the number of images with an estimate and their precision. Models that are incapable of estimating a confidence will have a flat precision curve.

5   Experiments

We first report experiments on the 7Scenes [62] dataset, demonstrating that our baselines are competitive with the state of the art when a large number of mapping images is available. We also show that as the number of mapping images reduces, map-free suitable methods degrade more gracefully than traditional approaches. Additional relative pose estimation experiments on ScanNet [19] are reported in the supplement, to allow comparison of our baselines against previous methods. Finally, we report performance on the new Niantic map-free relocalization dataset and identify areas for improvement.

5.1   7Scenes

First, we compare methods described in Sects. 3.1 and 3.2 against traditional methods when all mapping frames are available. Figure 6a shows impressive scores of structure-based DSAC* [14] (trained with depth from PlaneR-CNN [40]), hLoc [53] and ActiveSearch [56,57]. When 5 reference frames can be

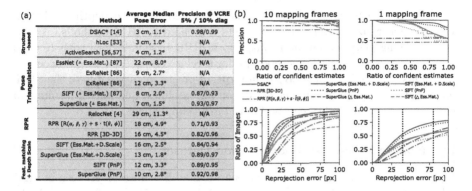

Fig. 6. 7Scenes results. (a) Using all mapping frames. Dataset-specific (7Scenes) methods in green, trained on SUNCG [65] in yellow, and trained on Scannet [19] in blue. (b) 10 and 1 mapping frame scenarios: precision curves (top), cumulative density of reprojection error (bottom). Dashed vertical lines indicate 1%, 5% and 10% of the image diagonal, correspondingly 8px, 40px and 80px. (Color figure online)

retrieved for each query using DenseVLAD [72] (following [87]), triangulation-based relative pose methods are competitive with structure-based methods, especially in average median rotation error. See results for EssNet [87], ExReNet [86] and our feature matching and triangulation baselines, denoted by △.

Closer to map-free relocalization, if for each query frame, we retrieve a single reference frame from the set of mapping images, the accuracy of metric relative pose estimation becomes more important, see the sections for relative pose regression (RPR) and *Feature Matching+D.Scale* in Fig. 6a. Unsurprisingly, methods in both families slightly degrade in performance, with Feature Matching + D.Scale methods beating RPR methods. However, all baselines remain competitive, despite depth [40] and RPR networks being trained on ScanNet [19] and evaluated on 7Scenes. High scores for all methods in Fig. 6a are partially explainable by the power of image retrieval and good coverage of the scene.

In map-free relocalization, the query and reference images could be far from each other. Thus, we evaluate the baselines on heavily sparsified maps, where metric relative pose accuracy is more important. We find the K most representative reference images of each scene by K-means clustering over DenseVLAD [72] descriptors of the mapping sequence. In Fig. 6b we show results for $K = 10$ and $K = 1$, where $K = 1$ corresponds to map-free relocalization. We show precision curves using a pose acceptance threshold of VCRE < 10% of the image diagonal (*i.e.*, 80px). We also plot the cumulative density of the VCRE. Unsurprisingly, pose triangulation methods fare well even when $K = 10$ but cannot provide estimates when $K = 1$. For $K = 1$, Feature Matching+D.Scale outperforms the competition. Specifically, SuperGlue (Ess.Mat.+D.Scale) recovers more than 50% of query images with a reprojection error below 40px.

DSAC* remains competitive in sparse regimes, but it requires training per scene, while the other baselines were trained on ScanNet. Both ScanNet and

7Scenes show very similar indoor scenes. Yet, single-image depth prediction seems to generalize better across datasets compared to RPR methods, as Feature Matching+D.Scale methods outperform RPR baselines both with $K = 10$ and $K = 1$ scenarios. RPR methods perform relatively well for larger accuracy thresholds but they perform poorly in terms of precision curves due to their lack of estimated confidence. Further details on all baselines, qualitative results and additional ablation studies can be found in the supplement.

5.2   Niantic Map-Free Relocalization Dataset

Figure 7 shows our main results on the Niantic map-free dataset. As seen in Fig 7 a, b and c, this dataset is much more challenging than 7Scenes for all methods. This is due to multiple factors: low overlap between query and reference images; the quality of feature matching is affected by variations in lighting, season, etc.; and the use of single-image depth prediction networks trained on KITTI for non-driving outdoor images.

In Fig. 7d and 7e we show results of the best methods in each family of baselines: RPR with 6D rotation and scaled translation parameterization and Super-Glue (Ess.Mat.+D.Scale). SuperGlue (Ess.Mat.+D.Scale) in Fig. 7e reports a median angular rotation below 10° for a large number of scenes. In these cases, the high variance of the median translation error is partly due to the variance of depth estimates. Further improvement of depth prediction methods in outdoor scenes should improve the metric accuracy of the translation error. Qualitative examples in Fig 7f shows where depth improvements could produce better results: both the angular pose and the absolute scale in the first row are accurate, while the second row has good angular pose and bad absolute scale.

The RPR method in Fig. 7d exhibits a different behavior: the average angular error is lower than for Feature Matching+D.Scale baselines, yet it rarely achieves high accuracy. This is also evident in Fig. 7 c, where Feature Matching+D.Scale methods outperform RPR methods for stricter thresholds, but degrade for broader thresholds. Indeed, when the geometric optimization fails due to poor feature matches, the estimated scaleless pose can be arbitrarily far from the ground truth. In contrast, RPR methods fail more gracefully due to adhering to the learned distribution of relative poses. For example, in Fig. 7c allowing for a coarser VCRE threshold of 10% of the image diagonal, the $[3D - 3D]$ and $[R(6D) + t]$ variants overtake all methods, including feature matching-based methods. Hence, RPR methods can be more accurate than feature matching at broad thresholds, but they offer lower precision in VCRE at practical thresholds.

RPR methods currently do not predict a confidence which prevents detecting spurious pose estimates, $e.g.$, when there is no visual overlap between images, as illustrated in the supplement. Although feature matching methods can estimate the confidence based on the number of inliers, the precision curves in Fig. 7a show that these confidences are not always reliable. Further research in modeling confidence of both families of methods could allow to combine their advantages.

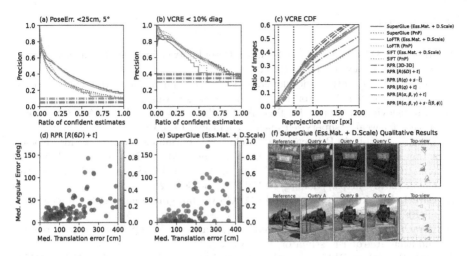

Fig. 7. Our dataset results. (a,b) Precision plots using pose error (a) and VCRE (b) thresholds. (c) VCRE CDF, vertical lines indicate 1, 5 and 10% of the image diagonal, corresp. 9px, 45px and 90px.(d,e) Scatter plot of median angular vs translation error for each scene, estimated using RPR $[R(6D) + t]$ (d) and SG [54] Ess.Mat.+D.scale (e). Each point represents a scene, and the colormap shows precision for pose error threshold 25 cm, 5°. (f) Qualitative results: the reference frame and three queries are shown for two scenes. The top view shows the ground truth (solid line, no fill) and estimated poses (dashed line, filled).

6   Conclusion and Future Work

We have proposed map-free relocalization, a new relocalization task. Through extensive experiments we demonstrate how existing methods for single-image depth prediction and relative pose regression can be used to address the task with some success. Our results suggest some directions for future research: improve the scale estimates by improving depth estimation in outdoor scenes; improve the accuracy of metric RPR methods; and derive a confidence for their estimates.

To facilitate further research, we have presented the Niantic map-free relocalization dataset and benchmark with a large number of diverse places of interest. We define an evaluation protocol to closely match AR use cases, and make the dataset and an evaluation service publicly available.

As methods for this task improve, we hope to evaluate at stricter pose error thresholds corresponding to visually more pleasing results. A version of map-free relocalization could use a burst of query frames rather than a single query frame to match some practical scenarios. Our dataset is already suitable for this variant of the task, so we hope to explore baselines for it in the future.

Acknowledgements. We thank Pooja Srinivas, Amy Duxbury, Camille François, Brian McClendon and their teams, for help with validating and anonymizing the dataset and Galen Han for his help in building the bundle adjustment pipeline. We also thank players of Niantic games for scanning places of interests.

References

1. Abouelnaga, Y., Bui, M., Ilic, S.: DistillPose: lightweight camera localization using auxiliary learning. In: IROS (2021)
2. Apple: ARKit. https://developer.apple.com/documentation/arkit/configuration_objects/understanding_world_tracking. Accessed 6 Mar 2022
3. Arandjelovic, R., Gronat, P., Torii, A., Pajdla, T., Sivic, J.: NetVLAD: CNN architecture for weakly supervised place recognition. In: CVPR (2016)
4. Balntas, V., Li, S., Prisacariu, V.: RelocNet: continuous metric learning relocalisation using neural nets. In: Ferrari, V., Hebert, M., Sminchisescu, C., Weiss, Y. (eds.) Computer Vision – ECCV 2018. LNCS, vol. 11218, pp. 782–799. Springer, Cham (2018). https://doi.org/10.1007/978-3-030-01264-9_46
5. Barath, D., Matas, J., Noskova, J.: MAGSAC: marginalizing sample consensus. In: CVPR (2019)
6. Barath, D., Noskova, J., Ivashechkin, M., Matas, J.: MAGSAC++, a fast, reliable and accurate robust estimator. In: CVPR (2020)
7. Besl, P., McKay, N.D.: A method for registration of 3-D shapes. IEEE TPAMI (1992)
8. Bhowmik, A., Gumhold, S., Rother, C., Brachmann, E.: Reinforced feature points: optimizing feature detection and description for a high-level task. In: CVPR (2020)
9. Brachmann, E., Humenberger, M., Rother, C., Sattler, T.: On the limits of pseudo ground truth in visual camera re-localisation. In: ICCV (2021)
10. Brachmann, E., et al.: DSAC-differentiable RANSAC for camera localization. In: CVPR (2017)
11. Brachmann, E., Rother, C.: Learning less is more - 6D camera localization via 3D surface regression. In: CVPR (2018)
12. Brachmann, E., Rother, C.: Expert sample consensus applied to camera re-localization. In: ICCV (2019)
13. Brachmann, E., Rother, C.: Neural-guided RANSAC: learning where to sample model hypotheses. In: ICCV (2019)
14. Brachmann, E., Rother, C.: Visual camera re-localization from RGB and RGB-D images using DSAC. IEEE TPAMI (2021)
15. Brahmbhatt, S., Gu, J., Kim, K., Hays, J., Kautz, J.: Geometry-aware learning of maps for camera localization. In: CVPR (2018)
16. Cai, R., Hariharan, B., Snavely, N., Averbuch-Elor, H.: Extreme rotation estimation using dense correlation volumes. In: CVPR (2021)
17. Campos, C., Tardós, J.D.: Scale-aware direct monocular odometry. In: ICUS (2021)
18. Chen, K., Snavely, N., Makadia, A.: Wide-baseline relative camera pose estimation with directional learning. In: CVPR (2021)
19. Dai, A., Chang, A.X., Savva, M., Halber, M., Funkhouser, T., Nießner, M.: ScanNet: Richly-annotated 3D reconstructions of indoor scenes. In: CVPR (2017)
20. Dai, A., Nießner, M., Zollhöfer, M., Izadi, S., Theobalt, C.: Bundlefusion: real-time globally consistent 3D reconstruction using on-the-fly surface reintegration. ACM TOG (2017)
21. Dusmanu, M., et al.: D2-net: a trainable CNN for joint detection and description of local features. In: CVPR (2019)
22. Eggert, D.W., Lorusso, A., Fisher, R.B.: Estimating 3-D rigid body transformations: a comparison of four major algorithms. Mach. Vision Appl. (1997)
23. Eigen, D., Puhrsch, C., Fergus, R.: Depth map prediction from a single image using a multi-scale deep network. In: NeurIPS (2014)

24. En, S., Lechervy, A., Jurie, F.: RPNet: an end-to-end network for relative camera pose estimation. In: Leal-Taixé, L., Roth, S. (eds.) ECCV 2018. LNCS, vol. 11129, pp. 738–745. Springer, Cham (2019). https://doi.org/10.1007/978-3-030-11009-3_46

25. Facil, J.M., Ummenhofer, B., Zhou, H., Montesano, L., Brox, T., Civera, J.: CAM-Convs: camera-aware multi-scale convolutions for single-view depth. In: CVPR (2019)

26. Fischler, M.A., Bolles, R.C.: Random sample consensus: a paradigm for model fitting with applications to image analysis and automated cartography. Commun. ACM (1981)

27. Gao, X.S., Hou, X.R., Tang, J., Cheng, H.F.: Complete solution classification for the perspective-three-point problem. IEEE TPAMI (2003)

28. Geiger, A., Lenz, P., Urtasun, R.: Are we ready for autonomous driving? The KITTI vision benchmark suite. In: CVPR (2012)

29. Godard, C., Mac Aodha, O., Brostow, G.J.: Unsupervised monocular depth estimation with left-right consistency. In: CVPR (2017)

30. Godard, C., Mac Aodha, O., Firman, M., Brostow, G.J.: Digging into self-supervised monocular depth prediction. In: ICCV (2019)

31. Google: ARCore. https://developers.google.com/ar/develop/fundamentals. Accessed 6 Mar 2022

32. Hartley, R., Zisserman, A.: Multiple View Geometry in Computer Vision. Cambridge University Press, Cambridge (2003)

33. Humenberger, M., et al.: Robust image retrieval-based visual localization using kapture (2020)

34. Izadi, S., et al.: Kinectfusion: real-time 3D reconstruction and interaction using a moving depth camera. In: ACM UIST (2011)

35. Jin, Y., Mishkin, D., Mishchuk, A., Matas, J., Fua, P., Yi, K.M., Trulls, E.: Image matching across wide baselines: from paper to practice. IJCV (2020)

36. Kendall, A., Cipolla, R.: Geometric loss functions for camera pose regression with deep learning. In: CVPR (2017)

37. Kendall, A., Grimes, M., Cipolla, R.: PoseNet: a convolutional network for real-time 6-DOF camera relocalization. In: CVPR (2015)

38. Li, X., Wang, S., Zhao, Y., Verbeek, J., Kannala, J.: Hierarchical scene coordinate classification and regression for visual localization. In: CVPR (2020)

39. Li, Z., Snavely, N.: MegaDepth: learning single-view depth prediction from internet photos. In: CVPR (2018)

40. Liu, C., Kim, K., Gu, J., Furukawa, Y., Kautz, J.: PlaneRCNN: 3D plane detection and reconstruction from a single image. In: CVPR (2019)

41. Lowe, D.G.: Distinctive image features from scale-invariant keypoints. IJCV (2004)

42. Melekhov, I., Ylioinas, J., Kannala, J., Rahtu, E.: Relative camera pose estimation using convolutional neural networks. In: Blanc-Talon, J., Penne, R., Philips, W., Popescu, D., Scheunders, P. (eds.) ACIVS 2017. LNCS, vol. 10617, pp. 675–687. Springer, Cham (2017). https://doi.org/10.1007/978-3-319-70353-4_57

43. Newcombe, R.A., et al.: KinectFusion: real-time dense surface mapping and tracking. In: ISMAR (2011)

44. Nister, D.: An efficient solution to the five-point relative pose problem. IEEE TPAMI (2004)

45. Pion, N., Humenberger, M., Csurka, G., Cabon, Y., Sattler, T.: Benchmarking image retrieval for visual localization. In: 3DV (2020)

46. Radenović, F., Tolias, G., Chum, O.: CNN image retrieval learns from BoW: unsupervised fine-tuning with hard examples. In: Leibe, B., Matas, J., Sebe, N., Welling, M. (eds.) ECCV 2016. LNCS, vol. 9905, pp. 3–20. Springer, Cham (2016). https://doi.org/10.1007/978-3-319-46448-0_1

47. Raguram, R., Frahm, J.-M., Pollefeys, M.: A comparative analysis of RANSAC techniques leading to adaptive real-time random sample consensus. In: Forsyth, D., Torr, P., Zisserman, A. (eds.) ECCV 2008. LNCS, vol. 5303, pp. 500–513. Springer, Heidelberg (2008). https://doi.org/10.1007/978-3-540-88688-4_37

48. Ranftl, R., Bochkovskiy, A., Koltun, V.: Vision transformers for dense prediction. In: ICCV (2021)

49. Ranftl, R., Koltun, V.: Deep fundamental matrix estimation. In: Ferrari, V., Hebert, M., Sminchisescu, C., Weiss, Y. (eds.) ECCV 2018. LNCS, vol. 11205, pp. 292–309. Springer, Cham (2018). https://doi.org/10.1007/978-3-030-01246-5_18

50. Ranftl, R., Lasinger, K., Hafner, D., Schindler, K., Koltun, V.: Towards robust monocular depth estimation: mixing datasets for zero-shot cross-dataset transfer. IEEE TPAMI (2020)

51. Rau, A., Garcia-Hernando, G., Stoyanov, D., Brostow, G.J., Turmukhambetov, D.: Predicting visual overlap of images through interpretable non-metric box embeddings. In: Vedaldi, A., Bischof, H., Brox, T., Frahm, J.-M. (eds.) ECCV 2020. LNCS, vol. 12350, pp. 629–646. Springer, Cham (2020). https://doi.org/10.1007/978-3-030-58558-7_37

52. Revaud, J., Weinzaepfel, P., de Souza, C.R., Humenberger, M.: R2D2: repeatable and reliable detector and descriptor. In: NeurIPS (2019)

53. Sarlin, P.E., Cadena, C., Siegwart, R., Dymczyk, M.: From coarse to fine: robust hierarchical localization at large scale. In: CVPR (2019)

54. Sarlin, P.E., DeTone, D., Malisiewicz, T., Rabinovich, A.: SuperGlue: learning feature matching with graph neural networks. In: CVPR (2020)

55. Sattler, T., Leibe, B., Kobbelt, L.: Fast image-based localization using direct 2D-to-3D matching. In: ICCV (2011)

56. Sattler, T., Leibe, B., Kobbelt, L.: Improving image-based localization by active correspondence search. In: Fitzgibbon, A., Lazebnik, S., Perona, P., Sato, Y., Schmid, C. (eds.) ECCV 2012. LNCS, vol. 7572, pp. 752–765. Springer, Heidelberg (2012). https://doi.org/10.1007/978-3-642-33718-5_54

57. Sattler, T., Leibe, B., Kobbelt, L.: Efficient & effective prioritized matching for large-scale image-based localization. IEEE TPAMI (2017)

58. Sattler, T., et al.: Benchmarking 6DOF outdoor visual localization in changing conditions. In: CVPR (2018)

59. Sattler, T., Weyand, T., Leibe, B., Kobbelt, L.: Image retrieval for image-based localization revisited. In: BMVC (2012)

60. Schönberger, J.L., Frahm, J.M.: Structure-from-motion revisited. In: CVPR (2016)

61. Shavit, Y., Ferens, R., Keller, Y.: Learning multi-scene absolute pose regression with transformers. In: ICCV (2021)

62. Shotton, J., Glocker, B., Zach, C., Izadi, S., Criminisi, A., Fitzgibbon, A.: Scene coordinate regression forests for camera relocalization in RGB-D images. In: CVPR (2013)

63. Silberman, N., Hoiem, D., Kohli, P., Fergus, R.: Indoor segmentation and support inference from RGBD images. In: Fitzgibbon, A., Lazebnik, S., Perona, P., Sato, Y., Schmid, C. (eds.) ECCV 2012. LNCS, vol. 7576, pp. 746–760. Springer, Heidelberg (2012). https://doi.org/10.1007/978-3-642-33715-4_54

64. Snavely, N., Seitz, S.M., Szeliski, R.: Photo tourism: exploring photo collections in 3D. In: ACM SIGGRAPH (2006)

65. Song, S., Yu, F., Zeng, A., Chang, A.X., Savva, M., Funkhouser, T.: Semantic scene completion from a single depth image. In: CVPR (2017)
66. Sun, J., Shen, Z., Wang, Y., Bao, H., Zhou, X.: LoFTR: detector-free local feature matching with transformers. In: CVPR (2021)
67. Sun, W., Jiang, W., Trulls, E., Tagliasacchi, A., Yi, K.M.: ACNe: attentive context normalization for robust permutation-equivariant learning. In: CVPR (2020)
68. Suwajanakorn, S., Snavely, N., Tompson, J., Norouzi, M.: Discovery of latent 3D keypoints via end-to-end geometric reasoning. In: NeurIPS (2018)
69. Tateno, K., Tombari, F., Laina, I., Navab, N.: CNN-SLAM: real-time dense monocular slam with learned depth prediction. In: CVPR (2017)
70. Tiwari, L., Ji, P., Tran, Q.-H., Zhuang, B., Anand, S., Chandraker, M.: Pseudo RGB-D for self-improving monocular SLAM and depth prediction. In: Vedaldi, A., Bischof, H., Brox, T., Frahm, J.-M. (eds.) ECCV 2020. LNCS, vol. 12356, pp. 437–455. Springer, Cham (2020). https://doi.org/10.1007/978-3-030-58621-8_26
71. Toft, C., Turmukhambetov, D., Sattler, T., Kahl, F., Brostow, G.J.: Single-image depth prediction makes feature matching easier. In: Vedaldi, A., Bischof, H., Brox, T., Frahm, J.-M. (eds.) ECCV 2020. LNCS, vol. 12361, pp. 473–492. Springer, Cham (2020). https://doi.org/10.1007/978-3-030-58517-4_28
72. Torii, A., Arandjelovic, R., Sivic, J., Okutomi, M., Pajdla, T.: 24/7 place recognition by view synthesis. In: CVPR (2015)
73. Türkoğlu, M.Ö., Brachmann, E., Schindler, K., Brostow, G., Monszpart, A.: Visual camera re-localization using graph neural networks and relative pose supervision. In: 3DV (2021)
74. Tyszkiewicz, M., Fua, P., Trulls, E.: Disk: learning local features with policy gradient. In: NeurIPS (2020)
75. Ummenhofer, B., et al.: Demon: depth and motion network for learning monocular stereo. In: CVPR (2017)
76. Valentin, J., et al.: Learning to navigate the energy landscape. In: 3DV (2016)
77. Walch, F., Hazirbas, C., Leal-Taixé, L., Sattler, T., Hilsenbeck, S., Cremers, D.: Image-based localization using LSTMs for structured feature correlation. In: ICCV (2017)
78. Wald, J., Sattler, T., Golodetz, S., Cavallari, T., Tombari, F.: Beyond controlled environments: 3D camera re-localization in changing indoor scenes. In: Vedaldi, A., Bischof, H., Brox, T., Frahm, J.-M. (eds.) ECCV 2020. LNCS, vol. 12352, pp. 467–487. Springer, Cham (2020). https://doi.org/10.1007/978-3-030-58571-6_28
79. Watson, J., Firman, M., Brostow, G.J., Turmukhambetov, D.: Self-supervised monocular depth hints. In: ICCV (2019)
80. Winkelbauer, D., Denninger, M., Triebel, R.: Learning to localize in new environments from synthetic training data. In: ICRA (2021)
81. Wu, C.: VisualSFM: a visual structure from motion system (2011). http://ccwu.me/vsfm/
82. Wu, C.: Towards linear-time incremental structure from motion. In: 3DV (2013)
83. Xian, K., et al.: Monocular relative depth perception with web stereo data supervision. In: CVPR (2018)
84. Yang, N., Wang, R., Stückler, J., Cremers, D.: Deep virtual stereo odometry: leveraging deep depth prediction for monocular direct sparse odometry. In: Ferrari, V., Hebert, M., Sminchisescu, C., Weiss, Y. (eds.) ECCV 2018. LNCS, vol. 11212, pp. 835–852. Springer, Cham (2018). https://doi.org/10.1007/978-3-030-01237-3_50
85. Yi, K.M., Trulls, E., Ono, Y., Lepetit, V., Salzmann, M., Fua, P.: Learning to find good correspondences. In: CVPR (2018)

86. Zhang, W., Kosecka, J.: Image based localization in urban environments. In: 3DV (2006)
87. Zhou, Q., Sattler, T., Pollefeys, M., Leal-Taixe, L.: To learn or not to learn: visual localization from essential matrices. In: ICRA (2020)
88. Zhou, T., Brown, M., Snavely, N., Lowe, D.G.: Unsupervised learning of depth and ego-motion from video. In: CVPR (2017)
89. Zhou, Y., Barnes, C., Lu, J., Yang, J., Li, H.: On the continuity of rotation representations in neural networks. In: CVPR (2019)

Self-distilled Feature Aggregation for Self-supervised Monocular Depth Estimation

Zhengming Zhou[1,2] and Qiulei Dong[1,2,3]

[1] National Laboratory of Pattern Recognition, Institute of Automation, Chinese Academy of Sciences, Beijing 100190, China
zhouzhengming2020@ia.ac.cn, qldong@nlpr.ia.ac.cn
[2] School of Artificial Intelligence, University of Chinese Academy of Sciences, Beijing 100049, China
[3] Center for Excellence in Brain Science and Intelligence Technology, Chinese Academy of Sciences, Beijing 100190, China

Abstract. Self-supervised monocular depth estimation has received much attention recently in computer vision. Most of the existing works in literature aggregate multi-scale features for depth prediction via either straightforward concatenation or element-wise addition, however, such feature aggregation operations generally neglect the contextual consistency between multi-scale features. Addressing this problem, we propose the Self-Distilled Feature Aggregation (SDFA) module for simultaneously aggregating a pair of low-scale and high-scale features and maintaining their contextual consistency. The SDFA employs three branches to learn three feature offset maps respectively: one offset map for refining the input low-scale feature and the other two for refining the input high-scale feature under a designed self-distillation manner. Then, we propose an SDFA-based network for self-supervised monocular depth estimation, and design a self-distilled training strategy to train the proposed network with the SDFA module. Experimental results on the KITTI dataset demonstrate that the proposed method outperforms the comparative state-of-the-art methods in most cases. The code is available at https://github.com/ZM-Zhou/SDFA-Net_pytorch.

Keywords: Monocular depth estimation · Self-supervised learning · Self-distilled feature aggregation

1 Introduction

Monocular depth estimation is a challenging topic in computer vision, which aims to predict pixel-wise scene depths from single images. Recently, self-supervised methods [12,14,16] for monocular depth learning have received much attention, due to the fact that they could be trained without ground truth depth labels.

Supplementary Information The online version contains supplementary material available at https://doi.org/10.1007/978-3-031-19769-7_41.

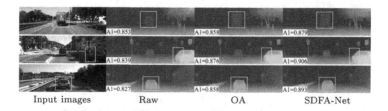

Input images   Raw   OA   SDFA-Net

Fig. 1. Depth map comparison by the networks (described in Sect. 4.3) with different feature aggregation techniques (the straightforward concatenation (Raw), the Offset-based Aggregation (OA) [25], and the proposed SDFA module) on KITTI [13], where 'A1' is an accuracy metric.

The existing methods for self-supervised monocular depth estimation could be generally categorized into two groups according to the types of training data: the methods which are trained with monocular video sequences [30,50,57] and the methods which are trained with stereo pairs [12,14,16]. Regardless of the types of training data, many existing methods [15,16,40–42,45,53] employ various encoder-decoder architectures for depth prediction, and the estimation processes could be considered as a general process that sequentially learns multi-scale features and predicts scene depths. In most of these works, their encoders extract multi-scale features from input images, and their decoders gradually aggregate the extracted multi-scale features via either straightforward concatenation or element-wise addition, however, although such feature aggregation operations have demonstrated their effectiveness to some extent in these existing works, they generally neglect the contextual consistency between the multi-scale features, i.e., the corresponding regions of the features from different scales should contain the contextual information for similar scenes. This problem might harm a further performance improvement of these works.

Addressing this problem, we firstly propose the Self-Distilled Feature Aggregation (SDFA) module for simultaneously aggregating a pair of low-scale and high-scale features and maintaining their contextual consistency, inspired by the success of a so-called 'feature-alignment' module that uses an offset-based aggregation technique for handling the image segmentation task [1,24,25,34,35]. It has to be pointed out that such a feature-alignment module could not guarantee its superior effectiveness in the self-supervised monocular depth estimation community. As shown in Fig. 1, when the Offset-based Aggregation (OA) technique is straightforwardly embedded into an encoder-decoder network for handling the monocular depth estimation task, although it performs better than the concatenation-based aggregation (Raw) under the metric 'A1' which is considered as an accuracy metric, it is prone to assign inaccurate depths into pixels on occlusion regions (e.g., the regions around the contour of the vehicles in the yellow boxes in Fig. 1) due to the calculated inaccurate feature alignment by the feature-alignment module for these pixels. In order to solve this problem, the proposed SDFA module employs three branches to learn three feature offset maps respectively: one offset map is used for refining the input low-scale feature, and the other two are jointly used for refining the input high-scale feature under a designed self-distillation manner.

Then, we propose an SDFA-based network for self-supervised monocular depth estimation, called SDFA-Net, which is trained with a set of stereo image pairs. The SDFA-Net employs an encoder-decoder architecture, which uses a modified version of tiny Swin-transformer [37] as its encoder and an ensemble of multiple SDFA modules as its decoder. In addition, a self-distilled training strategy is explored to train the SDFA-Net, which selects reliable depths by two principles from a raw depth prediction, and uses the selected depths to train the network by self-distillation.

In sum, our main contributions include:

- We propose the Self-Distilled Feature Aggregation (SDFA) module, which could effectively aggregate a low-scale feature with a high-scale feature under a self-distillation manner.
- We propose the SDFA-Net with the explored SDFA module for self-supervised monocular depth estimation, where an ensemble of SDFA modules is employed to both aggregate multi-scale features and maintain the contextual consistency among multi-scale features for depth prediction.
- We design the self-distilled training strategy for training the proposed network with the explored SDFA module. The proposed network achieves a better performance on the KITTI dataset [13] than the comparative state-of-the-art methods in most cases as demonstrated in Sect. 4.

2   Related Work

Here, we review the two groups of self-supervised monocular depth estimation methods, which are trained with monocular videos and stereo pairs respectively.

2.1   Self-supervised Monocular Training

The methods trained with monocular video sequences aim to simultaneously estimate the camera poses and predict the scene depths. An end-to-end method was proposed by Zhou et al. [57], which comprised two separate networks for predicting the depths and camera poses. Guizilini et al. [20] proposed PackNet where the up-sampling and down-sampling operations were re-implemented by 3D convolutions. Godard et al. [15] designed the per-pixel minimum reprojection loss, the auto-mask loss, and the full-resolution sampling in Monodepth2. Shu et al. [45] designed the feature-metric loss defined on feature maps for handling less discriminative regions in the images.

Additionally, the frameworks which learnt depths by jointly using monocular videos and extra semantic information were investigated in [4,21,31,33]. Some works [2,28,55] investigated jointly learning the optical flow, depth, and camera pose. Some methods were designed to handle self-supervised monocular depth estimation in the challenging environments, such as the indoor environments [27,56] and the nighttime environments [36,49].

2.2 Self-supervised Stereo Training

The methods trained with stereo image pairs generally estimate scene depths by predicting the disparity between an input pair of stereo images. A pioneering work was proposed by Garg et al. [12], which used the predicted disparities and one image of a stereo pair to synthesize the other image at the training stage. Godard et al. [14] proposed a left-right disparity consistency loss to improve the robustness of monocular depth estimation. Tosi et al. [47] proposed monoResMatch which employed three hourglass-structure networks for extracting features, predicting raw disparities and refining the disparities respectively. FAL-Net [16] was proposed to learn depths under an indirect way, where the disparity was represented by the weighted sum of a set of discrete disparities and the network predicted the probability map of each discrete disparity. Gonzalez and Kim [18] proposed the ambiguity boosting, which improved the accuracy and consistency of depth predictions.

Additionally, for further improving the performance of self-supervised monocular depth estimation, several methods used some extra information (e.g., disparities generated with either traditional algorithms [47,53,59] or extra networks [3,6,22], and semantic segmentation labels [59]). For example, Watson et al. [53] proposed depth hints which calculated with Semi Global Matching [23] and used to guide the network to learn accurate depths. Other methods employed knowledge distillation [19] for self-supervised depth estimation [40,41]. Peng et al. [40] generated an optimal depth map from the multi-scale outputs of a network, and trained the same network with this distilled depth map.

3   Methodology

In this section, we firstly introduce the architecture of the proposed network. Then, the Self-Distilled Feature Aggregation (SDFA) is described in detail. Finally, the designed self-distilled training strategy is given.

3.1 Network Architecture

The proposed SDFA-Net adopts an encoder-decoder architecture with skip connections for self-supervised monocular depth estimation, as shown in Fig. 2.

Encoder. Inspired by the success of vision transformers [5,10,44,51,54] in various visual tasks, we introduce the following modified version of tiny Swin-transformer [37] as the backbone encoder to extract multi-scale features from an input image $I^l \in \mathbb{R}^{3 \times H \times W}$. The original Swin-transformer contains four transformer stages, and it pre-processes the input image through a convolutional layer with stride $= 4$, resulting in 4 intermediate features $[C_1, C_2, C_3, C_4]$ with the resolutions of $[\frac{H}{4} \times \frac{W}{4}, \frac{H}{8} \times \frac{W}{8}, \frac{H}{16} \times \frac{W}{16}, \frac{H}{32} \times \frac{W}{32}]$. Considering rich spatial information is important for depth estimation, we change the convolutional layer with stride $= 4$ in the original Swin-transformer to that with stride $= 2$ in order to keep more high-resolution image information, and accordingly, the resolutions of the

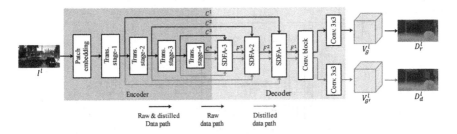

Fig. 2. Architecture of SDFA-Net. SDFA-Net is used to predict the volume-based depth representation V_g^l and the depth map D^l from the input image I^l. The features extracted by the encoder could be passed through the raw and distilled data paths to predict raw and distilled depth maps D_r^l, D_d^l respectively.

features extracted from the modified version of Swin-transformer are twice as they were, i.e. $[\frac{H}{2} \times \frac{W}{2}, \frac{H}{4} \times \frac{W}{4}, \frac{H}{8} \times \frac{W}{8}, \frac{H}{16} \times \frac{W}{16}]$.

Decoder. The decoder uses the multi-scale features $\{C^i\}_{i=1}^4$ extracted from the encoder as its input, and it outputs the disparity-logit volume V_g^l as the scene depth representation. The decoder is comprised of three SDFA modules (denoted as SDFA-1, SDFA-2, SDFA-3 as shown in Fig. 2), a convolutional block and two convolutional output layers. The SDFA module is proposed for adaptively aggregating the multi-scale features with learnable offset maps, which would be described in detail in Sect. 3.2. The convolutional block is used for restoring the spatial resolution of the aggregated feature to the size of the input image, consisting of a nearest up-sampling operation and two 3×3 convolutional layers with the ELU activation [7]. For training the SDFA modules under the self-distillation manner and avoiding training two networks, the encoded features could be passed through the decoder via two data paths, and be aggregated by different offset maps. The two 3×3 convolutional output layers are used to predict two depth representations at the training stage, which are defined as the raw and distilled depth representations respectively. Accordingly, the two data paths are defined as the raw data path and the distilled data path. Once the proposed network is trained, given an arbitrary test image, only the outputted distilled depth representation $V_{g'}^l$ is used as its final depth prediction at the inference stage.

3.2 Self-Distilled Feature Aggregation

The Self-Distilled Feature Aggregation (SDFA) module with learnable feature offset maps is proposed to adaptively aggregate the multi-scale features and maintain the contextual consistency between them. It jointly uses a low-scale decoded feature from the previous layer and its corresponding feature from the encoder as the input, and it outputs the aggregated feature as shown in Fig. 3. The two features are inputted from either raw or distilled data path.

As seen from Fig. 3, under the designed module SDFA-i ($i = 1, 2, 3$), the feature F^{i+1} (specially $F^4 = C^4$) from its previous layer is passed through a

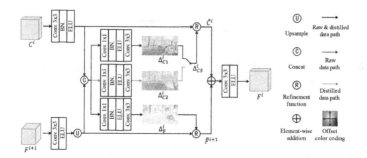

Fig. 3. Self-Distilled Feature Aggregation. Different offset maps are chosen in different data paths.

3×3 convolutional layer with the ELU activation [7] and up-sampled by the standard bilinear interpolation. Meanwhile, an additional 3×3 convolutional layer with the Batch Normalization (BN) [26] and ELU activation is used to adjust the channel dimensions of the corresponding feature C^i from the encoder to be the same as that of F^{i+1}. Then, the obtained two features via the above operations are concatenated together and passed through three branches for predicting the offset maps Δ^i_F, Δ^i_{C1}, and Δ^i_{C2}. The offset map Δ^i_F is used to refine the up-sampled F^{i+1}. According to the used data path P, a switch operation '$\mathcal{S}(\Delta^i_{C1}, \Delta^i_{C2}|P)$' is used to select an offset map from Δ^i_{C1} and Δ^i_{C2} for refining the adjusted C^i, which is formulated as:

$$\Delta^i_{CS} = \mathcal{S}(\Delta^i_{C1}, \Delta^i_{C2}|P) = \begin{cases} \Delta^i_{C1}, & P = \text{raw} \\ \Delta^i_{C2}, & P = \text{distilled} \end{cases} . \tag{1}$$

After obtaining the offset maps, a refinement function '$\mathcal{R}(F, \Delta)$' is designed to refine the feature F by the guidance of the offset map Δ, which is implemented by the bilinear interpolation kernel. Specifically, to generate a refined feature $\tilde{F}(p)$ on the position $p = [x, y]^\top$ from $F \in \mathbb{R}^{C \times H \times W}$ with an offset $\Delta(p)$, the refinement function is formulated as:

$$\tilde{F}(p) = \mathcal{R}(F, \Delta(p)) = \langle F(p + \Delta(p)) \rangle \quad , \tag{2}$$

where '$\langle \cdot \rangle$' denotes the bilinear sampling operation. Accordingly, the refined features $\tilde{F}^{i+1}$ and $\tilde{C}^i$ are generated with:

$$\tilde{F}^{i+1} = \mathcal{R}\left(\mathcal{U}\left(\text{conv}\left(F^{i+1}\right)\right), \Delta^i_F\right) \quad , \tag{3}$$

$$\tilde{C}^i = \mathcal{R}\left(\text{convb}\left(C^i\right), \Delta^i_{CS}\right) \quad , \tag{4}$$

where '$\mathcal{U}(\cdot)$' denotes the bilinear up-sample, 'conv$(\cdot)$' denotes the 3×3 convolutional layer with the ELU activation, and 'convb$(\cdot)$' denotes the 3×3 convolutional layer with the BN and ELU. Finally, the aggregated feature F^i is obtained with the two refined features as:

$$F^i = \text{conv}\left(\tilde{C}^i \oplus \tilde{F}^{i+1}\right) \quad , \tag{5}$$

where '⊕' denotes the element-wise addition.

Considering that (i) the offsets learnt with self-supervised training are generally suboptimal because of occlusions and image ambiguities and (ii) more effective offsets are expected to be learnt by utilizing extra clues (e.g., some reliable depths from the predicted depth map in Sect. 3.3), SDFA is trained in the following designed self-distillation manner.

3.3 Self-distilled Training Strategy

In order to train the proposed network with a set of SDFA modules, we design a self-distilled training strategy as shown in Fig. 4, which divides each training iteration into three sequential steps: the self-supervised forward propagation, the self-distilled forward propagation and the loss computation.

Self-supervised Forward Propagation. In this step, the network takes the left image I^l in a stereo pair as the input, outputting a left-disparity-logit volume $V_g^l \in \mathbb{R}^{N \times H \times W}$ via the raw data path. As done in [16], I^l and V_g^l are used to synthesize the right image $\hat{I}^r$, while the raw depth map D_r^l is obtained with V_g^l.

Specifically, given the minimum and maximum disparities $d_{\min}$ and $d_{\max}$, we firstly define the discrete disparity level d_n by the exponential quantization [16]:

$$d_n = d_{\max} \left(\frac{d_{\min}}{d_{\max}} \right)^{\frac{n}{N-1}} , \quad n = 0, 1, ..., N-1 , \tag{6}$$

where N is the total number of the disparity levels. For synthesizing the right images, each channel of V_g^l is shifted with the corresponding disparity d_n, generating the right-disparity-logit volume $\hat{V}_g^r$. A right-disparity-probability volume $\hat{V}_p^r$ is obtained by passing $\hat{V}_g^r$ through the softmax operation and is employed for stereoscopic image synthesis as:

$$\hat{I}^r = \sum_{n=0}^{N-1} \hat{V}_{p|n}^r \odot I_n^l , \tag{7}$$

where $\hat{V}_{p|n}^r$ is the nth channel of $\hat{V}_p^r$, '$\odot$' denotes the element-wise multiplication, and I_n^l is the left image shifted with d_n. For obtaining the raw depth map, V_g^l is passed through the softmax operation to generate the left-disparity-probability volume V_p^l. According to the stereoscopic image synthesis, the nth channel of V_p^l approximately equals the probability map of d_n. And a pseudo-disparity map d_+^l is obtained as:

$$d_+^l = \sum_{n=0}^{N-1} V_{p|n}^l \cdot d_n . \tag{8}$$

Given the baseline length B of the stereo pair and the horizontal focal length f_x of the left camera, the raw depth map D_r^l for the left image is calculated via $D_r^l = \frac{B f_x}{d_+^l}$.

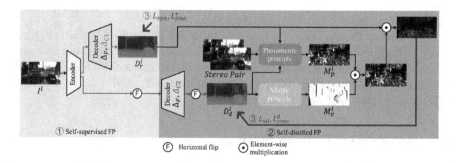

Fig. 4. Self-distilled training strategy. It comprises three steps: the self-supervised Forward Propagation (FP), the self-distilled Forward Propagation (FP), and the loss computation.

Self-distilled Forward Propagation. For alleviating the influence of occlusions and learning more accurate depths, the multi-scale features $\{C^i\}_{i=1}^4$ extracted by the encoder are horizontally flipped and passed through the decoder via the distilled data path, outputting a new left-disparity-logit volume $V_{g'}^l$. After flipping $V_{g'}^l$ back, the distilled disparity map d_d^l and distilled depth map D_d^l are obtained by the flipped $V_{g'}^l$ as done in Eq. (8). For training the SDFA-Net under the self-distillation manner, we employs two principles to select reliable depths from the raw depth map D_r^l: the photometric principle and the visible principle. As shown in Fig. 4, we implement the two principles with two binary masks, respectively.

The photometric principle is used to select the depths from D_r^l in pixels where they make for a superior reprojected left image. Specifically, for an arbitrary pixel coordinate p in the left image, its corresponding coordinate p' in the right image is obtained with a left depth map D^l as:

$$p' = p - \left[\frac{B f_x}{D^l(p)}, 0\right]^\top .\qquad(9)$$

Accordingly, the reprojected left image $\hat{I}^l$ is obtained by assigning the RGB value of the right image pixel p' to the pixel p of $\hat{I}^l$. Then, a weighted sum of the L_1 loss and the structural similarity (SSIM) loss [52] is employed to measure the photometric difference between the reprojected image $\hat{I}^l$ and raw image I^l:

$$l(D^l) = \left(\alpha \left\| \hat{I}^l - I^l \right\|_1 + (1-\alpha)\text{SSIM}(\hat{I}^l, I^l)\right) ,\qquad(10)$$

where α is a balance parameter. The photometric principle is formulated as:

$$M_p^l = \left[\left(l\left(D_r^l\right) - l\left(D_d^l\right)\right) < \epsilon\right) \cap l\left(D_r^l\right) < t_1\right] ,\qquad(11)$$

where '$[\cdot]$' is the iverson bracket, ϵ and t_1 are predefined thresholds. The second term in the iverson bracket is used to ignore the inaccurate depths with large photometric errors.

The visible principle is used to eliminate the potential inaccurate depths in the regions that are visible only in the left image, e.g., the pixels occluded in the right image or out of the edge of the right image. The pixels which are occluded in the right image could be found with the left disparity map d^l. Specifically, it is observed that for an arbitrary pixel location $p = [x, y]^\top$ and its horizontal right neighbor $p_i = [x + i, y]^\top (i = 1, 2, ..., K)$ in the left image, if the corresponding location of p is occluded by that of p_i in the right image, the difference between their disparities should be close or equal to the difference of their horizontal coordinates [59]. Accordingly, the occluded mask calculated with the distilled disparity map d_d^l is formulated as:

$$M_{occ}^l(p) = \left[\min_i \left(\left| d_d^l(p_i) - d_d^l(p) - i \right| \right) < t_2 \right] \quad , \tag{12}$$

where t_2 is a predefined threshold. Additionally, an out-of-edge mask M_{out}^l [38] is jointly utilized for filtering out the pixels whose corresponding locations are out of the right image. Accordingly, the visible principle is formulated as:

$$M_v^l = M_{occ}^l \odot M_{out}^l \quad . \tag{13}$$

Loss Computation. In this step, the total loss is calculated for training the SDFA-Net. It is noted that the raw depth map D_r^l is learnt by maximizing the similarity between the real image I^r and the synthesized image $\hat{I}^r$, while the distilled depth map D_d^l is learnt with self-distillation. The total loss function is comprised of four terms: the image synthesis loss L_{syn}, the self-distilled loss L_{sd}, the raw smoothness loss L_{smo}^+, and the distilled smoothness loss L_{smo}^d.

The image synthesis loss L_{syn} contains the L_1 loss and the perceptual loss [29] for reflecting the similarity between the real right image I^r and the synthesized right image $\hat{I}^r$:

$$L_{syn} = \left\| \hat{I}^r - I^r \right\|_1 + \beta \sum_{i=1,2,3} \left\| \phi_i(\hat{I}^r) - \phi_i(I^r) \right\|_2 \quad , \tag{14}$$

where '$\| \cdot \|_1$' and '$\| \cdot \|_2$' represent the L_1 and L_2 norm, $\phi_i(\cdot)$ denotes the output of ith pooling layer of a pretrained VGG19 [46], and β is a balance parameter.

The self-distilled loss L_{sd} adopts the L_1 loss to distill the accurate depths from the raw depth map D_r^l to the distilled depth map D_d^l, where the accurate depths are selected by the photometric and visible masks M_p^l and M_v^l:

$$L_{sd} = M_p^l \odot M_v^l \odot \left\| D_d^l - D_r^l \right\|_1 \quad . \tag{15}$$

The edge-aware smoothness loss L_{smo} is employed for constraining the continuity of the pseudo disparity map d_+^l and the distilled disparity map d_d^l:

$$L_{smo}^+ = \left\| \partial_x d_+^l \right\|_1 e^{-\gamma \left\| \partial_x I^l \right\|_1} + \left\| \partial_y d_+^l \right\|_1 e^{-\gamma \left\| \partial_y I^l \right\|_1} \quad , \tag{16}$$

$$L_{smo}^d = \left\| \partial_x d_d^l \right\|_1 e^{-\gamma \left\| \partial_x I^l \right\|_1} + \left\| \partial_y d_d^l \right\|_1 e^{-\gamma \left\| \partial_y I^l \right\|_1} \quad , \tag{17}$$

where '∂_x', '∂_y' are the differential operators in the horizontal and vertical directions respectively, and γ is a parameter for adjusting the degree of edge preservation.

Accordingly, the total loss is a weighted sum of the above four terms, which is formulated as:

$$L = L_{syn} + \lambda_1 L_{smo}^+ + \lambda_2 L_{sd} + \lambda_3 L_{smo}^d \quad , \tag{18}$$

where $\{\lambda_1, \lambda_2, \lambda_3\}$ are three preseted weight parameters. Considering that the depths learnt under the self-supervised manner are unreliable at the early training stage, λ_2 and λ_3 are set to zeros at these training iterations, while the self-distilled forward propagation is disabled.

4   Experiments

In this section, we evaluate the SDFA-Net as well as 15 state-of-the-art methods and perform ablation studies on the KITTI dataset [13]. The Eigen split [11] of KITTI which comprises 22600 stereo pairs for training and 679 images for testing is used for network training and testing, while the improved Eigen test set [48] is also employed for network testing. Additionally, 22972 stereo pairs from the Cityscapes dataset [8] are jointly used for training as done in [16]. At both the training and inference stages, the images are resized into the resolution of 1280 × 384. We utilize the crop proposed in [12] and the standard depth cap of 80 m in the evaluation. The following error and accuracy metrics are used as done in the existing works [15,16,33,40,45]: Abs Rel, Sq Rel, RMSE, logRMSE, A1 = $\delta < 1.25$, A2 = $\delta < 1.25^2$, and A3 = $\delta < 1.25^3$. Please see the supplemental material for more details about the datasets and metrics.

4.1   Implementation Details

We implement the SDFA-Net with PyTorch [39], and the modified version of tiny Swin-transformer is pretrained on the ImageNet1K dataset [9]. The Adam optimizer [32] with $\beta_1 = 0.5$ and $\beta_2 = 0.999$ is used to train the SDFA-Net for 50 epochs with a batch size of 12. The initial learning rate is firstly set to 10^{-4}, and is downgraded by half at epoch 30 and 40. The self-distilled forward propagation and the corresponding losses are used after training 25 epochs. For calculating disparities from the outputted volumes, we set the minimum and the maximum disparities to $d_{\min} = 2, d_{\max} = 300$, and the number of the discrete levels are set to $N = 49$. The weight parameters for the loss function are set to $\lambda_1 = 0.0008, \lambda_2 = 0.01$ and $\lambda_3 = 0.0016$, while we set $\beta = 0.01$ and $\gamma = 2$. For the two principles in the self-distilled forward propagation, we set $\alpha = 0.15, \epsilon = 1e - 5, t_1 = 0.2, t_2 = 0.5$ and $K = 61$. We employ random resizing (from 0.75 to 1.5) and cropping (192 × 640), random horizontal flipping, and random color augmentation as the data augmentations.

Table 1. Quantitative comparison on both the raw and improved KITTI Eigen test sets. ↓ / ↑ denotes that lower/higher is better. The best and the second best results are in **bold** and underlined in each metric.

Method	PP	Sup.	Data.	Resolution	Abs Rel ↓	Sq Rel ↓	RMSE ↓	logRMSE ↓	A1 ↑	A2 ↑	A3 ↑
Raw Eigen test set											
R-MSFM6 [58]		M	K	320 × 1024	0.108	0.748	4.470	0.185	0.889	0.963	0.982
PackNet [20]		M	K	384 × 1280	0.107	0.802	4.538	0.186	0.889	0.962	0.981
SGDepth [33]		M(s)	K	384 × 1280	0.107	0.768	4.468	0.186	0.891	0.963	0.982
FeatureNet [45]		M	K	320 × 1024	0.104	0.729	4.481	0.179	0.893	0.965	0.984
monoResMatch [47]	✓	S(d)	K	384 × 1280	0.111	0.867	4.714	0.199	0.864	0.954	0.979
Monodepth2 [15]	✓	S	K	320 × 1024	0.105	0.822	4.692	0.199	0.876	0.954	0.977
DepthHints [53]	✓	S(d)	K	320 × 1024	0.096	0.710	4.393	0.185	0.890	0.962	0.981
DBoosterNet-e [18]		S	K	384 × 1280	0.095	0.636	4.105	0.178	0.890	0.963	0.984
SingleNet [3]	✓	S	K	320 × 1024	0.094	0.681	4.392	0.185	0.892	0.962	0.981
FAL-Net [16]	✓	S	K	384 × 1280	0.093	0.564	3.973	0.174	0.898	0.967	**0.985**
Edge-of-depth [59]	✓	S(s,d)	K	320 × 1024	0.091	0.646	4.244	0.177	0.898	0.966	0.983
PLADE-Net [17]	✓	S	K	384 × 1280	**0.089**	0.590	4.008	0.172	0.900	0.967	**0.985**
EPCDepth [40]	✓	S(d)	K	320 × 1024	0.091	0.646	4.207	0.176	0.901	0.966	0.983
SDFA-Net (Ours)		S	K	384 × 1280	0.090	0.538	3.896	0.169	0.906	**0.969**	**0.985**
SDFA-Net (Ours)	✓	S	K	384 × 1280	**0.089**	0.531	3.864	0.168	**0.907**	**0.969**	**0.985**
PackNet [20]		M	CS+K	384 × 1280	0.104	0.758	4.386	0.182	0.895	0.964	0.982
SemanticGuide [21]		M(s)	CS+K	384 × 1280	0.100	0.761	4.270	0.175	0.902	0.965	0.982
monoResMatch [47]	✓	S(d)	CS+K	384 × 1280	0.096	0.673	4.351	0.184	0.890	0.961	0.981
DBoosterNet-e [18]	✓	S	CS+K	384 × 1280	0.086	0.538	3.852	0.168	0.905	0.969	**0.985**
FAL-Net [16]	✓	S	CS+K	384 × 1280	0.088	0.547	4.004	0.175	0.898	0.966	0.984
PLADE-Net [17]	✓	S	CS+K	384 × 1280	0.087	0.550	3.837	0.167	0.908	**0.970**	**0.985**
SDFA-Net (Ours)		S	CS+K	384 × 1280	0.085	0.531	3.888	0.167	0.911	0.969	**0.985**
SDFA-Net (Ours)	✓	S	CS+K	384 × 1280	**0.084**	0.523	3.833	0.166	**0.913**	**0.970**	**0.985**
Improved Eigen test set											
DepthHints [53]	✓	S(d)	K	320 × 1024	0.074	0.364	3.202	0.114	0.936	0.989	0.997
WaveletMonoDepth [43]	✓	S(d)	K	320 × 1024	0.074	0.357	3.170	0.114	0.936	0.989	0.997
DBoosterNet-e [18]		S	K	384 × 1280	0.070	0.298	2.916	0.109	0.940	0.991	0.998
FAL-Net [16]	✓	S	K	384 × 1280	0.071	0.281	2.912	0.108	0.943	0.991	0.998
PLADE-Net [17]	✓	S	K	384 × 1280	**0.066**	0.272	2.918	0.104	0.945	0.992	0.998
SDFA-Net (Ours)	✓	S	K	384 × 1280	0.074	**0.228**	**2.547**	**0.101**	**0.956**	**0.995**	**0.999**
PackNet [20]		M	CS+K	384 × 1280	0.071	0.359	3.153	0.109	0.944	0.990	0.997
DBoosterNet-e [18]	✓	S	CS+K	384 × 1280	**0.062**	0.242	2.617	**0.096**	0.955	0.994	0.998
FAL-Net [16]	✓	S	CS+K	384 × 1280	0.068	0.276	2.906	0.106	0.944	0.991	0.998
PLADE-Net [17]	✓	S	CS+K	384 × 1280	0.065	0.253	2.710	0.100	0.950	0.992	0.998
SDFA-Net (Ours)	✓	S	CS+K	384 × 1280	0.069	**0.207**	**2.472**	**0.096**	**0.963**	**0.995**	**0.999**

4.2   Comparative Evaluation

We firstly evaluate the SDFA-Net on the raw KITTI Eigen test set [11] in comparison to 15 existing methods listed in Table 1. The comparative methods are trained with either monocular video sequences (M) [20,21,33,45,58] or stereo image pairs (S) [3,15–18,40,43,47,53,59]. It is noted that some of them are trained with additional information, such as the semantic segmentation label (s) [21,33,59], and the offline computed disparity (d) [40,43,47,53,59]. Additionally, we also evaluate the SDFA-Net on the improved KITTI Eigen test set [48]. The corresponding results are reported in Table 1.

As seen from Lines 1–23 of Table 1, when only the KITTI dataset [13] is used for training (K), SDFA-Net performs best among all the comparative methods

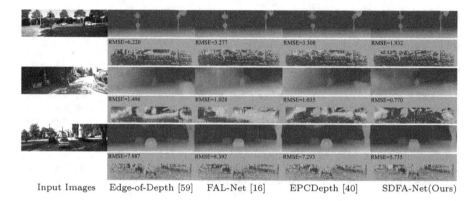

Input Images   Edge-of-Depth [59]   FAL-Net [16]   EPCDepth [40]   SDFA-Net(Ours)

Fig. 5. Visualization results of Edge-of-Depth [59], FAL-Net [16], EPCDepth [40] and our SDFA-Net on KITTI. The input images and predicted depth maps are shown in the odd rows, while we give the corresponding 'RMSE' error maps calculated with the improved Eigen test set in the even rows. For the error maps, red indicates larger error, and blue indicates smaller error. (Color figure online)

in most cases even without the post processing (PP.), and its performance is further improved when a post-processing operation step [14] is implemented on the raw Eigen test set. When both Cityscapes [8] and KITTI [13] are jointly used for training (CS + K), the performance of SDFA-Net is further improved, and it still performs best among all the comparative methods on the raw Eigen test set, especially under the metric 'Sq. Rel'.

As seen from Lines 24–34 of Table 1, when the improved Eigen test set is used for testing, the SDFA-Net still performs superior to the comparative methods in most cases. Although it performs relatively poorer under the metric 'Abs Rel.', its performances under the other metrics are significantly improved. These results demonstrate that the SDFA-Net could predict more accurate depths.

In Fig. 5, we give several visualization results of SDFA-Net as well as three typical methods whose codes have been released publicly: Edge-of-Depth [59], FAL-Net [16], and EPCDepth [40]. It can be seen that SDFA-Net does not only maintain delicate geometric details, but also predict more accurate depths in distant scene regions as shown in the error maps.

4.3   Ablation Studies

To verify the effectiveness of each element in SDFA-Net, we conduct ablation studies on the KITTI dataset [13]. We firstly train a baseline model only with the self-supervised forward propagation and the self-supervised losses (i.e. the image synthesis loss L_{syn} and the raw smoothness loss L_{smo}^{+}). The baseline model comprises an original Swin-transformer backbone (Swin) as the encoder and a raw decoder proposed in [15]. It is noted that in each block of the raw decoder, the multi-scale features are aggregated by the straightforward concatenation. Then we replace the backbone by the modified version of the Swin-transformer

Table 2. (a) Quantitative comparison on the raw Eigen test set in the ablation study. (b) The average norms (in pixels) of the offset vectors on the raw Eigen test set.

Method	Abs Rel↓	Sq Rel↓	RMSE↓	logRMSE↓	A1↑	A2↑	A3↑
Swin+Raw	0.102	0.615	4.323	0.184	0.887	0.963	0.983
Swin† + Raw	0.100	0.595	4.173	0.180	0.893	0.966	0.984
Swin† + SDFA-1	0.097	0.568	3.993	0.175	0.898	0.968	0.985
Swin† + SDFA-2	0.099	0.563	3.982	0.174	0.900	0.968	0.985
Swin† + SDFA-3	0.094	0.553	3.974	0.172	0.896	0.966	0.985
Swin† + OA	0.091	0.554	4.082	0.174	0.899	0.967	0.984
SDFA-Net	**0.090**	**0.538**	**3.896**	**0.169**	**0.906**	**0.969**	**0.985**

SDFA-i	Δ_F^i	Δ_{C1}^i	Δ_{C2}^i
1	1.39	0.14	0.59
2	5.85	0.53	0.56
3	14.89	0.65	0.65

(a)   (b)

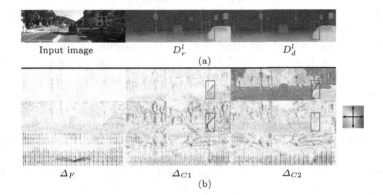

Input image D_r^l D_d^l
(a)

Δ_F Δ_{C1} Δ_{C2}
(b)

Fig. 6. (a) Visualization of the raw and distilled depth maps predicted with the SDFA-Net on KITTI. (b) Visualization of the corresponding offset maps learnt in SDFA modules. From the top to bottom, the offset maps are learnt in SDFA-1, SDFA-2, and SDFA-3. The color coding is visualized on the right part, where the hue and saturation denote the norm and direction of the offset vector respectively. The blue arrows in the bottom left offset map show the offset vectors in a 3×3 patch. (Color figure online)

(Swin†). We also sequentially replace the ith raw decoder block with the SDFA module, and our full model (SDFA-Net) employs three SDFA modules. The proposed self-distilled training strategy is used to train the models that contain the SDFA module(s). Moreover, an Offset-based Aggregation (OA) module which comprises only two offset branches and does not contain the 'Distilled data path' is employed for comparing to the SDFA module. We train the model which uses the OA modules in the decoder with the self-supervised losses only. The corresponding results are reported in Table 2(a).

From Lines 1–5 of Table 2(a), it is noted that using the modified version ('Swin†') of Swin transformer performs better than its original version for depth prediction, indicating that the high-resolution features with rich spatial details are important for predicting depths. By replacing different decoder blocks with the proposed SDFA modules, the performances of the models are improved in most cases. It demonstrates that the SDFA module could aggregate features more effectively for depth prediction. As seen from Lines 6–7 of Table 2(a), although 'Swin† + OA' achieves better quantitative performance compared to 'Swin† + Raw', our full model performs best under all the metrics, which indicates that the SDFA module is benefited from the offset maps learnt under the self-distillation.

Additionally, we visualize the depth maps predicted with the models denoted by 'Swin† + Raw', 'Swin† + OA', and our SDFA-Net in Fig. 1. Although the offset-based aggregation improves the quantitative results of depth prediction, 'Swin† + OA' is prone to predict inaccurate depths on occlusion regions. SDFA-Net does not only alleviate the inaccurate predictions in these regions but also further boost the performance of depth estimation. It demonstrates that the SDFA module maintains the contextual consistency between the learnt features for more accurate depth prediction. Please see the supplemental material for more detailed ablation studies.

To further understand the effect of the SDFA module, in Fig. 6, we visualize the predicted depth results of SDFA-Net, while the feature offset maps learnt by each SDFA module are visualized with the color coding. Moreover, the average norms of the offset vectors in these modules are shown in Table 2(b), which are calculated on the raw Eigen test set [13]. As seen from the first column in Table 2(b) and Fig. 6(b), the offsets in Δ_F reflect the scene information, and the norms of these offset vectors are relatively long. These offsets are used to refine the up-sampled decoded features with the captured information. Since the low-resolution features contain more global information and fewer scene details, the directions of the offset vectors learnt for these features (e.g., Δ_F^3 in bottom left of Fig. 6(b)) vary significantly. Specifically, considering that the features in an arbitrary 3×3 patch are processed by a convolutional kernel, the offsets in the patch could provide the adaptive non-local information for the process as shown with blue arrows in Fig. 6(b). As shown in the last two columns in Table 2(b) and Fig. 6(b), the norms of the offsets in Δ_{C1} and Δ_{C2} are significantly shorter than that in Δ_F. Considering that the encoded features maintain reliable spatial details, these offsets are used to refine the features in little local areas.

Moreover, it is noted that there are obvious differences between the Δ_{C1} and Δ_{C2}, which are learnt under the self-supervised and self-distillation manners respectively. Meanwhile, the depth map obtained with the features aggregated by Δ_F and Δ_{C2} are more accurate than that obtained with the features aggregated by Δ_F and Δ_{C1}, especially on occlusion regions (marked by red boxes in Fig. 6(a)). These visualization results indicate that the learnable offset maps are helpful for adaptively aggregating the contextual information, while the self-distillation is helpful for further maintaining the contextual consistency between the learnt feature, resulting in more accurate depth predictions.

5   Conclusion

In this paper, we propose the Self-Distilled Feature Aggregation (SDFA) module to adaptively aggregate the multi-scale features with the learnable offset maps. Based on SDFA, we propose the SDFA-Net for self-supervised monocular depth estimation, which is trained with the proposed self-distilled training strategy. Experimental results demonstrate the effectiveness of the SDFA-Net. In the future, we will investigate how to aggregate the features more effectively for further improving the accuracy of self-supervised monocular depth estimation.

Acknowledgements. This work was supported by the National Natural Science Foundation of China (Grant Nos. U1805264 and 61991423), the Strategic Priority Research Program of the Chinese Academy of Sciences (Grant No. XDB32050100), the Beijing Municipal Science and Technology Project (Grant No. Z211100011021004).

References

1. Cardace, A., Ramirez, P.Z., Salti, S., Di Stefano, L.: Shallow features guide unsupervised domain adaptation for semantic segmentation at class boundaries. In: Proceedings of the IEEE/CVF Winter Conference on Applications of Computer Vision, pp. 1160–1170 (2022)
2. Chen, Y., Schmid, C., Sminchisescu, C.: Self-supervised learning with geometric constraints in monocular video: connecting flow, depth, and camera. In: ICCV, pp. 7063–7072 (2019)
3. Chen, Z., et al.: Revealing the reciprocal relations between self-supervised stereo and monocular depth estimation. In: ICCV, pp. 15529–15538 (2021)
4. Cheng, B., Saggu, I.S., Shah, R., Bansal, G., Bharadia, D.: S^3Net: semantic-aware self-supervised depth estimation with monocular videos and synthetic data. In: Vedaldi, A., Bischof, H., Brox, T., Frahm, J.-M. (eds.) ECCV 2020. LNCS, vol. 12375, pp. 52–69. Springer, Cham (2020). https://doi.org/10.1007/978-3-030-58577-8_4
5. Cheng, Z., Zhang, Y., Tang, C.: Swin-depth: using transformers and multi-scale fusion for monocular-based depth estimation. IEEE Sens. J. **21**(23), 26912–26920 (2021)
6. Choi, H., et al.: Adaptive confidence thresholding for monocular depth estimation. In: ICCV, pp. 12808–12818 (2021)
7. Clevert, D.A., Unterthiner, T., Hochreiter, S.: Fast and accurate deep network learning by exponential linear units (ELUs). arXiv preprint arXiv:1511.07289 (2015)
8. Cordts, M., et al.: The cityscapes dataset for semantic urban scene understanding. In: CVPR, pp. 3213–3223 (2016)
9. Deng, J., Dong, W., Socher, R., Li, L.J., Li, K., Fei-Fei, L.: ImageNet: a large-scale hierarchical image database. In: CVPR, pp. 248–255 (2009)
10. Dosovitskiy, A., et al.: An image is worth 16×16 words: transformers for image recognition at scale. In: ICLR (2021)
11. Eigen, D., Puhrsch, C., Fergus, R.: Depth map prediction from a single image using a multi-scale deep network. In: Advances in Neural Information Processing Systems, pp. 2366–2374 (2014)
12. Garg, R., B.G., V.K., Carneiro, G., Reid, I.: Unsupervised CNN for single view depth estimation: geometry to the rescue. In: Leibe, B., Matas, J., Sebe, N., Welling, M. (eds.) ECCV 2016. LNCS, vol. 9912, pp. 740–756. Springer, Cham (2016). https://doi.org/10.1007/978-3-319-46484-8_45
13. Geiger, A., Lenz, P., Urtasun, R.: Are we ready for autonomous driving? The KITTI vision benchmark suite. In: CVPR, pp. 3354–3361 (2012)

14. Godard, C., Mac Aodha, O., Brostow, G.J.: Unsupervised monocular depth estimation with left-right consistency. In: CVPR, pp. 270–279 (2017)
15. Godard, C., Mac Aodha, O., Firman, M., Brostow, G.J.: Digging into self-supervised monocular depth estimation. In: ICCV, pp. 3828–3838 (2019)
16. GonzalezBello, J.L., Kim, M.: Forget about the lidar: self-supervised depth estimators with med probability volumes. Adv. Neural. Inf. Process. Syst. **33**, 12626–12637 (2020)
17. GonzalezBello, J.L., Kim, M.: PLADE-Net: towards pixel-level accuracy for self-supervised single-view depth estimation with neural positional encoding and distilled matting loss. In: CVPR, pp. 6851–6860 (2021)
18. GonzalezBello, J.L., Kim, M.: Self-supervised deep monocular depth estimation with ambiguity boosting. IEEE TPAMI (2021)
19. Gou, J., Yu, B., Maybank, S.J., Tao, D.: Knowledge distillation: a survey. IJCV **129**(6), 1789–1819 (2021)
20. Guizilini, V., Ambrus, R., Pillai, S., Raventos, A., Gaidon, A.: 3d packing for self-supervised monocular depth estimation. In: CVPR, pp. 2485–2494 (2020)
21. Guizilini, V., Hou, R., Li, J., Ambrus, R., Gaidon, A.: Semantically-guided representation learning for self-supervised monocular depth. In: International Conference on Learning Representations (ICLR) (2020)
22. Guo, X., Li, H., Yi, S., Ren, J., Wang, X.: Learning monocular depth by distilling cross-domain stereo networks. In: Ferrari, V., Hebert, M., Sminchisescu, C., Weiss, Y. (eds.) ECCV 2018. LNCS, vol. 11215, pp. 506–523. Springer, Cham (2018). https://doi.org/10.1007/978-3-030-01252-6_30
23. Hirschmuller, H.: Accurate and efficient stereo processing by semi-global matching and mutual information. In: CVPR, vol. 2, pp. 807–814. IEEE (2005)
24. Huang, S., Lu, Z., Cheng, R., He, C.: FaPN: feature-aligned pyramid network for dense image prediction. In: ICCV, pp. 864–873 (2021)
25. Huang, Z., Wei, Y., Wang, X., Liu, W., Huang, T.S., Shi, H.: AlignSeg: feature-aligned segmentation networks. IEEE TPAMI **44**(1), 550–557 (2021)
26. Ioffe, S., Szegedy, C.: Batch normalization: accelerating deep network training by reducing internal covariate shift. In: International Conference on Machine Learning, pp. 448–456 (2015)
27. Ji, P., Li, R., Bhanu, B., Xu, Y.: MonoIndoor: towards good practice of self-supervised monocular depth estimation for indoor environments. In: ICCV, pp. 12787–12796 (2021)
28. Jiao, Y., Tran, T.D., Shi, G.: EffiScene: efficient per-pixel rigidity inference for unsupervised joint learning of optical flow, depth, camera pose and motion segmentation. In: CVPR, pp. 5538–5547 (2021)
29. Johnson, J., Alahi, A., Fei-Fei, L.: Perceptual losses for real-time style transfer and super-resolution. In: Leibe, B., Matas, J., Sebe, N., Welling, M. (eds.) ECCV 2016. LNCS, vol. 9906, pp. 694–711. Springer, Cham (2016). https://doi.org/10.1007/978-3-319-46475-6_43
30. Johnston, A., Carneiro, G.: Self-supervised monocular trained depth estimation using self-attention and discrete disparity volume. In: CVPR, pp. 4756–4765 (2020)
31. Jung, H., Park, E., Yoo, S.: Fine-grained semantics-aware representation enhancement for self-supervised monocular depth estimation. In: ICCV, pp. 12642–12652 (2021)
32. Kingma, D.P., Ba, J.: Adam: a method for stochastic optimization. arXiv preprint arXiv:1412.6980 (2014)

33. Klingner, M., Termöhlen, J.-A., Mikolajczyk, J., Fingscheidt, T.: Self-supervised monocular depth estimation: solving the dynamic object problem by semantic guidance. In: Vedaldi, A., Bischof, H., Brox, T., Frahm, J.-M. (eds.) ECCV 2020. LNCS, vol. 12365, pp. 582–600. Springer, Cham (2020). https://doi.org/10.1007/978-3-030-58565-5_35

34. Li, X., et al.: Improving semantic segmentation via decoupled body and edge supervision. In: Vedaldi, A., Bischof, H., Brox, T., Frahm, J.-M. (eds.) ECCV 2020. LNCS, vol. 12362, pp. 435–452. Springer, Cham (2020). https://doi.org/10.1007/978-3-030-58520-4_26

35. Li, X., et al.: Semantic flow for fast and accurate scene parsing. In: Vedaldi, A., Bischof, H., Brox, T., Frahm, J.-M. (eds.) ECCV 2020. LNCS, vol. 12346, pp. 775–793. Springer, Cham (2020). https://doi.org/10.1007/978-3-030-58452-8_45

36. Liu, L., Song, X., Wang, M., Liu, Y., Zhang, L.: Self-supervised monocular depth estimation for all day images using domain separation. In: ICCV, pp. 12737–12746 (2021)

37. Liu, Z., et al.: Swin transformer: hierarchical vision transformer using shifted windows. In: ICCV, pp. 10012–10022 (2021)

38. Mahjourian, R., Wicke, M., Angelova, A.: Unsupervised learning of depth and ego-motion from monocular video using 3d geometric constraints. In: CVPR, pp. 5667–5675 (2018)

39. Paszke, A., et al.: PyTorch: an imperative style, high-performance deep learning library. Adv. Neural. Inf. Process. Syst. **32**, 8026–8037 (2019)

40. Peng, R., Wang, R., Lai, Y., Tang, L., Cai, Y.: Excavating the potential capacity of self-supervised monocular depth estimation. In: ICCV, pp. 15560–15569 (2021)

41. Pilzer, A., Lathuiliere, S., Sebe, N., Ricci, E.: Refine and distill: exploiting cycle-inconsistency and knowledge distillation for unsupervised monocular depth estimation. In: CVPR, pp. 9768–9777 (2019)

42. Poggi, M., Aleotti, F., Tosi, F., Mattoccia, S.: On the uncertainty of self-supervised monocular depth estimation. In: CVPR, pp. 3227–3237 (2020)

43. Ramamonjisoa, M., Firman, M., Watson, J., Lepetit, V., Turmukhambetov, D.: Single image depth prediction with wavelet decomposition. In: CVPR, pp. 11089–11098 (2021)

44. Ranftl, R., Bochkovskiy, A., Koltun, V.: Vision transformers for dense prediction. In: ICCV, pp. 12179–12188 (2021)

45. Shu, C., Yu, K., Duan, Z., Yang, K.: Feature-metric loss for self-supervised learning of depth and egomotion. In: Vedaldi, A., Bischof, H., Brox, T., Frahm, J.-M. (eds.) ECCV 2020. LNCS, vol. 12364, pp. 572–588. Springer, Cham (2020). https://doi.org/10.1007/978-3-030-58529-7_34

46. Simonyan, K., Zisserman, A.: Very deep convolutional networks for large-scale image recognition. arXiv preprint arXiv:1409.1556 (2014)

47. Tosi, F., Aleotti, F., Poggi, M., Mattoccia, S.: Learning monocular depth estimation infusing traditional stereo knowledge. In: CVPR, pp. 9799–9809 (2019)

48. Uhrig, J., Schneider, N., Schneider, L., Franke, U., Brox, T., Geiger, A.: Sparsity invariant CNNs. In: 2017 International Conference on 3D Vision (3DV), pp. 11–20 (2017)

49. Wang, K., et al.: Regularizing nighttime weirdness: efficient self-supervised monocular depth estimation in the dark. In: ICCV, pp. 16055–16064 (2021)

50. Wang, L., Wang, Y., Wang, L., Zhan, Y., Wang, Y., Lu, H.: Can scale-consistent monocular depth be learned in a self-supervised scale-invariant manner? In: ICCV, pp. 12727–12736 (2021)

51. Wang, W., et al.: Pyramid vision transformer: a versatile backbone for dense prediction without convolutions. In: ICCV, pp. 568–578 (2021)
52. Wang, Z., Bovik, A.C., Sheikh, H.R., Simoncelli, E.P.: Image quality assessment: from error visibility to structural similarity. IEEE TIP **13**(4), 600–612 (2004)
53. Watson, J., Firman, M., Brostow, G.J., Turmukhambetov, D.: Self-supervised monocular depth hints. In: ICCV (2019)
54. Yang, G., Tang, H., Ding, M., Sebe, N., Ricci, E.: Transformer-based attention networks for continuous pixel-wise prediction. In: ICCV, pp. 16269–16279 (2021)
55. Yin, Z., Shi, J.: GeoNet: unsupervised learning of dense depth, optical flow and camera pose. In: CVPR, pp. 1983–1992 (2018)
56. Zhou, J., Wang, Y., Qin, K., Zeng, W.: Moving indoor: unsupervised video depth learning in challenging environments. In: ICCV, pp. 8618–8627 (2019)
57. Zhou, T., Brown, M., Snavely, N., Lowe, D.G.: Unsupervised learning of depth and ego-motion from video. In: CVPR, pp. 1851–1858 (2017)
58. Zhou, Z., Fan, X., Shi, P., Xin, Y.: R-MSFM: recurrent multi-scale feature modulation for monocular depth estimating. In: ICCV, pp. 12777–12786 (2021)
59. Zhu, S., Brazil, G., Liu, X.: The edge of depth: explicit constraints between segmentation and depth. In: CVPR, pp. 13116–13125 (2020)

Planes vs. Chairs: Category-Guided 3D Shape Learning Without any 3D Cues

Zixuan Huang[1(✉)], Stefan Stojanov[1], Anh Thai[1], Varun Jampani[2], and James M. Rehg[1]

[1] Georgia Institute of Technology, Atlanta, USA
zhuang374@gatech.edu
[2] Google Research, Cambridge, USA

Abstract. We present a novel 3D shape reconstruction method which learns to predict an implicit 3D shape representation from a single RGB image. Our approach uses a set of single-view images of multiple object categories without viewpoint annotation, forcing the model to learn across multiple object categories without 3D supervision. To facilitate learning with such minimal supervision, we use category labels to guide shape learning with a novel categorical metric learning approach. We also utilize adversarial and viewpoint regularization techniques to further disentangle the effects of viewpoint and shape. We obtain the first results for large-scale (more than 50 categories) single-viewpoint shape prediction using a single model. We are also the first to examine and quantify the benefit of class information in single-view supervised 3D shape reconstruction. Our method achieves superior performance over state-of-the-art methods on ShapeNet-13, ShapeNet-55 and Pascal3D+.

1 Introduction

Reconstructing the 3D shape of objects from 2D images is a fundamental computer vision problem. Recent works have demonstrated that high-quality 3D shape reconstructions can be obtained from a single RGB input image [4,7,10,30]. In the most straightforward approach, explicit 3D supervision from ground-truth (GT) object shapes is used to train a shape reconstruction model [30,39,43,51]. While this approach yields strong performance, GT 3D shape is difficult to obtain on a large scale. This limitation can be addressed via multi-view supervision [17,21,24,26,37,45,52,53], in which the learner optimizes a reprojection loss over a set of viewpoints. This approach becomes more difficult when fewer viewpoints are available during training. The limiting case is single-view supervision, which is challenging due to the well-known entanglement between shape and pose.[1] Prior works (see Table 1) have addressed this

---

[1] When only a single view of each object instance is available for training, there are an infinite number of possible 3D shapes that could explain the image under appropriate camera viewpoints.

Supplementary Information The online version contains supplementary material available at https://doi.org/10.1007/978-3-031-19769-7_42.

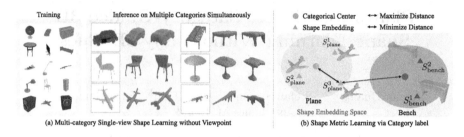

(a) Multi-category Single-view Shape Learning without Viewpoint

(b) Shape Metric Learning via Category label

Fig. 1. (a) We present the first method to learn 3D shape reconstruction from single-viewpoint images over *multiple object categories simultaneously*, without 3D supervision or viewpoint annotations. (b) To facilitate shape learning under such a challenging scenario, We leverage category labels to perform metric learning in the shape embedding space. For each category, we learn a shape center. Then with a given sample, we minimize the distance between its shape embedding and the corresponding category center, while contrasting this distance with other inter-category distances.

challenge in two ways. The first approach assumes that the camera viewpoint is known at training time [20,25]. While this strong assumption effectively reduces entanglement, it is not scalable as it requires pose annotations.

A second line of attack trains separate models for each category of 3D objects [8,14,15,18,23,36,44,49,55]. This single category approach provides the learner with a strong constraint – each model only needs to learn the "pattern" of shape and pose entanglement associated with a single object category. This approach has the disadvantage that data cannot be pooled across categories: pooling data can be beneficial to tasks such as generalization to unseen categories of objects (zero-shot [19,57] or few-shot [53]) and viewpoint learning. For example, tables and chairs have many features (e.g. legs) in common, and a multi-class reconstruction model could leverage such similarities in constructing shared feature representations.

We introduce a novel *multi-category, single-view (MCSV)* 3D shape learning method, shown in Fig. 1 (a), that does not use shape or viewpoint supervision, and can learn to represent multiple object categories within a single reconstruction model. Our method exploits the observation that shapes within a category (e.g. planes) are more similar to each other than they are to the shapes in other categories (e.g. planes vs. chairs). We use a shape metric learning approach, illustrated in Fig. 1 (b), which learns category shape embeddings (category centroids denoted as green/blue dots) together with instance shape embeddings (green/blue triangles) and minimizes the distance between shape instances and their corresponding category centers (solid arrow) while maximizing the distance to the other categories (dotted arrow). This method has two benefits. First, the learned shape manifold captures a strong prior knowledge about shape distance, which helps to eliminate erroneous shapes that might otherwise explain the input image. Second, similar shapes tend to be clustered together, enabling the supervision received by a particular shape to affect its neighbors, and helping to overcome the limitation of having only a single viewpoint. As a consequence of this approach, we are able to train a single model on all 55 object categories in ShapeNet-55, which has not been accomplished in prior work.

Another important aspect of our approach is that we represent object shape as an implicit signed distance field (SDF) instead of the more commonly used mesh representation, in order to accommodate objects with varying topologies. This is important for scaling MCSV models to handle large numbers of object categories. Specifically, our model (Fig. 1 (a)) takes a single input image and produces an implicit SDF function for the 3D object shape. In contrast to mesh and voxel representations, SDF representations create challenges in training because they are less constrained and require a more complex differentiable rendering pipeline. A beneficial aspect of our approach is the use of adversarial learning [15,55] and cycle-consistency [35,55] for regularization.

In summary, our 3D reconstruction learning approach has the following favorable properties:

- **Extremely weak supervision.** We use only single-view masked images with category labels during training (considered as 'self-supervised' in prior works).
- **Categorical shape metric learning.** We make effective use of class labels with our novel shape metric learning.
- **Single template-free multi-category model.** We learn a single model to reconstruct multiple categories without using any category-specific templates.
- **Implicit SDF representation.** We use implicit SDFs, instead of meshes, to represent diverse object topologies.
- **Order of magnitude increase in object categories learned in a single model.** Due to the scalability of our approach, we are the first to train a MCSV model on all 55 categories of ShapetNetCoreV2 [2].
- **State-of-the-art reconstructions.** Experiments on ShapeNet [2] and Pascal3D+ [50] show consistently better performance than previous methods.

To the best of our knowledge, our work is the first successful exploration of the challenging MCSV problem without using known viewpoints or supervised pretraining.

2 Related Work

Single-View Supervision. The body of work that is most closely-related to this paper are methods that use single-view supervision to learn 3D shape reconstruction with a back-projection loss [6,8,14,15,18,20,23,25,36,44,49,55]. These works can be organized as in Table 1 and largely differ in their choice of 1) learning a single multi-category (multi-class) vs multiple single-category models for reconstruction; 2) shape representation (e.g. implicit SDF vs explicit mesh); 3) known vs. unknown viewpoint assumption. We are the first to demonstrate the feasibility of single-view, multi-category learning of an implicit shape representation (SDF) under the unknown viewpoint condition via the use of category labels. We are also the first to provide results on all of the ShapeNet-55 classes in a single model, an order of magnitude more than prior works.

Table 1. Single-view supervised methods for shape reconstruction. For methods without supervision, they are mainly verified under a limited setting [14] or for specific objects [49]. K: keypoints, T: category templates, M: masks, m: mesh, vox: voxel, pc: pointcloud, im: implicit representation.

Model	[18]	[44]	[8]	[23]	[6]	[15]	[14]	[55]	[36]	[25]	[49]	[20]	[40]	Ours
Multi-Class Rec.	–	–	–	–	–	–	–	–	–	–	–	✓	✓	✓
Supervision	K, M	M, T	M, T	M	M	M	–	M	M	M, V	–	M, V	M, V	M
3D Rep.	m	m	m	m	vox	vox	m	m	pc	im	d	m	m	im

Within this body of work [20,25,55] and the concurrent work of [40] are the most closely related. Kato et al. [20] (VPL) and Simoni et al. [40] are the only prior works to address multi-category shape learning with a single model, but both assume access to ground truth (GT) viewpoint and use mesh representations. Similar to our work, Ye et al. [55] do not use viewpoint GT, but they train one model for each object category. Further, their mesh prediction is refined from a low-resolution voxel prediction by optimizing on each individual sample. The latter is less efficient than feedforward models and can potentially harm the prediction of concave shapes or details. In contrast with [21,40,55] we use an implicit shape representation. Further, as shown in Table 1, some earlier approaches use fixed meshes as deformable shape templates. In contrast with implicit functions, meshes with fixed topology restrict the possible 3D shapes that can be accurately represented.

Lin et al. [25] is the only prior work to use an implicit SDF representation for single view supervision, and we adopt their SDF-SRN network architecture and associated reprojection losses in our formulation. This work trains single-category models and assumes that the camera viewpoint is known, whereas one of the main goals of our work is to remove these assumptions and to demonstrate that the unconstrained SDF representation can be successful in the multi-category, unknown viewpoint case. We consistently outperform a version of SDF-SRN modified for our setting in Table 2, which highlights our innovations.

Shape Supervision. Prior works on single image 3D reconstruction with explicit 3D geometric supervision have achieved impressive results [4,7,10,30,51]. However, the requirement of 3D supervision limits the applicability of these methods. To overcome this, subsequent works employ multi-view images for supervision. The use of image-only supervision is enabled by differentiable rendering, which enables the generation of 2D images/masks from the predicted 3D shape and the comparison to the ground truth reference images as supervision. These methods can be grouped by their representation of shape, including voxels [45,52,53], pointclouds [17,24], meshes [21,26] and implicit representation [37]. In contrast to these works, we allow only *a single image per object instance* in the training dataset.

Category Information. Few prior works have explored using category-specific priors for few-shot shape reconstruction [31,46], where they assume voxel templates are available for each category. Our method uses category priors for shape

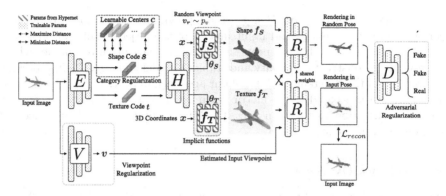

Fig. 2. Approach Overview. Given the input image, the encoders E and V predict the shape s and texture t codes and viewpoint v. The hypernetwork H takes both latent codes and outputs the parameters θ_S, θ_T of the implicit functions f_S and f_T. These functions are sampled with a learnable ray tracer R to render the image with the predicted viewpoint v and thus to compute the reconstruction loss. We also render an extra image from a random viewpoint for the adversarial and viewpoint regularizations. Our shape metric learning technique is performed over the shape latent code.

reconstruction, but by only leveraging the significantly weaker supervision of category labels.

Deep Metric Learning. To harness shape learning with category labels, our method also shares ideas with metric learning. The key idea of metric learning is to learn an embedding space where similar instances are together and dissimilar instances are far according to some distance metric. This paradigm has shown success in self-supervised [3,13] and supervised [22] learning, as well as many downstream tasks e.g. person re-identification [16]. We use the idea of metric learning for single-view shape reconstruction and demonstrate its effectiveness for the first time.

3 Approach

Given a collection of n images concatenated with their masks $\{I_i \in \mathbb{R}^{h \times w \times 4}\}_{i=1}^{n}$ and class labels $\{\mathbf{y}_i \in \{0, 1, \cdots, c\}\}$, our goal is to learn a single-view 3D reconstruction model without any 3D, viewpoint, or multi-view supervision. In contrast to most existing work, we learn a single network that works across all object categories. We represent shape using an implicit shape representation function $f_S : \mathbb{R}^3 \to \mathbb{R}$, a multi-layer perceptron (MLP) that maps 3D coordinates to signed distance function (SDF) values $s \in \mathbb{R}$. Following a standard framework for learning 3D shape via differentiable rendering, our model first infers shape, texture and viewpoints from the input images, which are then rendered into images. The rendered image is then compared to the input image, providing a training signal for model learning. However, in our challenging multi-category setting, without viewpoint or 3D supervision, providing supervision only by comparing the rendered image and input images of the object results in poor performance. To mitigate this, we propose a set of regularization techniques which improve the reconstruction quality and stabilize model training. We first present an overview of the different modules of our approach in Sect. 3.1, and then introduce our category-based shape metric learning in Sect. 3.2. Finally, we present other regularization methods in Sect. 3.3. Implementation details and licenses are described in the supplement.

3.1   Network Overview

Our model is trained end-to-end and consists of four main modules (Fig. 2).

Image Encoder. The image encoder E maps an image with mask channel $I \in \mathbb{R}^{h \times w \times 4}$ to a latent shape vector $s \in \mathbb{R}^l$ for the downstream shape predictor, and a latent texture vector $t \in \mathbb{R}^l$ for texture predictor.

Shape and Texture Prediction Module. Following the design of hypernetworks [11,25], the shape and texture prediction module uses latent codes s and t to predict the parameters of shape and texture implicit functions f_S and f_T that map 3D query points x to SDF and texture predictions:

$$\theta_S, \theta_T \leftarrow H(s, t; \theta_H) \tag{1}$$

$$f_S : (x; \theta_S) - \text{SDF prediction} \tag{2}$$

$$f_T : (x; \theta_T) - \text{Texture prediction} \tag{3}$$

f_S and f_T are MLPs with predefined structure and parameters estimated by the hypernet H with parameters θ_H.

Viewpoint Prediction Module. In our model, the viewpoint is represented as the trigonometric functions of Euler angles for continuity [1], i.e. $v = [\cos \gamma, \sin \gamma]$ with γ denoting 3 Euler angles. The viewpoint prediction network $V(I; \theta_V)$ predicts the viewpoint v of the object in the input image I with regard to a canonical pose (can be different from the human-defined canonical pose), where θ_V are the learnable parameters.

Differentiable Renderer. We use an SDF-based implicit differentiable renderer from [25] which takes the shape, texture and the viewpoint as inputs and renders the corresponding image. Formally, we denote it as a functional, $R(f_S, f_T, v; \theta_R)$, which maps shape SDF, texture implicit function and the viewpoint into 2D RGB image with an alpha channel, in $\mathbb{R}^{h \times w \times 4}$. The renderer itself is also learnable with θ_R parameters. Note that as the renderer in [25] cannot render masks, we modify it to render an extra alpha channel from the SDF field following [54].

Before we describe different regularizations, we briefly discuss the major challenge in learning our model. During training, the only supervision signals are the input images and the masks of the objects. The common approach is to minimize a reconstruction loss via differentiable rendering[2]

$$\mathcal{L}_{recon} = \|I - R(f_S, f_T, v; \theta_R)\|_2^2. \tag{4}$$

This learning scheme works well with multi-view supervision [37], or even single-view images with viewpoint ground truth [25]. However, when there is only single-view supervision without viewpoint ground truth, this is significantly more difficult. There are infinite combinations of shapes and viewpoints that perfectly render the given image, and the model does not have any guidance to identify the correct shape-viewpoint combination. To tackle this problem, we utilize category labels to guide the learning of shape via metric learning. We also make use of other regularizations including adversarial learning and cycle consistency to facilitate learning.

3.2 Shape Metric Learning with Category Guidance

Our novel shape metric learning approach is the key ingredient that enables us to train a single model containing 55 shape categories. The key idea is to learn a metric which maps shapes within the same category (e.g. chairs) to be close to each other, while mapping shapes in different categories (e.g. planes vs. chairs) to be farther apart. This approach leverages qualitative label information to obtain a continuous metric for shape similarity, resulting in two benefits. First, the shape manifold defined by the learned metric constrains the space of possible 3D shapes, helping to eliminate spurious shape solutions that would otherwise be consistent with the input image (see Fig. 3 (a) for an example). Second, the shape manifold tends to group similar shapes together, creating a neighborhood structure over the training samples. For example, without category guidance it can be difficult for the network to learn the relationship between two images of different chairs captured from different, unknown, camera viewpoints. However, with the learned shape manifold these instances are grouped together, facilitating the sharing of supervisory signals from the rendering-based losses. A qualitative reconstruction comparison in Fig. 3 (a) and t-SNE visualizations of shape embedding space in Fig. 3 (b) provide further insight into these benefits. Note that as the image discriminator used for adversarial regularization uses images and labels as input, category information is still available during training even without metric learning. Therefore, the current comparison demonstrates the effectiveness of metric learning for further utilizing category label information.

Formally, for a batch of n images $\{I_i\}_{i=1}^n$, we extract both shape latent codes and texture latent codes via an image encoder E. For the ith sample, we denote the latent shape code as s_i and the category label as $y_i \in C$ where C is the set of integer labels. There are two types of losses commonly used for metric learning: embedding losses [12,38,48] that compare different instances (e.g., triplet

---

[2] Masks are leveraged with additional losses that enforce consistency between the shape and the input mask. For conciseness, we omit them here and refer readers to [25] for more details.

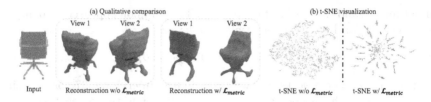

Fig. 3. (a) Category-guided metric learning is beneficial for eliminating erroneous shapes that can explain input image. In the first view (close to input viewpoint), both shapes looks plausible and can explain input image. But from another view, it is clear that our metric-learned shape is more reasonable as a chair, while the baseline can be closer to a table/nightstand. (b) t-SNE visualization of the shape embedding space on ShapeNet-55. Category-guided metric learning results in a more structured representation and facilitates joint learning.

loss), and proxy-based losses [27,47,56] which learn parameters representing categories (e.g. weight vectors in normalized softmax loss or ProxyNCA [34]). As the embedding losses usually require specific sampling strategies to work well, we follow the latter for metric learning in this work.

Specifically, for each category we define a learnable category center vector (or category embedding) $\{c_k\}_{k=1}^{C}$ and optimize a distance metric so that instances of the same category are close and instances of different categories are far apart, as illustrated in the green shaded region of Fig. 2 and Fig. 1(b). We optimize the following loss between learned shape instance embeddings s and category center embeddings c:

$$\mathcal{L}_{metric} = -\sum_{i=0}^{N} log \frac{exp(d(s_i, c_{y_i})/\tau)}{\sum_{k \in \mathcal{C}} exp(d(s_i, c_k)/\tau)}, \qquad (5)$$

where d is a similarity measure and τ is the temperature. Following normalized softmax loss [47] we use cosine similarity as a similarity measure. The centers are randomly initialized from a uniform distribution and updated directly through gradient descent. We set the temperature τ to be 0.3 across all our experiments.

3.3   Shape and Viewpoint Regularization

Adversarial regularization. To further facilitate learning, we use adversarial regularization to ensure that predicted shapes and texture fields will render realistic images from any viewpoints in addition to the input view. Given the training image collection, we estimate the data manifold via an image discriminator. If the rendered images also lie on the training data manifold, most erroneous shapes and texture predictions will be eliminated. To achieve this, similar to [15,55], we use a discriminator with the adversarial training [9]. Furthermore, because our multi-category setting results in a complex image distribution, we condition our discriminator on category labels for easier training following [33].

While our work is not the first work to use GANs for shape learning in general, we believe it still provides useful new insights by combining adversarial training with unconstrained implicit representation and class conditioning under our challenging MCSV setting.

Formally, suppose $I_{recon} = R(f_S, f_T, v)$ is the output of the renderer for estimated shape, texture and viewpoint given an input image. We sample another random viewpoint v' from a prior distribution $P_v(v)$ and render another image, $I_{rnd} = R(f_S, f_T, v')$ from this viewpoint. We match the distribution of I and I_{recon}, I_{rnd} with adversarial learning as shown in Fig. 2. Specifically, we optimize the objective $\mathcal{L}_{gan}$ below by alternatively updating our model (the E, H, R, V networks) and the discriminator $D(I; \theta_D)$

$$\min_{\theta_E, \theta_H, \theta_V, \theta_R} \max_{\theta_D} \mathcal{L}_{gan} = \mathbb{E}[\log D(I)] + \mathbb{E}[\log(1 - D([I_{recon}, I_{rnd}]))]. \quad (6)$$

Here, $[I_{recon}, I_{rnd}]$ is the stack of reconstructed and randomly rendered images (on batch dimension). For the update step of the reconstruction model, we follow a non-saturating scheme [9] where we maximize $\mathbb{E}[\log(D([I_{recon}, I_{rnd}]))]$ instead of minimizing $\mathbb{E}[\log(1 - D([I_{recon}, I_{rnd}]))]$. We also use R_1 regularizer [29] and spectral normalization [32] to stabilize the training process.

Viewpoint Regularization via Cycle-Consistency. We regularize the viewpoint prediction module via cycle-consistency [58], with a similar approach as the state-of-the-art at self-supervised viewpoint estimation [35] and Ye et al. [55]. However, they both rely on a strong shape symmetry assumptions to facilitate the view learning. In our approach, we do not use any such restrictive assumptions. The key idea of viewpoint regularization, $\mathcal{L}_{cam}$, is that we can render an arbitrary number of images by sampling viewpoints with a given shape and texture. This provides us many rendered image-viewpoint pairs to supervise the viewpoint predictor. Please see the supplement for a formal description of $\mathcal{L}_{cam}$.

4   Experiments

This section presents the experiments of applying our method to multiple synthetic and real datasets, as well as ablations of its individual components. Following an overview of the datasets, metrics and baselines, we give results on synthetic and real data.

An additional goal is to quantify the value of category labels in the MCSV setting. By comparing category label-based supervision with a standard two-view reconstruction approach, we find that the value of the category label is approximately equal to 20% of an additional view. We also show that models trained under MCSV setting can outperform and generalize better than category-specific models. See the supplement for details.

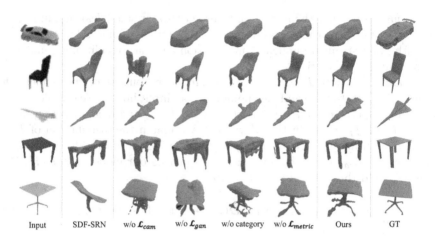

Input   SDF-SRN   w/o $\mathcal{L}_{cam}$ w/o $\mathcal{L}_{gan}$ w/o category   w/o $\mathcal{L}_{metric}$ Ours   GT

Fig. 4. Qualitative comparison on ShapeNet-13. Our method learns both better global structures and details on various categories.

4.1   Datasets

For synthetic data we use two splits of the ShapeNet [2] dataset, and for real data we use Pascal3D+ [50]. We provide additional results on Pix3D [41] in the supplement.

ShapeNet-13. This dataset consists of images from the 13 biggest categories of ShapeNet, originally used by Kato et al. [21] and following works learning 3D shape without explicit 3D supervision [25,37]. 24 views per object are generated by placing cameras at a fixed 30° elevation and with azimuthal increments around the object of 15°. Following SDF-SRN [25], we use a 70/10/20 split for training, validation and testing, resulting in 30643, 4378 and 8762 objects respectively. We use only one view per object instance, which is randomly pre-selected out of the 24 views for each object.

ShapeNet-55. To create a more challenging setting in comparison to prior work, we use all of the categories of ShapeNet.v2. This is challenging due to (1) approximately four times more shape categories; and (2) uniformly sampling at random the azimuth in the $[0°, 360°]$ range and elevation in $[20°, 40°]$. As a result, the image and shape distributions that the model needs to learn are significantly more complex. To reduce the data imbalance, we randomly sample at most 500 objects per category. Like ShapeNet13 we use only one image per object instance and a 70/10/20 split, with 15031, 1705, 3427 images respectively.

Pascal 3D+. This dataset contains real-world images with 3D ground truth obtained by CAD model alignment. Its challenges compared to ShapeNet come from (1) inaccurate object masks that include noisy backgrounds—an especially difficult setting for adversarial learning; and (2) the viewpoints vary in azimuth, elevation and tilt, creating challenges because different categories have different and unknown viewpoint distributions. We combine the commonly used motor-cycle and chair categories for a total of 2307 images, and as in [25] we use the ImageNet [5] subset of Pascal3D+ with 1132/1175 images for training/testing.

4.2 Evaluation Metrics

We use Chamfer Distance (CD) and F-score under different thresholds to evaluate shape reconstruction performance following [10,25,42,43]. We use the Marching Cubes algorithm [28] to convert the implicit representation to meshes prior to computing the metrics. Shapes are transformed into camera coordinates for evaluation. See the supplement for evaluation details.

Chamfer Distance. Following [25,37], CD is defined as an average of accuracy and completeness. Given two surface point sets S_1 and S_2, CD is given as:

$$d_{CD}(S_1, S_2) = \frac{1}{2|S_1|} \sum_{x \in S_1} \min_{y \in S_2} \|x - y\|_2 + \frac{1}{2|S_2|} \sum_{y \in S_2} \min_{x \in S_1} \|x - y\|_2 \quad (7)$$

F-score. Compared to CD which accumulates distances, F-score measures accuracy and completeness by thresholding the distances. For a threshold d (F-Score@d), precision is the percentage of predicted points that have neighboring ground truth points within distance d, recall is the percentage of ground truth points that have neighboring predicted points within d. F-score, the harmonic mean of precision and recall, can be intuitively interpreted as the percentage of surface that is correctly reconstructed.

4.3 Baselines

There are no directly comparable state-of-the-art methods under our MCSV setting. Instead, we slightly alter the setting of two recent methods SDF-SRN [25] and Ye et al. [55] to compare with our model.

SDF-SRN [25] learns to predict implicit shapes using single images and known camera poses in training. For comparison purposes, we attach our viewpoint prediction module to SDF-SRN and train it across multiple categories simultaneously. We compare to SDF-SRN on all datasets.

Ye et al. [55] learns category-specific voxel prediction without viewpoint GT.[3] We compare our method with Ye et al. on Pascal3D+ (see supplement for additional ShapeNet-13 comparison). For Ye et al., we train different models for different categories following their original setting and compute the average metric over categories. We use GT masks as supervision.

---

[3] Their method has an optional per-sample test-time optimization to further refine the voxels and convert into meshes. Using the authors' implementation of this optimization did not lead to improved results in our experiments, so we use the voxel prediction as the output for evaluation.

Table 2. Quantitative result measured by CD and F-score on ShapeNet-13. Our method performs favorably to baselines and other SOTA methods.

Methods	F-Score@1.0↑	F-Score@5.0↑	F-Score@10.0↑	CD↓
w/o category	0.1589	0.6261	0.8527	0.520
w/o $\mathcal{L}_{metric}$	0.1875	0.6864	0.8805	0.458
w/o $\mathcal{L}_{cam}$	0.1837	0.6741	0.8758	0.463
w/o $\mathcal{L}_{gan}$	0.1846	0.6437	0.8422	0.532
Ours	**0.2005**	**0.7168**	**0.8949**	**0.430**
SDF-SRN	0.1606	0.5441	0.7584	0.682

4.4   ShapeNet-13

We perform experiments on ShapeNet-13 and show quantitative and qualitative results in Table 2 and Fig. 4. See additional results in the supplement.

Ablation Study. We first analyze the results of ablating the multiple regularization techniques used in our approach. In Table 2, 'w/o category' shows the results of the model without any category information and 'w/o $\mathcal{L}_{metric}$' is the model with only GAN conditioning to leverage category labels, by removing the metric learning component (both use all losses except $\mathcal{L}_{metric}$). Comparing them to our full model, we clearly see our metric learning helps utilize category information in a more efficient way (w/o $\mathcal{L}_{metric}$ vs Ours). We also see the benefit of having category information for shape reconstruction tasks (w/o category vs Ours). We further ablate on our camera (w/o $\mathcal{L}_{cam}$ vs Ours) and GAN regularization (w/o $\mathcal{L}_{gan}$ vs Ours) and quantify their value. Our qualitative result verifies these findings as well. We include more quantitative findings about the value of category labels for shape learning in the supplement.

SOTA Comparison. Comparing with SDF-SRN in Table 2 we see that our approach improves on the adapted SDF-SRN by a 23.8% (w/o category) and 37.0% (with category, Ours) CD decrease. Qualitatively, our approach captures both the overall shape topology and details (legs of table in the bottom row) whereas SDF-SRN fails. These results demonstrate the effectiveness of our proposed model.

4.5   ShapeNet-55

Compared to ShapeNet-13, ShapeNet-55 is significantly more challenging due to both the number of categories and viewpoint variability. For results see Table 3 and Fig. 5 (a).

Ablation Study. On this dataset, we again ablate on the value of category labels and our proposed shape metric learning. Similar to ShapeNet-13, we clearly see our metric learning method utilizes category information more effectively than GAN conditioning only (w/o $\mathcal{L}_{metric}$ vs Ours in Table 3). Meanwhile, we demonstrate the huge benefit of having category information for shape reconstruction on this dataset (w/o category vs Ours, 32.4% CD decrease). This is much more significant than ShapeNet-13 because ShapeNet-55 has much higher data complexity, where a better structured shape latent space can be quite beneficial to shape learning. Qualitatively, without category regularization the model cannot even capture the global shape successfully, whereas without shape metric learning it only partially succeeds at highly symmetric objects (two vases in Fig. 5 (a)). For others (e.g. bench at row 2) its reconstruction can only explain the input view. This further strengthen the discussion in Fig. 3 (a), where shape metric learning can help eliminate erroneous shape predictions.

Table 3. Quantitative result measured by CD and F-score on ShapeNet-55. Our method performs favorably to baselines and other SOTA methods.

Methods	F-Score@1.0↑	F-Score@5.0↑	F-Score@10.0↑	CD↓
w/o category	0.0977	0.4365	0.6815	0.801
w/o $\mathcal{L}_{metric}$	0.1431	0.5758	0.8047	0.620
Ours	**0.1619**	**0.6164**	**0.8386**	**0.541**
SDF-SRN [25]	0.0707	0.2750	0.4806	1.172

SOTA Comparison. Comparing with row 4 of Table 3 we see that our approach improves on the adapted SDF-SRN by 31.6% (without category) and 53.8% (with category) CD decrease. The significant performance improvement justifies the effectiveness of our overall framework, as well as the effectiveness of the categorical shape metric learning. Qualitatively, SDF-SRN collapses and fails to capture the topology of most inputs, while our method with still demonstrate a great qualitative reconstruction performance. More qualitative results on various categories and reconstructed textures are included in the supplement.

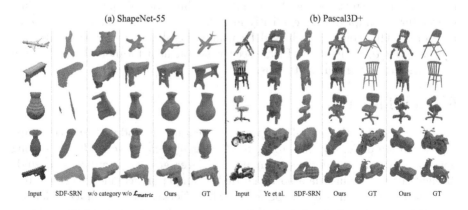

Fig. 5. Qualitative comparison on ShapeNet55 and Pascal3D+. Our method learns both better global 3D structure and shape details on various categories.

4.6   Pascal3D+

We compare our method to SDF-SRN [25] and Ye et al. [55] on Pascal3D+, as shown in Table 4 and Fig. 5 (b). It is clear that our method outperforms the SOTA on both metrics, and our metric learning method is critical to the good performance. We also compare reconstructions qualitatively. As shown in Fig. 5 (b), SDF-SRN collapses to a thin flake that can only explain input images, while Ye et al. lack some important details. This further verifies the effectiveness of our proposed method.

Table 4. Quantitative result measured by CD and F-score on Pascal3D+. Our method performs favorably to other SOTA methods.

Methods	F-Score@1.0↑	F-Score@5.0↑	F-Score@10.0↑	CD↓
SDF-SRN [25]	0.0954	0.4656	0.7474	0.777
Ye et al. [55]	**0.1195**	0.5429	0.8252	0.625
Ours w/o $\mathcal{L}_{metric}$	0.1038	0.5108	0.7799	0.673
Ours	0.1139	**0.5460**	**0.8455**	**0.580**

5   Limitation and Discussion

One limitation is that reconstruction accuracy decreases for some samples with large concavity. For example, for bowls or bathtubs on ShapeNet-55, our method cannot capture the concavity of the inner surface. Concavity is hard to model without explicit guidance, and this can potentially be improved by explicitly modeling lighting and shading. On the other hand, ShapeNet-55 represents a class imbalance challenge, where images in different categories range from 40 to 500 images. This makes learning on rare classes difficult.

We also observe a training instability of our model caused by the usage of adversarial regularization. Meanwhile, we think our shape metric learning can be further improved by explicitly modeling the multi-modal nature of some categories. This could be achieved by using several proxies for each category. We leave this to future work. Please see the supplement for more discussion and qualitative examples on limitations.

6   Conclusion

We have presented the first 3D shape reconstruction method with an SDF shape representation under the challenging *multi-category, single-view (MCSV)* setting without viewpoint supervision. Our method leverages category labels to guide implicit shape learning via a novel metric learning approach and additional regularization. Our results on ShapeNet-13, ShapeNet-55 and Pascal3D+, demonstrate the superior quantitative and qualitative performance of our method over prior works. Our findings are the first to quantify the benefit of category information in single-image 3D reconstruction.

Acknowledgments. We thank Miao Liu for helpful comments and discussions.

References

1. Beyer, L., Hermans, A., Leibe, B.: Biternion nets: continuous head pose regression from discrete training labels. In: Gall, J., Gehler, P., Leibe, B. (eds.) GCPR 2015. LNCS, vol. 9358, pp. 157–168. Springer, Cham (2015). https://doi.org/10.1007/978-3-319-24947-6_13
2. Chang, A.X., et al.: Shapenet: sn information-rich 3d model repository. arXiv preprint arXiv:1512.03012 (2015)
3. Chen, T., Kornblith, S., Norouzi, M., Hinton, G.: A simple framework for contrastive learning of visual representations. In: International Conference on Machine Learning, pp. 1597–1607. PMLR (2020)
4. Choy, C.B., Xu, D., Gwak, J.Y., Chen, K., Savarese, S.: 3D-R2N2: a unified approach for single and multi-view 3D object reconstruction. In: Leibe, B., Matas, J., Sebe, N., Welling, M. (eds.) ECCV 2016. LNCS, vol. 9912, pp. 628–644. Springer, Cham (2016). https://doi.org/10.1007/978-3-319-46484-8_38
5. Deng, J., Dong, W., Socher, R., Li, L.J., Li, K., Fei-Fei, L.: Imagenet: a large-scale hierarchical image database. In: 2009 IEEE Conference on Computer Vision and Pattern Recognition, pp. 248–255. IEEE (2009)
6. Gadelha, M., Maji, S., Wang, R.: 3D shape induction from 2D views of multiple objects. In: 2017 International Conference on 3D Vision (3DV), pp. 402–411. IEEE (2017)
7. Gkioxari, G., Malik, J., Johnson, J.: Mesh R-CNN. In: Proceedings of the IEEE/CVF International Conference on Computer Vision (ICCV), October 2019
8. Goel, S., Kanazawa, A., Malik, J.: Shape and viewpoint without keypoints. In: Vedaldi, A., Bischof, H., Brox, T., Frahm, J.-M. (eds.) ECCV 2020. LNCS, vol. 12360, pp. 88–104. Springer, Cham (2020). https://doi.org/10.1007/978-3-030-58555-6_6
9. Goodfellow, I.J., et al.: Generative adversarial networks. arXiv preprint arXiv:1406.2661 (2014)
10. Groueix, T., Fisher, M., Kim, V.G., Russell, B., Aubry, M.: AtlasNet: A Papier-Mâché approach to learning 3D surface generation. In: Proceedings IEEE Conference on Computer Vision and Pattern Recognition (CVPR) (2018)
11. Ha, D., Dai, A., Le, Q.V.: Hypernetworks. arXiv preprint arXiv:1609.09106 (2016)
12. Hadsell, R., Chopra, S., LeCun, Y.: Dimensionality reduction by learning an invariant mapping. In: 2006 IEEE Computer Society Conference on Computer Vision and Pattern Recognition (CVPR 2006), vol. 2, pp. 1735–1742. IEEE (2006)
13. He, K., Fan, H., Wu, Y., Xie, S., Girshick, R.: Momentum contrast for unsupervised visual representation learning. In: Proceedings of the IEEE/CVF Conference on Computer Vision and Pattern Recognition, pp. 9729–9738 (2020)
14. Henderson, P., Ferrari, V.: Learning single-image 3d reconstruction by generative modelling of shape, pose and shading. Int. J. Comput. Vis. **128**(4), 835–854 (2019). https://doi.org/10.1007/s11263-019-01219-8
15. Henzler, P., Mitra, N.J., Ritschel, T.: Escaping plato's cave: 3D shape from adversarial rendering. In: Proceedings of the IEEE/CVF International Conference on Computer Vision, pp. 9984–9993 (2019)
16. Hermans, A., Beyer, L., Leibe, B.: In defense of the triplet loss for person re-identification. arXiv preprint arXiv:1703.07737 (2017)

17. Insafutdinov, E., Dosovitskiy, A.: Unsupervised learning of shape and pose with differentiable point clouds. arXiv preprint arXiv:1810.09381 (2018)

18. Kanazawa, A., Tulsiani, S., Efros, A.A., Malik, J.: Learning category-specific mesh reconstruction from image collections. In: Ferrari, V., Hebert, M., Sminchisescu, C., Weiss, Y. (eds.) ECCV 2018. LNCS, vol. 11219, pp. 386–402. Springer, Cham (2018). https://doi.org/10.1007/978-3-030-01267-0_23

19. Kar, A., Häne, C., Malik, J.: Learning a multi-view stereo machine. arXiv preprint arXiv:1708.05375 (2017)

20. Kato, H., Harada, T.: Learning view priors for single-view 3D reconstruction. In: Proceedings of the IEEE/CVF Conference on Computer Vision and Pattern Recognition, pp. 9778–9787 (2019)

21. Kato, H., Ushiku, Y., Harada, T.: Neural 3d mesh renderer. In: Proceedings of the IEEE Conference on Computer Vision and Pattern Recognition, pp. 3907–3916 (2018)

22. Khosla, P., et al.: Supervised contrastive learning. arXiv preprint arXiv:2004.11362 (2020)

23. Li, X., et al.: Self-supervised single-view 3D reconstruction via semantic consistency. In: Vedaldi, A., Bischof, H., Brox, T., Frahm, J.-M. (eds.) ECCV 2020. LNCS, vol. 12359, pp. 677–693. Springer, Cham (2020). https://doi.org/10.1007/978-3-030-58568-6_40

24. Lin, C.H., Kong, C., Lucey, S.: Learning efficient point cloud generation for dense 3D object reconstruction. In: AAAI Conference on Artificial Intelligence (AAAI) (2018)

25. Lin, C.H., Wang, C., Lucey, S.: Sdf-srn: Learning signed distance 3D object reconstruction from static images. arXiv preprint arXiv:2010.10505 (2020)

26. Liu, S., Li, T., Chen, W., Li, H.: Soft rasterizer: a differentiable renderer for image-based 3D reasoning. In: Proceedings of the IEEE/CVF International Conference on Computer Vision, pp. 7708–7717 (2019)

27. Liu, W., Wen, Y., Yu, Z., Li, M., Raj, B., Song, L.: Sphereface: deep hypersphere embedding for face recognition. In: Proceedings of the IEEE Conference on Computer Vision and Pattern Recognition, pp. 212–220 (2017)

28. Lorensen, W.E., Cline, H.E.: Marching cubes: A high resolution 3d surface construction algorithm. ACM siggraph computer graphics **21**(4), 163–169 (1987)

29. Mescheder, L., Geiger, A., Nowozin, S.: Which training methods for gans do actually converge? In: International conference on machine learning. pp. 3481–3490. PMLR (2018)

30. Mescheder, L., Oechsle, M., Niemeyer, M., Nowozin, S., Geiger, A.: Occupancy networks: Learning 3d reconstruction in function space. In: Proceedings of the IEEE/CVF Conference on Computer Vision and Pattern Recognition. pp. 4460–4470 (2019)

31. Michalkiewicz, M., Parisot, S., Tsogkas, S., Baktashmotlagh, M., Eriksson, A., Belilovsky, E.: Few-shot single-view 3-D object reconstruction with compositional priors. In: Vedaldi, A., Bischof, H., Brox, T., Frahm, J.-M. (eds.) ECCV 2020. LNCS, vol. 12370, pp. 614–630. Springer, Cham (2020). https://doi.org/10.1007/978-3-030-58595-2_37

32. Miyato, T., Kataoka, T., Koyama, M., Yoshida, Y.: Spectral normalization for generative adversarial networks. arXiv preprint arXiv:1802.05957 (2018)

33. Miyato, T., Koyama, M.: CGANs with projection discriminator. arXiv preprint arXiv:1802.05637 (2018)

34. Movshovitz-Attias, Y., Toshev, A., Leung, T.K., Ioffe, S., Singh, S.: No fuss distance metric learning using proxies. In: Proceedings of the IEEE International Conference on Computer Vision, pp. 360–368 (2017)
35. Mustikovela, S.K., et al.: Self-supervised viewpoint learning from image collections. In: Proceedings of the IEEE/CVF Conference on Computer Vision and Pattern Recognition, pp. 3971–3981 (2020)
36. Navaneet, K., Mathew, A., Kashyap, S., Hung, W.C., Jampani, V., Babu, R.V.: From image collections to point clouds with self-supervised shape and pose networks. In: Proceedings of the IEEE/CVF Conference on Computer Vision and Pattern Recognition, pp. 1132–1140 (2020)
37. Niemeyer, M., Mescheder, L., Oechsle, M., Geiger, A.: Differentiable volumetric rendering: learning implicit 3d representations without 3D supervision. In: Proceedings of the IEEE/CVF Conference on Computer Vision and Pattern Recognition, pp. 3504–3515 (2020)
38. Oord, A.V.D., Li, Y., Vinyals, O.: Representation learning with contrastive predictive coding. arXiv preprint arXiv:1807.03748 (2018)
39. Saito, S., Huang, Z., Natsume, R., Morishima, S., Kanazawa, A., Li, H.: Pifu: pxel-aligned implicit function for high-resolution clothed human digitization. In: Proceedings of the IEEE/CVF International Conference on Computer Vision, pp. 2304–2314 (2019)
40. Simoni, A., Pini, S., Vezzani, R., Cucchiara, R.: Multi-category mesh reconstruction from image collections. In: 2021 International Conference on 3D Vision (3DV), pp. 1321–1330. IEEE (2021)
41. Sun, X., et al.: Pix3d: dataset and methods for single-image 3d shape modeling. In: IEEE Conference on Computer Vision and Pattern Recognition (CVPR) (2018)
42. Tatarchenko, M., Richter, S.R., Ranftl, R., Li, Z., Koltun, V., Brox, T.: What do single-view 3D reconstruction networks learn? In: Proceedings of the IEEE/CVF Conference on Computer Vision and Pattern Recognition, pp. 3405–3414 (2019)
43. Thai, A., Stojanov, S., Upadhya, V., Rehg, J.M.: 3D reconstruction of novel object shapes from single images. arXiv preprint arXiv:2006.07752 (2020)
44. Tulsiani, S., Kulkarni, N., Gupta, A.: Implicit mesh reconstruction from unannotated image collections. arXiv preprint arXiv:2007.08504 (2020)
45. Tulsiani, S., Zhou, T., Efros, A.A., Malik, J.: Multi-view supervision for single-view reconstruction via differentiable ray consistency. In: Proceedings of the IEEE Conference on Computer Vision and Pattern Recognition, pp. 2626–2634 (2017)
46. Wallace, B., Hariharan, B.: Few-shot generalization for single-image 3D reconstruction via priors. In: Proceedings of the IEEE/CVF International Conference on Computer Vision, pp. 3818–3827 (2019)
47. Wang, F., Xiang, X., Cheng, J., Yuille, A.L.: Normface: L2 hypersphere embedding for face verification. In: Proceedings of the 25th ACM international conference on Multimedia, pp. 1041–1049 (2017)
48. Weinberger, K.Q., Saul, L.K.: Distance metric learning for large margin nearest neighbor classification. J. Mach. Learn. Res. $10(2)$, 207–244 (2009)
49. Wu, S., Rupprecht, C., Vedaldi, A.: Unsupervised learning of probably symmetric deformable 3D objects from images in the wild. In: Proceedings of the IEEE/CVF Conference on Computer Vision and Pattern Recognition, pp. 1–10 (2020)
50. Xiang, Y., Mottaghi, R., Savarese, S.: Beyond pascal: a benchmark for 3D object detection in the wild. In: IEEE Winter Conference on Applications of Computer Vision, pp. 75–82. IEEE (2014)

51. Xu, Q., Wang, W., Ceylan, D., Mech, R., Neumann, U.: DISN: deep implicit surface network for high-quality single-view 3D reconstruction. arXiv preprint arXiv:1905.10711 (2019)
52. Yan, X., Yang, J., Yumer, E., Guo, Y., Lee, H.: Perspective transformer nets: learning single-view 3D object reconstruction without 3D supervision. arXiv preprint arXiv:1612.00814 (2016)
53. Yang, G., Cui, Y., Belongie, S., Hariharan, B.: Learning single-view 3D reconstruction with limited pose supervision. In: Ferrari, V., Hebert, M., Sminchisescu, C., Weiss, Y. (eds.) ECCV 2018. LNCS, vol. 11219, pp. 90–105. Springer, Cham (2018). https://doi.org/10.1007/978-3-030-01267-0_6
54. Yariv, L., et al.: Multiview neural surface reconstruction by disentangling geometry and appearance. arXiv preprint arXiv:2003.09852 (2020)
55. Ye, Y., Tulsiani, S., Gupta, A.: Shelf-supervised mesh prediction in the wild. arXiv preprint arXiv:2102.06195 (2021)
56. Zhai, A., Wu, H.Y.: Classification is a strong baseline for deep metric learning. arXiv preprint arXiv:1811.12649 (2018)
57. Zhang, X., Zhang, Z., Zhang, C., Tenenbaum, J.B., Freeman, W.T., Wu, J.: Learning to reconstruct shapes from unseen classes. arXiv preprint arXiv:1812.11166 (2018)
58. Zhu, J.Y., Park, T., Isola, P., Efros, A.A.: Unpaired image-to-image translation using cycle-consistent adversarial networks. In: Proceedings of the IEEE International Conference on Computer Vision, pp. 2223–2232 (2017)

Author Index

Printed in the United States
by Baker & Taylor Publisher Services